AF316999

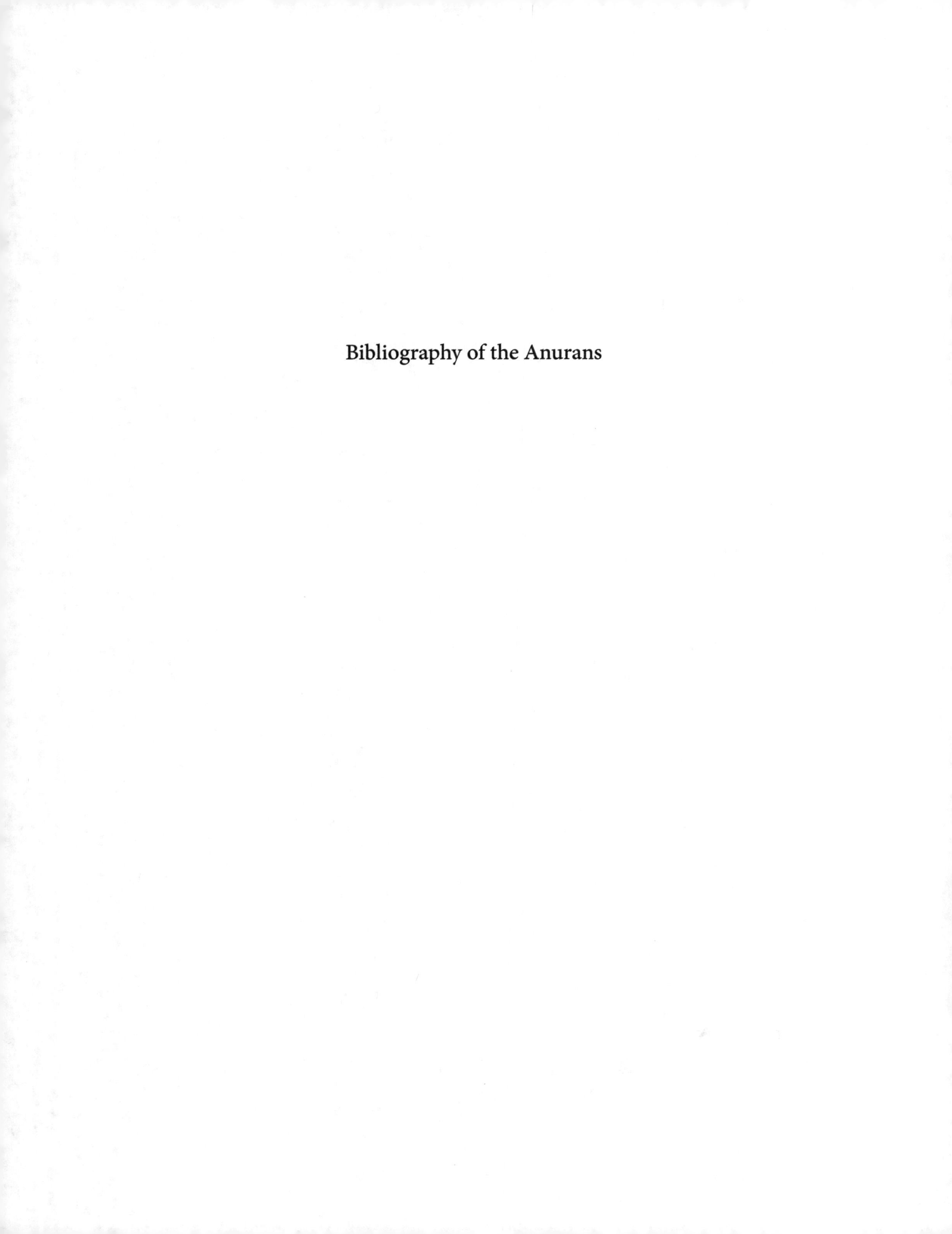

Bibliography of the Anurans

Bibliography of the Anurans of the United States and Canada

Part 1: 1698–2012. Part 2: 2013–2021

C. Kenneth Dodd, Jr.

International Society for the History
and Bibliography of Herpetology
www.ISHBH.com

2022

Wahlgreniana is named in honor of Richard Wahlgren (1946–2019) A founding member and first Chairman of the *International Society for the History and Bibliography of Herpetology*. Without Richard's tireless dedication to ISHBH, the society could not have made it through its early years. *Wahlgreniana* is a series of book length works complementing the ISHBH journal, *Bibliotheca Herpetologica*. Books in this series are published on an irregular basis and are sold separately from ISHBH subscriptions.

International Society for the History and Bibliography of Herpetology
© 2022 ISHBH & C. Kenneth Dodd, Jr.
All rights reserved. Published 1 October 2022
Printed on demand by IngramSpark global distribution network
Layout & design by Breck Bartholomew

ISBN: 979-8-218-06245-3 (hardcover)
ISBN: 979-8-218-06246-0 (ebook)

www.ISHBH.com

Cover and titlepage illustration: *Bufo americanus* by John H. Richard (1807–1881). Originally published in Baird, Spencer F. (1859) Reptiles of the Boundary Survey. Pp. 33–35, 41 plates *in:* Emory, W. H. (Ed.) *United States and Mexican Boundary Survey. Volume II, part 2.* United States Government, Washington, D.C.

Contents

Foreword

There is a lengthy pedigree of documented concerns about the proliferation of scientific literature and the consequent difficulty in 'keeping current,' especially for readers (and writers) with broad interests. Among my favorite of such lamentations is that of Charles Maynard (1889) decrying the need for his readers to subscribe to "half a dozen" society publications in order to keep track of his writings. His solution to the problem was the simple expedient of privately publishing his own journal to concentrate his writings and place them in aggregated form into the hands of his readers. Well and good for his contemporaries, but original copies of Maynard's *Contributions to Science* are now difficult (a possible euphemism for 'impossible') to find. His ideas and data have been rescued in our digital age by reprinted versions available, at least occasionally, via Amazon and other sources.

The passage of time has not ameliorated Maynard's concerns about the proliferation of outlets for scientific research. Every year brings more and more information into the world of scientific scholarship. By the turn of the 21st century it had become extraordinarily difficult to 'keep up' with information relevant to scholars pursuing even a focused research project. The problem becomes particularly acute for those with broad interests that extend across traditionally recognized boundaries. Almost 20 years ago Aaron Bauer (pers. com., August 2022) made an effort to estimate the number of post-Linneaus journals that at least occasionally carried articles pertaining to natural history; he landed on the jaw-dropping figure of approximately 250,000! Even for the (relatively) focused community of herpetologists there is now a plethora of serial publications that regularly contain content about amphibians and reptiles, ranging from titles exclusively dedicated to herpetology, to those with a regional or taxonomic focus, to titles dedicated to general natural history or science for which herpetology constitutes an important component of the content. Navigating all of this remains a challenge, despite the many tools now available to help with awareness of, and access to, the voluminous literature in herpetology.

Although private publication was, and to some degree still is, an option, more prevalent (and longer-lasting) solutions included the preparation of the annual and/or periodic index to contents of serial publications, and bibliographies of titles that were collated and proffered by dedicated scholars in the hopes of facilitating access to the literature pertinent to a particular topic. Apart from the collected works of an author presented in a Festschrift or obituary, printed and online bibliographies are no longer common, but they remain quite helpful. The increasingly prevalent attitude that 'it's all online' has been expressed in my hearing at least since the early 1990s when the Society of Vertebrate Paleontology moved to kill the nearly 70-year run of the invaluable *Bibliography of Fossil Vertebrates*. At that time there was certainly not anything even approximating truth in the claim that it was 'all online' but with every passing year, the claim more closely approaches some form of reality; much more is now online now than it was then. That reality facilitates *access* for some scholars, especially those from wealthy nations or institutions. But not all scholars are as well served because access to online resources remains a challenge in many countries, especially those without consistent or reliable access to electricity, and many academic institutions even in wealthy nations do not pay for access to content that is, technically, online. I am, therefore, delighted that this current compilation has options for both print and electronic access. To some degree online availability also helps with *awareness*, especially if the scholar is careful with keyword selection for searches, is attuned to various stumbling blocks that can impair a successful search and participates in the various options that provide notification of journal contents, or platforms that allow scholars to share their newly published works online.

But *awareness* is tricky. What if the target of interest is a detailed overview of a broad and complex topic encompassing over 300 years of scholarship and published literature? Needs such as those are best addressed by bibliographies such as this one, which remain a valuable starting point for broad overviews. As concentrated summations of information on a focused topic, works such as this remain valuable tools for serious established scholars as well as those with a budding interest in the topic. Such a need arose somewhat unexpectedly in the late 1980s with the recognition that amphibian populations were declining across the globe. That stimulated a surge in research on amphibians and a search for causes of declines. In addition to data on population status and basic information about the natural history of species, their population biology, their significance in ecosystems, and anthropogenic influences on populations were emphasized as important and relevant data (Wake and Morowitz, 1991). The need for those data certainly contributed to the increase in annual publications about anuran natural history noted in the introductory paragraphs herein.

Ken Dodd has drawn from decades of experience (who's counting how many?) with herpetological literature to compile a record of published material pertaining to the broad topic of 'the natural history of anurans in the United States and Canada.' This is not his first foray into bibliographic documentation. In addition to the excellent bibliographies in his books he has also produced bibliographic summations of anuran defensive mechanisms (Dodd, 1975), endangered and threatened amphibians and reptiles in the United States (Dodd, 1979, 1981), primary literature on the loggerhead sea turtle (Dodd, 1987), and the herpetofauna of Florida (Enge and Dodd, 1986, 1992). Much of the content of this current effort appeared in his 2013 compilation on anurans of the United States and Canada, but this is a much-expanded treatment. I confess I have a long-standing fascination with productions like this one. Over the last 25 years I have repeatedly turned to the bibliographies assembled by Domning (1996) and Zhao and Zhao (1994), simultaneously for my own research questions, or those of my students, or to meet a simple urge to satisfy my curiosity on various points. Those works garner few citations in the literature but are of inestimable value to me as sources of information, guideposts to the literature, and stimulants for new questions, thoughts, and explorations.

This compilation on anurans is destined to be another to which I refer often, and I have already benefitted from it. Over 280 pages of literature citations include references to numerous books and monographs, as well as references to papers from over 700 serial publications (ok, I counted). Without overlap, there are 123 *Journals*, 69 *Bulletins*, 38 *Proceedings*, 26 *Naturalists*, 19 *Transactions*, 15 *Occasional Papers*, eight *Magazines*, eight *Reviews*, seven *Annals*, five *Newsletters*, four *Leaflets*, three *Quarterlies*, and two *Applications*. As Dodd notes in his introductory paragraphs, many titles are drawn from journals with long histories of carrying herpetological content. But titles that are well off the beaten path for many herpetologists find a place here as well. As an example, Hall's monograph on Blowflies might not immediately jump to the mind of many students contemplating anuran biology, but an account of parasitism of *Rana catesbeiana* by the fly *Bufolucilia* (now *Lucilia*) *silvarum* appears on p. 218—I *certainly* would have missed this without Ken's guidance. In a separate vein, when I noted the citation to Harlan's 1835 *Medical and Physical Researches*, which I have normally consulted in reference to salamanders or mammals, I was curious about the anuran content… so, I looked. Two prominent examples of its relevance for inclusion here are *Genera of North American Reptilia, and a synopsis of the species* (pp. 84–160; mostly descriptions, but with brief notes on natural history for some species) and *Descriptions of several new species of batrachian reptiles, with observations on the larvae of frogs* (pp. 214–228). So, I learned a bit more about a volume with which I was inadequately familiarized, and that is never a bad thing!

Are there published titles that are not included here? I know the nature of our literature too well to doubt that there are. But that in no way diminishes this work and its significance. Everyone who acquaints themselves with it will find at least a few inducements to seek out particular references and read them. The serious scholar will find much here with which to engage. My somewhat-more-than-casual perusal yielded five citations to papers of immediate interest that I had never seen before, and many titles with which I was unfamiliar. I have been living without the *Proceedings of the Natural History Association of Miramichi* for far too long now, and I have set aside a space on my shelves for the *Proceedings of the Natural Science Association of Staten Island*. So, the hunt is on… and I love the hunt; it is one more gift from the mind, effort, energy, and keyboard of Ken Dodd. I hope that similar discoveries and enjoyment will await any reader of this brief introduction to a masterful compilation of interesting and relevant literature. Happy hunting!

Christopher J. Bell
21 August, 2022

REFERENCES

Dodd, C. K., Jr. 1975. A Bibliography of Anuran Defensive Mechanisms. Smithsonian Herpetological Information Service 37. 10 pp.

Dodd, C. K., Jr. 1979. A Bibliography of Endangered and Threatened Amphibians and Reptiles in the United States and its Territories (Conservation, Distribution, Natural History, Status). Smithsonian Herpetological Information Service 46. 35 pp.

Dodd, C. K., Jr. 1981. A Bibliography of Endangered and Threatened Amphibians and Reptiles in the United States and it Territories (Conservation, Distribution, Natural History, Status). Supplement. Smithsonian Herpetological Information Service 49:1–16.

Dodd, C. K., Jr. 1987. A Bibliography of the Loggerhead Sea Turtle *Caretta caretta* (Linnaeus), 1758. Endangered Species Report No. 16, U. S. Fish and Wildlife Service,

Albuquerque, New Mexico. 64 pp.

Dodd, C. K., Jr. 2013a. A bibliography of the anurans of the United States and Canada (1734–2012). Herpetological Conservation and Biology 8 (Monograph 4):1–202.

Domning, D. P. 1996. Bibliography and Index of the Sirenia and Desmostylia. Smithsonian Contributions to Paleobiology 80. iii + [1] + 611 pp.

Enge, K. N., and C. K. Dodd, Jr. 1986. A Bibliography of the Herpetofauna of Florida. Smithsonian Herpetological Information Service 72. 68 pp.

Enge, K. M., and C. K. Dodd, Jr. 1992. An Indexed Bibliography of the Herpetofauna of Florida. Florida Game and Fresh Water Fish Commission, Nongame Wildlife Program, Technical Report No. 11. iv + 231 pp.

Harlan, R. 1835. Medical and Physical Researches: Or Original Memoirs in Medicine, Surgery, Physiology, Geology, Zoology, and Comparative Anatomy. Printed by Lydia R. Bailey, Philadelphia, Pennsylvania. xxxix, 9-653, [2] pp., 1 p. errata.

Maynard, C. J. 1889. Prefatory. Contributions to Science [By Charles J. Maynard] 1(1):i–ii.

Wake, D. B., and H. J. Morowitz, co-chairs. 1991. Declining amphibian populations—a global phenomenon? Findings and recommendations. Alytes 9(2):33–42.

Zhao, E.-M., and H. H. Zhao. 1994. Chinese Herpetological Literature. Catalogue and Indices. Publishing House of the Chengdu University of Science and Technology, Chengdu, China. 399 pp.

Abstract

In 2013, I published an extensive bibliography (Dodd 2013a) of the anurans of the United States and Canada based largely upon background material from The Frogs of the United States and Canada (Dodd 2013b). The bibliography was produced with the objective of developing a comprehensive reference to publications on the natural history of North American anurans. It focused on life history, ecology, systematics, behavior, physiological ecology, diseases, parasites, and conservation biology, and included important references on distribution and other topics useful to understanding frogs in their natural environment. Strictly physiological, developmental, and genetics citations were excluded, as were routine new distribution records, especially when life history information was not included. Master's theses and doctoral dissertations were included as they were encountered, but I made no attempt to access all potentially citable graduate research, much of which remains unpublished or unavailable. I also excluded what is popularly termed "gray" literature, as well as most popular and hobbyist publications.

Since the original bibliography (Dodd 2013a), I have made an effort to update references previously omitted because of redundancy, triviality (short notes, especially in the early days of Copeia and Herpetologica), oversight, or because they were published before the end of 2012 but after my frog book had gone to press. I also examined much older and obscure natural histories for references to frogs. As such, I extended the bibliographic references to Gabriel Thomas' (1698) mention of the bullfrog, the earliest reference to a specific species that I encountered. Frogs were often mentioned in early travelogues, but rarely can the species be identified. As previously, I have limited the scope of this updated bibliography to topics that might be termed field ecology or natural history, but that does not mean that highly complex topics, such as molecular biology, population modelling, and laboratory research are excluded; they are all vital to telling us what frogs do in nature and how they do it. Publishing this updated and comprehensive bibliography will allow researchers, students, and naturalists access to a large amount of information that might not be readily available through traditional database searches. Naturally, some publications likely have escaped my notice, despite years of intensive search and cross–referencing. To those authors, I offer my sincere apology.

I especially thank Breck Bartholomew, Christopher Bell, and Aaron Bauer for their comments, suggestions, assistance and patience in making publication of this bibliography a reality.

Composite Bibliographic Trivia

Journals. — The most cited journals involving frog natural history, not surprisingly, are those that have been publishing the longest and that specialize in herpetology: Copeia (now Ichthyology and Herpetology), Herpetological Conservation and Biology, Herpetological Review, Herpetologica, and the Journal of Herpetology. Other journals with substantial numbers of North American frog papers include American Midland Naturalist, Animal Behaviour, Biological Conservation, Canadian Journal of Zoology, Canadian Field-Naturalist, Conservation Biology, Ecology, Environmental Toxicology and Chemistry, Evolution, Northeastern Naturalist, Oecologia, Southeastern Naturalist, Southwestern Naturalist, and Northwestern Naturalist.

Graduate research. — The combined bibliography contains citations for 227 doctoral dissertations and 295 Master's/Honors theses.

Most published authors. — Of course, most herpetologists are interested in taxa besides to frogs, particularly salamanders and non–avian reptiles; few specialize only on frogs. Further, not all citations are of equivalent page length, duration of study, rigor or importance. Given these caveats, the top 10 names in terms of numbers of authored and co–authored papers involving the life history of frogs are: A.N. Bragg, A.R. Blaustein, R.D. Semlitsch, H.C. Gerhardt, S.E. Trauth, R. Relyea, P.S. Corn, B.K. Sullivan, D.M. Green, and W.E. Meshaka, Jr. With the exception of Bragg (d. 1968) and Semlitsch (d. 2015), these biologists remain active in frog research.

Number of publications annually. — Not surprisingly, there has been an explosion in publishing on North American anuran life history since the early 1990s. For example, there were 78–90 papers published annually from 1990–1994, 101–112 from 1995–1997, 140–171 from 1998–2003, 183–217 from 2004–2012, 96–135 from 2013–2020, and 217 in 2021. No doubt, this pace continues.

Literature Cited

Dodd, C. K., Jr. 2013a. A bibliography of the anurans of the United States and Canada (1734–2012). Herpetological Conservation and Biology 8 (Monograph 4):1–202.

Dodd, C.K., Jr. 2013b. The Frogs of the United States and Canada. Johns Hopkins University Press, Baltimore, Maryland, 2 volumes, i–xxvii + 962 pp.

Part 1:1698–2012

Abbott, C.C. 1868. Catalogue of the Vertebrate Animals of New Jersey. Appendix E, *In* G.H. Cook, Geology of New Jersey, Newark, New Jersey. [frogs on pp. 804-805]

Abbott, C.C. 1882. Notes on the habits of the "savannah cricket frog." American Naturalist 16:707–711.

Abbott, C.C. 1884a. Recent studies of the spade–foot toad. American Naturalist 18:1075–1080.

Abbott, C.C. 1884b. Hibernation in the lower vertebrates. Science 4(75):34–39. [Reprinted in 1887 in The Swiss Cross 2(6):172–174.]

Abbott, C.C. 1885. A Naturalist's Rambles about Home. D. Appleton, New York.

Abbott, C.C. 1888. Cyclopedia of Natural History. Nims and Knight, Troy, New York.

Abbott, C.C. 1894. The intelligence of batrachians. Science 3:66–67.

Abbott, C.C. 1904. One explanation of reported showers of toads. Proceedings of the American Philosophical Society 43:163–164.

AbuBakr, S., and S.S. Crupper. 2010. Prevalence of cadmium resistance in *Staphylococcus sciuri* isolated from the gray treefrog, *Hyla chrysoscelis* (Anura: Hylidae). Phyllomedusa 9:141–146.

Acker, P.M., K.C. Kruse, and E.B. Krehbiel. 1986. Aging *Bufo americanus* by skeletochronology. Journal of Herpetology 20:570–574.

Ackleh, A.S., K. Deng, and Q. Huang. 2010. Existence–uniqueness results and difference for an amphibian juvenile–adult model. AMS Contemporay Mathematics Series 513:1–23.

Ackleh, A.S., K. Deng, and Q. Huang. 2011. Stochastic juvenile–adult models with application to a green tree frog population. Journal of Biological Dynamics 5:64–83.

Ackroyd, J.F., and R.L. Hoffman. 1946. An albinistic specimen of *Pseudacris feriarum*. Copeia 1946:257–258.

Adam, M.D., and M.J. Lacki. 1993. Factors affecting amphibian use of road–rut ponds in Daniel Boone National Forest. Transactions of the Kentucky Academy of Science 54:13–16.

Adama, D., K. Lansley, and M.–A. Beaucher. 2004. Northern leopard frog (*Rana pipiens*) recovery: captive rearing and reintroduction in southeast British Columbia, 2003. Columbia Basin Fish and Wildlife Compensation Program, Nelson, British Columbia.

Adams, C.K., and D. Saenz. 2012. Leaf litter of invasive Chinese tallow (*Triadica sebifera*) negatively affects hatching success of an aquatic breeding anuran, the southern leopard frog (*Lithobates sphenocephalus*). Canadian Journal of Zoology 90: 991–998.

Adams, C.K., D. Saenz, and E. Fucik. 2008. *Gastrophryne carolinensis* (Eastern Narrow–mouthed Toad). Hind limb malformation. Herpetological Review 39:460–461.

Adams, C.K., D. Saenz, and R.N. Connor. 2011. Palatability of twelve species of anuran larvae in eastern Texas. American Midland Naturalist 166:211–223.

Adams, L.W., T. M. Franklin, L. E. Dove, and J. M. Duffield. 1986. Design considerations for wildlife in urban stormwater management. *In* Transactions of the 51st North American Wildlife and Natural Resources Conference, pp. 249–259. Wildlife Management Institute Washington, D.C.

Adams, M.J. 1993. Summer nests of the tailed frog (*Ascaphus truei*) from the Oregon Coast Range. Northwestern Naturalist 74:15–18.

Adams, M.J. 1999. Correlated factors in amphibian decline: exotic species and habitat change in western Washington. Journal of Wildlife Management 63:1162–1171.

Adams, M.J. 2000. Pond permanence and the effects of exotic vertebrates on anurans. Ecological Applications 10:559–568.

Adams, M.J., and S. Claeson. 1998. Field response of tadpoles to conspecific and heterospecifics alarm. Ethology 104:955–961.

Adams, M.J., and R.B. Bury. 2002. The endemic headwater stream amphibians of the American Northwest: associations with environmental gradients in a large forested preserve. Global Ecology and Biogeography 11:169–178.

Adams, M.J., and C.A. Pearl. 2007. Problems and opportunities managing invasive bullfrogs: is there any hope? Pp. 679–693 *In* F. Gherardi (ed.). Biological Invaders in Inland Waters: Profiles, Distribution, and Threats. Springer, The Netherlands.

Adams, M.J., R.B. Bury, and S.A. Swarts. 1998. Amphibians of the Fort Lewis Military Reservation, Washington: sampling techniques and community patterns. Northwestern Naturalist 79:12–18.

Adams, M.J., S.D. West, and L. Kalmbach. 1999. Amphibian and reptile surveys of U.S. Navy lands on the Kitsap and Toandos peninsulas, Washington. Northwestern Naturalist 80:1–7.

Adams, M.J., D.E. Schindler, and R.B. Bury. 2001. Association of amphibians with attenuation of ultraviolet–B radiation in montane ponds. Oecologia 128:519–528.

Adams, M.J., C.A. Pearl, and R.B. Bury. 2003. Indirect facilitation of an anuran invasion by non–native fishes. Ecology Letters 6:343–351.

Adams, M.J., B.R. Hossack, R.A. Knapp, P.S. Corn, S.A. Diamond, P.C. Trenham, and D.B. Fagre. 2005. Distribution patterns of lentic–breeding amphibians in relation to ultraviolet radiation exposure in western North America. Ecosystems 8:488–500.

Adams, M.J., S.K. Galvan, D. Reinitz, R.A. Cole, S. Payre, M. Hahr, and P. Govindarajulu. 2007. Incidence of the fungus, *Batrachochytrium dendrobatidis*, in amphibian populations along the northwest coast of North America. Herpetological Review 38:430–431.

Adams, M.J., C.A. Pearl, B. McCreary, S.K. Galvan, S. Wessell–Kelly, W.H. Wente, C.W. Anderson, and A.B. Kuehl. 2009. Short–term effect of cattle exclosures on Columbia spotted frog (*Rana luteiventris*) populations and habitat in northeastern Oregon. Journal of Herpetology 43:132–138.

Adams, M.J., N.D. Chelgren, D. Reinitz, R.A. Cole, L.J. Rachowicz, S. Galvan, B. McCreary, C.A. Pearl, L.L. Bailey, J. Bettaso, E.L. Bull, and M. Leu. 2010. Using occupancy models to understand the distribution of an amphibian pathogen, *Batrachochytrium dendrobatidis*. Ecological Applications 20:289–302.

Adams, M.J., C.A. Pearl, S.K. Galvan, and B. McCreary. 2011. Non–native species impacts on pond occupancy by an anuran. Journal of Wildlife Management 75:30–35.

Adams, S.B., and C.A. Frissell. 2001. Thermal habitat use and evidence of seasonal migration by Rocky Mountain tailed frogs, *Ascaphus montanus*, in Montana. Canadian Field-Naturalist 115:251–256.

Adams, S.B., D.A. Schmetterling, and M.K. Young. 2005. Instream movements by boreal toads (*Bufo boreas boreas*). Herpetological Review 36:27–33.

Adamson, M.L. 1981a. Development and transmission of *Gyrinicola batrachiensis* (Walton, 1929) (Pharyngodonidae: Oxyuroidea). Canadian Journal of Zoology 59:1351–1367.

Adamson, M.L. 1981b. Seasonal changes in populations of *Gyrinicola batrachiensis* (Walton, 1929) in wild tadpoles. Canadian Journal of Zoology 59:1377–1386.

Adamson, M.L. 1981c. *Gyrinicola batrachiensis* (Walton, 1929) n. comb. (Oxyuroidea; Nematoda) from tadpoles in eastern and central Canada. Canadian Journal of Zoology 59:1344–1350.

Ade, C.M., M.D. Boone, and H.J. Puglis. 2010. Effects of an insecticide and potential predators on green frogs and northern cricket frogs. Journal of Herpetology 44:591–600.

Adkins–Giese, C.L. 2012. Protecting rare amphibians under the U.S. Endangered Species Act. Froglog (102):21–23.

Adler, K. 1970. The role of extraoptic photoreceptors in amphibian rhythms and orientation: a review. Journal of Herpetology 4:99–112.

Adler, K. 1980. Individuality in the use of orientation cues by green frogs. Animal Behaviour 28: 413–425.

Adler, K. 2003. America's first herpetological expedition: William Bartram's travels in the southeastern United States (1773–1776). Bonner zoologische Beiträge 52:275–295.

Adolph, E.F. 1931a. Body size as a factor in the metamorphosis of tadpoles. Biological Bulletin 61:376–386.

Adolph, E.F. 1931b. The size of the body and the size of the environment in the growth of tadpoles. Biological Bulletin 61:350–375.

AFCC. 1936. A future in frogs. American Frog Canning Company, New Orleans, Louisiana.

Agassiz, L. 1850. Lake Superior: its physical character, vegetation, and animals, compared with those of other and similar regions. With a narrative of the tour by J. Elliot Cabot and contributions by other scientific gentlemen. Part 7. Description of some new species of reptiles from the region of Lake Superior. Gould, Kendall and Lincoln, Boston, 378–382 + plate 6.

Aitchison, S.W., and D.S. Tomko. 1974. Amphibians and reptiles of Flagstaff, Arizona. Plateau 47:18–25.

Akers, E.C. 1997. Effects of predators and water color on growth, shape, and coloration of the tadpoles of *Hyla chrysoscelis* (Anura: Hylidae). M.S. thesis, Mississippi State University, Mississippi State.

Akers, E.C., C.M. Taylor, and R. Altig. 2008. Effects of clay–associated organic material on the growth of *Hyla chrysos-*

celis tadpoles. Journal of Herpetology 42:408–410.

Akin, G.C. 1966. Self–inhibition of growth in *Rana pipiens* tadpoles. Physiological Zoology 39:341–356.

Alexander, D.G. 1966. An ecological study of the swamp cricket frog, *Pseudacris nigrita feriarum* (Baird), with comparative notes on two other hylids of the Chapel Hill, North Carolina, region. Ph.D. Dissertation, University of North Carolina, Chapel Hill.

Alexander, L.G., J.H.K. Pechmann and P.J. DeVries. 2012. Effects of salinity on early life stages of the Gulf Coast toad, *Incilius nebulifer* (Anura: Bufonidae). Copeia 2012:106–114.

Alexander, T.R. 1965. Observations on the feeding behavior of *Bufo marinus* (Linne). Herpetologica 20:255–259.

Alford, R.A. 1981. Community organization and behavior of anuran larvae in a northern Florida temporary pond. M.S. thesis, University of Florida, Gainesville.

Alford, R.A. 1986a. Effects of parentage on competitive ability and vulnerability to predation in *Hyla chrysoscelis* tadpoles. Oecologia 68:199–204.

Alford, R.A. 1986b. Habitat use and positional behavior of anuran larvae in a northern Florida temporary pond. Copeia 1986:408–423.

Alford, R.A. 1989a. Variation in predator phenology affects predator performance and prey community composition. Ecology 70:206–219.

Alford, R.A. 1989b. Competition between larval *Rana palustris* and *Bufo americanus* is not affected by variation in reproductive phenology. Copeia 1989:993–1000.

Alford, R.A., and M.L. Crump. 1982. Habitat partitioning among size classes of larval southern leopard frogs, *Rana utricularia*. Copeia 1982:367–373.

Alford, R.A., and H.W. Wilbur. 1985. Priority effects in experimental pond communities: competition between *Bufo* and *Rana*. Ecology 66:1097–1105.

Alford, R.A., and R.N. Harris. 1988. Effects of larval growth history on anuran metamorphosis. American Naturalist 131:91–106.

Algar, A.C., J.T. Kerr, and D.J. Currie. 2011. Quantifying the importance of regional and local filters for community trait structure in tropical and temperate zones. Ecology 92:903–914.

Ali, M.F., K.R. Lips, F.C. Knoop, B. Fritzsch, C. Miller, and J.M. Conlon. 2002. Antimicrobial peptides and protease inhibitors in the skin secretions of the crawfish frog, *Rana areolata*. Biochimica et Biophysica Acta–Proteins and Proteomics 1601:55–63.

Allaback, M.L., D.M. Laabs, D.S. Keegan, and J.D. Harwayne. 2010. *Rana draytonii* (California Red–legged Frog). Dispersal. Herpetological Review 41:204–206.

Allan, D.M. 1973. Some relationships of vocalization to behavior in the Pacific treefrog, *Hyla regilla*. Herpetologica 29:366–371.

Allan, S.J., and A.M. Simmons. 1994. Temporal features mediating call recognition in the green treefrog, *Hyla cinerea*: amplitude modulation. Animal Behaviour 47:1073–1086.

Allard, H.A. 1907. Fowler's toad (*Bufo fowleri* Putnam). Science 26:383-384.

Allard, H.A. 1908. *Bufo fowleri* (Putnam) in northern Georgia. Science 28:655–656.

Allen, A.C. 1963. The amphibia of Wayne County, Ohio. Journal of the Ohio Herpetological Society 4:23–30.

Allen, D. 1937. Some notes on the amphibia of a waterfowl sanctuary, Kalamazoo, Michigan. Copeia 1937:190–191.

Allen, D.M. 1975. An analysis of vocalization and female discriminatory responses in the Pacific treefrog, *Hyla regilla*. M.S. thesis, California State University, Fullerton.

Allen, E.R. 1938. Notes on Wright's bullfrog, *Rana heckscheri* (Wright). Copeia 1938:50.

Allen, E.R. 1941. The value of toads. All–Pets Magazine. [reprinted by The Florida Reptile Institute, Silver Springs, Florida]

Allen, E.R., and W.T. Neill. 1953. The treefrog, *Hyla septentrionalis*, in Florida. Copeia 1953:127–128.

Allen, J.A. 1868. Catalogue of the reptiles and batrachians found in the vicinity of Springfield, Massachusetts, with notices of all other species known to inhabit the state. Proceedings of the Boston Society of Natural History 12:3–38.

Allen, L.S. 2006. Collaboration in the Borderlands: The Malpai Borderlands Group: After 10 years of efforts to preserve the open spaces and way of life of the Borderlands Region, the Malpai Borderlands Group is now internationally recognized as an outstanding example of collaborative planning and management of large landscapes. Rangelands 28:17–21.

Allen, M.J. 1932. A survey of the amphibians and reptiles of Harrison County, Mississippi. American Museum Novitates 542:1–20.

Allen, W.R. 1932. Further comment on the activity of the spade–foot toad. Copeia 1932:104.

Allran, J.W., and W.H. Karasov. 2001. Effects of atrazine on embryos, larvae, and adults of anuran amphibians. Environmental Toxicology and Chemistry 20:769–775.

Altig, R. 1970. A key to the tadpoles of the continental United States and Canada. Herpetologica 26:180–207.

Altig, R. 1971. Descriptive notes on the tadpoles of *Pseudacris ornata* and *Bufo alvarius*. Texas Journal of Science 23:301–303.

Altig, R. 1972a. Notes on the larvae and premetamorphic tadpoles of four *Hyla* and three *Rana* with notes on tadpole color patterns. Journal of the Elisha Mitchell Scientific Society 88:113–119.

Altig, R.A. 1972b. Defensive behavior in *Rana areolata* and *Hyla avivoca*. Quarterly Journal of the Florida Academy of Sciences 35:212–216.

Altig, R., and E.D. Brodie, Jr. 1968. Albinistic and cyanistic frogs from Oregon. The Wasmann Journal of Biology 26:241–242.

Altig, R., and E.D. Brodie, Jr. 1972. Laboratory behavior of *Ascaphus truei* tadpoles. Journal of Herpetology 6:21–24.

Altig, R., and M.T. Christensen. 1981. Behavioral characteristics of the tadpoles of *Rana heckscheri*. Journal of Herpetology 15:151–154.

Altig, R., and C.K. Dodd, Jr. 1987. The status of the Amargosa toad (*Bufo nelsoni*) in the Amargosa River drainage of Nevada. Southwestern Naturalist 32:276–278.

Altig, R., and R.W. McDiarmid. 1999. Body plan. Development and morphology. Pp. 24–51 *In* R.W. McDiarmid and R.G. Altig (eds.), Tadpoles. The Biology of Anuran Larvae. University of Chicago Press, Chicago.

Altig, R., M.R. Whiles, and C.L. Taylor. 2007. What do tadpoles really eat? Assessing the trophic status of an understudied and imperiled group of consumers in freshwater habitats. Freshwater Biology 52:386–395.

Alvarez, J.A. 2004. *Rana aurora draytonii* (California Red–legged Frog). Microhabitat. Herpetological Review 35:162–163.

Alvarez, J.A. 2011. *Bufo boreas* (Western Toad). Davian behavior. Herpetological Review 42:408–409.

Amburgey, S., W.C. Funk, M. Murphy and E. Muths. 2012. Effects of hydroperiod duration on survival, developmental rate, and size at metamorphosis in boreal chorus frog tadpoles (*Pseudacris maculata*). Herpetologica 68:456–467.

Andersen, M.L. 1973. Thermal relations in *Acris crepitans*. M.A. thesis, University of Kansas, Lawrence.

Anderson, A.L.,and W.D. Brown. 2009. plasticity of hatching in green frogs (*Rana clamitans*) to both egg and tadpole predators. Herpetologica 65:207–213.

Anderson, A.M., D.A. Haukos, and J.T. Anderson. 1999a. Habitat use by anurans emerging and breeding in playa wetlands. Wildlife Society Bulletin 27:759–769.

Anderson, A.M., D.A. Haukos, and J.T. Anderson. 1999b. Diet composition of three anurans from the playa wetlands of northwest Texas. Copeia 1999:515–520.

Anderson, A.R., and J.W. Petranka. 2003. Odonate predator does not affect hatching time or morphology of embryos of two amphibians. Journal of Herpetology 37:65–71.

Anderson, G.A., S.C. Schell, and I. Pratt. 1965. The life cycle of *Bunoderella metteri* (Allocreadidae: Bunoderinae), a trematode parasite of *Ascaphus truei*. Journal of Parasitology 51:579–582.

Anderson, J.D., K.A. Hawthorne, J.M. Galandak, and M.J. Ryan. 1978. A report on the status of the endangered reptiles and amphibians of New Jersey. Bulletin of the New Jersey Academy of Science 23:26–33.

Anderson, J.L., and B.W. Buttrey. 1962. Enteric protozoa of four species of frogs from the Lake Itasca region of Minnesota. Proceedings of the South Dakota Academy of Science 41:73–82.

Anderson, K. 1991. Chromosome evolution in Holarctic *Hyla* treefrogs. Pp. 299–331 *In* D.M. Green and S.K. Sessions (eds.), Amphibian Cytogenetics and Evolution. Academic Press, San Diego, California.

Anderson, K., and P.E. Moler. 1986. Natural hybrids of the Pine Barrens treefrog, *Hyla andersonii* with *H. cinerea* and *H. femoralis* (Anura: Hylidae): morphological and chromosomal evidence. Copeia 1986:70–76.

Anderson, L.R., and J.A. Arruda. 2006. Land use and anuran biodiversity in southeast Kansas, USA. Amphibian and Reptile Conservation 4:46–59.

Anderson, M.T., J.M. Kiesecker, D.P. Chivers and A.R. Blaustein. 2001. The direct and indirect effects of temperature on a predator–prey relationship. Canadian Journal of Zoology 79: 1834–1841.

Anderson, P.K. 1942. Amphibians and reptiles of Jackson County, Missouri. Bulletin of the Chicago Academy of Sciences 6:203–220.

Anderson, P.K. 1951. Albinism in tadpoles of *Microhyla carolinensis*. Herpetologica 7:56.

Anderson, P.K. 1954. Studies in the ecology of the narrow–mouthed toad, *Microhyla carolinensis carolinensis*. Tulane Studies in Zoology 2:15–46.

Anderson, P.K., E.A. Liner, and R.E. Etheridge. 1952. Notes on amphibian and reptile populations in a Louisiana pineland area. Ecology 33:274–278.

Anderson, R.C., and G.F. Bennett. 1963. Opthalmic myiasis in amphibians in Algonquin Park, Ontario, Canada. Canadian Journal of Zoology 41:1169–1170.

Anderson, T.R. 1965. Frog captures a fledgling eastern phoebe. Auk 82:285–286.

Andrews, E.A. 1928. Tree frogs and pitcher plants. Science 67:269–270.

Andrews, J.S. 2001. The Atlas of the Reptiles and Amphibians of Vermont. Privately Published, Middlebury, Vermont.

Andrews, K.D., R.L. Lampley, M.A. Gillman, D.T. Corey, S.R. Ballard, M.J. Blasczyk, and W.G. Dyer. 1992. Helminths of *Rana catesbeiana* in southern Illinois with a checklist of helminths in bullfrogs of North America.

Transactions of the Illinois State Academy of Science 85:147–172.

Angermann, J.E., G.M. Fellers, and F. Matsumura. 2002. Polychlorinated biphenyls and toxaphene in Pacific tree frog tadpoles (*Hyla regilla*) from the California Sierra Nevada, USA. Environmental Toxicology and Chemistry 21:2209–2215.

Anholt, B.R., D.K. Skelly, and E.E. Werner. 1996. Factors modifying antipredator behavior in larval toads. Herpetologica 52:301–313.

Anholt, B.R., E. Werner, and D.K. Skelly. 2000. Effect of food and predators on the activity of four larval ranid frogs. Ecology 81:3509–3521.

Ankley, G.T., J.E. Tietge, D.L. DeFoe, K.M. Jensen, G.W. Holcombe, E.J. Durhan, and S.A. Diamond. 1998. Effects of ultraviolet light and methoprene on survival and development of *Rana pipiens*. Environmental Toxicology and Chemistry 17:2530-2542.

Ankley, G.T., J.E. Tietge, G.W. Holcombe, D.L. DeFoe, S.A. Diamond, K.M. Jensen, and S.J. Degitz. 2000. Effects of laboratory ultraviolet radiation and natural sunlight on survival and development of *Rana pipiens*. Canadian Journal of Zoology 78:1092–1100.

Ankley, G.T., S.A. Diamond, J.E. Tietge, G.W. Holcombe, K.M. Jensen, D.L. DeFoe, and R. Peterson. 2002. Assessment of the risk of ultraviolet radiation to amphibians. I. Dose–dependent induction of hind limb malformations in the northern leopard frog (*Rana pipiens*). Environmental Science & Technology 36:2853–2858.

Ankley, G. T., S.J. Degitz, S.A. Diamond, and J. E. Tietge. 2004a. Assessment of environmental stressors potentially responsible for malformations in North American anuran amphibians. Ecotoxicology and Environmental Safety 58:7–16.

Ankley, G.T., D.W. Kuehl, M.D. Kahl, K.M. Jensen, B.C. Butterworth, and J.W. Nichols. 2004b. Partial life-cycle toxicity and bioconcentration modeling of perfluorooctanesulfonate in the northern leopard frog (*Rana pipiens*). Environmental Toxicology and Chemistry 23:2745-2755.

ANLFRT (Alberta Northern Leopard Frog Recovery Team). 2005. Alberta Northern Leopard Frog Recovery Plan, 2005–2010. Alberta Sustainable Resource Development, Fish and Wildlife Division, Alberta Species at Risk Recovery Plan No. 7, Edmonton.

Anonymous. Undated. The amphibians of Linn County, Iowa. Biology Club, Roosevelt High School, Cedar Rapids, Iowa.

Anonymous. 1892. Bullfrogs for market. The Evening Star, Washington, DC. [2 January].

Anonymous. 1895. The gopher frog. Natural Science News 1(29):115.

Anonymous. 1899. Miss Seldon's frog farm. The Evening Star, Washington, DC. [4 February].

Anonymous. 1901. Frogs for market: a big hatchery to be established in Massachusetts. Scientific American Supplement (1314):1057. [9 March].

Anonymous. 1903. In the early hours. Forest and Stream 60(June 6):441.

Anonymous. 1907. Frogs. Okoboji Protective Association Bulletin 3:5.

Anonymous. 1918. The batrachians of Rhode Island. Part 2. Toads and frogs. Roger Williams Park, Park Museum Bulletin 10(4):93–96.

Anonymous. 1922. Amphibia and Reptilia. Cornell Rural School Leaflet 15(4):303–346.

Anonymous. 1931. Frog industry in Louisiana. Louisiana Department of Conservation, Division of Fisheries Educational Pamphlet No. 2. 40 pp.

Anonymous. 1933. Raising Frogs for Profit. American Frog Industries, Houston, Texas. 40 pp.

Anonymous. 1935a. Frog industry in Louisiana. Louisiana Department of Conservation, Division of Fisheries Bulletin No. 26. 44 pp. [reprinted 1938 and 1939, 47 pp.]

Anonymous. 1935b. Frogs' legs: amphibians leap in as bivalves go out. Newsweek 5(18):26, 28. [May 04, 1935].

Anonymous. 1946. Tobacco prices affect quantity of frog legs. Science News Letter 50:184. [September 21, 1946].

Anonymous. 1975. How to Raise Bullfrogs for Fantastic Profits. A Complete Instruction Manual. Quaestar Publishing, Rochester, New York. 61 pp.

Anonymous. 2007. Rare blue frog discovered at Corkscrew Swamp Sanctuary. Florida Naturalist (Spring):11.

Anton, T.G. 1999. Current distribution and status of amphibians and reptiles in Cook County, Illinois. Transactions of the Illinois State Academy of Science 92:211–232.

Anzalone, C.R., L.B. Kats, and M.S. Gordon. 1998. Effects of solar UV–B radiation on embryonic development in *Hyla cadaverina*, *Hyla regilla*, and *Taricha torosa*. Conservation Biology 12:646–653.

Applegarth, J.S. 1979. Environmental implications of herpetofaunal remains from archeological sites west of Carlsbad, New Mexico. Pp. 159-167 *In* H.H. Genoways and R.J. Baker (eds.), Biological Investigations in the Guadalupe Mountains National Park, Texas. Proceedings and Transactions Series No. 4. National Park Service, Washington, D.C.

Applegate, R.D. 1990. *Rana catesbeiana, Rana palustris* (Bullfrog, Pickerel Frog). Predation. Herpetological Review 21:90–91.

Aresco, M.J. 1996. Geographic variation in the morphology and lateral stripe of the green treefrog (*Hyla cinerea*) in the southeastern United States. American Midland Naturalist 135:293–298.

Aresco, M.J. 2004. River frog *Rana heckscheri* (Wright). Pp. 17–18 *In* R.E. Mirarchi, M.A. Bailey, T.M. Haggerty, and T.L. Best (eds.), Alabama Wildlife. Vol. 3. Imperiled Amphibians, Reptiles, Birds, and Mammals. University of Alabama Press, Tuscaloosa.

Aresco, M.J., and R.N. Reed. 1998. *Rana capito sevosa* (Dusky Gopher Frog). Predation. Herpetological Review 29:40.

Aresco, M.J., and M.S. Gunzburger. 2004. Effects of large–scale sediment removal on herpetofauna in Florida wetlands. Journal of Herpetology 38:275–279.

Arndt, R.G. 1977. A blue variant of the green frog, *Rana clamitans melanota* (Amphibia, Anura, Ranidae) from Delaware. Journal of Herpetology 11:102–103.

Arnold, H.L. 1968. Poisonous Plants of Hawaii. Tong Publishing, Honolulu.

Arnold, S.J., and R.J. Wassersug. 1978. Differential predation on metamorphic anurans by garter snakes (*Thamnophis*): social behavior as a possible defense. Ecology 59:1014–1022.

Aronson, L.R. 1943a. The sexual behavior of Anura. 5. Oviposition in the green frog, *Rana clamitans*, and the bull frog, *Rana catesbeiana*. American Museum Novitates 1224:1–6.

Aronson, L.R. 1943b. The sexual behavior of anura. 4. Oviposition in the mink frog, *Rana septentrionalis* Baird. American Midland Naturalist 29:242–244.

Aronson, L.R. 1943c. The sexual behavior of Anura. 3. The "release" mechanism and sex recognition in *H. andersonii*. Copeia 1943:246–249.

Aronson, L.R. 1944. The sexual behavior of Anura. 6. The mating pattern of *Bufo americanus*, *Bufo fowleri*,and *Bufo terrestris*. American Museum Novitates 1250:1–15.

Asay, M.J., P.G. Harowicz, and L. Su. 2005. Chemically mediated mate recognition in the tailed frog (*Ascaphus truei*). Chemical Signals in Vertebrates 10:24–31.

Ashley, E.P., and J.T. Robinson. 1996. Road mortality of amphibians, reptiles and other wildlife on the Long Point causeway, Lake Erie, Ontario. Canadian Field-Naturalist 110:403–412.

Ashton, A.D., and F.C. Rabalais. 1978. Helminth parasites of some anurans of northwestern Ohio. Proceedings of the Helminthological Society of Washington 45:141–142.

Ashton, D.T. 2002. A comparison of abundance and assemblage of lotic amphibians in late–seral and second–growth redwood forests in Humboldt County, California. M.A. thesis, Humboldt State University, Arcata, California.

Ashton, D.T., and R.J. Nakamoto. 2007. *Rana boylii* (Foothill Yellow–legged Frog). Predation. Herpetological Review 38:442.

Ashton, D.T., S.B. Marks, and H.H. Welsh, Jr. 2006. Evidence of continued effects from timber harvesting on lotic amphibians in redwood forests of northwestern California. Forest Ecology and Management 221:183–193.

Ashton, K.G., and A.C.S. Knipps. 2011. Effects of fire history on amphibian and reptile assemblages in rosemary scrub. Journal of Herpetology 45:497–503.

Ashton, R.E., Jr., and P.S. Ashton. 1988. Handbook of Reptiles and Amphibians of Florida. Part 3. The Amphibians. Windward Publishing, Inc., Miami.

Ashton, R.E., Jr., S.I. Guttman, and P. Buckley. 1973. Notes on the distribution, coloration, and breeding of the Hudson Bay toad, *Bufo americanus copei* (Yarrow and Henshaw). Journal of Herpetology 7:17–20.

Asquith, A. 1986. Implications of variation in the mating call of the green treefrog, *Hyla cinerea*. M.S. thesis, Mississippi State University, Mississippi State.

Asquith, A., and P. Zimba. 1988. Geographic variation in the mating call of the green treefrog (*Hyla cinerea*). American Midland Naturalist 119:101–110.

Asquith, A., and R. Altig. 1990. Male call frequency as a criterion for female choice in *Hyla cinerea*. Journal of Herpetology 24:198–201.

Atkinson, C. 1896. Batrachia of Turkey Lake, Indiana. Proceedings of the Indiana Academy of Science 5:258-261.

Atlas, M. 1935. The effect of temperature on the development of *Rana pipiens*. Physiological Zoology 8:290–310.

Atwood, M.A. 2011. Effects of basal resources on the food web of temporary freshwater pools: implications for amphibians and restoration efforts. M.S. thesis, State University of New York, Syracuse.

Aubry, K.B., and P.A. Hall. 1991. Terrestrial amphibian communities in the southern Washington Cascade Range. Pp. 327–340 *In* L.F. Ruggiero, K.B. Aubry, A.B. Carey, and M.H. Huff (coordinators). Wildlife and Vegetation of Unmanaged Douglas–fir Forests. USDA Forest Service General Technical Report PNW–GTR–285, Portland, Oregon.

Auburn, J.S., and D.H. Taylor. 1979. Polarized light perception and orientation in larval bullfrogs *Rana catesbeiana*. Animal Behaviour 27:658–668.

Audo, M.C., T.M. Mann, T. L. Polk, C.M. Loudenslager, W.J. Diehl, and R. Altig. 1995. Food deprivation during different periods of tadpole (*Hyla chrysoscelis*) ontogeny affects metamorphic erformance differently. Oecologia 103:518–522.

Auffenberg, W. 1956. Remarks on some Miocene anurans from Florida, with a description of a new species of *Hyla*. Breviora (52):1–11.

Ault, K.K., J.E. Johnson, H.C Pinkart, and R.S. Wagner.

2012. Genetic comparison of water molds from embryos of amphibians *Rana cascadae*, *Bufo boreas* and *Pseudacris regilla*. Diseases of Aquatic Organisms 99:127–137.

Austin, D.F. 1973. Range expansion of the Cuban tree frog in Florida. Florida Naturalist 46(4):28.

Austin, D.F., and A. Schwartz. 1975. Another exotic amphibian in Florida, *Eleutherodactylus coqui*. Copeia 1975:188.

Austin, J.D., and K.R. Zamudio. 2008. Incongruence in the pattern and timing of intra–specific diversification in bronze frogs and bullfrogs (Ranidae). Molecular Phylogenetics and Evolution 48:1041–1053.

Austin, J.D., S.C. Lougheed, L. Niedrauer, A.A. Chek, and P.T. Boag. 2002. Cryptic lineages in a small frog: the post–glacial history of the spring peeper, *Pseudacris crucifer* (Anura: Hylidae). Molecular Phylogenetics and Evolution 25:316–329.

Austin, J.D., S.C. Lougheed, P.E. Moler, and P.T. Boag. 2003a. Phylogenetics, zoogeography, and the role of dispersal and vicariance in the evolution of the *Rana catesbeiana* (Anura: Ranidae) species group. Biological Journal of the Linnean Society 80:601–624.

Austin, J.D., J.A. Dávila, S.C. Lougheed, and P.T. Boag. 2003b. Genetic evidence for female–biased dispersal in the bullfrog, *Rana catesbeiana* (Ranidae). Molecular Ecology 12:3165–3172.

Austin, J.D., S.C. Lougheed, and P.T. Boag. 2004a. Discordant temporal and geographic patterns in maternal lineages of eastern North American frogs, *Rana catesbeiana* (Ranidae) and *Pseudacris crucifer* (Hylidae). Molecular Phylogenetics and Evolution 32:799–816.

Austin, J.D., S.C. Lougheed, and P.T. Boag. 2004b. Controlling for the effects of history and nonequilibrium conditions in gene flow estimates in northern bullfrog (*Rana catesbeiana*) populations. Genetics 168:1491–1506.

Austin, J.D., T.A. Gorman, and D. Bishop. 2011a. Assessing fine–scale genetic structure and relatedness in the micro–endemic Florida bog frog. Conservation Genetics 12:833–838.

Austin, J.D., T.A. Gorman, D. Bishop, and P. Moler. 2011b. Genetic evidence of contemporary hybridization in one of North America's rarest anurans, the Florida bog frog. Animal Conservation 14:553–561.

Avila, V.L., and P.G. Frye. 1978. Feeding behavior of the African clawed frog (*Xenopus laevis* Daudin): (Amphibia, Anura, Pipidae): effect of prey type. Journal of Herpetology 12:391–396.

Awan, A.R., and G.R. Smith. 2007a. The effect of group size on the responses of wood frog tadpoles to fish. American Midland Naturalist 158:79–84.

Awan, A.R., and G.R. Smith. 2007b. The effect of group size on the activity of leopard (*Rana pipiens*) tadpoles. Jour-nal of Freshwater Ecology 22:355–357.

Awbrey, F.T. 1963. Homing and home range in *Bufo valliceps*. Texas Journal of Science 15:127–141.

Awbrey, F.T. 1965. An experimental investigation of the effectiveness of anuran mating calls as isolating mechanisms. Ph.D. Dissertation, University of Texas, Austin.

Awbrey, F.T. 1968. Call discrimination in female *Scaphiopus couchi* and *Scaphiopus hurteri*. Copeia 1968:420–423.

Awbrey, F.T. 1972. "Mating call" of a *Bufo boreas* male. Copeia 1972:579–581.

Awbrey, F.T. 1978. Social interaction among chorusing Pacific tree frogs, *Hyla regilla*. Copeia 1978:208–214.

Axtell, C.B. 1976. Comparisons of morphology, lactate dehydrogenase, and distribution of *Rana blairi* and *Rana utricularia* in Illinois and Missouri. Transactions of the Illinois State Academy of Science 69:37–48.

Axtell, R.W. 1958. Female reaction to the male call in two anurans (Amphibia). Southwestern Naturalist 3:70–76.

Axtell, R.W. 1963. A reinterpretation of the distribution of *Bufo w. woodhousei* Girard especially on the southeastern margin of its range. Herpetologica 19:115–122.

Axtell, R.W., and N. Haskell. 1977. An interhiatal population of *Pseudacris streckeri* from Illinois, with an assessment of its postglacial dispersion history. Natural History Miscellanea (202):1–8.

Azous, A.L., and K.O. Richter. 1995. Amphibian and plant community responses to changing hydrology in urban wetlands. Pp. 156–162 *In* Puget Sound Research '95 Proceedings. Volume 1. Puget Sound Water Quality Authority, Olympia, Washington.

Babbitt, K.J. 1995. *Bufo terrestris* (Southern Toad). Oophagy. Herpetological Review 26:30.

Babbitt, K.J. 2001. Behaviour and growth of southern leopard frog (*Rana sphenocephala*) tadpoles: effects of food and predation risk. Canadian Journal of Zoology 79:809–814.

Babbitt, K.J., and F. Jordan. 1996. Predation on *Bufo terrestris* tadpoles: effects of cover and predatory identity. Copeia 1996:485–488.

Babbitt, K.J., and G.W. Tanner. 1997. Effects of cover and predator identity on predation of *Hyla squirella* tadpoles. Journal of Herpetology 31:128–130.

Babbitt, K.J., and G.W. Tanner. 1998. Effects of cover and predator size on survival and development of *Rana utricularia* tadpoles. Oecologia 114:258–262.

Babbitt, K.J., and G.W. Tanner. 2000. Use of temporary wetlands by anurans in a hydrologically modified landscape. Wetlands 20:313–322.

Babbitt, K.J., and W.E. Meshaka. 2000. Benefits of eating conspecifics: effects of background diet on survival and metamorphosis in the Cuban treefrog (*Osteopilus septen-*

trionalis). Copeia 2000:469–474.

Babbitt, K.J., M.J. Baber, and T.L. Tarr. 2003. Patterns of larval amphibian distribution along a wetland hydroperiod gradient. Canadian Journal of Zoology 81:1539–1552.

Babbitt, K.J., M.J. Baber, and G.W. Tanner. 2005. The impact of agriculture on temporary wetland amphibians in Florida. Pp. 48–55 *In* W.E. Meshaka, Jr. and K.J. Babbitt (eds.), Amphibians and Reptiles. Status and Conservation in Florida. Krieger Publishing, Malabar, Florida.

Babbitt, K.J., M.J. Baber, and L.A. Brandt. 2006. The effect of woodland proximity and wetland characteristics on larval anuran assemblages in an agricultural landscape. Canadian Journal of Zoology 84:510–519.

Babbitt, L.H. 1937. The amphibia of Connecticut. Connecticut State Geological and Natural History Survey Bulletin No. 57.

Babcock, H.L. 1921. Does the cricket frog occur in New England? Copeia (90):8.

Babcock, H.L. 1926. A time-table of New England frogs and toads. Bulletin of the Boston Society of Natural History 38:11-14.

Babcock, H.L., and I. Hoopes. 1940. An almanac of frogs and toads. New England Naturalist 6:7-9.

Baber, M.J. 2001. Understanding anuran community structure in temporary wetlands: the interaction and importance of landscape and biotic processes. Ph.D. Dissertation, Florida International University, Miami.

Baber, M.J., and K.J. Babbitt. 2003. The relative impacts of native and introduced predatory fish on a temporary wetland tadpole assemblage. Oecologia 136:289–295.

Baber, M.J., and K.J. Babbitt. 2004. Influence of habitat complexity on predator–prey interactions between the fish (*Gambusia holbrooki*) and tadpoles of *Hyla squirella* and *Gastrophryne carolinensis*. Copeia 2004:173–177.

Baber, M.J., E. Fleishman, K.J. Babbitt, and T.L. Tarr. 2004. The relationship between wetland hydroperiod and nestedness patterns in assemblages of larval amphibians and predatory macroinvertebrates. Oikos 107:16–27.

Baber, M.J., K.J. Babbitt, F. Jordan, H.L. Jelks, and W.M. Kitchens. 2005. Relationships among habitat type, hydrology, predator composition, and distribution of larval anurans in the Florida Everglades. Pp. 154–160 *In* W.E. Meshaka, Jr. and K.J. Babbitt (eds.), Amphibians and Reptiles. Status and Conservation in Florida. Krieger Publishing, Malabar, Florida.

Bachmann, K. 1969. Temperature adaptations of amphibian embryos. American Naturalist 103:115–130.

Backlund, D. 1994. The wood frog. South Dakota Conservation Digest 61(1):20-21.

Backlund, D. 1996. Gray treefrogs. South Dakota Conservation Digest 63(5):16-17.

Backlund, D. 2002. The northern cricket frog. South Dakota Conservation Digest 69(1):26-27.

Backlund, D. 2009. Spring chorus. South Dakota Conservation Digest 76(2):28-29.

Backus, R.H. 1954. Notes on the frogs and toads of Labrador. Copeia 1954:226–227.

Bacon, E.J., and Z.M. Anderson. 1976. Distributional records of amphibians and reptiles from Coastal Plain of Arkansas. Proceedings of the Arkansas Academy of Science 30:14–15.

Bagdonas, K.R., and D. Pettus. 1976. Genetic compatibility in wood frogs (Amphibia, Anura, Ranidae). Journal of Herpetology 10:105–112.

Bagnara, J.T., and J. J. Kollros. 1956. Abbreviated larval period of *Rana catesbeiana* in Iowa. Proceedings of the Iowa Academy of Science 63:729–731.

Bagnara, J.T., and J.S. Frost. 1977. Leopard frog supply. Science 197:106–107. [reprinted in Herpetological Review 8:118–119]

Bahlman, A. 2010. Variation in six local populations of the western toad (*Anaxyrus boreas*) in the Great Basin: A common garden experiment. M.S. thesis, University of Nevada, Reno.

Bailey, M.A. 1989. Migration of *Rana areolata sevosa* and associated winter–breeding amphibians at a temporary pond in the lower coastal plain of Alabama. M.S. thesis, Auburn University, Auburn, Alabama.

Bailey, M.A. 1990. Movement of the dusky gopher frog (*Rana areolata sevosa*) at a temporary pond in the lower Coastal Plain of Alabama. Pp. 27–43 *In* C.K. Dodd, Jr., R.E. Ashton, Jr., R. Franz, and E. Wester (eds.), Burrow Associates of the Gopher Tortoise. Proceedings of the 8th Annual Meeting, Gopher Tortoise Council, Gainesville, Florida.

Bailey, M.A. 1991. The dusky gopher frog in Alabama. Journal of the Alabama Academy of Science 62:28–34.

Bailey, M.A., and D.B. Means. 2004. Gopher frog *Rana capito* Leconte and Mississippi gopher frog *Rana sevosa* Goin and Netting. Pp. 15–17 *In* R.E. Mirarchi, M.A. Bailey, T.M. Haggerty, and T.L. Best (eds.), Alabama Wildlife. Vol. 3. Imperiled Amphibians, Reptiles, Birds, and Mammals. University of Alabama Press, Tuscaloosa.

Bailey, M.A., J.N. Holmes, K.A. Buhlmann, and J.C. Mitchell. 2006. Habitat Management Guidelines for Amphibians and Reptiles of the Southeastern United States. Partners in Amphibian and Reptile Conservation, Technical Publication HMG–2, Montgomery, Alabama.

Bailey, R.M. 1943. Four species new to the Iowa herpetofauna, with notes on their natural histories. Proceedings of the Iowa Academy of Science 50:347–352.

Bailey, R. M. 1944a. Iowa's Frogs and Toads. Part 1. Iowa Conservationist 3(3):17–20.

Bailey, R. M. 1944b. Iowa's Frogs and Toads. Part 2. Iowa Conservationist 3(4):25, 27–30.

Bailey, R. M., and M. K. Bailey. 1941. The distribution of Iowa toads. Iowa State College Journal of Science 15:169–177.

Bailey, V. 1933. Cave life of Kentucky. American Midland Naturalist 14:385–635.

Baird, S.F. 1854. Descriptions of new genera and species of North American frogs. Proceedings of the Academy of Natural Sciences of Philadelphia 7:59–62.

Baird, S.F. 1859. Reptiles of the boundary, with notes by the naturalists of the survey. Pp. 1–35 *In* United States and Mexican Boundary Survey, under the order of Lieut. Col. W.H. Emory. U.S. Government Printing Office, Washington, D.C.

Baird, S.F., and C. Girard. 1852. Descriptions on new species of reptiles, collected by the U.S. Exploring Expedition under the command of Capt. Charles Wilkes, U.S.N. Part 1.including the species from the west coast of North America. Proceedings of the Academy of Natural Sciences of Philadelphia 6:174–177.

Baird, S.F., and C.H. Girard. 1853. Descriptions of new species of reptiles collected by the U.S. Exploring Expedition under the command of Capt. Charles Wilkes, U.S.N. 2nd part. Proceedings of the Academy of Natural Sciences of Philadelphia 6:378–379.

Baird, T. 1983. Influence of social and predatory stimuli on the air–breathing behavior of the African clawed frog, *Xenopus laevis*. Copeia 1983:411–420.

Baker, B.J., and J.M.L. Richardson. 2006. The effect of artificial light on male breeding–season behaviour in green frogs, *Rana clamitans melanota*. Canadian Journal of Zoology 84: 1528–1532.

Baker, M.R. 1977. Redescription of *Oswaldocruzia pipiens* (Nematoda: Trichostrongylidae) from amphibians of eastern North America. Canadian Journal of Zoology 55:104–109.

Baker, M.R. 1978a. Morphology and taxonomy of *Rhabdias* sp. (Nematoda: Rhabdiasidae) from reptiles and amphibians of southern Ontario. Canadian Journal of Zoology 56:2127–2141.

Baker, M.R. 1978b. Transmission of *Cosmocercoides dukae* (Nematoda: Cosmocercoidea) to amphibians. Journal of Parasitology 64:765–766.

Baker, M.R. 1978c. Development and transmission of *Oswaldocruzia pipiens* Walton,1929 (Nematoda: Trichostrongylidae) in amphibians. Canadian Journal of Zoology 56:1026–1031.

Baker, M.R. 1979a. The free–living and parasitic development of *Rhabdias* sp. (Nematoda: Rhabdiassidae) in amphibians. Canadian Journal of Zoology 57:161–178.

Baker, M.R. 1979b. Seasonal population changes in *Rhab-dias ranae* Walton, 1929 (Nematoda: Rhabdiasidae) in *Rana sylvatica* of Ontario. Canadian Journal of Zoology 57:179–183.

Baker, M.R. 1987. Synopsis of the Nematoda parasitic in amphibians and reptiles. Memorial University of Newfoundland Occasional Papers in Biology No. 11.

Baker, R.H. 1942. The bullfrog, a Texas wildlife resource. Texas Game, Fish and Oyster Commission, Bulletin No. 23.

Balas, C.J. 2008. The effects of conservation programs on amphibians of the Prairie Pothole Region's glaciated plain. M.S. thesis, Humboldt State University, Arcata, California.

Balas, C.J., N.H. Euliss, Jr., and D.M. Mushet. 2012. Influence of conservation programs on amphibians using seasonal wetlands in the Prairie Pothole Region. Wetlands 32:333–345.

Baldauf, R.J. 1943. Hand list of Berks County amphibians and reptiles. Leaflets of the Mengel Natural History Society (1):1–8.

Baldauf, R.J. 1987. Houston invaded by frogs…and few people know it. Explorer 29(1):4-6.

Baldwin, R.F., and P.G. deMaynadier. 2009. Assessing threats to pool–breeding amphibian habitat in an urbanizing landscape. Biological Conservation 142:1628–1638.

Baldwin, R.F., A.J.K. Calhoun, and P.G. deMaynadier. 2006a. Conservation planning for amphibian species with complex habitat requirements: a case study using movements and habitat selection of the wood frog *Rana sylvatica*. Journal of Herpetology 40:442–453.

Baldwin, R.F., A.J.K. Calhoun and P.G. deMaynadier. 2006b. The significance of hydroperiod and stand maturity for pool–breeding amphibians in forested landscapes. Canadian Journal of Zoology 84: 1604–1615.

Baldwin, R.F., P.G. deMaynadier, and A.J.K. Calhoun. 2007. *Rana sylvatica* (Wood Frog). Predation. Herpetological Review 38:194–195.

Balfour, P.S., and J. Ranlett. 2006. *Spea hammondii* (Western Spadefoot). Predation. Herpetological Review 37:212.

Ball, R.W., and D.L. Jameson. 1966. Premating isolating mechanisms in sympatric and allopatric *Hyla regilla* and *Hyla californiae*. Evolution 20:533–551.

Ball, R.W., and D.L. Jameson. 1970. Biosystematics of the canyon tree frog *Hyla cadaverina* Cope (= *Hyla californiae* Gorman). Proceedings of the California Academy of Sciences 37:363–380.

Ball, S.C. 1933. The spade-foot toads (*Scaphiopus holbrookii*) in Connecticut in 1933. Anatomical Record (Supplement) 57(4):101.

Ball, S.C. 1936. The distribution and behavior of the spade-foot toad in Connecticut. Transactions of the Connecticut Academy of Arts and Science 32:351–379.

Ballard, A.S., P.S. Balfour, and C.J. Conway. 2008. *Rana catesbeiana* (American Bullfrog). Predation. Herpetological Review 39:462–463.

Ballard, S.R. 1994. Status of the herpetofauna in the LaRue-Pine Hills/Otter Pond Research Natural Area in Union County, Illinois. M.S. thesis, Southern Illinois University, Carbondale.

Ballinger, R.E. 1966. Natural hybridization of the toads *Bufo woodhousei* and *Bufo speciosus*. Copeia 1966:366–368.

Ballinger, R.E., and C.O. McKinney. 1966. Developmental temperature tolerance of certain anuran species. Journal of Experimental Zoology 161:21–28.

Ballinger, R.E., J.D. Lynch and P.H. Cole. 1979. Distribution and natural history of amphibians and reptiles in western Nebraska with ecological notes on the herptiles of Arapaho Prairie. Prairie Naturalist 11:65–74.

Ballinger, R.E., J.W. Meeker, and M. Thies. 2000. A checklist and distribution maps of the amphibians and reptiles of South Dakota. Transactions of the Nebraska Academy of Sciences 26:29–46.

Ballinger, R.E., J.D. Lynch, and G.R. Smith. 2010. Amphibians and Reptiles of Nebraska. Rusty Lizard Press, Oro Valley, Arizona.

Banas, J.A., W.J. Loesche, and G.W. Nace. 1988. Classification and distribution of large intestinal bacteria in non–hibernating and hibernating leopard frogs (*Rana pipiens*). Applied and Environmental Microbiology 54:2305–2310.

Bancroft, B.A., N.J. Baker, and A.R. Blaustein. 2007. A meta–analysis of the effects of ultraviolet B radiation and its synergistic interactions with pH, contaminants, and disease on amphibian survival. Conservation Biology 22:987–996.

Bancroft, B.A., N.J. Baker, C.L. Searle, T.S. Garcia, and A.R. Blaustein. 2008. Larval amphibians seek warm temperatures and do not avoid harmful UVB radiation. Behavioral Ecology 19:879–886.

Bancroft, B.A., B.A. Han, C.L. Searle, L.M. Biga, D.H. Olson, L.B. Kats, J.J. Lawler and A.R. Blaustein. 2011. Species–level correlates of susceptibility to the pathogenic amphibian fungus *Batrachochytrium dendrobatidis* in the United States. Biodiversity and Conservation 20:1911–1920.

Bancroft, G.T., J.S. Godley, D.T. Gross, N.N. Rojas, D.A. Sutphen, and R.W. McDiarmid. 1983. The herpetofauna of Lake Conway: species accounts. U.S. Army Corps of Engineers Aquatic Plant Control Research Program, Miscellaneous Paper A–83–5.

Bang, D., and V. Mack. 1998. Enriching the environment of the laboratory bullfrog (*Rana catesbeiana*). Lab Animal 27:41–42.

Bank, M.S., J. Crocker, B. Connery, and A. Amirbahman. 2007. Mercury bioaccumulation in green frog (*Rana clamitans*) and bullfrog (*Rana catesbeiana*) tadpoles from Acadia National Park, Maine, USA. Environmental Toxicology and Chemistry 26:118–125.

Banta, A.M. 1914. Sex recognition and the mating behavior of the wood frog, *Rana sylvatica*. Biological Bulletin 26:171–183.

Banta, B.H. 1961. On the occurrence of *Hyla regilla* in the lower Colorado River, Clark County, Nevada. Herpetologica 17:106–108.

Banta, B.H. 1962. A preliminary account of the herpetofauna of the Saline Valley Hydrographic Basin, Inyo County, California. The Wasmann Journal of Biology 20:161–251.

Banta, B.H. 1973. A supernumerary forelimb on a Spring Peeper, *Hyla crucifer crucifer* Wied (Amphibia: Salientia) from south central Michigan. Herpeton, Journal of the Southwestern Herpetologists Society 7(1):6-7.

Banta, B.H. 1974. Death–feigning in *Hyla regilla* on Santa Cruz Island, Santa Barbara County, California. Bulletin of the Maryland Herpetological Society 10:88.

Banta, B.H., and D. Morafka. 1966. An annotated check list of the recent amphibians and reptiles inhabiting the city and county of San Francisco, California. The Wasmann Journal of Biology 24:223–236.

Banta, B.H., and G. Carl. 1967. Death–feigning behavior in the eastern gray treefrog *Hyla versicolor versicolor*. Herpetologica 23:317–318.

Barbeau, T.R., and H.B. Lillywhite. 2005. Body wiping behaviors associated with cutaneous lipids in hylid tree frogs of Florida. Journal of Experimental Biology 208:2147–2156.

Barber, M.A., and C.H. King. 1927. The tadpole of the spadefoot toad an enemy of mosquito larvae. Public Health Reports 42(52):3189-3193.

Barber, P.H. 1999a. Phylogeography of the canyon treefrog, *Hyla arenicolor* (Cope) based on mitochondrial DNA sequence data. Molecular Ecology 8:547–562.

Barber, P.H. 1999b. Patterns of gene flow and population genetic structure in the canyon treefrog, *Hyla arenicolor* (Cope). Molecular Ecology 8:563–576.

Barbour, R.W. 1953. The amphibians of Big Black Mountain, Harlan County, Kentucky. Copeia 1953:84–89.

Barbour, R.W. 1956a. *Pseudacris brachyphona* in Tennessee. Copeia 1956:54.

Barbour, R.W. 1956b. The Salientia of Kentucky: identification and distribution. Transactions of the Kentucky Academy of Science 17:81–86.

Barbour, R.W. 1957a. Observations on the mountain chorus frog, *Pseudacris brachyphona* (Cope) in Kentucky. American Midland Naturalist 57:125–128.

Barbour, R.W. 1957b. A distributional study of *Pseudacris nigrita* in Kentucky. American Midland Naturalist 57:129–132.

Barbour, R.W. 1971. Amphibians & Reptiles of Kentucky. University Press of Kentucky, Lexington.

Barbour, R.W., and E.P. Walters. 1941. Notes on the breeding habits of *Pseudacris brachyphona*. Copeia 1941:116.

Barbour, R.W., and W.L. Gault. 1952. Notes on the spadefoot toad *Scaphiopus h. holbrooki*. Copeia 1952:192.

Barbour, T. 1910. *Eleutherodacylus ricordii* in Florida. Proceedings of the Biological Society of Washington 23:100.

Barbour, T. 1931. Another introduced frog in North America. Copeia 1931:140.

Barnett, H.K., and J.S. Richardson. 2002. Predation risk and competition effects on the life– history characteristics of larval Oregon spotted frog and larval red–legged frog. Oecologia 132:436–444.

Barnum, B.B. 1953. Like to be a frog farmer? The Reclamation ERA 39(6):118-119.

Barr, T.C., Jr. 1953. Notes on the occurrence of ranid frogs in caves. Copeia 1953:60–61.

Barrantine, C.D. 1991a. Survival of billbugs (*Sphenophorus* spp.) ingested by Western toads (*Bufo boreas*). Herpetological Review 22:5.

Barrantine, C.D. 1991b. Food habits of Western toads (*Bufo boreas halophilus*) foraging from a residential lawn. Herpetological Review 22:84–87.

Barrass, A.N. 1985. The effects of highway traffic noise on the phonotactic and associated reproductive behavior of selected anurans. Ph.D. Dissertation, Vanderbilt University, Nashville, Tennessee.

Barrett, K., C. Guyer, and D. Watson. 2010. Water from urban streams slows growth and speeds metamorphosis in Fowler's toad (*Bufo fowleri*) larvae. Journal of Herpetology 44:297–300.

Barron, M.G., and K.B. Woodburn. 1995. Ecotoxicology of chlorpyrifos. Reviews of Environmental Contamination and Toxicology 144:1–93.

Barry, D.S., T.K. Pauley, and J.C. Maerz. 2008. Amphibian use of man–made pools on clear–cuts in the Allegheny Mountains of West Virginia, USA. Applied Herpetology 5:121–128.

Barta, J.R., and S.S. Desser. 1984. Blood parasites of amphibians from Algonquin Park, Ontario. Journal of Wildlife Diseases 20:180–189.

Bartareau, T.M. 2004. PVC pipe diameter influences the species and sizes of treefrogs captured in a Florida coastal scrub community. Herpetological Review 35:150–152.

Bartareau, T.M., and W.E. Meshaka. 2007. *Osteopilus septen-trionalis* (Cuban Treefrog). Diet. Herpetological Review 38:324–325.

Bartelt, P.E. 1998. *Bufo boreas* (Western Toad). Mortality. Herpetological Review 29:96.

Bartelt, P.E. 2000. A biophysical analysis of habitat selection in western toads (*Bufo boreas*) in southwestern Idaho. Ph.D. Dissertation, Idaho State University, Pocatello.

Bartelt, P.E., and C.R. Peterson. 2005. Physically modeling operative temperatures and evaporative rates in amphibians. Journal of Thermal Biology 30:93–102.

Bartelt, P.E., C.R. Peterson, and R.W. Klaver. 2004. Sexual differences in the post–breeding movements and habitats selected by Western toads (*Bufo boreas*) in southeastern Idaho. Herpetologica 60:455–467.

Bartelt, P.E., R.W. Klaver, and W.P. Porter. 2010. Modeling amphibian energetics, habitat suitability, and movements of western toads, *Anaxyrus (= Bufo) boreas*, across present and future landscapes. Ecological Modelling 221:2675–2686.

Bartelt, P.E., A.L. Gallant, R.W. Klaver, C.K. Wright, D.A. Patla, and C.R. Peterson. 2011. Predicting breeding habitat for amphibians: a spatiotemporal analysis across Yellowstone National Park. Ecological Applications 21:2530–2547.

Bartlett, R.D. 1989. On the trail of the Florida bog frog. Vivarium 1(3):8–11.

Bartlett, R.D., and P.P. Bartlett. 1999. A Field Guide to Florida Reptiles and Amphibians. Gulf Publishing Company, Houston, Texas.

Bartlett, R.D., and P.P. Bartlett. 2006. Guide and Reference to the Amphibians of Eastern and Central North America (North of Mexico). University Press of Florida, Gainesville.

Bartlett, R.D., and P.P. Bartlett. 2009. Guide and Reference to the Amphibians of Western North America (North of Mexico) and Hawaii. University Press of Florida, Gainesville.

Bartlett, R.D., and P.P. Bartlett. 2011. Florida's Frogs, Toads, and Other Amphibians. A Guide to their Identification and Habits. University Press of Florida, Gainesville.

Bartlett–Healy, K., W. Crans, and R. Gaugler. 2008. Phonotaxis to amphibian vocalizations in *Culex territans* (Diptera: Culicidae). Annals of the Entomological Society of America 101:95–103.

Bartram, W. 1791. Travels through North and South Carolina, Georgia, East and West Florida, the Cherokee Country, the Extensive Territories of the Muscoculges, or Creek Confederation, and the Country of the Chactaws. James and Johnson, Philadelphia.

Basey, H.E. 1969. Sierra Nevada Amphibians. Sequoia Natural History Association, Three Rivers, California.

Basey, H.E. 1976. Discovering Sierra Reptiles and Amphibians. Yosemite Association, Yosemite National Park, California. [3rd printing in 1991 with different cover]

Bastien, H., and R. Leclair, Jr. 1992. Aging wood frogs (*Rana sylvatica*) by skeletochronology. Journal of Herpetology 26:222–225.

Bateman, H., M.J. Harner, and A. Chung-MacCoubrey. 2008. Abundance and reproduction of toads (*Bufo*) along a regulated river in the southwestern United States: importance of flooding in riparian ecosystems. Journal of Arid Environments 72:1613-1619.

Bates, M.E., B.F. Cropp, M. Gonchar, J. Knowles, J.A. Simmons, and A. Megela Simmons. 2010. Spatial location influences vocal interactions in bullfrog choruses. Journal of the Acoustical Society of America 127:2664–2677.

Batista, C.G. 2002. *Rana catesbeiana* (Bullfrog). Effects on native anuran community. Herpetological Review 33:131.

Batts, B.S. 1960. Distribution of *Pseudacris nigrita nigrita* and *Pseudacris nigrita feriarum* in the Piedmont and Coastal Plain regions of North Carolina. Herpetologica 16:45–47.

Baud, D.R., and M.L. Beck. 2005. Interactive effects of UV–B and copper on spring peeper tadpoles (*Pseudacris crucifer*). Southeastern Naturalist 4:15–22.

Bauer Dial, C.A., and N.A. Dial. 1995. Lethal effects of the consumption of field levels of paraquat–contaminated plants on frog tadpoles. Bulletin of Environmental Contamination and Toxicology 55:870–877.

Baughman, B., and B.D. Todd. 2007. The role of substrate cues in habitat selection by juvenile *Bufo terrestris* and *Scaphiopus holbrookii*. Journal of Herpetology 41:154–157.

Baxley, D., and C. Qualls. 2009. Habitat associations of reptile and amphibian communities in longleaf pine habitats of south Mississippi. Herpetological Conservation and Biology 4:295–305.

Baxter, G.T. 1946. A study of the amphibians and reptiles of Wyoming. M.S. thesis, University of Wyoming, Laramie.

Baxter, G.T. 1947. The amphibians and reptiles of Wyoming. Wyoming Wildlife 11:30–34.

Baxter, G.T. 1952a. The relation of temperature to the altitudinal distribution of frogs and toads in southeastern Wyoming. Ph.D. Dissertation, University of Michigan. Ann Arbor.

Baxter, G.T. 1952b. Notes on growth and the reproductive cycle of the leopard frog, *Rana pipiens* Schreber, in southern Wyoming. Journal of the Colorado–Wyoming Academy of Science 4:91.

Baxter, G.T., and M.D. Stone. 1985. Amphibians and Reptiles of Wyoming. 2nd ed. Wyoming Game and Fish Department, Cheyenne. [first edition, 1980]

Baxter, G.T., M.R. Stromberg, and C.K. Dodd, Jr. 1982. The status of the Wyoming toad (*Bufo hemiophrys baxteri*). Environmental Conservation 9:348, 338. [pagination is correct]

Bayless, L.E. 1966. Comparative ecology of two sympatric species of *Acris* (Anura; Hylidae) with emphasis on interspecific competition. Ph.D. Dissertation, Tulane University, New Orleans, Louisiana.

Bayless, L.E. 1969a. Post–metamorphic growth of *Acris crepitans*. American Midland Naturalist 81:590–592.

Bayless, L.E. 1969b. Ecological divergence and distribution of sympatric *Acris* populations (Anura: Hylidae). Herpetologica 25:181–187.

Bayne, K.A. 2004. The natural history and morphology of the eastern cricket frog, *Acris crepitans crepitans*, in West Virginia. M.S. thesis, Marshall University, Huntington, West Virginia.

Bazazi, S., K. S. Pfennig , N. O. Handegard, and I. D. Couzin. 2012. Vortex formation and foraging in polyphenic spadefoot toad tadpoles. Behavioral Ecology and Sociobiology 66:879–889.

Beane, J.C. 1990. *Rana palustris* (Pickerel Frog). Predation. Herpetological Review 21:59.

Beane, J.C. 1998. Status of the river frog, *Rana heckscheri* (Anura: Ranidae), in North Carolina. Brimleyana 25:69–79.

Beane, J.C., and L.T. Pusser. 2005. *Bufo terrestris* (Southern Toad). Diet and scavenging. Herpetological Review 36:432.

Beane, J.C., and L.T. Pusser. 2006. *Rana heckscheri* (River Frog). Predation. Herpetological Review 37:339.

Beane, J.C., T.J. Thorp, and D.A. Jackan. 1998. *Heterodon simus* (Southern Hognose Snake). Diet. Herpetological Review 29:44–45.

Beane, J.C., A.L. Braswell, J.C. Mitchell, W.M. Palmer, and J.R. Harrison III. 2010. Amphibians & Reptiles of the Carolinas and Virginia. University of North Carolina Press, Chapel Hill.

Beard, K.H. 2007. Diet of the invasive frog, *Eleutherodacylus coqui*, in Hawaii. Copeia 2007:281–291.

Beard, K.H., and J. Baillie. 1998. *Rana catesbeiana* (Bullfrog). Diet. Herpetological Review 29:40.

Beard, K.H., and E.M. O'Neill. 2005. Infection of an invasive frog *Eleutherodacylus coqui* by the chytrid fungus *Batrachochytrium dendrobatidis* in Hawaii. Biological Conservation 126:591–595.

Beard, K.H., and W.C. Pitt. 2005. Potential consequences of the coqui frog invasion in Hawaii. Diversity and Distributions 11:427–433.

Beard, K.H., and W.C. Pitt. 2006. Potential predators of an

invasive frog (*Eleutherodactylus coqui*) in Hawaiian forests. Journal of Tropical Ecology 22:345–347.

Beard, K.H., R. Al–Chokhachy, N.C. Tuttle and E. M. O'Neill. 2008. Population density and growth rates of *Eleutherodactylus coqui* in Hawaii. Journal of Herpetology 42:626–636.

Beard, K.H., E. A. Price and W. C. Pitt. 2009. Biology and impacts of Pacific island invasive species. 5. *Eleutherodactylus coqui*, the Coqui Frog (Anura: Leptodactylidae). Pacific Science 63:297–316.

Beard, K.H., C.A. Olson and W.C. Pitt. 2012. Biology and impacts of Pacific island invasive species: 8. *Eleutherodactylus planirostris*, the Greenhouse Frog (Anura: Eleutherodactylidae). Pacific Science 66:255–270.

Bergstrom, T. M. 2010. A natural history study of *Bufo a. americanus*, the eastern American toad, and the phenology of spring rreeders in southwest West Virginia. M.S. thesis, Marshall University, Huntington, West Virginia.

Beasley, V.R., S.A. Faeh, B. Wikoff, C. Staehle, J. Eisold, D. Nichols, R. Cole, A.M. Schotthoefer, M. Greenwell, and L.E. Brown. 2005. Risk factors and declines in northern cricket frogs (*Acris crepitans*). Pp. 75–86 *In* M.J. Lannoo (ed.), Amphibian Declines. The Conservation Status of United States Species. University of California Press, Berkeley.

Beauclerc, K.B., B. Johnson, and B.N. White. 2010. Distinctiveness of declining northern populations of Blanchard's cricket frog (*Acris blanchardi*) justifies recovery efforts. Canadian Journal of Zoology 88:553–566.

Beaudry, F. 2007. *Rana sylvatica* (Wood Frog). Predation. Herpetological Review 38:195.

Beauregard, N., and R. Leclair, Jr. 1988. Multivariate analysis of the summer habitat structure of *Rana pipiens* Schreber, in Lac Saint Pierre (Québec, Canada). Pp. 129–143 *In* R.C. Szaro, K.E. Severson, and D.R. Patton (technical coordinators), Management of Amphibians, Reptiles, and Small Mammals in North America. Proceedings of a Symposium. USDA Forest Service, General Technical Report RM–166.

Bechtel, H.B. 1995. Reptile and Amphibian Variants. Colors, Patterns, and Scales. Krieger Publishing, Malabar, Florida.

Beck, C.G., and J.D. Congdon. 2003. Energetics of metamorphic climax in the southern toad (*Bufo terrestris*). Oecologia 137:344–351.

Beck, C.W. 1997. Effect of changes in resource level on age and size at metamorphosis in *Hyla squirella*. Oecologia 112:187–192.

Beck, C.W., and J.D. Congdon. 1999. Effects of individual variation in age and size at metamorphosis on growth and survivorship of southern toads (*Bufo terrestris*) metamorphs. Canadian Journal of Zoology 77:944–951.

Beck, C.W., and J.D. Congdon. 2000. Effects of age and size at metamorphosis on performance and metabolic rates of Southern toad, *Bufo terrestris*, metamorphs. Functional Ecology 14:32–38.

Beck, W.M., Jr. 1948. An ecological study of the cold–blooded vertebrates of a north Florida lake. M.S. thesis, University of Florida, Gainesville.

Becker, C.G., C.R. Fonseca, C.F.B. Haddad, R.F. Batista, and P.I. Prado. 2007. Habitat split and the global decline of amphibians. Science 318:1775–1777.

Becker, C.G., D. Rodriguez, A.V. Longo, A.L. Talaba, and K.R. Zamudio. 2012. Disease risk in temperate amphibian populations is higher at closed–canopy sites. PLoS One 7:e48205.

Becker, S.N. 2009. Hanging out with the cool frogs: do operative and body temperatures explain population response to disease? M.S. thesis, Southern Illinois University, Carbondale.

Bednarz, J.C., and J.J. Dinsmore. 1985. Flexible dietary response and feeding ecology of the red–shouldered hawk (*Buteo lineatus*) in Iowa. Canadian Field-Naturalist 99:262–264.

Bee, M.A. 2001. Habituation and sensitization of aggression in bullfrogs (*Rana catesbeiana*): testing the dual–process theory of habituation. Journal of Comparative Psychology 115:307–316.

Bee, M.A. 2002. Territorial male bullfrogs (*Rana catesbeiana*) do not assess fighting ability based on size–related variation in acoustic signals. Behavioral Ecology 13:109–124.

Bee, M.A. 2003. Experience–based plasticity of acoustically evoked aggression in a territorial frog. Journal of Comparative Physiology 189A:485–496.

Bee, M.A. 2004. Within–individual variation in bullfrog vocalizations: implications for an acoustically mediated social recognition system. Journal of the Acoustical Society of America 116: 3770–3781.

Bee, M.A. 2007a. Sound source segregation in grey treefrogs: spatial release from masking by the sound of a chorus. Animal Behaviour 74:549–558.

Bee, M.A. 2007b. Selective phonotaxis by male wood frogs (*Rana sylvatica*) to the sound of a chorus. Behavioral Ecology and Sociobiology 61:955–966.

Bee, M.A. 2008a. Parallel female preferences for call duration in a diploid ancestor of an allotetraploid treefrog. Animal Behaviour 76:845–853.

Bee, M.A. 2008b. Finding a mate at a cocktail party: spatial release from masking improves acoustic mate recognition in grey treefrogs. Animal Behaviour 75:1781–1791.

Bee, M.A. 2010. Spectral preferences and the role of spatial coherence in simultaneous integration in gray treefrogs (*Hyla chrysoscelis*). Journal of Comparative Psychology 124:412–424.

Bee, M.A., and S.A. Perrill. 1996. Responses to conspecific advertisement calls in the green frog (*Rana clamitans*) and their role in male–male communication. Behaviour 133:283–301.

Bee, M.A., and T.R. Schachtman. 2000. Is habituation a mechanism of neighbor recognition in green frogs? Behavioral Ecology and Sociobiology 48:165–168.

Bee, M.A., and H.C. Gerhardt. 2001a. Neighbour–stranger discrimination by territorial male bullfrogs (*Rana catesbeiana*). 1. Acoustic basis. Animal Behaviour 62:1129–1140.

Bee, M.A., and H.C. Gerhardt. 2001b. Neighbour–stranger discrimination by territorial male bullfrogs (*Rana catesbeiana*). 2. Perceptual basis. Animal Behaviour 62:1141–1150.

Bee, M.A., and H.C. Gerhardt. 2001c. Habituation as a mechanism of reduced aggression between adjacently territorial male bullfrogs (*Rana catesbeiana*). Journal of Comparative Psychology 115:68–82.

Bee, M.A., and A.C. Bowling. 2002. Socially–mediated pitch alteration by territorial male bullfrogs, *Rana catesbeiana*. Journal of Herpetology 36:140–143.

Bee, M.A., and H.C. Gerhardt. 2002. Individual voice recognition in a territorial frog (*Rana catesbeiana*). Proceedings of the Royal Society B 269:1443–1448.

Bee, M.A., and E.M. Swanson. 2007. Auditory masking of anuran advertisement calls by road traffic noise. Animal Behaviour 74:1765–1776.

Bee, M.A., and K.K. Riemersma. 2008. Does common spatial origin promote the auditory grouping of temporally separated signal elements in grey treefrogs? Animal Behaviour 76:831–843.

Bee, M.A., and J.J. Schwartz. 2009. Behavioral measures of signal recognition thresholds in frogs in presence and absence of chorus–shaped noise. Journal of the Acoustical Society of America 126:2788–2801.

Bee, M.A., S.A. Perrill, and P.C. Owen. 1999. Size assessment in simulated territorial encounters between male green frogs (*Rana clamitans*). Behavioral Ecology and Sociobiology 45:177–184.

Bee, M.A., S.A. Perrill, and P.C. Owen. 2000. Male green frogs lower the pitch of acoustic signals in defense of territories: a possible dishonest signal of size? Behavioral Ecology 11:169–177.

Bee, M.A., C.E. Kozich, K.J. Blackwell, and H.C. Gerhardt. 2001. Individually distinct advertisement calls of territorial male green frogs, *Rana clamitans*: implications for individual discrimination. Ethology 107: 65–84.

Bee, M.A., J.M. Cook, E.K. Love, L.R. O'Bryan, B.A. Pettitt, K. Schrode and A. Vélez. 2010. Assessing acoustic signal variability and the potential for sexual selection and social recognition in boreal chorus frogs (*Pseudacris maculata*). Ethology 116: 564–576.

Bee, M.A., A. Vélez, and J.D. Forester. 2012. Sound level discrimination by gray treefrogs in the presence and absence of chorus–shaped noise. Journal of the Acoustical Society of America 131:4188–4195.

Behle, W.H., and R.J. Erwin. 1962. The green frog (*Rana clamitans*) established at West Ogden, Weber County, Utah. Proceedings of the Utah Academy of Sciences, Arts, and Letters 39:74–76.

Behler, J.L., and F.W. King. 1979. The Audubon Society Field Guide to North American Reptiles and Amphibians. Alfred A. Knopf, New York.

Beiswenger, R.E. 1972. Aggregative behavior of tadpoles of the American toad, *Bufo americanus*, in Michigan. Ph.D. Dissertation, University of Michigan, Ann Arbor.

Beiswenger, R.E. 1975. Structure and function in aggregations of tadpoles of the American toad, *Bufo americanus*. Herpetologica 31:222–233.

Beiswenger, R.E. 1977. Diel patterns of aggregative behavior in tadpoles of *Bufo americanus*, in relation to light and temperature. Ecology 58:98–108.

Beiswenger, R.E. 1978. Responses of *Bufo* tadpoles (Amphibia, Anura, Bufonidae) to laboratory gradients of temperature. Journal of Herpetology 12:499–504.

Beiswenger, R.E. 1981. Predation by grey jays on aggregating tadpoles of the boreal toad (*Bufo boreas*). Copeia 1981:274–276.

Beiswenger, R.E. 1986. An endangered species, the Wyoming toad *Bufo hemiophrys baxteri*—the importance of an early warning system. Biological Conservation 37:59–71.

Belden, J.B., S.T. McMurry, L.M. Smith, and P. Reilley. 2010. Acute toxicity of fungicide formulations to amphibians at environmental relevant concentrations. Environmental Toxicology and Chemistry 29:2477–2480.

Belden, L.K. 2006. Impact of eutrophication on wood frog, *Rana sylvatica*, tadpoles infected with *Echinostoma trivolvis* cercariae. Canadian Journal of Zoology 84:1315–1321.

Belden, L.K., and A.R. Blaustein. 2002. Exposure of red–legged frog embryos to ambient UV–B radiation in the field negatively affects larval growth and development. Oecologia 130:551–554.

Belden, L.K., and J.M. Wojdak. 2011. The combined influence of trematode parasites and predatory salamanders on wood frog (*Rana sylvatica*) tadpoles. Oecologia 166:1077–1086.

Belden, L.K., E.L. Wildy, A.C. Hatch, and A.R. Blaustein. 2000. Juvenile western toads, *Bufo boreas*, avoid chemical cues of snakes fed juvenile, but not larval, conspecifics. Animal Behaviour 59:871–875.

Belden, L.K., I.T. Moore, R.T. Mason, J.C. Wingfield, and

A.R. Blaustein. 2003. Survival, the hormonal stress response and UV–B avoidance in Cascades frog tadpoles (*Rana cascadae*) exposed to UV–B radiation. Functional Ecology 17:409–416.

Belden, L.K., I.T. Moore, J.C. Wingfield, and A.R. Blaustein. 2005. Corticosterone and growth in Pacific treefrog (*Hyla regilla*) tadpoles. Copeia 2005:424–430.

Bellis, E.D. 1953. The effects of temperature on the breeding calls of some Oklahoma salientians. M.S. thesis, University of Oklahoma, Norman.

Bellis, E.D. 1957a. An ecological study of the wood frog *Rana sylvatica* LeConte. Ph.D. Dissertation, University of Minnesota, Minneapolis.

Bellis, E.D. 1957b. The effects of temperature on salientian breeding calls. Copeia 1957:85–89.

Bellis, E.D. 1959. A study of movement of American toads in a Minnesota bog. Copeia 1959:173–174.

Bellis, E.D. 1961. Growth of the wood frog, *Rana sylvatica*. Copeia 1961:74–77.

Bellis, E.D. 1962a. The influence of humidity on wood frog activity. American Midland Naturalist 68:139–148.

Bellis, E.D. 1962b. Cover value and escape habits of the wood frog in a Minnesota bog. Herpetologica 17:228–231.

Bellis, E.D. 1965. Home range and movements of the wood frog in a northern bog. Ecology 46:89–98.

Bellocq, M.I., K. Kloosterman, and S.M. Smith. 2000. The diet of coexisting species of amphibians in Canadian jack pine forests. Herpetological Journal 10:63-68.

Beltz, E. 2007. Scientific and common names of the reptiles and amphibians of North America explained. http://ebeltz.net/herps/etymain.html.

Benard, M.F. 2007. Predators and mates: conflicting selection on the size of male Pacific treefrogs (*Pseudacris regilla*). Journal of Herpetology 41:317–320.

Benard, M.F., and J.M. Maher. 2011. Consequences on intraspecific niche variation: phenotypic similarity increases competition among recently metamorphosed frogs. Oecologia 166:585–592.

Bennett, D.W. 2008. Amphibian responses to forest management practices in southwestern Georgia. M.S. thesis, University of Florida, Gainesville.

Bennett, L. 2003. The miracle of the toads. Alberta Naturalist 33:72–73.

Bennett, R.S., E.E. Klaas, J.R. Coats, M.A. Mayse, and E.J. Kolbe. 1983. Fenvalerate residues in nontarget organisms from treated cotton fields. Bulletin of Environmental Contamination and Toxicology 31:61–65.

Bennett, S.H., J.W. Gibbons, and J. Glanville. 1980. Terrestrial activity, abundance and diversity of amphibians in differently managed forest types. American Midland Naturalist 103:412–416.

Benson, A.J. 2000. Documenting over a century of aquatic introductions in the United States. Pp. 1–31 *In* R. Claudi and J.H. Leach (eds.), Nonindigenous Freshwater Organisms. Vectors, Biology, and Impacts. Lewis Publishers, Boca Raton, Florida.

Bergeron, C.M., C.M. Bodinof, J.M. Unrine, and W.A. Hopkins. 2010a. Mercury accumulation along a contamination gradient and nondestructive indices of exposure in amphibians. Environmental Toxicology and Chemistry 29: 980–988.

Bergeron, C.M., C.M. Bodinof, J.M. Unrine, and W.A. Hopkins. 2010b. Bioaccumulation and maternal transfer of mercury and selenium in amphibians. Environmental Toxicology and Chemistry 29: 989–997.

Bergeron, C.M., W.A. Hopkins, B.D. Todd, M.J. Hepner, and J.M. Unrine. 2011a. Interactive effects of maternal and dietary mercury exposure have latent and lethal consequences for amphibian larvae. Environmental Science & Technology 45:3781–3787.

Bergeron, C. M., W.A. Hopkins, C.M. Bodinof, S.A. Budischak, H. Wada, and J.M. Unrine. 2011b. Counterbalancing effects of maternal mercury exposure during different stages of early ontogeny. Science of the Total Environment 409:4746–4752.

Bergman, C.M. 1999. Range extension of spring peepers, *Pseudacris crucifer*, in Labrador. Canadian Field-Naturalist 113:309-310.

Beringer, J., and T.R. Johnson. 1995. *Rana catesbeiana* (Bullfrog). Diet. Herpetological Review 26:98.

Bernal, X., and S.R. Ron. 2004. *Leptodactylus fragilis* (White–lipped Foamfrog). Courtship. Herpetological Review 35:372–373.

Berns, M.W. 1966. Some genetic and histochemical aspects of the variant blue frog in the genus *Rana*. M.S. thesis, Cornell University, Ithaca, New York.

Berns, M.W., and L.D. Uhler. 1966. Blue frogs of the genus *Rana*. Herpetologica 22:181–183.

Berns, M.W., and K.S. Narayan. 1970. An histochemical and ultrastructural analysis of the dermal chromatophores of the variant ranid blue frog. Journal of Morphology 132:169–180.

Berrill, M., S. Bertram, A. Wilson, S. Louis, and D. Brigham. 1993. Lethal and sublethal impacts of pyrethroid insecticides on amphibian embryos and tadpoles. Environmental Toxicology and Chemistry 12:525–539.

Berrill, M., S. Bertram, L. McGillivray, M. Kolohon, and B. Pauli. 1994. Effects of low concentrations of forest use pesticides on frog embryos and tadpoles. Environmental Toxicology and Chemistry 13:657–664.

Berrill, M., S. Bertram, L. McGillivray, M. Kolohon, and D. Ostrander. 1995. Comparative sensitivity of amphibian tadpoles to single and pulsed exposures of the forest–use insecticide fenitrothion. Environmental Toxicology and Chemistry 14:1011–1018.

Berrill, M., S. Bertram, and B. Pauli. 1997. Effects of pesticides on amphibian embryos and larvae. SSAR Herpetological Conservation 1:233–245.

Berrill, M., D. Coulson, L. McGillivary, and B. Pauli. 1998. Toxicity of endosulfan to aquatic stages of anuran amphibians. Environmental Toxicology and Chemistry 17:1738–1744.

Bertram, S., and M. Berrill. 1997. Fluctuations in a northern population of gray treefrogs, *Hyla versicolor*. SSAR Herpetological Conservation 1:57–63.

Bertram, S., M.Berrill, and E. Nol. 1996. Male mating success and variation in chorus attendance within and among breeding seasons in the gray treefrog (*Hyla versicolor*). Copeia 1996:729–734.

Berven, K.A. 1981. Mate choice in the wood–frog, *Rana sylvatica*. Evolution 35:707–722.

Berven, K.A. 1982a. The genetic basis of altitudinal variation in the wood frog, *Rana sylvatica*. 1. An experimental analysis of life history traits. Evolution 36:962–983.

Berven, K.A. 1982b. The genetic basis of altitudinal variation in the wood frog, *Rana sylvatica*. 2. An experimental analysis of larval development. Oecologia 52:360–369.

Berven, K.A. 1988. Factors affecting variation in reproductive traits within a population of wood frogs (*Rana sylvatica*). Copeia 1988:605–615.

Berven, K.A. 1990. Factors affecting population fluctuations in larval and adult stages of the wood frog (*Rana sylvatica*). Ecology 71:1599–1608.

Berven, K.A. 1995. Population regulation in the wood frog, *Rana sylvatica*, from three diverse geographic locations. Australian Journal of Ecology 20:385–392.

Berven, K.A. 2009. Density dependence in the terrestrial stage of wood frogs: evidence from a 21–year population study. Copeia 2009:328–338.

Berven, K.A., and B.G. Chadra. 1988. The relationship among egg size, density and food level on larval development in the wood frog (*Rana sylvatica*). Oecologia 75:67–72.

Berven, K.A., and T.A. Grudzien. 1990. Dispersal in the wood frog (*Rana sylvatica*): implications for genetic population structure. Evolution 44:2047–2056.

Berven, K.A., and R.S. Boltz. 2001. Interactive effects of leech (*Desserobdella picta*) infection on wood frog (*Rana sylvatica*) tadpole fitness traits. Copeia 2001:907–915.

Berven, K.A., D.E. Gill, and S.J. Smith–Gill. 1979. Countergradient selection in the green frog, *Rana clamitans*. Evolution 33:609–623.

Bettaso, J., A. Haggerty, and E. Russell. 2008. *Rana boylii* (Foothill Yellow–legged Frog). Necrogamy. Herpetolog-ical Review 39:462.

Bettaso, J., J. Carlson, B. Fahey and M. Thomas. 2011. *Rana boylii* (Foothill Yellow–legged Frog) and *Anaxyrus boreas* (Western Toad). Interspecific amplexus. Herpetological Review 42:589.

Beverley, R. 1705. The History of Virginia, in Four Parts. R. Parker, London. [frogs in Chapter XIX of Part 4]

Bevier, C.R., K. Larson, K. Reilly, and S. Tat. 2004. Vocal repertoire and calling activity of the mink frog, *Rana septentrionalis*. Amphibia–Reptilia 25:255–264.

Bevier, C.R., D.C. Tierney, L.E. Henderson, and H.E. Reid. 2006. Chorus attendance and site fidelity in the mink frog, *Rana septentrionalis*: are males territorial? Journal of Herpetology 40:160–164.

Beyer, G.E. 1900a. Louisiana herpetology. Proceedings of the Louisiana Society of Naturalists for 1897–99. Appendix 1, pp. 25–33.

Beyer, G.E. 1900b. A check–list of the batrachians and reptiles of Louisiana. Proceedings of the Louisiana Society of Naturalists for 1897–99. Appendix 1, pp. 33–46.

Bezy, K.B., R.L. Bezy, K. Bolles, and E.F. Enderson. 2004. Breeding behavior of the western chorus frog (*Pseudacris triseriata* complex) in Arizona: do chorus frogs call in the snow on the Colorado Plateau? Sonoran Herpetologist 17:82–85.

Bezy, R.L., W.C. Sherbrooke, and C.H. Lowe. 1966. Rediscovery of *Eleutherodactylus augusti* in Arizona. Herpetologica 22:221–225.

Bider, J.R., and K.A. Morrison. 1981. Changes in toad (*Bufo americanus*) responses to abiotic factors at the northern limit of their distribution. American Midland Naturalist 106:293–304.

Bider, J.R., and S. Matte. 1996. The Atlas of Amphibians and Reptiles of Québec. St. Lawrence Valley Natural History Society and the Ministère de l'Environnment et de la Faune Direction de la faune et des habitats Québec, Sainte–Anne–de–Bellevue, Québec.

Biek, R., L. S. Mills and R. B. Bury. 2002. Terrestrial and stream amphibians across clearcut-forest interfaces in the Siskiyou Mountains, Oregon. Northwest Science 76:129–140.

Biesterfeldt, J.M., J.W. Petranka, and S. Sherbondy. 1993. Prevalence of chemical interference competition in natural populations of wood frogs, *Rana sylvatica*. Copeia 1993:688–695.

Biklé, A. 1985. Amphibians of the Landels–Hill Big Creek Reserve. Environmental Field Program, University of California–Santa Cruz, Special Paper No. 4.

Billings, J. T. 1973. A partial definition of the fundamental niche of larval *Rana pipiens* (Ranidae, Amphibia). M.S. thesis, University of Nebraska-Lincoln.

Bimber, D.L., and R.A. Mitchell. 1978. Effects of diquat on amphibian embryo development. Ohio Journal of Science 78:50–51.

Binckley, C.A., and W.J. Resetarits, Jr. 2002. Reproductive divisions under threat of predation: squirrel treefrog (*Hyla squirella*) responses to banded sunfish (*Enneacanthus obesus*). Oecologia 130:157–161.

Binckley, C.A., and W.J. Resetarits. 2003. Functional equivalence of non–lethal effects: generalized fish avoidance determines distribution of gray treefrog, *Hyla chrysoscelis*, larvae. Oikos 102:623–629.

Binckley, C. A., and W.J. Resetarits, Jr. 2007. Effects of forest canopy on habitat selection in treefrogs and aquatic insects: implications for communities and metacommunities. Oecologia 153:951–958.

Binckley, C. A., and W.J. Resetarits, Jr. 2008. Oviposition behavior partitions aquatic landscapes along predation and nutrient gradients. Behavioral Ecology 19:552–557.

Birchfield, G.L. 2002. Green frog (*Rana clamitans*) movement behavior and terrestrial habitat use in fragmented landscapes in central Missouri. Ph.D. Dissertation, University of Missouri, Columbia.

Birchfield, G.L., and J.E. Deters. 2005. Movement paths of displaced northern green frogs (*Rana clamitans melanota*). Southeastern Naturalist 4:63–76.

Birdsall, C.W., C.E. Grue, and A. Anderson. 1986. Lead concentrations in bullfrog *Rana catesbeiana* and green frog *R. clamitans* tadpoles inhabiting highway drainages. Environmental Pollution (Series A) 40:233–247.

Birge, W.J. 1978. Aquatic toxicology of trace elements of coal and fly ash. Pp. 219–240 *In* J.H. Thorp and G.W. Gibbons (eds.), Energy and Environmental Stress in Aquatic Systems, U.S. Department of Energy Symposium Series No. 48.

Birge, W.J., and J.J. Just. 1973. Sensitivity of vertebrate embryos to heavy metals as a criterion of water quality. Water Resources Research Institute, University of Kentucky, Research Report 61.

Birge, W.J., J.A, Black, and A.G. Westerman. 1979. Evaluation of aquatic pollutants using fish and amphibian eggs as bioassay organisms. Pp. 108–118 *In* Animals as Monitors of Environmental Pollutants. National Academy of Sciences, Washington.

Birge, W.J., J.A. Black, and R.A. Kuehne. 1980. Effects of organic compounds on amphibian reproduction. Water Resources Research Institute, University of Kentucky, Research Report 121.

Birge, W.J., J.A. Black, A.G. Westerman, and B.A. Ramey. 1983. Fish and amphibian embryos: a model system for evaluating tetratogenicity. Fundamental and Applied Toxicology 3:237–242.

Birge, W.J., A.G. Westerman, and J.A. Spromberg. 2000. Comparative toxicology and risk assessment of amphibians. Pp. 727–791 *In* D.W. Sparling, G. Linder, and C.A. Bishop (eds.), Ecotoxicology of Amphibians and Reptiles. SETAC Press, Pensacola, Florida.

Birkhead, R.D., G.R. Mullen, and W.C. Welbourn. 2007. *Rana sphenocephala utricularia*. (Southern Leopard Frog). Ectoparasites. Herpetological Review 38:194.

Birx-Raybuck, D.A., S.J. Price, and M.E. Dorcas. 2010. Pond age and riparian zone proximity influence anuran occupancy of urban retention ponds. Urban Ecosystems 13:181-190.

Bishop, C.A., K.E. Pettit, M.E. Gartshore, and D.A. MacLeod. 1997. Extensive monitoring of anuran populations using call counts and road transects in Ontario (1992 to 1993). SSAR Herpetological Conservation 1:149–160.

Bishop, C.A., N.A. Mahony, J. Struger, P. Ng, and K.E. Pettitt. 1999. Anuran development, density and diversity in relation to agricultural activity in the Holland River watershed, Ontario, Canada (1990–1992). Environmental Monitoring and Assessment 57:21–43.

Bishop, D.C. 2003. *Rana okaloosae* (Florida Bog Frog). Predation. Herpetological Review 34:235.

Bishop, D.C. 2005. Ecology and distribution of the Florida bog frog and flatwoods salamander on Eglin Air Force Base. Ph.D. Dissertation, Virginia Polytechnic Institute and State University, Blacksburg.

Bishop, D.C., C.A. Haas, and L.O. Mahoney. 2012. Responses of *Lithobates okaloosae*, *L. clamitans* and *L. sphenocephalus* tadpoles to chemical cues of snake and fish predators. Florida Scientist 75:1–10.

Bishop, L.A., and T.M. Farrell. 1994. *Thamnophis sauritus sackenii* (Peninsula Ribbon Snake). Behavior. Herpetological Review 25:127.

Bishop, S.C. 1927. The amphibians and reptiles of Allegany State Park. New York State Museum Handbook No. 3.

Bisrat, S.A. 2010. Modeling bark beetle outbreak and fire interactions in western U.S. forests and the invasion potential of an invasive Puerto Rican frog in Hawaii using remote sensing data. Ph.D. Dissertation, Utah State University, Logan.

Bisrat, S.A., M.A. White, K.H. Beard, and D.R. Cutler. 2012. Predicting the distribution potential of an invasive frog using remotely sensed data in Hawaii. Diversity and Distributions 18:648–660.

Black, I.H., and K.L. Gosner. 1958. The barking tree frog, *Hyla gratiosa* in New Jersey. Herpetologica 13:254–255.

Black, J.D. 1938. Additional records of *Rana sylvatica* in Arkansas. Copeia 1938:48–49.

Black, J.D., and S.C. Dellinger. 1938. Herpetology of Arkansas. Part 2. The Amphibians. Occasional Papers of the University of Arkansas Museum No. 2:1–30.

Black, J.H. 1967. A blue leopard frog from Montana. Herpetologica 23:314–315.

Black, J.H. 1968. A possible stimulus for the formation of some aggregations in tadpoles of *Scaphipus bombifrons*. Proceedings of the Oklahoma Academy of Science 49:13–14.

Black, J.H. 1969. Ethoecology of tadpoles of *Bufo americanus charlesmithi* in a temporary pool in Oklahoma. Journal of Herpetology 3:198.

Black, J.H. 1970. Amphibians of Montana. Montana Fish and Game Department, Animals of Montana Series No. 1.

Black, J.H. 1971. The toad genus *Bufo* in Montana. Northwest Science 45:156–162.

Black, J.H. 1973a. A checklist of the cave fauna of Oklahoma: Amphibia. Proceedings of the Oklahoma Academy of Science 53:33–37.

Black, J.H. 1973b. Ethoecology of *Scaphiopus* (Pelobatidae) larvae in temporary pools in central and southwestern Oklahoma. Ph.D. Dissertation, University of Oklahoma, Norman.

Black, J.H. 1974a. Larval spadefoot survival. Journal of Herpetology 8:371–373.

Black, J.H. 1974b. Bullfrog eating a bird. Herpetological Review 5:105.

Black, J.H. 1975. The formation of "tadpole nests" by anuran larvae. Herpetologica 31:76–79.

Black, J.H. 1976. Oklahoma leopard frogs. Bulletin of the Oklahoma Herpetological Society 1:6–10.

Black, J.H., and J.N. Black. 1968. Frog with a tail. Montana Outdoors 3(3):not paginated (3 pp.).

Black, J.H., and J.N. Black. 1969. Postmetamorphic basking aggregations of the boreal toad, *Bufo boreas boreas*. Canadian Field-Naturalist 83:155–156.

Black, J.H., and R.B. Brunson. 1971. Breeding behavior of the boreal toad, *Bufo boreas boreas* (Baird and Girard), in western Montana. Great Basin Naturalist 31:109–113.

Black, J.H., and G. Sievert. 1989. A Field Guide to the Amphibians of Oklahoma. Oklahoma Department of Wildlife Conservation, Oklahoma City.

Black, J.H., T. Hunkapiller, and D. Dawson. 1976. Winter activity in Oklahoma frogs. Bulletin of the Oklahoma Herpetological Society 1:22.

Blackburn, L.M., C.D. Hoefler, and P. Nanjappa. 2002. *Acris crepitans blanchardi* (Blanchard's Cricket Frog). Predation. Herpetological Review 33:299.

Blair, A.P. 1941a. Variation, isolation mechanisms and hybridization in certain toads. Genetics 26:398–417.

Blair, A.P. 1941b. Isolating mechanisms in tree frogs. Proceedings of the National Academy of Sciences 27:14–17.

Blair, A.P. 1942. Isolating mechanisms in a complex of four species of toads. Biological Symposia 6:235–249.

Blair, A.P. 1943a. Geographical variation of ventral markings in toads. American Midland Naturalist 29:615–620.

Blair, A.P. 1943b. Population structure in toads. American Naturalist 77:563–568.

Blair, A.P. 1946. Description of a six–year–old hybrid toad. American Museum Novitates 1327:1–3.

Blair, A.P. 1947a. Variation in two characteristics in *Bufo fowleri* and *Bufo americanus*. American Museum Novitates 1343:1–5.

Blair, A.P. 1947b. The male warning vibration in *Bufo*. American Museum Novitates 1344:1–7.

Blair, A.P. 1947c. Defensive use of parotoid secretion by *Bufo marinus*. Copeia 1947:137.

Blair, A.P. 1950a. Notes on Oklahoma microhylid frogs. Copeia 1950:152.

Blair, A.P. 1950b. Skittering locomotion in *Acris crepitans*. Copeia 1950:237.

Blair, A.P. 1951a. Winter activity in Oklahoma frogs. Copeia 1951:178.

Blair, A.P. 1951b. Note on the herpetology of the Elk Mountains, Colorado. Copeia 1951:239–240.

Blair, A.P. 1955. Distribution, variation, and hybridization in a relict toad (*Bufo microscaphus*) in southwestern Utah. American Museum Novitates 1722:1–38.

Blair, A.P. 1963. Notes on anuran behavior especially *Rana catesbeiana*. Herpetologica 19:151.

Blair, A.P. 1967. Bullfrog attempts to catch dove. Southwestern Naturalist 12:201–202.

Blair, A.P., and H.L. Lindsay, Jr. 1961. *Hyla avivoca* (Hylidae) in Oklahoma. Southwestern Naturalist 6:202.

Blair, J., and R.J. Wassersug. 2000. Variation in the pattern of predator–induced damage to tadpole tails. Copeia 2000:390–401.

Blair, W.F. 1936. A note on the ecology of *Microhyla olivacea*. Copeia 1936:115.

Blair, W.F. 1949. Development of the solitary spadefoot toad in Texas. Copeia 1949:72.

Blair, W.F. 1951. Interbreeding of natural populations of vertebrates. American Naturalist 85:9–30.

Blair, W.F. 1953. Growth, dispersal and age at sexual maturity of the Mexican toad (*Bufo valliceps* Wiegmann). Copeia 1953:208–212.

Blair, W.F. 1955a. Size difference as a possible isolation mechanism in *Microhyla*. American Naturalist 89:297–302.

Blair, W.F. 1955b. Mating call and stage of speciation in the *Microhyla olivacea–M. carolinensis* complex. Evolution 9:469–480.

Blair, W.F. 1955c. Differentiation of mating call in spadefoots, genus *Scaphiopus*. Texas Journal of Science 7:183–188.

Blair, W.F. 1956a. The mating calls of hybrid toads. Texas Journal of Science 8:350–355.

Blair, W.F. 1956b. Call difference as an isolation mechanism in Southwestern toads (genus *Bufo*). Texas Journal of Science 8:87–106.

Blair, W.F. 1956c. Comparative survival of hybrid toads (*Bufo woodhousei* x *B. valliceps*) in nature. Copeia 1956:259–260.

Blair, W.F. 1956d. Mating call and possible stage of speciation in the Great Basin spadefoot. Texas Journal of Science 8:236–238.

Blair, W.F. 1957a. Mating call and relationships of *Bufo hemiophrys* Cope. Texas Journal of Science 9:99–108.

Blair, W.F. 1957b. Structure of the call and relationships of *Bufo microscaphus* Cope. Copeia 1957:208–212.

Blair, W.F. 1958a. Call difference as an isolation mechanism in Florida species of hylid frogs. Quarterly Journal of the Florida Academy of Sciences 21:32–48.

Blair, W.F. 1958b. Mating call and stage of speciation of two allopatric populations of spadefoots (*Scaphiopus*). Texas Journal of Science 10:484–488.

Blair, W.F. 1958c. Call structure and species groups in U.S. treefrogs (*Hyla*). Southwestern Naturalist 3:77–89.

Blair, W.F. 1958d. Response of a green treefrog (*Hyla cinerea*) to the call of the male. Copeia 1958:333–334.

Blair, W.F. 1959. Genetic compatibility and species groups in U.S. toads (*Bufo*). Texas Journal of Science 11:427–453.

Blair, W.F. 1960a. A breeding population of the Mexican toad (*Bufo valliceps*) in relation to its environment. Ecology 41:165–174.

Blair, W.F. 1960b. Mating call as evidence of relations in the *Hyla eximia* group. Southwestern Naturalist 5:129–135.

Blair, W.F. 1961a. Further evidence bearing on intergroup and intragroup genetic compatibility in toads (genus *Bufo*). Texas Journal of Science 13:163–175.

Blair, W.F. 1961b. Calling and spawning seasons in a mixed population of anurans. Ecology 42:99–110.

Blair, W.F. 1963a. Intragroup genetic compatibility in the *Bufo americanus* species group of toads. Texas Journal of Science 15:15–34.

Blair, W.F. 1963b. Evolutionary relationships of North American toads of the genus *Bufo*: a progress report. Evolution 17:1–16.

Blair, W.F. 1964. Evidence bearing on the relationships of the *Bufo boreas* group of toads. Texas Journal of Science 16:181–192.

Blair, W.F. 1966. Genetic compatability in the *Bufo valliceps* and closely related groups of toads. Texas Journal of Sci-

ence 18:333–351.

Blair, W.F. 1971. Structure of the call and relationships of *Bufo microscaphus* Cope. Copeia 1957:208–212.

Blair, W.F. 1972. Evidence from hybridization. Pp. 196–232 *In* W.F. Blair (ed.), Evolution in the Genus *Bufo*. University of Texas Press, Austin.

Blair, W.F. 1974. Character displacement in frogs. American Zoologist 14:1119–1125.

Blair, W.F., and D. Pettus. 1954. The mating call and its significance in the Colorado River toad (*Bufo alvarius* Girard). Texas Journal of Science 6:72–77.

Blair, W.F., and M.J. Littlejohn. 1960. Stage of speciation of two allopatric populations of chorus frogs (*Pseudacris*). Evolution 14:82–87.

Blair, W.R. 1964. Evidence bearing on the relationships of the *Bufo boreas* group of toads. Texas Journal of Science 16:181–192.

Blanchard, F. N. 1922. The amphibians and reptiles of western Tennessee. Occasional Papers of the Museum of Zoology, University of Michigan 117:1 – 18.

Blanchard, F.N. 1925. A collection of amphibians and reptiles from southern Indiana and adjacent Kentucky. Papers of the Michigan Academy of Science, Arts and Letters 5:367–388.

Blanchard, F.N. 1933. Late autumn collections and hibernating situations of the salamander *Hemidactylium scutatum* (Schlegel) in southern Michigan. Copeia 1933:216.

Blaney, R.M. 1971. An annotated check list and biogeographic analysis of the insular herpetofauna of the Apalachicola region, Florida. Herpetologica 27:406–430.

Blasus, R. E. 1997. Amphibian & reptile time table for Minnesota: a guide to the seasonal activity of Minnesota's amphibians & reptiles. Minnesota Herpetological Society Occasional Paper (4):1-30.

Blatchley, W.S. 1892. Notes on the batrachians and reptiles of Vigo County, Indiana. Journal of the Indiana Society of Natural History 14:27.

Blatchley, W.S. 1899. Notes on the batrachians and reptiles of Vigo County, Indiana. II. Annual Report of the Indiana Department of Geology and Natural Resources 24:537–552.

Blaustein, A.R., 1988. Ecological correlates and potential functions of kin recognition and kin association in anuran larvae. Behavior Genetics 18:449–464.

Blaustein, A.R. 1994. Amphibians in a bad light. Natural History 103(10):32–37.

Blaustein, A.R., and R.K. O'Hara. 1982a. Kin recognition in *Rana cascadae* tadpoles: maternal and paternal effects. Animal Behaviour 30:1151–1157.

Blaustein, A.R., and R.K. O'Hara. 1982b. Kin recognition cues in *Rana cascadae* tadpoles. Behavioral and Neural

Biology 36:77–87.

Blaustein, A.R., and R.K. O'Hara. 1986. An investigation of kin recognition in red–legged frogs (*Rana aurora*) tadpoles. Journal of Zoology 209:347–353.

Blaustein, A.R., and L.K. Belden. 1987. Aggregation behaviour in *Rana cascadae* tadpoles: association preferences among wild aggregations and responses to non–kin. Animal Behaviour 35:1549–1555.

Blaustein, A.R., and B. Waldman. 1992. Kin recognition in anuran amphibians. Animal Behaviour 44:207–221.

Blaustein, A.R., and L.K. Belden. 2003. Amphibian defenses against ultraviolet–B radiation. Evolution & Development 5:89–97.

Blaustein, A.R., and P.T.J. Johnson. 2003a. The complexity of deformed amphibians. Frontiers in Ecology and the Environment 1:87–94.

Blaustein, A.R., and P.T.J. Johnson. 2003b. Explaining frog deformities. Scientific American 288:60–65.

Blaustein, A.R., and P.T.J. Johnson. 2010. When an infection turns lethal. Nature 465: 881–882.

Blaustein, A.R., and L.B. Kats. 2003. Amphibians in a very bad light. Bioscience 53: 1028–1029.

Blaustein, A.R., R.K. O'Hara, and D.H. Olson. 1984. Kin preference behaviour is present after metamorphosis in *Rana cascadae* frogs. Animal Behaviour 32:445–450.

Blaustein, A.R., K.S. Chang, H.G. Lefcourt, and R.K. O'Hara. 1990. Toad tadpole kin recognition: recognition of half siblings and the role of maternal cues. Ethology Ecology & Evolution 2:215–226.

Blaustein, A.R., D.G. Hokit, R.K. O'Hara, and R.A. Holt. 1994a. Pathogenic fungus contributes to amphibian losses in the Pacific Northwest. Biological Conservation 67:251–254.

Blaustein, A.R., P.D. Hoffman, D.G. Hokit, J.M. Kiesecker, S.C. Walls, and J.B. Hays. 1994b. UV repair and resistance to solar UV–B in amphibian eggs: a link to population declines. Proceedings of the National Academy of Sciences of the United States of America 91:1791–1795.

Blaustein, A.R., D.B. Wake, and W.P. Sousa. 1994c. Amphibian declines: judging stability, persistence, and susceptibility of populations to local and global extinctions. Conservation Biology 8:60-71.

Blaustein, A. R., J. J. Beatty, D. H. Olson, and R. M. Storm. 1995a. The Biology of Amphibians and Reptiles in Old-Growth Forests in the Pacific Northwest. United States Department of Agriculture, Forest Service, Pacific Northwest Research Station General Technical Report PNW-GTR-337.

Blaustein, A.R., J.M. Kiesecker, D.G. Hokit, and S.C. Walls. 1995b. Amphibian declines and UV radiation. Bioscience 45:514–515

Blaustein, A.R., P.D. Hoffman, J.M. Kiesecker, and J.B. Hayes. 1996a. DNA repair activity and resistance to solar UV–B radiation in eggs of the red–legged frog. Conservation Biology 10:1398–1402.

Blaustein, A.R., J.M. Kiesecker, S.C. Walls, and D.G. Hokit. 1996b. Field experiments, amphibian mortality, and UV radiation. Bioscience 46: 386–388.

Blaustein, A.R., J.M. Kiesecker, D.P. Chivers, and R.G. Anthony. 1997. Ambient UV–B radiation causes deformities in amphibian embryos. Proceedings of the National Academy of Sciences of the United States of America 94:13735–13737.

Blaustein, A.R., J.M. Kiescecker, D.P. Chivers, D.G. Hokit, A. Marco, L.K. Belden, and A. Hatch. 1998. Effects of ultraviolet radiation on amphibians: field experiments. American Zoologist 38:799–812.

Blaustein, A.R., J.B. Hays, P.D. Hoffman, D.P. Chivers, J.M. Kiesecker, W.P. Leonard, A. Marco, D.H. Olson, J.K. Reaser, and R.G. Anthony. 1999. DNA repair and resistance to UV–B radiation in western spotted frogs. Ecological Applications 9:1100–1105.

Blaustein, A.R., J.M. Romansic, J.M. Kiesecker, and A.C. Hatch. 2003. Ultraviolet radiation, toxic chemicals and amphibian population declines. Diversity and Distributions 9:123–140.

Blaustein, A.R., B. Han, B. Fasy, J. Romansic, E.A. Scheessele, R.G. Anthony, A. Marco, D.P. Chivers, L.K. Belden, J.M. Kiesecker, T. Garcia, M. Lizana, and L.B. Kats. 2004. Variable breeding phenology affects the exposure of amphibian embryos to ultraviolet radiation and optical characteristics of natural waters protect amphibians from UV–B in the U.S. Pacific Northwest: comment. Ecology 85:1747–1754.

Blaustein, A.R., J.M. Romansic, and E.A. Scheessele. 2005a. Ambient levels of ultraviolet–B radiation cause mortality in juvenile western toads, *Bufo boreas*. American Midland Naturalist 154:375–382.

Blaustein, A.R., J.M. Romansic, E.A. Scheessele, B.A. Han, A.P. Pessier, and J.E. Longcore. 2005b. Interspecific variation in susceptibility of frog tadpoles to the pathogenic fungus *Batrachochytrium dendrobatidis*. Conservation Biology 19:1460-1468.

Blaustein, A.R., B.A. Han, R.A. Relyea, P.T.J. Johnson, J.C. Buck, S.S. Gervasi, and L.B. Kats. 2011. The complexity of amphibian population declines: understanding the role of cofactors in driving amphibian losses. Annals of the New York Academy of Sciences 1223:108–119.

Blaustein, A.R., S.S. Gervasi, P.T.J. Johnson, J.T. Hoverman, L.K. Belden, P.W. Bradley, and G.Y. Xie. 2012. Ecophysiology meets conservation: understanding the role of disease in amphibian population declines. Philosophical Transactions of the Royal Society B 367:1688–1707.

Bleakney, S. 1952. The amphibians and reptiles of Nova Scotia. Canadian Field-Naturalist 66:125–129.

Bleakney, S. 1954. Range extensions of amphibians in eastern Canada. Canadian Field-Naturalist 68:165–171.

Bleakney, J.S. 1958a. A zoogeographical study of the amphibians and reptiles of eastern Canada. National Museum of Canada Bulletin 155:1–119.

Bleakney, S. 1958b. Cannibalism in *Rana sylvatica* tadpoles. Herpetologica 14:34.

Bleakney, S. 1959. Postglacial dispersal of the western chorus frog in eastern Canada. Canadian Field-Naturalist 73:197–205.

Bleakney, S. 1963. First North American record of *Bufolucilia silvarum* (Meigen) (Diptera: Calliphoridae) parasitizing *Bufo terrestris americanus* Holbrook. Canadian Entomologist 95:107.

Blem, C.R., J.W. Steiner, and M.A. Miller. 1978. Comparison of jumping abilities of the cricket frogs *Acris gryllus* and *Acris crepitans*. Herpetologica 34:288–291.

Blem, C.R., C.A. Ragan, and L.S. Scott. 1986. The thermal physiology of two sympatric treefrogs *Hyla cinerea* and *Hyla chrysoscelis* (Anura: Hylidae). Comparative Biochemistry and Physiology A 85:563–570.

Blihovde, W.B. 2000a. *Rana capito aesopus* (Florida Gopher Frog). Ectoparasites. Herpetological Review 31:101.

Blihovde, W.B. 2000b. Terrestrial behavior of the Florida gopher frog (*Rana capito aesopus*). M.S. thesis, University of Central Florida, Orlando.

Blihovde, W.B. 2006. Terrestrial movements and upland habitat use of gopher frogs in central Florida. Southeastern Naturalist 5:265–276.

Blomquist, S.M. 2005. *Bufo retiformis* Sanders and Smith, 1951. Sonoran Green Toad. Pp. 433– 435 *In* M.J. Lannoo (ed.), Amphibian Declines. The Conservation Status of United States Species. University of California Press, Berkeley.

Blomquist, S.M., and J.C. Tull. 2002. *Rana luteiventris* (Columbia Spotted Frog). Burrow use. Herpetological Review 33:131.

Blomquist, S.M., and M.J. Hunter, Jr. 2007. Externally attached radio–transmitters have limited effects of the anti–predator behavior and vagility of *Rana pipiens* and *Rana sylvatica*. Journal of Herpetology 41:430–438.

Blomquist, S.M., and M.L. Hunter, Jr. 2009. A multi–scale assessment of habitat selection and movement patterns by northern leopard frogs (*Lithobates* [*Rana*] *pipiens*) in a managed forest. Herpetological Conservation and Biology 4:142–160.

Blosser, E. M., and L.P. Lonnibos. 2012. Florida's frog–biting mosquitoes. Wing Beats (Spring):5–8, 11–12. [University of Florida Entomology Department]

Blouin, M.S. 1989a. Life history correlates of a color polymorphism in the ornate chorus frog, *Pseudacris ornata*. Copeia 1989:319–325.

Blouin, M.S. 1989b. Inheritance of a naturally occurring color polymorphism in the ornate chorus frog, *Pseudacris ornata*. Copeia 1989:1056–1059.

Blouin, M.S. 1990. Evolution of palatability differences between closely–related treefrogs. Journal of Herpetology 24:309–311.

Blouin, M.S. 1991. Proximate development causes of limb length variation between *Hyla cinerea* and *Hyla gratiosa* (Anura: Hylidae). Journal of Morphology 209:305–310.

Blouin, M.S. 1992a. Comparing bivariate reaction norms among species: time and size at metamorphosis in three species of *Hyla* (Anura: Hylidae). Oecologia 90:288–293.

Blouin, M.S. 1992b. Genetic correlations among morphometric traits and rates of growth and differentiation in the green tree frog, *Hyla cinerea*. Evolution 46:735–744.

Blouin, M.S., and M.L.G. Loeb. 1991. Effects of environmentally induced development rate variation on head and limb morphology in the green tree frog, *Hyla cinerea*. American Naturalist 138:717–728.

Blouin, M.S., and S.T. Brown. 2000. Effects of temperature–induced variation in larval growth rate on head width and leg length at metamorphosis. Oecologia 125:358–361.

Blouin, M.S., I. C. Phillipsen, and K. J. Monsen. 2010. Population structure and conservation genetics of the Oregon spotted frog, *Rana pretiosa*. Conservation Genetics 11:2179–2194.

Bly, B.L. 2004. Population dynamics of three amphibian species across the Sheyenne National Grassland of southeastern South Dakota. M.S. thesis, University of North Dakota, Grand Forks.

Boatright–Horowitz, S.S., C.A. Cheney, and A.M. Simmons. 1999. Atmospheric and underwater propagation of bullfrog vocalizations. Bioacoustics 9:257–280.

Boatright–Horowitz, S. L., S. S. Horowitz, and A. M. Simmons. 2000. Patterns of vocal interactions in a bullfrog (*Rana catesbeiana*) chorus: preferential responding to far neighbors. Ethology 106: 701–712.

Bodnar, D.A. 1996. The separate and combined effects of harmonic structure, phase, and FM on female preferences in the barking treefrog (*Hyla gratiosa*). Journal of Comparative Physiology A 178:173–182.

Boenke, M. 2011. Terrestrial habitat and ecology of Fowler's toads, *Anaxyrus fowleri*. M.S. thesis, McGill University, Montreal, Québec.

Boenke, M., and D.M. Green. 2012. *Anaxyrus fowleri* (Fowler's Toad). Habitat. Herpetological Review 43:461.

Bogart, J.P., and C.E. Nelson. 1976. Evolutionary implications from karyotypic analysis of frogs of the families Microhylidae and Rhinophrynidae. Herpetologica 32:199–208.

Bogart, J.P., and A.P. Jaslow. 1979. Distribution and call parameters of *Hyla chrysoscelis* and *Hyla versicolor* in Michigan. Royal Ontario Museum Life Sciences Contribution No. 117.

Bogert, C.M. 1930. An annotated list of the amphibians and reptiles of Los Angeles County, California. Bulletin of the Southern California Academy of Sciences 29:3–14.

Bogert, C.M. 1933. Notes on Grand Canyon amphibians. Grand Canyon Nature Notes. [reprinted in S. Lamb (ed). 1994. The Best of Grand Canyon Nature Notes, 1926–1935. Grand Canyon Natural History Association]

Bogert, C.M. 1947. Results of the Archbold Expeditions. No. 57. A field study of homing in the Carolina toad. American Museum Novitates 1355:1–24.

Bogert, C.M. 1958. Sounds of North American frogs. The biological significance of voice in frogs. Folkway Records and Service Corporation, New York. 16 pp.

Bogert, C.M. 1960. The influence of sound on the behavior of amphibians and reptiles. Pp. 37–320 *In* W.E. Lanyon and W.N. Tavolga (eds.), Animal Sounds and Communication, American Institute of Biological Sciences Publication No. 7, Washington, D.C.

Bogert, C.M. 1962. Isolation mechanisms in toads of the *Bufo debilis* group in Arizona and western Mexico. American Museum Novitates 2100:1–37.

Bogosian, V., III, E.C. Hellgren and R.W. Moody. 2012. Assemblages of amphibians, reptiles, and mammals on an urban military base in Oklahoma. Southwestern Naturalist 57:277–284.

Bohnsack, K.K. 1951. Temperature data on the terrestrial hibernation of the greenfrog, *Rana clamitans*. Copeia 1951:236–239.

Bohnsack, K.K. 1952. Terrestrial hibernation of the bullfrog, *Rana catesbeiana* Shaw. Copeia 1952:114.

Boice, R., and D.W. Witter. 1969. Hierarchical feeding behaviour in the leopard frog (*Rana pipiens*). Animal Behaviour 17:474–479.

Boice, R., and C. Boice. 1970. Interspecific competition in captive *Bufo marinus* and *Bufo americanus* toads. Journal of Biological Psychology 12:32–36.

Boice, R., and R.C. Williams. 1971. Delay in onset of tonic immobility in *Rana pipiens*. Copeia 1971:747–748.

Boiteau, G., and P. C. McCarthy. 2010. Is there a role for stripes of adults and colour of larvae in determining the avoidance of the Colorado potato beetle by the American toad? Canadian Journal of Zoology 88: 468–478.

Boivin, E. 2012. *Batrachochytrium dendrobatidis* in the Adirondacks, New York, USA. Herpetological Review 43:610.

Bolek, M.G. 2006. The role of arthropod second intermediate hosts as avenues for and constraints on the transmission of frog lung flukes (Digenea: Haematoloechidae). Ph.D. Dissertation, University of Nebraska, Lincoln.

Bolek, M.G., and J.R. Coggins. 1998. Endoparasites of Cope's gray treefrog, *Hyla chrysoscelis*, and western chorus frog, *Pseudacris t. triseriata*, from southeastern Wisconsin. Journal of the Helminthological Society of Washington 65:212–218.

Bolek, M.G., and J.R. Coggins. 2000. Seasonal occurrence and community structure of helminth parasites from the eastern American toad, *Bufo americanus americanus*, from southeastern Wisconsin, U.S.A. Comparative Parasitology 67:202–209.

Bolek, M.G., and J.R. Coggins. 2001. Seasonal occurrence and community structure of helminth parasites in green frogs, *Rana clamintans melanota*, from southeastern Wisconsin, U.S.A. Comparative Parasitology 68:164–172.

Bolek, M.G., and J.R. Coggins. 2002. Observations on myiasis by the calliphorid, *Bufolucilia silvarum*, in the eastern American toad (*Bufo americanus americanus*) from southeastern Wisconsin. Journal of Wildlife Diseases 38:598–603.

Bolek, M.G., and J.R. Coggins. 2003. Helminth community structure of sympatric eastern American toad, *Bufo americanus americanus*, northern leopard frog, *Rana pipiens*, and blue–spotted salamander, *Ambystoma laterale*, from southeastern Wisconsin. Journal of Parasitology 89:673–680.

Bolek, M.G., and J. Janovy, Jr. 2004a. Observations on myiasis by the calliphorids, *Bufolucilia silvarum* and *Bufolucilia elongata*, in wood frogs, *Rana sylvatica*, from southeastern Wisconsin. Journal of Parasitology 90:1169–1171.

Bolek, M.G., and J. Janovy, Jr. 2004b. *Rana catesbeiana* (Bullfrog). Gigantic tadpole. Herpetological Review 35:376–377.

Bolek, M.G., and J. Janovy, Jr. 2004c. *Rana blairi* (Plains Leopard Frog). Prey. Herpetological Review 35:262.

Bolek, M.G., and J. Janovy, Jr. 2005. New host and distribution records for the amphibian leech *Desserobdella picta* (Rhynchobdellida: Glossiphoniidae) from Nebraska and Wisconsin. Journal of Freshwater Ecology 20:187–189.

Bolek, M.G., and J. Janovy, Jr. 2007. *Rana catesbeiana* (American Bullfrog). Diet. Herpetological Review 38:325–326.

Bolek, M.G., and J. Janovy, Jr. 2008. Alternative life cycle strategies of *Megalodiscus temperatus* in tadpoles and metamorphosed anurans. Parasite, Journal de la Societe Française de Parasitologie 15:396–401.

Bolek, M.G., K.K. Brotan, R.E. Rudolph, and J. Janovy, Jr. 2007. *Bufo woodhousii* (Woodhouse's Toad). Cannibalism. Herpetological Review 38:319.

Bolek, M.G., S. D. Snyder, and J. Janovy, Jr. 2009a. Alternative life cycle strategies and colonization of young anurans by *Gorgoderina attenuata* in Nebraska. Journal of Parasitology 95:604–616.

Bolek, M.G., S. D. Snyder, and J. Janovy, Jr. 2009b. Redescription of the frog bladder fluke *Gorgoderina attenuata* from the northern leopard frog, *Rana pipiens*. Journal of Parasitology 95:665–668.

Bollinger, R. R. 1966. A study of the biology of the trypanosomes *of Rana clamitans clamitans*. Ph.D. Dissertation, Tulane University, New Orleans, Louisiana.

Bollinger, R. R., J. R. Seed, and A. A. Gam. 1968. Studies on frog trypanosomiasis II. Seasonal variations in the parasitemia levels of *Trypanosoma rotatorium* in *Rana clamitans* from Louisiana. Tulane Studies in Zoology and Botany 15: 64–69.

Bonin, J., J.–L. DesGranges, J. Rodrigue, and M. Ouellet. 1997a. Anuran species richness in agricultural landscapes of Québec: foreseeing long–term results of road call surveys. SSAR Herpetological Conservation 1:141–149.

Bonin, J., M. Ouellet, J. Rodrigue, J.–L. DesGranges, F. Gagné, T.F. Sharbel, and L.A. Lowcock. 1997b. Measuring the health of frogs in agricultural habitats subjected to pesticides. SSAR Herpetological Conservation 1:246–257.

Bonine, K.E., G.H. Dayton, and R.E. Jung. 2001. Attempted predation of Couch's spadefoot (*Scaphiopus couchii*) juveniles by ants (*Aphaenogaster cockerelli*). Southwestern Naturalist 46:104–106.

Bonnaterre, M. l'A. 1789. Tableau Encyclopédique et Méthodique des Trois Règnes de la Nature. Erpétologie. Chez Panckoucke, Hôtel de Thou, Paris.

Boone, M.D. 2005. Juvenile frogs compensate for small metamorph size with terrestrial growth: overcoming the effects of larval density and insecticide exposure. Journal of Herpetology 39:416–423.

Boone, M.D. 2008. Examining the single and interactive effects of three insecticides on amphibian metamorphosis. Environmental Toxicology and Chemistry 27:1561–1568.

Boone, M.D., and C.M. Bridges. 1999. The effect of temperature on the potency of carbaryl for survival of tadpoles of the green frog (*Rana clamitans*). Environmental Toxicology and Chemistry 18:1482–1484.

Boone, M.D., and C.M. Bridges. 2003. Effects of carbaryl on green frog (*Rana clamitans*) tadpoles: Timing of exposure versus multiple exposures. Environmental Toxicology and Chemistry 22:2695–2702.

Boone, M.D., and R.D. Semlitsch. 2001. Interactions of an insecticide with larval density and predation in experimental amphibian communities. Conservation Biology 15:228–238.

Boone, M.D., and R.D. Semlitsch. 2002. Interactions of an insecticide with competition and pond drying in amphibian communities. Ecological Applications 12:307–316.

Boone, M.D., and R.D. Semlitsch. 2003. Interactions of bullfrog tadpole predators and an insecticide: predation release and facilitation. Oecologia 422:610–616.

Boone, M.D., and S.M. James. 2003. Interactions of an insecticide, herbicide, and natural stressors in amphibian community mesocosms. Ecological Applications 13:829–841.

Boone, M.D., and N.H. Sullivan. 2012. The impact of an insecticide changes with amount of leaf litter input: implications for amphibian populations. Environmental Toxicology and Chemistry 31:1518–1524.

Boone, M.D., C.M. Bridges, and B.B. Rothermel. 2001. Growth and development of larval green frogs (*Rana clamitans*) exposed to multiple doses of an insecticide. Oecologia 129:518–524.

Boone, M.D., E.E. Little, and R.D. Semlitsch. 2004a. Overwintered bullfrog tadpoles negatively affect salamanders and anurans in native amphibian communities. Copeia 2004:683–690.

Boone, M.D., R.D. Semlitsch, J.F. Fairchild, and B.B. Rothermel. 2004b. Effects of an insecticide on amphibians in large–scale experimental ponds. Ecological Applications 14:685–691.

Boone, M.D., R.D. Semlitsch, E.E. Little, and M.C. Doyle. 2007. Multiple stressors in amphibian communities: interactive effects of chemical contamination, bullfrog tadpoles, and bluegill sunfish. Ecological Applications 17:291–301

Boone, M.D., R.D. Semlitsch, and C. Mosby. 2008. Suitability of golf course ponds for amphibian metamorphosis when bullfrogs are removed. Conservation Biology 22:172–179.

Bos, D.H., and J.W. Sites. 2001. Phylogeography and conservation genetics of the Columbia spotted frog (*Rana luteiventris*; Amphibia, Ranidae). Molecular Ecology 10: 1499–1513.

Bosakowski, T. 1999. Amphibian macrohabitat associations on a private industrial forest in western Washington. Northwestern Naturalist 80:61–69.

Bosc, L.A.G. 1800. *In* F.M. Daudin, Histoire Naturelle des Quadrupèdes Ovipaires. Livraison 1:10, plate 5, Marchant et Cie, Paris.

Bosc, L.A.G., and F.M. Daudin. 1801. *Laraine ocularie, Hyla ocularis. In* C.S. Sonnini and P.A. Latreille, Histoire Naturelle des Reptiles, avec Figures dissinées d'après Nature. Vol. 2. Deterville, Paris.

Boschulte, D.S. 1993. Toxicity of six commonly used herbicides on larval bullfrogs (*Rana catesbeiana*). M.S. thesis, Illinois State University, Normal.

[1]Bossert, M., M. Draud, and T. Draud. 2003a. *Bufo fowleri* (Fowler's Toad) and *Malaclemys terrapin terrapin* (Northern Diamondback Terrapin). Refugia and nesting. Herpetological Review 34:49.

[1]Bossert, M., M. Draud, and T. Draud. 2003b. *Bufo fowleri* (Fowler's Toad) and *Malaclemys terrapin terrapin* (Northern Diamondback Terrapin). Refugia and nesting. Herpetological Review 34:135.

Botch, P.S., M.C. Beers, and T.M. Judd. 2007. The effects of calcium on the feeding preference of the tadpole of *Bufo americanus*. Bulletin of the Maryland Herpetological Society 43:162–166.

Bouchard, J.L. 1951. The platyhelminthes parasitizing some northern Maine amphibia. Transactions of the American Microscopical Society 70:245–250.

Bouchard, J.L. 1953. An ecological and taxonomic study of helminth parasites from Oklahoma amphibians. Ph.D. Dissertation, University of Oklahoma, Norman.

Bouchard, J., A.T. Ford, F.E. Eigenbrod, and L. Fahrig. 2009. Behavioral responses of northern leopard frogs (*Rana pipiens*) to roads and traffic: implications for population persistence. Ecology and Society 14(2):23.

Boughton, R.G. 1997. The use of PVC refugia as a trapping technique for hylid treefrogs. M.S. thesis, University of Florida, Gainesville.

Boughton, R.G., J. Staiger, and R. Franz. 2000. Use of PVC pipe refugia as a sampling technique for hylid treefrogs. American Midland Naturalist 144:168–177.

Boulenger, G.A. 1882. Descriptions of a new genus and species of frogs of the family Hylidae. The Annals and Magazine of Natural History, Series 5, 10:326–328.

Boulenger, G.A. 1899. On the American spade–foot (*Scaphiopus solitarius* Holbrook). Proceedings of the Zoological Society of London 1899(3):790–793.

Boulenger, G.A. 1917. Descriptions of new frogs of the genus *Rana*. The Annals and Magazine of Natural History 20:413–418.

Boundy, J., and T.G. Balgooyen. 1988. Record lengths of some amphibians and reptiles from the western United States. Herpetological Review 19:26–27.

Bourque, R.M. 2008. Spatial ecology of an inland population of the foothill yellow–legged frog (*Rana boylii*) in Tehama County, California. M.A. thesis, Humboldt State University, Arcata, California.

Bourque, R.M., and J.B. Bettaso. 2011. *Rana boylii* (Foothill Yellow–legged Frog). Reproduction. Herpetological Review 42:589.

Bovbjerg, R.V. 1965. Experimental studies on the dispersal of the frog, *Rana pipiens*. Proceedings of the Iowa Academy of Science 72:412–418.

Bovbjerg, R.V., and A.M. Bovbjerg. 1964. Summer emigrations of the frog *Rana pipiens* in northwestern Iowa. Proceedings of the Iowa Academy of Science 71:511–518.

Bowen, K.D., and E.A. Beever. 2010. Daytime amphibian surveys in three protected areas in the western Great Lakes. IRCF Reptiles & Amphibians 17:26–31, 34–35.

Bowerman, J. 2010. Submerged calling by Oregon spotted frogs (*Rana pretiosa*) remote from breeding aggregations. IRCF Reptiles & Amphibians 17:84–87.

Bowerman, J., and P.T.J. Johnson. 2003. Timing of trematode–related malformations in Oregon spotted frogs and Pacific treefrogs. Northwestern Naturalist 84:142–145.

Bowerman, J, P.T.J. Johnson, and T. Bowerman. 2010. Sublethal predators and their injured prey: linking aquatic predators and severe limb abnormalities in amphibians. Ecology 91:242–251.

Bowers, D.G., D.E. Andersen, and N.H. Euliss, Jr. 1998. Anurans as indicators of wetland condition in the Prairie Pothole Region of North Dakota: an environmental monitoring and assessment program pilot project. Pp. 369–378 *In* M.J. Lannoo (ed.), Status & Distribution of Midwestern Amphibians. University of Iowa Press, Iowa City.

Bowker, R.W., and B.K. Sullivan. 1991. *Bufo punctatus* x *B. retiformis* (Red–spotted Toad, Sonoran Green Toad). Natural hybridization. Herpetological Review 22:54.

Bowler, J.K. 1977. Longevity of reptiles and amphibians in North American collections. SSAR Herpetological Circular No. 6.

Boyd, C.E. 1963. Waif dispersal in toads. Herpetologica 18:269.

Boyd, C.E. 1964. The distribution of cricket frogs in Mississippi. Herpetologica 20:201–202.

Boyd, C.E., S.V. Vinson, and D.E. Ferguson. 1963. Possible DDT resistance in two species of frogs. Copeia 1963:426–429.

Boyd, C.E., and D.H. Vickers. 1963. Distribution of some Mississippi amphibians and reptiles. Herpetologica 19:202–205.

Boyd, S.H. 1975. Inhibition of fish reproduction by *Rana catesbeiana* larvae. Physiological Zoology 48:225–234.

Boykin, K.G. 2006. Multiscale analysis of habitat, vegetation change, and streamflow as ecological factors affecting population dynamics of *Rana chiricahuensis*. Ph.D. Dissertation, New Mexico State University, Las Cruces.

Boykin, K.G., and K.C. McDaniel. 2008. Simulated potential effects of ecological factors on a hypothetical population of Chiricahua leopard frog (*Rana chiricahuensis*). Ecological Modelling 18:175–181.

Brach, V. 1992. Discovery of the Rio Grande chirping frog in Smith County, Texas (Anura: Leptodactylidae). Texas Journal of Science 44:490.

Brach, V. 1998. The jewel of the reeds. Reptiles Magazine 6(3):10–13.

Bradford, D.F. 1982. Oxygen relations and water balance during hibernation, and temperature regulation during summer, in a high–elevation amphibian (*Rana muscosa*). Ph.D. Dissertation, University of California, Los Angeles.

Bradford, D.F. 1983. Winterkill, oxygen regulations, and energy metabolism of a submerged dormant amphibian, *Rana muscosa*. Ecology 64:1171–1183.

Bradford, D.F. 1984. Temperature modulation in a high–elevation amphibian, *Rana muscosa*. Copeia 1984:966–976.

Bradford, D.F. 1989. Allotopic distribution of native frogs and introduced fishes in high Sierra Nevada lakes of California: implication of the negative effect of fish introductions. Copeia 1989:775–778.

Bradford, D.F. 1991. Mass mortality and extinction in a high–elevation population of *Rana muscosa*. Journal of Herpetology 25:174–177.

Bradford, D.F., C. Swanson, and M.S. Gordon. 1992. Effects of low pH and aluminum on two declining species of amphibians in the Sierra Nevada, California. Journal of Herpetology 26:369–377.

Bradford, D.F., F. Tabatabai, and D.M. Graber. 1993. Isolation of remaining populations of the native frog, *Rana muscosa*, by introduced fishes in Sequoia and Kings Canyon National Parks, California. Conservation Biology 7:882–888.

Bradford, D.F., M.S. Gordon, D.F. Johnson, R.D. Andrews, and W.B. Jennings. 1994a. Acidic deposition as an unlikely cause for amphibian population declines in the Sierra Nevada, California. Biological Conservation 69:155–161.

Bradford, D.F., D.M. Graber, and F. Tabatabai. 1994b. Population declines of the native frog, *Rana muscosa*, in Sequoia and Kings Canyon National Parks, California. Southwestern Naturalist 39:323–327.

Bradford, D.F., S.D. Cooper, T.M. Jenkins, Jr., K. Kratz, O. Sarnelle, and A.D. Brown. 1998. Influences of natural acidity and introduced fish on faunal assemblages in California alpine lakes. Canadian Journal of Fisheries and Aquatic Sciences 55:2478–2491.

Bradford, D.F., A.C. Neale, M.S. Nash, D.W. Sada, and J.R. Jaeger. 2003. Habitat patch occupancy by toads (*Bufo punctatus*) in a naturally fragmented desert landscape. Ecology 84:1012–1023.

Bradford, D.F., J.R. Jaeger, and R.D. Jennings. 2004. Population status and distribution of a decimated amphibian, the relict leopard frog (*Rana onca*). Southwestern Naturalist 49: 218–228.

Bradford, D.F., J.R. Jaeger, and S.A. Shanahan. 2005a. Distributional changes and population status of amphibians in the eastern Mojave Desert. Western North American Naturalist 65:462–472.

Bradford, D.F., R.D. Jennings, and J.R. Jaeger. 2005b. *Rana onca* Cope, 1875(b). Relict Leopard Frog. Pp. 567–568 *In* M.J. Lannoo (ed.), Amphibian Declines. The Conservation Status of United States Species. University of California Press, Berkeley.

Bradford, D.F., R.A. Knapp, D.W. Sparling, M.S. Nash, K.A. Stanley, N.G. Tallent–Halsell, L.L. McConnell, and S.M. Simonich. 2011. Pesticide distributions and population declines of California, USA, alpine frogs, *Rana muscosa* and *Rana sierrae*. Environmental Toxicology and Chemistry 30:682–691.

Bradley, G.A., P.C. Rosen, M.J. Sredl, T.R. Jones, and J.E. Longcore. 2002. Chytridiomycosis in native Arizona frogs. Journal of Wildlife Diseases 38:206–212.

Brady, M.K. 1925. Notes on the herpetology of Hog Island. Copeia (137):110–111.

Brady, M. 1927. Notes on the amphibians and reptiles of the Dismal Swamp. Copeia (162):26–29.

Brady, M.K., and F. Harper. 1935. A Florida subspecies of *Pseudacris nigrita* (Hylidae). Proceedings of the Biological Society of Washington 48:107–110.

Bragg, A.N. 1936a. Notes on the breeding habits, eggs, and embryos of *Bufo cognatus* with a description of the tadpole. Copeia 1936:14–20.

Bragg, A.N. 1936b. The ecological distribution of some North American anura. The American Naturalist 70:459–466.

Bragg, A.N. 1937a. A note on the metamorphosis of the tadpoles of *Bufo cognatus*. Copeia 1937:227–228.

Bragg, A.N. 1937b. Observations on *Bufo cognatus* with special reference to the breeding habits and eggs. American Midland Naturalist 18:273–284.

Bragg, A.N. 1939. Possible hybridization of *Bufo cognatus* and *B.w. woodhousii*. Copeia 1939:173.

Bragg, A.N. 1940a. Observations on the ecology and natural history of Anura. 2. Habits, habitat and breeding of *Bufo woodhousii woodhousii* (Girard) in Oklahoma. American Midland Naturalist 24:306–335.

Bragg, A.N. 1940b. Observations on the ecology and natural history of Anura. 6. The ecological importance of the study of the habits of animals as illustrated by toads. The Wasmann Collector 4:6–16.

Bragg, A.N. 1940c. Observations on the ecology and natural history of Anura. 1. Habits, habitat and breeding of *Bufo cognatus* Say. American Naturalist 74:322–349, 424–438.

Bragg, A.N. 1941a. Tadpoles of *Scaphiopus bombifrons* and *Scaphiopus hammondii*. The Wasmann Collector 4:92–94.

Bragg, A.N. 1941b. Some observations on amphibia at and near Las Vegas, New Mexico. Great Basin Naturalist 2:109–117.

Bragg, A.N. 1941c. Observations on the ecology and natural history of Anura. 9. The invasion of the Canadian River flood plain by two prairie species. Proceedings of the Oklahoma Academy of Science 22:73–74.

Bragg, A.N. 1941d. A corrected list of the amphibia known in central Oklahoma. Proceedings of the Oklahoma Academy of Science 1941:16–17.

Bragg, A.N. 1941e. Observations on the ecology and natural history of anura VIII. Some factors in the initiation of breeding behavior. Turtox News 19:1–3.

Bragg, A.N. 1941f. Observations on the ecology and natural history of anura. XI. The invasion of the Canadian River flood plain by two prairie species. Proceedings of the Oklahoma Academy of Science 1941:73–74.

Bragg, A.N. 1942. Observations on the ecology and natural history of Anura. 10. The breeding habits of *Pseudacris streckeri* Wright and Wright in Oklahoma including a description of the eggs and tadpoles. The Wasmann Collector 5:47–62.

Bragg, A.N. 1943a. Observations on the ecology and natural history of Anura. 15. The hylids and microhylids in Oklahoma. Great Basin Naturalist 4:62–80.

Bragg, A.N. 1943b. Observations on the ecology and natural history of Anura. 16. Life–history of *Pseudacris clarkii* (Baird) in Oklahoma. The Wasmann Collector 5:129–140.

Bragg, A.N. 1944a. The spadefoot toads in Oklahoma with a summary of our knowledge of the group. American Naturalist 78:517–533.

Bragg, A.N. 1944b. Breeding habits, eggs, and tadpoles of *Scaphiopus hurterii*. Copeia 1944:230–241.

Bragg, A.N. 1945a. Notes on the psychology of frogs and toads. Journal of General Psychology 32:27–37.

Bragg, A.N. 1945b. The spadefoot toads in Oklahoma with a summary of our knowledge of the group. 2. American Naturalist 79:52–72.

Bragg, A.N. 1945c. Breeding and tadpole behavior in *Scaphiopus hurterii* near Norman, Oklahoma, spring, 1945. The Wasmann Collector 6: 69–78.

Bragg, A.N. 1946a. Aggregation with cannibalism in tadpoles of *Scaphiopus bombifrons* with some general remarks on the probable evolutionary significance of such phenomena. Herpetologica 3:89–97.

Bragg, A.N. 1946b. Some salientian adaptations. Great Basin Naturalist 7:11–15.

Bragg, A.N. 1947. Comment on Smith (Univ. Kans. Publ. Mus. Nat. Hist., 1946, 1, 93–96). Science 106:166.

Bragg, A.N. 1948a. Observations on *Hyla versicolor* in Oklahoma. Proceedings of the Oklahoma Academy of Science 28:31–35.

Bragg, A.N. 1948b. Observations on the life history of *Pseudacris triseriata* (Wied.) in Oklahoma. The Wasmann Collector 7:149–168.

Bragg, A.N. 1950a. Salientian breeding dates in Oklahoma. Pp. 35–38 *In* Researches on the Amphibia of Oklahoma. University of Oklahoma Press, Norman.

Bragg, A.N. 1950b. Observations on the ecology and natural history of Anura. 17. Adaptations and distribution in accordance with habitats in Oklahoma. Pp. 59–100 *In* Researches on the Amphibia of Oklahoma. University of Oklahoma Press, Norman.

Bragg, A.N. 1950c. Observations on *Microhyla* (Salientia: Microhylidae). The Wasmann Journal of Biology 8:113–118.

Bragg, A.N. 1950d. The identification of salientia in Oklahoma. Pp. 9–29 *In* Researches on the Amphibia of Oklahoma. University of Oklahoma Press, Norman.

Bragg, A.N. 1950e. Observations on the ecology and natural history of Anura. 14. Growth rates and age at sexual maturity of *Bufo cognatus* under natural conditions in central Oklahoma. Pp. 47–58 *In* Researches on the Amphibia of Oklahoma. University of Oklahoma Press, Norman.

Bragg, A.N. 1950f. Size range in adults of the toad *Bufo cognatus*. Copeia 1951:153.

Bragg, A.N. 1952. Decline in toad populations in central Oklahoma. Proceedings of the Oklahoma Academy of Science 33:70.

Bragg, A.N. 1953. A study of *Rana areolata* in Oklahoma. The Wasmann Journal of Biology 11:273–318.

Bragg, A.N. 1954a. A second season with *Rana areolata* in Oklahoma. The Wasmann Journal of Biology 12: 337–343.

Bragg, A.N. 1954b. *Bufo terrestris charlesmithi*, a new subspecies from Oklahoma. The Wasmann Journal of Biology 12:245–254.

Bragg, A.N. 1954c. Aggregational behavior and feeding reactions in tadpoles of the savannah spadefoot. Herpetologica 10:97–102.

Bragg, A.N. 1955a. Supplementary notes on the young of *Rana areolata*. Herpetologica 11:87–88.

Bragg, A.N. 1955b. The tadpole of *Bufo debilis debilis*. Herpetologica 11:211–212.

Bragg, A.N. 1955c. Taxonomic and physiological factors in

the embryonic development of certain toads. Copeia 1955:62.

Bragg, A.N. 1955d. In quest of spadefoots. New Mexico Quarterly 25(4):345–358.

Bragg, A.N. 1956a. Dimorphism and cannibalism in tadpoles of *Scaphiopus bombifrons* (Amphibia, Salientia). Southwestern Naturalist 1:105–108.

Bragg, A.N. 1956b. Further observations on spadefoot toads. Herpetologica 12:201–204.

Bragg, A.N. 1957a. Some factors in the feeding of toads. Herpetologica 13:189–191.

[2]Bragg, A.N. 1957b. Parasitism of spadefoot tadpoles by *Saprolegnia*. Herpetologica 13:191.

Bragg, A.N. 1957c. Amphibian eggs produced singly or in masses. Herpetologica 13:212.

Bragg, A.N. 1957d. Variation in colors and color patterns in tadpoles in Oklahoma. Copeia 1957:36–39.

Bragg, A.N. 1958a. On metamorphic and postmetamorphic aggregations in spadefoots. Herpetologica 13:273.

[2]Bragg, A.N. 1958b. Parasitism of spadefoot tadpoles by *Saprolegnia*. Herpetologica 14:34.

Bragg, A.N. 1958c. Taxonomic status of the gray tree frog in Oklahoma. Herpetologica 14:141–147.

Bragg, A.N. 1958d. The eggs of the dwarf American toad. Herpetologica 13:273–275.

Bragg, A.N. 1958e. A melanistic tendency in the Great Plains toad, *Bufo cognatus*. Southwestern Naturalist 3:229–230.

Bragg, A.N. 1958f. The clasping reflex observed in juvenile spadefoots (*Scaphiopus*). Southwestern Naturalist 3:229.

Bragg, A.N. 1959a. Response of a female *Pseudacris nigrita triseriata* to the call of a male. Copeia 1959:341.

Bragg, A.N. 1959b. Voice in Salientia. I. Observations on *Hyla versicolor* x *chrysoscelis* intergrades in central Oklahoma with a few data from Arkansas. The Wasmann Journal of Biology 17:231–243.

Bragg, A.N. 1959c. Behavior of tadpoles of Hurter's spadefoot during an exceptionally rainy season. The Wasmann Journal of Biology 17:23–42.

Bragg, A.N. 1960a. Ovulation and breeding patterns in Salientia. Herpetologica 16:124.

Bragg, A.N. 1960b. Population fluctuation in the amphibian fauna of Cleveland County, Oklahoma during the past twenty–five years. Southwestern Naturalist 5:165–169.

Bragg, A.N. 1960c. Experimental observations on the feeding of spadefoot tadpoles. Southwestern Naturalist 5:201–207.

Bragg, A.N. 1960d. Feeding in the Houston toad. Southwestern Naturalist 5:106.

Bragg, A.N. 1961a. The behavior and comparative developmental rates in nature of tadpoles of a spadefoot, a toad, and a frog. Herpetologica 17:80–84.

Bragg, A.N. 1961b. A theory of the origin of spade–footed toads deduced principally by a study of their habits. Animal Behaviour 9: 178–186.

Bragg, A.N. 1962a. *Saprolegnia* on tadpoles again in Oklahoma. Southwestern Naturalist 7:79–80.

Bragg, A.N. 1962b. Predator–prey relationship in two species of spadefoot tadpoles with notes on some other features of their behavior. The Wasmann Journal of Biology 20:81–97.

Bragg, A.N. 1962c. Predation on arthropods by spadefoot tadpoles. Herpetologica 18:144.

Bragg, A.N. 1964a. A hypothesis to explain the almost exclusive use of temporary water by breeding spadefoot toads. Proceedings of the Oklahoma Academy of Science 44:24–25.

Bragg, A.N. 1964b. Mass movements resulting in aggregations of tadpoles of the Plains spadefoot, some of them in response to light and temperature (Amphibia: Salientia). The Wasmann Journal of Biology 22:299–305.

Bragg, A.N. 1964c. Further study of predation and cannibalism in spadefoot tadpoles. Herpetologica 20:17–24.

Bragg, A.N. 1964d. Lens action by jelly of frog's eggs. Proceedings of the Oklahoma Academy of Science 44:23.

Bragg, A.N. 1965. Gnomes of the Night. The Spadefoot Toads. University of Pennsylvania Press, Philadelphia.

Bragg, A.N. 1966. Longevity of the tadpole stage in the Plains spadefoot (Amphibia: Salientia). The Wasmann Journal of Biology 24:71–73.

Bragg, A.N. 1967. Recent studies on the spadefoot toads. Bios 38:75–84.

Bragg, A.N. 1968. The formation of feeding schools in tadpoles of spadefoots. The Wasmann Journal of Biology 26:11–16.

Bragg, A.N., and C.C. Smith. 1942. Observations on the ecology and natural history of Anura. 9. Notes on breeding behavior in Oklahoma. Great Basin Naturalist 3:33–50.

Bragg, A.N., and C.C. Smith. 1943. Observations on the ecology and natural history of Anura. 4. The ecological distribution of toads in Oklahoma. Ecology 24:285–309.

Bragg, A.N., and H.A. Dundee. 1948. Salientian collections in Oklahoma, 1948. Proceedings of the Oklahoma Academy of Science 1948:24–25.

Bragg, A.N., and O. Sanders. 1951. A new subspecies of the

---

[2]Identical in content and must have been mistakenly printed twice.

Bufo woodhousii group of toads (Salientia: Bufonidae: *Bufo woodhousii velatus* subsp. nov.). The Wasmann Journal of Biology 9:363–378.

Bragg, A.N., and V.E. Dowell. 1954. Leopard frog eggs in October. Proceedings of the Oklahoma Academy of Science 35:41.

Bragg, A.N., and W.N. Bragg. 1958a. Parasitism of spadefoot tadpoles by *Saprolegnia*. Herpetologica 14:34.

Bragg, A.N., and W.N. Bragg. 1958b. Variations in the mouth parts in tadpoles of *Scaphiopus* (*Spea*) *bombifrons* Cope (Amphibia: Salientia). Southwestern Naturalist 3:55–69.

Bragg, A.N., and M. Brooks. 1958. Social behavior in juveniles of *Bufo cognatus* Say. Herpetologica 14:141–147.

Bragg, A.N., and O.M. King. 1960. Aggregational and associated behavior in tadpoles of the Plains spadefoot. The Wasmann Journal of Biology 18:273–289.

Bragg, A.N., and S. Hayes. 1963. A study of labial teeth rows in tadpoles of Couch's spadefoot. The Wasmann Journal of Biology 21:149–154.

Bragg, A.N., and J.H. Black. 1970. Four cases of probable albinism in tadpoles. Proceeedings of the Oklahoma Academy of Science 49:14–15.

Bragg, A.N., and J. Nelson. 1966. Further notes on predation by tadpoles of the plains spadefoot. Proceedings of the Oklahoma Academy of Science 46:25–26.

Bragg, A.N., R. Matthews, and R. Kingsinger, Jr. 1964. The mouth parts of tadpoles of Hurter's spadefoot. Herpetologica 19:284–285.

Bragg, A.N., N. Taylor, and R.J. Taylor. 1968. A range extension of the pickerel frog, *Rana palustris palustris*, in Oklahoma with indications of an extension of the Austroriparian life province. Southwestern Naturalist 13:372–374.

Braid, M.R., C.B. Raymond, and W.S. Sanders. 1994. Feeding trials with the dusky gopher frog, *Rana capito sevosa*, in a recirculating water system and other aspects of their culture as part of a "headstarting" effort. Journal of the Alabama Academy of Science 65:249–262.

Braid, M.R., K.N. Butler, A.F. Blankenship III, and A. Schultz–Cantley. 2000. *Rana capito sevosa* (Dusky Gopher Frog). Metamorph captures. Herpetological Review 31:101–102.

Branch, L.C., and D.G. Hokit. 2000. A comparison of scrub herpetofauna on two central Florida sand ridges. Florida Scientist 63:108–117.

Brand, A.B, and J.W. Snodgrass. 2010. Value of artificial habitats for amphibian reproduction in altered landscapes. Conservation Biology 24:295–301.

Brand, A.B., J.W. Snodgrass, M.T. Gallagher, R.E. Casey, and R. Van Meter. 2010. Lethal and sublethal effects of embryonic and larval exposure of *Hyla versicolor* to stormwater pond sediments. Archives of Environmental Contamination and Toxicology 58:325–331.

Brander, S.M., J.A. Royle, and M. Eames. 2007. Evaluation of the status of anurans on a refuge in suburban Maryland. Journal of Herpetology 41:52–60.

Brandon, R.A., and S.R. Ballard. 1998. Status of Illinois chorus frogs in southern Illinois.Pp.102–112 *In* M.J. Lannoo (ed.), Status & Conservation of Midwestern Amphibians. University of Iowa Press, Iowa City.

Brandt, B.B. 1936a. Parasites of certain North Carolina salientia. Ecological Monographs 6:493–532.

Brandt, B.B. 1936b. The frogs and toads of eastern North Carolina. Copeia 1936:215–223.

Brandt, B.B. 1954. Salientia of Bleckley County, Georgia, and vicinity. Herpetologica 9:141–145.

Brandt, B.B., and C.F. Walker. 1933. A new species of *Pseudacris* from the southeastern United States. Occasional Papers of the Museum of Zoology, University of Michigan No. 272.

Brannelly, L.A., M. W. H. Chatfield and C. L. Richards–Zawacki. 2012. Field and laboratory studies of the susceptibility of the green treefrog (*Hyla cinerea*) to *Batrachochytrium dendrobatidis* infection. PLoS One 7(6): e38473.

Brannon, M.P. 2006. *Bufo a. americanus* (Eastern American Toad). Leucism. Herpetological Review 37:333–334.

Branson, B.A. 1975. Claude who? Another unwanted exotic species. National Parks and Conservation Magazine (June):17–18.

Brashears, V. Sr., and V. Brashears, Jr. 1950. Frog Raising. Brashears Printing Co., Berryville, Arkansas.

Brashears, V. III, and K.B. Brashears. 1982. The Frog Manual. Raising to Recipes. Brashears, Houston, Texas.

Braswell, A. 1988. A survey of amphibians and reptiles of Nags Head Woods Ecological Preserve. Association of Southeastern Biologists Bulletin 35:199–217.

Braswell, A.L. 1993. Status report on *Rana capito capito* Leconte, the Carolina gopher frog in North Carolina. Nongame and Endangered Wildlife Program, North Carolina Wildlife Resources Commission, Raleigh.

Brattstrom, B.H. 1953. The amphibians and reptiles from Rancho La Brea. Transactions of the San Diego Society of Natural History 11:365–392.

Brattstrom, B.H. 1958. New records of Cenozoic amphibians and reptiles from California. Bulletin of the Southern California Academy of Sciences 54:1–4.

Brattstrom, B.H. 1962. Thermal control of aggregation behavior in tadpoles. Herpetologica 18:38–46.

Brattstrom, B.H. 1963. A preliminary review of the thermal requirements of amphibians. Ecology 44:238–255.

Brattstrom, B.H. 1968. Thermal acclimation in anuran amphibians as a function of latitude and altitude. Comparative Biochemistry and Physiology 24:93–111.

Brattstrom, B.H. 1979. Amphibian temperature regulation studies in the field and laboratory. American Zoologist 19:345–356.

Brattstrom, B.H., and J.W. Warren. 1955. Observations on the ecology and behavior of the Pacific treefrog, *Hyla regilla*. Copeia 1955:181–191.

Brattstrom, B.H., and M.C. Bondello. 2012. Effects of off–road vehicle noise on desert vertebrates. Pp. 167–206 *In* R.H. Webb and H.G. Wilshire (eds.), Environmental Effects of Off–Road Vehicles. Springer–Verlag, New York.

Breckenridge, W.J. 1944. Reptiles and Amphibians of Minnesota. University of Minnesota Press, Minneapolis.

Breckenridge, W.J., and J.R. Tester. 1961. Growth, local movements and hibernation of the Manitoba toad, *Bufo hemiophrys*. Ecology 42:637–646.

Breden, F. 1987. The effect of post–metamorphic dispersal on the population genetic structure of Fowler's toad, *Bufo woodhousei fowleri*. Copeia 1987:386–395.

Breden, F. 1988. Natural history and ecology of Fowler's Toad, *Bufo woodhousei fowleri* (Amphibia: Bufonidae), in the Indiana Dunes National Seashore. Fieldiana Zoology, new series (18):1–16.

Breden, F., and C.H. Kelly. 1982. The effect of conspecific interactions on metamorphosis in *Bufo americanus*. Ecology 63:1682–1689.

Breden, F., A. Lum, and R. Wassersug. 1982. Body size and orientation in aggregates of toad tadpoles *Bufo woodhousei*. Copeia 1982:672–680.

Breder, C.M., Jr. 1927. Frog tagging: a method of studying anuran life habits. Zoologica 9:201–229.

Brehaut, L. 2012. Local and landscape variables influencing the use of ponds by Wood Frogs (*Lithobates sylvaticus*) in the Shakwak Valley, Yukon. Undergraduate thesis, Queen's University, Kingston, Ontario.

Brekke, D.R., S.D. Hillyard, and R.M. Winokur. 1991. Behavior associated with the water absorption response by the toad, *Bufo punctatus*. Copeia 1991:393–401.

Brem, F.M. 2012. Factors affecting infectivity and pathogenicity of *Batrachochytrium dendrobatidis* in amphibians. Ph.D. Dissertation, University of Memphis, Memphis, Tennessee.

Brennan, T.C., and A.T. Holycross. 2006. Amphibians and Reptiles in Arizona. Arizona Game and Fish Department, Phoenix.

Brenner, F.J. 1969. Role of temperature and fat deposition in hibernation and reproduction in two species of frogs. Herpetologica 25:105–113.

Brenowitz, E.E., and G.J. Rose. 1994. Behavioural plasticity mediates aggression in choruses of the Pacific treefrog. Animal Behaviour 47:633–641.

Brenowitz, E.A., and G.J. Rose. 1999. Female choice and plasticity of male calling behaviour in the Pacific treefrog. Animal Behaviour 57:1337–1342.

Brenowitz, E.A., W. Wilczynski, and H.H. Zakon. 1984. Acoustic communication in spring peepers: environmental and behavioral aspects. Journal of Comparative Physiology 155A:585–592.

Bresler, J.B. 1963. Pigmentation characteristics of *Rana pipiens*: dorsal region. American Midland Naturalist 70:197–207.

Bresler, J.B. 1964. Pigmentation characteristics of *Rana pipiens*: tympanum spot, line on upper jaw, and spots on upper eyelids. American Midland Naturalist 72:382–389.

Bresler, J., and A.N. Bragg. 1954. Variations in the rows of labial teeth in tadpoles. Copeia 1954:255–257.

Brickell, J. 1737. The Natural History of North Carolina. With an Account of the Trade, Manners, and Customes of the Christian and Indian Inhabitants. James Carson, Dublin.

Bridges, A.S., and M.E. Dorcas. 2000. Temporal variation in anuran calling behavior: implications for surveys and monitoring programs. Copeia 2000:587–592.

Bridges, C.M. 1997. Tadpole swimming performance and activity affected by acute exposure to sublethal levels of carbaryl. Environmental Toxicology and Chemistry 16:1935–1939.

Bridges, C.M. 1999a. Effects of a pesticide on tadpole activity and predator avoidance behavior. Journal of Herpetology 33:303–306.

Bridges, C.M. 1999b. Predator–prey interactions between two amphibian species: effects of insecticide exposure. Aquatic Ecology 33:205–211.

Bridges, C.M. 2000. Long–term effects of pesticide exposure at various life stages of the southern leopard frog (*Rana sphenocephala*). Archives of Environmental Contamination and Toxicology 39:91–96.

Bridges, C.M. 2002. Tadpoles balance foraging and predator avoidance: effects of predation, pond drying, and hunger. Journal of Herpetology 36:627–634.

Bridges, C.M., and R.D. Semlitsch. 2000. Variation in pesticide tolerance of tadpoles among and within species of Ranidae and patterns of amphibian decline. Conservation Biology 14:1490–1499.

Bridges, C. M., and R.D. Semlitsch. 2001. Genetic variation in insecticide tolerance in a population of southern leopard frogs (*Rana sphenocephala*): implications for amphibian conservation. Copeia 2001:7–13.

Bridges, C.M., and M.D. Boone. 2003. The interactive effects of UV–B and insecticide exposure on tadpole survival, growth and development. Biological Conservation 113:49–54.

Briggler, J.T. 1998. Amphibian use of constructed woodland ponds in the Ouachita National Forest. M.S. thesis, University of Arkansas, Fayetteville.

Briggler, J.T. 2000. *Rana utricularia* (Southern Leopard Frog). Predation. Herpetological Review 31:171.

Briggler, J.T., and T.R. Johnson. 2008. Missouri's Toads and Frogs. Missouri Department of Conservation, Jefferson City.

Briggler, J.T., K.M. Lohraff, and G.L. Adams. 2001. Amphibian parasitism by the leech *Desserobdella picta* at a small pasture pond in northwest Arkansas. Journal of Freshwater Ecology 16:105–111.

Briggs, C.J., V.T. Vredenburg, R.A. Knapp, and L.J. Rachowicz. 2005. Investigating the population–level effects of chytridiomycosis: an emerging infectious disease of amphibians. Ecology 86:3149–3159.

Briggs, C.J., R.A. Knapp, and V.T. Vredenburg. 2010. Enzootic and epizootic dynamics of the chytrid fungal pathogen of amphibians. Proceedings of the National Academy of Sciences of the United States of America 107:9695–9700.

Briggs, J.L. 1975. A case of *Bufolucilla elongata* Shannon 1924 (Diptera: Calliphoridae) myiasis in the American toad, *Bufo americanus* Holbrook 1836. Journal of Parasitology 61:412.

Briggs, J.L. 1978. An asymptotic growth model allowing seasonal variation in growth rates, with application to a population of the Cascade frog, *Rana cascadae* (Amphibia, Anura, Ranidae). Journal of Herpetology 12:559–564.

Briggs, J.L. 1987. Breeding biology of the Cascades frog, *Rana cascadae*, with comparisons to *R. aurora* and *R. pretiosa*. Copeia 1987:241–245.

Briggs, J.L., and R.M. Storm. 1970. Growth and population structure of the Cascade frog, *Rana cascadae* Slater. Herpetologica 26:283–300.

Brill, T.M. 1993. Postmetamorphic population structure and the effects of salinity on larval development and size at metamorphosis of *Bufo woodhousei fowleri* Hinckley (Anura: Bufonidae). M.S. thesis, Bloomsburg University, Bloomsburg, Pennsylvania.

Brimley, C.S. 1926. Revised key and list of the amphibians and reptiles of North Carolina. Journal of the Elisha Mitchell Scientific Society 42:75–93.

Brimley, C.S. 1944. Amphibians and Reptiles of North Carolina. Carolina Biological Supply Company, Elon College, North Carolina. [Issued in 6 supplements to Volumes 3 and 4 of Carolina Tips: No. 10 (April, 1940) 10-11; No. 11 (June, 1940) 14-15; No. 12 (August, 1940) 18-19; No. 13 (September, 1940) 22-23; No. 14 (November, 1940) 26-27; No. 15 (January, 1941) 2-3.]

Brindle, A.A., S.B. Karr, G.R. Smith, and J.A. Rettig. 2009. Within–pond oviposition site selection in the wood frog (*Rana sylvatica*). Bulletin of the Maryland Herpetological Society 45:57–60.

Britson, C.A., and R.E. Kissell, Jr. 1996. Effects of food type on developmental characteristics of an ephemeral pond–breeding anuran, *Pseudacris triseriata feriarum*. Herpetologica 52:374–382.

Britson, C.A., and S.T. Threlkeld. 1998. Abundance, metamorphosis, developmental and behavioral abnormalities in *Hyla chrysoscelis* tadpoles following exposure to three agrichemicals and methyl mercury in outdoor mesocosms. Bulletin of Environmental Contamination and Toxicology 61:154–161.

Britson, C.A., and S.T. Threlkeld. 2000. Interactive effects of anthropogenic, environmental, and biotic stressors on multiple endpoints in *Hyla chrysoscelis*. Journal of the Iowa Academy of Science 107:61–66.

Brocchi, P. 1877. Sur quelques batraciens Raniformes et Bufoniformes d l'Amérique Centrale. Bulletin de la Société Philomathique de Paris, Série 7, 1:175–197.

Brocchi, M.P. 1879. Sur divers Batraciens anoures de l'Amérique Centrale. Bulletin de la Société Philomathique de Paris, Série 7, 3(1):19–24.

Brockelman, W.Y. 1968. Natural regulation of density in tadpoles of *Bufo americanus*. Ph.D. Dissertation, University of Michigan, Ann Arbor.

Brockelman, W.Y. 1969. An analysis of density effects and predation in *Bufo americanus* tadpoles. Ecology 50:632–644.

Brode, W.E. 1958. The occurrence of the pickerel frog, three salamanders and two snakes in Mississippi caves. Copeia 1958:47–48.

Brode, W.E. 1959. Territoriality in *Rana clamitans*. Herpetologica 15:140.

Brode, W.E. 1960. Some notes on the occurrence of *Rana palustris* in caves. Journal of the Mississippi Academy of Science 6:233.

Brodie, E.D., Jr. 1968. A case of interbreeding between *Bufo boreas* and *Rana cascadae*. Herpetologica 24: 86.

Brodie, E.D., Jr., and D.R. Formanowicz. 1983. Prey size preference of predators: differential vulnerability of larval anurans. Herpetologica 39:67–75.

Brodie, E.D., Jr., and D.R. Formanowicz. 1987. Antipredator mechanisms of larval anurans: protection of palatable individuals. Herpetologica 43:369–373.

Brodie, E.D., Jr., D.R. Formanowicz, Jr., and E.D. Brodie III. 1978. The development of noxiousness of *Bufo americanus* tadpoles to aquatic insect predators. Herpetologica 34:302–306.

Brodkin, M.A., H. Madhoun, M. Rameswaran, and I. Vatnick. 2007. Atrazine is an immune disruptor in adult northern leopard frogs (*Rana pipiens*). Environmental Toxicology and Chemistry 26:80–84.

Brodman, R. 2003. Amphibians and reptiles from twenty–three counties of Indiana. Proceedings of the Indiana Academy of Science 112:43–54.

Brodman, R. 2009. A 14–year study of amphibian populations and metacommunities. Herpetological Conservation and Biology 4:106–119.

Brodman, R., and M. Kilmurry. 1998. Status of amphibians in northwestern Indiana. Pp. 125–136 *In* M.J. Lannoo (ed.), Status & Distribution of Midwestern Amphibians. University of Iowa Press, Iowa City.

Brodman, R., S. Cortwright, and A. Resetar. 2002. Historical changes of reptiles and amphibians of northwest Indiana Fish and Wildlife properties. American Midland Naturalist 147:135– 144.

Brodman, R., M. Parrish, H. Kraus, and S. Cortwright. 2006. Amphibian biodiversity recovery in a large–scale ecosystem restoration. Herpetological Conservation and Biology 1:101–108.

Broel, A. 1937. Frog Raising. American Frog Canning Co., New Orleans, Louisiana.

Broel, A. 1943. Frog Raising. Marlboro House, Detroit. Spiral Bound. [other copies dated 1944, 1945]

Broel, A. 1950. Frog Raising for Pleasure and Profit. Marlboro House, New Orleans, Louisiana. [additional copies dated 1951, 1953, 1954, 1960 editions published in Detroit]

Brooks, D.R. 1975. A review of the genus *Allassostomoides* (Trematoda: Paramphistomidae) with a redescription of *Allassostomoides chelydrae*. Journal of Parasitology 61:882–885.

Brooks, D.R. 1976. Parasites of amphibians of the Great Plains. Part 2. Platyhelminths of amphibians in Nebraska. Bulletin of the University of Nebraska State Museum 10:65–92.

Brooks, D.R. 1979. New records for amphibian and reptile trematodes. Proceedings of the Helminthological Society of Washington 46:286–289.

Brooks, D.R., and N.J. Welch. 1976. Parasites of amphibians of the Great Plains. 1. The cercariae of *Cephalogonimus brevicirrus* Ingles, 1932 (Trematoda:Cephalogonimidae). Proceedings of the Helminthological Society of Washington 43:92–93.

Brooks, G.R., Jr. 1959. The food habits of *Rana catesbeiana* Shaw and other ranids from five different pond types. M.S. thesis, University of Richmond, Richmond, Virginia.

Brooks, G.R., Jr. 1964. An analysis of the food habits of the bullfrog, *Rana catesbeiana*, by body size, sex, month, and habitat. Virginia Journal of Science 15:173–186.

Brothers, D.R. 1965. An annotated list of the amphibians and reptiles of northeastern North Carolina. Journal of the Elisha Mitchell Scientific Society 81:119–124.

Brothers, D.R. 1994. *Bufo boreas* (Western Toad). Predation. Herpetological Review 25:117.

Browder, L.W. 1972. Genetic and embryological studies of albinism in *Rana pipiens*. Journal of Experimental Zoology 180:149–156.

Browder, L.W., J.C. Underhill, and D.J. Merrell. 1966. Mid–dorsal stripe in the wood frog. Journal of Heredity 57:65–67.

Brower, J. van Zant, and L.P. Brower. 1962. Experimental studies of mimicry. 6. The reaction of toads (*Bufo terrestris*) to honeybees (*Apis mellifera*) and their dronefly mimics (*Eristalis vinetorum*). American Midland Naturalist 96:297–307.

Brower, J. van Zant, and L.P. Brower. 1965. Experimental studies of mimicry. 8. Further investigations of honeybees (*Apis mellifera*) and their dronefly mimics (*Eristalis* spp.). American Naturalist 99:173–187.

Brower, J. van Zant, L.P. Brower, and P.W. Westcott. 1960. Experimental studies of mimicry. 5. The reaction of toads (*Bufo terrestris*) to bumblebees (*Bombus americanorum*) and their robberfly mimics (*Mallophora bomboides*), with a discussion of aggressive mimicry. American Naturalist 94:343–355.

Brown, A., B. Christensen, T.D. Schwaner, and M. Nickerson. 2000. *Bufo microscaphus* (ArizonaToad). Reproduction. Herpetological Review 31:168.

Brown, B.C. 1950. An annotated check list of the reptiles and amphibians of Texas. Baylor University, Waco, Texas.

Brown, C. 1997. Habitat structure and occupancy patterns of the montane frog, *Rana cascadae*, in the Cascades Range, Oregon, at multiple scales: implications for population dynamics in patchy landscapes. M.S. thesis, Oregon State University, Corvallis.

Brown, C., K. Kiehl, and L. Wilkinson. 2012. Advantages of long–term, multi–scale monitoring: assessing the current status of the Yosemite toad (*Anaxyrus* [*Bufo*] *canorus*) in the Sierra Nevada, California, USA. Herpetological Conservation and Biology 7:115–131.

Brown, C.J., B. Blossey, J.C. Maerz, and S.J. Joule. 2006. Invasive plant and experimental venue affect tadpole performance. Biological Invasions 8:327–338.

Brown, C.S. 1995. Rear leg amputation and subsequent adaptive behavior during reintroduction of a green treefrog, *Hyla cinerea*. Bulletin of the Association of Reptile and Amphibian Veterinarians 5(2):6–7.

Brown, D. 2009. Amphibian populations at the Celery Bog Nature Area and the potential effects of land use and water quality. M.S. thesis, Purdue University, West Lafayette, Indiana.

Brown, D.J., G.M. Street, R.W. Nairn, and M.R. Forstner. 2012. A place to call home: amphibian use of created and restored wetlands. International Journal of Ecology doi:10.1155/2012/989872.

Brown, D.J., M.C. Jones, J. Bell, and M.R.J. Forstner. 2012 (2015). Feral hog damage to endangered Houston toad (*Bufo houstonensis*) habitat in the Lost Pines of Texas. Texas Journal of Science 64:73-88.

Brown, D.R. 1984. *Rana palustris* (Pickerel Frog). Oviposition. Herpetological Review 15:110–111.

Brown, E.E. 1979. Some snake food records from the Carolinas. Brimleyana 1:113–124.

Brown, E.E. 1980. Some historical data bearing on the Pine Barrens treefrog, *Hyla andersoni*, in South Carolina. Brimleyana 3:113–117.

Brown, E.E. 1992. Notes on amphibians and reptiles of the western Piedmont of North Carolina. Journal of the Elisha Mitchell Scientific Society 108:38–54.

Brown, H.A. 1967a. Embryonic temperature adaptations and genetic compatibility in two allopatric populations of the spadefoot toad, *Scaphiopus hammondi*. Evolution 21:742–761.

Brown, H.A. 1967b. High temperature tolerance of the eggs of a desert anuran, *Scaphiopus hammondii*. Copeia 1967:365–370.

Brown, H.A. 1969. The heat resistance of some anuran tadpoles (Hylidae and Pelobatidae). Copeia 1969:138–147.

Brown, H.A. 1975a. Embryonic temperature adaptations of the Pacific treefrog, *Hyla regilla*. Comparative Biochemistry and Physiology 51A:863–873.

Brown, H.A. 1975b. Reproduction and development of the red–legged frog, *Rana aurora*, in northwestern Washington. Northwest Science 49:241–252.

Brown, H.A. 1975c. Temperature and development of the tailed frog, *Ascaphus truei*. Comparative Biochemistry and Physiology 50A:397–405.

Brown, H.A. 1976. The status of California and Arizona populations of the western spadefoot toads (genus *Scaphiopus*). Natural History Museum of Los Angeles County Contributions in Science No. 286:1–15.

Brown, H.A. 1977. A case of interbreeding between *Rana aurora* and *Bufo boreas* (Amphibia, Anura). Journal of Herpetology 11:92–94.

Brown, H.A. 1989. Tadpole development and growth of the Great Basin spadefoot toad, *Scaphiopus intermontanus*, from central Washington. Canadian Field-Naturalist 103:531–534.

Brown, H.A. 1990. Morphological variation and age–class determination in overwintering tadpoles of the tailed frog, *Ascaphus truei*. Journal of Zoology 220:171–184.

Brown, J.S. 1956. The frogs and toads of Alabama. Ph.D. Dissertation, University of Alabama, Tuscaloosa.

Brown, J.S., and H.T. Boschung, Jr. 1954. *Rana palustris* in Alabama. Copeia 1954:226.

Brown, L.E. 1964. An electrophoretic study of variation in the blood proteins of the toads *Bufo americanus* and *Bufo woodhousei*. Systematic Zoology 13:92–95.

Brown, L.E. 1969. Natural hybrids between two toad species in Alabama. Quarterly Journal of the Florida Academy of Sciences 32:285–290.

Brown, L.E. 1970. Interspecies interactions as possible causes of racial size differences in the toads *Bufo americanus* and *Bufo woodhousei*. Texas Journal of Science 21:261–267.

Brown, L.E. 1971a. Natural hybridization and reproductive ecology of two toad species in a disturbed environment. American Midland Naturalist 86:78–85.

Brown, L.E. 1971b. Natural hybridization and trend toward extinction in some relict Texas toad populations. Southwestern Naturalist 16:185–199.

Brown, L.E. 1973. Speciation in the *Rana pipiens* complex. American Zoologist 13:73–79.

Brown, L.E. 1974. Behavioral reactions of bullfrogs while attempting to eat toads. Southwestern Naturalist 19:329–340.

Brown, L.E. 1975. The status of the near–extinct Houston toad (*Bufo houstonensis*) with recommendations for its conservation. Herpetological Review 6:37–40.

Brown, L.E. 1978. Subterranean feeding by the chorus frog *Pseudacris streckeri* (Anura: Hylidae). Herpetologica 34:212–216.

Brown, L.E. 2012. *Pseudacris streckeri illinoensis* (Illinois Chorus Frog). Large wound in holotype. Herpetological Bulletin (112):43–44.

Brown, L.E., and J.R. Pierce. 1965. Observations on the breeding behavior of certain anuran amphibians. Texas Journal of Science 26:313–317.

Brown, L.E., and J.R. Pierce. 1967. Male–male interactions and chorusing intensities of the Great Plains toad, *Bufo cognatus*. Copeia 1967:149–154.

Brown, L.E., and M.A. Ewert. 1971. A natural hybrid between the toads *Bufo hemiophrys* and *Bufo cognatus* in Minnesota. Journal of Herpetology 5:78–82.

Brown, L.E., and J.R. Brown. 1972a. Call types of the *Rana pipiens* complex in Illinois. Science 176:928–929.

Brown, L.E., and J.R. Brown. 1972b. Mating calls and distributional records of treefrogs of the *Hyla versicolor* complex in Illinois. Journal of Herpetology 6:233–234.

Brown, L.E., and M.J. Littlejohn. 1972. Male release call in the *Bufo americanus* group. Pp. 310–323 *In* W.F. Blair (ed.), Evolution in the Genus *Bufo*. University of Texas Press, Austin.

Brown, L.E., and J.R. Brown. 1973. Notes on breeding choruses of two anurans (*Scaphiopus holbrookii,Pseudacris streckeri*) in southern Illinois. Natural History Miscellanea (192):1–3.

Brown, L.E., and J.R. Brown. 1977. Comparison of environmental and body temperatures as predictors of mating call parameters of spring peepers. American Midland Naturalist 97:209–211.

Brown, L.E., and R.S. Funk. 1977. Absence of dorsal spotting in two species of leopard frogs (Anura: Ranidae). Herpetologica 33:290–293.

Brown, L.E., and D.B. Means. 1984. Fossorial behavior and ecology of the chorus frog *Pseudacris ornata*. Amphibia–Reptilia 5:261–273.

Brown, L.E., and G.B. Rose. 1988. Distribution, habitat, and calling season of the Illinois chorus frog (*Pseudacris streckeri illinoensis*) along the lower Illinois River. Illinois Natural History Survey Biological Notes 132:1–13.

Brown, L.E., and M.A. Morris. 1990. Distribution, habitat, and zoogeography of the Plains leopard frog (*Rana blairi*) in Illinois. Illinois Natural History Survey Biological Notes 136:1-6.

Brown, L.E., and J.E. Cima. 1998. Illinois chorus frogs and the Sand Lake dilemma. Pp. 301–311 *In* M.J. Lannoo (ed.), Status & Conservation of Midwestern Amphibians. University of Iowa Press, Iowa City.

Brown, L.E., and A. Mesrobian. 2005. Houston toads and Texas politics. Pp. 150–167 *In* M.J. Lannoo (ed.), Amphibian Declines. The Conservation Status of United States Species. University of California Press, Berkeley.

Brown, L.E., H.O. Jackson, and J.R. Brown. 1972. Burrowing behavior of the chorus frog, *Pseudacris streckeri*. Herpetologica 28:325–328.

Brown, L.E., H.M. Smith, and R.S. Funk. 1977. Request for the conservation of *Rana sphenocephala* Cope, 1886, and the suppression of *Rana utricularia* Harlan, 1826 and *Rana virescens* Cope, 1889 (Amphibia: Salientia). Bulletin of Zoological Nomenclature 33:195–203.

Brown, L.E., M.A. Morris, and T.R. Johnson. 1993. Zoogeography of the plains leopard frog (*Rana blairi*). Bulletin of the Chicago Academy of Sciences 15:1–13.

Brown, M.B., and C.R. Brown. 2009. *Lithobates catesbeianus* (American Bullfrog). Predation on cliff swallows. Herpetological Review 40:206.

Brown, M.G., E. K. Dobbs, J. W. Snodgrass, and D. R. Ownby. 2012. Ameliorative effects of sodium chloride on acute copper toxicity among Cope's gray tree frog (*Hyla chrysoscelis*) and green frog (*Rana clamitans*) embryos. Environmental Toxicology and Chemistry 31:836–842.

Brown, R.E. 1972. Size variation and food habits of larval bullfrogs, *Rana catesbeiana* Shaw, in western Oregon. Ph.D. Dissertation, Oregon State University, Corvallis.

Brown, R.L. 1974. Diets and habitat preferences of selected anurans in southeast Arkansas. American Midland Naturalist 91:468–473.

Brown, R.M., and D.H. Taylor. 1995. Compensatory escape mode trade–offs between swimming performance and maneuvering behavior through larval ontogeny of the wood frog, *Rana sylvatica*. Copeia 1995:1–7.

Brown, W. 1953. Brolo Frog Farm. Privately published, West Frankfort, Illinois. 45 pp.

Brown, W.C., and J.R. Slater. 1939. The amphibians and reptiles of the islands of the state of Washington. Occasional Papers, Department of Biology, College of Puget Sound No. 4:6–31.

Browne, C.L., and C.A. Paszkowski. 2010. Factors affecting the timing of movements to hibernation sites by western toads (*Anaxyrus boreas*). Herpetologica 66:250–258.

Browne, C.L., C.A. Paszkowski, A.L. Foote, A. Moenting, and S.M. Boss. 2009. The relationship of amphibian abundance to habitat features across spatial scales in the Boreal Plains. Ecoscience 16:209–223.

Brues, C.T. 1932. Further studies on the fauna of North American hot springs. Proceedings of the American Academy of Arts and Science 67:185–303.

Brugger, K.E. 1984. Aspects of reproduction in the squirrel treefrog, *Hyla squirella*. M.S. thesis, University of Florida, Gainesville.

Bruggers, R.L. 1973. Food habits of bullfrogs in northwest Ohio. Ohio Journal of Science 73:185–188.

Bruggers, R.L., and W.B. Jackson. 1974. Eye–lens weight of the bullfrog (*Rana catesbeiana*) related to larval development, transformation, and age of adults. Ohio Journal of Science 74:282–286.

Bruneau, M., and E. Magnin. 1980a. Vie larvaire des ouaouarons *Rana catesbeiana* Shaw (Amphibia Anura) des Laurentides, Québec. Canadian Journal of Zoology 58:169–174.

Bruneau, M., and E. Magnin. 1980b. Croissance, nutrition, et reproduction des ouaouarons *Rana catesbeiana* Shaw (Amphibia Anura) des Laurentides au nord du Montréal. Canadian Journal of Zoology 58:175–183.

Brunner, J.L., K.E. Barnett, C.J. Gosier, S.A. McNulty, M.J. Rubbo, and M.B. Kolozsvary. 2011. *Ranavirus* infection in die–offs of vernal pool amphibians in New York, USA. Herpetological Review 42:76–79.

Bryan, E.H., Jr. 1932. Frogs in Hawaii. Mid-Pacific Magazine 43(1):61-64.

Bryan, W.A. 1915. Natural History of Hawaii. Hawaiian Gazette Co., Ltd., Honolulu.

Bryce, F.D. 1975. Reptiles and amphibians of the Wichitas. Great Plains Journal 15:65–78.

Bryer, P.J. 2002. The embryonic world of wood frogs, *Rana*

sylvatica: natal pond learning and anti–predator behaviors. M.S. thesis, University of Maine, Orono.

Bryson, R.W., J. R. Jaeger, J.A. Lemos–Espinal, and D. Lazcano. 2012. A multilocus perspective on the speciation history of a North American aridland toad (*Anaxyrus punctatus*). Molecular Phylogenetics and Evolution 64:393–400.

Buchanan, B. W. 1988. Territoriality in the squirrel treefrog, *Hyla squirella*: competition for diurnal retreat sites. M.S. thesis, University of Southwestern Louisiana, Lafayette.

Buchanan, B.W. 1992. Bimodal nocturnal activity pattern of *Hyla squirella*. Journal of Herpetology 26:521–522.

Buchanan, B.W. 1993. Effects of enhanced lighting on the behaviour of nocturnal frogs. Animal Behaviour 45:893–899.

Buchanan, B.W. 1994. Sexual dimorphism in *Hyla squirella*: chromatic and pattern variation between the sexes. Copeia 1994:797–796.

Buchanan, B.W. 1998. Low–illumination prey detection by squirrel treefrogs. Journal of Herpetology 32:270–274.

Buchanan, B.W., and R.C. Taylor. 1996. Lightening the load: micturition enhances jumping performance of squirrel treefrogs. Journal of Herpetology 30:410–413.

Buchholz, D.R., and T.B. Hayes. 2000. Larval period comparison for the spadefoot toads *Scaphiopus couchii* and *Spea multiplicata* (Pelobatidae: Anura). Herpetologica 56:455–468.

Buchholz, D.R., and T.B. Hayes. 2002. Evolutionary patterns of diversity in spadefoot toad metamorphosis (Anura: Pelobatidae). Copeia 2002:180–189.

Buck, J.C., E.A. Scheessele, R.A. Relyea and A.R. Blaustein. 2012. The effects of multiple stressors on wetland communities: pesticides, pathogens and competing amphibians. Freshwater Biology 57: 61–73.

Buckle, J. 1971. A recent introduction of frogs to Newfoundland. Canadian Field-Naturalist 85:72–74.

Budischak, S.A., L.K. Belden and W.A. Hopkins. 2008. The effects of malathion on embryonic development and latent susceptibility to trematode parasites in ranid tadpoles. Environmental Toxicology and Chemistry 27:2496–2500.

Budischak S.A., L.K. Belden, and W.A. Hopkins. 2009. Relative toxicity of malathion to trematode–infected and noninfected *Rana palustris* tadpoles. Archives of Environmental Contamination and Toxicology 56:123–128.

Buhlmann, K.A. 2001. A biological inventory of eight caves in northwestern Georgia with conservation implications. Journal of Cave and Karst Studies 63:91–98.

Buhlmann, K.A., and J.C. Mitchell. 1997. Ecological notes on the amphibians and reptiles of the Naval Surface Warfare Center, Dahlgren Laboratory, King George County, Virginia. Banisteria 9:45–51.

Buhlmann, K.A., J.C. Mitchell, and C.A. Pague. 1993. Amphibian and small mammal abundance and diversity in saturated forested wetlands and adjacent uplands of southeastern Virginia. Pp. 1–7 *In* S.D. Eckles, A. Jennings, A. Spingarn, and C. Wienhold (eds.), Proceedings of a Workshop on Saturated Forested Wetlands in the Mid–Atlantic Region. The State of the Science. U.S. Fish and Wildlife Service, Annapolis, Maryland.

Buhlmann, K.A., J.C. Mitchell, and L.R. Smith. 1999. Descriptive ecology of the Shenandoah Valley sinkhole pond system in Virginia. Banisteria 13:23–51.

Bulen, B.J., and C.A. Distel. 2011. Carbaryl concentration gradients in realistic environments and their influence on our understanding of the tadpole food web. Archives of Environmental Contamination and Toxicology 60:343–350.

Bulger, J.B., N.J. Scott, Jr., and R.B. Seymour. 2003. Terrestrial activity and conservation of adult California red–legged frogs *Rana aurora draytonii* in coastal forests and grasslands. Biological Conservation 110:85–95.

Bull, E.L. 2003. Diet and prey availability of Columbia spotted frogs in northeastern Oregon. Northwest Science 77:349–356.

Bull, E.L. 2006. Sexual differences in the ecology and habitat selection of western toads (*Bufo boreas*) in northeastern Oregon. Herpetological Conservation and Biology 1:27–38.

Bull, E.L. 2009. Dispersal of newly metamorphosed and juvenile Western toads (*Anaxyrus boreas*) in northeastern Oregon, USA. Herpetological Conservation and Biology 4:236–247.

Bull, E.L., and B.E. Carter. 1996a. Winter observations of tailed frogs in northeastern Oregon. Northwestern Naturalist 77:45–47.

Bull, E.L., and B.E. Carter. 1996b. Tailed frogs: distribution, ecology, and association with timber harvest in northeastern Oregon. Research Paper PNW–RP–497, U.S. Forest Service, Pacific Northwest Research Station, Portland, Oregon.

Bull, E.L., and M.P. Hayes. 2000. Livestock effects on reproduction of the Columbia spotted frog. Journal of Range Management 53:291–294.

Bull, E.L., and M.P. Hayes. 2001. Post–breeding season movements of Columbia spotted frogs (*Rana luteiventris*) in northeastern Oregon. Western North American Naturalist 61:119–123.

Bull, E.L., and D.B. Marx. 2002. Influence of fish and habitat on amphibian communities in high elevation lakes in northwestern Oregon. Northwest Science 76:240–248.

Bull, E.L., and C. Carey. 2008. Breeding frequency of western toads (*Bufo boreas*) in northeastern Oregon. Herpetological Conservation and Biology 3:282–288.

Bull, E.L., and J.L. Hayes. 2009. Selection of diet by metamorphic and juvenile western toads (*Bufo boreas*) in northwestern Oregon. Herpetological Conservation and Biology 4:85–95.

Bullard, A.J. 1965. Additional records of the tree frog *Hyla andersonii* from the Coastal Plain of North Carolina. Herpetologica 21:154–155.

Buller, C. R. 1928. Methods employed in producing the Bullfrog (*Rana catesbeiana*) tadpoles at the Pennsylvania State Hatcheries. Bulletin of the Commonwealth of Pennsylvania Board of Fish Commissioners (6): 1-13.

Bumpus, H.C. 1885a. Reptiles and batrachians of Rhode Island. X. Class Batrachia. Random Notes on Natural History 2(7):52–53.

Bumpus, H.C. 1885b. Reptiles and batrachians of Rhode Island. XI. Random Notes on Natural History 2(8):59–60.

Bumpus, H.C. 1885c. Reptiles and batrachians of Rhode Island. XII. Random Notes on Natural History 2(9):68–69.

Bumpus, H.C. 1885d. Reptiles and batrachians of Rhode Island. XIII. Random Notes on Natural History 2(10):75.

Bumpus, H.C. 1885e. Reptiles and batrachians of Rhode Island. XIV. Random Notes on Natural History 2(11):83–84.

Bumpus, H.C. 1885f. Reptiles and batrachians of Rhode Island. XV. Random Notes on Natural History 2(12):93.

Bumpus, H.C. 1886a. Reptiles and batrachians of Rhode Island. XVI. Random Notes on Natural History 3(1):7–8.

Bumpus, H.C. 1886b. Reptiles and batrachians of Rhode Island. XVII. Random Notes on Natural History 3(2):13.

Bundy, D., and C.R. Tracy. 1977. Behavioral response of American toads (*Bufo americanus*) to stressful thermal and hydric environments. Herpetologica 33:455–458.

Bunnell, J.F., and R.A. Zampella. 1999. Acid water anuran pond communities along a regional forest to agro–urban ecotone. Copeia 1999:614–627.

Bunnell, J.F., and R.A. Zampella. 2008. Native fish and anuran assemblages differ between impoundments with and without non–native centrarchids and bullfrogs. Copeia 2008:931–939.

Burbrink, F.T., C.A. Phillips, and E.J. Heske. 1998. A riparian zone in southern Illinois as a potential dispersal corridor for reptiles and amphibians.Biological Conservation 86:107–115.

Burdick, S.L. 2008. Blanchard's Cricket Frog seasonal status and distribution in southeastern South Dakota. M.S. thesis, University of South Dakota, Vermillion.

Burdick, S.L., and D.L. Swanson. 2010. Status, distribution and microhabitats of Blanchard's Cricket Frog *Acris blanchardi* in South Dakota. Herpetological Conservation and Biology 5:9–16.

Burger, W.L., Jr., and A.N. Bragg. 1947. Notes on *Bufo boreas* (B. and G.) from the Gothic region of Colorado. Proceedings of the Oklahoma Academy of Science 27:61–65.

Burger, W.L., P.W. Smith, and H.M. Smith. 1949. Notable records of reptiles and amphibians in Oklahoma, Arkansas, and Texas. Journal of the Tennessee Academy of Science 14:130–134.

Burgess, R.C., Jr. 1950. Development of spade–foot toad larvae under laboratory conditions. Copeia 1950:49–53.

Burgett, A.A., C.D. Wright, G.R. Smith, D.T. Fortune, and S.L Johnson. 2007. Impact of ammonium nitrate on wood frog (*Rana sylvatica*) tadpoles: effects on survivorship and behavior. Herpetological Conservation and Biology 2:29–34.

Burke, V. 1933. Bacteria as food for vertebrates. Science 78:194–195.

Burkett, R.D. 1969. An ecological study of the cricket frog, *Acris crepitans*, in northeastern Kansas. Ph.D. Dissertation, University of Kansas, Lawrence.

Burkett, R.D. 1984. An ecological study of the cricket frog, *Acris crepitans*. Pp. 89–103 *In* R.A. Seigel, L.E. Hunt, J.L. Knight, L. Malaret, and N.L. Zuschlag (eds.), Vertebrate Ecology and Systematics. A Tribute to Henry S. Fitch. Museum of Natural History, University of Kansas, Lawrence.

Burkett, R.D. 1989. Status of diploid/tetraploid gray treefrogs (*Hyla chrysoscelis–Hyla versicolor*) in the mid–South. Pp. 51–57 *In* A.F. Scott (ed.), Proceedings of the Second Annual Symposium on the Natural History of Lower Tennessee and Cumberland Valleys. Center for Field Biology, Austin Peay State University, Clarksville, Tennessee.

Burkett, R.D. 1991. Ecological comparisons of some closely related species of amphibians in Land Between the Lakes and the mid–South. Journal of the Tennessee Academy of Science 66:161–164.

Burkholder, G. 1998. *Hyla chrysoscelis* (Gray Treefrog). Hibernacula. Herpetological Review 29:231.

Burkholder, L.L., and L.V. Diller. 2007. Life history of postmetamorphic coastal tailed frogs (*Ascaphus truei*) in northwestern California. Journal of Herpetology 41:251–262.

Burley, L.A., A.T. Moyer, and J.W. Petranka. 2006. Density of an intraguild predator mediates feeding group size, intraguild egg predation, and intra– and interspecific competition. Oecologia 148:641–649.

Burmeister, S., and W. Wilczynski. 2000. Social signals influence hormones independently of calling behavior in the treefrog, *Hyla cinerea*. Hormones and Behavior

38:201–209.

Burmeister, S., and W. Wilczynski. 2001. Social context influences androgenic effects on calling in the green treefrog (*Hyla cinerea*). Hormones and Behavior 40:550–558.

Burmeister, S., W. Wilczynski, and M.J. Ryan. 1999a. Temporal call changes and prior experience affect graded signalling in the cricket frog. Animal Behaviour 57:611–618.

Burmeister, S., J. Konieczka, and W. Wilczynski. 1999b. Agonistic encounters in a cricket frog (*Acris crepitans*) chorus: behavioral outcomes vary with local competition and within breeding season. Ethology 105:335–347.

Burmeister, S., A.G. Ophir, M.J. Ryan, and W. Wilczynski. 2002. Information transfer during cricket frog contests. Animal Behaviour 64:715–725.

Burne M.R. 2000. Conservation of vernal pool–breeding amphibian communities: habitat and landscape association with community richness. M.S. thesis, University of Massachusetts, Amherst.

Burne, M.R., and C.R. Griffin. 2005. Habitat associations of pool–breeding amphibians in eastern Massachusetts, USA. Wetlands Ecology and Management 13:247–259.

Burnley, J.M. 1973. Eastern spadefoots, *Scaphiopus holbrooki*, found on the south fork of Long Island during 1973. Engelhardtia 6:10.

Burnley, J.M. 1979. Photographic glimpses of the Long Island Pine Barrens. Privately published, Massapequa Park, New York. 11 pp.

Bumzahem, C.B. 1953. Digital trepidation in the subspecies of the toad *Bufo woodhousei*. Copeia 1953:181.

Burress, R.M. 1982. Effects of synergized rotenone on non-target organisms in ponds. U.S. Fish and Wildlife Service, Investigations in Fish Control No 91:1–7.

Burroughs, J. 1913. The Writings of John Burroughs. 17. The Summit of the Years. Houghton Mifflin Co., Boston, Massachusetts.

Bursey, C.R., and W.F. DeWolf II. 1998. Helminths of the frogs, *Rana catesbeiana*, *Rana clamitans*, and *Rana palustris*, from Coshocton County, Ohio. Ohio Journal of Science 98:28–29.

Bursey, C.R., and S.R. Goldberg. 2001. *Falcaustra lowei* n. sp. and other helminths from the Tarahumara frog, *Rana tarahumarae* (Anura: Ranidae), from Sonora, Mexico. Journal of Parasitology 87:340–344.

Bursey, C.R., S.R. Goldberg, and J.B. Bettaso. 2010. Persistence and stability of the component helminth community of the foothill yellow–legged frog, *Rana boylii* (Ranidae), from Humboldt County, California, 1964–1965, versus 2004–2007. American Midland Naturalist 163:476–482.

Burt, C.E. 1932. Records of amphibians and reptiles from the eastern and central United States (1931). American Midland Naturalist 13:75–85.

Burt, C.E. 1933a. Amphibians from the Great Basin of the West and adjacent areas (1932). American Midland Naturalist 14:350–354.

Burt, C.E. 1933b. A contribution to the herpetology of Kentucky. American Midland Naturalist 14:669–679.

Burt, C.E. 1935. Further records of the ecology and distribution of amphibians and reptiles in the Middle West. American Midland Naturalist 16:311–336.

Burt, C.E. 1936. Contributions to the herpetology of Texas. 1. Frogs of the genus *Pseudacris*. American Midland Naturalist 17:770–775.

Burt, C.E. 1937 (1938). Contributions to Texan herpetology. 6. Narrow–mouthed froglike toads (*Microhyla* and *Hypopachus*). Papers of the Michigan Academy of Science, Arts and Letters 23:607–610.

Burton, E.C., M.J. Gray, and A.C. Schmutzer. 2006. Comparison of anuran call survey durations in Tennessee wetlands. Proceedings of the Annual Conference of Southeastern Fish and Wildlife Agencies, pp. 15–18.

Burton, E.C., D.L. Miller, E.L. Styer, and M.J. Gray. 2008. Amphibian ocular malformation associated with Frog Virus 3. Veterinary Journal 177:442–444.

Burton, E.C., M.J. Gray, A.C. Schmutzer, and D.L. Miller. 2009. Differential responses of postmetamorphic amphibians to cattle grazing in wetlands. Journal of Wildlife Management 73:269–277.

Bury, R.B. 1968. The distribution of *Ascaphus truei* in California. Herpetologica 24:39–46.

Bury, R.B. 1970. Food similarities in the tailed frog, *Ascaphus truei*, and the Olympic salamander, *Rhyacotriton olympicus*. Copeia 1970:170–171.

Bury, R.B. 1973. The Cascade frog, *Rana cascadae*, in the North Coast Range of California. Northwest Science 47:228–229.

Bury, R.B. 1983. Differences in amphibian populations in logged and old growth redwood forest. Northwest Science 57:167–178.

Bury, R.B. 1988. Habitat relationships and ecological importance of amphibians and reptiles. Pp. 61–76 *In* K.J. Raedeke (ed.), Streamside Management. Riparian Wildlife and Forestry Interactions. Institute of Forest Resources, University of Washington, Contribution No. 59.

Bury, R.B. 2004. Wildfire, fuel reduction, and herpetofaunas across diverse landscape mosaics in northwestern forests. Conservation Biology 18:968-975.

Bury, R.B. 2008. Amphibians in the Willamette Valley, Oregon: survival in a rapidly urbanizing and developing region. Pp. 435–444 *In* J.C. Mitchell, R.E Jung Brown, and B. Bartholomew (eds.), Urban Herpetology. Herpetological Conservation 3. Society for the Study of Amphibians and Reptiles, Salt Lake City, Utah.

Bury, R.B., and R.A. Luckenbach. 1976. Introduced amphibians and reptiles in California. Biological Conservation 10:1–14.

Bury, R.B., and J.A. Whelan. 1984. Ecology and management of the bullfrog. U.S. Fish and Wildlife Service Resource Publication 155.

Bury, R.B., and P.S. Corn. 1988a. Responses of aquatic and streamside amphibians to timber harvest: a review. Pp. 165–181 *In* K.J. Raedeke (ed.), Streamside Management. Riparian Wildlife and Forestry Interactions. Institute of Forest Resources, University of Washington, Contribution No. 59.

Bury, R.B., and P.S. Corn. 1988b. Douglas–fir forests in the Oregon and Washington Cascades: relation of the herpetofauna to stand age and moisture. Pp. 11–22 *In* R.C. Szaro, K.E. Severson, and D.R. Patton (eds.), Management of Amphibians, Reptiles, and Small Mammals in North America. USDA Forest Service General and Technical Report RM–166.

Bury, R.B., and M.J. Adams. 1999. Variation in age at metamorphosis across a latitudinal gradient for the tailed frog, *Ascaphus truei*. Herpetologica 55:283–291.

Bury, R.B., C.K. Dodd, Jr., and G.M. Fellers. 1980. Conservation of the Amphibia of the United States. A Review. U.S. Fish and Wildlife Service Resource Publication 134.

Bury, R.B., P.S. Corn, and K.B. Aubry. 1991a. Regional patterns of terrestrial amphibian communities in Oregon and Washington. Pp. 341–350 *In* L.F. Ruggiero, K.B. Aubry, A.B. Carey, and M.H. Huff (coordinators), Wildlife and Vegetation of Unmanaged Douglas–fir Forests. USDA Forest Service General Technical Report PNW–GTR–285, Portland, Oregon.

Bury, R.B., P.S. Corn, K.B. Aubrey, F.F. Gilbert, and L.L.C. Jones. 1991b. Aquatic amphibian communities in Oregon and Washington. Pp. 353–362 *In* L.F. Ruggiero, K.B. Aubry, A.B. Carey, and M.H. Huff (coordinators), Wildlife and Vegetation of Unmanaged Douglas–fir Forests. USDA Forest Service General Technical Report PNW–GTR–285, Portland, Oregon.

Bury, R.B., P. Loafman, D. Rofkar, and K.I. Mike. 2001. Clutch sizes and nests of tailed frogs from the Olympic Peninsula, Washington. Northwest Science 75:419–422.

Busack, S.D., and R.B. Bury. 1975. Toad in exile. National Parks and Conservation Magazine (March):15–16.

Busby, W.H., and W.R. Brecheisen. 1997. Chorusing phenology and habitat associations of the crawfish frog, *Rana areolata* (Anura: Ranidae), in Kansas. Southwestern Naturalist 42:210–217.

Busby, W.H., and J.R. Parmelee. 1996. Historical changes in a herpetofaunal assemblage in the Flint Hills of Kansas. American Midland Naturalist 135:81–91.

Busby, W.H., J.T. Collins, and G. Suleiman. 2005. The Snakes, Lizards, Turtles, and Amphibians of Fort Riley and Vicinity. 2nd rev. ed. Kansas Biological Survey, Lawrence. [first edition, 1996]

Buseck, R.S., D.A. Keinath, and M. Geraud. 2005. Species assessment for Great Basin spadefoot toad (*Spea intermontana*) in Wyoming. Wyoming Natural Diversity Database, University of Wyoming, Laramie.

Bush, F.M. 1959. Foods of some Kentucky herptiles. Herpetologica 15:73–77.

Bush, F.M., and E.F. Menhinick. 1962. The food of *Bufo woodhousei fowleri* Hinckley. Herpetologica 18:110–114.

Bush, S.L., H.C. Gerhardt, and J. Schul. 2002. Pattern recognition and call preferences in treefrogs (Anura: Hylidae): a quantitative analysis using a no–choice paradigm. Animal Behaviour 63:7–14.

Bushnell, R.J., E.P. Bushnell, and M.V. Parker. 1939. A chromosome study of five members of the Family Hylidae. Journal of the Tennessee Academy of Science 14:209–215.

Butler, J.M. 2007. *Rana sphenocephala* (Southern Leopard Frog) and *Scaphiopus holbrookii* (Eastern Spadefoot). Reproductive behavior. Herpetological Review 38:444.

Butterfield, B.P. 1988. Age structure and reproductive biology of the Illinois chorus frog (*Pseudacris streckeri illinoensis*) from northeastern Arkansas. M.S. thesis, Arkansas State University, Jonesboro.

Butterfield, B.P., W.E. Meshaka, and S.E. Trauth. 1989. Fecundity and egg mass size of the Illinois chorus frog, *Pseudacris streckeri illinoensis* (Hylidae), from northeastern Arkansas. Southwestern Naturalist 34:556–557.

Butterfield, B.P., H.A. Messer, L.D. Bennie, and J.W. Stanley. 2009. Distribution *Pseudacris crucifer* (Spring Peeper). Herpetological Review 40:234–235.

Buttermore, K.F., P.N. Litkenhaus, D.C. Torpey, G.R. Smith, and J.E. Rettig. 2011. Effects of mosquitofish (*Gambusia affinis*) cues on wood frog (*Lithobates sylvaticus*) tadpole activity. Acta Herpetologica 6:81–85.

Buzo, D. 2008. A GIS model for identifying potential breeding habitat for the Houston toad (*Bufo houstonensis*). M.S. thesis, Texas State University–San Marcos, Texas.

Cabarle, K.C. 2006. Geographic distribution: *Rana sylvatica* (Wood Frog). Herpetological Review 37:359.

Cagle, F.R. 1941. Key to the reptiles and amphibians of Illinois. Museum of Natural and Social Sciences Southern Illinois Normal University, Contribution No. 5: 1–32.

Cagle, F.R. 1942. Herpetological fauna of Jackson and Union counties, Illinois. American Midland Naturalist 28:164–200.

Cagle, F.R. 1954. A Texas population of the cricket frog, *Acris*. Copeia 1954:227–228.

Cahn, A.R. 1926. A set of albino frog eggs. Copeia (151):107–109.

Cahn, A.R. 1939. The barking frog, *Hyla gratiosa*, in northern Alabama. Copeia 1939:52–53.

Cain, B.W., and S.R. Utesch. 1976. An unusual color pattern of the green tree frog *Hyla cinerea*. Southwestern Naturalist 21:235–236.

Caldwell, J.P. 1982. Disruptive selection: a tail color polymorphism in *Acris* tadpoles in response to differential predation. Canadian Journal of Zoology 60:2818–2827.

Caldwell, J.P. 1986. Selection of egg deposition sites: a seasonal shift in the southern leopard frog, *Rana sphenocephala*. Copeia 1986:249–253.

Caldwell, J.P. 1987. Demography and life history of two species of chorus frogs (Anura: Hylidae) in South Carolina. Copeia 1987:114–127.

Caldwell, J.P., J.H. Thorp, and T.O. Jervey. 1980. Predator–prey relationships among larval dragonflies, salamanders, and frogs. Oecologia 46:285–289.

Calef, G.W. 1973a. Natural mortality of tadpoles in a population of *Rana aurora*. Ecology 54:741–758.

Calef, G.W. 1973b. Spatial distribution and "effective" breeding population of red–legged frogs (*Rana aurora*) in Marion Lake, British Columbia. Canadian Field-Naturalist 87:279–284.

Calfee, R.D., and E.E. Little. 2003. The effects of ultraviolet–B radiation on the toxicity of fire–fighting chemicals. Environmental Toxicology and Chemistry 22:1525–1531.

Calhoon, R.E., and D.L. Jameson. 1970. Canonical correlation between variation in weather and variation in size in the Pacific tree frog in southern California. Copeia 1970:124–134.

Calhoun, A.J.K., T.E. Walls, S.S. Stockwell, and M.M. McCollough. 2003. Evaluating several vernal pools as a basis for conservation strategies: a Maine case study. Wetlands 23:70–81.

Calhoun, A.J., N.A. Miller, and M.W. Klemens. 2005. Conserving pool–breeding amphibians in human–dominated landscapes through local implementation of Best Development Practices. Wetlands Ecology and Management 13:291–304.

Calsbeek, R., and S.R. Kuchta. 2011. Predator mediated selection and the impact of developmental stage on viability in wood frog tadpoles (*Rana sylvatica*). BMC Evolutionary Biology 11:e353.

Camara, J., and B.W. Buttrey. 1961. Intestinal protozoa from tadpoles and adults of the mink frog, *Rana septentrionalis* Baird. Proceedings of the South Dakota Academy of Science 40:59–66.

Cameron, A.W., and A.J. Tomlinson. 1962. Dispersal of the introduced green frog in Newfoundland. Bulletin of the National Museum of Canada (183):104–110.

Cameron, J.A. 1940. Effect of fluorine on hatching time and hatching stage in *Rana pipiens*. Ecology 21:288–292.

Camp, C.D., and R.A. Pyron. 2003. Geographic distribution: *Rana sylvatica* (Wood Frog). Herpetological Review 34:382.

Camp, C.D., C.E. Condee, and D.G. Lovell. 1990. Oviposition, larval development, and metamorphosis in the wood frog, *Rana sylvatica* (Anura: Ranidae), in Georgia. Brimleyana 16:17–21.

Camp, C.L. 1915. *Batrachoseps major* and *Bufo cognatus californicus*, new amphibia from southern California. University of California Publications in Zoology 12:327–334.

Camp, C.L. 1916. Description of *Bufo canorus*, a new toad from the Yosemite National Park. University of California Publications in Zoology 17:59–62.

Camp, C.L. 1917. Notes on the systematic status of the toads and frogs of California. University of California Publications in Zoology 17:115–125.

Campbell, B. 1931. *Rana tarahumarae*, a frog new to the United States. Copeia 1931:164.

Campbell, B. 1933. Further notes on spadefoots (*Scaphiopus hammondii*). Copeia 1933:100.

Campbell, B. 1934a. Notes on three amphibians. Crater Lake National Park Nature Notes 2:8.

Campbell, B. 1934b. Report on a collection of reptiles and amphibians made in Arizona during the summer of 1933. Occasional Papers of the Museum of Zoology, University of Michigan No. 289.

Campbell, C.A. 1969. Who cares for the Fowler's toad? Ontario Naturalist 7(4):24–27.

Campbell, C.E., I.G. Warkentin, and K.G. Powell. 2004. Factors influencing the distribution and potential spread on introduced anurans in western Newfoundland. Northeastern Naturalist 11:151–162.

Campbell, G.R. 1978. The Nature of Things on Sanibel. Press Printing Co., Fort Myers, Florida.

Campbell, J.B. 1970a. Life history of *Bufo boreas boreas* in the Colorado Front Range. Ph.D. Dissertation, University of Colorado, Boulder.

Campbell, J.B. 1970b. Food habits of the boreal toad, *Bufo boreas boreas*, in the Colorado Front Range. Journal of Herpetology 4:83–85.

Campbell, J.B. 1970c. Hibernacula of a population of *Bufo boreas boreas* in the Colorado Front Range. Herpetologica 26:278–282.

Campbell, J.B. 1970d. New elevational records for the boreal toad (*Bufo boreas boreas*). Arctic and Alpine Research 2:157–159.

Campbell, J.B. 1976. Environmental controls on boreal toad populations in the San Juan Mountains. Pp. 289–295 *In* H.W. Steinhoff and J.D. Ives (eds.), Ecological Impacts of Snowpack Augmentation in the San Juan Mountains, Colorado. San Juan Ecology Project, Colorado State University Publishing, Fort Collins.

Campbell, J.B., and W.G. Degenhardt. 1971. *Bufo boreas boreas* in New Mexico. Southwestern Naturalist 16:209–220.

Campbell, K.R., T.S. Campbell, and S.A. Johnson. 2010. The use of PVC pipe refugia to evaluate spatial and temporal distributions of native and introduced treefrogs. Florida Scientist 73:78–88.

Campbell, P.M., and W.K. Davis. 1968. Vertebrates in the stomachs of *Bufo valliceps*. Herpetologica 24:327–328.

Campbell, R.A. 1968. A comparative study of the parasites of certain Salientia from Pocohontas State Park, Virginia. Virginia Journal of Science 19:13–20.

Campbell, R.W. 2007. Turkey vultures feeding on live Western toad toadlets. Wildlife Afield 4: 275–277.

Campbell, T.S. 2007. *Osteopilus septentrionalis* (Cuban Treefrog). Saurophagy. Herpetological Review 38:440.

Camponelli, K.M., R.E. Casey, J.W. Snodgrass, S.M. Lev, and E.R. Landa. 2009. Impacts of weathered tire debris on the development of *Rana sylvatica* larvae. Chemosphere 74:717–722.

Capranica, R.R. 1965. The Evoked Vocal Response of the Bullfrog. MIT Press, Cambridge, Massachusetts.

Capranica, R.R. 1968. The vocal repertoire of the bullfrog (*Rana catesbeiana*). Behaviour 31:302–325.

Capranica, R., L. Frishkopf, and M.H. Goldstein, Jr. 1967. Experiments on speech perception: voice and hearing in the bullfrog. Pp. 244–246 *In* Proceedings of a Symposium on Models for the Perception of Speech and Visual Form. MIT Press, Cambridge, Massachusetts.

Capranica, R.R., L.S. Frishkopf, and E. Nevo. 1973. Encoding of geographic dialects in the auditory system of the cricket frog. Science 182:1272–1275.

Carcagno, E. 2009. Orientation of vernal pool amphibians in an industrial forest landscape. M.S. thesis, University of New Hampshire, Durham.

Carey, C. 1976. Thermal physiology and energetics of boreal toads, *Bufo boreas boreas*. Ph.D. Dissertation, University of Michigan, Ann Arbor.

Carey, C. 1993. Hypothesis concerning the causes of the disappearance of boreal toads from the mountains of Colorado. Conservation Biology 7:355–362.

Carey, C., and C.J. Bryant. 1995. Possible interrelations among environmental toxicants, amphibian development, and decline of amphibian populations. Environmental Health Perspectives 103 (Supplement 4):13–17.

Carey, C., and L.J. Livo. 2009. Chytridiomycosis in Woodhouse's toad (*Anaxyrus woodhousii*) in Colorado. Herpetological Review 40:50–52.

Carey, C., G.D. Maniero, and J.F. Stinn. 1996. Effect of cold on immune function and susceptibility to bacterial infection in toads (*Bufo marinus*). Pp. 123–129 *In* F. Geiser, A.J. Hulbert, and S.C. Nicol (eds.), Adaptations to Cold. Tenth International Hibernation Symposium. University of New England Press, Armidale, New South Wales, Australia.

Carey, C., N. Cohen, and L. Rollins–Smith. 1999. Amphibian declines: an immunological perspective. Developmental and Comparative Immunology 23:459–472.

Carey, C., P.S. Corn, M.S. Jones, L.J. Livo, E. Muths, and C.W. Loeffler. 2005. Factors limiting the recovery of boreal toads (*Bufo b. boreas*). Pp. 222–236 *In* M.J. Lannoo (ed.), Amphibian Declines. The Conservation Status of United States Species. University of California Press, Berkeley.

Carey, C., J.E. Bruzgul, L.J. Livo, M.L. Walling, K.A. Kuehl, B.F. Dixon, A.P. Pessier, R.A. Alford, and K.B. Rogers. 2006. Experimental exposures of boreal toads (*Bufo boreas*) to a pathogenic chytrid fungus (*Batrachochytrium dendrobatidis*). EcoHealth 3:5–21.

Carey, S.D. 1982. Geographic distribution. *Eleutherodactylus planirostris*. Herpetological Review 13:130.

Carfagno, G.L.F., J.M. Carithers, L.J. Mycoff, and R.M. Lehtinen. 2011. How the cricket frog lost its spot: the inducible defense hypothesis. Herpetologica 67:386–396.

Carl, G.C. 1942. The western spadefoot toad in British Columbia. Copeia 1942:129.

Carl, G.C. 1943. The amphibians of British Columbia. British Columbia Provincial Museum, Bulletin No. 2.

Carl, G.C. 1949. Extensions of known ranges of some amphibians in British Columbia. Herpetologica 5:139.

Carl, G.C., and G.A. Hardy. 1943. Report on a collecting trip to the Lac La Hache area, British Columbia. Report of the British Columbia Provincial Museum for 1942:25–49.

Carl, G.C., and I. M. Cowan. 1945. Notes on some frogs and toads of British Columbia. Copeia 1945:52–53.

Carpenter, C.C. 1953a. A study of hibernacula and hibernating associations of snakes and amphibians in Michigan. Ecology 34:74–80.

Carpenter, C.C. 1953b. Aggregation behavior of tadpoles of *Rana p. pretiosa*. Herpetologica 9:77–78.

Carpenter, C.C. 1953c. An ecological survey of the herpetofauna of the Grand Teton–Jackson Hole area of Wyoming. Copeia 1953:170–174.

Carpenter, C.C. 1954a. A study of amphibian movement in the Jackson Hole Wildlife Park. Copeia 1954:197–200.

Carpenter, C.C. 1954b. Feeding aggregation of narrow mouth toads (*Microhyla carolinensis olivacea*). Proceedings of the Oklahoma Academy of Science 35:45.

Carpenter, C.C. 1955. The amphibians and reptiles of the University of Oklahoma Biological Station area in south central Oklahoma. Proceedings of the Oklahoma Academy of Science 36:39–46.

Carpenter, C.C. 1957. Michigan's frogs and toads. The Jack-Pine Warbler 35:64–67.

Carpenter, C.C., and D.E. Delzell. 1951. Road records as indicators of differential spring migrations of amphibians. Herpetologica 7:63–64.

Carpenter, H.L., and E.O. Morrison. 1973. Feeding behavior of the bullfrog, *Rana catesbeiana*, in northcentral Texas. Bios 44:188–193.

Carpenter, N.M., M.L. Casazza, and G.D. Wylie. 2002. *Rana catesbeiana* (Bullfrog). Diet. Herpetological Review 33:130.

Carr, A.F., Jr. 1934. A key to the breeding–songs of Florida frogs. Florida Naturalist 7(2):19–23.

Carr, A.F., Jr. 1937. The geographic and ecological distribution of the reptiles and amphibians of Florida. Ph.D. Dissertation, University of Florida, Gainesville.

Carr, A.F., Jr. 1940a. A contribution to the herpetology of Florida. University of Florida Publication, Biological Science Series 3(1):1–118.

Carr, A.F., Jr. 1940b. Dates of frog choruses in Florida. Copeia 1940:55.

Carr, L.W., and L. Fahrig. 2001. Effect of road traffic on two amphibian species of differing vagility. Conservation Biology 15:1071–1078.

Carr, L.W., L. Fahrig, and S.E. Pope. 2000. Impacts of landscape transformation by roads. Pp. 225–243 *In* K.J. Gutzwiller (ed.), Applying Landscape Ecology in Biological Conservation. Springer–Verlag, New York.

Cartwright, M.E., S.E. Trauth, and J.D. Wilhide. 1998. Wood frog (*Rana sylvatica*) use of wildlife ponds in northcentral Arkansas. Journal of the Arkansas Academy of Science 52:32–34.

Cary, J.A. 2010. Determining habitat characteristics that predict oviposition site selection for pond–breeding northern red–legged frogs (*Rana aurora*) in Humboldt County, California. M.A. thesis, Humboldt State University, Arcata, California.

Case, S.M. 1976. Evolutionary studies in selected North American frogs of the genus *Rana* (Amphibia, Anura). Ph.D. Dissertation, University of California, Berkeley.

Case, S.M. 1978a. Electrophoretic variation in two species of ranid frogs, *Rana boylei* and *R. muscosa*. Copeia 1978:311–320.

Case, S.M. 1978b. Biochemical systematics of members of the genus *Rana* native to western North America. Systematic Zoology 27:299–311.

Case, S.M., P.G. Haneline, and M.F. Smith. 1975. Protein variation in several species of *Hyla*. Systematic Zoology 24:281–295.

Cash, M.N., and J.P. Bogart. 1978. Cytological differentiation of the diploid–tetraploid species pair of North American treefrogs (Amphibia, Anura, Reptilia). Journal of Herpetology 12:555–558.

Cash, W.B. 1994. Herpetofaunal diversity of a temporary wetland in the Southeast Atlantic Coastal Plain. M.S. thesis, Georgia Southern University, Statesboro.

Casper, G.S. 1996. Geographic distributions of the amphibians and reptiles of Wisconsin. An interim report of the Wisconsin Herpetological Atlas Project. Milwaukee Public Museum, Milwaukee, Wisconsin.

Casper, G.S. 1998. Review of the status of Wisconsin amphibians. Pp. 199–205 *In* M.J. Lannoo (ed.), Status & Distribution of Midwestern Amphibians. University of Iowa Press, Iowa City.

Castano, B., S. Miely, G.R. Smith, and J.E. Rettig. 2010. Interactive effects of food availability and temperature on wood frog (*Rana sylvatica*) tadpoles. Herpetological Journal 20:209–211.

Catalano, P.A., and A.M. White. 1977. New host records for *Haematoloechus complex* (Seely) Krull, 1933 from *Hyla crucifer* and *Rana sylvatica*. Ohio Journal of Science 77:99.

Catesby, M. 1731–1747. The Natural History of Carolina, Florida and the Bahama Islands. Benjamin White, London. [Frogs appear on Plates 69-72 in Volume 2. These were included in Part 9 of Catesby, issued on 7 June 1739.]

Cecil, S.G., and J.J. Just. 1979. Survival rate, population density and development of a naturally occurring anuran larvae (*Rana catesbeiana*). Copeia 1979:447–453.

Cely, J.E., and J.A. Sorrow, Jr. 1986. Distribution and habitat of *Hyla andersonii* in South Carolina. Journal of Herpetology 20:102–104.

Chamberlain, E.B. 1939. Frogs and toads of South Carolina. Charleston Museum Leaflet No. 12.

Chamberlain, F.M. 1897. Notes on the edible frogs of the United States and their artificial propagation. U.S. Bureau of Fisheries for 1897:249–261.

Chambers, D.L. 2011. Increased conductivity affects corticosterone levels and prey consumption in larval amphibians. Journal of Herpetology 45:219–223.

Chan, L.M., and K.R. Zamudio. 2009. Population differentiation of temperate amphibians in unpredictable environments. Molecular Ecology 18:3185–3200.

Chan–McLeod, A. C. A. 2003. Factors affecting the permeability of clearcuts to red–legged frogs. Journal of Wildlife Management 67:663–671.

Chan–McLeod, A.A., and B. Wheeldon. 2004. *Rana aurora* (Northern Red–legged Frog). Habitat and movement. Herpetological Review 35:375.

Chan–McLeod, A.C.A., and A. Moy. 2007. Evaluating residual tree patches as stepping stones and short–term refugia for red–legged frogs. Journal of Wildlife Management 71:1836–1844.

Chance, K.B. 2002. A telemetric study of winter habitat selection by the American bullfrog, *Rana catesbeiana*, in east–central Kansas. M.S. thesis, Emporia State University, Emporia, Kansas.

Chandler, J.H., and L.L. Marking. 1975. Toxicity of the lampricide 3–trifluoromethyl–4–nitrophenol (TFM) to selected aquatic invertebrates and frog larvae. U.S. Fish and Wildlife Service, Investigations in Fish Control No. 62:1–7.

Chandler, J.H., and L.L. Marking. 1982. Toxicity of rotenone to selected aquatic invertebrates and frog larvae. Progressive Fish Culturist 44:78–80.

Chantell, C.J. 1970. Upper Pliocene frogs from Idaho. Copeia 1970:654–664.

Chantell, C.J. 1971. Fossil amphibians from the Egelhoff Local Fauna of northcentral Nebraska. Contributions from the Museum of Paleontology, University of Michigan 23:239–246.

Chapel, W.L. 1939. Field notes on *Hyla wrightorum* Taylor. Copeia 1939:225–227.

Chapman, B.R., and S.D. Castro. 1972. Additional vertebrate prey of the loggerhead shrike. Wilson Bulletin 84:496–497.

Chapman, R.M. 1966. Light wavelength and energy preferences of the bullfrog: evidence for color vision. Journal of Comparative Physiology and Psychology 61:429–435.

Chatfield, M.W.H., B.B. Rothermel, C.S. Brooks, and J.B. Kay. 2009. Detection of *Batrachochytrium dendrobatidis* in amphibians from the Great Smoky Mountains of North Carolina and Tennessee. Herpetological Review 40:176–179.

Cheek, A. O., C. F. Ide, J. E. Bollinger, C. V. Rider, and J. A. McLachlan. 1999. Alteration of leopard frog (*Rana pipiens*) metamorphosis by the herbicide acetochlor. Archives of Environmental Contamination and Toxicology 37:70–77.

Chelgren, N.D., D.K. Rosenberg, S.S. Heppell, and A.I. Gitelman. 2006. Carryover aquatic effects on survival of metamorphic frogs during pond emigration. Ecological Applications 16:250–261.

Chelgren, N.D., C.A. Pearl, M.J. Adams, and J. Bowerman. 2008a. Demography and movement in a relocated population of Oregon spotted frogs (*Rana pretiosa*): influence of season and gender. Copeia 2008:742–751.

Chelgren, N.D., D.K. Rosenberg, S.S. Heppell, and A. I. Gitelman. 2008b. Individual variation affects departure rate from the natal pond in an ephemeral pond–breeding anuran. Canadian Journal of Zoology 86: 260–267.

Chelgren, N.D., B. Samora, M.J. Adams, and B. McCreary. 2011. Using spatiotemporal models and distance sampling to map the space use and abundance of newly metamorphosed western toads (*Anaxyrus boreas*). Herpetological Conservation and Biology 6:175–190.

Chen, C.Y., K.M. Hathaway, and C.L. Folt. 2004. Multiple stress effects of Vision™ herbicide, pH, and food on zooplankton and larval amphibian species from forest wetlands. Environmental Toxicology and Chemistry 23:823–831.

Chen, C.Y., K.M. Hathaway, D.G. Thompson, and C.L. Folt. 2008. Multiple stressor effects of herbicide, pH, and food on wetland zooplankton and a larval amphibian. Ecotoxicology and Environmental Safety 71:209–218.

Chen, G.J., and S.S. Desser. 1989. The Coccidia (Apicomplexa: Eimeriidae) of frogs from Algonquian Park, with descriptions of two new species. Canadian Journal of Zoology 67:1686–1689.

Chen, T.–H., J.A. Gross, and W.H. Karasov. 2006. Sublethal effects of lead on northern leopard frog (*Rana pipiens*) tadpoles. Environmental Toxicology and Chemistry 25:1383–1389.

Cheng, T.C. 1961. Description, life history and developmental pattern of *Glypthelmins pennsylvaniensis* n. sp. (Trematoda: Brachycoeliidae), new parasite of frogs. Journal of Parasitology 47:469–477.

Cheng, T.–H. 1929a. Intersexuality in *Rana cantabrigensis*. Journal of Morphology and Physiology 48:345–369.

Cheng, T.–H. 1929b. A new case of intersexuality in *Rana cantabrigensis*. Biological Bulletin 57:412–421.

Cheng, T.–H. 1930. Intersexuality in tadpoles of *Rana cantabrigensis*. Papers of the Michigan Academy of Science, Arts and Letters 11:353–368.

Chermock, R.L. 1952. A key to the amphibians and reptiles of Alabama. Geological Survey of Alabama, Museum Paper 23.

Chestnut, T., J.E. Johnson, and R.S. Wagner. 2008. Results of amphibian chytrid (*Batrachochytrium dendrobatidis*) sampling in Denali National Park, Alaska, USA. Herpetological Review 39:202–204.

Childs, C. 2000. The Secret Knowledge of Water. Little, Brown and Co., New York.

Childs, H.E., Jr. 1953. Selection by predation on albino and normal spadefoot toads. Evolution 7:228–233.

Childs, H.E., Jr. 1955. An incidence of albinism in yellow–legged frogs. Bulletin of the Southern California Academy of Sciences 54:126–127.

Chivers, D.P., and R.S. Mirza. 2001. Importance of predator diet cues in responses of larval wood frogs to fish and invertebrate predators. Journal of Chemical Ecology 27:45–51.

Chivers, D.P., J.M. Kiesecker, E.L. Wildy, L.K. Belden, L.B. Kats, and A.R. Blaustein. 1999a. Avoidance response of post–metamorphic anurans to cues of injured conspecifics and predators. Journal of Herpetology 33:472–476.

Chivers, D.P., J.M. Kiesecker, A. Marco, E.L. Wildy, and A.R. Blaustein. 1999b. Shifts in life history as a response to predation in Western toads (*Bufo boreas*). Journal of Chemical Ecology 25:2455–2463.

Chivers, D.P., J.M. Kiesecker, A. Marco, J. DeVito, M.T. Anderson, and A.R. Blaustein. 2001a. Predator–induced life history changes in amphibians: egg predation induces hatching. Oikos 92: 135–142.

Chivers, D.P., E.L. Wildy, J.M. Kiesecker and A.R. Blaustein. 2001. Avoidance response of juvenile Pacific treefrogs to chemical cues of introduced predatory bullfrogs. Journal of Chemical Ecology 27: 1667–1676.

Chrapliwy, P.S. 1953. Taxonomy and distribution of the spadefoot toads of North America (Salientia: Pelobatidae). M.S. thesis, University of Kansas, Lawrence.

Chrapliwy, P.S., and K.L. Williams. 1957. A species of frog new to the fauna of the United States: *Pternohyla fodiens* Boulenger. Natural History Miscellanea (160):1–2.

Chrapliwy, P.S., and E. V. Malnate. 1961. The systematic status of the spadefoot toad *Spea laticeps* Cope. Texas Journal of Science 13:160–162.

Christein, D., and D.H. Taylor. 1978. Population dynamics in breeding aggregations of the American toad, *Bufo americanus* (Amphibia, Anura, Bufonidae). Journal of Herpetology 12:17–24.

Christein, D., S.I. Guttman, and D.H. Taylor. 1979. Heterozygote deficiencies in a breeding population of *Bufo americanus* (Bufonidae: Anura): the test of a hypothesis. Copeia 1979:498–502.

Christian, K.A. 1976. Ontogeny of the food niche of *Pseudacris triseriata*. M.S. thesis, Colorado State University, Fort Collins.

Christian, K.A. 1982. Changes in the food niche during postmetamorphic ontogeny of the frog *Pseudacris triseriata*. Copeia 1982:73–80.

Christiansen, J.L. 1981. Population trends among Iowa's amphibians and reptiles. Proceedings of the Iowa Academy of Science 88:24–27.

Christiansen, J.L. 1998. Perspectives on Iowa's declining amphibians and reptiles. Journal of the Iowa Academy of Science 105:109–114.

Christiansen, J.L. 2001. Non–native amphibians and reptiles in Iowa. Journal of the Iowa Academy of Science 108:210–211.

Christiansen, J.L., and R.R. Burken. 1978. The endangered and uncommon reptiles and amphibians of Iowa. Iowa Science Teachers Journal, Special Issue, 26 pp.

Christiansen, J.L., and C.M. Mabry. 1985. The amphibians and reptiles of Iowa's Loess Hills. Proceedings of the Iowa Academy of Science 92:159–163.

Christiansen, J.L., and R.M. Bailey. 1991. The salamanders and frogs of Iowa. Iowa Department of Natural Resources, Nongame Technical Series 3.

Christiansen, J.L., and H. Feltman. 2000. A relationship between trematode metacercariae and bullfrog limb abnormalities. Journal of the Iowa Academy of Science 107:79–85.

Christie, K., J. Schul, and A.S. Feng. 2010. Phonotaxis to male's calls embedded within a chorus by female gray treefrogs, *Hyla versicolor*. Journal of Comparative Physiology A 196:569–579.

Christie, P. 1997. Reptiles and Amphibians of Prince Edward County Ontario. Natural Heritage Books, Toronto.

Christin, M.–S., A.D. Gendron, P. Brousseau, L. Ménard, D.J. Marcogliese, D. Cyr, S. Ruby, and M. Fournier. 2003. Effects of agricultural pesticides on the immune system of *Rana pipiens* and on its resistance to parasitic infection. Environmental Toxicology and Chemistry 22:1127–1133.

Christin, M.–S., L. Ménard, A.D. Gendron, S. Ruby, D. Cyr, D.J. Marcogliese, L. Rollins–Smith, and M. Fournier. 2004. Effects of agricultural pesticides on the immune system of *Xenopus laevis* and *Rana pipiens*. Aquatic Toxicology 67:33–43.

Christman, B., K.B. Malmos, A.T. Holycross, and R.C. Averill–Murray. 2000. *Bufo punctatus* and *Bufo woodhousii* (Red–spotted Toad and Woodhouse's Toad). Hybridization. Herpetological Review 31:99–100.

Christman, S.P. 1970. *Hyla andersoni* in Florida. Quarterly Journal of the Florida Academy of Sciences 33:80.

Christman, S.P. 1974. Geographic variation for salt water tolerance in the frog *Rana sphenocephala*. Copeia 1974:773–778.

Christoffel, R., R. Hay and M. Wolfgram. 2001. Amphibians of Wisconsin. Wisconsin Department of Natural Resources PUB–ER–105.

Chubbs, T.E., and F.R. Phillips. 1998. Distribution of the wood frog, *Rana sylvatica*, in Labrador. Canadian Field-Naturalist 112:329-331.

Church, D.R. 2008. Role of current versus historical hydrology in amphibian species turnover within local pond communities. Copeia 2008:115–125.

Church, D.R., H.M. Wilbur, S.M. Roble, F.C. Huber, and M.W. Donahue. 2002. Observations on breeding by eastern spadefoots (*Scaphiopus holbrookii*) in Augusta County, Virginia. Banisteria 20:71–72.

Churchill, T.A., and K.B. Storey. 1996. Organ metabolism and cryoprotectant synthesis during freezing in spring peepers *Pseudacris crucifer*. Copeia 1996:517–525.

Claflin, W.J. 1962. Larval growth and development of *Pseudacris nigrita triseriata* and *Rana pipiens* in semi–permanent ponds in south–eastern South Dakota. M.A. thesis, State University of South Dakota, Vermillion.

Clark, C.B., and A.N. Bragg. 1950. A comparison of the ovaries of two species of *Bufo* with different ecological requirements. Pp. 143–152 *In* Researches on the Amphibia of Oklahoma, University of Oklahoma Press, Norman.

Clark, G.W., J. Bradford, and R. Nussbaum. 1969. Blood parasites of some Pacific Northwest amphibians. Bulletin of the Wildlife Disease Association 5:117–118.

Clark, H.O., Jr. 2008. *Spea hammondii* (Western Spadefoot). Predation and use as burrow decorations. Herpetological Review 39:80–81.

Clark, K.C., and R.J. Hall. 1985. Effects of elevated hydrogen ion and aluminum concentration on the survival of amphibian embryos and larvae. Canadian Journal of Zoology 63:116–123.

Clark, K.C., and B.D. LaZerte. 1985. A laboratory study of the effects of aluminum and pH on amphibian eggs and tadpoles. Canadian Journal of Fisheries and Aquatic Sciences 42:1544–1551.

Clark, K.C., and B.D. LaZerte. 1987. Intraspecific variation in hydrogen ion and aluminum toxicity in *Bufo americanus* and *Ambystoma maculatum*. Canadian Journal of Fisheries and Aquatic Sciences 44:1622–1628.

Clark, P.J., J.M. Reed, B.G. Tavernia, B.S. Windmiller, and J.V. Regosin. 2008. Urbanization effects on spotted salamander and wood frog presence and abundance. Pp. 67–75 *In* J.C. Mitchell, R.E. Jung Brown, and B. Bartholomew (eds.), Urban Herpetology. Herpetological Conservation 3. Society for the Study of Amphibians and Reptiles, Salt Lake City, Utah.

Clark, R.F. 1949. Snakes of the hill parishes of Louisiana. Journal of the Tennessee Academy of Science 24:244–261.

Clark, V.R. 2001. *Rana heckscheri* (River Frog). Ectoparasites. Herpetological Review 32:36.

Clarke, R.D. 1972. The effect of toe clipping on survival in Fowler's toad (*Bufo woodhousei fowleri*). Copeia 1972:182–185.

Clarke, R.D. 1974a. Activity and movement patterns in a population of Fowler's toad, *Bufo woodhousei fowleri*. American Midland Naturalist 92:257–274.

Clarke, R.D. 1974b. Food habits of the toads, genus *Bufo* (Amphibia, Bufonidae). American Midland Naturalist 91:140–147.

Clarke, R.D. 1974c. Postmetamorphic growth rates in a natural population of Fowler's toad, *Bufo woodhousei fowleri*. Canadian Journal of Zoology 52:1489–1498.

Clarke, R.D. 1977. Postmetamorphic survivorship of Fowler's toad, *Bufo woodhousei fowleri*. Copeia 1977:594–597.

Clarke, R.F. 1958. An ecological study of reptiles and amphibians in Osage County, Kansas. Emporia State Research Studies 7:1–52.

Clarkson, R.W., and J.C. DeVos, Jr. 1986. The bullfrog, *Rana catesbeiana* Shaw, in the lower Colorado River, Arizona–California. Journal of Herpetology 20:42–49.

Clarkson, R.W., and J.C. Rorabaugh. 1989. Status of leopard frogs (*Rana pipiens* complex: Ranidae) in Arizona and southeastern California. Southwestern Naturalist 34:531–538.

Claussen, D.L. 1973a. The water relations of the tailed frog, *Ascaphus truei*, and the Pacific treefrog, *Hyla regilla*. Comparative Biochemistry and Physiology 44A:155–171.

Claussen, D.L. 1973b. The thermal relations of the tailed frog, *Ascaphus truei*, and the Pacific treefrog, *Hyla regilla*. Comparative Biochemistry and Physiology 44A:137–153.

Claussen, D.L. 1974. Urinary bladder water reserves in the terrestrial toad, *Bufo fowleri*, and the aquatic frog, *Rana clamitans*. Herpetologica 30:360–367.

Claussen, D.L., and J.R. Layne, Jr. 1983. Growth and survival of juvenile toads, *Bufo woodhousei*, maintained on different diets. Journal of Herpetology 17:107–112.

Clawson, M.E., and T.S. Baskett. 1982. Herpetofauna of the Ashland Wildlife Area, Boone County, Missouri. Transactions, Missouri Academy of Science 16:5–16.

Clawson, R.G., B.G. Lockley, and R.H. Jones. 1997. Amphibian responses to helicopter harvesting in forested floodplains of low order, blackwater streams. Forest Ecology and Management 90:225–235.

Clemmer, J.H., E.Z. Miller, L. Wolgamott, G.R. Smith, and J.E. Rettig. 2011. The effects of temperature and salinity on wood frog (*Lithobates sylvaticus*) tadpole growth and survival. Bulletin of the Maryland Herpetological Society 47:38–41.

Cobb, J.N. 1904. The commercial fisheries of the Hawaiian Islands in 1903. Pp. 433–512 *In* Report of the U.S. Bureau of Fisheries for 1904. U.S. Bureau of Fisheries Document 590. U.S. Government Printing Office, Washington, DC. [see pp. 482-483, 487]

Cochran, D.M. 1932. Our friend the frog. National Geographic Magazine 61(5):628-654.

Cochran, D.M. 1961. Type specimens of reptiles and amphibians in the United States National Museum. Bulletin of the United States National Museum (220): xv + 291 pp.

Cochran, D.M., and C.J. Goin. 1970. The New Field Book of Reptiles and Amphibians. G.P. Putnam's Sons, New York.

Cochran, P.A. 1999. *Rana sylvatica* (Wood Frog). Predation. Herpetological Review 30:94.

Cochran, P.A. 2008. An unusual microhabitat for an American toad (*Anaxyrus americanus*). Bulletin of the Chicago Herpetological Society 43:62.

Cochran, P.A., and J.A. Cochran. 2003. *Pseudacris crucifer* (Spring Peeper). Predation. Herpetological Review 34:360.

Cochran, P.M. 2007. *Bufo americanus* (American Toad). Predation. Herpetological Review 38:178.

Cockerell, T. 1927. Zoology of Colorado. University of Colorado, Boulder, Colorado.

Cocroft, R.B. 1994. A cladistic analysis of chorus frog phylogeny (Hylidae: *Pseudacris*). Herpetologica 50:420–437.

Cocroft, R.B., and M.J. Ryan. 1995. Patterns of advertisement call evolution in toads and chorus frogs. Animal Behaviour 49:283–303.

Cogger, H.G., and R.G. Zweifel (eds.). 1992. Reptiles & Amphibians. Smithmark Publishers, New York.

Cogger, H.G., and R.G. Zweifel (eds.). 1998. Encyclopedia of Reptiles & Amphibians. Second edition, Smithmark Publishers, New York.

Coggins, J.R., and R.A. Sajdak. 1982. A survey of helminth parasites in the salamanders and certain anurans from Wisconsin. Proceedings of the Helminthological Society of Washington 49:99–102.

Cohen, J.S., S. Ng and B. Blossey. 2012a. Quantity counts: amount of litter determines tadpole performance in experimental mesocosms. Journal of Herpetology 46:85–90.

Cohen, J.S., J.C. Maerz, and B. Blossey. 2012b. Traits, not origin, explain impacts of plants on larval amphibians. Ecological Applications 22:218–222.

Cohen, N.W., and W.E. Howard. 1958. Bullfrog food and growth at the San Joaquin Experimental Range, California. Copeia 1958:223–225.

Colbert, N., R.F. Baldwin, and R. Thiet. 2011. A developer-initiated conservation plan for pool–breeding amphibians in Maine, USA: a case study. Journal of Conservation Planning 7: 27–38.

Cole, C.J. 1962. Notes on the distribution and food habits of *Bufo alvarius* at the eastern edge of its range. Herpetologica 18:172–175.

Cole, C.J. 1966. The chromosomes of *Acris crepitans blanchardi*. Copeia 1966:578–580.

Cole, C.J., C.H. Lowe, and J.W. Wright. 1968. Karyotypes of eight species of toads (genus *Bufo*) in North America. Copeia 1968:96–100.

Cole, E.C., W.C. McComb, M. Newton, C.L. Chambers, and J.P. Leeming. 1997. Response of amphibians to clearcutting, burning, and glyphosate application in the Oregon Coast Range. Journal of Wildlife Management 61:656–664.

Cole, L.M., and J.E. Casida. 1983. Pyrethroid toxicology in the frog. Pesticide Biochemistry and Physiology 20:217–224.

Coleman, J.L., N.B. Ford and K. Herriman. 2008. A road survey of amphibians and reptiles in a bottomland hardwood forest. Southeastern Naturalist 7:339–348.

Collier, A., J.B. Keiper, and L.P. Orr. 1998. The invertebrate prey of the northern leopard frog, *Rana pipiens*, in a northeastern Ohio population. Ohio Journal of Science 98:39–41.

Collins, J.P. 1975. A comparative study of the life history strategies in a community of frogs. Ph.D. Dissertation, University of Michigan, Ann Arbor.

Collins, J.P., and M.A. Lewis. 1979. Overwintering tadpoles and breeding season variation in the *Rana pipiens* complex in Arizona. Southwestern Naturalist 24:371–373.

Collins, J.P., and H.M. Wilbur. 1979. Breeding habits and habitats of the amphibians of the Edwin S. George Reserve, Michigan, with notes on the local distribution of fishes. Occasional Papers of the Museum of Zoology, University of Michigan No. 686.

Collins, J.T. 1993. Amphibians & Reptiles in Kansas. 3rd rev. ed. University of Kansas, Museum of Natural History, Lawrence. [first edition, 1974; second edition, 1982]

Collins, J.T., and S.L. Collins. 2006. Amphibians, turtles and reptiles of Cheyenne Bottoms. Sternberg Museum of Natural History, Fort Hays State University, Fort Hays, Kansas . [first edition, 1993]

Collins, J.T., S.L. Collins, and T.W. Taggart. 2010. Amphibians, Reptiles, and Turtles in Kansas. Eagle Mountain Publishing, Eagle Mountain, Utah.

Collins, J.T., S.L. Collins, and T.W. Taggart. 2011. Amphibians, Reptiles, and Turtles of the Cimarron National Grassland Kansas. U.S. Forest Service, Cimarron National Grassland, Elkhart, Kansas. [first edition, 1991]

Collopy, M.W., and J.R. Koplin. 1983. Diet, capture success, and mode of hunting by female American kestrels in winter. Condor 85:369–371.

Coman, D.R. 1972. The native mammals, reptiles and amphibians of Mount Desert Island, Maine. Privately published, Bar Harbor, Maine.

Combs, A., G. Hess, C. Chapman, A. Stovall, Z. Burns,

S. Matthews, and J.D. Goins. 2005. *Rana catesbeiana* (American Bullfrog). Diet. Herpetological Review 36:439.

Conant, R. 1940. *Rana virgatipes* in Delaware. Herpetologica 1:176–177.

Conant, R. 1945. An annotated check list of the amphibians and reptiles of the Del–Mar–Va Peninsula. Society of Natural History of Delaware, pp. 1–8.

Conant, R. 1947. The carpenter frog in Maryland. Maryland, A Journal of Natural History 17:72–73.

Conant, R. 1957. Reptiles and amphibians of the northeastern states. Zoological Society of Philadephia, 3rd edition. [first edition, 1947; 2nd edition, 1952]

Conant, R. 1962. Notes on the distribution of reptiles and amphibians in the Pine Barrens of southern New Jersey. New Jersey Nature Notes (New Jersey Audubon Society) 17:16–21.

Conant, R. 1977 (1978). Semiaquatic reptiles and amphibians of the Chihuahuan Desert and their relationships to drainage patterns of the region. Pp. 455–491 *In* R.H. Wauer and D.H. Riskind (eds.), Transactions of the Symposium on the Biological Resources of the Chihuahuan Desert Region, United States and Mexico. U.S. National Park Service Transactions and Proceedings Series No. 3.

Conant, R. 1979. A zoogeographical review of the amphibians and reptiles of southern New Jersey, with emphasis on the Pine Barrens. Pp. 467–488 *In* R.T.T. Forman (ed.), Pine Barrens: Ecosystem and Landscape. Academic Press, New York.

Conant, R., and J.T. Collins. 1998. A Field Guide to the Reptiles and Amphibians. Eastern and Central North America. 3rd exp. ed. Houghton Mifflin Co., Boston, Massachusetts [first edition, 1958; second edition, 1991]

Conant, R., J.C. Mitchell, and C.A. Pague. 1990. Herpetofauna of the Virginia barrier islands. Virginia Journal of Science 41:364–380.

Conaway, C.H., and D.E. Metter. 1967. Skin glands associated with breeding in *Microhyla carolinensis*. Copeia 1967:672–673.

Conger, P. 1922. Distribution of batrachians in Wisconsin. M.S. thesis, University of Wisconsin, Madison.

Conlon, J.M. 2004. The therapeutic potential of antimicrobial peptides from frog skin. Reviews in Medical Microbiology 15:17–25.

Conlon, J.M., T. Halverson, J. Dulka, J.E. Platz, and F.C. Knoop. 1999. Peptides with antimicrobial activity of the brevinin–1 family isolated from the skin secretions of the southern leopard frog, *Rana sphenocephala*. Journal of Peptide Research 54:522–527.

Conlon, J.M., A. Sonnevend, C. Davidson, D.D. Smith, and P.F. Nielsen. 2004a. The ascaphins: a family of antimicrobial peptides from the skin secretions of the most primitive extant frog, *Ascaphus truei*. Biochemical and Biophysical Research Communications 320:170–175.

Conlon, J.M., J. Kolodziejek, and N. Nowotny. 2004b. Antimicrobial peptides from ranid frogs: taxonomic and phylogenetic markers and a potential source of new therapeutic agents. Biochimica et Biophysica Acta 1696:1–14.

Conlon, J.M., A. Sonnevend, A. Davidson, A. Demandt, and T. Jouenne. 2005a. Host–defense peptides isolated from the skin secretions of the northern red–legged frog *Rana aurora aurora*. Developmental and Comparative Immunology 29:83–90.

Conlon, J.M., T. Jouenne, P. Cosette, D. Cosquer, H. Vaudry, C.K. Taylor, and P.W. Abel. 2005b. Bradykinin–related peptides and tryptophyllins in the skin secretions of the most primitive extant frog, *Ascaphus truei*. General and Comparative Endocrinology 143:193–199.

Conlon, J.M., B. Abraham, A. Sonnevend, T. Jouenne, P. Cosette, J. Leprince, H. Vaudry, and C.A. Bevier. 2005c. Purification and characterization of antimicrobial peptides from the skin secretions of the carpenter frog *Rana virgatipes* (Ranidae, Aquarana). Regulatory Peptides 131:38–45.

Conlon, J.M., N. Al–Ghafari, L. Coquet, J. Leprince, T. Jouenne, H. Vaudry, and C. Davidson. 2006. Evidence from peptidomic analysis of skin secretions that the red–legged frogs, *Rana aurora draytonii* and *Rana aurora aurora*, are distinct species. Peptides 27:1305–1312.

Conlon, J.M., L. Coquet, J. Leprince, T. Jouenne, H. Vaudry, J. Kolodziejek, N. Nowotny, C.R. Bevier, and P.E. Moler. 2007a. Peptidomic analysis of skin secretions from *Rana heckscheri* and *Rana okaloosae* provides insight into phylogenetic relationships among frogs of the *Aquarana* species group. Regulatory Peptides 138:87–93.

Conlon, J.M., C.R. Bevier, L. Coquet, J. Leprince, T. Jouenne, H. Vaudry, and B.R. Hossack. 2007b. Peptidomic analysis of skin secretions supports separate species status for the tailed frogs, *Ascaphus truei* and *Ascaphus montanus*. Comparative Biochemistry and Physiology 2D:121–125.

Conlon, J.M., E. Ahmed, L. Coquet, T. Jouenne, J. Leprince, H. Vaudry, and J.D. King. 2009. Peptides with potent cytolytic activity from the skin secretions of the North American leopard frogs, *Lithobates blairi* and *Lithobates yavapaiensis*. Toxicon 53:699–705.

Conlon, J.M., L. Coquet, J. Leprince, T. Jouenne, H. Vaudry, and J.D. King. 2010. Primary structures of skin antimicrobial peptides indicate a close, but not conspecific, phylogenetic relationship between the leopard frogs *Lithobates onca* and *Lithobates yavapaiensis* (Ranidae). Comparative Biochemistry and Physiology 151C:313–317.

Conlon, J.M., M. Mechkarska, E. Ahmed, L. Coquet, T.

Jouenne, J. Leprince, H. Vaudry, M.P. Hayes, and G. Padgett–Flohr. 2011. Host defense peptides in skin secretions of the Oregon spotted frog *Rana pretiosa*: implications for species resistance to chytridiomycosis. Developmental and Comparative Immunology 35:644–649.

Connell, J. 2006. Mercury concentrations in the muscles of the North American bullfrogs (*Rana catesbeiana*). Ph.D. Dissertation, University of Nevada–Las Vegas.

Connolly, J.C., B.L. Kress, G.R. Smith, J.E. Rettig. 2011. Possible behavioral avoidance of UV–B radiation and sunlight in wood frog (*Lithobates sylvaticus*) tadpoles. Current Herpetology 30:1–5.

Constible, J.M., P.T. Gregory, and B.R. Anholt. 2001. Patterns of distribution, relative abundance, and microhabitat use of anurans in a boreal landscape influenced by fire and timber harvest. Ecoscience 8:462–470.

Constible, J.M., P.T. Gregory, and K.W. Larsen. 2010. The pitfalls of extrapolation in conservation: movements and habitat use of a threatened toad are different in the boreal forest. Animal Conservation 13:43–52.

Converse, K.A., J. Mattsson, and L. Eaton–Poole. 2000. Field surveys of Midwestern and Northeastern Fish and Wildlife Service lands for the presence of abnormal frogs and toads. Journal of the Iowa Academy of Science 107:160–167.

Cook, D. 1997a. Biology of the California red–legged frog: a synopsis. Transactions of the Western Section of the Wildlife Society 33:79–82.

Cook, D. 1997b. Microhabitat use and reproductive success of the California red–legged frog (*Rana aurora draytonii*) and bullfrog (*Rana catesbeiana*) in an ephemeral marsh. M.A. thesis, Sonoma State University, Sonoma, California.

Cook, D. 2002. *Rana aurora draytonii* (California Red–legged Frog). Predation. Herpetological Review 33:303.

Cook, D., and M.R. Jennings. 2001. *Rana aurora draytonii* (California Red–legged Frog). Predation. Herpetological Review 32:182–183.

Cook, D.G., and M.R. Jennings. 2007. Microhabitat use of the California red–legged frog and introduced bullfrog in an ephemeral marsh. Herpetologica 63:430–440.

Cook, D.G., S. White, P. White and E. White. 2012. *Rana boylii* (Foothill Yellow–legged Frog). Upland movement. Herpetological Review 43:325–326.

Cook, F.R. 1960. New localities for the Plains spadefoot toad, tiger salamander and the Great Plains toad in the Canadian prairies. Copeia 1960:363–364.

Cook, F.R. 1963. The rediscovery of the mink frog in Manitoba. Canadian Field-Naturalist 77:129–130.

Cook, F.R. 1964a. Additional records and a correction of the type locality for the boreal chorus frog in northwestern Ontario. Canadian Field-Naturalist 78:186–192.

Cook, F.R. 1964b. The rusty colour phase of the Canadian toad, *Bufo hemiophrys*. Canadian Field-Naturalist 78:263–267.

Cook, F. R. 1964c. A northern range extension for *Bufo americanus* with notes on *B. americanus* and *Rana sylvatica*. Canadian Field-Naturalist 78:65–66.

Cook, F. R. 1964d. A range extension for the wood frog in northeastern Saskatchewan. Canadian Field–Naturalist. 78:268–269.

Cook, F. R. 1964e. The status of records of the western spotted frog, *Rana pretiosa*, in Saskatchewan. Copeia 1964: 219.

Cook, F.R. 1965. Additions to the known range of some amphibians and reptiles in Saskatchewan. Canadian Field-Naturalist 79:112–120.

Cook, F.R. 1966. A guide to the amphibians and reptiles of Saskatchewan. Saskatchewan Museum of Natural History, popular series 13:1–40.

Cook, F.R. 1967. An analysis of the herpetofauna of Prince Edward Island. National Museum of Canada Bulletin 212.

Cook, F.R. 1977. Records of the boreal toad from the Yukon and northern British Columbia. Canadian Field-Naturalist 91:185–186.

Cook, F.R. 1978. Amphibians and reptiles of Saskatchewan. Saskatchewan Museum of Natural History, popular series, No. 13.

Cook, F.R. 1981. Amphibians and reptiles of the Ottawa District, revised edition. Trail & Landscape (Ottawa Field Naturalist's Club) 15:75-109.

Cook, F.R. 1983. An analysis of toads of the *Bufo americanus* group in a contact zone in central northern North America. National Museums of Canada, Publications in Natural Sciences 3:i–89.

Cook, F.R., and D.R.M. Hatch. 1964. A spadefoot toad from Manitoba. Canadian Field-Naturalist 78:60–61.

Cook, F.R., and J.C. Cook. 1981. Attempted avian predation by a Canadian toad, *Bufo americanus hemiophrys*. Canadian Field-Naturalist 95:346–347.

Cook, J.C. 2008. Transmission and occurrence of *Dermomycoides* sp. in *Rana sevosa* and other ranids in the north central Gulf of Mexico states. M.S. thesis, University of Southern Mississippi, Hattiesburg.

Cook, R.P. 1989. And the voice of the grey tree frog was heard again in the land. Park Science 9(3):6–7.

Cook, R.P. 2008. Potential and limitations of herpetofaunal restoration in an urban landscape. Pp. 465–478 *In* J.C. Mitchell, R.E. Jung Brown, and B. Bartholomew (eds.), Urban Herpetology. Herpetological Conservation 3, Society for the Study of Amphibians and Reptiles, Salt Lake City, Utah.

Cook, R.P., T.A. Tupper, P.W.C. Paton, and B.C. Timm. 2011. Effects of temperature and temporal factors on anuran detection probabilities at Cape Cod National Seashore, Massachusetts, USA: implications for long–term monitoring. Herpetological Conservation and Biology 6:25–39.

Cooper, J.E. 1948. Maryland frogs and toads. Maryland Naturalist 18:37-40.

Cooper, J.E. 1950. Maryland tree frogs. Maryland Naturalist 20:77-80.

Cooper, J.E. 1953. Notes on the amphibians and reptiles of southern Maryland. Maryland Naturalist 23:90–100.

Cooper, J.E. 1958. Some albino reptiles and polydactylous frogs. Herpetologica 14:54–56.

Cooper, J.E. 1970. *Hyla femoralis* in Maryland, revisited. Bulletin of the Maryland Herpetological Society 6:14–15.

Cooper, J.E., S.S. Robinson, and J.B. Funderburg (eds.). 1977. Endangered and Threatened Plants and Animals of North Carolina. North Carolina State Museum of Natural History, Raleigh.

Cooper, J.G. 1860. Report upon the reptiles collected on the survey. Pp. 292–306 + plates XII–XXI *In* Explorations and Surveys for a Railroad Route from the Mississippi River to the Pacific Ocean. Route near the Forty–seventh and Forty–ninth Parallels, Explored by I.I. Stevens, Governor of Washington Territory, in 1853–'55. No. 4. War Department, U.S. Government Printing Office, Washington, D.C.

Cooper, J.G. 1869. The naturalist in California. American Naturalist 3:470–481.

Cooper, W.E., Jr. 2011. Escape strategy and vocalization during escape by American bullfrogs (*Lithobates catesbeianus*). Amphibia–Reptilia 32:213–221.

Cope, E.D. 1862. On some new and little known American Anura. Proceedings of the Academy of Natural Sciences of Philadelphia 14:151–159.

Cope, E.D. 1863. On *Trachycephalus, Scaphiopus* and other American Batrachia. Proceedings of the Academy of Natural Sciences of Philadelphia 15:43–54.

Cope, E.D. 1864. On a collection of reptiles from Owen's Valley, California, made by Dr. G.H. Horn, with remarks on the origin of species. Proceedings of the Academy of Natural Sciences of Philadelphia 19:85–86.

Cope, E.D. 1866. On the structures and distribution of the genera of the arciferous anura. Journal of the Academy of Natural Sciences of Philadelphia, series 2, 6:67–112.

[3]Cope, E.D. 1866 (1867). On the reptilia and Batrachia of the Sonoran Province of the Nearctic Region. Proceedings of the Academy of Natural Sciences of Philadelphia 18:300–314.

Cope, E.D. 1871. Catalogue of Reptilia and Batrachia obtained by C.J. Maynard in Florida. Second and Third Annual Report of the Trustees, Peabody Academy of Science for 1869–1870:82–85.

Cope, E.D. 1875. *Rana onca*, Cope, sp. nov. Pp. 528–529 and Plate 15 *In* H.C. Yarrow, Report upon the collections of batrachians and reptiles made in portions of Nevada, Utah, California, Colorado, New Mexico, and Arizona during 1871, 1872, 1873, and 1874. Ch. 4, pp. 509–584 *In* Report Upon Geographical and Geological Surveys West of the One Hundredth Meridian in Charge of First Lieut. Geo. M. Wheeler. Vol. 5. Zoology. U.S. Government Printing Office, Washington, D.C.

Cope, E.D. 1877a. On some new and little known reptiles and fishes from the Austroriparian Region. Proceedings of the American Philosophical Society 17:63–68.

Cope, E.D. 1877b (1878). A new genus of Cystignathidae from Texas. American Naturalist 12:253.

Cope, E.D. 1879a. A contribution to the herpetology of Montana. American Naturalist 13:432–441.

Cope, E.D. 1879b. Eleventh contribution to the herpetology of tropical America. Proceedings of the American Philosophical Society 18:261–277.

Cope, E.D. 1880. On the zoological position of Texas. Bulletin of the United States National Museum 17:1–51.

Cope, E.D. 1883. Notes on the geographic distribution of Batrachia and Reptilia inwestern North America. Proceedings of the Academy of Natural Sciences of Philadelphia 35:10–35.

Cope, E.D. 1886. Synonymic list of the North American species of *Bufo* and *Rana*, with descriptions of some Batrachia, from specimens in the National Museum. Proceedings of the American Philosophical Society 23:514–526.

Cope, E.D. 1888a. On a new species of *Bufo* from Texas. Proceedings of the United States National Museum 11:317–318.

Cope, E.D. 1888b. Catalogue of the Batrachia and Reptilia bought by William Taylor from San Diego, Texas. Proceedings of the United States National Museum 11:395–398.

Cope, E.D. 1889. The Batrachia of North America. Bulletin of the United States National Museum 34:1–525.

Cope, E.D. 1890. Dr. Leonhard Stejneger on *Bufo lentiginosus woodhousei*. American Naturalist 24:1204–1205.

Cope, E.D. 1891. A new species of frog from New Jersey. American Naturalist 25:1017–1019.

Cope, E.D. 1892. The Batrachia and Reptilia of northwestern Texas. Proceedings of the Academy of Natural Sciences of Philadelphia 44:331–337.

---

[3]Dated 1866 but published in 1867. The date attributed to newly described species is thus 1867. Also see Cope (1877).

Cope, E.D. 1893a. A contribution to the herpetology of British Columbia. Proceedings of the Academy of Natural Sciences of Philadelphia 46:181–184.

Cope, E.D. 1893b. On a collection of Batrachia and Reptilia from southwest Missouri. Proceedings of the Academy of Natural Sciences of Philadelphia 45:383–385.

Cope, E.D. 1893c. On the Batrachia and Reptilia of the Plains at Latitude 36°30'. Proceedings of the Academy of Natural Sciences of Philadelphia 45:386–387.

Cope, E.D. 1893d. On a new spade–foot from Texas. American Naturalist 27:155–156.

Corcoran, M.F., and J.R. Travis. 1980. A comparison of the karyotypes of the frogs *Rana areolata*, *Rana sphenocephala*, and *Rana pipiens*. Herpetologica 36:296–300.

Corkran, C.C., and C. Thoms. 1996. Amphibians of Oregon, Washington and British Columbia. Lone Pine Publishing, Edmonton, Alberta.

Corkran, C.C., and C. Thoms. 2006. Amphibians of Oregon, Washington and British Columbia. 2nd ed. Lone Pine Publishing, Edmonton, Alberta.

Corn, P.S. 1980a. Polymorphic reproductive behavior in male chorus frogs (*Pseudacris triseriata*). Journal of the Colorado–Wyoming Academy of Sciences 12:6–7.

Corn, P.S. 1980b. Comment on the occurrence of *Pseudacris clarki* in Montana. Bulletin of the Chicago Herpetological Society 15:77–78.

Corn, P.S. 1981. Field evidence for a relationship between color and developmental rate in the northern leopard frog (*Rana pipiens*). Herpetologica 37:155–160.

Corn, P.S. 1986. Genetic and developmental studies of albino chorus frogs. Journal of Heredity 77:164–168.

Corn, P.S. 1993. *Bufo boreas* (Boreal Toad). Predation. Herpetological Review 24:57.

Corn, P.S. 1994. What we know and don't know about amphibian declines in the West. Pp. 59–67 *In* W.W. Covington and L.F. DeBano (eds.), Sustainable Ecological Systems: Implementing and Ecological Approach to Land Management. USDA Forest Service, General Technical Report RM–247, Ft. Collins, Colorado.

Corn, P.S. 1998. Effects of ultraviolet radiation on boreal toads in Colorado. Ecological Applications 8:18–26.

Corn, P.S. 2000. Amphibian declines: review of some current hypotheses. Pp. 663–696 *In* D.W. Sparling, G. Linder, and C.A. Bishop (eds.), Ecotoxicology of Amphibians and Reptiles. SETAC Press, Pensacola, Florida.

Corn, P.S. 2003. Endangered toads in the Rockies. Pp. 43–51 *In* L. Taylor, K. Martin, D. Hik, and A. Ryall (eds.), Ecological and Earth Sciences in Mountain Areas. The Banff Center, Banff, Alberta.

Corn, P.S. 2005. Climate change and amphibians. Animal Biodiversity and Conservation 28:59–67.

Corn, P.S. 2007. Amphibians and disease: implications for conservation in the Greater Yellowstone Ecosystem. Yellowstone Science 15(2):11–16.

Corn, P.S., and J.C. Fogleman. 1984. Extinction of montane populations of the Northern Leopard Frog (*Rana pipiens*) in Colorado. Journal of Herpetology 18:147–152.

Corn, P.S., and R.B. Bury. 1989. Logging in western Oregon: responses of headwater habitats and stream amphibians. Forest Ecology and Management 29:39–57.

Corn, P.S., and L.J. Livo. 1989. Leopard frog and wood frog reproduction in Colorado and Wyoming. Northwestern Naturalist 70:1–9.

Corn, P.S., and F.A. Vertucci. 1992. Descriptive risk assessment of the effects of acidic deposition on Rocky Mountain amphibians. Journal of Herpetology 26:361–369.

Corn, P.S., and E. Muths. 2002. Variable breeding phenology affects the exposure of amphibian embryos to ultraviolet radiation. Ecology 83:2958–2963.

Corn, P.S., and E. Muths. 2004. Variable breeding phenology affects the exposure of amphibians to ultraviolet radiation: reply. Ecology 85:1759–1763.

Corn, P.S., W. Stolzenburg, and R.B. Bury. 1989. Acid precipitation studies in Colorado and Wyoming: interim report of surveys of montane amphibians and water chemistry. U.S. Fish and Wildlife Service Biological Report 80(40.26).

Corn, P.S., M.L. Jennings, and E. Muths. 1997. Survey and assessment of amphibian populations in Rocky Mountain National Park. Northwestern Naturalist 78:34–55.

Corn, P.S., E. Muths, and W.M. Iko. 2000. A comparison in Colorado of three methods to monitor breeding amphibians. Northwestern Naturalist 81:22–30.

Corn, P.S., B.R. Hossack, E. Muths, D.A. Patla, C.R. Peterson, and A.L. Gallant. 2005. Status of amphibians on the Continental Divide: surveys on a transect from Montana to Colorado, USA. Alytes 22:85–94.

Corn, P.S., E. Muths, A.M. Kissel, and R.D. Scherer. 2011. Breeing chorus indices are weakly related to estimated abundance of boreal chorus frogs. Copeia 2011:365–371.

Cornell, T.J., K.A. Berven, and G.J. Gamboa. 1989. Kin recognition by tadpoles and froglets of the wood frog *Rana sylvatica*. Oecologia 78:312–316.

Corrington, J.D. 1929. Herpetology of the Columbia, South Carolina, region. Copeia (172):58–83.

Corse, W.A., and D.E. Metter. 1980. Economics, adult feeding and larval growth of *Rana catesbeiana* on a fish hatchery. Journal of Herpetology 14:231–238.

Corser, J.D. 2008. The Cumberland Plateau disjunct paradox and the biogeography and conservation of pond–breeding amphibians. American Midland Naturalist

159:498–503.

Cort, W.W. 1919. A new distome from *Rana aurora*. University of California Publications in Zoology 19:283–298.

Corum, S. 2003. Effects of Sacramento pikeminnow on foothill yellow–legged frogs in coastal streams. M.S. thesis, Humboldt State University, Arcata, California.

Cortwright, S.A. 1998. Ten– to eleven–year population trends of two pond–breeding amphibians species, red–spotted newts and green frogs. Pp. 61–71 *In* M.J. Lannoo (ed.), Status & Distribution of Midwestern Amphibians. University of Iowa Press, Iowa City.

Cortwright, S.A., and C.E. Nelson. 1990. An examination of multiple factors affecting community structure in an aquatic amphibian community. Oecologia 83:123–131.

Cory, L., and J.J. Manion. 1953. Predation on eggs of the woodfrog *Rana sylvatica*, by leeches. Copeia 1953:66.

Cory, L., and J.J. Manion. 1955. Ecology and hybridization in the genus *Bufo* in the Michigan–Indiana region. Evolution 9:42–51.

Costanzo, J.P., and R.E. Lee, Jr. 1993. Cryoprotectant production capacity of the freeze–tolerant wood frog, *Rana sylvatica*. Canadian Journal of Zoology 71:71–75.

Costanzo, J.P., and R. E. Lee. 2005. Cryoprotection by urea in a terrestrially hibernating frog. Journal of Experimental Biology 208:4079–4089.

Costanzo, J.P., R.E. Lee, Jr., and M.F. Wright. 1991. Effect of cooling rate on the survival of frozen wood frogs. Journal of Comparative Physiology B 161:225–229.

Costanzo, J.P., R.E. Lee, Jr., and M.F. Wright. 1992a. Cooling rate influences cryoprotectants distribution and organ dehydration in freezing wood frogs. Journal of Experimental Zoology 261:373–378.

Costanzo, J.P., M.F. Wright, and R.E. Lee, Jr. 1992b. Freeze tolerance as an overwintering adaptation in Cope's gray treefrog (*Hyla chrysoscelis*). Copeia 1992:565–569.

Costanzo, J.P., R.E. Lee, Jr., and P.H. Lortz. 1993. Physiological responses of freeze tolerant and –intolerant frogs: clues to evolution of anuran freeze tolerance. American Journal of Physiology 265:R721–R725.

Costanzo, J.P., J. T. Irwin, and R. E. Lee. 1997. Freezing impairment of male reproductive behaviors of the freeze–tolerant wood frog, *Rana sylvatica*. Physiological Zoology 70: 158–166.

Cotten, T.B., M.A. Kwiatkowski, D. Saenz, and M. Collyer. 2012. Effects of an invasive plant, Chinese tallow (*Triadica sebifera*), on development and survival of anuran larvae. Journal of Herpetology 46:186–193.

Coues, E., and H.C. Yarrow. 1878. Notes on the herpetology of Dakota and Montana. Bulletin of the United States Geological and Geographical Survey 4:259–291.

Counts, C.L. III, and R.W. Taylor. 1977. A xanthoma of indeterminate origin in *Bufo americanus* (Amphibia, Anura, Bufonidae). Journal of Herpetology 11:235–236.

Courtois, D., R. Leclair, Jr., S. Lacasse, and P. Magnan. 1995. Habitats préférentiels d'amphibiens ranidés dans les lacs oligotrophes du Bouclier laurentien, Québec. Canadian Journal of Zoology 73:1744–1753.

Cousineau, M., and K. Rogers. 1991. Observations on sympatric *Rana pipiens*, *R. blairi*, and their hybrids in eastern Colorado. Journal of Herpetology 25:114–116.

Cowan, I. M. 1936. A review of the reptiles and amphibians of British Columbia. British Columbia Museum, pp. K15–K25.

Cowan, I.M. 1939. The vertebrate fauna of the Peace River District of British Columbia. Occasional Papers of the British Provincial Museum 1:1–102.

Cowan, I.M. 1941. Longevity of the red–legged frog. Copeia 1941:48.

Cowles, R.B. 1924. Notes regarding the breeding habits of *Scaphiopus hammondii* Baird. Pomona College Journal of Entomology and Zoology 16:108–110.

Cowles, R.B., and C.M. Bogert. 1936. The herpetology of the Boulder Dam region (Nev., Ariz., Utah). Herpetologica 1:33–42.

Cowman, D.F. 2005. Pesticides and amphibian declines in the Sierra Nevada Mountains, California. Ph.D. Dissertation, Texas A&M University, College Station.

Cowman, D.F., and L.E. Mazanti. 2000. Ecotoxicology of "new generation" pesticides to amphibians. Pp. 233–268 *In* D.W. Sparling, G. Linder, and C.A. Bishop (eds.), Ecotoxicology of Amphibians and Reptiles, SETAC Press, Pensacola, Florida.

Cox, P. 1898. Batrachia of New Brunswick. Bulletin of the Natural History Society of New Brunswick 4:64–66.

Cox, P. 1899. The anoura of New Brunswick. Proceedings of the Natural History Association of Miramichi 1:9–19.

Cragin, F.W. 1881. A preliminary catalogue of Kansas reptiles and batrachians. Transactions of the Kansas Academy of Science 7:112–120.

Crawford, J.A., D.B. Shepard, and C.A. Conner. 2009. Diet composition and overlap between recently metamorphosed *Rana areolata* and *Rana sphenocephala*: implications for a frog of conservation concern. Copeia 2009:642–646.

Crawshaw, G.J. 1997. Diseases in Canadian amphibian populations. SSAR Herpetological Conservation 1:258–270.

Crawshaw, L.I., R.N. Rausch, L.P. Wollmuth, and E.J. Bauer. 1992. Seasonal rhythms of development and temperature selection in larval bullfrogs, *Rana catesbeiana* Shaw. Physiological Zoology 65:346–359.

Crayon, J.J. 1998. *Rana catesbeiana* (Bullfrog). Diet. Herpetological Review 29:232.

Crayon, J.J., and R.L. Hothem. 1998. *Xenopus laevis* (African Clawed Frog). Predation. Herpetological Review 29:165–166.

Creel, G.C. 1963. Bat as a food item of *Rana pipiens*. Texas Journal of Science 15:104–106.

Creel, T.L., G.W. Foster, D.J. Forrester, D. Sabrina–Osorio, L. Garcia–Prieto, J.M. Martinez–Cruz, S.P. Diaz–Camacho, and K. Noda. 2000. Parasites of the green treefrog, *Hyla cinerea*, from Orange Lake, Alachua County, Florida, U.S.A. Comparative Parasitology 67:255–258.

Crenshaw, J.W., Jr. 1958. The role of hybridization in the development of biological (species) reproductive barriers with special reference to the Amphibia and Reptilia. Yearbook of the American Philosophical Society 1958:250–252.

Crenshaw, J.W., and W.F. Blair. 1959. Relationships in the *Pseudacris nigrita* complex in southwestern Georgia. Copeia 1959:215–222.

Creusere, F.M., and W.G. Whitford. 1976. Ecological relationships in a desert anuran community. Herpetologica 32:7–18.

Crisafulli, C.M., L.S. Trippe, C.P. Hawkins, and J.A. MacMahon. 2005. Amphibian responses to the 1980 eruption of Mount St. Helens. Pp. 183–197 *In* V.H. Dale, F.J. Swanson, and C.M. Crisafulli (eds.), Ecological Responses to the 1980 Eruption of Mount St. Helens. Springer, New York.

Crockett, M.E. 2001. Survey and comparison of amphibian assemblages in two physiographic regions of northeast Tennessee. M.S. thesis, East Tennessee State University, Johnson City, Tennessee.

Croes, S.A., and R.E. Thomas. 2000. Freeze tolerance and cryoprotectant synthesis of the Pacific tree frog *Hyla regilla*. Copeia 2000:863–868.

Cromer, R.B., J.D. Lanham, and H.H. Hanlin. 2002. Herpetofaunal response to gap and skidder–rut wetland creation in a southern bottomland hardwood forest. Forest Science 48:407–413.

Cronin, J.T., and J. Travis. 1986. Size–limited predation on larval *Rana areolata* (Anura: Ranidae) by two species of backswimmer (Insecta: Notonectidae). Herpetologica 42:171–174.

Crosby, C.R., and S.C. Bishop. 1925. A new genus and two new species of spiders collected by *Bufo quercicus*. Florida Entomologist 9:33–36.

Crosby, M.K., L.E. Licht and J.Z. Fu. 2009. The effect of habitat fragmentation on finescale population structure of wood frogs (*Rana sylvatica*). Conservation Genetics 10:1707–1718.

Croshaw, D.A. 2005. Cryptic behavior is independent of dorsal color polymorphism in juvenile northern leopard frogs (*Rana pipiens*). Journal of Herpetology 39:125–129.

Cross, C.L., and S.L. Gerstenberger. 2002. *Rana catesbeiana* (American Bullfrog). Diet. Herpetological Review 33:129–130.

Cross, K., and J.M. Hranitz. 1999. *Bufo americanus* (American Toad). Endoparasite. Herpetological Review 31:39.

Crossland, M.R., G.P. Brown, M. Anstis, C. Shilton, and R. Shine. 2008. Mass mortality of native anuran tadpoles in tropical Australia due to the invasive cane toad (*Bufo marinus*). Biological Conservation 141:2387–2394.

Crosswhite, E., and M. Wyman. 1920. [No title: figure and notes on an abnormal toad, presumably *A. boreas*]. Pomona College Journal of Entomology and Zoology 12:78.

Croteau, M.C. 2009. Chronic exposure to UV–B radiation and 4–tert–octylphenol disrupt metamorphosis and the thyroid system of northern leopard frog (*Rana pipiens*) tadpoles. Ph.D. Dissertation, University of Ottawa, Ottawa, Ontario.

Croteau, M.C., N. Hogan, J.C. Gibson, D. Lean, and V.L. Trudeau. 2008a. Toxicological threats to amphibians and reptiles in urban environments. Pp. 197–209 *In* J.C. Mitchell, R.E. Jung Brown, and B. Bartholomew (eds.), Urban Herpetology. Herpetological Conservation 3. Society for the Study of Amphibians and Reptiles, Salt Lake City, Utah.

Croteau, M.C., M.A. Davidson, D.R.S. Lean, and V.L. Trudeau. 2008b. Global increases in ultraviolet B radiation: potential impacts on amphibian development and metamorphosis. Physiological and Biochemical Zoology 81:743–761.

Crother, B.I. (Committee Chair). 2012. Scientific and Standard English Names of Amphibians and Reptiles on North America North of Mexico, with Comments Regarding Confidence in Our Understanding. 7th ed. Society for the Study of Amphibians and Reptiles, Herpetological Circular No. 39.

Crouch, W.B., and P.W.C. Paton. 2000. Using egg–mass counts to monitor wood frog populations. Wildlife Society Bulletin 28:895–901.

Crouch, W.B. III, and P.W.C. Paton. 2002. Assessing the use of call surveys to monitor breeding anurans in Rhode Island. Journal of Herpetology 36:185–192.

Crump, D., M. Berrill, D. Coulson, D. Lean, L. McGillivray, and A. Smith. 1999. Sensitivity of amphibian embryos, tadpoles, and larvae to enhanced UV–B radiation in natural pond conditions. Canadian Journal of Zoology 77:1956–1966.

Crump, M.L. 1981a. Energy accumulation and amphibian metamorphosis. Oecologia 49:167–169.

Crump, M.L. 1981b. Intraclutch egg size variability in *Hyla crucifer* (Anura: Hylidae). Copeia 1981:302–308.

Crump, M.L. 1986. Cannibalism by younger tadpoles: another hazard of metamorphosis. Copeia 1986:1007–1009.

Cuellar, H.S. 1968. Genetic compatibility of *Rana areolata* with southwestern members of the *Rana pipiens* complex: (Anura: Ranidae). M.S. thesis, Texas Tech University, Lubbock.

Cuellar, H.S. 1971. Levels of genetic compatibility of *Rana areolata* with southwestern members of the *Rana pipiens* complex (Anura: Ranidae). Evolution 25:399–409.

Cuellar, O. 1994. Ecological observations on *Rana pretiosa* in western Utah. Alytes 12:109–121.

Culley, D.D., Jr. 1973. Use of bullfrogs in biological research. American Zoologist 13:85–90.

Culley, D.D., Jr. 1981. Have we turned the corner on bullfrog culture? Aquaculture Magazine 7:20-24.

Culley, D.D., and C. Gravois. 1970. A new look at an old problem. The American Fish Farmer 1(5):5-10.

Culley, D.D., and C. Gravois. 1971. Recent developments in frog culture. Proceedings of the 25th Annual Conference of the Southeastern Association of Game and Fish Commissioners, pp. 583-597.

Culley Jr, D.D., N.D. Horseman, R.L. Amborski, and S.P. Meyers. 1978. Current status of amphibian culture with emphasis on nutrition, diseases, and reproduction of the bullfrog, *Rana catesbeiana*. Proceedings of the Annual Meeting-World Mariculture Society 9(1-4):653-669.

Culp, C.E., J.O. Falkinham III, and L.K. Belden. 2007. Identification of the natural bacterial microflora on the skin of eastern newts, bullfrog tadpoles and redback salamanders. Herpetologica 63:66–71.

Cummins, H. 1920. The role of voice and coloration in spring migration and sex recognition in frogs. Journal of Experimental Zoology 30:325–343.

Cunjak, R.A. 1986. Winter habitat of northern leopard frogs, *Rana pipiens*, in a southern Ontario stream. Canadian Journal of Zoology 64:255–257.

Cunningham, J. 2003. Pond–breeding amphibian species distributions in a beaver–modified landscape, Acadia National Park, Mount Desert Island, Maine. M.S. thesis, University of Maine, Orono.

Cunningham, J., A.J.K. Calhoun, and W.E. Glanz. 2007. Pond–breeding amphibian species richness and breeding habitat selection in a beaver–modified landscape. Journal of Wildlife Management 71:2517–2526.

Cunningham, J.D. 1955a. A case of cannibalism in the toad *Bufo boreas*. Herpetologica 10:166.

Cunningham, J.D. 1955b. Observations on the ecology of the canyon treefrog, *Hyla californiae*. Herpetologica 20:55–61.

Cunningham, J.D. 1955c. Observations on the natural history of the California toad, *Bufo californicus* Camp. Herpetologica 17:255–260.

Cunningham, J.D. 1955d. Notes on abnormal *Rana aurora draytoni*. Herpetologica 11:149.

Cunningham, J.D. 1959. Reproduction and food of some California snakes. Herpetologica 15:17–19.

Cunningham, J.D. 1962. Observations on the natural history of the California toad, *Bufo californicus* Camp. Herpetologica 17:255–260.

Cunningham, J.D. 1963. Additional observations on the ecology of the Yosemite toad, *Bufo canorus*. Herpetologica 19:56–61.

Cunningham, J.D., and D.P. Mullally. 1956. Thermal factors in the ecology of the Pacific treefrog. Herpetologica 12:68–79.

Cunnington, G.M., and L. Fahrig. 2010. Plasticity in the vocalizations of anurans in response to traffic noise. Acta Oecologica 36:463-470.

Cupp, E.W., D. Zhang, X. Yue, M.S. Cupp, C. Guyer, T.R. Sprenger, and T.R. Unnasch. 2004. Identification of reptilian and amphibian blood meals from mosquitoes in an eastern equine encephalomyelitis virus focus in central Alabama. American Journal of Tropical Medicine and Hygiene 71:272-276.

Cupp, P.V., Jr. 1974. Thermal tolerances and acclimation during anuran development and metamorphosis. Ph.D. Dissertation, Clemson University, Clemson, South Carolina.

Cupp, P.V., Jr. 1980. Thermal tolerance of five salientian amphibians during development and metamorphosis. Herpetologica 36:234–244.

Currie, W., and E.D. Bellis. 1969. Home range and movements of the bullfrog, *Rana catesbeiana* Shaw, in an Ontario pond. Copeia 1969:688–692.

Curry, T.R., and M.P. Hayes. 2009. *Rana aurora* (Northern Red–legged Frog). Egg mass disturbance. Herpetological Review 40:208–209.

Cushman, S.A. 2006. Effects of habitat loss and fragmentation on amphibians: a review and prospectus. Biological Conservation 128:231–240.

Cutler, W.H., II. 1989. *Rana sylvatica* (Wood Frog). Attempted predation. Herpetological Review 20:69.

Daigle, C. 1997. Distribution and abundance of the chorus frog *Pseudacris triseriata* in Québec. SSAR Herpetological Conservation 1:73–77.

Dailey, M.D., and S.R. Goldberg. 2000. *Langeronia burseyi* sp. n. (Trematoda: Lecithodendriidae) from the California treefrog, *Hyla cadaverina* (Anura: Hylidae), with revision of the genus *Langeronia* Caballero and Bravo–Hollis, 1949. Comparative Parasitology 67:165–168.

Dale, J.M., B. Freedman, and J. Kerekes. 1985a. Acidity

and associated water chemistry of amphibian habitats in Nova Scotia. Canadian Journal of Zoology 63:97–105.

Dale, J.M., B. Freedman, and J. Kerekes. 1985b. Experimental studies of the effects of acidity and associated water chemistry on amphibians. Proceedings of the Nova Scotian Institute of Science 35:35–54.

Daly, E.W., and P.T.J. Johnson. 2011. Beyond immunity: quantifying the effects of host anti–parasite behavior on parasite transmission. Oecologia 165:1043–1050.

Damm, S.H. 2003. Effects of aluminum, copper, and lead on the development and survival of gray treefrog (*Hyla versicolor*) larvae. M.S. thesis, University of Akron, Akron, Ohio.

Daniel, R., and B. Edmond. 2006. Atlas of Missouri amphibians and reptiles for 2005. http://atlas.moherp.org/pubs/atlas05.pdf.

D'Aoust–Messier, A.–M. 2012. Colonizing northern landscapes: population genetics and phylogeography of Wood Frogs (*Lithobates sylvaticus*) in the James Bay area. M.S. thesis, Laurentian University, Sudbury, Ontario.

Dapson, R.W., and L. Kaplan. 1975. Biological half–life and distribution of radiocesium in a contaminated population of green treefrogs *Hyla cinerea*. Oikos 26:39–42.

Dare, O.K., and M.R. Forbes. 2009. Patterns of trematode and nematode lungworm infections in northern leopard frogs and wood frogs from Ontario, Canada. Journal of Helminthology 83:339–343.

Da Silva, E.T., E. P. dos Reis, P.S. Santos, and R.N. Feio. 2010. *Lithobates catesbeianus* (American Bullfrog). Diet. Herpetological Review 41:475–476.

Da Silva, H.R. 1997. Two character states new for hylines and the taxonomy of the genus *Pseudacris*. Journal of Herpetology 31:609–613.

Daszak, P., A.A. Cunningham, and A.D. Hyatt. 2003. Infectious disease and amphibian population declines. Diversity and Distributions 9:141–150.

Daszak, P., A. Strieby, A.A. Cunningham, J.E. Longcore, C. Brown, and D. Porter. 2004. Experimental evidence that the bullfrog (*Rana catesbeiana*) is a potential carrier of chytridiomycosis, an emerging fungal disease of amphibians. Herpetological Journal 14:201–207.

Daszak, P., D.E. Scott, A.M. Kilpatrick, C. Faggioni, J.W. Gibbons, and D. Porter. 2005. Amphibian population declines at Savannah River Site are linked to climate, not chytridiomycosis. Ecology 86:3232–3237.

Datta, S., L. Hansen, L. McConnell, J. Baker, J. LeNoir, and J.N. Selber. 1998. Pesticides and PCB contaminants in fish and tadpoles from the Kaweah River Basin, California. Bulletin of Environmental Contamination and Toxicology 60:829–836.

Daudin, F.M. 1802. Histoire Naturelle des Rainettes, des Grenouilles, et des Crapauds. Bertrandet, Libraire Levrault, Paris.

Daugherty, C.H. 1979. Population ecology and genetics of *Ascaphus truei*: an examination of gene flow and natural selection. Ph.D. Dissertation, University of Montana, Missoula.

Daugherty, C.H., and A.L. Sheldon. 1982a. Age–determination, growth, and life history of a Montana population of the tailed frog (*Ascaphus truei*). Herpetologica 38:461–468.

Daugherty, C.H., and A.L. Sheldon. 1982b. Age–specific movement patterns of the frog *Ascaphus truei*. Herpetologica 38:468–474.

Davidson, C. 2004. Declining downwind: amphibian population declines in California and historic pesticide use. Ecological Applications 14:1892–1902.

Davidson, C. 2010. *Rana draytonii* (California red–legged frog). Prey. Herpetological Review 41:66.

Davidson, C., and R.A. Knapp. 2007. Multiple stressors and amphibian declines: dual impacts of pesticides and fish on yellow–legged frogs. Ecological Applications 17:587–597.

Davidson, C., H.B. Shaffer, and M.R. Jennings. 2001. Declines of the California red–legged frog: climate, UV–B, habitat, and pesticides hypotheses. Ecological Applications 11:464–479.

Davidson, C., H.B. Shaffer, and M.R. Jennings. 2002. Spatial tests of the pesticide drift, habitat destruction, UV–B, and climate–change hypotheses for California amphibian declines. Conservation Biology 16:1588–1601.

Davidson, C., M.F. Bernard, H.B. Shaffer, J.M. Parker, C. O'Leary, J.M. Conlon, and L.A. Rollins–Smith. 2007. Effects of chytrid and carbaryl exposure on survival, growth and skin peptide defenses in foothill yellow–legged frogs. Environmental Science & Technology 41:1771–1776.

Davidson, C., K. Stanley, and S.M. Simonich. 2012. Contaminant residues and declines of the Cascades frog (*Rana cascadae*) in the California Cascades, USA. Environmental Toxicology and Chemistry 31:1895–1902.

Davidson, S.A., and D.L. Chambers. 2011a. Occurrence of *Batrachochytrium dendrobatidis* in amphibians of Wise County, Virginia, USA. Herpetological Review 42:214–215.

Davidson, S.R.A., and D.L. Chambers. 2011b. Ranavirus prevalence in amphibian populations of Wise County, Virginia. Herpetological Review 42:540–542.

Davis, A.B., and P.A. Verrell. 2005. Demography and reproductive ecology of the Columbia spotted frog (*Rana luteiventris*) across the Palouse. Canadian Journal of Zoology 83:702–711.

Davis, A.K., M.J. Yabsley, M.K. Keel, and J.C. Maerz. 2007. Discovery of a novel alveolate pathogen affecting southern leopard frogs in Georgia: description of the disease and host effects. EcoHealth 4:310–317.

Davis, A.K., M.K. Keel, A. Ferreira, and J.C. Maerz. 2010. Effects of chytridiomycosis on circulating white blood cell distributions of bullfrog larvae (*Rana catesbeiana*). Comparative Clinical Pathology 19:49–55.

Davis, B.J., and N. Hollenback. 1978. New records of the bird–voiced treefrog, *Hyla avivoca* (Hylidae), from Arkansas and Louisiana. Southwestern Naturalist 23:161–162.

Davis, D.D. 1933. Unusual behavior in a leopard frog. Copeia 1933:223–224.

Davis, J.G., and S.A. Menze. 2000. Ohio Frog and Toad Atlas. Ohio Biological Survey Miscellaneous Contribution No. 6.

Davis, J.G., and S.A. Menze. 2002. In Ohio's Backyard: Frogs and Toads. Ohio Biological Survey, Backyard Series No. 3.

Davis, J.G., P.J. Krusling, and J.W. Ferner. 1998. Status of amphibian populations in Hamilton County, Ohio. Pp. 155–165 *In* M.J. Lannoo (ed.), Status & Distribution of Midwestern Amphibians. University of Iowa Press, Iowa City.

Davis, J.R., D.T. Eastlake, A.J. Kouba, and C.K. Vance. 2012. *Batrachochytrium dendrobatidis* detected in Fowler's toad (*Anaxyrus fowleri*) populations in Memphis, Tennessee. Herpetological Review 43:81–83.

Davis, L.C. 1973. The herpetofauna of Peccary Cave, Arkansas. M.S. thesis, University of Arkansas, Fayetteville.

Davis, M.J., P. Kleinhenz, and M.D. Boone. 2011. Juvenile green frog (*Rana clamitans*) predatory ability not affected by exposure to carbaryl at different times during larval development. Environmental Toxicology and Chemistry 30:1618–1620.

Davis, M.J., J.L. Purrenhage, and M.D. Boone. 2012. Elucidating predator–prey interactions using aquatic microcosms: complex effects of a crayfish predator, vegetation, and atrazine on tadpole survival and behavior. Journal of Herpetology 46:527–534.

Davis, M.S. 1980. Observations on the life history of the wood frog, *Rana sylvatica* LeConte, in Alabama. M.S. thesis, Auburn University, Auburn, Alabama.

Davis, M.S., and G.W. Folkerts. 1980. A beetle (*Eutheola rugiceps* LeC.: Scarabaeidae) penetrates the stomach wall of its predator (*Rana sylvatica* LeC.: Amphibia). Coleopterists Bulletin 34:396.

Davis, M.S., and T.R. Jones. 1982. Reassessment of the distribution of three amphibians in Alabama. Journal of the Alabama Academy of Science 53:10–16.

Davis, M.S., and G.W. Folkerts. 1986. Life history of the wood frog, *Rana sylvatica* LeConte (Amphibia: Ranidae), in Alabama. Brimleyana 12:29–50.

Davis, N.S., and F.L. Rice. 1883. List of the batrachia and reptilia of Illinois. Bulletin of the Chicago Academy of Sciences 1(3):25 – 32.

Davis, S.L. 1958. Notes on the Amphibia in Acadia National Park, Maine. M.S. thesis, Cornell University, Ithaca, New York.

Davis, T.M., and P.T. Gregory. 2003. Decline and local extinction of the western toad, *Bufo boreas,* on southern Vancouver Island, British Columbia, Canada. Herpetological Review 34:350–352.

Davis, W.T. 1884. The reptiles and batrachians of Staten Island. Proceedings of the Natural Science Association of Staten Island, Extra No. 1: 13.

Davis, W.T. 1904. *Hyla andersonii* and *Rana virgatipes* at Lakehurst, New Jersey. American Naturalist 38:893.

Davis, W.T. 1907. Additional observations on *Hyla andersonii* and *Rana virgatipes* in New Jersey. American Naturalist 41:49–51.

Davis, W.T. 1910. An addition to the list of Staten Island frogs. Proceedings of the Staten Island Association of Arts and Sciences 3(2):66–67.

Dawson, D.E., and M.E. Hostetler. 2007. Herpetofaunal use of edge and interior habitats in urban forest remnants. Urban Habitats 5:103–125.

Dawson, J.T. 1982. Kin recognition and schooling in the American toad (*Bufo americanus*). Ph.D. Dissertation, State University of New York, Albany.

Dayton, G.H. 2000. *Gastrophryne olivacea* (Narrow–mouthed Toad). Vocalization. Herpetological Review 31:40.

Dayton, G.H., and R.E. Jung. 1999. *Scaphiopus couchii* (Couch's Spadefoot). Predation. Herpetological Review 30:164.

Dayton, G.H., and L.A. Fitzgerald. 2001. Competition, predation, and the distributions of four desert anurans. Oecologia 129:430–435.

Dayton, G.H., and S.D. Wapo. 2002. Cannibalistic behavior in *Scaphiopus couchii*: more evidence for larval anuran oophagy. Journal of Herpetology 36:531–532.

Dayton, G.H., and C.W. Painter. 2005. *Bufo speciosus* Girard, 1854. Texas Toad. Pp. 435–436 *In* M.J. Lannoo (ed.), Amphibian Declines. The Conservation Status of United States Species. University of California Press, Berkeley.

Dayton, G.H., and L.A. Fitzgerald. 2006. Habitat suitability models for desert amphibians. Biological Conservation 132:40–49.

Dayton, G.H., and L.A. Fitzgerald. 2011. The advantage of no defense: predation enhances cohort survival in a desert amphibian. Aquatic Ecology 45:325–333.

Dayton, G.H., R.E. Jung,, and S. Droege. 2004. Large–scale habitat associations of four desert anurans in Big Bend National Park, Texas. Journal of Herpetology 38:619–

627.

Dayton, G.H., R. Skiles,, and L. Dayton. 2007. Frogs & Toads of Big Bend National Park. Texas A&M University Press, College Station.

Dean, J. 1980. Encounters between bombardier beetles and two species of toads (*Bufo americanus, B. marinus*): speed of prey-capture does not determine success. Journal of Comparative Physiology A 135:41-50.

Dean, R.A. 1966. High temperature tolerances of anuran amphibians. M.S. thesis, University of North Dakota, Grand Forks.

Dearolf, K. 1956. Survey of North American cave vertebrates. Proceedings of the Pennsylvania Academy of Science 30:201–210.

Deban, S.M., and K.C. Nishikawa. 1992. Kinematics of prey capture and the mechanism of tongue protraction in the green tree frog *Hyla cinerea*. Journal of Experimental Biology 170:235–256.

DeBenedictis, P.A. 1970. Interspecific competition between tadpoles of *Rana pipiens* and *Rana sylvatica*: an experimental field study. Ph.D. Dissertation, University of Michigan, Ann Arbor.

DeBenedictis, P.A. 1974. Interspecific competition between larvae of *Rana pipiens* and *Rana sylvatica*: an experimental field study. Ecological Monographs 44:129–151.

Deckert, R.F. 1914a. List of salientia from Jacksonville, Fla. Copeia (3):3.

Deckert, R.F. 1914b. Further notes on the salientia of Jacksonville, Fla. Copeia (5):2–4.

Deckert, R.F. 1914c. Further notes on the salientia of Jacksonville, Fla. Copeia (9):1–3.

Deckert, R.F. 1915a. An albino pond frog. Copeia (24):53–54.

Deckert, R.F. 1915b. Further notes on the salientia of Jacksonville, Fla. Copeia (18):3–5.

Deckert, R.F. 1915c. Concluding notes on the salientia of Jacksonville, Fla. Copeia (20):21–24.

Deckert, R.F. 1921. Amphibian notes from Dade Co., Florida. Copeia (92):20–23.

Deckert, R.F. 1922. Notes on Dade County salientia. Copeia (112):88.

DeFauw, S.L., and P.J. English. 1994. Geographic distribution. *Scaphiopus holbrookii holbrookii* (Eastern Spadefoot). Herpetological Review 25:162.

De Garady, C.J., and R.S. Halbrook. 2006. Using anurans as bioindicators of PCB contaminated streams. Journal of Herpetology 40:127–130.

Degenhardt, W.G., C.W. Painter, and A.H. Price. 1996. Amphibians & Reptiles of New Mexico. University of New Mexico Press, Albuquerque.

Degitz, S.J., P.A. Kosian, E.A. Makynen, K.M. Jensen, and G.T. Ankley. 2000. Stage-and species-specific developmental toxicity of all-trans retinoic acid in four native North American ranids and *Xenopus laevis*. Toxicological Sciences 57:264-274.

Degner, J.F. 2007. Genetic and phenotypic evolution in the ornate chorus frog (*Pseudacris ornata*): testing the relative roles of natural selection. M.S. thesis, University of Central Florida, Orlando.

Degner, J.F., D.M. Silva, T.D. Hether, J.M. Daza, and E.A. Hoffman. 2010. Fat frogs, mobile genes: unexpected phylogeographic patterns for the ornate chorus frog (*Pseudacris ornata*). Molecular Ecology 19:2501–2515.

DeGraaf, J.D., and D.G. Nein. 2010. Predation of spotted turtle (*Clemmys guttata*) hatchling by green frog (*Rana clamitans*). Northeastern Naturalist 17:667–670.

DeGraaf, R.M., and D.D. Rudis. 1983. Amphibians and Reptiles of New England. Habitats and Natural History. University of Massachusetts Press, Amherst.

DeGraaf, R.M., and D.D. Rudis. 1986. New England Wildlife: Habitat, Natural History, and Distribution. USDA Forest Service, General Technical Report NE–108, Broomall, Pennsylvania.

DeGraaf, R.M., and D.D. Rudis. 1990. Herpetofaunal species composition and relative abundance among three New England forest types. Forest Ecology and Management 32:155–165.

Deguise, I., and J.S. Richardson. 2009a. Prevalence of the chytrid fungus (*Batrachochytrium dendrobatidis*) in western toads in southwestern British Columbia, Canada. Northwestern Naturalist 90:35–38.

Deguise, I., and J.S. Richardson. 2009b. Movement behaviour of adult western toads in a fragmented, forest landscape. Canadian Journal of Zoology 87: 1184–1194.

Deichman, J.L., W.E. Duellman, and G.B. Williamson. 2008. Predicting biomass from snout–vent length in New World frogs. Journal of Herpetology 42:238–245.

De Kay, J.E. 1842. Zoology of New York. Part III Reptiles and Amphibia. W & A White and J. Visscher, Albany, New York

Delis, P.R. 1993. Effects of urbanization on the community of anurans of a pine flatwood habitat in west central Florida. M.S. thesis, University of South Florida, Tampa.

Delis, P.R. 2001. *Hyla gratiosa* and *H. femoralis* (Anura: Hylidae) in west central Florida: a comparative study of rarity and commonness. Ph.D. Dissertation, University of South Florida, Tampa.

Delis, P.R., and A.P. Summers. 1996. *Hyla gratiosa* (Barking Treefrog). Reproduction. Herpetological Review 27:18–19.

Delis, P.R., H.R. Mushinsky, and E.D. McCoy. 1996. Decline of some west–central Florida anuran populations in response to habitat degradation. Biodiversity and Conservation 5:1579–1595.

Delis, P.R., C. Kindlin, and R.L. Stewart. 2010. The herpetofauna of Letterkenny Army Depot, south–central Pennsylvania: a starting point to the long–term monitoring and management of amphibians and reptiles. Journal of Kansas Herpetology 34:11–16.

Delnicki, D., and E. Bolen. 1977. Use of black–bellied tree duck nest sites by other species. Southwestern Naturalist 22:275–277.

Delvinquier, B.L.J., and S.S. Desser. 1996. Opalinidae (Sarcomastigophora) in North American amphibia: genus *Opalina* Purkinje and Valentin, 1835. Systematic Parasitology 33:33–51.

Delzell, D.E. 1958. Spatial movement and growth of *Hyla crucifer*. Ph.D. Dissertation, University of Michigan, Ann Arbor.

Delzell, D.E. 1979. A provisional checklist of amphibians and reptiles in the Dismal Swamp area, with comments on their range of distribution. Pp. 244–260 *In* P.W. Kirk, Jr. (ed.), The Great Dismal Swamp. University Press of Virginia, Charlottesville.

deMaynadier, P.G., and M.L. Hunter, Jr. 1998. Effects of silvicultural edges on the distribution and abundance of amphibians in Maine. Conservation Biology 12:340–352.

deMaynadier, P.G., and M.L. Hunter, Jr. 1999. Forest canopy closure and juvenile emigration by pool–breeding amphibians in Maine. Journal of Wildlife Management 63:441–450.

deMaynadier, P.G., and M.L. Hunter, Jr. 2000. Road effects on amphibian movements in a forested landscape. Natural Areas Journal 20:56–65.

Demlong, M.J. 1997. Head–starting *Rana subaquavocalis* in captivity. Reptiles Magazine (January):24–28, 30, 32–33.

Denman, N.S., and L. Denman. 1985. Geographic distribution. *Bufo americanus copei* (Hudson Bay Toad). Herpetological Review 16:114.

DeNoyelles, F., W.D. Kettle, C.H. Fromm, M.F. Moffett, and S.L. Dewey. 1989. Use of experimental ponds to assess the effects of a pesticide on the aquatic environment. Pp. 41–56 *In* J.R. Voshell (ed.), Using Mesocosms to Assess the Aquatic Ecological Risk of Pesticides: Theory and Practice. Entomological Society of America, Lanham, Maryland.

Dentel, J.A., A. Reist, O. Hornacek, J. McBride, C. White, T.D. Schwaner and J. Spurgat. 2011. *Lithobates catesbeianus* (American Bullfrog). Predation. Herpetological Review 42:89.

Denton, R.D. 2011. Amphibian community similarity between natural ponds and constructed ponds of multiple types in Daniel Boone National Forest, Kentucky. M.S. thesis, Eastern Kentucky University, Richmond.

Denton, R.D., and S.C. Richter. 2012. A quantitative comparison of two common amphibian sampling techniques for wetlands. Herpetological Review 43:44–47.

Denver, R.J., N. Mirhadi, and M. Phillips. 1998. Adaptive plasticity in amphibian metamorphosis: response of *Scaphiopus hammondii* tadpoles to habitat desiccation. Ecology 79:1859–1872.

Depkin, F.C., M.C. Coulter, and A.L. Bryan, Jr. 1992. Food of nestling wood storks in east central Georgia. Waterbirds 15:219–225.

Depoe, C.E., J.H. Funderberg, and T.L. Quay. 1961. The reptiles and amphibians of North Carolina: (A preliminary check list and bibliography). Journal of the Elisha Mitchell Scientific Society 17:125–139.

Dernehl, P.H. 1902. Place–modes of *Acris Gryllus* for Madison, Wis. Bulletin of the Wisconsin Natural History Society 2:75–83.

de Solla, S.R., C.A. Bishop, K.E. Pettit, and J.E. Elliott. 2002a. Organochlorine pesticides and polychlorinated biphenyls (PCBs) in eggs of red–legged frogs (*Rana aurora*) and northwestern salamanders (*Ambystoma gracile*) in an agricultural landscape. Chemosphere 46:1027–1032.

de Solla, S.R., K.E. Pettit, C.A. Bishop, K.M. Cheng, and J.E. Elliott. 2002b. Effects of agricultural runoff on native amphibians in the lower Fraser River valley, British Columbia, Canada. Environmental Toxicology and Chemistry 21:353–360.

de Solla, S.R., L.J. Shirose, K.J. Fernie, G.C. Barrett, C.S. Brousseau, and C.A. Bishop. 2005. Effect of sampling effort and species detectability on volunteer based anuran monitoring programs. Biological Conservation 121:585–594.

Desroches, J.–F., and D. Rodrigue. 2004. Amphibiens et Reptiles du Québec et des Maritimes. Éditions Michel Quintin, Waterloo, Québec.

Desroches, J.–F., and I. Picard. 2006. *Rana sylvatica* (Wood Frog). Tadpole maximum size. Herpetological Review 37:449–450.

Dessauer, H.C., and E. Nevo. 1969. Geographic variation of blood and liver proteins in cricket frogs. Biochemical Genetics 3:171–188.

Desser, S., and J.R. Barta. 1984a. An intraerythrocytic virus and rickettsia of frogs from Algonquin Park, Ontario. Canadian Journal of Zoology 62:1521–1524.

Desser, S., and J.R. Barta. 1984b. *Thrombocytozoons ranarum* Tchacarof 1863, a prokaryotic parasite in thrombocytes of the mink frog *Rana septentrionalis* in Ontario. Journal of Parasitology 70:454–456.

Desser, S., and S. Jones. 1985. *Hexamita intestinalis* Dujardin in the blood of frogs from southern and central Ontario.

Journal of Parasitology 71:841.

Desser, S., J. Lom, and I. Dykova. 1986. Developmental stages of *Sphaerospora ohlmacheri* (Whinery, 1893) n. comb. (Myxozoa: Myxosporea) in the renal tubules of bullfrog tadpoles, *Rana catesbeiana* from Lake of Two Rivers, Algonquin Park, Ontario. Canadian Journal of Zoology 64:2344–2347.

Desser, S., M.E. Siddall, and J.R. Barta. 1989. Ultrastructural observations of the developmental stages of *Lankesterella minima* (Apicomplexa) in experimentally infected *Rana catesbeiana* tadpoles. Journal of Parasitology 76:97–103.

Desser, S., H. Hong, M.E. Siddall, and J.R. Barta. 1993. An ultrastructural study of *Brugerolleia algonquinensis* gen. nov., sp. nov. (Diplomonadina: Diplomonadida), a flagellate parasite in the blood of frogs from Ontario, Canada. European Journal of Protistology 29:72–80.

Dever, J.A. 2007. Fine scale genetic structure in the threatened foothill yellow–legged frog (*Rana boylii*). Journal of Herpetology 41:168–173.

Devito, J., and J. Markow. 1998. Amphibians and reptiles of the Connecticut College Arboretum. Connecticut College Arboretum Bulletin No. 36, 57 pp.

DeVito, J., D.P. Chivers, J.M. Kiesecker, A. Marco, E.L. Wildy, and A.R. Blaustein. 1998. The effects of snake predation on metamorphosis of western toads, *Bufo boreas* (Amphibia, Bufonidae). Ethology 104:185–193.

DeVito, J., D.P. Chivers, J.M. Kiesecker, L.K. Belden, and A.R. Blaustein. 1999. Effects of snake predation on aggregation and metamorphosis of Pacific treefrog (*Hyla regilla*) larvae. Journal of Herpetology 33:504–507.

de Vlaming, V.L., and R.B. Bury. 1970. Thermal selection in tadpoles of the tailed frog, *Ascaphus truei*. Journal of Herpetology 4:179–189.

Dexter, R.W. 1973. Nomenclatural history and status of Fowler's toad (*Bufo fowleri*). HISS News-Journal1:154–157.

Diakow, C. 1977. Initiation and inhibition of the release croak of *Rana pipiens*. Physiology & Behavior 19:607–610.

Dial, N.A., and C.A. Bauer. 1984. Teratogenic and lethal effects of paraquat on developing frog embryos (*Rana pipiens*). Bulletin of Environmental Contamination and Toxicology 33:592–597.

Dial, N.A., and C.A. Bauer Dial. 1987. Lethal effects of diquat and paraquat on developing frog embryos and 15–day–old tadpoles, *Rana pipiens*. Bulletin of Environmental Contamination and Toxicolgy 38:1006–1011.

Diana, S.G., W.J. Resetarits, Jr., D.J. Schaeffer, K. B. Beckmen, and V.R. Beasley. 2000. Effects of atrazine on amphibian growth, development, and survival in artificial aquatic communities. Environmental Toxicology and Chemistry 19:2961–2967.

Dibble, C.J., J.E. Kauffman, E.M. Zuzik, G.R. Smith, and J.E. Rettig. 2009. Effects of potential predator and competitor cues and sibship on wood frog (*Rana sylvatica*) embryos. Amphibia–Reptilia 30:294–298.

Di Berardino, M.A. 1962. The karyotype of *Rana pipiens* and investigation of its stability during embryonic differentiation. Developmental Biology 5:101–126.

Di Candia, M.R., and E.J. Routman. 2007. Cytonuclear discordance across a leopard frog contact zone. Molecular Phylogenetics and Evolution 45:564–575.

Dickerson, M.C. 1906. The Frog Book. North American Toads and Frogs with a Study of the Habits and Life Histories of those of the Northeastern States. Doubleday, Page & Co., New York.

Dickey, L.B. 1921. A new amphibian cestode. Journal of Parasitology 7:129–137.

Dickinson, W.E. 1956. Common tree toad. Lore 6(2):67.

Dickinson, W.E. 1965. Handbook of Amphibians and Turtles of Wisconsin. Milwaukee Public Museum, Milwaukee.

Dickinson, W.E. 1972. The amphibians and reptiles of Forest, Florence and Marinette counties with special reference to the Pine, Popple and Pike watersheds. Transactions of the Wisconsin Academy of Sciences, Arts and Letters 60: 303-308.

Dickman, M. 1968. The effect of grazing by tadpoles on the structure of a periphyton community. Ecology 49:1188–1190.

Dickson, N.J. 2002. The natural history and possible extirpation of Blanchard's cricket frog, *Acris crepitans blanchardi*, in West Virginia. M.S. thesis, Marshall University, Huntington, West Virginia.

Didiuk, A. 1997. Status of amphibians in Saskatchewan. SSAR Herpetological Conservation 1:110–116.

Diekamp, B., and H.C. Gerhardt. 1995. Neurophysiological and behavioral studies of vocal communication in the gray treefrog, *Hyla versicolor*. Journal of Comparative Physiology 177A:173–190.

Diener, R.A. 1965. The occurrence of tadpoles of the green treefrog, *Hyla cinerea cinerea* (Schneider), in Trinity Bay, Texas. British Journal of Herpetology 3:198–199.

Dill, L.M. 1977. "Handedness" in the Pacific tree frog (*Hyla regilla*). Canadian Journal of Zoology 55:1926–1929.

Diller, L.V., and R.L. Wallace. 1999. Distribution and habitat of *Ascaphus truei* in streams on managed, young growth forests in north coastal California. Journal of Herpetology 33:71– 79.

DiMauro, D., and M.L. Hunter, Jr. 2002. Reproduction of amphibians in natural and anthropogenic temporary pools in managed forests. Forest Science 48:397–406.

Dimmitt, M.A., and R. Ruibal. 1980a. Exploitation of food resources by spadefoot toads (*Scaphiopus*). Copeia

1980:854–862.

Dimmitt, M.A., and R. Ruibal. 1980b. Environmental correlates of emergence in spadefoot toads (*Scaphiopus*). Journal of Herpetology 14:21–29.

Dinehart, S.K., L.M. Smith, S.T. McMurry, T.A. Anderson, P.N. Smith, and D.A. Haukos. 2009. Toxicity of a glufosinate–and several glyphosate–based herbicides to juvenile amphibians from the Southern High Plains, USA. Science of the Total Environment 407:1065–1071.

Dinehart, S.K., L. M. Smith. S. T. McMurry, P. N. Smith, T. A. Anderson, and D. A. Haukos. 2010. Acute and chronic toxicity of Roundup WeatherMAX® and Ignite® 280 SL to larval *Spea multiplicata* and *S. bombifrons* from the Southern High Plains, USA. Environmental Pollution 158: 2610–2617.

Dinsmore, A. 2004. Geographic distribution. *Eleutherodactylus planirostris*. Herpetological Review 35:403.

Dinsmore, S.C., II, and D.L. Swanson. 2008. Temporal patterns of tissue glycogen, glucose, and glycogen phosphorylase activity prior to hibernation in freeze-tolerant chorus frogs, *Pseudacris triseriata*. Canadian Journal of Zoology 86:1095-110.

Dirig, R. 1978. A large diurnal breeding assemblage of American toads, *Bufo americanus americanus*, in the Catskill Mountains. Pitch Pine Naturalist 4:1–2.

Distel, C.A. 2010. Effects of an insecticide on competition in anurans: could pesticide–induced competitive exclusion be a mechanism for amphibian declines? Ph.D. Dissertation, Miami University, Oxford, Ohio.

Distel, C.A., and M.D. Boone. 2009. Effects of aquatic exposure to the insecticide carbaryl and density on aquatic and terrestrial growth and survival in American toads. Environmental Toxicology and Chemistry 28:1963–1969.

Distel, C.A., and M.D. Boone. 2010. Effects of aquatic exposure to the insecticide carbaryl are species–specific across life stages and mediated by heterospecific competitors in anurans. Functional Ecology 24:1342–1352.

Distel, C.A., and M.D. Boone. 2011a. Pesticide has asymmetric effects on two tadpole species across density gradient. Environmental Toxicology and Chemistry 30:650–658.

Distel, C.A., and M.D. Boone. 2011b. Insecticide has asymmetric effects on two tadpole species despite priority effects. Ecotoxicology 20:875–884.

Ditmars, R.L. 1905. The batracians of the vicinity of New York City. American Museum Journal 5(4):161–206. [republished as American Museum of Natural History, Guide Leaflet No. 20]

Dixon, J.R. 1967. The amphibians and reptiles of Los Angeles County California. Los Angeles County Museum, Science Series 23, Zoology No. 10.

Dixon, J.R. 2000. Amphibians and Reptiles of Texas, with Keys, Taxonomic Synopses, Bibliography, and Distribution Maps. 2nd ed. Texas A & M University Press, College Station. [first edition, 1987]

Dobbs, E.K., M.G. Brown, J.W. Snodgrass, and D.R. Ownby. 2012. Salt toxicity to treefrogs (*Hyla chrysoscelis*) depends on depth. Herpetologica 68:22–30.

Dodd, C.K., Jr. 1977. Immobility in juvenile *Bufo woodhousei fowleri*. Journal of the Mississippi Academy of Sciences 22:90–94.

Dodd, C.K., Jr. 1979. A photographic technique to study tadpole populations. Brimleyana 2:131–136.

Dodd, C.K., Jr. 1991. Drift fence–associated sampling bias of amphibians at a Florida sandhills temporary pond. Journal of Herpetology 25:296–301.

Dodd, C.K., Jr. 1992. Biological diversity of a temporary pond herpetofauna in north Florida sandhills. Biodiversity and Conservation 1:125–142.

Dodd, C.K., Jr. 1994. The effects of drought on population structure, activity, and orientation of toads (*Bufo quercicus* and *B. terrestris*) at a temporary pond. Ethology Ecology & Evolution 6:331–349.

Dodd, C.K., Jr. 1995. The ecology of a sandhills population of the eastern narrow–mouthed toad, *Gastrophryne carolinensis*, during a drought. Bulletin of the Florida Museum of Natural History 38:11–41.

Dodd, C.K., Jr. 1996. Use of terrestrial habitats by amphibians in the sandhill uplands of north–central Florida. Alytes 14:42–52.

Dodd, C.K., Jr. 2003. Monitoring amphibians in Great Smoky Mountains National Park. U.S. Geological Survey, Circular No. 1258.

Dodd, C.K., Jr. 2004. The Amphibians of Great Smoky Mountains National Park. University of Tennessee Press, Knoxville.

Dodd, C.K., Jr. (ed.). 2010. Amphibian Ecology and Conservation. A Handbook of Techniques. Oxford University Press, Oxford, United Kingdom.

Dodd, C.K., Jr., and P.V. Cupp. 1978. The effect of temperature and stage of development on the duration of immobility in selected anurans. British Journal of Herpetology 5:783–788.

Dodd, C.K., Jr. and B.G. Charest. 1988. The herpetofaunal community of temporary ponds in north Florida sandhills: species composition, temporal use, and management implications. pp. 87-97 *In* R.C. Szaro, K.E. Severson, and D.R. Patton (eds.), Management of Amphibians, Reptiles, and Small Mammals in North America, USDA Forest Service General & Technical Report RM-166.

Dodd, C.K., Jr., and R.A. Seigel. 1991. Relocation, repatriation and translocation of amphibians and reptiles: are they conservation strategies that work? Herpetologica

47:335–350.

Dodd, C.K., Jr., and B.S. Cade. 1998. Movement patterns and the conservation of amphibians breeding in small, temporary wetlands. Conservation Biology 12:331–339.

Dodd, C.K., Jr., and L.L. Smith. 2003. Habitat destruction and alteration: historical trends and future prospects for amphibians. Pp. 94–112 *In* R.D. Semlitsch (ed.), Amphibian Conservation. Smithsonian Books, Washington, D.C.

Dodd, C.K., Jr., and W.J. Barichivich. 2007. Establishing a baseline and faunal history in amphibian monitoring programs: the amphibians of Harris Neck, GA. Southeastern Naturalist 6:125–134.

Dodd, C.K., Jr., M.L. Griffey, and J.D. Corser. 2001. The cave associated amphibians of Great Smoky Mountains National Park: review and monitoring. Journal of the Elisha Mitchell Scientific Society 117:139–149.

Dodd, C.K., Jr., W.J. Barichivich, and L.L. Smith. 2004. Effectiveness of a barrier wall and culverts in reducing wildlife mortality on a heavily traveled highway in Florida. Biological Conservation 118:619–631.

Dodd, C.K., Jr., W.J. Barichivich, S.A. Johnson, and J.S. Staiger. 2007. Changes in a northwestern Florida Gulf Coast herpetofaunal community over a 28–y period. American Midland Naturalist 158:29–48.

Dodd, C. K., Jr., J. Loman, D. Cogălniceanu, and M. Puky. 2012. Monitoring amphibian populations. Pp. 3577–3635 *In* H.H. Heatwole and J. W. Wilkenson (eds.), Conservation and Decline of Amphibians: Ecological Aspects, Effect of Humans and Management. Amphibian Biology, Volume 10, Surrey Beatty & Sons, Chipping Norton, New South Wales, Australia.

Dodge, N.N. 1938. Amphibians and reptiles of Grand Canyon National Park. Grand Canyon Natural History Association, Natural History Bulletin No. 9.

Doherty, J.A., and H.C. Gerhardt. 1984. Evolutionary and neurobiological implications of selective phonotaxis in the spring peeper (*Hyla crucifer*). Animal Behaviour 32:875–881.

Dole, J.W. 1965a. Summer movements of adult leopard frogs, *Rana pipiens* Schreber, in northern Michigan. Ecology 46:236–255.

Dole, J.W. 1965b. Spatial relations in natural populations of the leopard frog, *Rana pipiens* Schreber, in northern Michigan. American Midland Naturalist 74:464–478.

Dole, J.W. 1967a. The role of substrate moisture and dew in the water economy of leopard frogs, *Rana pipiens*. Copeia 1967:141–149.

Dole, J.W. 1967b. Spring movements of leopard frogs, *Rana pipiens* Schreber, in northern Michigan. American Midland Naturalist 78:167–181.

Dole, J.W. 1968. Homing in leopard frogs, *Rana pipiens*.

Ecology 49:386–399.

Dole, J.W. 1971. Dispersal of recently metamorphosed leopard frogs, *Rana pipiens*. Copeia 1971:221–228.

Dole, J.W. 1972a. Homing and orientation of displaced toads, *Bufo americanus*, to their home sites. Copeia 1972:151–158.

Dole, J.W. 1972b. The role of olfaction and audition in the orientation of leopard frogs, *Rana pipiens*. Herpetologica 28:258–260.

Dole, J.W. 1972c. Evidence of celestial orientation in newly–metamorphosed *Rana pipiens*. Herpetologica 28:273–276.

Dole, J.W. 1973. Celestial orientation in recently metamorphosed *Bufo americanus*. Herpetologica 29:59–62.

Dole, J.W. 1974. Home range in the canyon tree frog (*Hyla cadaverina*). Southwestern Naturalist 19:105–107.

Dole, J.W., B.B. Rose, and K.H. Tachiki. 1981. Western toads (*Bufo boreas*) learn odor of prey insects. Herpetologica 37:63–68.

Dole, J.W., B.B. Rose, and C.F. Baxter. 1985. Hyperosmotic saline environment alters feeding behavior in the western toad, *Bufo boreas*. Copeia 1985:645–648.

Donald, D.B., W. T. Aitken, C. Paquette, and S.S. Wulff. 2011. Winter snowfall determines the occupancy of northern prairie wetlands by tadpoles of the Wood Frog (*Lithobates sylvaticus*). Canadian Journal of Zoology 89:1063–1073.

Donnelly, M.A., C.J. Farrell, M.J. Baber, and J.L. Glenn. 2001. The amphibians and reptiles of the Kissimmee River. 1. Patterns of abundance and occurrence in altered floodplain habitats. Herpetological Natural History 8:161–170.

Doody, J.S., and J.E. Young. 1995. Temporal variation in reproduction and clutch mortality of leopard frogs (*Rana utricularia*) in south Mississippi. Journal of Herpetology 29:614–616.

Doody, J.S., J.E. Young, and G.N. Johnson. 1995. *Rana capito* (Gopher Frog). Combat. Herpetological Review 26:202–203.

Dorcas, M., and W. Gibbons. 2011. Frogs. The Animal Answer Guide. Johns Hopkins University Press, Baltimore, Maryland.

Dorcas, M.E., S.J. Price, and G.E. Vaughan. 2006. Amphibians and reptiles of the Great Falls bypassed reaches in South Carolina. Journal of the North Carolina Academy of Science 122:1–9.

Dorcas, M.E., S.J. Price, J.C. Beane, and S.C. Owen. 2007. The Frogs and Toads of North Carolina. Field Guide and Recorded Calls. North Carolina Wildlife Resources Commission, Raleigh, North Carolina.

Doty, T.L. 1978. A study of larval amphibian population dynamics in a Rhode Island vernal pond. Ph.D. Dissertation, University of Rhode Island, Kingston.

Doubledee, R.A., E.B. Miller, and R.M. Nisbet. 2003. Bullfrogs, disturbance regimes, and the persistence of California red–legged frogs. Journal of Wildlife Management 67:424–438.

Dougherty, C.K., and G.R. Smith. 2006. Acute effects of road de–icers on the tadpoles of three anurans. Applied Herpetology 3:87–93.

Douglas, C.L. 1966. Amphibians and reptiles of Mesa Verde National Park, Colorado. University of Kansas Publications, Museum of Natural History 15:711–744.

Dowe, B.J. 1979. The effect of time of oviposition and micro-environment on growth of larval bullfrogs (*Rana catesbeiana*) in Arizona. M.S. thesis, Arizona State University, Tempe.

Dowler, R.C., J.K. McCoy, and L.J. Fohn. 2010. *Scaphiopus couchii* (Couch's Spadefoot). Predation. Herpetological Review 41:480.

Dowling, H.G. 1957. A review of the amphibians and reptiles of Arkansas. University of Arkansas Museum, Occasional Papers No. 3.

Doyle, J.M. 2011a. Evaluating the phenotype–linked fertility hypothesis: Do exaggerated male phenotypes advertise potential fecundity benefits to females? Ph.D. Dissertation, Purdue University, West Lafayette, Indiana.

Doyle, J.M. 2011b. Sperm depletion and a test of the phenotype–linked fertility hypothesis in gray treefrogs (*Hyla versicolor*). Canadian Journal of Zoology 89: 853–858.

Drake, C.J. 1914. The food of *Rana pipiens* Shreber. The Ohio Naturalist 14:257–269.

Drake, D.L. 2010. *Lithobates onca* (Relict Leopard Frog). Cannibalistic oophagy. Herpetological Review 41:198–199.

Drake, D.L., and S.E. Trauth. 2010. *Spea intermontana* (Great Basin Spadefoot). Algal symbiosis. Herpetological Review 41:481–482.

Drake, D.L., A. Drayer, and S.E. Trauth. 2007a. *Bufo americanus* (American Toad). Algal symbiosis. Herpetological Review 38:435–436.

Drake, D.L., R. Altig, J.B. Grace, and S.C. Walls. 2007b. Occurrence of oral deformities in larval anurans. Copeia 2007:449–458.

Drayer, A.N. 2011. Efficacy of constructed wetlands of various depths for natural amphibian community conservation. M.S. thesis, Eastern Kentucky University, Richmond.

Dreitz, V.J. 2006. Issues in species recovery: an example based on the Wyoming toad. Bioscience 56:765–771.

Dreslik, M.J., and J.M. Mui. 2005. *Pseudacris crucifer crucifer* (Northern Spring Peeper). Aggressive behavior. Herpeto-

logical Review 36:54–55.

Driver, E.C. 1936. Observations on *Scaphiopus holbrooki* (Harlan). Copeia 1936:67–69.

Drost, C.A., and G.M. Fellers. 1996. Collapse of a regional frog fauna in the Yosemite area of the California Sierra Nevada. Conservation Biology 10:414–425.

Duarte, A., D.J. Brown, and M.R.J. Forstner. 2011. Estimating abundance of the endangered Houston toad on a primary recovery site. Journal of Fish and Wildlife Management 2:207–215.

Dubois, A. 2006. Naming taxa from cladograms: a cautionary tale. Molecular Phylogenetics and Evolution 42:317–330.

DuBois, R.B., and F.M. Stoll. 1995. Downstream movement of leopard frogs in a Lake Superior tributary exemplifies the concept of a lotic macrodrift. Journal of Freshwater Ecology 10:135–139.

Ducey, P.K., W. Newman, K.D. Cameron, and M. Messere. 1998. Herpetofauna of the highly–polluted Onondaga Lake ecosystem, Onondaga County, New York. Herpetological Review 29:118–119.

Dudley, R., V.A. King, and R. Wassersug. 1991. The implications of shape and metamorphosis for drag forces on a generalized pond tadpole (*Rana catesbeiana*). Copeia 1991:252–257.

Duellman, W.E. 1951. Notes on the reptiles and amphibians of Greene County, Ohio. Ohio Journal of Science 51:335–341.

Duellman, W.E. 1955. Systematic status of the KeyWest spadefoot toad, *Scaphiopus holbrookii albus*. Copeia 1955:141–143.

Duellman, W.E. 1968. The taxonomic status of some American hylid frogs. Herpetologica 24:194–209.

Duellman, W.E. 1986. Biology of Amphibians. McGraw Hill, New York.

Duellman, W.E., and L.N. Bell. 1955. The frogs and toads of the Everglades National Park. Everglades Natural History 3(2):102–115.

Duellman, W.E., and A. Schwartz. 1958. Amphibians and reptiles of southern Florida. Bulletin of the Florida State Museum, Biological Sciences 3:181–324.

Duffitt, A.D., and M.S. Finkler. 2011. Sex–related differences in somatic stored energy reserves of *Pseudacris crucifer* and *Pseudacris triseriata* during the early breeding season. Journal of Herpetology 45:224–229.

Duffus, A.L.J., and D.H. Ireland. 2006. *Rana sylvatica* (Wood Frog). Terrestrial amplexing pairs. Herpetological Review 37:449.

Duffus, A.L.J., B.D. Pauli, K. Wozney, C.R. Brunetti, and M. Berrill. 2008. Frog Virus 3–like infections in aquatic amphibian communities. Journal of Wildlife Diseases

44:109–120.

Dumas, P.C. 1957. *Rana sylvatica* Le Conte in Idaho. Copeia 1957:150–151.

Dumas, P.C. 1966. Studies of the *Rana* species complex in the Pacific Northwest. Copeia 1966:60–74.

Duméril, A.M.C., and G. Bibron. 1841. Erpetologie General, ou Histoire Naturelle Complete des Reptiles, Vol. 8., Roret, Paris.

Dunbar, N.J. 1949. Wood frogs at Chimo. The Arctic Circular 2(1):9–10.

Dundee, H.A. 1974. Recognition characters for *Rana grylio*. Journal of Herpetology 8:275–276.

Dundee, H.A. 1999. *Gastrophryne olivacea* (Great Plains Narrowmouth Toad). Aggregation with tarantula. Herpetological Review 30:91–92.

Dundee, H.A., and D.A. Rossman. 1989. The Amphibians and Reptiles of Louisiana. Louisiana State University Press, Baton Rouge.

Dundee, H.A., C. Shillington, and C.M. Yeary. 2012. Interactions between tarantulas (*Aphonopelma hentzi*) and narrow–mouthed toads (*Gastrophryne olivacea*): support for a symbiotic relationship. Tulane Studies in Zoology and Botany 32:31–38.

Dunlap, D.G. 1955. Inter– and intraspecific variation in Oregon frogs of the genus *Rana*. American Midland Naturalist 54:314–331.

Dunlap, D.G. 1959. Notes on the amphibians and reptiles of Deschutes County, Oregon. Herpetologica 15:173–177.

Dunlap, D.G. 1963. The status if [sic] the gray treefrog, *Hyla versicolor* in South Dakota. Proceedings of the South Dakota Academy of Science 42:136–139.

Dunlap, D.G. 1967. Selected records of amphibians and reptiles from South Dakota. Proceedings of the South Dakota Academy of Science 46:100–106.

Dunlap, D.G. 1969. Evidence for a daily rhythm of heat resistance in the cricket frog, *Acris crepitans*. Copeia 1969:852–854.

Dunlap, D.G. 1977. Wood and western spotted frogs (Amphibia, Anura, Ranidae) in the Big Horn Mountains of Wyoming. Journal of Herpetology 11:85–87.

Dunlap, D.G., and R.M. Storm. 1951. The Cascade frog in Oregon. Copeia 1951:81.

Dunlap, D.G., and K.C. Kruse. 1976. Frogs of the *Rana pipiens* complex in the northern and central Plains states. Southwestern Naturalist 20:559–571.

Dunlap, D.G., and J.E. Platz. 1981. Geographic variation of proteins and call in *Rana pipiens* from the northcentral United States. Copeia 1981:876–879.

Dunlap, D.G., and C.K. Satterfield. 1985. Habitat selection in larval anurans: early experience and substrate pattern selection in *Rana pipiens*. Developmental Psychobiology 18:37–58.

Dunn, E.R. 1917. Reptile and amphibian collections from the North Carolina mountains, with especial reference to salamanders. Bulletin of the American Museum of Natural History 37:593–634.

Dunn, E.R. 1930. Reptiles and amphibians of Northampton and vicinity. Bulletin of the Boston Society of Natural History 57:3–8.

Dunn, E.R. 1935. The survival value of specific characters. Copeia 1935:85–98.

Dunn, E.R. 1937. The status of *Hyla evittata* Miller. Transactions of the Biological Society of Washington 50:9–10.

Dunn, E.R. 1938. Notes on frogs of the genus *Acris*. Proceedings of the Academy of Natural Sciences of Philadelphia 90:153–154.

Dunn, H.H. 1932. Frogs by the million raised on odd farm. Popular Science Monthly 120(2):54-55, 134.

Dupré, R.K., and J.W. Petranka. 1985. Ontogeny of temperature selection in larval amphibians. Copeia 1985:462–467.

Dupuis, L., and D. Steventon. 1999. Riparian management and the tailed frog in northern coastal forests. Forest Ecology and Management 124:35–43.

Dupuis, L., and P. Friele. 2006. The distribution of the Rocky Mountain tailed frog (*Ascaphus montanus*) in relation to the fluvial system: implications for management and conservation. Ecological Research 21:489–502.

Durham, F.E. 1956. Amphibians and reptiles of the North Rim, Grand Canyon, Arizona. Herpetologica 12:220–224.

Durham, L., and G.W. Bennett. 1963. Age, growth, and homing in the bullfrog. Journal of Wildlife Management 27:107–123.

Dury, R., and R.S. Williams. 1933. Notes on some Kentucky amphibians and reptiles. The Baker–Hunt Foundation Museum, Williams Natural History Collection Bulletin No. 1.

Dusi, J.L. 1949. The natural occurrence of "redleg," *Pseudomonas hydrophila*, in a population of American toads, *Bufo americanus*. Ohio Journal of Science 49:70–71.

Dyche, L.L. 1914. Ponds, pond fish, and pond fish culture. Kansas Department of Fish and Game, Bulletin 1.

Dyer, W.G. 1991. Helminth parasites of amphibians from Illinois and adjacent Midwestern states. Transactions of the Illinois State Academy of Science 84:125–143.

Dyrkacz, S. 1981. Recent instances of albinism in North American amphibians and reptiles. SSAR Herpetological Circular No. 11.

Earl, J.E. 2012. Effects of spatial subsidies and canopy cover on pond communities and multiple life stages in amphibians. Ph.D. Dissertation, University of Missouri, Columbia.

Earl, J.E., and H.H. Whiteman. 2009. Effects of pulsed nitrate exposure on amphibian development. Environmental Toxicology and Chemistry 28:1331–1337.

Earl, J.E., and H.H. Whiteman. 2010. Evaluation of phosphate toxicity in Cope's Gray Treefrog (*Hyla chrysoscelis*) tadpoles. Journal of Herpetology 44:201–208.

Earl, J.E., and R.D. Semlitsch. 2012. Reciprocal subsidies in ponds: does leaf litter input increase frog biomass export? Oecologia 170:1077–1087.

Earl, J.E., T.M. Luhring, B.K. Williams, and R.D. Semlitsch. 2011. Biomass export of salamanders and anurans from ponds is affected differentially by changes in canopy cover. Freshwater Biology 56:2473–2482.

Earl, J.E., K.E. Cohagen, and R.D. Semlitsch. 2012. Effects of leachate from tree leaves and grass litter on tadpoles. Environmental Toxicology and Chemistry 31:1511–1517.

Eason, G.W., Jr., and J.E. Fauth. 2001. Ecological correlates of anuran species richness in temporary pools: a field study in South Carolina, USA. Israel Journal of Zoology 47:347–365.

Easteal, S. 1981. The history of introductions of *Bufo marinus* (Amphibia: Anura): a natural experiment in evolution. Biological Journal of the Linnaean Society 16:93–113.

Eaton, B.R. 2004. Ecology of anurans in boreal Alberta. Ph.D. Dissertation, University of Alberta, Edmonton.

Eaton, B.R., and C.A. Paszkowski. 1999. *Rana sylvatica* (Wood Frog). Predation. Herpetological Review 30:164.

Eaton, B.R., and Z.C. Eaton. 2001. An observation of a mallard, *Anas platyrhynchos*, feeding on a wood frog, *Rana sylvatica*. Canadian Field-Naturalist 115:499–500.

Eaton, B.R., C. Grekul, and C.A. Paszkowski. 1999. An observation of interspecific amplexus between boreal, *Bufo boreas*, and Canadian, *Bufo hemiophrys*, toads with a range extension for boreal toad in central Alberta. Canadian Field-Naturalist 113:512–513.

Eaton, B.R., C.A. Paszkowski, and R. Chapman. 2003. *Rana sylvatica* (Wood Frog). Parasite. Herpetological Review 34:55.

Eaton, B.R., S. Eaves, C. Stevens, A. Puchniak, and C.A. Paszkowski. 2004. Deformity levels in wild populations of the wood frog (*Rana sylvatica*) in three ecoregions of western Canada. Journal of Herpetology 38:283–287.

Eaton, B.R., W.W. Tonn, C.A. Paszkowski, A.J. Danylchuk, and S.M. Boss. 2005a. Indirect effects of fish winterkills on amphibian populations in boreal lakes. Canadian Journal of Zoology 83:1532–1539.

Eaton, B.R., C.A. Paszkowski, K. Kristensen, and M. Hiltz. 2005b. Life–history variation among populations of Canadian toads in Alberta, Canada. Canadian Journal of Zoology 83:1421–1430.

Eaton, B.R., C.L. Browne, C.A. Paszlowski, Z.C. Eaton, and R. Chapman. 2005c. *Bufo boreas* (Western Toad). Seed dispersal. Herpetological Review 36:52–53.

Eaton, B.R., A.E. Moenting, C.A. Paszkowski, and D. Shpeley. 2008. Myiasis by *Lucilla silvarum* (Calliphoridae) in amphibian species in boreal Alberta, Canada. Journal of Parasitology 94:949–952.

Eaton, T.H., Jr. 1935. Report on amphibians and reptiles of the Navajo country. Rainbow Bridge–Monument Valley Expedition Bulletin No. 3, 18 pp.

Eaton, T.H., Jr., and R.M. Imagawa. 1948. Early development of *Pseudacris clarkii*. Copeia 1948:263–265.

Eaves, S.E. 2004. Amphibian distribution in the aspen parkland. M.S. thesis, University of Alberta, Edmonton.

Echaubard, P., K. Little, B. Pauli, and D. Lesbarrères. 2010. Context–dependent effects of ranaviral infection on northern leopard frog life history traits. PLoS One 5(10):e13723.

Echternacht, A.C. 1964. Albino specimens from Arizona. Herpetologica 20:211–212.

Eckert, J.C. 1934. The California toad in relation to the hive bee. Copeia 1934:92–93.

Eddy, S.B. 1976. Population ecology of the leopard frog, *Rana pipiens pipiens* Schreber, at Delta Marsh, Manitoba. M.S. thesis, University of Manitoba, Winnipeg.

Edge, C.B. D. G. Thompson, C. Hao, J. E. Houlahan. 2012. A silviculture application of the glyphosate–based herbicide VisionMAX to wetlands has limited direct effects on amphibian larvae. Environmental Toxicology and Chemistry 31:2375–2383.

Edginton, A.N., P.M. Sheridan, G.R. Stephenson, D.G. Thompson, and H.J. Boermans. 2004. Comparative effects of pH and Vision™ herbicide on two life stages of four anuran amphibian species. Environmental Toxicology and Chemistry 23:815–822.

Edgren, R.A., Jr. 1944. Notes on amphibians and reptiles from Wisconsin. American Midland Naturalist 32:495–498.

Edgren, R.A. 1954. Factors controlling color change in the tree frog, *Hyla versicolor* Wied. Proceedings of the Society for Experimental Biology and Medicine 87:20–23.

Edwards, J. 1999. Time course for cryoprotectant synthesis in the freeze-tolerant chorus frog (*Pseudacris triseriata*). M.A. thesis, University of South Dakota, Vermillion.

Edwards, J.R., K.L. Koster, and D.L. Swanson. 2000. Time course for cryoprotectant synthesis in the freeze–tolerant chorus frog, *Pseudacris triseriata*. Comparative Biochemistry and Physiology 125A:367–375.

Edwards, T.M., K.A. McCoy, T. Barbeau, M.W. McCoy, J.M.

Thro, and L.J. Guillette, Jr. 2006. Environmental context determines nitrate toxicity in southern toad (*Bufo terrestris*) tadpoles. Aquatic Toxicology 78:50–58.

Egan, R.S. 2001. Within–pond and landscape–level factors influencing the breeding effort of *Rana sylvatica* and *Ambystoma maculatum*. M.S. thesis, University of Rhode Island, Kingston.

Egan, R.S., and P.W.C. Paton. 2004. Within–pond parameters affecting oviposition by wood frogs and spotted salamanders. Wetlands 24:1–13.

Egan, R.S., and P.W.C. Paton. 2008. Multiple scale habitat characteristics of pond–breeding amphibians across a rural–urban gradient. Pp. 53–65 *In* J.C. Mitchell, R.E. Jung Brown, and B. Bartholomew (eds.), Urban Herpetology. Herpetological Conservation 3. Society for the Study of Amphibians and Reptiles, Salt Lake City, Utah.

Ehret, G., and H.C. Gerhardt. 1980. Auditory masking and effects of noise on responses of the green treefrog (*Hyla cinerea*) to synthetic mating calls. Journal of Comparative Physiology A 141:13–18.

Ehrlich, D. 1979. Predation by bullfrog tadpoles (*Rana catesbeiana*) on eggs and newly hatched larvae of the Plains leopard frog (*Rana blairi*). Bulletin of the Maryland Herpetological Society 15:25–26.

Eidietis, L. 2005a. A biomechanical description of the anuran tadpole startle response and some implications of anatomical diversity. Ph.D. Dissertation, University of Michigan, Ann Arbor.

Eidietis, L. 2005b. Size-related performance variation in the wood frog (*Rana sylvatica*) tadpole tactile-stimulated startle response. Canadian Journal of Zoology 83:1117-1127.

Eidietis, L. 2006. The tactile-stimulated startle response of tadpoles: acceleration performance and its relationship to the anatomy of wood frog (*Rana sylvatica*), bullfrog (*Rana catesbeiana*), and American toad (*Bufo americanus*) tadpoles. Journal of Experimental Zoology 305A:348-362.

Eigenbrod, F., S.J. Hecnar, and L. Fahrig. 2008. The relative effects of road traffic and forest cover on anuran populations. Biological Conservation 141:35–46.

Eigenbrod, F., S.J. Hecnar, and L. Fahrig. 2009. Quantifying the road–effect zone: threshold effects of a motorway on anuran populations in Ontario, Canada. Ecology and Society 14(1):24.

Einem, G.E., and L.D. Ober. 1956. The seasonal behavior of certain Floridian Salientia. Herpetologica 12:205–212.

Eklöv, P. 2000. Chemical cues from multiple predator–prey interactions induce changes in behavior and growth of anuran larvae. Oecologia 123:192–199.

El Balaa, R, and G. Blouin–Demers. 2011. Unpalatability of northern leopard frog *Lithobates pipiens* Schreber, 1782 tadpoles. Herpetology Notes 4:159.

Elinson, R.P. 1974. A block to cross–fertilization located in the egg jelly of the frog *Rana clamitans*. Journal of Embryology and Experimental Morphology 32:325–335.

Elinson, R.P. 1975a. Viable amphibian hybrids produced by circumventing a block to cross–fertilization (*Rana clamitans* X *Rana catesbeiana*). Journal of Experimental Zoology 192:323–329.

Elinson, R.P. 1975b. Fertilization of green frog (*Rana clamitans*) eggs in their native jelly by bullfrog (*Rana catesbeiana*) sperm. Journal of Experimental Zoology 193:419–423.

Elinson, R.P. 1977. Macrocephaly and microcephaly in hybrids between the bullfrog *Rana catesbeiana* and the mink frog *Rana septentrionalis* (Amphibia, Anura, Ranidae). Journal of Herpetology 11:94–96.

Elinson, R.P. 1981a. Have you seen a bullfrog–green frog hybrid? Herpetological Review 12:104.

Elinson, R.P. 1981b. Genetic analysis of developmental arrest in an amphibian hybrid (*Rana catesbeiana, Rana clamitans*). Developmental Biology 81:167–176.

Elinson, R.P. 1993. Viable triploid hybrids between a arge frog, *Rana catesbeiana*, and a small one, *Rana septentrionalis*. Herpetological Review 24:46–47.

Elliott, L., C. Gerhardt, and C. Davidson. 2009. The Frogs and Toads of North America. Houghton Mifflin Harcourt, Boston.

Ellis, M.M. 1917. Amphibians and reptiles of the Douglas Lake (Michigan) Region. 19th Michigan Academy of Science Reports, pp. 45–63.

Ellis, M.M., and J. Henderson. 1913. Amphibia and reptilia of Colorado. Part 1. University of Colorado Studies 10:39–129.

Ellis, M.M., and J. Henderson. 1915. Amphibia and reptilia of Colorado. Part 2. University of Colorado Studies 11:253–263.

Ely, C. 1944. Development of *Bufo marinus* larvae in dilute sea water. Copeia 1944:256.

Emerson, S.B., J. Travis, and M. Blouin. 1988. Evaluating a hypothesis about heterochrony: larval life–history traits and juvenile hind–limb morphology in *Hyla crucifer*. Evolution 42: 68–78.

Emery, A.R., A.H. Berst, and K. Kodaira. 1972. Under–ice observations of wintering sites of leopard frogs. Copeia 1972:123–126.

Emlen, S.T. 1968. Territoriality in the bullfrog, *Rana catesbeiana*. Copeia 1968:240–243.

Emlen, S.T. 1976. Lek organization and mating strategies in the bullfrog. Behavioral Ecology and Sociobiology 1:283–313.

Emlen, S.T. 1977. "Double–clutching" and its possible significance in the bullfrog. Copeia 1977:749–751.

Emmerson, F.H. 1971. Agonistic behavior in some captive frogs (*Rana*) from Nevada. Herpetological Review 3:29.

Emmett, A.D., and F.P. Allen. 1919. Nutritional studies on the growth of frog larvae (*Rana pipiens*). Journal of Biological Chemistry 38:325–344.

Enderson, E.F., and R.L. Bezy. 2000. Geographic distribution: *Pternohyla fodiens* (Lowland burrowing treefrog). Herpetological Review 31:251–252.

Engbrecht, N. J. 2010. The status of Crawfish Frogs (*Lithobates areolatus*) in Indiana, and a tool to assess populations. M.S. thesis, Indiana State University, Terre Haute.

Engbrecht, N.J., and J.L. Heemeyer. 2010. *Lithobates areolatus circulosus* (Northern Crawfish Frog). *Heterodon platirhinos* (Eastern Hog–nosed Snake). Predation. Herpetological Review 41:168–170.

Engbrecht, N.J., and M.J. Lannoo. 2010. A review of the status and distribution of crawfish frogs (*Lithobates areolatus*) in Indiana. Proceedings of the Indiana Academy of Science 119:64–73.

Engbrecht, N.J., and M.J. Lannoo. 2012. Crawfish frog behavioral differences in postburned and vegetated grasslands. Fire Ecology 8:63-76.

Engbrecht, N.J., S.J. Lannoo, J.O. Whitaker and M.J. Lannoo. 2011. Comparative morphometrics in ranid frogs (subgenus *Nenirana*): are apomorphic elongation and a blunt snout responses to small–bore burrow dwelling in crawfish frogs (*Lithobates areolatus*)? Copeia 2011:285–295.

Engbrecht, N.J., J.L. Heemeyer, V.C. Kinney and M.J. Lannoo. 2012. *Lithobates areolatus circulosus* (Northern Crawfish Frog). Thwarted predation. Herpetological Review 43:323–324.

Enge, K.M. 1984. Effects of clearcutting and site preparation on the herpetofauna of a north Florida flatwoods. M.S. thesis, University of Florida, Gainesville.

Enge, K.M. 1998a. Herpetofaunal survey of an upland hardwood forest in Gadsden County, Florida. Florida Scientist 61:141–159.

Enge, K.M. 1998b. Herpetofaunal drift–fence survey of steephead ravines in 2 river drainages. Proceedings of the Annual Conference of Southeastern Association of Fish and Wildlife Agencies 1998:336–348.

Enge, K.M. 2002. Herpetofaunal drift–fence survey of two seepage bogs in Okaloosa County, Florida. Florida Scientist 65:67–82.

Enge, K.M. 2005a. Commercial harvest of amphibians and reptiles in Florida for the pet trade. Pp. 198–211 *In* W.E. Meshaka, Jr. and K.J. Babbitt (eds.), Amphibians and Reptiles. Status and Conservation in Florida. Krieger Publishing, Malabar, Florida.

Enge, K.M. 2005b. Herpetofaunal drift–fence surveys of steephead ravines in the Florida Panhandle. Southeastern Naturalist 4:657–678.

Enge, K.M., and W.R. Marion. 1986. Effects of clearcutting and site preparation on herpetofauna of a north Florida flatwoods. Forest Ecology and Management 14:177–192.

Enge, K.M., and K.N. Wood. 1998. Herpetofaunal surveys of the Big Bend Wildlife Management Area, Taylor County, Florida. Florida Scientist 61:61–87.

Enge, K.M., and K.N. Wood. 2000. A herpetofaunal survey of Chassahowitzka Wildlife Management Area, Hernando County, Florida. Herpetological Natural History 7:117–144.

Enge, K.M., and K.N. Wood. 2001. Herpetofauna of Chinsegut Nature Center, Hernando County, Florida. Florida Scientist 64:283–305.

Enge, K.M., D.T. Cobb, G.L. Sprandel, and D.L. Francis. 1996. Wildlife captures in a pipeline trench in Gadsden County, Florida. Florida Scientist 59:1–11.

Engelbert, J., M.Patrick, and S.E. Trauth. 2008. *Rana catesbeiana* (American Bullfrog). Lithophagy. Herpetological Review 39:80.

Engelhardt, G.P. 1917. Grand Canyon notes. Copeia (39):5–7.

Engelhardt, G.P. 1918. Batrachians from southwestern Utah. Copeia (60):77–80.

Engels, W.L. 1942. Vertebrate fauna of North Carolina coastal islands. A study in the dynamics of animal distribution. 1. Ocracoke Island. American Midland Naturalist 28:273–304.

Engels, W.L. 1952. Vertebrate fauna of North Carolina coastal islands. 2. Shackleford Banks. American Midland Naturalist 47:702–742.

Engeman, R.M., and E.M. Engeman. 1996. Longevity of Woodhouse's toad in Colorado. Northwestern Naturalist 77:23.

Engeman, R.M., and M.A. Engeman. 2006. *Bufo woodhousii* (Woodhouse's Toad). Survival. Herpetological Review 37:442–443.

Engle, J.E. 2001. Population biology and natural history of Columbia spotted frogs (*Rana luteiventris*) in the Owyhee Uplands of southwest Idaho: implications for monitoring and management. M.S. thesis, Boise State University, Boise, Idaho.

Epple, A.O. 1983. The Amphibians of New England. Down East Books, Camden, Maine.

Ercole, J.A. 1970. Mass dieoff of bullfrogs and leopard frogs by freezing in central Arizona. Journal of the Arizona Academy of Science 6:139.

Ernst, C.H. 2001. Some ecological parameters of the wood

turtle, *Clemmys insculpta*, in southeastern Pennsylvania. Chelonian Conservation and Biology 4:94–99.

Ernst, C.H., S.C. Belfit, S.W. Sekscienski, and A.F. Laemmerzahl. 2011. The amphibians and reptiles of Ft. Belvoir and northern Virginia. Bulletin of the Maryland Herpetological Society 33:1-62.

Ernst, J.A. 2001. Assessment of the *Rana pipiens* complex in southwestern South Dakota. M.S. thesis, University of Wisconsin–Stevens Point.

Ernst, S.G. 1945. The food of the red–shouldered hawk in New York State. Auk 62:452–453.

Errington, P.L., and W.J. Breckenridge. 1938. Food habits of buteo hawks in the north–central United States. Wilson Bulletin 50:113–121.

Ervin, E.L. 2005. *Pseudacris cadaverina* (Cope, 1866). California Treefrog. Pp. 467–470 *In* M.J. Lannoo (ed.), Amphibian Declines. The Conservation Status of United States Species. University of California Press, Berkeley.

Ervin, E.L., and R.N. Fisher. 2001. *Thamnophis hammondii* (Two–striped Garter Snake). Prey. Herpetological Review 32:265–266.

Ervin, E.L., and T.L. Cass. 2007. *Spea hammondii* (Western Spadefoot). Reproductive pattern. Herpetological Review 38:196–197.

Ervin, E.L., and R.N. Fisher. 2007. *Thamnophis hammondii* (Two–striped Gartersnake). Foraging behavior. Herpetological Review 38:345–346.

Ervin, E.L., R.N. Fisher, and K. Madden. 2000. *Hyla cadaverina* (California Treefrog). Predation. Herpetological Review 31:234.

Ervin, E.L., A.E. Anderson, T.L. Cas, and R.E. Murcia. 2001. *Spea hammondii* (Western Spadefoot Toad). Elevation record. Herpetological Review 32:36.

Ervin, E.L., S.J. Mullin, M.L. Warburton, and R.N. Fisher. 2003. *Thamnophis hammondii* (Two–striped Garter Snake). Prey. Herpetological Review 34:74–75.

Ervin, E.L., C.D. Smith, and S.V. Christopher. 2005. *Spea hammondii* (Western Spadefoot). Reproduction. Herpetological Review 36:309–310.

Ervin, E.L., D.A. Kisner, and R.N. Fisher. 2006. *Bufo californicus* (Arroyo Toad). Mortality. Herpetological Review 37:199.

Ervin, E.L., A.K. Gonzales, D.S. Johnston, and J.P. Pittman. 2007. *Spea hammondii* (Western Spadefoot). Predation. Herpetological Review 38:197–198.

Etges, W.J. 1979. Ecological genetic relationships in selected anurans of the southeastern United States. M.S. thesis, University of Georgia, Athens.

Etges, W.J. 1987. Call site choice in male anurans. Copeia 1987:910–923.

Evans, H.E., and R.M. Roecker. 1951. Notes on the herpe-

tology of Ontario, Canada. Herpetologica 7:69–71.

Evenden, F.G., Jr. 1943. Notes on amphibia of the Cascade Mountains in Oregon. Copeia 1943:251–252.

Evenden, F.G., Jr. 1948. Food habits of *Triturus granulosus* in western Oregon. Copeia 1948:219–220.

Evermann, B.W., and H.W. Clark. 1916. The turtles and batrachians of the Lake Maxinkuckee region. Proceedings of the Indiana Academy of Science 26:472–518.

Evermann, B.W., and H.W. Clark. 1920. Lake Maxinkuckee. A Physical and Biological Survey. Volume 1. Indiana Department of Conservation, Publication No. 7. [amphibians on pages 620–644]

Evrard, J.O., and S.R. Hoffman. 2000. Amphibians and reptiles captured in drift fences in northwest Wisconsin pine barrens. Journal of the Iowa Academy of Science 107:182–186.

Ewert, M.A. 1969. Seasonal movements of the toads *Bufo americanus* and *B. cognatus* in northwestern Minnesota. Ph.D. Dissertation, University of Minnesota, Minneapolis.

Faeh, S.A., D.K. Nichols, and V.R. Beasley. 1998. Infectious diseases of amphibians. Pp. 259– 265 *In* M.J. Lannoo (ed.), Status & Distribution of Midwestern Amphibians. University of Iowa Press, Iowa City.

Fair, J.W. 1969. Survival of weevils through the digestive tract of an amphibian. Bulletin of the Southern California Academy of Sciences 68:260–261.

Fair, J.W. 1970. Comparative rates of rehydration from soil in two species of toads, *Bufo boreas* and *Bufo punctatus*. Comparative Biochemistry and Physiology 34:281–287.

Fairchild, L. 1981. Mate selection and behavioral thermoregulation in Fowler's toads. Science 212:950–951.

Fairchild, L. 1984. Male reproductive tactics in an explosive breeding toad population. American Zoologist 24:407–418.

Faivovich, J., C.F.B. Haddad, P.C.A. Garcia, D.R. Frost, J.A. Campbell, and W.C. Wheeler. 2005. Systematic review of the frog family Hylidae, with special reference to Hylinae: phylogenetic analysis and taxonomic revision. Bulletin of the American Museum of Natural History 294:1–240.

Farmer, A.L., L.L. Smith, S.B. Castleberry, and J.W. Gibbons. 2009. A comparison of techniques for sampling amphibians in isolated wetlands in Georgia, USA. Applied Herpetology 6:327–341.

Farner, D.S., and J. Kezer. 1953. Notes on the amphibians and reptiles of Crater Lake National Park. American Midland Naturalist 50:448–462.

Farrar, E.S., and J.D. Hey. 1997. Carnivorous spadefoot (*Spea bombifrons* Cope) tadpoles and fairy shrimp in western Iowa. Journal of the Iowa Academy of Science 104:4-7.

Farrell, M.P. 1971. Effect of temperature and photoperiod acclimations on the water economy of *Hyla crucifer*. Herpetologica 27:41–48.

Farrell, T.M., M.A. Pilgrim, P.G. May, and W.B. Blihovde. 2011. The herpetofauna of Lake Woodruff National Wildlife Refuge, Florida. Southeastern Naturalist 10:647–658.

Farringer, J.E. 1972. The determination of acute toxicity of rotenone and bayer 73 to selected aquatic organisms. M.S. thesis, University of Wisconsin–La Crosse.

Farris, J.S., A. Kluge, and M.F. Margoliash. 1979. Paraphyly of the *Rana boylii* species group. Systematic Zoology 28:627–634.

Farris, J.S., A. Kluge, and M.F. Margoliash. 1982. Immunological distance and the phylogenetic relationships of the *Rana boylii* species group. Systematic Zoology 31:479–491.

Fashingbauer, B.A. 1957. The effect of aerial spraying with DDT on a population of wood frogs. Flicker 29(4):160.

Fauth, J.E. 1999. Identifying keystone species from field data – an example from 20 temporary ponds. Ecology Letters 2:36–43.

Feaver, P.E. 1971. Breeding pool selection and larval mortality of three California amphibians: *Ambystoma tigrinum californiense* Gray, *Hyla regilla* Baird and Girard, and *Scaphiopus hammondii* Girard. M.A. thesis, Fresno State College, Fresno, California.

Feder, J.H. 1977. Genetic variation and biochemical systematics in western *Bufo*. M.A. thesis, University of California, Berkeley.

Feder, J.H. 1979. Natural hybridization and genetic divergence between the toads *Bufo boreas* and *Bufo punctatus*. Evolution 33:1089–1097.

Feder, M.E. 1983. The relation of air breathing and locomotion to predation on tadpoles, *Rana berlandieri*, by turtles. Physiological Zoology 56:522–531.

Fedewa, L. 2003. Microbial ecology of developing southern toads, spring peepers, and narrow–mouthed toads. M.S. thesis, University of Georgia, Athens.

Fedewa, L.A. 2006. Fluctuating gram–negative microflora in developing anura. Journal of Herpetology 40:131–135.

Federighi, H. 1938. Albinism in *Rana pipiens* (Schreber). Ohio Journal of Science 38:37–40.

Feinberg, J.A., and T.M. Green. 2004. *Scaphiopus holbrookii* (Eastern Spadefoot). Diurnal breeding. Herpetological Review 35:53.

Feinberg, J.A., and K. Hoffmann. 2004. *Scaphiopus holbrookii* (Eastern Spadefoot). Coloration and behavior. Herpetological Review 35:377.

Feldhammer, G.A., M.B. Epstein, and W.B. Taliaferro. 1987. Prey remains in barn owl pellets from a South Carolina barrier island. Georgia Journal of Science 45:148–151.

Feldman, C.R., and J.A. Wilkinson. 2000. *Rana muscosa* (Mountain Yellow–legged Frog). Predation. Herpetological Review 31:102.

Feliciano, S.L. 2000. A comparison of two northern cricket frog (*Acris crepitans*) populations and their habitats during drought conditions in southeastern New York and the Florida Panhandle. M.S. thesis, University of Florida, Gainesville.

Felix, Z.I., Y. Wang, and C.J. Schweitzer. 2010. Effects of experimental canopy manipulation on amphibian egg deposition. Journal of Wildlife Management 74:496–503.

Fellers, G.M. 1975. Behavioral interactions in North American treefrogs (Hylidae). Chesapeake Science 16:218–219.

Fellers, G.M. 1976. Social interactions in North American treefrogs. Ph.D. Dissertation, University of Maryland, College Park.

Fellers, G.M. 1979a. Aggression, territoriality, and mating behaviour in North American treefrogs. Animal Behaviour 27:107–119.

Fellers, G.M. 1979b. Mate selection in the gray treefrog, *Hyla versicolor*. Copeia 1979:286–290.

Fellers, G.M., and C.A. Drost. 1993. Disappearance of the Cascades frog *Rana cascadae* at the southern end of its range, California, USA. Biological Conservation 65:177–181.

Fellers, G.M., and L.L. Wood. 2004. *Rana aurora draytonii* (California Red–legged Frog). Predation. Herpetological Review 35:163.

Fellers, G.M., and P.M. Kleeman. 2007. California red–legged frog (*Rana draytonii*) movement and habitat use: implications for conservation. Journal of Herpetology 41:276–286.

Fellers, G.M., D.E. Green, and J.E. Longcore. 2001a. Oral chytridiomycosis in the mountain yellow–legged frog (*Rana muscosa*). Copeia 2001:945–953.

Fellers, G.M., A.E. Launer, G. Rathbun, S. Bobzien, J. Alvarez, D. Sterner, R.B. Seymour, and M. Westphal. 2001b. Overwintering tadpoles in the California Red–legged frog (*Rana aurora draytonii*). Herpetological Review 32:156–157.

Fellers, G.M., L.L. McConnell, D. Pratt, and S. Datta. 2004. Pesticides in mountain yellow–legged frogs (*Rana muscosa*) from the Sierra Nevada Mountains of California, USA. Environmental Toxicology and Chemistry 23:2170–2177.

Fellers, G.M., D.F. Bradford, D. Pratt, and L.L. Wood. 2007a. Demise of repatriated populations of mountain yellow–legged frogs (*Rana muscosa*) in the Sierra Nevada of California. Herpetological Conservation and Biology

2:5–21.

Fellers, G.M., K.L. Pope, J.E. Stead, M.S. Koo, and H.H. Welsh, Jr. 2007b. Turning population trend monitoring into active conservation: can we save the Cascades frog (*Rana cascadae*) in the Lassen Region of California. Herpetological Conservation and Biology 3:28–39.

Fellers, G.M., R.A. Cole, D.M. Rentz, and P.M. Kleeman. 2011. Amphibian chytrid fungus (*Batrachochytrium dendrobatidis*) in coastal and montane California, USA anurans. Herpetological Conservation and Biology 6:383–394.

Felson, A.J. 2010. Suburbanization and amphibians: designed ecological solutions. Ph.D. Dissertation, Rutgers University, New Brunswick, New Jersey.

Feminella, J.W., and C.P. Hawkins. 1994. Tailed frog tadpoles differentially alter their feeding behavior in response to non–visual cues from four predators. Journal of the North American Benthological Society 13:310–320.

Feng, A.S., H.C. Gerhardt, and R.R. Capranica. 1976. Sound localization behavior of the green treefrog (*Hyla cinerea*) and the barking treefrog (*H. gratiosa*). Journal of Comparative Physiology A 107:241–252.

Fenoilio, D.B., G.O. Graening, and J.F. Stout. 2005. Seasonal movement patterns of pickerel frogs (*Rana palustris*) in an Ozark cave and trophic implications supported by stabile isotope evidence. Southwestern Naturalist 50:385–389.

Fenton, M.H. undated [1928]. Scientific Method of Bullfrog Culture in Connection with Muskrat Farming. M.H. Fenton Publisher, Pickerel, Ontario. 53 pp. [other copies say Forest Green, Missouri]

Fenton, M.H. 1932. Raising Jumbo Bullfrogs. Charilon Fur and Frog Farm Inc., Chicago, Illinois. 80 pp. [second edition, 1933]

Ferguson, D.E. 1952. The distribution of amphibians and reptiles of Wallowa County, Oregon. Herpetologica 8:66–68.

Ferguson, D.E. 1954. An interesting factor influencing *Bufo boreas* reproduction at high elevations. Herpetologica 10:199.

Ferguson, D.E. 1960. Observations on movements and behavior of *Bufo fowleri* in residential areas. Herpetologica 16:112–114.

Ferguson, D.E. 1961. The herpetofauna of Tishomingo County, Mississippi, with comments on its zoogeographic affinities. Copeia 1961:391–396.

Ferguson, D.E. 1963. Orientation in three species of anuran amphibians. Ergebnisse der Biologie 26:128–134.

Ferguson, D.E. 1966a. Sun–compass orientation in anurans. Pp. 21–33 *In* R.M. Storm (ed.), Animal Orientation and Navigation. Oregon State University Press, Corvallis.

Ferguson, D.E. 1966b. Evidence of sun–compass orientation in the chorus frog, *Pseudacris triseriata*. Herpetologica 22:106–112.

Ferguson, D.E., and H.F. Landreth. 1966. Celestial orientation of Fowler's toad *Bufo fowleri*. Behaviour 26:107–123.

Ferguson, D.E., and C.C. Gilbert. 1968. Tolerances of three species of anuran amphibians to five chlorinated hydrocarbon insecticides. Journal of the Mississippi Academy of Sciences 13:135–138.

Ferguson, D.E., K.E. Payne, and R.M. Storm. 1958. Notes on the herpetology of Baker County, Oregon. Great Basin Naturalist 18:63–65.

Ferguson, D.E., H.F. Landreth, and M.R. Turnipseed. 1965. Astronomical orientation of the southern cricket frog, *Acris gryllus*. Copeia 1965:58–66.

Ferguson, D.E., H.F. Landreth, and J.P. McKeown. 1967. Sun compass orientation of the Northern Cricket Frog, *Acris crepitans*. Animal Behaviour 15:45–53.

Ferguson, D.E., J.P. McKeown, O.S. Bosarge, and H.F. Landreth. 1968. Sun–compass orientation of bullfrogs. Copeia 1968:230–235.

Ferguson, J.H., and C.H. Lowe. 1969. Evolutionary relationships in the *Bufo punctatus* group. American Midland Naturalist 81:435–466.

Ferner, J.W., T. Rice, and K.S. Neltner. 1992. Predation on birds, at a birdfeeder, by bullfrogs, *Rana catesbeiana* Shaw. Pp. 61–62 *In* P.D. Strimple and J.L. Strimple (eds.), Contributions in Herpetology. Greater Cincinnati Herpetological Society, Cincinnati, Ohio.

Ferrari, M.C.O., and D.P. Chivers. 2008. Cultural learning of predator recognition in mixed–species assemblages of frogs: the effect of tutor–to–observer ratio. Animal Behaviour 75: 1921–1925.

Ferrari, M.C.O., and D. P. Chivers. 2009a. Sophisticated early life lessons: threat–sensitive generalization of predator recognition by embryonic amphibians. Behavioral Ecology 20: 1295–1298.

Ferrari, M.C.O., and D. P. Chivers. 2009b. Temporal variability, threat sensitivity and conflicting information about the nature of risk: understanding the dynamics of tadpole antipredator behaviour. Animal Behaviour 78:11–16.

Ferrari, M.C.O., and D.P. Chivers. 2010. The ghost of predation future: threat–sensitive and temporal assessment of risk by embryonic woodfrogs. Behavioral Ecology and Sociobiology 64:549–555.

Ferrari, M.C.O., F. Messier, and D.P. Chivers. 2007a. First documentation of cultural transmission of predator recognition by larval amphibians. Ethology 113:621–627.

Ferrari, M.C.O., F. Messier, and D.P. Chivers. 2007b. Degradation of chemical alarm cues under natural conditions:

risk assessment by larval woodfrogs. Chemoecology 17:263–266.

Ferrari, M.C.O., F. Messier, and D. P. Chivers. 2008. Larval amphibians learn to match antipredator response intensity to temporal patterns of risk. Behavioral Ecology 19:980–983.

Ferrari, M.C.O., G.E. Brown, F. Messier, and D.P. Chivers. 2009. Threat–sensitive generalization of predator recognition by larval amphibians. Behavioral Ecology and Sociobiology 63: 1369–1375.

Ferrari, M.C.O., A.K. Manek, and D.P. Chivers. 2010. Temporal learning of predation risk by embryonic amphibians. Biology Letters 6:308–310.

Ferrari, M.C.O., G. E. Brown, G. R. Bortolotti, and D. P. Chivers. 2010. Linking predator risk and uncertainty to adaptive forgetting: a theoretical framework and empirical test using tadpoles. Proceedings of the Royal Society B 277:2205–2210.

Ferrari, M.C.O., G.E. Brown, G.R. Bortolotti, and D.P. Chivers. 2011. Prey behaviour across antipredator adaptation types: how does growth trajectory influence learning of predators? Animal Cognition 14:809–816.

Ferrari, M.C.O., J. Vrtĕlová, G.E. Brown, and D. P. Chivers. 2012. Understanding the role of uncertainty on learning and retention of predator information. Animal Cognition 15:807–813.

Fetkavich, C., and L.J. Livo. 1998. Late–season boreal toad tadpoles. Northwestern Naturalist 79:120–121.

Ficetola, G.F., C. Miaud, F. Pompanon, and P. Taberlet. 2009. Species detection using environmental DNA from water samples. Biology Letters 4:423–425.

Fichter, E., and A.D. Linder. 1964. The amphibians of Idaho. Idaho State University Museum, Pocatello.

Fidenci, P. 2004. The California red–legged frog, *Rana aurora draytonii*, along the Arroyo Santo Domingo, northern Baja California, Mexico. Herpetological Bulletin 88:27–31.

Fidenci, P. 2006. *Rana boylii* (Foothill Yellow–legged Frog). Predation. Herpetological Review 37:208.

Field, K.J., and D. Groebner. 2005. *Rana chiricahuensis* (Chiricahua Leopard Frog). Reproduction. Herpetological Review 36:306–307.

Field, K.J., T.L. Beatty, Sr., and T.L. Beatty, Jr. 2003. *Rana subaquavocalis* (Ramsey Canyon Leopard Frog). Diet. Herpetological Review 34:235.

Figiel, C.R., and R.D. Semlitsch. 1991. Effects of nonlethal injury and habitat complexity on predation in tadpole populations. Canadian Journal of Zoology 69:830–834.

Filho, C.B.C., H.C. Costa, E.T Da Silva, O.P. Ribeiro Filho, and R.N. Feio. 2008. *Lithobates catesbeianus* (American Bullfrog). Prey. Herpetological Review 39:338.

Fincel, E.A. 2006. Urea: a natural cryoprotectant in the freeze–tolerant wood frog (*Rana sylvatica*). M.S. thesis, Eastern Illinois University, Charleston.

Findlay, C.S., and J. Houlahan. 1997. Anthropogenic correlates of species richness in southeastern Ontario wetlands. Conservation Biology 11:1000–1009.

Findlay, C.S., L. Lenton, and L. Zheng. 2001. Land–use correlates of anuran community richness and composition in southeastern Ontario wetlands. Ecoscience 8:336–343.

Findley, J.S. 1964. Verification of the occurrence of *Bufo microscaphus* Cope in New Mexico. Southwestern Naturalist 9:102–109.

Finlay, J.C., and V.T. Vredenburg. 2007. Introduced trout sever trophic connections in watersheds: consequences for a declining amphibian. Ecology 88:2187–2198.

Finley, R.B., Jr. 1953. A northern record of *Hyla arenicolor* in western Colorado. Copeia 1953:180.

Fischer, T.D. 1998. Anura of eastern South Dakota: their distribution and characteristics of their wetland habitats, 1997–1998. M.S. thesis, South Dakota State University, Brookings.

Fischer, T.D., D.C. Backlund, K.F. Higgins, and D.E. Naugle. 1999. Field Guide to South Dakota Amphibians. South Dakota Agricultural Experiment Station Bulletin 733, Brookings, South Dakota.

Fishbeck, D.W., and J.C. Underhill. 1960. Amphibians of eastern South Dakota. Herpetologica 16:131–136.

Fishbeck, D.W., and J.C. Underhill. 1971. Distribution of stripe polymorphism in wood frogs, *Rana sylvatica* LeConte, from Minnesota. Copeia 1971:253–259.

Fisher, A.K. 1893. Hawks and owls of the United States in their relation to agriculture. U.S. Department of Agriculture Bulletin No. 3, Washington, D.C.

Fisher, C., A. Joynt, and R.J. Brooks. 2007. Reptiles and Amphibians of Canada. Lone Pine Publishing, Edmonton, Alberta.

Fisher, H.T., and A. Richards. 1950. The annual ovarian cycle of *Acris crepitans* Baird. Pp. 129–142 *In* Researches on the Amphibia of Oklahoma, University of Oklahoma Press, Norman.

Fisher, M.C., and T.W.J. Garner. 2007. The relationship between the emergence of *Batrachochytrium dendrobatidis*, the international trade in amphibians and introduced amphibian species. Fungal Biology Reviews 21:2–9.

Fisher, R.N., and H.B. Shaffer. 1996. The decline of amphibians in California's Great Central Valley. Conservation Biology 10:1387–1397.

Fishwild, T.G., R.A. Schemidt, K.M. Jankens, K.A. Berven, G.J. Gamboa, and C.M. Richards. 1990. Sibling recognition by larval frogs (*Rana pipiens, R. sylvatica*, and

Pseudacris crucifer). Journal of Herpetology 24:40–44.

Fitch, H.S. 1936. Amphibians and reptiles of the Rogue River Basin, Oregon. American Midland Naturalist 17:634–652.

Fitch, H.S. 1956a. Temperature responses in free–living amphibians and reptiles of northeastern Kansas. University of Kansas Publications, Museum of Natural History 8:417–476.

Fitch, H.S. 1956b. A field study of the Kansas ant–eating frog, *Gastrophryne olivacea*. University of Kansas Publications, Museum of Natural History 8:275–306.

Fitch, H.S. 1956c. Early sexual maturity and longevity under natural conditions in the Great Plains narrow–mouthed frog. Herpetologica 12:281–282.

Fitch, H.S. 1958. Home ranges, territories, and seasonal movements of vertebrates of the Natural History Reservation. University of Kansas Publications, Museum of Natural History 11:63–326.

Fitch, H.S. 1960. Autecology of the copperhead. University of Kansas Publications, Museum of Natural History 13:85–288.

Fite, K.V. 1973. The visual fields of the frog and toad: a comparative study. Behavioral Biology 9:707–718.

Fite, K.V., A. Blaustein, L. Bengston, and H.E. Hewitt. 1998. Evidence of retinal light damage in *Rana cascadae*: a declining amphibian species. Copeia 1998:906–914.

Fitzgerald, G.J. 1974. Influence of moon phase and weather factors on locomotory activity in *Bufo americanus*. Oikos 25:338–340.

Fitzgerald, G.J., and J.R. Bider. 1974a. Seasonal activity of the toad *Bufo americanus* in southern Quebec as revealed by a sand transect technique. Canadian Journal of Zoology 52:1–5.

Fitzgerald, G.J., and J.R. Bider. 1974b. Evidence for a relationship between geotaxis and seasonal movements in the toad *Bufo americanus*. Oecologia 17:277–280.

Fitzgerald, G.J., and J.R. Bider. 1974c. Evidence of a relationship between age and activity in the toad *Bufo americanus*. Canadian Field-Naturalist 88:499–501.

Fitzgerald, K.T., H.M. Smith, and L.J. Guillette, Jr. 1981. Nomenclature of the diploid species of the diploid–tetraploid *Hyla versicolor* complex. Journal of Herpetology 15:355–356.

Fleet, R., and B. Autry. 1999. Herpetofaunal assemblages of four forest types from the Caddo Lake area of northeastern Texas. Texas Journal of Science 51:297–308.

Fleming, P.L. 1976. A study of the distribution and ecology of *Rana clamitans* Latreille. Ph.D. Dissertation, University of Minnesota, Minneapolis.

Flesch, A.D., D.E. Swann, D.S. Turner, and B.F. Powell. 2010. Herpetofauna of the Rincon Mountains, Arizona. Southwestern Naturalist 55:240–253.

Flores, G. 1978. The reproductive ecology of the northern spring peeper, *Hyla crucifer crucifer*, in southern Connecticut. HERP, Bulletin of the New York Herpetological Society 14:22–25.

Florey, J., and S.J. Mullin. 2005. A survey of anuran breeding activity in Illinois, 1986–1989. Transactions of the Illinois State Academy of Science 98:39–47.

Florida Department of Agriculture. 1936. Bullfrog farming and frogging in Florida. Bulletin (new series) No. 56. [reprinted 1952]

Flowers, M.A. 1994. Feeding ecology and habitat use of juvenile Great Plains toads (*Bufo cognatus*) and Woodhouse's toads (*B. woodhousei*). M.A. thesis, University of South Dakota, Vermillion.

Flowers, M.A., and B.M. Graves. 1995. Prey selectivity and size–specific diet changes in *Bufo cognatus* and *B. woodhousii* during early postmetamorphic ontogeny. Journal of Herpetology 29:608–612.

Flowers, M.A., and B.M. Graves. 1997. Juvenile toads avoid chemical cues from snake predators. Animal Behaviour 53:641–646.

Floyd, T.M., and E.S. Kilpatrick. 2002. *Pseudacris brachyphona* (Mountain Chorus Frog). Verification of historical occurrence. Herpetological Review 33:48.

Floyd, T.M., K.R. Russell, C.E. Moorman, D.H. Van Lear, D.C. Guynn, Jr., and J.D. Lanham. 2002. Effects of prescribed fire on herpetofauna within hardwood forests of the upper Piedmont of South Carolina: a preliminary analysis. Pp. 123–127 *In* K.W. Outcalt (ed.), Proceedings of the 11th Biennial Southern Silvicultural Research Conference. USDA Forest Service, Southern Research Station, General, and Technical Report SRS–48, Asheville, North Carolina.

Flury, A.G. 1951. Variations in some local populations of *Hyla versicolor* in central Texas. M.A. thesis, University of Texas, Austin.

Flury, A.G. 1957. The leopard frog. Texas Game and Fish 15(4):4–5, 22–23.

Fogell, D.D. 2010. A Field Guide to the Amphibians and Reptiles of Nebraska. Institute of Agriculture and Natural Resources, University of Nebraska–Lincoln.

Fogleman, J.C., P.S. Corn, and D. Pettus. 1980. The genetic basis for a dorsal color polymorphism in *Rana pipiens*. Journal of Heredity 71:439–440.

Folt, B.T. 2012. Geographic distribution: *Lithobates heckscheri* (River Frog). Herpetological Review 43:99.

Fontenot, B.E., and P.M. Hampton. 2009. *Scaphiopus couchii* (Couch's Spadefoot). Predation. Herpetological Review 40:73.

Fontenot, C.L., Jr. 2003. *Hyla versicolor* (Gray Treefrog). Behavior. Herpetological Review 34:358.

Fontenot, C.L., Jr. 2011. *Hyla cinerea* (Green Treefrog).Winter aggregation. Herpetological Review 42:84–85.

Fontenot, L.W., G.P. Noblet, and S.G. Platt. 1994. Rotenone hazards to amphibians and reptiles. Herpetological Review 25:150–153, 156.

Fontenot, L.W., G.P. Noblet, J.M. Akins, M.D. Stephens, and G.P. Cobb. 2000. Bioaccumulation of polychlorinated biphenyls in ranid frogs and northern water snakes from a hazardous waste site and a contaminated watershed. Chemosphere 40:803-809.

Force, E.R. 1925. Notes on reptiles and amphibians of Okmulgee County, Oklahoma. Copeia (141):25–27.

Force, E.R. 1930. The amphibians and reptiles of Tulsa County, Oklahoma, and vicinity. Copeia 1930:25–39.

Force, E.R. 1933. The age of attainment of sexual maturity of the leopard frog *Rana pipiens* (Scheber) in northern Michigan. Copeia 1933:128–131.

Ford, L.S., and D.C. Cannatella. 1993. The major clades of frogs. Herpetological Monographs 7:94–117.

Forester, D.C. 1973. Mating call as a reproductive isolating mechanism between *Scaphiopus bombifrons* and *S. hammondii*. Copeia 1973:60–67.

Forester, D.C. 1975. Laboratory evidence for potential gene flow between two species of spadefoot toads, *Scaphiopus bombifrons* and *Scaphiopus hammondii*. Herpetologica 31:282–286.

Forester, D.C., and R. Czarnowsky. 1985. Sexual selection in the spring peeper, *Hyla crucifer* (Amphibia, Anura): role of the advertisement call. Behaviour 92:112–128.

Forester, D.C., and R. Daniel. 1986. Observations on the social behavior of the southern cricket frog, *Acris gryllus* (Anura: Hylidae). Brimleyana 12:5–11.

Forester, D.C., and D.V. Lykens. 1986. Significance of satellite males in a population of spring peepers (*Hyla crucifer*). Copeia 1986:719–724.

Forester, D. C., and W. K. Harrison. 1987. The significance of antiphonal vocalisation by the spring peeper, *Pseudacris crucifer* (Amphibia, Anura). Behaviour 103:1–15.

Forester, D.C., and D.V. Lykens. 1988. The ability of wood frog eggs to withstand prolonged terrestrial stranding: an empirical study. Canadian Journal of Zoology 66:1733–1735.

Forester, D.C., and K.J. Thompson. 1998. Gauntlet behaviour as a male sexual tactic in the American toad (Amphibia: Bufonidae). Behaviour 135:99–119.

Forester, D. C., D. V. Lykens, and W. K. Harrison. 1989. The significance of persistent vocalisation by the spring peeper, *Pseudacris crucifer* (Anura: Hylidae). Behavior 108: 197–208.

Forester, D.C., S. Knoedler, and R. Sanders. 2003. Life history and status of the mountain chorus frog (*Pseudacris brachyphona*) in Maryland. Maryland Naturalist 46:1–15.

Forester, D.C., J.W. Snodgrass, K. Marsalek, and Z. Lanham. 2006. Post–breeding dispersal and summer home range of female American toads (*Bufo americanus*). Northeastern Naturalist 13:59–72.

Formanowicz, D.R. 1981. Palatability and antipredator behavior of the treefrog *Hyla versicolor* to the shrew *Blarina brevicauda*. Journal of Herpetology 15:235–236.

Formanowicz, D.R., Jr. 1986. Anuran tadpole/ aquatic insect predator–prey interactions: tadpole size and predator capture success. Herpetologica 42:367–373.

Formanowicz, D.R., Jr., and E.D. Brodie, Jr. 1979. Palatability and antipredator behavior of selected *Rana* to the shrew *Blarina*. American Midland Naturalist 101:456–458.

Formanowicz, D.R., Jr., and E.D. Brodie, Jr. 1982. Relative palatabilities of members of a larval amphibian community. Copeia 1982:91–97.

Formanowicz, D.R., Jr., and M.S. Bobka. 1989. Predation risk and microhabitat preference: an experimental study of the behavioral responses of prey and predator. American Midland Naturalist 121:379–386.

Forsman, E.D., and J.K. Swingle. 2007. Use of arboreal nests of tree voles (*Arborimus* spp.) by amphibians. Herpetological Conservation and Biology 2:113–118.

Forstner, J.M., M.R.J. Forstner, and J.R. Dixon. 1998. Ontogenetic effects on prey selection and food habits of two sympatric east Texas ranids: the southern leopard frog, *Rana sphenocephala*, and the bronze frog, *Rana clamitans clamitans*. Herpetological Review 29:208–211.

Forstner, M.R.J., and T.L. Ahlbrandt. 2003. Abiotic pond characteristics potentially influencing breeding of Houston Toads (*Bufo houstonensis*). Texas Journal of Science 55:315–322.

Forstner, M.R.J., and P. Crump. 2011. Houston toad population supplementation in Texas, USA. Pp. 71-76 *In* P.S. Soorae (ed.), Global Re-introduction Perspectives: 2011. More Case Studies from Around the Globe. IUCN/SSC Re-introduction Specialist Group, Gland, Switzerland.

Fort, D.J., R.L. Rogers, J.H. Thomas, W.A. Hopkins, and C. Schlekat. 2006. Comparative developmental toxicity of nickel to *Gastrophryne carolinensis*, *Bufo terrestris*, and *Xenopus laevis*. Archives of Environmental Contamination and Toxicology 51:703–710.

Fortin, C., M. Ouellet, and M.–J. Grimard. 2003. La Rainette faux–grillon boréale (*Pseudacris maculata*): présence officiellement validée au Québec. Le naturaliste canadien 127(2):71–75.

Fortin, C., M. Ouellet, and P. Galois. 2004a. Les amphibians et les reptiles des îles de l'estuaire du Saint–Laurent:

mieux connaître pour mieux conserver. Le naturaliste canadien 128(1):61–67.

Fortin, C., P. Galois, M. Ouellet, and G.J. Doucet. 2004b. Utilisation des emprises de lignes de transport d'énergie électrique par les amphibians et les reptiles en forêt décidue au Québec. Le naturaliste canadien 128(1):68–75.

Fortman, J.R. 1973. Characters of F$_1$ hybrid tadpoles and adults between six species of *Hyla* (Hylidae, Anura). Ph.D. Dissertation, Mississippi State University, Mississippi State.

Fortman, J., and R. Altig. 1973. Characters of F$_1$ hybrid tadpoles between six species of *Hyla*. Copeia 1973:411–416.

Fortman, J., and R. Altig. 1974. Characters of F1 hybrid frogs from six species of *Hyla*. Herpetologica 30:221–234.

Foster, B.J., D.W. Sparks, and J.E. Duchamp. 2004. Urban herpetology. 2. Amphibians and reptiles of the Indianapolis Airport conservation lands. Proceedings of the Indiana Academy of Science 113:53–59.

Foster, C.D. 2007. *Pseudacris regilla* (Pacific Treefrog). Death feigning. Herpetological Review 38:192.

Foster, C.D., E. Sandhaus, A. Boinay, and V. Hubbartt. 2007. *Bufo boreas halophilus* (California Toad). Cannibalism. Herpetological Review 38:178–179.

Foster, C.D., and S.J. Mullin. 2008. Speed and endurance of *Thamnophis hammondii* are not affected by consuming the toxic frog *Xenopus laevis*. Southwestern Naturalist 53:370–373.

Foster, W.A. 1967. Chorus structure and vocal response in the Pacific tree frog, *Hyla regilla*. Herpetologica 23:100–104.

Fouquette, M.J., Jr. 1954. Food competition among four sympatric species of garter snakes, genus *Thamnophis*. Texas Journal of Science 6:172–188.

Fouquette, M.J., Jr. 1960. Call structure in frogs of the Family Leptodactylidae. Texas Journal of Science 12:201–215.

Fouquette, M.J., Jr. 1968. Remarks on the type specimen of *Bufo alvarius* Girard. Great Basin Naturalist 28:70–72.

Fouquette, M.J., Jr. 1975. Speciation in chorus frogs.1. Reproductive character displacement in the *Pseudacris nigrita* complex. Systematic Zoology 24:16–23.

Fouquette, M.J., Jr. 1980. Effect of environmental temperatures on body temperature of aquatic–calling anurans. Journal of Herpetology 14: 347–352.

Fouquette, M.J., Jr. 2005. *Rhinophrynus dorsalis* Dumeril and Bibron, 1841. Burrowing Toad (Sapo Borracho). Pp. 599–600 *In* M.J. Lannoo (ed.), Amphibian Declines. The Conservation Status of United States Species. University of California Press, Berkeley.

Fouquette, M.J., and M.J. Littlejohn. 1960. Patterns of oviposition in two species of hylid frogs. Southwestern Naturalist 5:92–96.

Fournier, M.A. 1997. Amphibians in the Northwest Territories. SSAR Herpetological Conservation 1:100–106.

Fournier, M.A. undated. Amphibians & reptiles in the Northwest Territories. Ecology North, Yellowknife, Northwest Territories. http://www.enr.gov. nt.ca/_live /documents/documentManagerUpload/amphibians_reptiles_pamphlet.pdf.

Fowler, H.W. 1905. The sphagnum frog of New Jersey— *Rana virgatipes* Cope. Proceedings of the Academy of Natural Sciences of Philadelphia 1905:662–664.

Fowler, H.W. 1906. Some cold–blooded vertebrates of the Florida Keys. Proceedings of the Academy of Natural Sciences of Philadelphia 58:77–113.

Fowler, H.W. 1918. An albino spring frog in winter. Copeia (61):84.

Fowler, J.A. 1960. Amphibian calling. Cranbrook Institute Newsletter 29:78–85.

Fowler, J.A., and G. Orton. 1947. The occurrence of *Hyla femoralis* in Maryland. Maryland, A Journal of Natural History 17:6–7.

Fowler, J.A., and R.L. Hoffman. 1951. *Gastrophryne carolinensis carolinensis* (Holbrook) in southwestern Virginia. Virginia Journal of Science 2:101.

Fox, S. 2008. Oophagy and larval cannibalism without polyphenism in tadpoles of the Great Basin spadefoot (*Spea intermontana*). Herpetological Review 39:151–154.

Fox, S.F., P.A. Shipman, R.E. Thill, J.P. Phelps, and D.M. Leslie, Jr. 2004. Amphibian communities under diverse forest management in Ouachita Mountains, Arkansas. Pp. 164–173 *In* J.M. Guldin (compiler), Ouachita and Ozark Mountains Symposium. Ecosystem Management Research. USDA Forest Service General and Technical Report SRS–74, Asheville, North Carolina.

Fracz, A. 2012. Importance of hydrologic connectivity for coastal wetlands to open water of eastern Georgian Bay. M.S. thesis, McMaster University, Hamilton, Ontario.

Fraker, M.E. 2007. Predation risk assessment and the anti–predator behavioral dynamics of larval anurans. Ph.D. Dissertation, University of Michigan, Ann Arbor.

Fraker, M.E. 2008a. The dynamics of predation risk assessment: responses of anuran larvae to chemical cues of predators. Journal of Animal Ecology 77:638–645.

Fraker, M.E. 2008b. The effect of hunger on the strength and duration of the antipredator behavioral response of green frog (*Rana clamitans*) tadpoles. Behavioral Ecology and Sociobiology 62:1201–1205.

Fraker, M.E. 2008c. The influence of the circadian rhythm of green frog (*Rana clamitans*) tadpoles on their antipredator behavior and the strength of the nonlethal effects of predators. American Naturalist 171:545–552.

Fraker, M.E. 2009a. Predation risk assessment by green frog

(*Rana clamitans*) tadpoles through chemical cues produced by multiple prey. Behavioral Ecology and Sociobiology 63:1397–1402.

Fraker, M.E. 2009b. The effect of prior experience on a prey's current perceived risk. Oecologia 158:765–774.

Fraker, M.E. 2009c. Perceptual limits to predation risk assessment in green frog (*Rana clamitans*) tadpoles. Behaviour 146:1025–1036.

Fraker, M.E. 2010. Risk assessment and anti–predator behavior of wood frog (*Rana sylvatica*) tadpoles: a comparison with green frog (*Rana clamitans*) tadpoles. Journal of Herpetology 44:390–398.

Fraker, M.E., and B. Luttbeg. 2012. Predator–prey space use and the spatial distribution of predation events. Behaviour 149:555–574.

Fraker, M., E. F. Hu, V. Cuddapah, S. A. McCollum, R. A. Relyea, J. Hempel, and R. J. Denver. 2009. Characterization of an alarm pheromone secreted by amphibian tadpoles that induces behavioral inhibition and suppression of the neuroendocrine stress axis. Hormones and Behavior 55:520–529.

Frandsen, J.C., and A.W. Grundmann. 1960. The parasites of some amphibians of Utah. Journal of Parasitology 46:678.

Franz, R. 1966. Red leg in a natural population of Maryland amphibians. Bulletin of the Maryland Herpetological Society 2:7.

Franz, R. 1967. Annotated checklist of cave–associated reptiles and amphibians of Maryland. Contribution No. 6 of The Maryland Cave Survey. Bulletin of the Maryland Herpetological Society 3:50–53.

Franz, R. 1969. The Maryland Cave Survey. Contribution No. 6. Reptiles and amphibians in Maryland caves. Baltimore Grotto News 10(2):31–35.

Franz, R. 1970a. Food of the larval tailed frog. Bulletin of the Maryland Herpetological Society 6: 49–51.

Franz, R. 1970b. Egg development of the tailed frog under natural conditions. Bulletin of the Maryland Herpetological Society 6:27–30.

Franz, R. 1971. Notes on the distribution and ecology of the herpetofauna of northwestern Montana. Bulletin of the Maryland Herpetological Society 7:1–10.

Franz, R. 1972. Feeding in Florida cricket frogs. Bulletin of the Maryland Herpetological Society 8:89–90.

Franz, R. 1986. The Florida gopher frog and the Florida pine snake as burrow associates of the gopher tortoise in northern Florida. Pp. 16–20 *In* D.R. Jackson and R.J. Bryant (eds.), The Gopher Tortoise and its Community. Proceedings of the 5th Annual Meeting of the Gopher Tortoise Council, Florida State Museum, Gainesville.

Franz, R. 2005. *Hyla gratiosa* (Barking Treefrog). Winter retreat. Herpetological Review 36:434–435.

Franz, R., and D.S. Lee. 1970. The ecological and biogeographical distribution of the tailed frog, *Ascaphus truei*, in the Flathead River Drainage of northwestern Montana. Bulletin of the Maryland Herpetological Society 6:62–73.

Franz, R., C.K. Dodd, Jr., and C. Jones. 1988. *Rana areolata aesopus* (Florida Gopher Frog). Movement. Herpetological Review 19:82.

Franz, R., C.K. Dodd, Jr., and A.M. Bard. 1992. The non–marine herpetofauna of Egmont Key, Hillsborough County, Florida. Florida Scientist 55:179–183.

Freake, M.J., S.C. Borland, and J.B. Phillips. 2002. Use of a magnetic compass for Y–axis orientation in larval bullfrogs, *Rana catesbeiana*. Copeia 2002:466–471.

Freda, J. 1986. The influence of acidic pond water on amphibians: a review. Water, Air, & Soil Pollution 30:439–450.

Freda, J. 1990. Effects of acidification on amphibians. Pp. 114–129 *In* J.P. Baker, D.P. Bernard, S.W. Christensen, M.J. Sale, J. Freda, K. Heltcher, L. Rowe, P. Scanion, P. Stokes, G. Suter, and W. Warren–Hicks (eds.), Biological Effects of Changes in Surface Water Acid–base Chemistry. State of Science. Technology Report 13, National Acid Precipitation Assessment Program, Washington, D.C.

Freda, J., and W.A. Dunson. 1984. Sodium balance of amphibian larvae exposed to low environmental pH. Physiological Zoology 57:435–443.

Freda, J., and W.A. Dunson. 1985. Field and laboratory studies of ion balance and growth rates of ranid tadpoles chronically exposed to low pH. Copeia 1985:415–423.

Freda, J., and W.A. Dunson. 1986a. Effects of low pH and other chemical variables on the local distribution of amphibians. Copeia 1986:454–466.

Freda, J., and R.J. Gonzale 1986b. Daily movements of the treefrog, *Hyla andersoni*. Journal of Herpetology 20:469–471.

Freda, J., and D.G. McDonald. 1990. Effects of aluminum on the leopard frog, *Rana pipiens*: life stage comparisons and aluminum uptake. Canadian Journal of Fisheries and Aquatic Sciences 47:210–216.

Freda, J., and D.H. Taylor. 1992. Behavioral response of amphibian larvae to acidic water. Journal of Herpetology 26:429–433.

Freda, J., and D.G. McDonald. 1993. Toxicity of amphibian breeding ponds in the Sudbury Region. Canadian Journal of Fisheries and Aquatic Sciences 50:1497–1503.

Freda, J., V. Cavdek, and D.G. McDonald. 1990. Role of organic complexation in the toxicity of aluminum to *Rana pipiens* embryos and *Bufo americanus* tadpoles. Canadian Journal of Fisheries and Aquatic Sciences 47:217–224.

Freda, J., W.J. Sadinski, and W.A. Dunson. 1991. Long term monitoring of amphibian populations with respect to the effects of acidic precipitation. Water, Air, & Soil Pollution 55:445–462.

Freed, A.N. 1980a. Prey selection and feeding behavior of the green treefrog (*Hyla cinerea*). Ecology 61:461–465.

Freed, A.N. 1980b. An adaptive advantage of basking behavior in an anuran amphibian. Physiological Zoology 53:433–444.

Freed, A.N. 1982a. A treefrog's menu: selection for an evening's meal. Oecologia 53:20–26.

Freed, A.N. 1982b. To eat or not to eat? That is the question of treefrog prey selection. Ph.D. Dissertation, University of Florida, Gainesville.

Freed, A.N. 1988. The use of visual cues for prey–selection by foraging treefrogs (*Hyla cinerea*). Herpetologica 44:18–24.

Freed, P.S., and K. Neitman. 1988. Notes on predation on the endangered Houston toad, *Bufo houstonensis*. Texas Journal of Science 40:454–456.

Freiburg, R.E. 1951a. An ecological study of the narrow–mouthed toad (*Microhyla*) in northeastern Kansas. M.A. thesis, University of Kansas, Lawrence.

Freiburg, R.E. 1951b. An ecological study of the narrow–mouthed toad (*Microhyla*) in northeastern Kansas. Transactions of the Kansas Academy of Science 54:374–386.

Freidenburg, L.K. 2003. Spatial ecology of the wood frog, *Rana sylvatica*. Ph.D. Dissertation, University of Connecticut, Storrs.

Freidenburg, L.K., and D.K. Skelly. 2004. Microgeographic variation in thermal preference by an amphibian. Ecology Letters 7:369–373.

Freidenfelds, N.A. 2012. *Acris crepitans* (Northern Cricket Frog). Fire ant envenomation. Herpetological Review 43:117.

Freidenfelds, N.A., J.L. Purrenhage, and K.J. Babbitt. 2011. Effects of clearcuts and forest buffer size on post–breeding emigration of wood frogs (*Lithobates sylvaticus*). Forest Ecology and Management. 261:2115–2122.

Frese, P.W., and A.M. Sullivan. 2000. *Rana areolata circulosa* (Northern Crawfish Frog). Vocalization. Herpetological Review 31:101.

Friebele, E., and J. Zambo. 2004. A Guide to the Amphibians and Reptiles of Jug Bay. Jug Bay Wetlands Sanctuary, Lothian, Maryland.

Fried, B., P. L. Pane, and A. Reddy. 1997. Experimental infection of *Rana pipiens* tadpoles with *Echinostoma trivolvis* cercariae. Parasitology Research 83:666–669.

Friet, S.C. 1993. Aquatic overwintering of the bullfrog (*Rana catesbeiana*) during natural hypoxia in an ice–covered pond. M.S. thesis, Dalhousie University, Halifax, Nova Scotia.

Frisbie, M.P., J. P. Costanzo, and R. E. Lee. 2000. Physiological and ecological aspects of low–temperature tolerance in embryos of the wood frog, *Rana sylvatica*. Canadian Journal of Zoology 78: 1032–1041.

Froiland, S.G. 1990. Natural History of the Black Hills and Badlands. Center for Western Studies, Augustana College, Sioux Falls, South Dakota.

Frost, D. 1983. Past occurrence of *Acris crepitans* (Hylidae) in Arizona. Southwestern Naturalist 28:105.

Frost, D.R. (ed.). 1985. Amphibian Species of the World. Allen Press and the Association of Systematics Collections, Lawrence, Kansas.

Frost, D.R., T. Grant, J. Faivovich, R.H. Bain, A. Haas, C.F.B. Haddad, R.O. de Sá, A. Channing, M. Wilkinson, S.C. Donnellan, C.J. Raxworthy, J.A. Campbell, B.L. Blotto, P. Moler, R.C. Drewes, R.A. Nussbaum, J.D. Lynch, D.M. Green, and W.C. Wheeler. 2006a. The amphibian tree of life. Bulletin of the American Museum of Natural History 297:1–370.

Frost, D.R., T. Grant, and J.R. Mendelson III. 2006b. *Ollotis* Cope, 1875 is the oldest name for the genus currently referred to as *Cranopsis* Cope, 1875 (Anura: Hyloides: Bufonidae). Copeia 2006:558.

Frost, D.R., T. Grant, J. Faivovich, R.H. Bain, A. Haas, C.F.B. Haddad, R.O. de Sá, A. Channing, M. Wilkinson, S.C. Donnellan, C.J. Raxworthy, J.A. Campbell, B.L. Blotto, P. Moler, R.C. Drewes, R.A. Nussbaum, J.D. Lynch, D.M. Green, and W.C. Wheeler. 2007. Is *The Amphibian Tree of Life* really fatally flawed? Cladistics 24:385–395.

Frost, D.R., J.R. Mendelson III, and J. Pramuk. 2009. Further notes on the nomenclature of Middle American toads (Bufonidae). Copeia 2009:418.

Frost, J.S., and E.W. Martin. 1971. A comparison of distribution and high temperature tolerance in *Bufo americanus* and *Bufo woodhousii fowleri*. Copeia 1971:750–751.

Frost, J.S., and J.T. Bagnara. 1976. A new species of leopard frog (*Rana pipiens* complex) from northwestern Mexico. Copeia 1976:332–338.

Frost, J.S., and J.T. Bagnara. 1977. Sympatry between *Rana blairi* and the southern form of leopard frog in southeastern Arizona (Anura: Ranidae). Southwestern Naturalist 22: 443–453.

Frost, J.S., and J.E. Platz. 1983. Comparative assessment of modes of reproductive isolation among four species of leopard frogs (*Rana pipiens* complex). Evolution 37:66–78.

Frost, S.W. 1924. Frogs as insect collectors. Bulletin of the New York Entomological Society 32:174–185.

Frost, S.W. 1935. The food of *Rana catesbeiana* Shaw. Copeia 1935:15–18.

Fucik, E.M. 2011. Interactions between invasive species and climate change: the effect on an East Texas anuran. M.S. thesis, Stephen F. Austin State University, Nacogdoches, Texas.

Fulk, F.D., and J.O. Whitaker, Jr. 1969. The food of *Rana catesbeiana* in three habitats in Owen County, Indiana. Proceedings of the Indiana Academy of Science 78:491–496.

Fuller, T.E., K.L. Pope, D.T. Ashton, and H.H. Welsh, Jr. 2011. Linking the distribution of an invasive amphibian (*Rana catesbeiana*) to habitat conditions in a managed river system in northern California. Restoration Ecology 19:204–213.

Fulmer, T., and R. Tumlison. 2004. Important records of the bird–voiced treefrog (*Hyla avivoca*) in the headwaters of the Ouachita River drainage of southwestern Arkansas. Southeastern Naturalist 3:259–266.

Fulton, M.H., and J.E. Chambers. 1985. The toxic and teratogenic effects of selected organophosphorus compounds on the embryos of three species of amphibians. Toxicology Letters 26:175–180.

Funasaki, G.Y., P.Y. Lai, and L.M. Nakahara. 1988. A review of the biological control introductions in Hawai'i: 1890 to 1985. Proceedings of the Hawaiian Entomological Society 28:105-158.

Funderburg, J.B., C.H. Hotchkiss, and P. Hertl. 1974a. The wood frog, *Rana sylvatica* LeConte, in the Virginia coastal plain. Bulletin of the Maryland Herpetological Society 10:58.

Funderburg, J.B., P. Hertl, and W.M. Kerfoot. 1974b. A range extension for the carpenter frog, *Rana virgatipes* Cope, in the Chesapeake Bay region. Bulletin of the Maryland Herpetological Society 10:77–79.

Funk, R.S. 1964. *Hyla wrightorum* (Salientia: Hylidae) in western Arizona. Southwestern Naturalist 9:206.

Funk, W.C. 2004. Patterns and consequences of dispersal in Columbia spotted frogs (*Rana luteiventris*). Ph.D. Dissertation, University of Montana, Missoula.

Funk, W.C., M.S. Blouin, P.S. Corn, B.A. Maxwell, D.S. Pilliod, S. Amish, and F.W. Allendorf. 2005a. Population structure of Columbia spotted frogs (*Rana luteiventris*) is strongly affected by the landscape. Molecular Ecology 14:483–496.

Funk, W.C., A.E. Greene, P.S. Corn, and F.W. Allendorf. 2005b. High dispersal in a frog species suggests that it is vulnerable to habitat fragmentation. Biology Letters 1:13–16.

Funk, W.C., C.A. Pearl, H.M. Draheim, M.J. Adams, T.D. Mullins, and S.M. Haig. 2008. Range–wide phylogeographic analysis of the spotted frog complex (*Rana luteiventris* and *Rana pretiosa*) in northwestern North America. Molecular Phylogenetics and Evolution 49:198–210.

Funk, W.C., T.S. Garcia, G.A. Cortina, and R.H. Hill. 2011. Population genetics of introduced bullfrogs, *Rana (Lithobates) catesbeianus* [sic], in the Willamette Valley, Oregon, USA. Biological Invasions 13:651–658.

Funkhouser, A. 1976. Observations on pancreas: body weight ratio change during development of the bullfrog, *Rana catesbeiana*. Herpetologica 32:370–371.

Funkhouser, J.W. 1949. Adventures with park Amphibia. Crater Lake National Park Nature Notes 15:12–13.

Gaertner, J.P., M.A. Gaston, D. Spontak, M.R. J. Forstner, and D. Hahn. 2009. Seasonal variation in the detection of *Batrachochytrium dendrobatidis* in a Texas population of Blanchard's cricket frog (*Acris crepitans blanchardi*). Herpetological Review 40:184–187.

Gaertner, J.P., D. McHenry, M.R.J. Forstner, and D. Hahn. 2010. Annual variation of *Batrachochytrium dendrobatidis* in the Houston toad (*Bufo houstonensis*) and a sympatric congener (*Bufo nebulifer*). Herpetological Review 41:456–459.

Gaertner, J.P., D.J. Brown, J.A. Mendoza, M.R.J. Forstner, T. Bonner, and D. Hahn. 2012. Geographic variation in *Batrachochytrium dendrobatidis* occurrence among populations of *Acris crepitans blanchardi* in Texas, USA. Herpetological Review 43:274–278.

Gage, M.W. 2011. Biodiversity in an isolated suburban reservation: amphibians in the Middlesex Fells. M.S. thesis, University of Massachusetts, Boston.

Gage, S.H. 1897. Life history of the toad. Cornell Nature Leaflet No. 9. [reprinted in Cornel Nature–Study Leaflets, New York Department of Agriculture Nature–Study Bulletin No. 1, pp. 185–206, 1904]

Gahl, M.K. 2007. Spatial and temporal patterns of amphibian disease in Acadia National Park wetlands. Ph.D. Dissertation, University of Maine, Orono.

Gahl, M.K., and A.J.K. Calhoun. 2010. The role of multiple stressors in ranavirus–caused amphibian mortalities in Acadia National Park wetlands. Canadian Journal of Zoology 88:108–121.

Gahl, M.K., J.E. Longcore, and J.E. Houlahan. 2011a. Varying responses of northeastern North American amphibians to the chytrid pathogen *Batrachochytrium dendrobatidis*. Conservation Biology 26:135–141.

Gahl, M. K., B. D. Pauli, and J. E. Houlahan. 2011b. Effects of chytrid fungus and a glyphosate–based herbicide on survival and growth of wood frogs (*Lithobates sylvaticus*) Ecological Applications 21:2521–2529.

Gaige, H.T. 1920. Observations upon the habits of *Ascaphus truei* Stejneger. Occasional Papers of the Museum of Zoology, University of Michigan 84:1–9.

Gaige, H.T. 1931. Notes on *Syrrhophus marnockii* Cope. Copeia 1931:63.

Gaige, H.T. 1932. The status of *Bufo copei*. Copeia 1932:134.

Gallie, J.A., R.L. Mumme, and S.A. Wissinger. 2001. Experience has no effect on the development of chemosensory recognition of predators by tadpoles of the American toad, *Bufo americanus*. Herpetologica 57:376–383.

Galois, P., and M. Ouellet. 2005. Le Grand Bois de Saint–Grégoire, un refuge pour l'herpétofaune dans la plaine montérégienne. Le naturaliste canadien 129(2):37–43.

Gamble, T., P.B. Berendzen, H.B. Shaffer, D.E. Starkey, and A.M. Simons. 2008. Species limits and phylogeography of North American cricket frogs (*Acris*: Hylidae). Molecular Phylogenetics and Evolution 48:112–125.

Gamboa, G.J., K.A. Berven, R.A. Schemidt, G. Fishwild, and K.M. Jankens. 1991. Kin recognition by larval wood frogs (*Rana sylvatica*): effects of diet and prior exposure to conspecifics. Oecologia 86:319–324.

Gambs, R.D., and M.J. Littlejohn. 1979. Acoustic behavior of males of the Rio Grande leopard frog (*Rana berlandieri*): an experimental analysis through field playback trials. Copeia 1979:643–650.

Garcia, T.S., J.M. Romansic, and A.R. Blaustein. 2006. Survival of three species of anuran metamorphs exposed to UV–B radiation and the pathogenic fungus *Batrachochytrium dendrobatidis*. Diseases of Aquatic Organisms 72:163–169.

Garcia, T.S., D.J. Paoletti, and A.R Blaustein. 2009. Correlated trait responses: comparing defense strategies across a stress gradient. Canadian Journal of Zoology 87: 41–49.

García–París, M., D.R. Buchholz, and G. Parra–Olea. 2003. Phylogenetic relationships of Pelobatoidea re–examined using mtDNA. Molecular Phylogenetics and Evolution 28:12–23.

Gardiner, D.M., and D.M. Hoppe. 1999. Environmentally induced limb malformations in mink frogs (*Rana septentrionalis*). Journal of Experimental Zoology 284:207–216.

Gardner, J.D. 1995. *Pseudacris regilla* (Pacific Chorus Frog). Reproduction. Herpetological Review 26:32.

Garman, H. 1892. A synopsis of the reptiles and amphibians of Illinois. Illinois Laboratory of Natural History Bulletin 3(13):215–388.

Garman, H. 1901. The food of the toad. Kentucky Agricultural Experiment Station, Bulletin No. 91:60–68.

Garman, S. 1877. On a variation in the colors of animals. Proceedings of the American Association for the Advancement of Science 25:194.

Garman, S. 1884. The North American reptiles and batrachians. Bulletin of the Essex Institute 16:3–64.

Garman, S. 1887. Reptiles and batrachians of Texas and Mexico. Bulletin of the Essex Institute 19:1–20.

Garner, T.W.J., M.W. Perkins, P. Govindarajulu, D. Seglie, S. Walker, A.A. Cunningham, and M.C. Fisher. 2006. The emerging amphibian pathogen *Batrachochytrium dendrobatidis* globally infects introduced populations of the North American bullfrog, *Rana catesbeiana*. Biology Letters 2:455–459.

Garnier, J.H. 1883. The mink or hoosier frog. American Naturalist 17:945–954.

Garrett, J.M., and D.G. Barker. 1987. A Field Guide to the Reptiles and Amphibians of Texas. Texas Monthly Press, Austin.

Garton, E.R., F. Grady, and S.D. Carey. 1993. The vertebrate fauna of West Virginia caves. West Virginia Speleological Survey Bulletin 11.

Garton, J.D., and H.R. Mushinsky. 1979. Integumentary toxicity and unpalatability as an antipredator mechamism in the narrow mouthed toad, *Gastrophryne carolinensis*. Canadian Journal of Zoology 57:1965–1973.

Garton, J.S., and R.A. Brandon. 1975. Reproductive ecology of the green treefrog, *Hyla cinerea*, in southern Illinois (Anura: Hylidae). Herpetologica 31:150–161.

Gartside, D.F. 1980. Analysis of a hybrid zone between chorus frogs of the *Pseudacris nigrita* complex in the southern United States. Copeia 1980:56–66.

Garwood, J.M. 2006. *Rana cascadae* (Cascades Frog). Predation. Herpetological Review 37:76.

Garwood, J.M. 2009. Spatial ecology of the Cascades frog: identifying dispersal, migration, and resource uses at multiple spatial scales. M.S. thesis, Humboldt State University, Arcata, California.

Garwood, J.M., and H.H. Welsh, Jr. 2005. *Rana cascadae* (Cascades Frog). Predation. Herpetological Review 36:165–166.

Garwood, J.M., and C.A. Wheeler. 2007. *Rana cascadae* (Cascades Frog). Predation. Herpetological Review 38:193–194.

Garwood, J.M., and C.W. Anderson. 2010. *Rana cascadae* (Cascades Frog). Necrogamy. Herpetological Review 41:204.

Garwood, J.M., C.A. Wheeler, R.M. Bourque, M.D. Larson, and H.H. Welsh, Jr. 2007. Egg mass drift increases vulnerability during early development of Cascades frogs (*Rana cascadae*). Northwestern Naturalist 88:95–97.

Garwood, J.M., C. W. Anderson, and S.J. Ricker. 2010. Bullfrog predation on a juvenile Coho salmon in Humboldt County, California. Northwestern Naturalist 91:99–101.

Gascon, C., and J.R. Rider. 1985. The effect of pH on bullfrog, *Rana catesbeiana*, and green frog, *Rana clamitans melanota*, tadpoles. Canadian Field-Naturalist 99:259–261.

Gascon, C., and D. Planas. 1986. Spring pond water chemistry and the reproduction of the wood frog, *Rana sylvatica*. Canadian Journal of Zoology 64:543–550.

Gascon, C., and J. Travis. 1992. Does the spatial scale of experimentation matter? A test with tadpoles and dragonflies. Ecology 73:2237–2243.

Gascon, C., D. Planas, and G. Moreau. 1987. The interaction of pH, calcium and aluminum concentrations on the survival and development of wood frog (*Rana sylvatica*) eggs and tadpoles. Annals of the Royal Society of Belgium 117:189–199.

Gaston, M.A., A. Fuji, F.W. Weckerly, and M.R.J. Forstner. 2010. Potential component Allee effects and their impact on wetland management in the conservation of endangered anurans. PLoS One 5:e10102.

Gates, G.O. 1957. A study of the herpetofauna in the vicinity of Wickenburg, Maricopa County, Arizona. Transactions of the Kansas Academy of Science 60:403–418.

Gatten, R.E., Jr., and C.J. Hill. 1984. Social influence on thermal selection by *Hyla crucifer*. Journal of Herpetology 18:87–88.

Gatten, R.E., Jr., and R.M. Clark. 1989. Locomotor performance of hydrated and dehydrated frogs: recovery following exhaustive exercise. Copeia 1989:451–455.

Gatz, A.J., Jr. 1981a. Non–random mating by size in American toads, *Bufo americanus*. Animal Behaviour 29:1004–1012.

Gatz, A.J., Jr. 1981b. Selective size mating in *Hyla versicolor* and *Hyla crucifer*. Journal of Herpetology 15:113–114.

Gatz, A.J., Jr. 1981c. Size selective mating in *Hyla versicolor* and *Hyla crucifer*. Journal of Herpetology 15:114–116.

Gaudin, A.J. 1964. The tadpole of *Hyla californiae* Gorman. Texas Journal of Science 16:80–84.

Gaudin, A.J. 1965. Larval development of the tree frogs *Hyla regilla* and *Hyla californiae*. Herpetologica 21:117–130.

Gaudin, A.J. 1969. A comparative study of the osteology and evolution of the Holarctic treefrogs: *Hyla, Pseudacris, Acris* and *Limnaeodus*. Ph.D. Dissertation, University of Southern California, Los Angeles.

Gaul, R.W., Jr., and J.C. Mitchell. 2007. The herpetofauna of Dare County, North Carolina: history, natural history, and biogeography. Journal of the North Carolina Academy of Science 123:65–109.

Gaulke, C.A., R.S. Wagner, J.E. Johnson, and J.T. Irwin. 2010. Northern leopard frogs (*Rana pipiens*) infected with *Batrachochytrium dendrobatidis* in the amphibian trade. Herpetological Review 41:322–323.

Gaulke, C.A., J.T. Irwin, and R.S. Wagner. 2011. Prevalence and distribution of *Batrachochytrium dendrobatidis* at montane sites in central Washington, USA. Herpetological Review 42:209–211.

Gayou, D.G. 1984. Effects of temperature on the mating call of *Hyla versicolor*. Copeia 1984:733–738.

Gee, J.H., and R.C. Waldick. 1995. Ontogenetic buoyancy changes and hydrostatic control in larval anurans. Copeia 1995:861–870.

Gehlbach, F.R. 1965. Herpetology of the Zuni Mountains region, northwestern New Mexico. Proceedings of the United States National Museum 116:243–332.

Gehlbach, F.R., and B.B. Collette. 1959. Distributional and biological notes on the Nebraskka [sic] herpetofauna. Herpetologica 15:141–143.

Gelczis, L., and C. Drost. 2004. A surprise in Surprise Canyon: a new amphibian species for Grand Canyon National Park. [Grand Canyon National Park] Nature Notes (Fall):1–3.

Geluso, K., and G.D. Wright. 2010. Geographic distribution. *Gastrophryne olivacea* (Western narrow–mouthed toad). Herpetological Review 41:103.

Gendron, A.D., D.J. Marcogliese, S. Barbeau, M.–S. Christin, P. Brousseau, S. Ruby, D. Cyr, and M. Fournier. 2003. Exposure of leopard frogs to a pesticide mixture affects life history characteristics of the lungworm *Rhabdias ranae*. Oecologia 135:469–476.

Gentry, G. 1955. An annonated [sic] check list of the amphibians and reptiles of Tennessee. Journal of the Tennessee Academy of Science 30:168–176.

Gentry, J.B., and M.H. Smith. 1968. Food habits and burrow associates of *Peromyscus polionotus*. Journal of Mammalogy 49:562–565.

George, I.D. 1940. A study of the bullfrog, *Rana catesbeiana* Shaw, at Baton Rouge, Louisiana. Ph.D. Dissertation, University of Michigan, Ann Arbor.

Georgel, C.T.W. 2001. Activity, diet, and microhabitat of the carpenter frog (*Rana virgatipes* Cope) in Virginia. M.S. thesis, Christopher Newport University, Newport News, Virginia.

Gergus, E.W.A. 1998. Systematics of the *Bufo microscaphus* complex: allozyme evidence. Herpetologica 54:317–325.

Gergus, E.W.A., B.K. Sullivan, and K.B. Malmos. 1997. Call variation in the *Bufo microscaphus* complex: implications for species boundaries and the evolution of mate recognition. Ethology 103:979–989.

Gergus, E.W.A., K.B. Malmos, and B.K. Sullivan. 1999. Natural hybridization among distantly related toads (*Bufo alvarius, Bufo cognatus, Bufo woodhousii*) in central Arizona. Copeia 1999:281–286.

Gergus, E.W.A., T.W. Reeder, and B.K. Sullivan. 2004. Geographic variation in *Hyla wrightorum*: advertisement calls, allozymes, mtDNA, and morphology. Copeia 2004:758–769.

Gerhardt, H.C. 1968. Evolutionary aspects of vocal communication and reproductive behavior in the *Hyla cinerea* species group. M.A. thesis, University of Texas, Austin.

Gerhardt, H.C. 1973. Reproductive interactions between *Hyla crucifer* and *Pseudacris ornata* (Anura: Hylidae). American Midland Naturalist 89:81–88.

Gerhardt, H.C. 1974a. Mating call differences between eastern and western populations of the treefrog *Hyla chrysoscelis*. Copeia 1974:534–536.

Gerhardt, H.C. 1974b. Behavioral isolation of the tree frogs, *Hyla cinerea* and *Hyla andersonii*. American Midland Naturalist 91:424–433.

Gerhardt, H.C. 1974c. The significance of some spectral features in mating call recognition in the green treefrog (*Hyla cinerea*). Journal of Experimental Biology 61:229–241.

Gerhardt, H.C. 1975. Sound pressure levels and radiation patterns of the vocalizations of some North American frogs and toads. Journal of Comparative Physiology 102:1–12.

Gerhardt, H.C. 1978a. Temperature coupling in the vocal communication system of the gray tree frog, *Hyla versicolor*. Science 199:992–994.

Gerhardt, H.C. 1978b. Discrimination of intermediate sounds in a synthetic call continuum by female green tree frogs. Science 199:1089–1091.

Gerhardt, H.C. 1978c. Mating call recognition in the green tree frog (*Hyla cinerea*): the significance of some fine-temporal properties. Journal of Experimental Biology 74:59–73.

Gerhardt, H.C. 1981a. Mating call recognition in the green treefrog (*Hyla cinerea*): importance of two frequency bands as a function of sound pressure level. Journal of Comparative Physiology 144:17–25.

Gerhardt, H.C. 1981b. Mating call recognition in the barking treefrog (*Hyla gratiosa*): responses to synthetic calls and comparisons with the green treefrog (*Hyla cinerea*). Journal of Comparative Physiology 144:17–25.

Gerhardt, H.C. 1982. Sound pattern recognition in some North American treefrogs (Anura: Hylidae): implications for mate choice. American Zoologist 22:581–595.

Gerhardt, H.C. 1983. Acoustic communication in treefrogs. Verhandlungen der Deutschen Zoologischen Gesellschaft 76:25–35.

Gerhardt, H.C. 1986. Recognition of spectral patterns in the green treefrog: neurobiology and evolution. Experimental Biology 45:167–178.

Gerhardt, H.C. 1987. Evolutionary and neurobiological implications of selective phonotaxis in the green treefrog, *Hyla cinerea*. Animal Behaviour 35:1479–1489.

Gerhardt, H.C. 1991. Female mate choice in treefrogs: static and dynamic acoustic criteria. Animal Behaviour 42:615–635.

Gerhardt, H.C. 1994. The evolution of vocalization in frogs and toads. Annual Review of Ecology and Systematics 25: 293–324.

Gerhardt, H.C. 2001. Acoustic communication in two groups of closely related treefrogs. Advances in the Study of Behavior 30:99–167.

Gerhardt, H.C., and K.M. Mudry. 1980. Temperature effects of frequency preferences and mating call frequencies in the green tree frog, *Hyla cinerea* (Anura: Hylidae). Journal of Comparative Physiology 137:1–6.

Gerhardt, H.C., and J. Rheinlaender. 1982. Localization of an elevated sound source by the green tree frog. Science 217:662–663.

Gerhardt, H.C., and G.M. Klump. 1988a. Phonotactic responses and selectivity of barking treefrogs (*Hyla gratiosa*) to chorus sounds. Journal of Comparative Physiology A 163:795–802.

Gerhardt, H.C., and G.M. Klump. 1988b. Masking of acoustic signals by the chorus background noise in the green tree frog: a limitation on mate choice. Animal Behaviour 36:1247–1249.

Gerhardt, H.C., and G.M. Klump. 1991. Female mate choice in treefrogs: static and acoustic criteria. Animal Behaviour 42:615–635.

Gerhardt, H.C, and J. Schul. 1999. A quantitative analysis of behavioral selectivity for pulse rise–time in the gray treefrog, *Hyla versicolor*. Journal of Comparative Physiology A 185:33–40.

Gerhardt, H.C., and M.A. Bee. 2007. Recognition and localization of acoustic signals. Pp. 113–146 *In* A.S. Feng, P.M. Narins, R.R. Fay, and A.N. Popper (eds.), Hearing and Sound Communication in Amphibians. Springer Handbook of Auditory Research. Springer, New York.

Gerhardt, H.C., S.I. Guttman, and A.A. Karlin. 1980. Natural hybrids between *Hyla cinerea* and *Hyla gratiosa*: morphology, vocalization and electrophoretic analysis. Copeia 1980:577–584.

Gerhardt, H.C., R.E. Daniel, S.A. Perrill, and S. Schramm. 1987. Mating behaviour and male mate success in the green treefrog. Animal Behaviour 35:1490–1503.

Gerhardt, H.C., B. Diekamp, and M. Ptacek. 1989. Intermale spacing in choruses of the spring peeper, *Pseudacris* (*Hyla*) *crucifer*. Animal Behaviour 38:1012–1024.

Gerhardt, H.C., S. Allan, and J.J. Schwartz. 1990. Female green treefrogs (*Hyla cinerea*) do not selectively respond to signals with a harmonic structure in noise. Journal of Comparative Physiology A 166:791–794.

Gerhardt, H.C., M.B. Ptacek, L. Barnett, and K.G. Torke. 1994. Hybridization in the diploid–tetraploid treefrogs *Hyla chrysoscelis* and *Hyla versicolor*. Copeia 1994:51–59.

Gerhardt, H.C., M.L. Dyson, and S.D. Tanner. 1996.Dynamic properties of the advertisement calls of gray tree frogs: patterns of variability and female choice. Behavioral Ecology 7: 7–18.

Gerhardt, H.C, S.D. Tanner, C.M. Corrigan, and H.C. Walton. 2000. Female preference functions based on call duration in the gray tree frog (*Hyla versicolor*). Behavioral Ecology 11:663–669.

Gerhardt, H.C., C.C. Martínez–Rivera, J.J. Schwartz, V.T. Marshall, and C.G. Murphy. 2007. Preferences based on spectral differences in acoustic signals in four species of treefrogs (Anura: Hylidae). Journal of Experimental Biology 210:2990–2998.

Gerlanc, N.M. 1999. Effects of breeding pool permanence on developmental rate of western chorus frogs, *Pseudacris triseriata*, in tallgrass prairie. M.S. thesis, Kansas State University, Manhattan.

Gerlanc, N.M., and G.A. Kaufman. 2003. Use of bison wallows by anurans on Konza Prairie. American Midland Naturalist 150:158–168.

Gerlanc, N.M., and G.A. Kaufman. 2005. Habitat of origin and changes in water chemistry influence development of western chorus frogs. Journal of Herpetology 39:254–265.

Germaine, S.S., and D.W. Hays. 2009. Distribution and postbreeding environmental relationships of northern leopard frogs (*Rana* [*Lithobates*] *pipiens*) in Washington. Western North American Naturalist 69:537–547.

Gerow, K., N.C. Cline, D.E. Swann, and M. Pokorny. 2010. Estimating annual vertebrate mortality on roads at Saguaro National Park, Arizona. Human–Wildlife Interactions 4:283–292.

Gets, L.L. 1958. The winter activities of *Rana clamitans* tadpoles. Copeia 1958:219.

Ghioca, D.M. 2005. Effect of land use on larval amphibian communities in playa wetlands of the Southern High Plains. Ph.D. Dissertation, Texas Tech University, Lubbock.

Ghioca, D.M., and L.M. Smith. 2007. Biases in trapping larval amphibians in Playa wetlands. Journal of Wildlife Management 71:991–995.

Ghioca–Robrecht, D.M., and L.M. Smith. 2011. The role of spadefoot toad tadpoles in wetland trophic structure as influenced by environmental and morphological factors. Canadian Journal of Zoology 89: 47–59.

Ghioca–Robrecht, D.M., L. M. Smith, and L. D. Densmore. 2009. Ecological correlates of trophic polyphenism in spadefoot tadpoles inhabiting playas. Canadian Journal of Zoology 87: 229–238.

Gibbons, J.W., and D.H. Bennett. 1974. Determination of anuran terrestrial activity patterns by a drift fence. Copeia 1974:236–243.

Gibbons, J.W., and J.W. Coker. 1978. Herpetofaunal colonization patterns of Atlantic Coast barrier islands. American Midland Naturalist 99:219–233.

Gibbons, J.W., and J.R. Harrison III. 1981. Reptiles and amphibians of Kiawah and Capers Islands, South Carolina. Brimleyana 5:145–162.

Gibbons, J.W., and R.D. Semlitsch. 1982. Terrestrial drift fences with pitfall traps: an effective technique for quantitative sampling of animal populations. Brimleyana 7:1–16.

Gibbons, J.W., and R.D. Semlitsch. 1991. Guide to the Reptiles and Amphibians of the Savannah River Site. University of Georgia Press, Athens.

Gibbons, J.W., V.J. Burke, J.E. Lovich, R.D. Semlitsch, T.D. Tuberville, J.R. Bodie, J.L. Greene, P.H. Niewiarowski, H.H. Whiteman, D.E. Scott, J.H.K. Pechmann, C.R. Harrison, S.H. Bennett, J.D. Krenz, M.S. Mills, K.A. Buhlmann, J.R. Lee, R.A. Seigel, A.D. Tucker, T.M. Mills, T. Lamb, M.E. Dorcas, J.D. Congdon, M.H. Smith, D.H. Nelson, M.B. Dietsch, H.G. Hanlin, J.A. Ott, and D.J. Karapatakis. 1997. Perceptions of species abundance, distribution, and diversity: lessons from four decades of sampling on a government–managed reserve. Environmental Management 21:259–268.

Gibbons, J.W., C.T. Winne, D.E. Scott, J.D. Willson, X. Glaudas, K.M. Andrews, B.D. Todd, L.A. Fedewa, L. Wilkinson, R.N. Tsaliagos, S.J. Harper, J.L. Greene, T. Tuberville, B.S. Metts, M.E. Dorcas, J.P. Nestor, C.A. Young, T. Akre, R.N. Reed, K.A. Buhlman, J. Norman, D.A. Croshaw, C. Hagen, and B.B. Rothermel. 2006. Remarkable amphibian biomass and abundance in an isolated wetland: implications for wetland conservation. Conservation Biology 20:1457–1465.

Gibbs, E.L., G.W. Nace, and M.B. Emmons. 1971. The live frog is almost dead. Bioscience 21:1027–1034.

Gibbs, J.P. 1998a. Amphibian movements in response to forest edges, roads, and streambeds in southern New England. Journal of Wildlife Management 62:584–589.

Gibbs, J.P. 1998b. Distribution of woodland amphibians along a forest fragmentation gradient. Landscape Ecology 13:263–268.

Gibbs, J.P., and W.G. Shriver. 2005. Can road mortality limit populations of pond–breeding amphibians? Wetlands Ecology and Management 13:281–289.

Gibbs, J.P., and J.M. Reed. 2008. Population and genetic linkages of vernal pool–breeding amphibians. Pp. 149–167 *In* A.J.K. Calhoun and P. deMaynadier (eds.). Science and Conservation of Vernal Pools in Northeastern North America. CRC Press, Boca Raton, Florida.

Gibbs, J.P., K.K. Whiteleather, and F.W. Schueler. 2005. Changes in frog and toad populations over 30 years in New York State. Ecological Applications 15:1148–1157.

Gibbs, J.P., A.R. Breisch, P.K. Ducey, G. Johnson, J.L. Behler, and R.C. Bothner. 2007. The Amphibians and Reptiles of New York State. Oxford University Press, Oxford, United Kingdom.

Gibbs, R.H., Jr. 1957. The chorus frog, *Pseudacris nigrita*, at Plattsburgh, New York. Copeia 1957:311–312.

Gibson, J.D., and P. Sattler. 2010. An unusual breeding event in the Urban Park in Danville, Virginia with specific notes on the eastern spadefoot (*Scaphiopus holbrookii*). Catesbeiana 30:73-81.

Gilbert, M., R. Leclair, Jr., and R. Fortin. 1994. Reproduction of the northern leopard frog (*Rana pipiens*) in floodplain habitat in the Richelieu River, P. Quebec, Canada. Journal of Herpetology 28:465–470.

Gilbert, P.W. 1942. Abnormally inflated lymph sacs in the wood frog. Copeia 1942:177.

Gilbertson, M.–K. 2001. Pesticide–induced immunosuppression in northern leopard frogs (*Rana pipiens*). Ph.D. Dissertation, University of Windsor, Windsor, Ontario.

Gilbertson, M.–K., G.D. Haffner, K.G. Drouillard, A. Albert, and B. Dixon. 2003. Immunosuppression in the northern leopard frog (*Rana pipiens*) induced by pesticide exposure. Environmental Toxicology and Chemistry 22:101–110.

Gilhen, J. 1984. Amphibians and Reptiles of Nova Scotia. Nova Scotia Museum, Halifax.

Gillilland, C.D., C.L. Summer, M.G. Gillilland, K. Kannan, D.L. Villeneuve, K.K. Coady, P. Muzzall, C. Mehne, and J.P. Giesy. 2001. Organochlorine insecticides, polychlorinated biphenyls, and metals in water, sediment, and green frogs from southwestern Michigan. Chemosphere 44:327-339.

Gillilland, M. G., III, and P. M. Muzzall. 1999. Helminths infecting froglets of the northern leopard frog (*Rana pipiens*) from Foggy Bottom Marsh, Michigan. Journal of the Helminthological Society of Washington 66:73–77.

Gillis, J.E. 1975. Characterization of a hybridizing complex of leopard frogs. Ph.D. Dissertation, Colorado State University, Fort Collins.

Gillis, J.E. 1979. Adaptive differences in the water economies of two species of leopard frogs from eastern Colorado (Amphibia, Anura, Ranidae). Journal of Herpetology 13:445–450.

Gilmore, R.J. 1924. Notes on the life history and feeding habits of the spadefoot toad of the western plains. Colorado College Publication, Science Series 13:1–12.

Girard, C. 1854. A list of the North American bufonids, with diagnoses of new species. Proceedings of the Academy of Natural Sciences of Philadelphia 7:86–88.

Girard, C. 1855. Abstract of a report to Lieutenant James McGillis, U.S.N., upon the reptiles collected during the U.S.N. Astronomical Expedition to Chili. Proceedings of the Academy of Natural Sciences of Philadelphia 7:226–227.

Girard, C. 1859. *In* S.F. Baird, Reptiles of the Boundary. United States and Mexican Boundary Survey, under the order of Lieut. Col. W.H. Emory. United States Government Printing Office, Washington, D.C.

Given, M.F. 1987. Vocalizations and acoustic interactions of the carpenter frog, *Rana virgatipes*. Herpetologica 43:467–481.

Given, M.F. 1988a. Growth rate and the cost of calling activity in male carpenter frogs, *Rana virgatipes*. Behavioral Ecology and Sociobiology 22:153–160.

Given, M.F. 1988b. Territoriality and aggressive interactions of male carpenter frogs, *Rana virgatipes*. Copeia 1988:411–421.

Given, M.F. 1990. Spatial distribution and vocal interaction in *Rana clamitans* and *R. virgatipes*. Journal of Herpetology 24:377–382.

Given, M.F. 1993a. Male response to female vocalizations in the carpenter frog, *Rana virgatipes*. Animal Behaviour 46:1139–1149.

Given, M.F. 1993b. Vocal interactions in *Bufo woodhousii fowleri*. Journal of Herpetology 27:447–452.

Given, M.F. 1996. Intensity modulation of advertisement calls in *Bufo woodhousii fowleri*. Copeia 1996:970–977.

Given, M.F. 1999a. Distribution records of *Rana virgatipes* and associated anuran species along Maryland's Eastern Shore. Herpetological Review 30:144–146.

Given, M.F. 1999b. Frequency alteration of the advertisement call in the carpenter frog, *Rana virgatipes*. Herpetologica 55:304–317.

Given, M.F. 2002. Interrelationships among calling effort, growth rate, and chorus tenure in *Bufo fowleri*. Copeia 2002:979–987.

Given, M.F. 2005. Vocalizations and reproductive behavior of male pickerel frogs, *Rana palustris*. Journal of Herpetology 39:223–233.

Given, M.F. 2008. Does physical or acoustical disturbance cause male pickerel frogs, *Rana palustris*, to vocalize underwater? Amphibia–Reptilia 29:177–184.

Glaser, H.S.R. 1970. The distribution of amphibians and reptiles in Riverside County, California. Riverside Museum Press, Natural History Series No. 1, 40 pp.

Glass, B.P. 1946. A new *Hyla* from south Texas. Herpetologica 3:101–103.

Gleason, R.L., and T.H. Craig. 1979. Food habits of burrowing owls in southeastern Idaho. Great Basin Naturalist 39:274–276.

Glennemeier, K.A. 2000. Role of corticosterone in the development and physiological ecology of *Rana pipiens* tadpoles and the disruption of this endocrine system by organochlorine contamination. Ph.D. Dissertation, University of Michigan, Ann Arbor.

Glennemeier, K.A., and L.J. Begnoche. 2002. Impact of organochlorine contamination on amphibian populations in southwestern Michigan. Journal of Herpetology 36:233–244.

Glennemeier, K.A., and R.J. Denver. 2002. Role for corticoids in mediating the response of *Rana pipiens* tadpoles to intraspecific competition. Journal of Experimental Zoology 292:32–40.

Glista, D.J, T.L. DeVault, and J.A. DeWoody. 2008. Vertebrate road mortality predominantly impacts amphibians. Herpetological Conservation and Biology 3:77–87.

Glooschenko, V., W.F. Weller, P.G.R. Smith, R. Alvo, and J.H.G. Archbold. 1992. Amphibian distribution with respect to pond water chemistry near Sudbury, Ontario. Canadian Journal of Fisheries and Aquatic Science 49 (Suppl. 1):114–121.

Glorioso, B.M., J.H. Waddle, M.E. Crockett, K.G. Rice, and H.F. Percival. 2012. Diet of the invasive Cuban Treefrog (*Osteopilus septentrionalis*) in pine rockland and mangrove habitats in South Florida. Caribbean Journal of Science 46:346–355.

Gloyd, H.K. 1937. A herpetological consideration of faunal areas in southern Arizona. Bulletin of the Chicago Academy of Sciences 5:79–136.

Gloyd, H.K., and R. Conant. 1990. Snakes of the *Agkistrodon* Complex. A Monographic Review. Society for the Study of Amphibians and Reptiles, Contributions in Herpetology No. 6.

Goater, C.P., and R.E. Vandenbos. 1997. Effects of larval history and lungworm infection on the growth and survival of juvenile wood frogs (*Rana sylvatica*). Herpetologica 53:331–338.

Goater, C.P., M. Mulvey, and G.W. Esch. 1990. Electrophoretic differentiation of two *Halipegus* (Trematoda: Hemiuridae) congeners in an amphibian population. Journal of Parasitology 76:431–434.

Godley, J.S., R.W. McDiarmid, and G.T. Bancroft. 1981. Large–scale operations management test of use of the White Amur for control of problem aquatic plants. Report 1, Volume 5. The herpetofauna of Lake Conway, Florida. U.S. Army Corps of Engineers, Technical Report A–78–2, Waterways Experiment Station, Vicksburg, Mississippi.

Godwin, G.J., and S.M. Roble. 1983. Mating success in male treefrogs, *Hyla chrysoscelis* (Anura: Hylidae). Herpetologica 39:141–146.

Goebel, A.M. 1996. Systematics and conservation of bufonids in North America and the *Bufo boreas* species group. Ph.D. Dissertation, University of Colorado, Boulder.

Goebel, A.M. 2005. Conservation systematics: the *Bufo boreas* species group. Pp. 210–221 *In* M.J. Lannoo (ed.), Amphibian Declines. The Conservation Status of United States Species. University of California Press, Berkeley.

Goebel, A.M., T.A. Ranker, P.S. Corn, and R.G. Olmstead. 2009. Mitochondrial DNA evolution in the *Anaxyrus boreas* species group. Molecular Phylogenetics and Evolution 50:209–225.

Goin, C.J. 1938. A large chorus of *Hyla gratiosa*. Copeia 1938:48.

Goin, C.J. 1943. The lower vertebrate fauna of the water hyacinth community in northern Florida. Proceedings of the Florida Academy of Sciences 6:143–154.

Goin, C.J. 1944. *Eleutherodactylus ricordii* at Jacksonville, Florida. Copeia 1944:192.

Goin, C.J. 1946. Studies on the life history of *Eleutherodactylus ricordii planirostris* (Cope) in Florida, with special reference to the local distribution of an allelomorphic color pattern. Ph.D. Dissertation, University of Florida, Gainesville.

Goin, C.J. 1947a. A note on the food of *Heterodon simus*. Copeia 1947:275.

Goin, C.J. 1947b. Studies on the life history of *Eleutherodactylus ricordii planirostris* (Cope) in Florida. University of Florida Studies, Biological Science Series 4:1–66.

Goin, C.J. 1948. The peep order in peepers: a swamp water serenade. Quarterly Journal of the Florida Academy of Sciences 11:59–61.

Goin, C.J. 1950. Color pattern inheritance in some frogs of the genus *Eleutherodactylus*. Bulletin of the Chicago Academy of Sciences 9:1–13.

Goin, C.J., and M.G. Netting. 1940. A new gopher frog from the Gulf Coast, with comments upon the *Rana areolata* group. Annals of the Carnegie Museum 28:137–168.

Goin, C.J., and O.B. Goin. 1953. Temporal variation in a small community of amphibians and reptiles. Ecology 34:406–408.

Goin, C.J., and O.B. Goin. 1957. Remarks on the behavior of the squirrel treefrog, *Hyla squirella*. Annals of the Carnegie Museum 35:27–36.

Goin, C.J., and L.H. Ogren. 1956. Parasitic copepods (Argulidae) on amphibians. Journal of Parasitology 42:154.

Goin, O.B. 1955. The World Outside My Door. Macmillan Press, New York.

Goin, O.B. 1958. A comparison of the nonbreeding habits of two treefrogs, *Hyla squirella* and *Hyla cinerea*. Quarterly Journal of the Florida Academy of Sciences 21:49–60.

Goin, O.B., and C.J. Goin. 1960. Return of the toad, *Bufo terrestris americanus*, to the breeding site. Herpetologica 16:276.

Goldberg, C.S. 2002. Habitat, spatial population structure, and methods for monitoring barking frogs (*Eleutherodactylus augusti*) in southern Arizona. M.S. thesis, University of Arizona, Tucson.

Goldberg, C.S., and C.R. Schwalbe. 2004a. Habitat use and spatial structure of a barking frog (*Eleutherodactylus augusti*) population in southeastern Arizona. Journal of Herpetology 38:305–312.

Goldberg, C.S., and C.R. Schwalbe. 2004b. Considerations for monitoring a rare anuran (*Eleutherodactylus augusti*). Southwestern Naturalist 49:442–448.

Goldberg, C.S., and L.P. Waits. 2009. Using habitat models to determine conservation priorities for pond–breeding amphibians in a privately–owned landscape of northern Idaho, USA. Biological Conservation 142:1096–1104.

Goldberg, C.S., and L.P. Waits. 2010. Comparative landscape genetics of two pond–breeding amphibian species in a highly modified agricultural area. Molecular Ecology 19:3650–3663.

Goldberg, C.S., K.J. Field, and M.J. Sredl. 2004a. Mitochondrial DNA sequences do not support species status of the Ramsey Canyon leopard frog (*Rana subaquavocalis*). Journal of Herpetology 38:313–319.

Goldberg, C.S., B.K. Sullivan, J.H. Malone, and C.R. Schwalbe. 2004b. Divergence among barking frogs (*Eleutherodactylus augusti*) in the southwestern United States. Herpetologica 60:312–320.

Goldberg, C.S., D.S. Pilliod, R.S. Arkle, and L.P. Waits. 2011. Molecular detection of vertebrates in stream waters: a demonstration using Rocky Mountain tailed frogs and Idaho giant salamanders. PLoS One 6(7):e22746.

Goldberg, S.R., and C.R. Bursey. 1991a. Helminths from the red–spotted toad, *Bufo punctatus* (Anura: Bufonidae), from southern Arizona. Journal of the Helminthological Society of Washington 58:267–269.

Goldberg, S.R., and C.R. Bursey. 1991b. Helminths of three toads, *Bufo alvarius*, *Bufo cognatus* (Bufonidae), and *Scaphiopus couchii* (Pelobatidae), from southern Arizona. Journal of the Helminthological Society of Washington 58:142–146.

Goldberg, S.R., and C.R. Bursey. 1996. Helminths of the oak toad (*Bufo quercicus*, Bufonidae) from Florida (USA). Alytes 14:122–126.

Goldberg, S.R., and C.R. Bursey. 2001a. Helminths of the California treefrog, *Hyla cadaverina* (Hylidae) from southern California. Bulletin of the Southern California Academy of Sciences 100:117–122.

Goldberg, S.R., and C.R. Bursey. 2001b. Persistence of the nematode *Ozwaldocruzia pipiens* (Molineidae), in the Pacific treefrog, *Hyla regilla* (Hylidae), from California. Bulletin of the Southern California Academy of Sciences 100:44–50.

Goldberg, S.R., and C.R. Bursey. 2002a. Helminths of the bullfrog, *Rana catesbeiana* (Ranidae), in California with revisions to the California anuran helminth list. Bulletin of the Southern California Academy of Sciences 101:118–130.

Goldberg, S.R., and C.R. Bursey. 2002b. Helminths of the Plains spadefoot, *Spea bombifrons*, the western spadefoot, *Spea hammondii*, and the Great Basin spadefoot, *Spea intermontana* (Pelobatidae). Western North American Naturalist 62:491–495.

Goldberg, S.R., and C.R. Bursey. 2005. *Scaphiopus couchii* (Couch's Spadefoot). Endoparasites. Herpetological Review 36:165–166.

Goldberg, S.R., C.R. Bursey, and I. Ramos. 1995. The component parasite community of three sympatric toad species, *Bufo cognatus*, *Bufo debilis* (Bufonidae), and *Spea multiplicata* (Pelobatidae)from New Mexico. Journal of the Helminthological Society of Washington 62:57–61.

Goldberg, S.R., C.R. Bursey, B.K. Sullivan, and Q.A. Truong. 1996a. Helminths of the Sonoran green toad, *Bufo retiformis* (Bufonidae), from southern Arizona. Journal of the Helminthological Society of Washington 63:120–122.

Goldberg, S.R., C.R. Bursey, K.B. Malmos, B.K. Sullivan, and H. Cheam. 1996b. Helminths of the southwestern toad, *Bufo microscaphus*, Woodhouse's toad, *Bufo woodhousii* (Bufonidae), and their hybrids from central Arizona. Great Basin Naturalist 56:369–374.

Goldberg, S.R., C.R. Bursey, E.W.A. Gergus, B.K. Sullivan, and Q.A. Truong. 1996c. Helminths of three treefrogs *Hyla arenicolor*, *Hyla wrightorum*, and *Pseudacris triseriata* (Hylidae) from Arizona. Journal of Parasitology 82:833–835.

Goldberg, S.R., C. R. Bursey, and H. Cheam. 1998a. Helminths in two native frog species (*Rana chiricahuensis*, *Rana yavapaiensis*) and one introduced frog species (*Rana catesbeiana*) (Ranidae) from Arizona. Journal of Parasitology 84:175–177.

Goldberg, S.R., C.R. Bursey, and H. Cheam. 1998b. Nematodes of the Great Plains narrow–mouthed toad, *Gastrophryne olivacea* (Microhylidae), from southern Arizona. Journal of the Helminthological Society of Washington 65:102–104.

Goldberg, S.R., C.R. Bursey, and S. Hernandez. 1999a. Helminths of the western toad, *Bufo boreas* (Bufonidae) from southern California. Bulletin of the Southern California Academy of Sciences 98:39–44.

Goldberg, S.R., C.R. Bursey, and G. Galindo. 1999b. Helminths of the lowland burrowing treefrog, *Pternohyla fodiens* (Hylidae), from southern Arizona. Great Basin Naturalist 59:195–197.

Goldberg, S.R., C.R. Bursey, and J.E. Platz. 2000. Helminths of the plains leopard frog, *Rana blairi* (Ranidae). Southwestern Naturalist 45:362–366.

Goldberg, S.R., C.R. Bursey, R.G. McKinnell, and I.S. Tan. 2001. Helminths of northern leopard frogs, *Rana pipi-*

ens (Ranidae), from North Dakota and South Dakota. Western North American Naturalist 61:248–251.

Goldberg, S.R., C.R. Bursey, and C. Wong. 2002. Helminths of the western chorus frog from eastern Alberta, Canada. Northwest Science 76:77–79.

Goldberg, S.R., C.R. Bursey, N. Nieto, and J. Bettaso. 2004. *Rana aurora* (Northern Red–legged Frog). Endoparasites. Herpetological Review 35:161–162.

Golden, D.M., and J.D. Kleopfer. 2011. *Scaphiopus holbrookii* (Eastern Spadefoot). Locomotion. Herpetological Review 42:263.

Golden, D.R., G.R. Smith, and J.E. Rettig. 2001. Effects of age and group size on habitat selection and activity level in *Rana pipiens* tadpoles. Herpetological Journal 11:69–73.

Goldstein, J.A. 2007. The effect of temperature on the development and behavior of Relict Leopard Frog tadpoles (*Rana onca*). M.S. thesis, University of Nevada, Las Vegas.

Goldsworthy, M.O. 2007. *Ascaphus truei* (Tailed Frog). Nest site. Herpetological Review 38:68–69.

Gomez, D.M., and R.G. Anthony. 1996. Amphibian and reptile abundance in riparian and upslope areas of five forest types in western Oregon. Northwest Science 70:109–119.

Gomez–Mestre, I., and D.R. Buchholz. 2006. Developmental plasticity mirrors differences among taxa in spadefoot toads linking plasticity and diversity. Proceedings of the National Academy of Sciences of the United States of America 103:19021–19026.

Gonsolin, T.E. 2010. Ecology of Foothill Yellow–legged Frogs in Upper Coyote Creek, Santa Clara County, CA. M.S. thesis, San Jose State University, San Jose, California.

Gonzalez, S.M. 2004. Biological indicators of wetland health: comparing qualitative and quantitative vegetation measures with anuran measures. M.S. thesis, University of South Florida, Tampa.

Goodman, B.A., and P.T.J. Johnson. 2011. Ecomorphology and disease: cryptic effects of parasitism on host habitat use, thermoregulation, and predator avoidance. Ecology 92:542–548.

Goodman, J.D. 1989. *Langeronia brenesi* n. sp. (Trematoda: Lecithodendriidae) in the mountain yellow–legged frog *Rana muscosa* from southern California. Transactions of the American Microscopical Society 108:387–393.

Goodman, R.M., and Y.T. Ararso. 2012. Survey of ranavirus and the fungus *Batrachochytrium dendrobatidis* in frogs of central Virginia, USA. Herpetological Review 43:78–80.

Goodsell, J.A., and L.B. Kats. 1999. Effect of introduced mosquitofish on Pacific treefrogs and the role of alternate prey. Conservation Biology 13:921–924.

Goodyear, C.P. 1971. Y–axis orientation of the oak toad, *Bufo quercicus*. Herpetologica 27:320–323.

Goodyear, C.P., and R. Altig. 1971. Orientation of bullfrogs (*Rana catesbeiana*) during metamorphosis. Copeia 1971:362–364.

Goraya, J., Y. Wang, Z. Li, M. O'Flaherty, F.C. Knoop, J.E. Platz, and J.M. Conlon. 2000. Peptides with antimicrobial activity from four different families isolated from the skins of the North American frogs *Rana luteiventris, Rana berlandieri* and *Rana pipiens*. European Journal of Biochemistry 267:894–900.

Gordon, K. 1939. The amphibia and reptilia of Oregon. Oregon State Monographs, Studies in Zoology No. 1.

Gordon, R.E. 1955. Additional remarks on albinism in *Microhyla carolinensis*. Herpetologica 11:240.

Gorham, S.W. 1963. The comparative number of species of amphibians in Canada and other countries III. Summary of species of anurans. Canadian Field-Naturalist 77:13–48.

Gorham, S.W. 1964. Notes on the amphibians of Browns Flat Area, New Brunswick. Canadian Field-Naturalist 78:154–160.

Gorham, S.W. 1970. The Amphibians and Reptiles of New Brunswick. New Brunswick Museum, Monographic Series No. 6.

Gorham, S.W. 1974. Checklist of World Amphibians/Listedes Amphibiens du Monde. The New Brunswick Museum, Saint John, New Brunswick.

Gorman, J. 1960. Treetoad studies, 1. *Hyla californiae*, new species. Herpetologica 16:214–222.

Gorman, R.R., and J.H. Ferguson. 1970. Sun–compass orientation in the Western toad, *Bufo boreas*. Herpetologica 26:34–45.

Gorman, T.A. 2009. Ecology of two rare amphibians of the Gulf Coastal Plain. Ph.D. Dissertation, Virginia Polytechnic Institute and State University, Blacksburg.

Gorman, T.A., and C.A. Haas. 2011. Seasonal microhabitat selection and use of syntopicpopulations of *Lithobates okaloosae* and *Lithobates clamitans clamitans*. Journal of Herpetology 45:313–318.

Gorman, T.A., and C.A. Haas. 2012. Tadpole competition: the influence of a common tadpole on the growth of a rare congener. Florida Scientist 75:11–24.

Gorman, T.A., D.C. Bishop, and C.A. Haas. 2009. Spatial interactions between two species of frogs: *Rana okaloosae* and *R. clamitans clamitans*. Copeia 2009:138–141.

Gorman, T.A., J.A. Austin, and C.A. Haas. 2012. Conservation of the Florida bog frog, one of North America's rarest amphibians. Froglog (102):24–25.

Gorman, W.L. 1986. Patterns of color polymorphism in the cricket frog, *Acris crepitans*, in Kansas. Copeia

1986:995–999.

Gorman, W.L. 1983. Patterns of genic variation in the cricket frog, *Acris crepitans*, in Kansas. Ph.D. Dissertation, University of Kansas, Lawrence.

Gorman, W.L., and M.S. Gaines. 1987. Patterns of genetic variation in the cricket frog, *Acris crepitans*, in Kansas. Copeia 1987:352–360.

Gosner, K.L. 1956. Experimental hybridization between two North American tree frogs. Herpetologica 12:285–289.

Gosner, K.L. 1959. Systematic variations in tadpole teeth with notes on food. Herpetologica 15:203–210.

Gosner, K.L. 1960. A simplified table for staging anuran embryos and larvae with notes on identification. Herpetologica 16:183–190.

Gosner, K.L., and I.H. Black. 1954. Larval development in *Bufo woodhousei fowleri* and *Scaphiopus holbrooki holbrooki*. Copeia 1954:251–255.

Gosner, K.L., and I.H. Black. 1955. The effects of temperature and moisture on the reproductive cycle of *Scaphiopus h. holbrooki*. American Midland Naturalist 54:192–203.

Gosner, K.L., and I.H. Black. 1956. Notes on amphibians from the upper Coastal Plain of North Carolina. Journal of the Elisha Mitchell Scientific Society 72:40–47.

Gosner, K.L., and I.H. Black. 1957a. The effects of acidity on the development and hatching of New Jersey frogs. Ecology 38:256–262.

Gosner, K.L., and I.H. Black. 1957b. Larval development in New Jersey Hylidae. Copeia 1957:31–36.

Gosner, K.L., and I.H. Black. 1958a. Notes on the life history of Brimley's chorus frog. Herpetologica 13:249–254.

Gosner, K.L., and I.H. Black. 1958b. Notes on larval toads in the eastern United States with special reference to natural hybridization. Herpetologica 14:133–140.

Gosner, K.L., and D.A. Rossman. 1959. Observations on the reproductive cycle of the swamp chorus frog, *Pseudacris nigrita*. Copeia 1959:263–266.

Gosner, K.L., and D.A. Rossman. 1960. Eggs and larval development of the treefrogs *Hyla crucifer* and *Hyla ocularis*. Herpetologica 16:225–232.

Gossling, J., W.J. Loesche, and G.W. Nace. 1982. Large intestine bacteria flora of nonhibernating and hibernating leopard frogs (*Rana pipiens*). Applied and Environmental Microbiology 44:59–66.

Gould, W.R., D.A. Patla, R. Daley, P.S. Corn, B.R. Hossack, R. Bennetts and C.R. Peterson. 2012. Estimating occupancy in large landscapes: evaluation of amphibian monitoring in the Greater Yellowstone Ecosystem. Wetlands 32:379–389.

Govindarajulu, P. 2004. Introduced bullfrogs (*Rana catesbeiana*) in British Columbia: impacts on native Pacific treefrogs (*Hyla regilla*) and red–legged frogs (*Rana auro-*

ra). Ph.D. Dissertation, University of Victoria, Victoria, British Columbia.

Govindarajulu, P., R. Altwegg, and B.R. Anholt. 2005. Matrix model investigation of invasive species control: bullfrogs on Vancouver Island. Ecological Applications 15:2161–2170.

Govindarajulu, P., W.S. Price, and B.R. Anholt. 2006. Introduced bullfrogs (*Rana catesbeiana*) in Canada: has their ecology diverged? Journal of Herpetology 40:249–260.

Graber, R.R., J.W. Graber, and E.L. Kirk. 1973. Illinois birds: Laniidae. Illinois Natural History Survey, Biological Note 83, 18 pp.

Gradwell, N. 1973. On the functional morphology of suction and gill irrigation in the tadpole of *Ascaphus*, and notes on hibernation. Herpetologica 29:84–93.

Graeter, G.J. 2005. Habitat selection and movement patterns of amphibians in altered forest habitats. M.S. thesis, University of Georgia, Athens.

Graeter, G.J., B.B. Rothermel, and J.W. Gibbons. 2008. Habitat selection and movement of pond–breeding amphibians in experimentally fragmented pine forests. Journal of Wildlife Management 72:473–482.

Grafe, U. 1997. Use of metabolic substrates in the gray treefrog *Hyla versicolor*: implications for calling behavior. Copeia 1997:356–362.

Graham, C., S.C. Richter, S. McClean, E.O'Kane, P.R. Flatt, and C. Shaw. 2006. Histamine–releasing and antimicrobial peptides from the skin secretions of the dusky gopher frog, *Rana sevosa*. Peptides 27:1313–1319.

Graham, S.P. 2009. *Acris gryllus* (Southern Cricket Frog). Predation. Herpetological Review 40:198.

Graham, S.P. 2010. Geographic distribution. *Pseudacris brachyphona* (Mountain Chorus Frog). Herpetological Review 41:241.

Graham, S.P., E.K. Timpe, and L. Giovanetto. 2007. Significant new records for Georgia herpetofauna. Herpetological Review 38:494–495.

Graham, S. P., D. A. Steen, R. D. Birkhead, and C. Guyer. 2012. The amphibians and reptiles of Tuskegee National Forest, Macon County, Alabama. Alabama Museum of Natural History Bulletin 29:1–59.

Graham, T.E. 1978. Massachusetts frogs and toads, part 1. Massachusetts Wildlife 29(5):12–14.

Graham, T.E. 1978. Massachusetts frogs and toads, part 2— conclusion. Massachusetts Wildlife 29(6):12–19.

Granatosky, M.C., and K.L. Krysko. 2011. Ontogenetic behavioral shifts in habitat utilization of treefrogs (Hylidae) in north–central Florida. IRCF Reptiles & Amphibians 18:194–201.

Granatosky, M.C., L.M. Wagner, and K.L. Krysko. 2011. *Osteopilus septentrionalis* (Cuban Treefrog). Prey. Herpetological Review 42:90.

Granoff, A. 1969. Viruses of amphibia. Current Topics in Microbiology and Immunology 50:107–137.

Granoff, A., P.E. Came, and K.A. Rafferty. 1965. The isolation and properties of viruses from *Rana pipiens*: their possible relationship to the renal adenocarcinoma of the leopard frog. Annals of the New York Academy of Sciences 126:237–255.

Grant, E.H.C., I. Chellman, P. Nanjappa, and R.E. Jung. 2004. *Rana* spp. (Multiple Ranid Species). Hibernacula. Herpetological Review 35:262–263.

Grant, J. 1961. The tailed toad in southeastern British Columbia. Canadian Field-Naturalist 75:165.

Grant, J.B. 2001. *Rana palustris* (Pickerel Frog). Production of odor. Herpetological Review 32:183.

Grant, K.P., and L.E. Licht. 1993. Acid tolerance of anuran embryos and larvae from central Ontario. Journal of Herpetology 27:1–6.

Grant, K.P., and L.E. Licht. 1995. Effects of ultraviolet radiation on life–history stages of anurans from Ontario, Canada. Canadian Journal of Zoology 73:2292–2301.

Grasso, R.L., R.M. Coleman, and C. Davidson. 2010. Palatability and antipredator response of Yosemite toads (*Anaxyrus canorus*) to nonnative trout (*Salvelinus fontinalis*) in the Sierra Nevada Mountains of California. Copeia 2010:457–462.

Gratwicke, B., T.E. Lovejoy, and D.E. Wildt. 2012. Will amphibians croak under the Endangered Species Act? Bioscience 62:197–202.

Gravel, M., M.J. Mazerolle, and M.–A. Villard. 2012. Interactive effects of roads and weather on juvenile amphibian movements. Amphibia–Reptilia 33:113–127.

Graves, B.M., and S.H. Anderson. 1987. Habitat Suitability Index Models. Bullfrog. U.S. Fish and Wildlife Service, Biological Report 82(10.138), 22 pp.

Graves, B.M., and J.J. Krupa. 2005. *Bufo cognatus* Say, 1823. Great Plains Toad. Pp. 401–404 *In* M.J. Lannoo (ed.), Amphibian Declines. The Conservation Status of United States Species. University of California Press, Berkeley.

Graves, B.M., C.H. Summers, and K.L. Olmstead. 1993. Sensory mediation of aggregation among postmetamorphic *Bufo cognatus*. Journal of Herpetology 27:315–319.

Gray, B.S., and M. Lethaby. 2010. Observations of limb abnormalities in amphibians from Erie County, Pennsylvania. Journal of Kansas Herpetology 35:14–16.

Gray, E.P., S. Nunziata, J.W. Snodgrass, D.R. Ownby, and J.E. Havel. 2010. Predation on green frog eggs (*Rana clamitans*) by ostracoda. Copeia 2010:452–456.

Gray, I.E. 1941. Amphibians and reptiles of the Duke Forest and vicinity. American Midland Naturalist 25:652–658.

Gray, L.A. 1992. Larval growth and age at completion of metamorphosis of the tailed frog, *Ascaphus truei*. M.S. thesis. Central Washington University, Ellensburg, Washington.

Gray, M.J. 2002. Effect of anthropogenic disturbance and landscape structure on body size, demographics, and chaotic dynamics of Southern High Plains amphibians. Ph.D. Dissertation, Texas Tech University, Lubbock.

Gray, M. J., and L. M. Smith. 2005. Influence of land use on postmetamorphic body size of playa lake amphibians. Journal of Wildlife Management 69: 515–524.

Gray, M.J., L.M. Smith, and R.I. Leyva. 2004a. Influence of agricultural landscape on a Southern High Plains, USA, amphibian assemblage. Landscape Ecology 19:719–729.

Gray, M.J., L.M. Smith, and R. Brenes. 2004b. Effects of agricultural cultivation on demographics of Southern High Plains amphibians. Conservation Biology 18:1368–1377.

Gray, M.J., D.L. Miller, A.C. Schmutzer, and C.A. Baldwin. 2007a. Frog Virus 3 prevalence in tadpole populations inhabiting cattle–access and non–access wetlands in Tennessee. Diseases of Aquatic Organisms 77:97–103.

Gray, M.J., S. Rajeev, D.L. Miller, A.C. Schmutzer, E.C. Burton, E.D. Rogers, and G.J. Hickling. 2007b. Preliminary evidence that American Bullfrogs (*Rana catesbeiana*) are suitable hosts for *Escherichia coli* O157:H7. Applied and Environmental Microbiology 73:4066–4068.

Gray, M.J., L.M. Smith, D.L. Miller, and C.R. Bursey. 2007c. Influences of agricultural land use on *Clinostomum attenuatum* metacercariae prevalence in Southern Great Plains amphibians, U.S.A. Herpetological Conservation and Biology 2:23–28.

Gray, M.J., D.L. Miller, and J.T. Hoverman. 2009. Ecology and pathology of amphibian ranaviruses. Diseases of Aquatic Organisms 87:243–266.

Gray, M.J., D.L. Miller, and J.T. Hoverman. 2012. Reliability of non–lethal surveillance methods for detecting ranavirus infection. Diseases of Aquatic Organisms 99:1–6.

Gray, P., and E. Stegall. 1986. Distribution and status of Strecker's chorus frog (*Pseudacris streckeri streckeri*) in Kansas. Transactions of the Kansas Academy of Science 89:81–85.

Gray, R.H. 1971. Fall activity and overwintering of the cricket frog, *Acris crepitans*, in central Illinois. Copeia 1971:748–750.

Gray, R.H. 1972. Metachrosis of the vertebral stripe in the cricket frog, *Acris crepitans*. American Midland Naturalist 87:549–551.

Gray, R.H. 1977. Lack of physiological differentiation in three color morphs of the cricket frog (*Acris crepitans*) in Illinois. Transactions of the Illinois State Academy of Science 70:73–79.

Gray, R.H. 1978. Nondifferential predation susceptibility

and behavioral selection in three color morphs of Illinois cricket frogs, *Acris crepitans*. Transactions of the Illinois State Academy of Science 71:356–360.

Gray, R.H. 1983. Seasonal, annual and geographic variation in color morph frequencies of the cricket frog, *Acris crepitans*, in Illinois. Copeia 1983:300–311.

Gray, R.H. 1984. Effective breeding size and the adaptive significance of color polymorphism in the cricket frog (*Acris crepitans*) in Illinois, U.S.A. Amphibia–Reptilia 5:101–107.

Gray, R.H. 2000a. Historical occurrence of malformations in the cricket frog, *Acris crepitans*, in Illinois. Transactions of the Illinois State Academy of Science 93:279–284.

Gray, R.H. 2000b. Morphological abnormalities in Illinois cricket frogs, *Acris crepitans*, 1968–1971. Journal of the Iowa Academy of Science 107:92–95.

Gray, R.H., and L.E. Brown. 2005. Decline of northern cricket frogs (*Acris crepitans*). Pp. 47–54 *In* M.J. Lannoo (ed.), Amphibian Declines. The Conservation Status of United States Species. University of California Press, Berkeley.

Graybeal, A. 1993. The phylogenetic utility of cytochrome *b*: lessons from bufonid frogs. Molecular and Phylogenetic Evolution 2:256–269.

Graybeal, A. 1997. Phylogenetic relationships of Bufonid frogs and tests of alternate macroevolutionary hypotheses characterizing their radiation. Zoological Journal of the Linnean Society 119:297–338.

Greckhamer, A. 1992a. Bermerkungen uber die Zucht des amerikanischen Laubfrosches *Hyla cinerea* (Schneider, 1799). Herpetofauna 14:17–20.

Greckhamer, A. 1992b. Observations on the breeding of the American green treefrog *Hyla cinerea* (Schneider, 1799). Nordisk Herpetologisk Forening 35:82–85.

Green, D.E., and C. Kagarise Sherman. 2001. Diagnostic histological findings in Yosemite toads (*Bufo canorus*) from a die–off in the 1970s. Journal of Herpetology 35:92–103.

Green, D.E., and E. Muths. 2005. Health evaluation of amphibians in and near Rocky Mountain National Park (Colorado, USA). Alytes 22:109–129.

Green, D.E., and C.K. Dodd, Jr. 2007. Presence of amphibian chytrid fungus *Batrachochytrium dendrobatidis* and other amphibian pathogens at warm–water fish hatcheries in southeastern North America. Herpetological Conservation and Biology 2:43–47.

Green, D.E., K.A. Converse, and A.K. Schrader. 2002. Epizootiology of sixty–four amphibian morbidity and mortality events in the USA, 1996–2001. Annals of the New York Academy of Sciences 969:232–339.

[4]Green, D.M. 1981a (1982). Mating call characteristics of hybrid toads (*Bufo americanus* x *B. fowleri*) at Long Point, Ontario. Canadian Journal of Zoology 60:3293–3297.

Green, D.M. 1981b. Adhesion and the toe pads of treefrogs. Copeia 1981:790–796.

Green, D.M. 1981c. Theoretical analysis of hybrid zones derived from an examination of two dissimilar zones of hybridization in toads (genus *Bufo*). Ph.D. Dissertation, University of Guelph, Guelph, Ontario.

Green, D.M. 1983. Allozyme variation through a clinal hybrid zone between the toads *Bufo americanus* and *B. hemiophrys* in southeastern Manitoba. Herpetologica 39:28–40.

Green, D.M. 1984. Sympatric hybridization and allozyme variation in the toads *Bufo americanus* and *B. fowleri* in southern Ontario. Copeia 1984:18–26.

Green, D.M. 1985a. Differentiation in amount of centromeric heterochromatin between subspecies of the red–legged frog, *Rana aurora*. Copeia 1985:1071–1074.

Green, D.M. 1985b. Biochemical identification of red–legged frogs, *Rana aurora draytoni* (Ranidae) at Duckwater, Nevada. Southwestern Naturalist 30:614–616.

Green, D.M. 1985c. Natural hybrids between the frogs *Rana cascadae* and *Rana pretiosa* (Anura: Ranidae). Herpetologica 41:262–267.

Green, D.M. 1986a. Systematics and evolution of western North American frogs allied to *Rana aurora* and *Rana boylii*: karyological evidence. Systematic Zoology 35:273–282.

Green, D.M. 1986b. Systematics and evolution of western North American frogs allied to *Rana aurora* and *Rana boylii*: electrophoretic evidence. Systematic Zoology 35:283–296.

Green, D.M. 1989. Fowler's toad, *Bufo woodhousii fowleri*, in Canada: biology and population status. Canadian Field-Naturalist 103:486–496.

Green, D.M. 1992. Fowler's toad (*Bufo woodhousei fowleri*) at Long Point, Ontario: changing abundance and implications for conservation. Canadian Wildlife Service Occasional Paper No. 76:37–43.

Green, D.M. 1996. The bounds of species: hybridization in the *Bufo americanus* group of North American toads. Israel Journal of Zoology 42:95–109.

Green, D.M. 1997. Temporal variation in abundance and age structure in Fowler's toads, *Bufo fowleri*, at Long Point, Ontario. SSAR Herpetological Conservation 1:45–56.

Green, D.M. 2003. The ecology of extinction: population

---

[4]This publication is variously cited as 1981 or 1982. Vol. 60 was published in 1981, although issue 12 did not appear until January 1982. The online journal is dated 1981, but the reprints are dated 1982.

fluctuation and decline in amphibians. Biological Conservation 111: 331–343.

Green, D.M. 2005. *Bufo americanus* Holbrook, 1836. American Toad. Pp. 386–390 *In* M.J. Lannoo (ed.), Amphibian Declines. The Conservation Status of United States Species. University of California Press, Berkeley.

Green, D.M., and R.W. Campbell. 1984. The amphibians of British Columbia. British Columbia Provincial Museum, Handbook No. 45.

Green, D.M., and D.M. Delisle. 1985. Allotriploidy in natural hybrid frogs, *Rana chiricahuensis* x *R. pipiens*, from Arizona: chromosomes and electrophoretic evidence. Journal of Herpetology 19:385–390.

Green, D.M., and C. Pustowka. 1997. Correlated morphological and allozyme variation in the hybridizing toads *Bufo americanus* and *Bufo hemiophrys*. Herpetologica 53:218–228.

Green, D.M., and C. Parent. 2003. Variable and asymmetric introgression in a hybrid zone in the toads, *Bufo americanus* and *Bufo fowleri*. Copeia 2003:34–43.

Green, D.M., C.H. Daugherty, and J.P. Bogart. 1980. Karyology and systematic relationships of the tailed frog *Ascaphus truei*. Herpetologica 36:346–352.

Green, D.M., A.O. Wasserman, and J.P. Bogart. 1981. Karyotypes of the frogs, *Rana septentrionalis* and *R. virgatipes*. Copeia 1981:879–882.

Green, D.M., T. F. Sharbel, J. Kearsley, and H. Kaiser. 1996. Postglacial range fluctuation, genetic subdivision and speciation in the western North American spotted frog complex, *Rana pretiosa*. Evolution 50:374–390.

Green, D.M., H. Kaiser, T.F. Sharbel, J. Kearsley, and K.R. McAllister. 1997. Cryptic species of spotted frogs, *Rana pretiosa* complex, in western North America. Copeia 1997:1–8.

Green, D.M., A.R. Yagi, and S.E. Hamill. 2011. Recovery strategy for the Fowler's Toad (*Anaxyrus fowleri*) in Ontario. Ontario Recovery Strategy Series. Ontario Ministry of Natural Resources, Peterborough, Ontario. vi + 21pp.

Green, N.B. 1938. The breeding habits of *Pseudacris brachyphona* (Cope) with a description of the eggs and tadpole. Copeia 1938:79–82.

Green, N.B. 1948. The spade–foot toad, *Scaphiopus h. holbrookii*, breeding in southern Ohio. Copeia 1948:65.

Green, N.B. 1952. A study of the life history of *Pseudacris brachyphona* (Cope) in West Virginia with special reference to behavior and growth of marked individuals. Ph.D. Dissertation, The Ohio State University, Columbus.

Green, N.B. 1963. The eastern spadefoot toad, *Scaphiopus holbrookii* Harlan, in West Virginia. Proceedings of the West Virginia Academy of Science 35:15–19.

Green, N.B. 1964. Postmetamorphic growth in the mountain chorus frog, *Pseudacris brachyphona* Cope. Proceedings of the West Virginia Academy of Science 36:34–38.

Green, N.B. 1965. Amphibians and Reptiles in West Virginia. Privately published, Marshall University, Huntington, West Virginia. 47 pp.

Green, N.B. 1969. The ratio of crescent and cruciform patterns in populations of the mountain chorus frog, *Pseudacris brachyphona*, in West Virginia. Proceedings of the West Virginia Academy of Science 41:142–144.

Green, N.B., and B. Dowler. 1967. Amphibians and reptiles of the Little Kanawha River basin. Proceedings of the West Virginia Academy of Science 38:50–57.

Green, N.B., and T.K. Pauley. 1987. Amphibians & Reptiles in West Virginia. University of Pittsburgh Press, Pittsburgh, Pennsylvania.

Greenberg, C.H. 1993. Effect of high-intensity wildfire and silvicultural treatments on biotic communities of sand pine scrub. Ph.D. Dissertation, University of Florida, Gainesville.

Greenberg, C.H. 2001a. Response of reptile and amphibian communities to canopy gaps created by wind disturbance in the southern Appalachians. Forest Ecology and Management 148:135–144.

Greenberg, C.H. 2001b. Spatio–temporal dynamics of pond use and recruitment in Florida gopher frogs (*Rana capito aesopus*). Journal of Herpetology 35:74–85.

Greenberg, C.H., and G.W. Tanner. 2004. Breeding pond selection and movement patterns by eastern spadefoot toads (*Scaphiopus holbrookii*) in relation to weather and edaphic conditions. Journal of Herpetology 38:569–577.

Greenberg, C.H., and G.W. Tanner. 2005a. Spatial and temporal ecology of eastern spadefoot toads on a Florida landscape. Herpetologica 61:20–28.

Greenberg, C.H., and G.W. Tanner. 2005b. Chaos and continuity: the role of isolated ephemeral wetlands on amphibian populations in xeric sand hills. Pp.79–90 *In* W.E. Meshaka, Jr. and K.J. Babbitt (eds.), Amphibians and Reptiles. Status and Conservation in Florida. Krieger Publishing, Malabar, Florida.

Greene, A.E., and W.C. Funk. 2009. Sexual selection on morphology in an explosive breeding amphibian, the Columbian spotted frog (*Rana luteiventris*). Journal of Herpetology 43:244–251.

Greenspan, S.E. 2011. Establishment of *Batrachochytrium dendrobatidis* in anuran epidermis and experimental transmission from bullfrogs to wood frogs. M.S. thesis, University of Maine, Orono.

Greenspan, S.E., J.L. Longcore, and A.J.K. Calhoun. 2012. Host invasion by *Batrachochytrium dendrobatidis*: fungal and epidermal ultrastructure in model anurans. Diseases of Aquatic Organisms 100:201–210.

Greenspan, S.E., A.J.K. Calhoun, J.E. Longcore, and M.G. Levy. 2012. Transmission of *Batrachochytrium dendrobatidis* to wood frogs (*Lithobates sylvaticus*) via a bullfrog (*L. catesbeianus*) vector. Journal of Wildlife Diseases 48:575-582.

Greenwell, M., V. Beasley, and L.E. Brown. 1996. The mysterious decline of the cricket frog. Aquaticus 26:48–54.

Greer, A.L., M. Berrill, and P.J.Wilson. 2005. Five amphibian mortality events associated with ranavirus infection in south central Ontario, Canada. Diseases of Aquatic Organisms 67: 9–14.

Gregoire, D.R. 2005. Tadpoles of the southeastern United States coastal plain. U.S. Geological Survey Report, Southeast Ecological Science Center, Gainesville. http://fl.biology.usgs.gov/armi/Guide_to_Tadpoles/guide_to_tadpoles.html

Gregoire, D.R., and M.S. Gunzburger. 2008. Effects of predatory fish on survival and behavior of larval gopher frogs (*Rana capito*) and southern leopard frogs (*Rana sphenocephala*). Journal of Herpetology 42:97–103.

Gregory, P.T. 1979. Predator avoidance behavior of the red–legged frog (*Rana aurora*). Herpetologica 35:175–184.

Greuter, K.L. 2004. Early juvenile ecology of the endangered Houston toad, *Bufo houstonensis* (Anura: Bufonidae). M.S. thesis, Texas State University, San Marcos.

Greuter, K.L., and M.R.J. Forstner. 2003. *Bufo houstonensis* (Houston Toad). Growth. Herpetological Review 34:355–356.

Gridi–Papp, M., and C.O. Gridi–Papp. 2005. Abnormal digits in Strecker's chorus frogs (*Pseudacris streckeri*, Hylidae) from central Texas. Southwestern Naturalist 50:490–494.

Griffin, D.R. 1976. The audibility of frog choruses to migrating birds. Animal Behaviour 24:421–427.

Griffin, P.C. 1999. *Bufo californicus*, arroyo toad movement patterns and habitat preferences. M.S. thesis, University of California, San Diego.

Griffin, P.C., and T.J. Case. 2002. *Bufo californicus* (Arroyo Toad). Predation. Herpetological Review 33:301.

Griffis–Kyle, K.L. 2005a. The effects of mineral nitrogen on embryonic and larval amphibians. Ph.D. Dissertation, Syracuse University, Syracuse, New York.

Griffis–Kyle, K.L. 2005b. Ontogenetic delays in effects of nitrite exposure on tiger salamanders (*Ambystoma tigrinum tigrinum*) and wood frogs (*Rana sylvatica*). Environmental Toxicology and Chemistry 24:1523–1527.

Griffis–Kyle, K.L. 2007. Sublethal effects of nitrite on tiger salamander (*Ambystoma tigrinum tigrinum*) and wood frog (*Rana sylvatica*) embryos and larvae: implications for field populations. Aquatic Ecology 41:119–127.

Griffis–Kyle, K.L. 2009. *Bufo debilis* (Green Toad). Breeding habitat selection. Herpetological Review 40:199–200.

Griffis–Kyle, K.L., and M.E. Ritchie. 2007. Amphibian survival, growth and development in response to mineral nitrogen exposure and predator cues in the field: an experimental approach. Oecologia 152:633–642.

Griffis–Kyle, K.L. and L.M. Navarrete. 2010. *Bufo debilis* (Green Toad). Mortality at a breeding site. Herpetological Review 41:334.

Griffis–Kyle, K.L., S.C. Kyle, and J.M. Jungels. 2011. Use of breeding sites by arid–land toads in rangelands: landscape level factors. Southwestern Naturalist 56:252–255.

Grimké, S., and R.G. Jaeger. 1998. Tadpole bullies: examining mechanisms of competition in a community of larval amphibians. Canadian Journal of Zoology 76:144–153.

Grinnell, J., and C. L. Camp. 1917. A distributional list of the amphibians and reptiles of California. University of California Publications in Zoology 17 (10):127–208.

Grinnell, J., and T.I. Storer. 1921. Reptiles and amphibians of Yosemite National Park. Pp. 175-182 *In* A. F. Hall (ed.) Handbook of Yosemite National Park. G.P. Putnam›s Sons, New York.

Grinnell, J., and T. I. Storer. 1924. Animal Life in the Yosemite. University of California Press, Berkeley.

Grinnell, J., J. Dixon, and J.M. Linsdale. 1930. Vertebrate natural history of a section of northern California through the Lassen Park region. University of California Publications in Zoology 35:1–594.

Grismer, L.L. 2002. Amphibians and Reptiles of Baja California, including its Pacific islands and the islands in the Sea of Cortés. University of California Press, Berkeley.

Groff, L.A. 2011. A species distribution model for guiding Oregon spotted frog (*Rana pretiosa*) surveys near the southern extent of its geographic range M.A. thesis, Humboldt State University, Arcata, California.

Grogan, C.B., and W.L. Grogan, Jr. 2011. *Ollotis alvaria* (Sonoran Desert Toad). Reproduction. Herpetological Review 42:89–90.

Grogan, W.L., Jr. 1974. Notes on *Lampropeltis calligaster rhombomaculata* and *Rana virgatipes*. Bulletin of the Maryland Herpetological Society 10:33–34.

Grogan, W.L., Jr., and P.G. Bystrak. 1973a. Early breeding activity of *Rana sphenocephala* and *Bufo woodhousei fowleri* in Maryland. Bulletin of the Maryland Herpetological Society 9:106.

Grogan, W.L., Jr., and P.G. Bystrak. 1973b. The amphibians and reptiles of Kent Island, Maryland. Bulletin of the Maryland Herpetological Society 9:115–118.

Gromko, M.H., F.S. Mason, and S.J. Smith–Gill. 1973. Analysis of the crowding effect in *Rana pipiens* tadpoles. Journal of Experimental Zoology 186:63–72.

Groner, M. L., and R. A. Relyea. 2011. A tale of two pesticides: How common insecticides affect aquatic communities. Freshwater Biology 56:2391–2404.

Gross, J.A., P.T.J. Johnson, L.K. Prahl, and W.H. Karasov. 2009. Critical period of sensitivity for effects of cadmium on frog growth and development. Environmental Toxicology and Chemistry 28:1227–1232.

Groves, J.D. 1980. Mass predation on a population of the American toad, *Bufo americanus*. American Midland Naturalist 103:202–203.

Groves, J.D., and F. Groves. 1978. Spider predation on amphibians and reptiles. Bulletin of the Maryland Herpetological Society 14:44–46.

Grubb, J.C. 1970. Orientation in post–reproductive Mexican toads, *Bufo valliceps*. Copeia 1970:674–680.

Grubb, J.C. 1972. Differential predation by *Gambusia affinis* on the eggs of seven species of anuran amphibians. American Midland Naturalist 88:103–108.

Grubb, J.C. 1973a. Olfactory orientation in *Bufo woodhousei fowleri*, *Pseudacris clarki* and *Pseudacris streckeri*. Animal Behaviour 21:726–732.

Grubb, J.C. 1973b. Orientation in newly metamorphosed Mexican toads, *Bufo valliceps*. Herpetologica 29:95–100.

Grubb, J.C. 1973c. Olfactory orientation in breeding Mexican toads, *Bufo valliceps*. Copeia 1973:490–497.

Grubb, J.C. 1975. Olfactory orientation in southern leopard frogs, *Rana utricularia*. Herpetologica 31:219–221.

Grubb, J.C. 1976. Maze orientation by the Mexican toads, *Bufo valliceps* (Amphibia, Anura, Bufonidae), using olfactory and configurational cues. Journal of Herpetology 10:97–104.

Gruia–Gray, J., and S. Desser. 1992. Cytopathological and epizootiology of frog erythrocytic virus in bullfrogs (*Rana catesbeiana*). Journal of Wildlife Diseases 28:34–41.

Gruia–Gray, J., M. Petric, and S. Desser. 1989. Ultrastructural, biochemical and biophysical properties of an erythrocytic virus of frogs from Ontario, Canada. Journal of Wildlife Diseases 25:497–506.

Guderyahn, L. 2006. Nationwide assessment of morphological abnormalities observed in amphibians collected from United States National Wildlife Refuges. CBFO–C0601. U.S. Fish and Wildlife Service, Annapolis, Maryland.

Guderyahn, L., S.B. Hager, and L. Scott. 2004. Evidence to support the presence of Cope's gray treefrog (*Hyla chrysoscelis*) at Green Wing Environmental Laboratory in northcentral Illinois. Transactions of the Illinois State Academy of Science 97:219–225.

Guerry, A.D., and M.L. Hunter, Jr. 2002. Amphibian distributions in a landscape of forests and agriculture: an examination of landscape composition and configuration. Conservation Biology 16:745–754.

Gunter, G. 1941. A plague of toads. Copeia 1941:266.

Gunzburger, M.S. 2004. The role of tadpole predation in the habitat distribution of the green treefrog (*Hyla cinerea*). Ph.D. Dissertation, Florida State University, Tallahassee.

Gunzburger, M.S. 2005. Differential predation on tadpoles influences the potential effects of hybridization between *Hyla cinerea* and *Hyla gratiosa*. Journal of Herpetology 39:682–687.

Gunzburger, M.S. 2006. Reproductive ecology of the green treefrog (*Hyla cinerea*) in northwestern Florida. American Midland Naturalist 155:321–328.

Gunzburger, M.S. 2007a. Habitat segregation in two sister taxa of hylid treefrogs. Herpetologica 63:301–310.

Gunzburger, M. S. 2007b. Evaluation of seven aquatic sampling methods for amphibians and other aquatic fauna. Applied Herpetology 4:47–63.

Gunzburger, M.S., and J. Travis. 2004. Evaluating predation pressure on green treefrog larvae across a habitat gradient. Oecologia 140:422–429.

Gunzburger, M.S., and J. Travis. 2005a. Effects of multiple predator species on green treefrog (*Hyla cinerea*) tadpoles. Canadian Journal of Zoology 83:996–1002.

Gunzburger, M.S., and J. Travis. 2005b. Critical literature review of the evidence for unpalatability of amphibian eggs and larvae. Journal of Herpetology 39:547–571.

Gunzburger, M.S., and J. Travis. 2007. Egg clutch characteristics of the barking treefrog, *Hyla gratiosa*, from North Carolina and Florida. Herpetological Review 38:22–24.

Gunzburger, M.S., W.B. Hughes, W.J. Barichivich, and J.S. Staiger. 2010. Hurricane storm surge and amphibian communities in coastal wetlands of northwestern Florida. Wetlands Ecology and Management 18:651–663.

Gupta, N. 2009. Effects of oil sands process–affected water and substrates on wood frog (*Rana sylvatica*) eggs and tadpoles. M.S. thesis, University of Saskatchewan, Saskatoon.

Guscio, C.G., B.R. Hossack, L.A. Eby, and P.S. Corn. 2007. Post–breeding habitat use by adult boreal toads (*Bufo boreas*) after wildfire in Glacier National Park, USA. Herpetological Conservation and Biology 3:55–62.

Guttman, D., J.E. Bramble, and O.J. Sexton. 1991. Observations on the breeding immigration of wood frogs *Rana sylvatica* reintroduced in east–central Missouri. American Midland Naturalist 125:269–274.

Guttman, S.I. 1972. [Cover photo of yellow saddle–backed *Rana clamitans*]. Bioscience 22(4).

Guttman, S.I. 1975. Genetic variation in the genus *Bufo*. Part 2. Isozymes in northern allopatric populations of the American toad *Bufo americanus*. Pp. 679–697 *In* C.L. Markert (ed.), Isozymes. 4. Genetics and Evolution, Academic Press, New York.

Guttman, S.I. 1985. Biochemical studies of anuran evolution. Copeia 1985:292–309.

Guzy, J.C., T.S. Campbell, and K.R. Campbell. 2006. Effects of hydrological alterations on frog and toad populations at Morris Bridge wellfield, Hillsborough County, Florida. Florida Scientist 69:276–287.

Guzy, J.C., E.D. McCoy, A.C. Deyle, S.M. Gonzalez, N. Halstead, and H.R. Mushinsky. 2012. Urbanization interferes with the use of amphibians as indicators of ecological integrity of wetlands. Journal of Applied Ecology 49:941–952.

Haag, H.B. 1931. Toxicological studies of *Derris elliptica* and its constituents. I. Rotenone. Journal of Pharmacology and Experimental Therapeutics 43:193–208.

Haber, V.R. 1926. The food of the Carolina treefrog, *Hyla cinerea* Schneider. Journal of Comparative Psychology 6:189–220.

Hadfield, S. 1966. Observations on body temperatures and activity in the toad *Bufo woodhousei fowleri*. Copeia 1966:581–582.

Haertel, J.D., and R.M. Storm. 1970. Experimental hybridization between *Rana pretiosa* and *Rana cascadae*. Herpetologica 26:436–446.

Haertel, J.D., A. Owczarzak, and R.M. Storm. 1974. A comparative study of the chromosomes from five species of the genus *Rana* (Amphibia: Salientia). Copeia 1974:109–114.

Haggerty, C. 2010. Anuran and tree community structure of cypress domes in Tampa, Florida, relative to time since incorporation within the urban landscape. M.S. thesis, University of South Florida, Tampa.

Hahn, D. 1968. A biogeographic analysis of the herpetofauna of the San Luis Valley, Colorado. M.S. thesis, Louisiana State University, Baton Rouge.

Haig, J.A. 1937. Climbing to success on frog legs. Nation's Business 25:24-27, 178-181. [May, 1937].

Hailman, J.P. 1982. Extremely low ambient light levels of *Ascaphus truei*. Journal of Herpetology 16:83–84.

Hailman, J.P. 1984. Bimodal nocturnal activity of the western toad (*Bufo boreas*) in relation to ambient illumination. Copeia 1984:283–290.

Hailman, J.P., and R.G. Jaeger. 1974. Phototactic responses to spectrally dominant stimuli and use of colour vision by adult anuran amphibians: a comparative survey. Animal Behaviour 22:757–795.

Hailman, J.P., and R.G. Jaeger. 1978. Phototactic responses of anuran amphibians to monochromatic stimuli of equal quantum intensity. Animal Behaviour 26:274–281.

Haislip, N.A., M.J. Gray, J.T. Hoverman, and D.L. Miller. 2011. Development and disease: how susceptibility to an emerging pathogen changes through anuran development. PLoS One 6(7): e22307.

Haislip, N. A., J. T. Hoverman, D. L. Miller, and M. J. Gray. 2012. Natural stressors and disease risk: does the threat of predation increase amphibian susceptibility to ranavirus? Canadian Journal of Zoology 90:893–902.

Hale, S.F., F. Retes, and T.R. Van Devender. 1977. New populations of *Rana tarahumarae* (Tarahumara frog) in Arizona. Journal of the Arizona Academy of Sciences 11:133–134.

Hale, S.F., C.R. Schwalbe, J.L. Jarchow, C.J. May, C.H. Lowe, and T.B. Johnson. 1995. Disappearance of the Tarahumara frog. Pp. 138–140 *In* E.T. LaRoe, G.S. Farris, C.E. Puckett, P.D. Doran, and M.J. Mac (eds.), Our Living Resources. A Report to the Nation on the Distribution, Abundance, and Health of U.S. Plants, Animals, and Ecosystems. National Biological Service, Washington, D.C.

Hale, S.F., P.C. Rosen, J.L. Jarchow, and G.A. Bradley. 2005. Effects of the chytrid fungus on the Tarahumara frog (*Rana tarahumarae*) in Arizona and Sonora, Mexico. Pp. 407–411 *In* G.J. Gottfried, B.S. Gebow, L.G. Eskew, C.B. Edminster, and B. Carelton (compilers), Connecting Mountain Islands and Desert Seas. Biodiversity and Management of the Madrean Archipelago. 2. USDA Forest Service Proceedings RMRS–P–36, Fort Collins, Colorado.

Hall, D.G. 1948. The Blowflies of North America. The Thomas Say Foundation, Baltimore.

Hall, J.A. 1993. Post–embryonic ontogeny and larval behavior of the spadefoot toad, *Scaphiopus intermontanus* (Anura: Pelobatidae). Ph.D. Dissertation, Washington State University, Pullman.

Hall, J.A., J.H. Larsen, Jr., D.E. Miller, and R.E. Fitzner. 1995. Discrimination of kin– and diet–based cues by larval spadefoot toads, *Scaphiopus intermontanus* (Anura: Peolobatidae), under laboratory conditions. Journal of Herpetology 29:233–243.

Hall, J.A., J.H. Larsen, Jr., and R.E. Fitzner. 1997. Postembryonic ontogeny of the spadefoot toad, *Scaphiopus intermontanus* (Anura: Pelobatidae): external morphology. Herpetological Monographs 11:124–178.

Hall, J.A., J.H. Larsen, Jr., and R.E. Fitzner. 2002. Morphology of the premetamorphic larva of the spadefoot toad, *Scaphiopus intermontanus* (Anura: Pelobatidae), with an emphasis on the lateral line system and mouthparts. Journal of Morphology 252:114–130.

Hall, J.G. 2006. Herpetofaunal sampling in the North Carolina coastal plain: a comparison between techniques across habitats. M.S. thesis, East Carolina University, Greenville.

Hall, R.J. 1990. Accumulation, metabolism and toxicity of parathion in tadpoles. Bulletin of Environmental Contamination and Toxicology 44:629–635.

Hall, R.J. 1994. Herpetofaunal diversity of the Four Holes Swamp, South Carolina. USDI National Biological Survey, Resource Publication 198.

Hall, R.J., and D. Swineford. 1979. Uptake of methoxychlor from food and water by the American toad (*Bufo americanus*). Bulletin of Environmental Contamination and Toxicology 23:335–337.

Hall, R.J., and D. Swineford. 1980. Toxic effects of endrin and toxaphene on the southern leopard frog *Rana sphenocephala*. Environmental Pollution (Series A) 23:53–65.

Hall, R.J., and E. Kolbe. 1980. Bioconcentration of organophosphorus pesticides to hazardous levels by amphibians. Journal of Toxicological and Environmental Health 6:853–860.

Hall, R.J., and D. Swineford. 1981. Acute toxicities of toxaphene and endrin to larvae of seven species of amphibians. Toxicology Letters 8:331–336.

Hall, R.J., and B.M. Mulhern. 1984. Are anuran amphibians heavy metal accumulators? Pp. 123–133 *In* R.A. Seigel, L.E. Hunt, J.L. Knight, L. Malaret, and N.L. Zuschlag (eds.), Vertebrate Ecology and Systematics. A Tribute to Henry S. Fitch. Museum of Natural History, University of Kansas, Lawrence, Kansas.

Hall, R.J., and P.F.P. Henry. 1992. Assessing effcts of pesticides on amphibians and reptiles: status and needs. Herpetological Journal 2:65–71.

Hallinan, T. 1923. Observations made in Duval County, northern Florida, on the gopher tortoise (*Gopherus polyphemus*). Copeia (115):11–20.

Hallowell, E. 1857 ("1856"). Notice of a collection of reptiles from Kansas and Nebraska, presented to the Academy of Natural Sciences by Dr. Hammond. Proceedings of the Academy of Natural Sciences of Philadelphia 8:238–253.

Hallowell, E. 1859. Explorations and surveys for a railroad route from the Mississippi River to the Pacific Ocean. Report in California for railroad routes to connect with the routes near the 35th and 32d parallels of north latitude, explored by Lieutenant R.S. Williamson, Corps of Topographical Engineers, 1853. Part 4. Zoological Report. No. 1. Report upon reptiles of the route. War Department, Washington, D.C. Pp. 1–27 + 10 plates.

Halverson, M.A. 2004. Relatedness and the ecology and evolution of the wood frog. Ph.D. Dissertation, Yale University, New Haven, Connecticut.

Halverson, M.A., D.K. Skelly, J.M. Kiesecker, and L.K. Freidenburg. 2003. Forest mediated light regime linked to amphibian distribution and performance. Oecologia 134:360–364.

Halverson, M. A., D. K. Skelly, and A. Caccone. 2006. Kin distribution of amphibian larvae in the wild. Molecular Ecology 15:1139–1145.

Hamel, J. 2009. *Anaxyrus americanus* (American Toad). Egg cannibalism. Herpetological Review 40:67–68.

Hamer, A.J., and M.J. McDonnell. 2008. Amphibian ecology and conservation in the urbanizing world: a review. Biological Conservation 141:2432–2449.

Hamerstrom, F.M., Jr., and F. Hamerstrom. 1951. Food of young raptors on the Edwin S. George Reserve. Wilson Bulletin 63:16–25.

Hamilton, H.L. 1941. The biological action of rotenone on freshwater animals. Proceedings of the Iowa Academy of Science 48:467–479.

Hamilton, W.J., Jr. 1930. Notes on the food of the American toad. Copeia 1930:45.

Hamilton, W.J., Jr. 1934. The rate of growth of the toad (*Bufo americanus americanus* Holbrook) under natural conditions. Copeia 1934:88–90.

Hamilton, W.J. 1948. The food and feeding behavior of the green frog, *Rana clamitans* Latreille, in New York State. Copeia 1948:203–207.

Hamilton, W.J., Jr. 1954. The economic status of the toad. Herpetologica 10:37–40.

Hamilton, W.J., Jr. 1955. Notes on the ecology of the oak toad in Florida. Herpetologica 11:205–210.

Hamilton, W.J., and J.A. Pollack. 1956. The food of some colubrid snakes from Fort Benning, Georgia. Ecology 37:519–526.

Hammerson, G.A. 1982a. Bullfrog eliminating leopard frogs in Colorado? Herpetological Review 13:115–116.

Hammerson, G.A. 1982b. The first record of *Rana sylvatica* from the southern Rocky Mountains and other early collections of A.E. Beardsley. Herpetological Review 13:10.

Hammerson, G.A. 1999. Amphibians and Reptiles in Colorado. A Colorado Field Guide. 2nd ed. University Press of Colorado, Niwot. [first edition, 1982]

Hammerson, G.A., and L.J. Livo. 1999. Conservation status of the northern cricket frog (*Acris crepitans*) in Colorado and adjacent areas at the northwestern extent of the range. Herpetological Review 30:78–80.

Hammond, J. I., D. K. Jones, P. R. Stephens, and R. A. Relyea. 2012. Phylogeny meets ecotoxicology: Evolutionary patterns in sensitivity to a common insecticide among North American amphibians. Evolutionary Applications 5:593–606.

Hampton, P. M., N. B. Ford, and K. Herriman. 2010. Impacts of active oil pumps and deer feed plots on amphibian and reptile assemblages in a floodplain. American Midland Naturalist 163:44–53.

Hampton, S.H., and E.P. Volpe. 1963. Development and interpopulation variability of the mouthparts of *Scaphiopus holbrooki*. American Midland Naturalist 70:319–328.

Han, B.A. 2008. The effects of an emerging pathogen on amphibian host behaviors and interactions. Ph.D. Dissertation, Oregon State University, Corvallis.

Han, B.A., P.W. Bradley, and A.R. Blaustein. 2008. Ancient behaviors of larval amphibians in response to an emerging fungal pathogen, *Batrachochytrium dendrobatidis*. Behavioral Ecology and Sociobiology 63:241–250.

Han, B.A., C.L. Searle and A.R. Blaustein. 2011. Effects of an infectious fungus, *Batrachochytrium dendrobatidis*, on amphibian predator–prey interactions. PLoS One 6(2): e16675.

Hankinson, T. L. 1917. Amphibians and reptiles of the Charleston region. Transactions of the Illinois State Academy of Science 10: 322–329.

Hanley, G.H. 1993. Enrichment options – California Toad. Animal Keeper's Forum 20:178.

Hanlin, H.G., F.D. Martin, L.D. Wike, and S.H. Bennett. 2000. Terrestrial activity, abundance and species richness of amphibians in managed forests in South Carolina. American Midland Naturalist 143:70–83.

Hannaca, W.L. 1933. The Frog Industry. Past, Present and Future. Chariton Corporation, Chicago, Illinois. 94 pp.

Hannon, S.J., C.A. Paszkowski, S. Boutin, J. DeGroot, S.E. Macdonald, M. Wheatley, and B.R. Eaton. 2002. Abundance and species composition of amphibians, small mammals and songbirds in riparian forest buffer strips of varying widths in the boreal mixedwood of Alberta. Canadian Journal of Forest Research 32:1784–1800.

Hanson, J.A., and J.L. Vial. 1956. Defensive behavior and effects of toxins in *Bufo alvarius*. Herpetologica 12:141–149.

Hansen, K.L. 1957. Movements, area of activity, and growth of *Rana heckscheri*. Copeia 1957:274–277.

Hansen, K.L. 1958. Breeding pattern of the eastern spadefoot toad. Herpetologica 14:57–67.

Hansen, K.L. 1965. Breeding records of the eastern spadefoot toad. Herpetologica 21:303–305.

Hansen, K.L. 1977. The jump of a frog. Florida Naturalist 53(1):6–8.

Harding, J.H. 1997. Amphibians and Reptiles of the Great Lakes Region. University of Michigan Press, Ann Arbor, Michigan.

Harding, J.H., and J.A. Holman. 1992. Michigan Frogs, Toads, and Salamanders. A Field Guide and Pocket Reference. Cooperative Extension Service, Michigan State University, East Lansing, Michigan.

Hardy, D.F. 1959. Chorus structure in the striped chorus frog, *Pseudacris nigrita*. Herpetologica 15:14–16.

Hardy, D.G. 1974. Some population dynamics of *Rana pipiens* in an area of southeastern South Dakota. M.S. thesis, University of South Dakota, Vermillion.

Hardy, J.D. 1953. Notes on the distribution of *Mycrohyla* [sic] *carolinensis* in southern Maryland. Herpetologica 8:162–166.

Hardy, J.D., Jr. 1964a. A new frog, *Rana palustris mansuetii*, subsp. nov. from the Atlantic Coastal Plain. Chesapeake Science 5:91–100.

Hardy, J.D., Jr. 1964b. The spontaneous occurrence of scoliosis in tadpoles of the leopard frog, *Rana pipiens*. Chesapeake Science 5:101–102.

Hardy, J.D. 2009. Eastern Narrow–mouthed Frog *Gastrophryne carolinensis*. Bulletin of the Maryland Herpetological Society 45:118–121.

Hardy, J.D., Jr. 1972a. Amphibians of the Chesapeake Bay region. Chesapeake Science 13:123–128.

Hardy, J.D., Jr. 1972b. Tentative outline for inventory of amphibians: *Hyla cinerea* (Green Treefrog). Chesapeake Science 13:186–190.

Hardy, J.D., Jr., and J.H. Gillespie. 1976. Hybridization between *Rana pipiens* and *Rana palustris* in a modified natural environment. Bulletin of the Maryland Herpetological Society 12:41–55.

Hardy, J.D., and R.J. Burroughs. 1986. Systematic status of the spring peeper, *Hyla crucifer* (Amphibia: Hylidae). Bulletin of the Maryland Herpetological Society 22:68–89.

Hardy, L.M. 1995. Checklist of the amphibians and reptiles of the Caddo Lake watershed in Texas and Louisiana. Bulletin of the Museum of Life Sciences, Louisiana State University in Shreveport, No. 10.

Hardy, L.M. 2004. Genus *Syrrhophus* (Anura: Leptodactylidae) in Louisiana. Southwestern Naturalist 49:263–266.

Hardy, L.M., and L.R. Raymond. 1991. Observations on the activity of the pickerel frog, *Rana palustris* (Anura: Ranidae), in northern Louisiana. Journal of Herpetology 25:220–222.

Hardy, R. 1947. Introduction of the bullfrog into Washington County, Utah. Herpetologica 3:169.

Harestad, A.S. 1985. *Scaphiopus intermontanus* (Great Basin Spadefoot Toad). Mortality. Herpetological Review 16:24.

Harfenist, A., T. Power, K.L. Clark, and D.B. Peakall. 1989. A review and evaluation of the amphibian toxicological literature. Canadian Wildlife Service Technical Report Series No. 61.

Hargis, S.E., M.–K. Harr, C.J. Henderson, W.J. Kim, and G.R. Smith. 2008. Factors influencing the distribution of overwintered bullfrog tadpoles (*Rana catesbeiana*) in two small ponds. Bulletin of the Maryland Herpetological Society 44:39–41.

Hargitt, C.W. 1888. Recent notes on *Scaphiopus holbrookii*. American Naturalist 22:535–537.

Harima, H. 1969. A survey of the herpetofauna of northwestern Florida. M.S. thesis, University of Alabama, Tuscaloosa.

Harkey, G.A., and R.D. Semlitsch. 1988. Effects of temperature on growth, development, and color polymorphism in the ornate chorus frog *Pseudacris ornata*. Copeia 1988:1001–1007.

Harlan, R. 1826. Descriptions of several new species of batrachian reptiles with observations on the larvae of frogs. American Journal of Science and Arts 10:53–66.

Harlan, R. 1835. Medical and Physical Researches: Or Original Memoirs in Medicine, Surgery, Physiology, Geology, Zoology, and Comparative Anatomy, Philadelphia. Lydia R. Bailey, Philadelphia, Pennsylvania.

Harless, M.L., C.J. Huckins, J.B. Grant and T.G. Pypker. 2011. Effects of six chemical deicers on larval wood frogs (*Rana sylvatica*). Environmental Toxicology and Chemistry 30:1637–1641.

Harner, M.J., A.J. Nelson, K. Geluso, and D.M. Simon. 2011. Chytrid fungus in American bullfrogs (*Lithobates catesbeianus*) along the Platte River, Nebraska, USA. Herpetological Review 42:549–551.

Harman, W.J., and A.R. Lawler. 1975. *Dero (Allodero) hylidae*, an oligochaete symbiont in hylid frogs in Mississippi. Transactions of the American Microscopical Society 94:38–42.

Harp, E.M., and J.W. Petranka. 2006. Ranavirus in wood frogs (*Rana sylvatica*): potential sources of transmission within and between ponds. Journal of Wildlife Diseases 42:307-318.

Harper, E.B. 2007. The role of terrestrial habitat in the population dynamics and conservation of pond breeding amphibians. Ph.D. Dissertation, University of Missouri, Columbia.

Harper, E.B., and R.D. Semlitsch. 2007. Density–dependence in the terrestrial life history stage of two anurans. Oecologia 153: 879–889.

Harper, E.B., T.A.G. Rittenhouse, and R.D. Semlitsch. 2008. Demographic consequences of terrestrial habitat loss for pool–breeding amphibians: predicting extinction risks associated with inadequate size of buffer zones. Conservation Biology 22:1205–1215.

Harper, F. 1928. Voices of New England frogs. Bulletin of the Boston Society of Natural History 46:3-9.

Harper, F. 1931a. Amphibians and reptiles of the Athabaska and Great Slave Lakes region. Canadian Field-Naturalist 45:68–70.

Harper, F. 1931b. Notes on two Georgia species of *Pseudacris*. Copeia 1931:159–161.

Harper, F. 1931c. A dweller in the piney woods. Scientific Monthly 32:176–181.

Harper, F. 1932. A voice from the pines. Natural History 32:280–288.

Harper, F. 1933. A tree–frog new to the Atlantic Coastal Plain. Journal of the Elisha Mitchell Scientific Society 48:228–231.

Harper, F. 1935a. Records of amphibians in the southeastern states. American Midland Naturalist 16:275–310.

Harper, F. 1935b. The name of the gopher frog. Proceedings of the Biological Society of Washington 48:79–82.

Harper, F. 1937. A season with Holbrook's chorus frog (*Pseudacris ornata*). American Midland Naturalist 18:260–272.

Harper, F. 1939a. A southern subspecies of the spring peeper (*Hyla crucifera*). Notulae Naturae 27:1–4.

Harper, F. 1939b. Distribution, taxonomy, nomenclature, and habits of the little tree–frog (*Hyla ocularis*). American Midland Naturalist 22:134–149.

Harper, F. 1940. Some works of Bartram, Daudin, Latreille, and Sonnini, and their bearing upon North American herpetological nomenclature. American Midland Naturalist 23:692–723.

Harper, F. 1947. A new cricket frog (*Acris*) from the middle western states. Proceedings of the Biological Society of Washington 60:39–40.

Harper, F. 1955. A new chorus frog (*Pseudacris*) from the eastern United States. Natural History Miscellanea (150):1–6.

Harper, F. 1956. Amphibians and reptiles of the Ungava Peninsula. Proceedings of the Biological Society of Washington 69:93–104.

Harper, F. 1963. Amphibians and reptiles of Keewatin and northern Manitoba. Proceedings of the Biological Society of Washington 76:159–168.

Harris, A.H. 1963. Ecological Distribution of some Vertebrates in the San Juan Basin, New Mexico. Museum of New Mexico Press, Papers in Anthropology 8:1-53, 62-63.

Harris, H.S., Jr. 1968. A survey of albinism in Maryland amphibians and reptiles. Bulletin of the Maryland Herpetological Society 4:57–60.

Harris, H.S., Jr. 1969. Distributional survey: Maryland and the District of Columbia. Bulletin of the Maryland Herpetological Society 5:97–161.

Harris, H.S., Jr. 1975. Distributional survey (Amphibia/Reptilia): Maryland and the District of Columbia. Bulletin of the Maryland Herpetological Society 11:73–167.

Harris, H.S., Jr., and K. Crocetti. 2008. The eastern spadefoot toad, *Scaphiopus holbrookii*, in Maryland. Bulletin of the Maryland Herpetological Society 44:107–110.

Harris, L.D., and C.R. Vickers. 1984. Some faunal community characteristics of cypress ponds and the changes induced by perturbations. Pp. 171–185 *In* K.C. Ewel and H.T. Odum (eds.), Cypress Swamps. University Presses of Florida, Gainesville.

Harris, M.L., C.A. Bishop, J. Struger, M.R. van den Heuvel, G.J. van der Kraak, D.G. Dixon, B. Ripley, and J.P. Bogart. 1998a. The functional integrity of northern leopard frog (*Rana pipiens*) and green frog (*Rana clamitans*) populations in orchard wetlands. 1. Genetics, physiology, and biochemistry of breeding adults and young–of–the–year. Environmental Toxicology and Chemistry 17:1338–1350.

Harris, M.L., C.A. Bishop, J. Struger, B. Ripley, and J.P. Bogart. 1998b. The functional integrity of northern leopard frog (*Rana pipiens*) and green frog (*Rana clamitans*) populations in orchard wetlands. 2. Effects of pesticides and eutrophic conditions on early life stage development. Environmental Toxicology and Chemistry 17:1351–1363.

Harris, M.L., L. Chora, C.A. Bishop, and J.P. Bogart. 2000. Species– and age–related differences in susceptibility to pesticide exposure for two amphibians, *Rana pipiens* and *Bufo americanus*. Bulletin of Environmental Contamination and Toxicology 64:263–270.

Harris, R.N., R.M. Brucker, J.B. Walke, M.H. Becker, C.R. Schwantes, D.C. Flaherty, B.A. Lam, D.C. Woodhams, C.J. Briggs, V.T. Vredenburg, and K.P.C. Minbiole. 2009. Skin microbes on frogs prevent morbidity and mortality caused by a lethal skin fungus. ISME Journal 3:818–824.

Harris, R.T. 1975. Seasonal activity and microhabitat utilization in *Hyla cadaverina* (Anura: Hylidae). Herpetologica 31:236–239.

Harris, S.M. 2006. Habitat selection by the relict leopard frog (*Rana onca*): assessment of vegetation use at two scales. M.S. thesis, University of Nevada, Las Vegas.

Harrison, J.R., J.W. Gibbons, D.H. Nelson, and C.L. Abercrombie III. 1979. Status report: amphibians. Pp. 73–78 *In* D.M. Forsythe, and W.B. Ezell, Jr. (eds.), Proceedings of the First South Carolina Endangered Species Symposium. South Carolina Wildlife and Marine Resources Department.

Hartman, F.A. 1906. Food habits of Kansas lizards and batrachians. Transactions of the Kansas Academy of Science 20:225–229.

Hartson, R.D., S.A. Orlofske, V. Melin, R.T. Dillon Jr., and P.T.J. Johnson. 2011. Land use and wetland spatial position jointly determine amphibian parasite communities. EcoHealth 8:485–500.

Harwood, P.D. 1931. The helminths parasitic in the Amphibia and Reptilia of Houston, Texas, and vicinity. Ph.D. Dissertation, Rice University, Houston, Texas.

Harwood, P.D. 1932. The helminths parasitic in the Amphibia and Reptilia of Houston, Texas, and vicinity. Proceedings of the United States National Museum 81(2940):1-71.

Hasken, J., J.L. Newby, A.M. Grelle, J. Boling, J. Estes, L.K. Garey, T. Wilmes, R. McKee, D. Gomez, T. Jackson, N. Gibson, E. Davinroy, C.E. Montgomery, and M.I. Kelrick. 2009. Evaluation of chytrid infection level in a newly discovered population of *Anaxyrus boreas* in the Rio Grande National Forest, Colorado, USA. Herpetological Review 40:426–428.

Hassinger, D.D. 1970. Notes on the thermal properties of frog eggs. Herpetologica 26:49–51.

Hassinger, D.D. 1972. Early life history and ecology of three congeneric species of *Rana* in New Jersey. Ph.D. Dissertation, Rutgers University, Newark, New Jersey.

Hatch, A.C., and A.R. Blaustein. 2000. Combined effects of UV–B, nitrate, and low pH reduce the survival and activity level of larval Cascades Frogs (*Rana cascadae*). Archives of Environmental Contamination and Toxicology 39:494–499.

Hatch, A.C., and A.R. Blaustein. 2003. Combined effects of UV–B radiation and nitrate fertilizer on larval amphibians. Ecological Applications 13:1083–1093.

Hatch, A.C., L.K. Belden, E. Scheessele, and A.R. Blaustein. 2001. Juvenile amphibians do not avoid potentially lethal levels of urea on soil substrate. Environmental Toxicology and Chemistry 20:2328–2335.

Hatfield, J.S., A.H. Price, D.D. Diamond, and C.D. True. 2004. Houston toad (*Bufo houstonensis*) in Bastrop County, Texas: need for protecting multiple subpopulations. Pp. 292-298 *In* H.R. Akcakaya, M.A. Burgman, O. Kindvall, C.C. Wood, P. Sjogren–Gulve, J.S. Hatfield, and M.A. McCarthy (eds.), Species Conservation and Management: Case Studies. Oxford University Press, Oxford, United Kingdom.

Hausfater, G., H.C. Gerhardt, and G.M. Klump. 1990. Parasites and mate choice in gray treefrogs, *Hyla versicolor*. American Zoologist 30:299–311.

Hawkes, V.C., and P.T. Gregory. 2012. Temporal changes in the relative abundance of amphibians relative to riparian buffer width in western Washington, USA. Forest Ecology and Management 274:67–80.

Hawkins, C.P., L.J. Gottschalk, and S.S. Brown. 1988. Densities and habitat of tailed frog tadpoles in small streams near Mt. St. Helens following the 1980 eruption. Journal of the North American Benthological Society 7:246–252.

Hay, O.P. 1887. The amphibians and reptiles of Indiana. Annual Report of the Indiana State Board of Agriculture 28:201-227.

Hay, O.P. 1889. Notes on the life–history of *Chorophilus triseriatus*. American Naturalist 23:770–774.

Hay, O.P. 1892. The batrachians and reptiles of the State of Indiana. Indiana Department of Geology and Natural

Resources, Seventeeth Annual Report, Indianapolis. Pp. 409–609.

Hay, R. 1998. Blanchard's cricket frogs in Wisconsin: a status report. Pp. 79–83 *In* M.J. Lannoo (ed.), Status & Conservation of Midwestern Amphibians. University of Iowa Press, Iowa City.

Hay, W.P. 1908. A list of the batrachians and reptiles of the District of Columbia and vicinity. Proceedings of the Biological Society of Washington 15:121-145.

Hayes, F.E. 1989. Antipredator behavior of recently metamorphosed toads (*Bufo a. americanus*) during encounters with garter snakes (*Thamnophis s. sirtalis*). Copeia 1989:1011–1015.

Hayes, M.P., and F.S. Cliff. 1982. A checklist of the herpetofauna of Butte County, the Butte Sink, and Sutter Buttes, California. Herpetological Review 13:85–87.

Hayes, M.P., and M.M. Miyamoto. 1984. Biochemical, behavioral and body size differences between the red–legged frogs, *Rana aurora aurora* and *Rana aurora draytonii*. Copeia 1984:1018–1022.

Hayes, M.P., and M.R. Tennant. 1985. Diet and feeding behavior of the California red–legged frog, *Rana aurora draytonii* (Ranidae). Southwestern Naturalist 30:601–605.

Hayes, M.P., and M.R. Jennings. 1986. Decline of ranid frog species in western North America: are bullfrogs (*Rana catesbeiana*) responsible? Journal of Herpetology 20:490–509.

Hayes, M.P., and M.R. Jennings. 1988. Habitat correlates of distribution of the California red–legged frog (*Rana aurora draytonii*) and the foothill yellow–legged frog (*Rana boylii*): implications for management. Pp. 144–158 *In* R.C. Szaro, K.E. Severson, and D.R. Patton (eds.), Management of Amphibians, Reptiles, and Small Mammals in North America. USDA Forest Service General and Technical Report RM–166.

Hayes, M.P., and D.M. Krempels. 1986. Vocal sac variation among frogs of the genus *Rana* from western North America. Copeia 1986:927–936.

Hayes, M.P., and C.B. Hayes. 2003. *Rana aurora aurora* (Northern Red–legged Frog). Juvenile growth. Male size at maturity. Herpetological Review 34:233–234.

Hayes, M.P., and C.B. Hayes. 2004. *Bufo boreas boreas* (Boreal Toad). Scats and behavior. Herpetological Review 35:369–370.

Hayes, M.P., and C.J. Rombough. 2004. *Rana aurora* (Northern Red–legged Frog). Predation. Herpetological Review 35:375–376.

Hayes, M.P., and R.F. Price. 2009. *Bufo boreas boreas* (Western Toad). Predation. Herpetological Review 40:68–69.

Hayes, M.P., C.A. Pearl, and C.J. Rombough. 2001. *Rana aurora aurora* (Northern Red–legged Frog). Movement. Herpetological Review 32:35–36.

Hayes, M.P., C.B. Hayes, and J.P. Schuett–Hames. 2004. *Rana aurora aurora* (Northern Red– legged Frog). Vocalization. Herpetological Review 35:52–53.

Hayes, M.P., C.J. Rombough, C.B. Hayes, and J.D. Engler. 2005. *Rana pretiosa* (Oregon Spotted Frog). Predation. Herpetological Review 36:307.

Hayes, M.P., J.D. Engler, and C.J. Rombaugh. 2006a. *Rana pretiosa* (Oregon Spotted Frog). Predation. Herpetological Review 37:209–210.

Hayes, M.P., M.R. Jennings, and G.B. Rathbun. 2006b. *Rana draytonii* (California Red–legged Frog). Prey. Herpetological Review 37:449.

Hayes, M.P., T. Quinn, D.J. Dugger, T.L. Hicks, M.A. Melchiors, and D.E. Runde. 2006c. Dispersion of coastal tailed frog (*Ascaphus truei*): an hypothesis relating occurrence of frogs in non–fish–bearing headwater basins to their seasonal movements. Journal of Herpetology 40:531–543.

Hayes, M.P., C.J. Rombough, and C.B. Hayes. 2007. *Rana aurora* (Northern Red–legged Frog). Movement. Herpetological Review 38:192–193.

Hayes, M.P., C.J. Rombough, G.E. Padgett–Flohr, L.A. Hallock, J.E. Johnson, R.S. Wagner, and J.D. Engler. 2009. Amphibian chytridiomycosis in the Oregon spotted frog (*Rana pretiosa*) in Washington state, USA. Northwestern Naturalist 90:148–151.

Hayes, T.B. 2004. There is no denying this: defusing the confusion about atrazine. Bioscience 54:1138–1149.

Hayes, T.B., K. Haston, M. Tsui, A. Hoang, C. Haeffele, and A. Vonk. 2002. Feminization of male frogs in the wild. Nature 419:895–896.

Hayes, T., K. Haston, M. Tsui, A. Hoang, C. Haeffele, and A. Vonk. 2003. Atrazine–induced hermaphroditism at 0.1 ppb in American leopard frogs (*Rana pipiens*): laboratory and field evidence. Environmental Health Perspectives 111:568–575.

Hayes, T.B., P. Falso, S. Gallipeau, and M. Stice. 2010. The cause of global amphibian declines: a developmental endocrinologist's perspective. Journal of Experimental Biology 213:921–933.

Hayes–Odum, L.A. 1990. Observations on reproduction and embryonic development in *Syrrhophus cystignathoides campi* (Anura: Leptodactylidae). Southwestern Naturalist 35:358–361.

Haynes, C.M., and S.D. Aird. 1981. The distribution and habitat requirements of the wood frog (Ranidae: *Rana sylvatica* Le Conte) in Colorado. Colorado Division of Wildlife, Special Report 50, Fort Collins.

Hays, M.R. 1955. Ultragulosity in the frog *Rana aurora draytoni*. Herpetologica 11:153.

Heard, M. 1904. A California frog ranch. Out West 21:20–27.

Heath, D.R., D.A. Saugey, and G.A. Heidt. 1986. Abandoned mine fauna of the Ouachita Mountains, Arkansas: vertebrate taxa. Proceedings of the Arkansas Academy of Science 40:33–36.

Heatwole, H. 1961. Habitat selection and activity of the wood frog, *Rana sylvatica* Le Conte. American Midland Naturalist 66:301–313.

Heatwole, H., and L.L. Getz. 1960. Studies on the amphibians and reptiles of Mud Lake Bog in southern Michigan. The Jack-Pine Warbler 38:107–112.

Heatwole, H., and A. Heatwole. 1968. Motivational aspects of feeding behavior in toads. Copeia 1968:692–698.

Heatwole, H., N. Poran, and P. King. 1999. Ontogenetic changes in the resistance of bullfrogs (*Rana catesbeiana*) to the venom of copperheads (*Agkistrodon contortrix contortrix*) and cottonmouths (*Agkistrodon piscivorus piscivorus*). Copeia 1999:808–814.

Hebard, W.B., and R.B. Brunson. 1963. Hind limb anomalies of a western Montana population of the Pacific tree frog, *Hyla regilla* Baird and Girard. Copeia 1963:570–572.

Hecht, M.K., and B.L. Matalas. 1946. A review of middle North American toads of the genus *Microhyla*. American Museum Novitates 1315:1–21.

Hecht–Kardasz, K.A. 2012. Geographic distribution: *Eleutherodactylus planirostris* (Greenhouse Frog). Herpetological Review 43:613.

Hecnar, S.J. 1995. Acute and chronic toxicity of ammonium nitrate fertilizer to amphibians from southern Ontario. Environmental Toxicology and Chemistry 14:2131–2137.

Hecnar, S.J. 1997. Amphibian pond communities in southwestern Ontario. SSAR Herpetological Conservation 1:1–15.

Hecnar, S.J., and R.T. M'Closkey. 1996a. Regional dynamics and the status of amphibians. Ecology 77:2091–2097.

Hecnar, S.J., and R.T. M'Closkey. 1996b. Amphibian species richness and distribution in relation to pond water chemistry in south–western Ontario, Canada. Freshwater Biology 36:7–15.

Hecnar, S.J., and R.T. M'Closkey. 1997a. Patterns of nestedness and species association in a pond–dwelling amphibian fauna. Oikos 80:371–381.

Hecnar, S.J., and R.T. M'Closkey. 1997b. The effects of predatory fish on amphibian species richness and distribution. Biological Conservation 79:123–131.

Hecnar, S.J., and R.T. M'Closkey. 1997c. Changes in the composition of a ranid frog community following bullfrog extinction. American Midland Naturalist 137:145–150.

Hecnar, S.J., and R.T. M'Closkey. 1998. Species richness patterns of amphibians in southwestern Ontario ponds. Journal of Biogeography 25:763–772.

Hecnar, S.J., and D.R. Hecnar. 1999. *Pseudacris triseriata* (Western Chorus Frog). Reproduction. Herpetological Review 30:38.

Hecnar, S.J., G.S. Casper, R.W. Russell, D.R. Hecnar, and J.N. Robinson. 2002. Nested species assemblages of amphibians and reptiles on islands in the Laurentian Great Lakes. Journal of Biogeography 29:475–489.

Hedeen, S.E. 1967. Feeding behavior of the great blue heron in Itasca State Park, Minnesota. Loon 39:116–120.

Hedeen, S.E. 1971a. Growth of the tadpoles of the mink frog, *Rana septentrionalis*. Herpetologica 27:160–165.

Hedeen, S.E. 1971b. Body temperatures of the mink frog, *Rana septentrionalis* Baird. Journal of Herpetology 5:211–212.

Hedeen, S.E. 1972a. Postmetamorphic growth and reproduction of the mink frog, *Rana septentrionalis* Baird. Copeia 1972:169–175.

Hedeen, S.E. 1972b. Escape behavior and causes of death of the mink frog, *Rana septentrionalis*. Herpetologica 28:261–262.

Hedeen, S.E. 1972c. Food and feeding behavior of the mink frog, *Rana septentrionalis* Baird, in Minnesota. American Midland Naturalist 88:291–300.

Hedeen, S.E. 1976. Scoliosis and hydrops in a *Rana catesbeiana* (Amphibia, Anura, Ranidae) tadpole population. Journal of Herpetology 10:261–262.

Hedeen, S.E. 1986. The southern geographic limit of the mink frog, *Rana septentrionalis*. Copeia 1986:239–244.

Hedge, T.A., K.E. Saunders, and C.A. Ross. 2002. Innovative housing and environmental enrichment for bullfrogs (*Rana catesbeiana*). Contemporary Topics in Laboratory Animal Science 41:120–121.

Hedges, S.B. 1986. An electrophoretic analysis of Holarctic hylid frog evolution. Systematic Zoology 35:1–21.

Hedges, S.B. 1989. Evolution and biogeography of West Indian frogs of the genus *Eleutherodactylus*: slow–evolving loci and the major groups. Pp. 305–370 *In* C.A. Woods (ed.), Biogeography of the West Indies. Past, Present, and Future. Sandhill Crane Press, Gainesville, Florida.

Hedges, S.B., W.E. Duellman, and M.P. Heinicke. 2008. New World direct–developing frogs (Anura: Terrarana): molecular phylogeny, classification, biogeography, and conservation. Zootaxa 1737:1–182.

Hedtke, S.F., and F.A. Puglisi. 1982. Short–term toxicity of five oils to four freshwater species. Archives of Environmental Contamination and Toxicology 11:425–430.

Heemeyer, J.L. 2011. Breeding migrations, survivorship, and obligate crayfish burrow use by adult crawfish frogs (*Lithobates areolatus*). M.S. thesis, Indiana State University, Terre Haute.

Heemeyer, J.L., and M.J. Lannoo. 2011. *Lithobates areolatus circulosus* (Northern Crawfish Frog). Winterkill. Herpetological Review 42:261–262.

Heemeyer, J.L., and M.J. Lannoo. 2012. Breeding migrations in crawfish frogs (*Lithobates areolatus*): long–distance movements, burrow philopatry, and mortality in a near–threatened species. Copeia 2012:440–450.

Heemeyer, J.L., J.G. Palis, and M.J. Lannoo. 2010. *Lithobates areolatus circulosus* (Northern Crawfish Frog). Predation. Herpetological Review 41:475.

Heemeyer, J.L., P.J. Williams, and M.J. Lannoo. 2012. Obligate crayfish burrow use and core habitat requirements of crawfish frogs. Journal of Wildlife Management 76:1081–1091.

Hefleck, S.K. 2001. Ecology of the exotic Cuban tree frog, *Osteopilus septentrionalis*, within Brevard County, Florida. M.S. thesis, Florida Institute of Technology, Melbourne.

Heilman, B. 1963. The frogs of spring are springing for their lives. Sports Illustrated 18:46–48; 62–64.

Heinen, J.T. 1985. Cryptic behavior in juvenile toads. Journal of Herpetology 19:524–527.

Heinen, J.T. 1992. Behavioral anti-predator strategies in newly-metamorphosed American toads (*Bufo americanus*) in response to predation risk by eastern garter snakes (*Thamnophis sirtalis*). Ph.D. Dissertation, University of Michigan, Ann Arbor.

Heinen, J.T. 1993a. Aggregations of newly metamorphosed *Bufo americanus*: tests of two hypotheses. Canadian Journal of Zoology 71:334–338.

Heinen, J.T. 1993b. Substrate choice and predation risk in newly metamorphosed American toads *Bufo americanus*: an experimental analysis. American Midland Naturalist 130:184–192.

Heinen, J.T. 1994a. Antipredator behavior of newly metamorphosed American toads (*Bufo a. americanus*), and mechanisms of hunting by eastern garter snakes (*Thamnophis s. sirtalis*). Herpetologica 50:137–145.

Heinen, J.T. 1994b. The significance of color change in newly metamorphosed American toads. Journal of Herpetology 28:87–93.

Heinen, J.T. 1995. Predator cues and prey responses: a test using eastern garter snakes (*Thamnophis s. sirtalis*) and American toads (*Bufo a. americanus*). Copeia 1995:738–741.

Heinen, J.T., and G. Hammond. 1997. Antipredator behaviors of newly metamorphosed green frogs (*Rana clamitans*) and leopard frogs (*R. pipiens*) in encounters with eastern garter snakes (*Thamnophis s. sirtalis*). American Midland Naturalist 137:136–144.

Heinen, J.T., and J. Abdella. 2005. On the advantages of putative cannibalism in American toad tadpoles (*Bufo a. americanus*): is it active or passive and why? American Midland Naturalist 153:338–347.

Heinicke, M.P., W.E. Duellman, and S.B. Hedges. 2007. Major Caribbean and Central American frog faunas originated by ancient oceanic dispersal. Proceedings of the National Academy of Sciences of the United States of America 104:10092–10097.

Heinicke, M.P., L.M. Diaz, and S.B. Hedges. 2011. Origin of invasive Florida frogs traced to Cuba. Biology Letters 7:407–410.

Heinrich, M.L. 1985. *Pseudacris triseriata triseriata* (Western Chorus Frog). Reproduction. Herpetological Review 16:24.

Heinrich, M.L., and D.W. Kaufman. 1985. Herpetofauna of the Konza Prairie Research Natural Area. Prairie Naturalist 17:101–112.

Hekkala, E.R., R.A. Saumure, J.R. Jaeger, H.–W. Herrman, M.J. Sredl, D.F. Bradford, D. Drabeck, and M.J. Blum. 2011. Resurrecting an extinct species: archival DNA, taxonomy, and conservation of the Vegas Valley leopard frog. Conservation Genetics 12:1379–1385.

Helbing, C.C., K. Ovaska, and L. Ji. 2006. Evaluation of the effect of acetochlor on thyroid hormone receptor gene expression in the brain and behavior of *Rana catesbeiana* tadpoles. Aquatic Toxicology 80:42–51.

Helgen, J. 2012. Peril in the Ponds. University of Massachusetts Press, Amherst.

Helgen, J.C., R.G. McKinnell, and M.C. Gernes. 1998. Investigation of malformed northern leopard frogs in Minnesota. Pp. 288–297 *In* M.J. Lannoo (ed.), Status & Conservation of Midwestern Amphibians. University of Iowa Press, Iowa City.

Helgen, J.C., M.C. Gernes, S.M. Kersten, J.W. Chirhart, J.T. Canfield, D. Bowers, J. Haferman, R.G. McKinnell, and D.M. Hoppe. 2000. Field investigations of malformed frogs in Minnesota 1993–97. Journal of the Iowa Academy of Science 107:96–112.

Heller, C.L. 1960. The Sierra yellow–legged frog. Yosemite Nature Notes 34:126–128.

Heller, J.A. 1927. Brewer's mole as food of the bullfrog. Copeia (165):116.

Hellman, R.E. 1953. A comparative study of the eggs and tadpoles of *Hyla phaeocrypta* and *Hyla versicolor* in Florida. Publications of the Research Division, Ross Allen's Reptile Institute 1:61–74.

Hemesath, L.M. 1998. Iowa's frog and toad survey, 1991–1994. Pp. 206–216 *In* M.J. Lannoo (ed.), Status & Conservation of Midwestern Amphibians. University of

Iowa Press, Iowa City.

Hemmings, V. 2010. Phylogeography of two closely related anurans, the Relict Leopard Frog (*Rana onca*) and Lowland Leopard Frog (*Rana yavapaiensis*). M.S. thesis, University of Nevada–Las Vegas.

Henderson, G.G., Jr. 1961. Reproductive potential of *Microhyla olivacea*. Texas Journal of Science 13:355–356.

Hendricks, F. S. 1973. Intestinal contents of *Rana pipiens* Schreber (Ranidae) larvae. Southwestern Naturalist 18:99–101.

Hennen, S. 1964. The karyotype of *Rana sylvatica* and its comparison with the karyotype of *Rana pipiens*. Journal of Heredity 55:124–128.

Henning, B.M., and A.J. Remsburg. 2009. Lakeshore vegetation effects on avian and anuran populations. American Midland Naturalist 161:123–133.

Henning, J.A., and G. Schirato. 2006. Amphibian use of Chehalis River floodplain wetlands. Northwestern Naturalist 87:209–214.

Henrich, T.W. 1968. Morphological evidence of secondary intergradation between *Bufo hemiophrys* Cope and *Bufo americanus* Holbrook in eastern South Dakota. Herpetologica 24:1–13.

Henshaw, S. 1904. Fauna of New England. 2. List of the batrachia. Occasional Papers of the Boston Society of Natural History No. 7:1–10.

Hensley, F.R. 1993. Ontogenetic loss of phenotypic plasticity of age at metamorphosis in tadpoles.Ecology 74:2405–2412.

Hensley, M. 1959. Albinism in North American amphibians and reptiles. Publications of the Museum, Michigan State University, Biological Series 1:133–159.

Hensley, M. 1962. Another snake recorded in the diet of the bullfrog. Herpetologica 18:141.

Hernandez, J.P. 2010. Priority effects of overwintered *Rana* tadpoles of larval southern toad (*Bufo terrestris* Bonnaterre). M.S. thesis, East Carolina University, Greenville, North Carolina.

Herreid, C.F. II. 1963. Range extension for *Bufo boreas boreas*. Herpetologica 19:218.

Herreid, C.F., and S. Kinney. 1966. Survival of Alaskan wood-frog (*Rana sylvatica*) larvae. Ecology 47:1039–1041.

Herreid, C.F., and S. Kinney. 1967. Temperature and development of the wood frog, *Rana sylvatica*, in Alaska. Ecology 48:579–590.

Herriman, M.W. 1933. Commercial Frog Raising. West Coast Frog Industries, North Hollywood, California. 52 pp.

Herrmann, H.L., K.J. Babbitt, M.J. Baber, and R.G. Congalton. 2005. Effects of landscape characteristics on amphibian distribution in a forest–dominated landscape.

Biological Conservation 123:139–149.

Hersikorn, B.D. 2010. In–situ caged wood frog (*Rana sylvatica*) survival and development in wetlands formed from oil sands process–affected materials. M.S. thesis, University of Saskatchewan, Saskatoon.

Hersikorn, B.D., and J.E.G. Smits. 2011. Compromised metamorphosis and thyroid hormone changes in wood frogs (*Lithobates sylvaticus*) raised on reclaimed wetlands on the Athabasca oil sands. Environmental Pollution 159:596–601.

Hersikorn, B.D., J. J.C. Ciborowski, and J. E.G. Smits. 2010. The effects of oil sands wetlands on wood frogs (*Rana sylvatica*). Toxicology & Environmental Chemistry 92:1513–1527.

Hetherington, T. E. 1985. Role of the opercularis muscle in seismic sensitivity in the bullfrog *Rana catesbeiana*. Journal of Experimental Zoology 235: 27–34.

Hewatt, W.G. 1934. A case of hermaphroditism in *Rana catesbeiana* Shaw. Copeia 1934:185.

Hewitt, H. 1950. The bullfrog as a predator on ducklings. Journal of Wildlife Management 14:244.

Hews, D.K. 1988. Alarm response in larval western toads, *Bufo boreas*: release of larval chemicals by a natural predator and its effect on predator capture efficiency. Animal Behaviour 36:125–133.

Hews, D.K., and A.R. Blaustein. 1985. An investigation of the alarm responses in *Bufo boreas* and *Rana cascadae* tadpoles. Behavioral and Neural Biology 43:47–57.

Heyer, W.R. 1978. Systematics of the fuscus group of the frog genus *Leptodactylus* (Amphibia, Leptodactylidae). Natural History Museum of Los Angeles County, Science Bulletin 29:1–85.

Heyer, W.R. 2002. *Leptodactylus fragilis*, the valid name for the Middle American and northern South American white-lipped frog (Amphibia: Leptodactylidae). Proceedings of the Biological Society of Washington 115:321–322.

Heymann, M.M. 1975. Reptiles and Amphibians of the American Southwest. Doubleshoe Publishers, Scottsdale, Arizona.

Hibbard, C.W., and A.B. Leonard. 1936. The occurrence of *Bufo punctatus* in Kansas. Copeia 1936:114.

Higginbotham, A.C. 1939. Studies on amphibian activity. 1. Preliminary report on the rhythmic activity of *Bufo americanus americanus* Holbrook and *Bufo fowleri* Hinckley. Ecology 20:58–70.

Higgins, C., and C. Sheard. 1926. Effects of ultraviolet radiation on the early development of *Rana pipiens*. Journal of Experimental Zoology 46:333–343.

Higgins, S.A. 2011. Analysis of urea and glycerol as cryoprotectants in the boreal chorus frog (*Pseudacris maculata*). M.S. thesis, University of South Dakota, Vermillion.

Higley, W. K. 1889. Reptilia and Batrachia of Wisconsin. Transactions of the Wisconsin Academy of Sciences, Arts and Letters 7:155–176.

Hildebrand, H. 1949. Notes on *Rana sylvatica* in the Labrador Peninsula. Copeia 1949:168–172.

Hildebrand, W.G. 2009a. Barking Treefrog *Hyla gratiosa*. Bulletin of the Maryland Herpetological Society 45:81–84.

Hildebrand, W.G. 2009b. Gray Treefrog. Bulletin of the Maryland Herpetological Society 45:85–88.

Hildebrand, W.G. 2009c. Green Treefrog *Hyla cinerea*. Bulletin of the Maryland Herpetological Society 45:89–92.

Hildebrand, W.G. 2009d. American Toad *Anaxyrus americanus*. Bulletin of the Maryland Herpetological Society 45:93–95.

Hildebrand, W.G. 2009e. Fowler's Toad *Anaxyrus fowleri*. Bulletin of the Maryland Herpetological Society 45:96–98.

Hildebrand, W.G. 2009f. Northern Cricket Frog *Acris crepitans* Bulletin of the Maryland Herpetological Society 45:99–102.

Hill, H.R. 1948. Amphibians and reptiles of Los Angeles County. Los Angeles County Museum, Science Series 12, Zoology 5:1–30.

Hill, R.L., M.G. Levy, E.K. Tempe, and J.B. Kaylock. 2011. Additional reports of the amphibian chytrid fungus *Batrachochytrium dendrobatidis* from Georgia, USA. Herpetological Review 42:376–378.

Hillery, C.M. 1984. Seasonality of two midbrain auditory responses in the treefrog, *Hyla chrysoscelis*. Copeia 1984:844–852.

Hillis, D.M. 1981. Premating isolating mechanisms among three species of the *Rana pipiens* complex in Texas and southern Oklahoma. Copeia 1981:312–319.

Hillis, D.M. 1982. Morphological differentiation and adaptation of the larvae of *Rana berlandieri* and *Rana sphenocephala* (*Rana pipiens* complex) in sympatry. Copeia 1982:168–174.

Hillis, D.M. 1988. Systematics of the *Rana pipiens* complex: puzzle and paradigm. Annual Review of Ecology and Systematics 19:39–63.

Hillis, D.M. 2007. Constraints in naming parts of the Tree of Life. Molecular Phylogenetics and Evolution 42:331–338.

Hillis, D.M., and S.K. Davis. 1986. Evolution of ribosomal DNA: fifty million years of recorded history in the frog genus *Rana*. Evolution 40:1275–1288.

Hillis, D.M., and T.P. Wilcox. 2005. Phylogeny of the New World true frogs (*Rana*). Molecular Phylogenetics and Evolution 34:299–314.

Hillis, D.M., J.S. Frost, and D.A. Wright. 1983. Phylogeny and biogeography of the *Rana pipiens* complex: a biochemical evaluation. Systematic Zoology 32:132–143.

Hillis, D.M., A.M. Hillis, and R.F. Martin. 1984. Reproductive ecology and hybridization of the endangered Houston toad (*Bufo houstonensis*). Journal of Herpetology 18:56–72.

Hillis, D.M., J.T. Collins, and J.P. Bogart. 1987. Distribution of diploid and tetraploid species of gray tree frogs (*Hyla chrysoscelis* and *Hyla versicolor*) in Kansas. American Midland Naturalist 117:214–217.

Hillman, S.S., P.C. Withers, R.C. Drewes, and S.D. Hillyard. 2009. Ecological and Environmental Physiology of Amphibians. Oxford University Press, Oxford, United Kingdom.

Himes, J.G., and T.W. Bryan. 1998. Geographic distribution, *Bufo americanus charlesmithi* (Dwarf American Toad). Herpetological Review 29:246.

Himmelman, J. 2006. Discovering Amphibians. Frogs and Salamanders of the Northeast. Down East Books, Camden, Maine.

Hinckley, M.H. 1880. Notes on eggs and tadpoles of *Hyla versicolor*. Proceedings of the Boston Society of Natural History 21:104–107.

Hinckley, M.H. 1881. On some differences in the mouth structure of tadpoles of the anourous batrachians found in Milton, Mass. Proceedings of the Boston Society of Natural History 21:307–315.

Hinckley, M.H. 1882. Notes on the development of *Rana sylvatica* LeConte. Proceedings of the Boston Society of Natural History 22:85–95.

Hinckley, M.H. 1884. Notes on the peeping frog, *Hyla pickeringii*, Leconte. Memoirs of the Boston Society of Natural History 3:311–318.

Hine, J.S. 1911. Beneficial habits of snakes, toads and related animals. The Agricultural College Extension Bulletin, The Ohio State University 6(7):3–14.

Hine, R.L., B.L. Les, and B.F. Hellmich. 1981. Leopard frog populations and mortality in Wisconsin, 1974–1976. Wisconsin Department of Natural Resources, Technical Bulletin No. 122.

Hinshaw, S.H., and B.K. Sullivan. 1990. Predation on *Hyla versicolor* and *Pseudacris crucifer* during reproduction. Journal of Herpetology 24:196–197.

Hinther, A., T.M. Edwards, L. J. Guillette, and C. C. Helbing. 2012. Influence of nitrate and nitrite on thyroid hormone responsive and stress–associated gene expression in cultured *Rana catesbeiana* tadpole tail fin tissue. Frontiers in Genetics 3(2012): 51.

Hinty, H.W. 1963. Hydropsic tadpoles of *Rana catesbeiana* from Ohio. Journal of the Ohio Herpetological Society 4:48.

Hird, D.W., S.L. Diesch, R.G. McKinnell, E. Gorham, F.B. Martin, S.W. Kurtz, and C. Dubrovolny. 1981. *Aeromonas hydrophila* in wild–caught frogs and tadpoles (*Rana pipiens*). Laboratory Animal Science 31:166–169.

Hird, D.W., S.L. Dietsch, R.G. McKinnell, E. Gorham, F.B. Martin, C.A. Meadows, and M. Gasiorowski. 1983. Enterobacteriaceae and *Aeromonas hydrophila* in Minnesota frogs and tadpoles (*Rana pipiens*). Applied and Environmental Microbiology 46:1423–1425.

Hirschfeld, C.J., and J.T. Collins. 1961. Notes on Ohio herpetology. Journal of the Ohio Herpetological Society 3:1–4.

Hirschfield, S.E. 1969. Vertebrate fauna of Nichol's Hammock, a natural trap. Quarterly Journal of the Florida Academy of Sciences 31:177–189.

Hisaoka, K.K., and J.C. List. 1957. The spontaneous occurrence of scoliosis in larvae of *Rana sylvatica*. Transactions of the American Microscopical Society 76:381–387.

Höbel, G. 2005a. *Rana palustris* (Pickerel Frog) and *Ambystoma maculatum* (Spotted Salamander). Reproductive behavior. Herpetological Review 36:55–56.

Höbel, G. 2005b. *Rana clamitans* (Green Frog) and *Rana catesbeiana* (American Bullfrog). Reproduction. Herpetological Review 36:439–440.

Höbel, G. 2011. *Hyla cinerea* (Green Treefrog). Cannibalism and defensive posture. Herpetological Review 42:85–86.

Höbel, G., and A. Slocum. 2010. *Hyla cinerea* (Green Treefrog). Morphology. Herpetological Review 41:335–336.

Hoberg, T., and C. Gause. 1992. Reptiles & amphibians of North Dakota. North Dakota Outdoors 55(1):7–19.

Hobson, C.S., and E.C. Moriarty. 2003. Geographic distribution. *Pseudacris nigrita nigrita*. Herpetological Review 34:259–260.

Hobson, J.A. 1966. Electrographic correlates of behavior in the frog with special reference to sleep. Electroencephalography and Clinical Neurophysiology 22:113–121.

Hobson, J.A., C.J. Goin, and O.B. Goin. 1967. Sleep behavior of frogs. Quarterly Journal of the Florida Academy of Sciences 30:184–186.

Hobson, J.A., C.J. Goin, and O.B. Goin. 1968. Electrographic correlates of behavior in treefrogs. Nature 220:386–387.

Hock, R. 1957. Alaskan zoogeography and Alaskan amphibia. Proceedings of the Fourth Alaskan Science Conference 1953, pp. 201–206. Privately Published, Fairbanks.

Hocking, D.J. 2010. *Hyla squirella* (Squirrel Treefrog). Reproduction. Herpetological Review 41:64.

Hocking, D.J., and R.D. Semlitsch. 2007. Effects of timber harvest on breeding–site selection by gray treefrogs (*Hyla versicolor*). Biological Conservation 138:506–513.

Hocking, D.J., and R.D. Semlitsch. 2008. Effects of experimental clearcut logging on gray treefrog (*Hyla versicolor*) tadpole performance. Journal of Herpetology 42:689–698.

Hocking, D.J., T.A.G. Rittenhouse, B.B. Rothermel, J.R. Johnson, C.A. Conner, E.B. Harper, and R.D. Semlitsch. 2008. Breeding and recruitment phenology of amphibians in Missouri oak–hickory woodlands. American Midland Naturalist 160:41–60.

Hodge, R.P. 1976. Amphibians & Reptiles in Alaska, the Yukon & Northwest Territories. Alaska Northwest Publishing Co., Anchorage.

Hoff, J.G., and S.A. Moss. 1974. A distress call in the bullfrog, *Rana catesbeiana*. Copeia 1974:533–534.

Hoffman, A.S., J.L. Heemeyer, P.J. Williams, J.R. Robb, D.R. Karns, V.C. Kinney, N.J. Engbrecht, and M.J. Lannoo. 2010. Strong site fidelity and a variety of imaging techniques reveal around–the–clock and extended activity patterns in crawfish frogs (*Lithobates areolatus*). Bioscience 60:829–834.

Hoffman, E.A., and M.S. Blouin. 2000. A review of color and pattern polymorphisms in anurans. Biological Journal of the Linnean Society 70:633–665.

Hoffman, E.A., and M.S. Blouin. 2004. Evolutionary history of the northern leopard frog: reconstruction of phylogeny, phylogeography, and historical changes in population demography from mitochondrial DNA. Evolution 58:145–159.

Hoffman, E.A., F.W. Schueler, and M.S. Blouin. 2004. Effective population sizes and temporal stability of genetic structure in *Rana pipiens*, the northern leopard frog. Evolution 58:2536–2545.

Hoffman, E.A., F.W. Schueler, A.G. Jones, and M.S. Blouin. 2006. An assessment of selection on a color polymorphism in the northern leopard frog. Molecular Ecology 15:2627–2641.

Hoffman, K.E. 2007. Testing the influence of Cuban treefrogs (*Osteopilus septentrionalis*) on native treefrog detection and abundance. M.S. thesis, University of Florida, Gainesville.

Hoffman, K.E., and S.A. Johnson. 2008. *Osteopilus septentrionalis* (Cuban Treefrog). Diet. Herpetological Review 39:339.

Hoffman, K.E., S.A. Johnson, and M.E. McGarrity. 2009. Interspecific variation in use of polyvinyl chloride (PVC) pipe refuges by hylid treefrogs: a potential source of capture bias. Herpetological Review 40:423–426.

Hoffman, N. 2003. Frog fence along Vermont Rt. 2 in Sandbar Wildlife Management Area. Collaboration,between Vermont Agency of Transportation and Vermont Agency of Natural Resources. International Conference on Ecology and Transportation 2003 Proceedings, pp. 431–432.

Hoffman, R.L. 1946. The voice of *Hyla versicolor* in Virginia. Herpetologica 3:141–142.

Hoffman, R.L. 1955. On the occurrence of two species of hylid frogs in Virginia. Herpetologica 11:30–32.

Hoffman, R.L. 1981. On the occurrence of *Pseudacris brachyphona* (Cope) in Virginia. Catesbeiana 1:9–13.

Hoffman, R.L. 1996. *Hyla chrysoscelis* also crosses the Blue Ridge: sic juvat transcendere montes. Catesbeiana 16:3–8.

Hoffman, R.L., and J.C. Mitchell. 1996. Records of anurans from Greensville County, Virginia. Banisteria 8:29-36.

Hoffpauir, C., and E.O. Morrison. 1966. *Rhabdias ranae* from *Bufo valliceps*. Southwestern Naturalist 11:302.

Hohn, S. 2003. A survey of New York State pet stores to investigate trade in native herpetofauna. Herpetological Review 34:23–27.

Hokit, D.G., and A.R. Blaustein. 1994. The effects of kinship on growth and development in tadpoles of *Rana cascadae*. Evolution 48:1383–1388.

Hokit, D.G., and A.R. Blaustein. 1995. Predator avoidance and alarm–response behaviour in kin–discriminating tadpoles (*Rana cascadae*). Ethology 101:280–290.

Hokit, D.G., and A.R. Blaustein. 1997. The effects of kinship on interactions between tadpoles of *Rana cascadae*. Ecology 78:1722–1735.

Hokit, D.G., and A. Brown. 2006. Distribution patterns of wood frogs (*Rana sylvatica*) in Denali National Park. Northwestern Naturalist 87:128–137.

Holbrook, C.T., and J.W. Petranka. 2004. Ecological interactions between *Rana sylvatica* and *Ambystoma maculatum*: evidence of interspecific competition and facultative intraguild predation. Copeia 2004:932–939.

Holbrook, J. E. 1835. American Herpetology; or a Description of the Reptiles Inhabiting the United States [Preprint sales brochure]. E. J. Van Brunt, Charleston, South Carolina.

Holbrook, J.E. 1836. North American Herpetology, or a Description of the Reptiles Inhabiting the United States. 1st ed., Vol. 1. J. Dobson and Son, Philadelphia, Pennsylvania.

Holbrook, J.E. "1836" [probably 1839]. North American Herpetology, or a Description of the Reptiles Inhabiting the United States. 1st ed., Vol. 1. Second version. J. Dobson and Son, Philadelphia, Pennsylvania.

Holbrook, J. E. 1838. North American Herpetology, or a Description of the Reptiles Inhabiting the United States. 1st ed., Vol. 3. J. Dobson and Son, Philadelphia, Pennsylvania.

Holbrook, J.E. 1840. North American Herpetology, or a Description of the Reptiles Inhabiting the United States. 1st ed. Vol. 4. J. Dobson and Son, Philadelphia, Pennsylvania.

Holbrook, J.E. 1842. North American Herpetology, or a Description of the Reptiles Inhabiting the United States. 2nd ed., Vol. 4. J. Dobson and Son, Philadelphia, Pennsylvania.

Holbrook, J.E. 1849. Reptiles. pp. 13–15 [in Appendix] *In* G. White, Statistics of the State of Georgia: including an account of its Natural, Civil, and Ecclesiastical History. W. Thorne Williams, Savannah.

Holland, A.A. 2002. Evaluating boreal toad (*Bufo boreas*) breeding habitat suitability. M.S. thesis, Colorado State University, Fort Collins.

Holland, A.A., K.R. Wilson, and M.S. Jones. 2006. Characteristics of boreal toad (*Bufo boreas*) breeding habitat in Colorado. Herpetological Review 37:157–159.

Holland, M. P., D. K. Skelly, M. Kashgarian, S. R. Bolden, L. M. Harrison and M. Cappello. 2007. Echinostome infection in green frogs (*Rana clamitans*) is stage and age dependent. Journal of Zoology 271:455–462.

Hollenbeck, R.R. 1974. Growth rates and movements within a population of *Rana pretiosa pretiosa* Baird and Girard in south central Montana. Ph.D. Dissertation, Montana State University, Bozeman.

Hollingsworth, B. 2002. Amphibians and reptiles in the headwaters of the San Diego River. San Diego Natural History Museum, Field Notes 16(2):18-20.

Hollis, D.M. 1998. Acoustic Relationships of the western toad, *Bufo boreas*, and the Yosemite toad, *Bufo canorus*: vocalization and Its role in natural hybridization. M.S. thesis, Fresno State University, Fresno, California.

Hollis, P. 1972. A survey of the parasites of the bullfrog, *Rana catesbeiana* Shaw, in central east Texas. Southwestern Naturalist 17:198–200.

Holloway, A.K., D.C. Canatella, H.C. Gerhardt, and D.M. Hillis. 2006. Polyploids with different origins and ancestors form a single sexual polyploid species. American Naturalist 167:E88–E101.

Holman, J.A. 1957. Bullfrog predation on the eastern spadefoot, *Scaphiopus holbrooki*. Copeia 1957:229.

Holman, J.A. 1958. Notes on reptiles and amphibians from Florida caves. Herpetologica 14:179–180.

Holman, J.A. 1961. Amphibians and reptiles of the Howard College Natural Area. Journal of the Alabama Academy of Science 32:77-87.

Holman, J.A. 1967. Additional Miocene anurans from Florida. Quarterly Journal of the Florida Academy of Sciences 30:121–140.

Holman, J.A. 1969. The Pleistocene amphibians and reptiles of Texas. Publications of the Museum, Michigan State University, Biological Series 4:161–192.

Holman, J.A. 2003. Fossil Frogs and Toads of North America. Indiana University Press, Bloomington.

Holman, J.A. 2012. The Amphibians and Reptiles of Michigan. Wayne State University Press, Detroit, Michigan.

Holman, J.A., and H.W. Campbell. 1958. An interesting feeding activity of *Microhyla carolinensis.* Herpetologica 14:205.

Holman, J.A., H.O. Jackson, and W.H. Hill. 1964. *Pseudacris streckeri illinoensis* Smith from extreme southern Illinois. Herpetologica 20:205.

Holomuzki, J.R. 1995. Oviposition sites and fish–deterrent mechanisms of two stream anurans. Copeia 1995:607–613.

Holomuzki, J.R. 1997. Habitat–specific life–histories and foraging by stream–dwelling American toads. Herpetologica 53:445–453.

Holomuzki, J.R., and N. Hemphill. 1996. Snail–tadpole interactions in streamside pools. American Midland Naturalist 136:315–327.

Holycross, A.T., B.G. Fedorko, and O. Fourie. 1999. *Bufo alvarius* (Colorado River Toad). Habitat. Herpetological Review 30:90.

Holzapfel, R.A. 1935. The cyclic character of hibernation of frogs. Ph.D. Dissertation, University of Oklahoma, Norman.

Holzapfel, R.A. 1937. Cyclic character of hibernation of frogs. Quarterly Review of Biology 12:65–84.

Holzman, N., and J.J. McManus. 1973. Effects of acclimation on metaboloic rate and thermal tolerance in the carpenter frog, *Rana vergatipes* [sic]. Comparative Biochemistry and Physiology 45A:833–842.

Holzwart, J., and K.D. Hall. 1984. The absence of glycerol in the hibernating American toad (*Bufo americanus*). Bios 55:31–36.

Horn, S., and M.D. Ulyshen. 2004. *Hyla cinerea* (Green Tree Frog). Diet. Herpetological Review 35:372.

Homan, R.N., B.S. Windmiller, and J.M. Reid. 2004. Critical thresholds associated with habitat loss for two vernal pool–breeding amphibians. Ecological Applications 14:1547–1553.

Homyack, J.D., and W. M. Giuliano. 2002. Effect of streambank fencing on herpetofauna in pasture stream zones. Wildlife Society Bulletin 30: 361–369.

Hoopes, I. 1930. *Bufo* in New England. Bulletin of the Boston Society of Natural History 57:13–20.

Hoopes, I. 1937a. The spadefoot toad in Essex County, Mass. Bulletin of the New England Museum of Natural History 85:12-13.

Hoopes, I. 1937b. The care of tadpoles. Bulletin of the New England Museum of Natural History No. 84:9–11.

Hoopes, I. 1937c. The care of tadpoles. New England Museum of Natural History Museum Leaflet No. 4. 4 pp. [text same as above but layout different]

Hoopes, I. 1938. Do you know the mink frog? New England Naturalist 1:4–6.

Hoopes, I. 1947. Notes on the spadefoot toad in captivity. Copeia 1947:138–139.

Hopey, M.E., and J.W. Petranka. 1994. Restriction of wood frogs to fish–free habitats: how important is adult choice versus direct predation? Copeia 1994:1023–1025.

Hopkins, W.A., M.T. Mendonca, and J.D. Congdon. 1997. Increased circulating levels of testosterone and corticosterone in Southern toads, *Bufo terrestris*, exposed to coal combustion waste. General and Comparative Endocrinology 108:237–246.

Hopkins, W.A., M.T. Mendonca, C.L. Rowe, and J.D. Congdon. 1998. Elevated trace element concentrations in Southern toads, *Bufo terrestris*, exposed to coal combustion waste. Archives of Environmental Contamination and Toxicology 35:325–329.

Hopkins, W.A., J.K. Ray, and J.D. Congdon. 2000. Incidence and impact of axial malformations in bullfrog larvae (*Rana catesbeiana*) developing in sites polluted by a coal burning power plant. Environmental Toxicology and Contamination 19:862–868.

Hopkins, W.A., S.E. DuRant, B.P. Staub, C.L. Rowe and B.P. Jackson. 2006. Reproduction, embryonic development, and maternal transfer of contaminants in the amphibian *Gastrophryne carolinensis.* Environmental Health Perspectives 114:661–666.

Hoppe, D.M. 1975. Selection pressures affecting color polymorphism in montane populations of the boreal chorus frog, *Pseudacris triseriata.* Ph.D. Dissertation, Colorado State University, Ft. Collins.

Hoppe, D.M. 1978. Thermal tolerance in tadpoles of the chorus frog *Pseudacris triseriata.* Herpetologica 34:318–321.

Hoppe, D.M. 1981. Chorus frogs and their colors. Pp. 7–8 *In* L. Ewell, K. Cram, and C. Johnson (eds.), Ecology of Reptiles and Amphibians in Minnesota. Proceedings of a Symposium. Bald Eagle Outdoor Learning Center, Cass Lake, Minnesota.

Hoppe, D.M. 2000. History of Minnesota frog abnormalities: do recent findings represent a new phenomenon? Journal of the Iowa Academy of Science 107:86–89.

Hoppe, D.M. 2005. Malformed frogs in Minnesota: history and interspecific differences. Pp. 103–108 *In* M.J. Lannoo (ed.), Amphibian Declines. The Conservation Status of United States Species. University of California Press, Berkeley.

Hoppe, D.M., and D. Pettus. 1984. Developmental features influencing color polymorphism in chorus frogs. Journal of Herpetology 18:113–120.

Hoppe, D.M., and R.G. McKinnell. 1997. Observations on the status of Minnesota leopard frog populations. Pp. 38–42 *In* J.J. Moriarty and D. Jones (eds.), Minnesota's Amphibians and Reptiles. Their Conservation and Status. Minnesota Herpetological Society, Serpent's Tale Natural History Book Distributors, Lanesboro, Minnesota.

Horn, S., J.L. Hanula, M.D. Ulyshen, and J.C. Kilgo. 2005. Abundance of green tree frogs and insects in artificial canopy gaps in a bottomland hardwood forest. American Midland Naturalist 153:321–326.

Horne, M.T., and W.A. Dunson. 1995a. Effects of low pH, metals, and water hardness on larval amphibians. Archives of Environmental Contamination and Toxicology 29:500–505.

Horne, M.T., and W.A. Dunson. 1995b. The interactive effects of low pH, toxic metals, and DOC on a simulated temporary pond community. Environmental Pollution 89:155–161.

Hossack, B.R. 2006. *Rana luteiventris* (Columbia Spotted Frog). Reproduction. Herpetological Review 37:208–209.

Hossack, B.R. 2011. Interactive effects of wildfire and disturbance history on amphibians and their parasites. Ph.D. Dissertation, University of Montana, Missoula.

Hossack, B.R., and P.S. Corn. 2007a. Wildfire effects on water temperature and selection of breeding sites by the boreal toad (*Bufo boreas*) in seasonal wetlands. Herpetological Conservation and Biology 3:46–54.

Hossack, B.R., and P.S. Corn. 2007b. Responses of pond–breeding amphibians to wildfire: short–term patterns in occupancy and colonization. Ecological Applications 17:1403–1410.

Hossack, B.R., P.S. Corn, and D.S. Pilliod. 2005. Lack of significant changes in the herpetofauna of Theodore Roosevelt National Park, North Dakota, since the 1920s. American Midland Naturalist 154:423–432.

Hossack, B.R., S.A. Diamond, and P.S. Corn. 2006a. Distribution of boreal toad populations in relation to estimated UV–B dose in Glacier National Park, Montana, USA. Canadian Journal of Zoology 84:98–107.

Hossack, B.R., P.S. Corn, and D.B. Fagre. 2006b. Divergent patterns of abundance and age–class structure of headwater stream tadpoles in burned and unburned watersheds. Canadian Journal of Zoology 84:1482–1488.

Hossack, B.R., L.A. Eby, C.G. Guscio, and P.S. Corn. 2009. Thermal characteristics of amphibian microhabitats in a fire–disturbed landscape. Forest Ecology and Management 258:1414–1421.

Hossack, B.R., M.J. Adams, E.H.C. Grant, C.A. Pearl, J.B. Bettaso, W.J. Barichivich, W.H. Lowe, K. True, J.L. Ware, and P.S. Corn. 2010. Low prevalence of chytrid fungus (*Batrachochytrium dendrobatidis*) in amphibians of U.S. headwater streams. Journal of Herpetology 44:253–260.

Hothem, R.L., A.M. Meckstroth, K.E. Wegner, M.R. Jennings, and J.J. Crayon. 2009. Diet of three species of anurans from the Cache Creek watershed, California, USA. Journal of Herpetology 43:275–283.

Hothem, R.L., M.R. Jennings, and J.J. Crayon. 2010. Mercury contamination in three species of anuran amphibians from the Cache Creek watershed, California, USA. Environmental Monitoring and Assessment 163:433–448.

Houck, W.J., and C. Henderson. 1953. A multiple appendage anomaly in the tadpole of *Rana catesbeiana*. Herpetologica 9:76–77.

Houlahan, J.E., and C.S. Findlay. 2003. The effects of adjacent land use on wetland amphibian species richness and community composition. Canadian Journal of Fisheries and Aquatic Sciences 60:1078–1094.

Houser, H.L., Jr., and D.A. Sutton. 1969. Morphologic and karyotypic differentiation of the California frogs *Rana muscosa* and *R. boylii*. Copeia 1969:184–188.

Hoverman, J.T., and R. A. Relyea. 2008. Temporal variation in predation risk: a mechanism underlying priority effects. Oikos 117:23–32.

Hoverman, J.T., M.J. Gray, and D.L. Miller. 2010. Anuran susceptibilies to ranaviruses: role of species identity, exposure route, and a novel virus isolate. Diseases of Aquatic Organisms 89:97–107.

Hoverman, J. T., M. J. Gray, N. A. Haislip, and D. L. Miller. 2011. Phylogeny, life history, and ecology contribute to differences in amphibian susceptibility to ranaviruses. EcoHealth 8:301–319.

Hoverman, J. T., M. J. Gray, D. L. Miller, and N. A Haislip. 2012. Widespread occurrence of ranavirus in pond–breeding amphibian populations. EcoHealth 9:36–48.

Hoverman, J.T., J.R. Mihaljevic, K.L.D. Richgels, J.L. Kerby, and P.T.J. Johnson. 2012. Widespread co–occurrence of virulent pathogens within California amphibian communities. EcoHealth 9:288–292.

Hovey, T.E., and D.R. Begen. 2003. *Rana catesbeiana* (Bullfrog). Predation. Herpetological Review 34:360–361.

Hovey, T.E., and E.L. Ervin. 2005. *Pseudacris cadaverina* (California Treefrog). Predation. Herpetological Review 36:304–305.

Hovingh, P. 1993a. Zoogeography and paleozoology of leeches, mollusks, and amphibians in Western Bonneville Basin, Utah, USA. Journal of Paleolimnology 9:41–54.

Hovingh, P. 1993b. Aquatic habitats, life history observations, and zoogeographic considerations of the spotted frog (*Rana pretiosa*) in Tule Valley, Utah. Great Basin Naturalist 53:168–179.

Hovingh, P. 1997. Amphibians in the eastern Great Basin (Nevada and Utah, USA): a geographical study with paleozoological models and conservation implications. Herpetological Natural History 5:97–134.

Hovingh, P., B. Benton, and D. Bornholdt. 1985. Aquatic parameters and life history observations of the Great Basin spadefoot toad in Utah. Great Basin Naturalist 45:22–30.

Howard, R.D. 1977. The evolution of mating strategies and resource utilization in bullfrogs, *Rana catesbeiana*. Ph.D. Dissertation, University of Michigan, Ann Arbor.

Howard, R.D. 1978a. The influence of male–defended oviposition sites on early embryo mortality in bullfrogs. Ecology 59:789–798.

Howard, R.D. 1978b. The evolution of mating strategies in bullfrogs, *Rana catesbeiana*. Evolution 32:850–871.

Howard, R.D. 1979. Estimating reproductive success in natural populations. American Naturalist 114:221–231.

Howard, R.D. 1980. Mating behaviour and mating success in woodfrogs, *Rana sylvatica*. Animal Behaviour 28:705–716.

Howard, R.D. 1981a. Sexual dimorphism in bullfrogs. Ecology 62:303–310.

Howard, R.D. 1981b. Male age–size distribution and male mating success in bullfrogs. Pp. 61–77 *In* R.D. Alexander and D.W. Tinkle (eds.), Natural Selection and Social Behavior. Chiron Press, New York.

Howard, R.D. 1983. Sexual selection and variation in reproductive success in a long–lived organism. American Naturalist 122:301–325.

Howard, R.D. 1984. Alternative mating behaviors of young male bullfrogs. American Zoologist 24:397–406.

Howard, R.D. 1988a. Sexual selection on male body size and mating behaviour in American toads, *Bufo americanus*. Animal Behaviour 36:1796–1808.

Howard, R.D. 1988b. Reproductive success in two species of anurans. Pp. 99–113 *In* T.H. Clutton–Brock (ed.), Reproductive Success. Studies of Individual Variation in Controlling Breeding Systems. University of Chicago Press, Chicago, Illinois.

Howard, R.D., and A.G. Kluge. 1985. Proximate mechanisms of sexual selection in wood frogs. Evolution 39:260–277.

Howard, R.D., and J.G. Palmer. 1995. Female choice in *Bufo americanus*: effects of dominant frequency and call order. Copeia 1995:212–217.

Howard, R.D., and J.R. Young. 1998. Individual variation in male vocal traits and female mating preferences in *Bufo americanus*. Animal Behaviour 55:1165–1179.

Howard, R.D., H.H. Whiteman, and T.I. Schueller. 1994. Sexual selection in American toads: a test of a good–genes hypothesis. Evolution 48:1286–1300.

Howard, W.E. 1950. Birds as bullfrog food. Copeia 1950:152.

Howe, C.M., M. Berrill, B.D. Pauli, C.C. Helbring, K. Werry, and N. Veldhoen. 2004. Toxicity of glyphosate–based pesticides to four North American frog species. Environmental Toxicology and Chemistry 23:1928–1938.

Howe, G.E., R. Gillis, and R.C. Mowbray. 1998. Effect of chemical synergy and larval stage on the toxicity of atrazine and alachlor to amphibian larvae. Environmental Toxicology and Chemistry 17:519–525.

Howe, R.H., Jr. 1899. North American wood frogs. Proceedings of the Boston Society of Natural History 28:369–374.

Hoyt, D.L. 1960. Mating behavior and eggs of the Plains spadefoot. Herpetologica 14:57–67.

Hranitz, J.M. 1989. Ecological and genetic variation between demes of *Bufo woodhousei fowleri* (Hinckley) (Chordata: Amphibia) on Assateague Island and the adjacent Delmar–va peninsula. M.S. thesis, Bloomsburg University, Bloomsburg, Pennsylvania.

Hranitz, J.M. 1993. Ontogenetic changes in the expression of isozymes, population genetic structure and heterozygosity–growth relationships in *Bufo woodhousii fowleri* (Anura: Bufonidae). Ph.D. Dissertation, Mississippi State University, Mississippi State.

Hranitz, J.M., and W.J. Diehl. 2000. Allozyme variation and population genetic structure during the life history of *Bufo woodhousii fowleri* (Amphibia: Anura). Biochemical Systematics and Ecology 28: 15–27.

Hranitz, J.M., T.S. Klinger, F.C. Hill, R.G. Sagar, T. Mencken, and J. Carr. 1993. Morphometric variation between *Bufo woodhousii fowleri* Hinckley (Anura: Bufonidae) on Assateague Island, Virginia, and the adjacent mainland. Brimleyana 19:65–75.

Hua, J., and R. A. Relyea. 2012. East Coast versus West Coast: Effects of an insecticide in communities containing different amphibian assemblages. Freshwater Science 21:787–799.

Huang, Q. 2011. Deterministic and stochastic structured-population models with application to green tree frogs. Ph.D. Dissertation, University of Louisiana–Lafayette.

Huang, R. 2011. Detecting the presence of *Batrachochytrium dendrobatidis* in amphibians in Piedmont vs Blue Ridge habitats in northern Georgia. M.S. thesis, Emory University, Atlanta, Georgia.

Huang, Y.–W., W.H. Karasov, K.A. Patnode, and C.R. Jefcoate. 1999. Exposure of northern leopard frogs in the Green Bay ecosystem to polychlorinated biphenyls, polychlorinated dibenzo–*p*–dioxins, and polychlorinated bibenzofurans is measured by direct chemistry but not hepatic ethoxyresorufin–*o*–deethylase activity. Environmental Toxicology and Chemistry 18:2123–2130.

Hubbs, C.L. 1917. A one–legged cricket frog. Copeia (50):99.

Hubbs, C.L. 1918. *Bufo fowleri* in Michigan, Indiana and Illinois. Copeia (55):40–43.

Hubbs, C., and N.E. Armstrong. 1961. Minimum developmental temperature tolerance of two anurans, *Scaphiopus couchi* and *Microhyla olivacea*. Texas Journal of Science 13:358–362.

Hubbs, C., and F.D. Martin. 1967. *Bufo valliceps* breeding in artificial pools. Southwestern Naturalist 12:105–106.

Hubbs, C., T. Wright, and O. Cuellar. 1963. Developmental temperature tolerance of central Texas populations of two anuran amphibians *Bufo valliceps* and *Pseudacris streckeri*. Southwestern Naturalist 8:142–149.

Hudson, G.E. 1942. The amphibians and reptiles of Nebraska. University of Nebraska, Nebraska Conservation Bulletin 24. [reprinted 1972]

Hudson, R.G. 1950. A possible hibernating site for the spring peeper. Copeia 1950:59.

Hudson, R.G. 1952. Observations on cricket frog locomotion. Copeia 1952:185.

Hudson, R.G. 1954. An annotated list of the reptiles and amphibians of the Unami Valley, Pennsylvania. Herpetologica 10:67–72.

Hudson, R.G. 1956. The leopard frog *Rana pipiens sphenocephala* in southeastern Pennsylvania. Herpetologica 12:182–183.

Huey, L.M. 1942. A vertebrate faunal survey of the Organ Pipe Cactus National Monument, Arizona. Transactions of the San Diego Society of Natural History 9(32):353-376.

Huey, R.B. 1980. Sprint velocity of tadpoles (*Bufo boreas*) through metamorphosis. Copeia 1980:537–540.

Hughes, N. 1962. The number and form of chromosomes in the genus *Scaphiopus*. Texas Journal of Science 14:225–228.

Hughes, N. 1963. Notes on two partial albino (?) toads, *Scaphiopus bombifrons*. Herpetologica 19:139–140.

Hughes, N. 1965. Comparison of frontoparietal bones of *Scaphiopus bombifrons* and *S. hammondii* as evidence of interspecific hybridization. Herpetologica 21:196–201.

Huheey, J.E. 1965a. Unusual behavior in juvenile toads, *Bufo americanus*. Journal of the Ohio Herpetological Society 5:33.

Huheey, J.E. 1965b. A toad in the diet of a frog. Journal of the Ohio Herpetological Society 5:33.

Huheey, J.E. 1980. Studies in warning coloration and mimicry VIII. Further evidence for a frequency–dependent model of predation. Journal of Herpetology 14:223–230.

Huheey, J.E., and A. Stupka. 1967. Amphibians and Reptiles of Great Smoky Mountains National Park. University of Tennessee Press, Knoxville.

Hulse, A.C., C.J. McCoy, and E.J. Censky. 2001. Amphibians and Reptiles of Pennsylvania and the Northeast. Cornell University Press, Ithaca, New York.

Humfeld, S.C., V.T. Marshall, and M.A. Bee. 2009. Context–dependent plasticity of aggressive signalling in a dynamic social environment. Animal Behaviour 78:915–924.

Humphries, W.J., and M.A. Sisson. 2012. Long distance migrations, landscape use, and vulnerability to prescribed fire of the gopher frog (*Lithobates capito*). Journal of Herpetology 46:665–670.

Hunka, D.L. 1974. Physiological aspects of overwintering in the boreal chorus frog, *Pseudacris triseriata maculata*. M.S. thesis, University of Manitoba, Winnipeg.

Hunsaker, D., II, and F.E. Potter, Jr. 1960. "Red leg" in a natural population of amphibians. Herpetologica 16:285–286.

Hunsaker, D. II, and P. Breese. 1967. Herpetofauna of the Hawaiian Islands. Pacific Science 21:423–428.

Hunter, M.L., Jr., J. Albright, and J. Arbuckle. 1992. The Amphibians and Reptiles of Maine. Maine Agricultural Experiment Station, Bulletin 838.

Hunt, R.H. 1980. Toad sanctuary in a tarantula burrow. Natural History 89(3):49–53.

Hunter, M.L., Jr., A.J.K. Calhoun, and M. McCollough. 1999. Maine Amphibians and Reptiles. University of Maine Press, Orono.

Hunziker, R. 1999. *Bufo punctatus*, a fine little toad. Reptile & Amphibian Hobbyist 4(11):34–38.

Hurst, B. 1944. The amphibians of Prince Edward Island. Acadian Naturalist 1(3):111–117.

Hurter, J. 1911. Herpetology of Missouri. Transactions of the Academy of Science of St. Louis 20(5):59–274.

Hutchison, V.H., and L.G. Hill. 1978. Thermal selection of bullfrog tadpoles (*Rana catesbeiana*) at different stages of development and acclimation temperatures. Journal of Thermal Biology 3:57–60.

Ideker, J. 1968. Secondary intergradation between *Bufo americanus americanus* Holbrook and *Bufo woodhousei woodhousei* Girard in southeastern South Dakota. M.A. thesis, University of South Dakota, Vermillion.

Ideker, J. 1976. Tadpole thermoregulatory behavior facilitates grackle predation. Texas Journal of Science 27:244–245.

Ideker, J. 1979. Adult *Cybister fimbriolatus* are predaceous (Coleoptera: Dytiscidae). Coleopterists Bulletin 33:41–44.

Illingworth, J.F. 1941. Feeding habits of *Bufo marinus*. Proceedings of the Hawaiian Entomological Society 11:51.

Inger, R.F. 1949. Trail of the Texas barking frog. Chicago Natural History Museum Bulletin 19 (7): 4-5.

Ingles, L.G. 1932a. *Cephalogonimus brevicirrus*, a new species of trematode from the intestine of *Rana aurora* from California. University of California Publications in Zoology 37:203–210.

Ingles, L.G. 1932b. Four new species of *Haematoloechus* (Trematoda) from *Rana aurora draytoni* from California. University of California Publications in Zoology 37:189–201.

Ingles, L.G. 1936. Worm parasites of California amphibia. Transactions of the American Microscopical Society 55:73–92.

Ingles, L.G., and M.P. Langston. 1933. A new species of bladder fluke from California frogs. Transactions of the American Microscopical Society 52:243–245.

Ingram, W.M., and E.C. Raney. 1943. Additional studies on the movement of tagged bullfrogs, *Rana catesbeiana* Shaw. American Midland Naturalist 29:239–241.

Ireland, D.H., A.J. Wirsing, and D.L. Murray. 2007. Phenotypically plastic responses of green frog embryos to conflicting predation risk. Oecologia 152:162–168.

Ireland, T.T., G.L. Wolters, and S.D. Schemnitz. 1994. Recolonization of wildlife on a coal strip–mine in northwestern New Mexico. Southwestern Naturalist 39:53–57.

Irwin, J.T. 2005. Overwintering in northern cricket frogs (*Acris crepitans*). Pp. 55–58 *In* M.J. Lannoo (ed.), Amphibian Declines. The Conservation Status of United States Species. University of California Press, Berkeley.

Irwin, J.T., J.P. Costanzo, and R.E. Lee, Jr. 1999. Terrestrial hibernation in the northern cricket frog, *Acris crepitans*. Canadian Journal of Zoology 77:1240–1246.

Irwin, J. T., J. P. Costanzo, and R. E. Lee. 2003. Postfreeze reduction of locomotor endurance in the freeze–tolerant wood frog, *Rana sylvatica*. Physiological and Biochemical Zoology 76: 331–338.

Irwin, K.J., C.K. Dodd, Jr., and M.L. Griffey. 1999. Geographic distribution. *Scaphiopus holbrooki* (Eastern Spadefoot Toad). Herpetological Review 30:232.

Irwin, K.J., L. Irwin, T.W. Taggart, J.T. Collins, and S.L. Collins. 2001. An herpetofaunal survey of the Apalachicola barrier islands, Florida: the 2000–2001 season. Kansas Herpetological Society Newsletter (123):12–15.

Irwin, M.S. 1929. A new lung fluke from *Rana clamitans* Latreille. Transactions of the American Microscopical Society 48:74–79.

Isaacs, J.S. 1971. Temporal stability of vertebral stripe color in a cricket frog population. Copeia 1971:551–552.

Isola, M. 2011. Does the matrix matter? A comparison on phenology and habitat utilization of two treefrog species in the Big Cypress National Preserve. M.S. thesis, Florida International University, Miami.

Iverson, J.B. 1973. *Hyla cinerea* (Green Treefrog). HISS News-Journal 1:153.

Jackman, R., S. Nowicki, D.J. Aneshansley, and T. Eisner. 1983. Predatory capture of toads by fly larvae. Science 222:515–516.

Jackson, A.W. 1952. The effect of temperature, humidity, and barometric pressure on the rate of call in *Acris crepitans* Baird in Brazos County, Texas. Herpetologica 8:18–20.

Jackson, C.G., Jr., and M.M. Jackson. 1970. Herpetofauna of Dauphin Island, Alabama. Quarterly Journal of the Florida Academy of Sciences 33:281–287.

Jackson, D.R., and E.G. Milstrey. 1989. The fauna of gopher tortoise burrows. Pp. 86–98 *In* J.E. Diemer, D.R. Jackson, J.L. Landers, J.N. Layne, and D.A. Wood (eds.), Gopher Tortoise Relocation Symposium Proceedings, Florida Game and Fresh Water Fish Commission, Nongame Wildlife Program Technical Report No.5.

Jackson, H.H.T. 1914. The land vertebrates of Ridgeway Bog, Wisconsin: their ecological succession and source of ingression. Bulletin of the Wisconsin Natural History Society 12:4–54.

Jackson, J.T., F.W. Weckerly, T.M. Swannack, and M.R.J. Forstner. 2006. Inferring absence of Houston toads given imperfect detection probabilities. Journal of Wildlife Management 70:1461–1463.

Jacobs, D.L. 1950. *Pseudacris nigrita triseriata* on the north shore of Lake Superior. Copeia 1950:154.

Jacobs, L., and J.E. Houlahan. 2011. Adjacent land–use affects amphibian community composition and species richness in managed forests in New Brunswick, Canada. Canadian Journal of Forest Research 41:1687–1697.

Jacobson, N.L. 1989. Breeding dynamics of the Houston toad. Southwestern Naturalist 34:374–380.

Jaeger, J.R., B.R. Riddle, R.D. Jennings, and D.F. Bradford. 2001. Rediscovering *Rana onca*: evidence for phylogenetically distinct leopard frogs from the border region of Nevada, Utah, and Arizona. Copeia 2001:339–354.

Jaeger, R.G., and J.P. Hailman. 1971. Two types of phototactic behaviour in anuran amphibians. Nature 230:189–190.

Jaeger, R.G., and J.P. Hailman. 1973. Effects of intensity on the phototactic responses of adult anuran amphibians: a comparative survey. Zeitschrift für Tierpsychologie 33:352–407.

Jaeger, R.G., and J.P. Hailman. 1976a. Ontogenetic shift of spectral phototactic preferences in anuran tadpoles. Journal of Comparative Physiology and Psychology 90:930–945.

Jaeger, R.G., and J.P. Hailman. 1976b. Phototaxis in anurans:

relation between intensity and spectral preferences. Copeia 1976:92–98.

James, H.A., and M.J. Ulmer. 1967. New amphibian host records for *Mesocestoides* sp. (Cestoda: Cyclophyllidea). Journal of Parasitology 53:59.

James, M.T., and T.P. Maslin. 1947. Notes on myiasis of the toad, *Bufo boreas boreas* Baird and Girard. Journal of the Washington Academy of Sciences 37:366–368.

James, P. 1966. The Mexican burrowing toad, *Rhinophrynus dorsalis*, an addition to the vertebrate fauna of the United States. Texas Journal of Science 18:272–276.

James, S.M. 2005. Amphibian metamorphosis and juvenile terrestrial performance following chronic cadmium exposure in the aquatic environment. Ph.D. Dissertation, University of Missouri, Columbia.

James, S.M., and R.D. Semlitsch. 2011. Terrestrial performance of juvenile frogs in two habitat types after chronic exposure to a contaminant. Journal of Herpetology 45:186–194.

James, S.M., E.E. Little, and R.D. Semlitsch. 2004a. Effects of multiple routes of cadmium exposure on the hibernation success of the American Toad (*Bufo americanus*). Archives of Environmental Contamination and Toxicology 46:518–527.

James, S.M., E.E. Little, and R.D. Semlitsch. 2004b. The effect of soil composition and hydration on the bioavailability and toxicity of cadmium to hibernating juvenile American toads (*Bufo americanus*). Environmental Pollution 132:523–532.

James, S.M., E.E. Little, and R.D. Semlitsch. 2005. Metamorphosis of two amphibian species after chronic cadmium exposure in outdoor aquatic mesocosms. Environmental Toxicology and Chemistry 24:1994–2001.

Jameson, D.L. 1950a. The development of *Eleutherodactylus latrans*. Copeia 1950:44–46.

Jameson, D.L. 1950b. The breeding and development of Strecker's chorus frog in central Texas. Copeia 1950:61.

Jameson, D.L. 1954. Social patterns in the leptodactylid frogs *Syrrhophus* and *Eleutherodactylus*. Copeia 1954:36–38.

Jameson, D.L. 1955. The population dynamics of the cliff frog, *Syrrhophus marnocki*. American Midland Naturalist 54:342–381.

Jameson, D.L. 1956a. Growth, dispersal and survival of the Pacific tree frog. Copeia 1956:25–29.

Jameson, D. L. 1956b. Survival of some central Texas frogs under natural conditions. Copeia 1956: 55–57.

Jameson, D.L. 1957. Population structure and homing responses in the Pacific tree frog. Copeia 1957:221–228.

Jameson, D.L., and A.G. Flury. 1949. The reptiles and amphibians of the Sierra Vieja range of southwestern Texas. Texas Journal of Science 1:54–79.

Jameson, D.L., and R.M. Myers. 1956. Food habits in juvenile frogs. Copeia 1956:261.

Jameson, D.L., and R.M. Myers. 1957. Albino Pacific tree frogs. Herpetologica 13:74.

Jameson, D.L., and S. Pequegnat. 1971. Estimation of relative viability and fecundity of color polymorphism in anurans. Evolution 25:180–194.

Jameson, D.L., and R.C. Richmond. 1971. Parallelism and convergence in the evolution of size and shape in Holarctic *Hyla*. Evolution 25:497–508.

Jameson, D.L., J.P. Mackey, and R.C. Richmond. 1966. The systematics of the Pacific tree frog, *Hyla regilla*. Proceedings of the California Academy of Sciences 33(19):551–620.

Jameson, D.L., W. Taylor, and J. Mountjoy. 1970. Metabolic and morphological adaptation to heterogenous environments by the Pacific tree toad, *Hyla regilla*. Evolution 24:75–89.

Jameson, D.L., J.P. Mackey, and M. Anderson. 1973. Weather, climate, and the external morphology of Pacific tree toads. Evolution 27:285–302.

Jameson, E.W., Jr. 1947. The food of the western cricket frog. Copeia 1947:212.

Jamieson, D.H., and S.E. Trauth. 1996. Dietary diversity and overlap between two subspecies of spadefoot toads (*Scaphiopus holbrookii holbrookii* and *S. h. hurterii*) in Arkansas. Proceedings of the Arkansas Academy of Science 50:75–78.

Jamieson, D.H., S.E. Trauth, and C.T. McAllister. 1993. Food habits of male bird–voiced treefrogs, *Hyla avivoca* (Anura: Hylidae), in Arkansas. Texas Journal of Science 45:45–49.

Jansen, K.P., A.P. Summers, and P.R. Delis. 2001. Spadefoot toads (*Scaphiopus holbrookii holbrookii*) in an urban landscape: effects of nonnatural substrates on burrowing in adults and juveniles. Journal of Herpetology 35:141–145.

Janson, H.S. 1953. Skittering locomotion in the frog *Hyla cinerea cinerea*. Copeia 1953:62.

Jasienski, M. 1992. Kinship effects and plasticity of growth in amphibian larvae. Ph.D. Dissertation, Harvard University, Cambridge, Massachusetts.

Jaslow, A.P., and R.C. Vogt. 1977. Identification and distribution of *Hyla versicolor* and *Hyla chrysoscelis* in Wisconsin. Herpetologica 33:201–205.

Jay, J.M., and W.J. Pohley. 1981. *Dermosporidium penneri* sp. n. from the skin of the American toad, *Bufo americanus* (Amphibia: Bufonidae). Journal of Parasitology 67:108–110.

Jefferson, D.M., and R.W. Russell. 2008. Ontogenetic and fertilizer effects on stable isotopes in the green frog (*Rana clamitans*). Applied Herpetology 5:189–196.

Jeffery, B.M., J.H. Waddle, and J.A. Maskell. 2004. *Hyla cinerea* (Green Treefrog). Predation. Herpetological Review 35:158.

Jenkins, J.R. 2000. Bioenergetics of freeze–thaw cycles in the chorus frog (*Pseudacris triseriata*). M.S. thesis, University of South Dakota, Vermillion.

Jennette, M.A. 2010. *Lithobates sylvaticus* (Wood Frog). Egg predation. Herpetological Review 41:476–477.

Jenni, D.A. 1969. A study of the ecology of four species of herons during the breeding season at Lake Alice, Alachua County, Florida. Ecological Monographs 39:245–270.

Jennings, D.T., H.S. Crawford, Jr., and M.L. Hunter, Jr. 1991. Predation by amphibians and small mammals on the spruce budworm (Lepidoptera: Tortricidae). The Great Lakes Entomologist 24(2):2.

Jennings, M.R. 1987. *Rana catesbeiana* (Bullfrog). Feeding. Herpetological Review 18:33.

Jennings, M.R. 1988. Natural history and decline of native ranids in California. Pp. 61–72 *In* H.P. De Lisle, P.R. Brown, B. Kaufman, and B.M. McGurty (eds.), Proceedings of the Conference on California Herpetology. Southwestern Herpetologists Society, Special Publication No. 4, Van Nuys, California.

Jennings, M.R. 1988. Origin of the population of *Rana aurora draytonii* on Santa Cruz Island, California. Herpetological Review 19:76.

Jennings, M.R. 1996. Status of amphibians. Pp. 921–944 *In* Sierra Nevada Ecosystem Project: Final Report to Congress. Vol. II. Assessements and Scientific Basis for Management Options. Center for Water and Wildland Resources, University of California, Davis, California.

Jennings, M.R. 2004. An annotated check list of the amphibians and reptiles of California and adjacent waters. California Fish and Game 90:161–213.

Jennings, M.R. 2005. *Rana fisheri* Stejneger, 1893.Vegas Valley Leopard Frog. Pp. 554–555 *In* M.J. Lannoo (ed.), Amphibian Declines. The Conservation Status of United States Species. University of California Press, Berkeley.

Jennings, M.R., and M.P. Hayes. 1984. The frogs of Tulare. Outdoor California 45(6):17–19.

Jennings, M.R., and M.P. Hayes. 1985. Pre–1900 overharvest of California red–legged frogs (*Rana aurora draytonii*): the inducement for bullfrog (*Rana catesbeiana*) introduction. Herpetologica 41:94–103.

Jennings, M.R., and M.P. Hayes. 1994a. Decline of native ranid frogs in the desert Southwest. Pp. 183–211 *In* P.R. Brown and J.W. Wright (eds.), Herpetology of the North American Deserts. Proceedings of a Symposium. Southwestern Herpetologists Society, Special Publication No. 5.

Jennings, M.R., and M.P. Hayes. 1994b. Amphibian and reptile species of special concern in California. California Department of Fish and Game, Inland Fisheries Division, Rancho Cordova, California.

Jennings, M.R., and M.P. Hayes. 2005. *Rana boylii* (Foothill Yellow–legged Frog). Coloration. Herpetological Review 36:438.

Jennings, M.R., D.F. Bradford, and D.F. Johnson. 1992. Dependence of the garter snake *Thamnophis elegans* on amphibians in the Sierra Nevada of California. Journal of Herpetology 26:503–505.

Jennings, M.R., J.C. Crayon, and R.L. Hothem. 2005. *Bufo boreas halophilus* (California Toad) and *Rana catesbeiana* (American Bullfrog). Amplexus. Herpetological Review 36:53.

Jennings, R.D., and N.J. Scott, Jr. 1993. Ecologically correlated morphological variation in tadpoles of the leopard frog, *Rana chiricahuensis*. Journal of Herpetology 27:285–293.

Jense, G.K., and R.L. Linder. 1970. Food habits of badgers in eastern South Dakota. Proceedings of the South Dakota Academy of Science 49:37–41.

Jensen, J.B. 1996. *Bufo terrestris* (Southern toad). Egg toxicity. Herpetological Review 27:138–139.

Jensen, J.B. 2000. *Rana capito* (Gopher Frog). Predation. Herpetological Review 31:42.

Jensen, J.B. 2004. Little grass frog *Pseudacris ocularis* (Bosc and Daudin). P. 27 *In* R.E. Mirarchi, M.A. Bailey, T.M. Haggerty, and T.L. Best (eds.), Alabama Wildlife. Vol. 3. Imperiled Amphibians, Reptiles, Birds, and Mammals. University of Alabama Press, Tuscaloosa.

Jensen, J.B., J.G. Palis, and M.A. Bailey. 1995. *Rana capito sevosa* (Dusky Gopher Frog). Submerged vocalization. Herpetological Review 26:98.

Jensen, J.B., M.A. Bailey, E.L. Blankenship, and C.D. Camp. 2003. The relationship between breeding by the gopher frog, *Rana capito* (Amphibia: Ranidae) and rainfall. American Midland Naturalist 150:185–190.

Jensen, J.B., C.D. Camp, W. Gibbons, and M.J. Elliott (eds.). 2008. Amphibians and Reptiles of Georgia. University of Georgia Press, Athens.

Jenssen, T.A. 1967. Food habits of the green frog, *Rana clamitans*, before and during metamorphosis. Copeia 1967:214–218.

Jenssen, T.A. 1968. Some morphological and behavioral characteristics of an intergrade population of the green frog, *Rana clamitans*, in southern Illinois. Transactions of the Illinois State Academy of Science 61:252–259.

Jenssen, T.A. 1972. Seasonal organ weights of the green frog, *Rana clamitans* (Anura: Ranidae), under natural conditions. Transactions of the Illinois State Academy of Science 65:15-24.

Jenssen, T.A., and W.D. Klimstra. 1966. Food habits of the green frog, *Rana clamitans*, in southern Illinois. American Midland Naturalist 76:169–182.

Jenssen, T.A., and W.B. Preston. 1968. Behavioral responses of the male green frog, *Rana clamitans*, to its recorded call. Herpetologica 24:181–182.

Jilek, R., and R. Wolff. 1978. Occurrence of *Spinitectus gracilis* (Nematoda, Spiuroidea) in the toad *Bufo woodhousei fowleri*. Journal of Parasitology 64:619.

Jirků, M., M.G. Bolek, C.M. Whipps, J. Janovy, Jr., M.L. Kent, and D. Modrý. 2006. A new species of *Myxidium* (Myxosporea: Myxidiidae), from the western chorus frog, *Pseudacris triseriata triseriata,* and Blanchard's cricket frog, *Acris crepitans blanchardi* (Hylidae), from eastern Nebraska: morphology, phylogeny, and critical comments on amphibian *Myxidium* taxonomy. Journal of Parasitology 92:611–619.

Jofré, M.B., and W.H. Karasov. 1999. Direct effect of ammonia on three species of North American anuran amphibians. Environmental Toxicology and Chemistry 18:1806–1812.

Jofré, M.B., M.L. Rosenshield, and W.H. Karasov. 2000. Effects of PCB 126 and ammonia, alone and in combination, on green frog (*Rana clamitans*) and leopard frog (*R. pipiens*) hatching success, development, and metamorphosis. Journal of the Iowa Academy of Science 107:113–122.

Johansen, F. 1926. Occurrences of frogs on Anticosti Island and Newfoundland. Canadian Field-Naturalist 40:16.

Johansen, K. 1962. Observations on the wood frog *Rana sylvatica* in Alaska. Ecology 43:146–147.

John, K.R., and D. Fenster. 1975. The effects of partitions on the growth rates of crowded *Rana pipiens* tadpoles. American Midland Naturalist 93:123–130.

John, K.R., and J.M. Fusaro. 1981. Growth and metamorphosis of solitary *Rana pipiens* tadpoles in confined space. Copeia 1981:737–741.

John–Alder, H.B., and P.J. Morin. 1990. Effects of larval density on jumping ability and stamina in newly metamorphosed *Bufo woodhousii fowleri*. Copeia 1990:856–860.

John–Alder, H.B., P.J. Morin, and S.P. Lawler. 1988. Thermal physiology, phenology, and distribution of tree frogs. American Naturalist 132:506–520.

John–Alder, H.B., M.C. Barnhart, and A.F. Bennett. 1989. Thermal sensitivity of swimming performance and muscle contraction in northern and southern populations of tree frogs (*Hyla crucifer*). Journal of Experimental Biology 142:357–372.

Johnson, B. 1989. Familiar Amphibians and Reptiles of Ontario. Natural Heritage/Natural History, Inc., Toronto, Ontario.

Johnson, B.K., and J.L. Christiansen. 1976. The food and food habits of Blanchard's cricket frog, *Acris crepitans blanchardi* (Amphibia, Anura, Hylidae), in Iowa. Journal of Herpetology 10:63–74.

Johnson, C. 1959. Genetic incompatibility in the call races of *Hyla versicolor* Le Conte in Texas. Copeia 1959:327–335.

Johnson, C. 1963. Additional evidence of sterility between the call–types in the *Hyla versicolor* complex. Copeia 1963:139–143.

Johnson, C. 1966. Species recognition in the *Hyla versicolor* complex. Texas Journal of Science 18:361–364.

Johnson, C.E. 1933. Absence of one eye in a frog. Copeia 1933:79–80.

Johnson, C.R. 1980. The effects of five organophosphorus insecticides on thermal stress in tadpoles of the Pacific treefrog, *Hyla regilla*. Zoological Journal of the Linnean Society 69:143–147.

Johnson, C.R., and R.B. Bury. 1965. Food of the Pacific treefrog, *Hyla regilla* Baird and Girard, in northern California. Herpetologica 21:56–58.

Johnson, C.R., and J.E. Prine. 1976. The effects of sublethal concentrations of organophosphorus insecticides and an insect growth regulator on temperature tolerance in hydrated and dehydrated juvenile Western toads, *Bufo boreas*. Comparative Biochemistry and Physiology 53A:147–149.

Johnson, D.H., M.D. Bryant, and A.H. Miller. 1948. Vertebrate animals of the Providence Mountains area of California. University of California Publications in Zoology 48:221–376.

Johnson, D.H., S.C. Fowle, and J.A. Jundt. 2000. The North American reporting center for amphibian malformations. Journal of the Iowa Academy of Science 107:123–127.

Johnson, D.H., and R.D. Batie. 2001. Surveys of calling amphibians in North Dakota. Prairie Naturalist 33:227–247.

Johnson, J.R. 2004. Fall breeding of the southern leopard frog (*Rana sphenocephala*) in central Missouri. Missouri Herpetological Association Newsletter 17:14-16.

Johnson, J.R. 2005a. Multi–scale investigations of gray treefrog movements: patterns of migration, dispersal, and gene flow. Ph.D. Dissertation, University of Missouri, Columbia.

Johnson, J.R. 2005b. *Hyla versicolor* (Gray Treefrog). Record size. Herpetological Review 36:302.

Johnson, J.R., and R.D. Semlitsch. 2003. Defining core habitat of local populations of the gray treefrog (*Hyla versicolor*) based on choice of oviposition site. Oecologia 137:205–210.

Johnson, J.R., J.H. Knouft, and R.D. Semlitsch. 2007. Sex

and seasonal differences in the spatial terrestrial distribution of gray treefrog (*Hyla versicolor*) populations. Biological Conservation 140:250–258.

Johnson, J.R., R.D. Mahan and R.D. Semlitsch. 2008. Seasonal terrestrial microhabitat use by gray treefrogs (*Hyla versicolor*) in Missouri oak–hickory forests. Herpetologica 64:259–269.

Johnson, K.A. 2003. Abiotic factors influencing the breeding, movement, and foraging of the eastern spadefoot (*Scaphiopus holbrookii*) in West Virginia. M.S. thesis, Marshall University, Huntington, West Virginia.

Johnson, K.E., and L. Eidietis. 2005. Tadpole body zones differ with regard to strike frequencies and kill rates by dragonfly naiads. Copeia 2005:909–913.

Johnson, L.M. 1988. Habitat selection in relation to food and feeding strategies of larval northern cricket frogs (*Acris crepitans*). M.S. thesis, Illinois State University, Normal.

Johnson, L.M. 1991. Growth and development of larval northern cricket frogs (*Acris crepitans*) in relation to phytoplankton abundance. Freshwater Biology 25:51–59.

Johnson, O.W. 1965. Early development, embryonic temperature tolerance and rate of development in *Rana pretiosa luteiventris* Thompson. Ph.D. Dissertation, Oregon State University, Corvallis.

Johnson, P.T.J., and D.R. Sutherland. 2003. Amphibian deformities and *Ribeiroia* infection: an emerging helminthiasis. Trends in Parasitology 19:332–335.

Johnson, P.T.J., and J.M. Chase. 2004. Parasites in the food web: linking amphibian malformations and aquatic eutrophication. Ecology Letters 7:521–526.

Johnson, P.T.J., K.B. Lunde, E.G. Ritchie, and A.E. Launer. 1999. The effect of trematode infection on amphibian limb development and survivorship. Science 284:802–804.

Johnson, P.T.J., K.B. Lunde, R.W. Haight, J. Bowerman, and A.R. Blaustein. 2001a. *Rebeiroia ondatrae* (Trematoda: Digenea) infection induces severe limb malformations in Western toads (*Bufo boreas*). Canadian Journal of Zoology 79:370–379.

Johnson, P.T.J., K.B. Lunde, E.G. Ritchie, J.K. Reaser, and A.E. Launer. 2001b. Morphological abnormality patterns in a California amphibian community. Herpetologica 57:336–352.

Johnson, P.T.J., K.B. Lunde, E.M. Thurman, E.G. Ritchie, S.N. Wray, D.R. Sutherland, J.M. Kapfer, T.J. Frest, J. Bowerman, and A.R. Blaustein. 2002. Parasite (*Riberoia ondatrae*) infection linked to amphibian malformations in the western United States. Ecological Monographs 72:151–168.

Johnson, P.T.J., D.R. Sutherland, J.M. Kinsella, and K.B. Lunde. 2004. Review of the trematode genus *Ribeiroia* (Psilostomidae): ecology, life history and pathogenesis with special emphasis on the amphibian malformation problem. Advances in Parasitology 57:191–253.

Johnson, P.T.J., E.R. Preu , D.R. Sutherland, J.M. Romansic, B. Han, and A.R. Blaustein. 2006. Adding infection to injury: synergistic effects of predation and parasitism on amphibian malformations. Ecology 87:2227–2235

Johnson, P.T.J., V.J. McKenzie, A.C. Peterson, J.L. Kerby, J. Brown, A.R. Blaustein, and T. Jackson. 2011. Regional decline of an iconic amphibian associated with elevation, land–use change, and invasive species. Conservation Biology 25:556–566.

Johnson, P.T.J., J.R. Rohr, J.T. Hoverman, E. Kellermanns, J. Bowerman, and K.B. Lunde. 2012. Living fast and dying of infection: host life history drives interspecific variation in infection and disease risk. Ecology Letters 15:235–242.

Johnson, R.N., D.G. Young, and J.F. Butler. 1993. Trypanosome transmission by *Corethrella wirthi* (Diptera: Chaoboridae) to the green treefrog *Hyla cinerea* (Anurra: Hylidae). Journal of Medical Entomology 30:918–921.

Johnson, S.A. 1996. *Hyla femoralis* (Pine Woods Treefrog). Predation. Herpetological Review 27:140.

Johnson, S.A., and R.C. Means. 2000. *Hyla squirella* (Squirrel Tree Frog). Reproduction. Herpetological Review 31:100.

Johnson, S.A., and M. McGarrity. 2010. Identification Guide to the Frogs of Florida. University of Florida Extension Service, Gainesville.

Johnson, S.A., M.E. McGarrity, and C.L. Staudhammer. 2010. An effective chemical deterrent for invasive Cuban treefrogs. Human–Wildlife Interactions 4:112–117.

Johnson, S.R. 2003. *Acris crepitans blanchardi* (Blanchard's Cricket Frog). Behavior. Herpetological Review 34:355.

Johnson, T.R. 1975. Fall choruses of *Hyla crucifer* in Missouri. St. Louis Herpetological Society Newsletter 2(12):5–6.

Johnson, T.R. 1977. The amphibians of Missouri. University of Kansas Museum of Natural History, Public Education Series No. 6:1-134.

Johnson, T.R. 2000. The Amphibians and Reptiles of Missouri. Missouri Department of Conservation, Jefferson City.

Johnson, V.O. 1939. A supplementary note on the larvae of *Bufo woodhousii woodhousii*. Herpetologica 1:160–164.

Johnsson, R.G., and P.C. Shelton. 1950. The vertebrates of Isle Royale National Park. Isle Royale Natural History Association, Isle Royale National Park, Michigan. [Revised by P.C. Shelton, 1964]

Johnston, G.F., and R. Altig. 1986. Identification characteristics of anuran tadpoles. Herpetological Review 17:36–

37.

Johnston, G.R., and J.C. Johnston. 2006. *Scaphiopus holbrookii holbrookii* (Eastern Spadefoot). Albinism. Herpetological Review 37:211–212.

Jolley, D.B., S.S. Ditchkoff, B.D. Sparklin, L.B. Hanson, M.S. Mitchell, and J.B. Grand. 2010. Estimate of herpetofauna depredation by a population of wild pigs. Journal of Mammalogy 91:519–524.

Jones, D. 2004. Aquatic and terrestrial use of habitat by the Amargosa Toad (*Bufo nelsoni*). M.S. thesis, University of Nevada, Reno.

Jones, D., E.T. Simandle, C.R. Tracy, and B. Hobbs. 2003. *Bufo nelsoni* (Amargosa Toad). Predation. Herpetological Review 34:229.

Jones, D. K., J. I. Hammond, and R. A. Relyea. 2009. Very highly toxic effects of endosulfan across nine species of tadpoles: Lag effects and family–level sensitivity. Environmental Toxicology and Chemistry 28:1939–1945.

Jones, D. K., J. R. Hammond, and R. A. Relyea. 2011. Competitive stress can make the herbicide Roundup® more deadly to larval amphibians. Environmental Toxicology and Chemistry 30:446–454.

Jones, H.N. 2011. *Bufo terrestris* (Southern Toad). Diet. Herpetological Review 42:409.

Jones, J., and B.H. Brattstrom. 1962. The call of the spring peeper, *Hyla crucifer*, in response to a recording of its own voice. Herpetologica 17: 246–250.

Jones, J.M. 1865. Contributions to the natural history of Nova Scotia: Reptilia. Transactions of the Nova Scotian Institute of Science 1:114-128.

Jones, J.M. 1973. Effects of thirty years hybridization on the toads *Bufo americanus* and *Bufo woodhousei fowleri* at Bloomington, Indiana. Evolution 27:435–448.

Jones, K., B. Rincon, and D.A. Steen. 2010. *Bufo terrestris* (Southern Toad). Predation. Herpetological Review 41:334–335.

Jones, K.B. 1981. Distribution, ecology, and habitat management of the reptiles and amphibians of the Hualapai–Aquarius Planning Area, Mohave and Yavapai counties, Arizona. U.S. Department of the Interior, Bureau of Land Management, Technical Note 353.

Jones, K.B. 1990. Habitat use and predatory behavior of *Thamnophis cyrtopsis* (Serpentes: Colubridae) in a seasonally variable aquatic environment. Southwestern Naturalist 35: 115–122.

Jones, L., D.R. Gossett, S.W. Banks, and M.L. McCallum. 2010. Antioxidant defense system in tadpoles of the American bullfrog (*Lithobates catesbeianus*) exposed to paraquat. Journal of Herpetology 44:222–228.

Jones, L.L.C., and M.G. Raphael. 1998. *Ascaphus truei* (Tailed Frog). Predation. Herpetological Review 29:39.

Jones, L.L.C., and M.J. Sredl. 2005. Chiricahua leopard frog status in the Galiuro Mountains, Arizona, with a monitoring framework for the species' entire range. Pp. 88–91 *In* G.J. Gottfried, B.S. Gebow, L.G. Eskew, C.B. Edminster, and B. Carelton (compilers), Connecting Mountain Islands and Desert Seas: Biodiversity and Management of the Madrean Archipelago. 2. USDA Forest Service Proceedings RMRS–P–36, Fort Collins, Colorado.

Jones, L.L.C., W.P. Leonard, and D.H. Olson (eds.). 2005. Amphibians of the Pacific Northwest. Seattle Audubon Society, Seattle, Washington.

Jones, M.C., J.R. Dixon, and M.R.J. Forstner. 2011. Is bigger always better? Mate selection in the Houston toad (*Bufo houstonensis*). Journal of Herpetology 45:455–456.

Jones, M.C., D.J. Brown, I. Mali, A. McKinney, and M.R. Forstner. 2012. Assessment of public knowledge and support for recovery of the endangered Houston toad (*Bufo houstonensis*) in Bastrop, Texas. Human Dimensions of Wildlife 17:220–224.

Jones, M.E.B., A.G. Armién, B.B. Rothermel, and A.P. Pessier. 2012. Granulomatous myositis associated with a novel alveolate pathogen in an adult southern leopard frog (*Lithobates sphenocephalus*). Diseases of Aquatic Organisms 102:163–167.

Jones, M.S., and B. Stiles. 2000. *Bufo boreas* (Boreal Toad). Predation. Herpetological Review 31:99.

Jones, M.S., J.P. Goettl, and L.J. Livo. 1999. *Bufo boreas* (Boreal Toad). Predation. Herpetological Review 30:91.

Jones, R.M. 1980. Nitrogen excretion by *Scaphiopus* tadpoles in ephemeral ponds. Physiological Zoology 53:26–31.

Jones, S.M., R.E. Ballinger, and J.W. Nietfeldt. 1981. Herpetofauna of Mormon Island Preserve, Hall County, Nebraska. Prairie Naturalist 13:33-41.

Jones, T.R., and R.J. Timmons. 2010. *Hyla wrightorum* (Arizona Treefrog). Predation. Herpetological Review 41:473–474.

Jordan, D.S. 1876. Manual of the Vertebrates of the Northern United States, including the District North of North Carolina and Tennessee, Exclusive of Marine Species. First Edition. Jansen, McClurg & Co., Chicago. [multiple editions, titles, and publishers, two others listed below]

Jordan, D.S. 1890. Manual of the Vertebrates of the Northern United States, including the District North and East of the Ozark Mountains, South of the Laurentian Hills, North of the Southern Boundary of Virginia, and East of the Missouri River. Fifth Edition. A.C. McClurg & Co., Chicago.

Jordan, D.S. 1929. Manual of the Vertebrates of the Northeastern United States Inclusive of Marine Species. 13th Edition. World Book Co., Yonkers-on-Hudson, New York.

Jordan, O.R., Jr., W.W. Byrd, and D.E. Ferguson. 1968a. Sun–compass orientation in *Rana pipiens*. Herpetologica 24:335–336.

Jordan, O.R., J.S. Garton, and R.F. Ellis. 1968b. The amphibians and reptiles of a middle Tennessee cedar glade. Journal of the Tennessee Academy of Science 43:72–78.

Jørgensen, M.B., and H.C. Gerhardt. 1991. Directional hearing in the gray tree frog *Hyla versicolor*: eardrum vibrations and phonotaxis. Journal of Comparative Physiology 169A: 177–183.

Joseph, M., J. Piovia–Scott, S.P. Lawler, and K.L. Pope. 2011. Indirect effects of introduced trout on Cascades frog (*Rana cascadae*) via shared aquatic prey. Freshwater Biology 56:828–838.

Joy, J.E., and B.T. Dowell. 1994. *Glypthelmins pennsylvaniensis* (Trematode: Digenea) in the spring peeper, *Pseudacris c. crucifer* (Anura: Hylidae), from southwestern West Virginia. Journal of the Helminthological Society of Washington 61:227–229.

Joy, J.E., and C.A. Bunten. 1997. *Cosmocercoides variabilis* (Nematoda, Cosmocercoidea) populations in the eastern American toad, *Bufo a. americanus* (Salientia, Bufonidae), from western West Virginia. Journal of the Helminthological Society of Washington 64:102–105.

Joy, J.E., E.M. Walker, S.G. Koh, J.M. Bentley, and A.G. Crank. 1996. Intrahepatic larval nematode infection in the northern spring peeper, *Pseudacris crucifer crucifer* (Anura: Hylidae), in West Virginia. Journal of Wildlife Diseases 32:340–343.

Judd, F.W. 1977. Toxicity of monosodium methanearsonate herbicide to Couch's spadefoot toad, *Scaphiopus couchi*. Herpetologica 33:44–46.

Judd, F.W., and K.J. Irwin. 2005. *Hypopachus variolosus* (Cope, 1866). Sheep Frog. Pp. 506–508 *In* M.J. Lannoo (ed.), Amphibian Declines. The Conservation Status of United States Species. University of California Press, Berkeley.

Judge, K.A., and R.J. Brooks. 2001. Chorus participation by male bullfrogs, *Rana catesbeiana*: a test of the energetic constraint hypothesis. Animal Behaviour 62:849–861.

Judge, K.A., S.J. Swanson, and R.J. Brooks. 2000. *Rana catesbeiana* (Bullfrog). Female vocalization. Herpetological Review 31:236–237.

Jung, R.E. 1993. Blanchard's cricket frogs (*Acris crepitans blanchardi*) in southwest Wisconsin. Transactions of the Wisconsin Academy of Sciences, Arts and Letters 81:79–87.

Jung, R.E., and C.H. Jagoe. 1995. Effects of low pH and aluminum on body size, swimming performance, and susceptibility to predation of green tree frog (*Hyla cinerea*) tadpoles. Canadian Journal of Zoology 73:2171–2183.

Jung, R.E., S. Claeson, J.E. Wallace, and W.C. Welbourn, Jr. 2001. *Eleutherodactylus guttilatus* (Spotted Chirping Frog), *Bufo punctatus* (Red–spotted Toad), *Hyla arenicolor* (Canyon Tree Frog), and *Rana berlandieri* (Rio Grande Leopard Frog). Mite infestation. Herpetological Review 32:33–34.

Jung, R.E., K.E. Bonine, M.L. Rosenshield, A. de la Rieza, S. Raimondo, and S. Droege. 2002a. Evaluation of canoe surveys for anurans along the Rio Grande in Big Bend National Park, Texas. Journal of Herpetology 36:390–397.

Jung, R.E., G.H. Dayton, S.J. Williamson, J.R. Sauer, and S. Droege. 2002b. An evaluation of population index and estimation techniques for tadpoles in desert pools. Journal of Herpetology 36:465–472.

Jungels, J.M., K.L. Griffis–Kyle, and W.J. Boeing. 2010. Low genetic differentiation among populations of the Great Plains toad (*Bufo cognatus*) in southern New Mexico. Copeia 2010:388–396.

Jurzenski, J., and W.W. Hoback. 2011. Opossums and leopard frogs consume the federally endangered American burying beetle (Coleoptera: Silphidae). Coleopterists Bulletin 65: 88–90.

Justis, C.S., and D.H. Taylor. 1976. Extraocular photoreception and compass orientation in bullfrogs, *Rana catesbeiana*. Copeia 1976:98–105.

Justus, J.T., M. Sandomir, T. Urquhart, and B.O. Ewan. 1977. Developmental rates of two species of toads from the desert Southwest. Copeia 1977:592–594.

Kaess, W., and F. Kaess. 1960. Perception of apparent motion in the common toad. Science 132:953.

Kagarise Sherman, C. 1980. A comparison of the natural history and mating system of two anurans: Yosemite toads (*Bufo canorus*) and black toads (*Bufo exsul*). Ph.D. Dissertation, University of Michigan, Ann Arbor.

Kagarise Sherman, C., and M.L. Morton. 1984. The toad that stays on its toes. Natural History 93(3):73–78.

Kagarise Sherman, C., and M.L. Morton. 1993. Population declines of Yosemite toads in the eastern Sierra Nevada of California. Journal of Herpetology 27:186–198.

Kaiser, B., and K. Burnett. 2006. Economic impacts of *E. coqui* frogs in Hawaii. Interdisciplinary Environmental Review 8:1-11.

Kalb, H.J., and G.R. Zug. 1990. Age estimates for a population of American toads, *Bufo americanus* (Salientia: Bufonidae), in northern Virginia. Brimleyana 16:79–86.

Kalm, P. 1753-1761. En Resa til Norra America. På Kong. Swenska Vetenskaps-academiens befallning, och Publici kostnad, I-III, Stockholm.

Kalm, P. 1770-1771. Travels into North America; containing its natural history, and a circumstantial Account of its Plantations and Agriculture in general, etc. Vols. I-III. Warrington.

Kaminsky, S. 1997. *Bufo americanus* (American Toad). Reproduction. Herpetological Review 28:84.

Kapfer, J.M., and T.C. Muehlfeld. 2006. *Rana sylvatica* (Wood Frog) and *Ambystoma maculatum* (Spotted Salamander). Behavior. Herpetological Review 37:210–211.

Kaplan, H.M., and L. Yoh. 1961. Toxicity of copper for frogs. Herpetologica 17:131–135.

Kaplan, H.M., and J.G. Overpeck. 1964. Toxicity of halogenated hydrocarbon insecticides for the frog, *Rana pipiens*. Herpetologica 20:163–169.

Kaplan, H.M., and S.S. Glaczenski. 1965. Hematological effects of organophosphate insecticides in the frog (*Rana pipiens*). Life Sciences 4:1213–1219.

Karlin, A.A., D.B. Means, S.I. Guttman, and D.D. Lambright. 1982. Systematics and the status of *Hyla andersonii* (Anura: Hylidae) in Florida. Copeia 1982:175–178.

Karlstrom, E.L. 1954. On robins and tadpoles. Yosemite Nature Notes 33:185–188.

Karlstrom, E.L. 1958. Sympatry of the Yosemite and Western toads in California. Copeia 1958:152–153.

Karlstrom, E.L. 1962. The toad genus *Bufo* in the Sierra Nevada of California: ecological and systematic relationships. University of California Publications in Zoology 62:1–104.

Karlstrom, E.L. 1986. Amphibian recovery in the North Fork Toutle River debris avalanche area of Mount St. Helens. Pp. 334–344 *In* S.A.C. Keller (ed.), Mount St. Helens. Five Years Later. Eastern Washington University Press, Cheney.

Karlstrom, E.L., and R.L. Livezey. 1955. The eggs and larvae of the Yosemite toad *Bufo canorus* Camp. Herpetologica 11:221–227.

Karns, D.R. 1983. Toxic bog water in northern Minnesota peatlands: ecological and evolutionary consequences for breeding amphibians. Ph.D. Dissertation, University of Minnesota, Minneapolis.

Karns, D.R. 1992a. Effects of acidic bog habitats on amphibian reproduction in a northern Minnesota peatland. Journal of Herpetology 26:401–412.

Karns, D.R. 1992b. Amphibians and reptiles. Pp. 131-150 *In* H.E. Wright, Jr., B. Coffin, and N. Aaseng (eds.), The Patterned Peatlands of Minnesota. University of Minnesota Press, Minneapolis.

Karns, D.R. 1994. The herpetofauna of Jefferson County: analysis of an amphibian and reptile community in southeastern Indiana. Proceedings of the Indiana Academy of Science 98:535-552.

Karraker, N.E. 2001. *Ascaphus truei* (Tailed Frog). Predation. Herpetological Review 32:100.

Karraker, N.E. 2007. Are embryonic and larval green frogs (*Rana clamitans*) insensitive to road deicing salt? Herpe-tological Conservation and Biology 2:35–41.

Karraker, N.E. 2008. Impacts of road deicing salts on amphibians and their habitats. Pp. 211– 223 *In* J.C. Mitchell, R.E. Jung Brown, and B. Bartholomew (eds.), Urban Herpetology. Herpetological Conservation 3. Society for the Study of Amphibians and Reptiles, Salt Lake City, Utah.

Karraker, N.E., and G.S. Beyersdorf. 1997. A tailed frog (*Ascaphus truei*) nest site in northwestern California. Northwestern Naturalist 78:110–111.

Karraker, N.E., and J.P. Gibbs. 2009. Amphibian production in forested landscapes in relation to wetland hydroperiod: a case study of vernal pools and beaver ponds. Biological Conservation 142:2293–2302.

Karraker, N.E., and G.R. Ruthig. 2009. Effect of road deicing salt on the susceptibility of amphibian embryos to infection by water molds. Environmental Research 109:40–45.

Karraker, N.E., and J.P. Gibbs. 2011. Contrasting road effect signals in reproduction of long–versus short–lived amphibians. Hydrobiologia 664:213–218.

Karraker, N.E., D.S. Pilliod, M.J. Adams, E.L. Bull, P.S. Corn, L.V. Diller, L.A. Dupuis, M.P. Hayes, B.R. Hossack, G.R. Hodgson, E.J. Hyde, K. Lohman, B.R. Norman, L.M. Ollivier, C.A. Pearl, and C.R. Peterson. 2006. Taxonomic variation in oviposition by tailed frogs (*Ascaphus* spp.). Northwestern Naturalist 87:87–97.

Kats, L.B., and R.G. Van Dragt. 1986. Background color–matching in the spring peeper, *Hyla crucifer*. Copeia 1986:109–115.

Kats, L.B., and R. P. Ferrer. 2003. Alien predators and amphibian declines: review of two decades of science and the transition to conservation. Diversity and Distributions 9: 99–110.

Kats, L.B., J.W. Petranka, and A. Sih. 1988. Antipredator defenses and the persistence of amphibian larvae with fishes. Ecology 69:1865–1870.

Kats, L.B., J.M. Kiesecker, D.P. Chivers, and A.R. Blaustein. 2000. Effects of UV–B radiation on anti–predator behavior in three species of amphibians. Ethology 106: 921–931.

Kapust, H.Q.W., K.R. McAllister, and M.P. Hayes. 2012. Oregon spotted frog (*Rana pretiosa*) response to enhancement of oviposition habitat degraded by invasive reed grass (*Phalaris arundinacea*). Herpetological Conservation and Biology 7:358–366.

Kauffeld, C.F. 1936. New York the type locality of *Rana pipiens* Schreber. Herpetologica 1:11.

Kauffeld, C.F. 1937. The status of the leopard frogs, *Rana brachycephala* and *Rana pipiens*. Herpetologica 1:84–87.

Kawamura, T., M. Nishioka, and H. Ueda. 1980. Inter-and intraspecific, European and American Toads. Scientific

Report of the Laboratory for Amphibian Biology, Hiroshima University 4 (1): 1-125.

Kawamura, T., M. Nishioka, and H. Ueda. 1981. Interspecific hybrids among Japanese, Formosan, European and American Brown Frogs. Scientific Report of the Laboratory for Amphibian Biology, Hiroshima University 5 (9): 195-323.

Kay, D.W. 1989. Movements and homing in the canyon tree frog (*Hyla cadaverina*). Southwestern Naturalist 34:293–295.

Kay, F.R. 1969. An albino Pacifc tree frog, *Hyla regilla*, from Death Valley, California. Great Basin Naturalist 29:111.

Keel, M.K., A.M. Ruiz, A.T. Fisk, W. Rumbeiha, A.K. Davis, and J.C. Maerz. 2010. Soft–tissue mineralization of bullfrog larvae (*Rana catesbeiana*) at a wastewater treatment facility. Journal of Veterinary Diagnostic Investigation 22:655–689.

Keeton, D., and C.C. Carpenter. 1955. A study of the variations of dorsal markings of *B. w. woodhousei*. Proceedings of the Oklahoma Academy of Science 36:81–83.

Kehr, A.I. 1997. Stage–frequency and habitat selection of a cohort of *Pseudacris ocularis* tadpoles (Hylidae: Anura) in a Florida temporary pond. Herpetological Journal 7:103–109.

Keinath, D., and J. Bennett. 2000. Distribution and status of the boreal toad (*Bufo boreas boreas*) in Wyoming. Wyoming Natural Diversity Database, University of Wyoming, Laramie.

Keiser, E.D., Jr., and L.D. Wilson. 1979. Checklist and key to the amphibians and reptiles of Louisiana. Second edition. Lafayette Natural History Museum Bulletin 1.

Kelleher, K.E., and J.R. Tester. 1969. Homing and survival in the Manitoba toad, *Bufo hemiophrys*, in Minnesota. Ecology 50:1040–1048.

Keller, M.J., and H.C. Gerhardt. 2001. Polyploidy alters advertisement call structure in gray treefrogs. Proceedings of the Royal Society B 268:341–345.

Kellner, A., and D.M. Green. 1995. Age structure and age at maturity in Fowler's toads, *Bufo woodhousii fowleri*, at their northern range limit. Journal of Herpetology 29:485–489.

Kellogg, R. 1932. Notes on the spadefoot of the western plains (*Scaphiopus hammondii*). Copeia 1932:36.

Kelly, A.G. 1933. Amphibians. Pp. 35–42 *In* A.G. Kelly (ed.). Nature Notes by the Staff of the Regional Museums. Palisades Interstate Park, New York and New Jersey.

Kelly, O.V. 2010. Automated digital individual identification system with an application to the northern leopard frog *Lithobates pipiens*. Ph.D. Dissertation, Idaho State University, Pocatello.

Kelsey, K.A. 1994. Responses of headwater stream amphibians to forest practices in western Washington. Northwest Science 68:133.

Kelsey, K.A. 1995. Responses of headwater stream amphibians to forest practices in western Washington. Ph.D. Dissertation, University of Washington, Seattle.

Kempema, S.L.F. 2008. Woodhouse's toad: a lesson in learning. South Dakota Conservation Digest 75(4):28-29.

Kennedy, J.P. 1962. Spawning season and experimental hybridization of the Houston toad, *Bufo houstonensis*. Herpetologica 17:239–245.

Kennedy, J.P. 1964. Experimental hybridization of the green treefrog *Hyla cinerea* Schneider (Hylidae). Zoologica 49:211–218.

Kenney, G., K. McKean, J. Martin, and C. Stearns. 2012. Identification of terrestrial wintering habitat of *Acris crepitans* (Northern Cricket Frog). Northeastern Naturalist 19:698–700.

Kent, D.M., M.A. Langston, and D.W. Hanf. 1997. Observations of vertebrates associated with gopher tortoise burrows in Orange County, Florida. Florida Scientist 60:197–201.

Kephart, D.G. 1982. Microgeographic variation in the diets of garter snakes. Oecologia 52:287–291.

Kerby, J.L., A. Wehrmann, and A. Sih. 2012. Impacts of the insecticide diazinon on the behavior of predatory fish and amphibian prey. Journal of Herpetology 46:171–176.

Kern, M.M., J.C. Guzy, S.J. Price, S.D. Hunt, E.A. Eskew, and M.E. Dorcas. 2012. Riparian–zone amphibians and reptiles within the Broad River Basin of South Carolina. Journal of the North Carolina Academy of Science 128:81–87.

Kessel, B. 1965. Breeding dates of *Rana sylvatica* at College, Alaska. Ecology 46:206–208.

Keyser, P.D., D.J. Sausville, W.M. Ford, D.J. Schwab, and P.H. Brose. 2004. Prescribed fire impacts to amphibians and reptiles on shelterwood-harvested oak-dominated forests. Virginia Journal of Science 55:159-168.

Khaner, M. 1935. Some entozoan found in certain amphibia in Quebec Province. M.A. thesis, McGill University, Montreal.

Kiesecker, J.M. 1996. pH–mediated predator–prey interactions between *Ambystoma tigrinum* and *Pseudacris triseriata*. Ecological Applications 6:1325–1331.

Kiesecker, J.M. 2002. Synergism between trematode infection and pesticide exposure: a link to amphibian limb deformities in nature? Proceedings of the National Academy of Sciences of the United States of America 99:9900–9904.

Kiesecker, J.M., and A.R. Blaustein. 1995. Synergism between UV–B radiation and a pathogen magnifies amphibian embryo mortality in nature. Proceedings of the National Academy of Sciences of the United States of America 92:11049–11052.

Kiesecker, J.M., and A.R. Blaustein. 1997a. Population differences in responses of red–legged frogs (*Rana aurora*) to introduced bullfrogs. Ecology 78:1752–1760.

Kiesecker, J.M., and A.R. Blaustein. 1997b. Influences of egg laying behavior on pathogenic infection of amphibian eggs. Conservation Biology 11:214–220.

Kiesecker, J.M., and A.R. Blaustein. 1998. Effects of introduced bullfrogs and smallmouth bass on microhabitat use, growth, and survival of native red–legged frogs (*Rana aurora*). Conservation Biology 12:776–787.

Kiesecker, J.M., and D.K. Skelly. 2000. Choice of oviposition site by gray treefrogs: the role of potential parasitic infection. Ecology 81:2939–2943.

Kiesecker, J.M., D.P. Chivers, and A.R. Blaustein. 1996. The use of chemical cues in predator recognition by western toad tadpoles. Animal Behaviour 32:1237–1245.

Kiesecker, J.M., D.P. Chivers, A. Marco, C. Quilchano, M.T. Anderson, and A.R. Blaustein. 1999. Identification of a disturbance signal in larval red–legged frogs, *Rana aurora*. Animal Behaviour 57:1295–1300.

Kiesecker, J.M., A.R. Blaustein, and C.L. Miller. 2001a. Potential mechanisms underlying the displacement of native red–legged frogs by introduced bullfrogs. Ecology 82:1964–1970.

Kiesecker, J.M., A.R. Blaustein, and L.K. Belden. 2001b. Complex causes of amphibian population declines. Nature 410:681–684.

Kiesecker, J.M., A.R. Blaustein, and C.L. Miller. 2001c. Transfer of a pathogen from fish to amphibians. Conservation Biology 15:1064–1070.

Kiesow, A.M. 2006a. South Dakota's wood frogs rediscovered. South Dakota Conservation Digest 73(2):26-27.

Kiesow, A.M. 2006b. Field Guide to Amphibians and Reptiles of South Dakota. South Dakota Department of Game, Fish and Parks, Pierre, South Dakota.

Kiffney, P.M., and J.S. Richardson. 2001. Interactions among nutrients, periphyton, and invertebrate and vertebrate (*Ascaphus truei*) grazers in experimental channels. Copeia 2001:422–429.

Kilby, J.D. 1936. A biological analysis of the food and feeding habits of *Rana sphenocephala* (Cope) and *Hyla cinerea* (Schneider). M.S. thesis, University of Florida, Gainesville.

Kilby, J.D. 1945. A biological analysis of the food and feeding habits of two frogs. Quarterly Journal of the Florida Academy of Sciences 8:71–104.

Killebrew, F.C., and R.C. Stone, Jr. 1978. An unusual color pattern in Couch's spadefoot, *Scaphiopus couchi* (Anura: Pelobatidae). Texas Journal of Science 39:389.

Kirn, A.J. 1949. Cannibalism among *Rana pipiens berlandieri*, and possibly by *Rana catesbeiana*, near Somerset, Texas. Herpetologica 5:84.

Kim, B., T.G. Smith, and S.S. Dresser. 1998. Life history and host specificity of *Hepatozoon clamitae* (Apicomplexia: Adeleorina) and ITS–1 nucleotide sequence variation of *Hepatozoon* species of frogs and mosquitoes from Ontario. Journal of Parasitology 84:789–797.

Kim, J.B., T. Halverson, Y.J. Basir, J. Dulka, F.C. Knoop, P.W. Abel, and J.M. Conlon. 2000. Purification and characterization of antimicrobial and vasorelaxant peptides from skin extracts and skin secretions of the North American pig frog *Rana grylio*. Regulatory Peptides 90:53–60.

Kimberling, D. N., A. R. Ferreira, S. M. Shuster, and P. Keim. 1996. RAPD marker estimation of genetic structure among isolated northern leopard frog populations in the south-western USA. Molecular Ecology 5:521–529.

Kime, N.M. 2001. Female mate choice for socially–variable advertisement calls in the cricket frog, *Acris crepitans*. Ph.D. Dissertation, University of Texas, Austin.

Kime, N.M., S.S. Burmeister, and M.J. Ryan. 2004. Female preferences for socially variable call characters in the cricket frog, *Acris crepitans*. Animal Behaviour 68:1391–1399.

King, F.W. 1932. Herpetological records and notes from the vicinity of Tuscon, Arizona, July and August, 1930. Copeia 1932:175–177.

King, F.W., and T. Krakauer. 1966. The exotic herpetofauna of Florida. Quarterly Journal of the Florida Academy of Sciences 29:144–154.

King, J.J., and R.S. Wagner. 2010. Toxic effects of the herbicide Roundup® Regular on Pacific Northwest amphibians. Northwestern Naturalist 91:318–324.

King, K.A., J.C. Rorabaugh, and J.A. Humphrey. 2002. *Rana catesbeiana* (Bullfrog). Diet. Herpetological Review 33:130–131.

King, K.C., J.D. McLaughlin, A.D. Gendron, B.D. Pauli, I. Giroux, B. Rondeau, M. Boily, P. Juneau, and D.J. Marcogliese. 2007. Impacts of agriculture on the parasite communities of northern leopard frogs (*Rana pipiens*) in southern Quebec, Canada. Parasitology 134:2063–2080.

King, K.C., A.D. Gendron, J.D. McLaughlin, I. Giroux, P. Brousseau, D. Cyr, S.M. Ruby, M. Fournier, and D.J. Marcogliese. 2008. Short–term seasonal changes in parasite community structure in northern leopard froglets (*Rana pipiens*) inhabiting agricultural wetlands. Journal of Parasitology 94:13–22.

King, K.C., J.D. Mclaughlin, M. Boily, and D.J. Marcogliese. 2010. Effects of agricultural landscape and pesticides on parasitism in native bullfrogs. Biological Conservation 143:302–310.

King, O.M. 1960. Observations on Oklahoma toads. Southwestern Naturalist 5:103.

King, R.B., and B. King. 1991. Sexual differences in color and color change in wood frogs. Canadian Journal of Zoology 69:1963–1968.

King, R.B., S. Hauff, and J.B. Phillips. 1994. Physiological color change in the green treefrog: responses to background brightness and temperature. Copeia 1994:422–432.

King, R.B., M.J. Oldham, W.F. Weller, and D. Wynn. 1997. Historic and current amphibian and reptile distributions in the island region of western Lake Erie. American Midland Naturalist 138:153–173.

King, W. 1939. A survey of the herpetology of Great Smoky Mountains National Park. American Midland Naturalist 21:531–582.

Kingsbury, B., and J. Gibson (eds.). 2002. Habitat Management Guidelines for Amphibians and Reptiles of the Midwest. Partners in Amphibian and Reptile Conservation, Technical Publication HMG–1, Montgomery, Alabama.

Kingsbury, B., and J. Gibson (eds.). 2012. Habitat Management Guidelines for Amphibians and Reptiles of the Midwest. Partners in Amphibian and Reptile Conservation, Technical Publication HMG–1, 2nd Edition, Montgomery, Alabama.

Kinney, V. C. 2011. Adult survivorship and juvenile recruitment in populations of Crawfish Frogs (*Lithobates areolatus*) with additional consideration of the population sizes of associated pond breeding species. M.S. thesis, Indiana State University, Terre Haute.

Kinney, V.C., and M.J. Lannoo. 2010. *Lithobates areolatus circulosus* (Northern Crawfish Frog). Breeding. Herpetological Review 41:197–198.

Kinney, V.C., J.L. Heemeyer, A.P. Pessier, and M.J. Lannoo. 2011. Seasonal pattern of *Batrachochytrium dendrobatidis* infection and mortality in *Lithobates areolatus*: affirmation of Vredenburg's "10,000 zoospore rule." PLoS One 6:e16708.

Kirk, J.J. 1988. Western spotted frog (*Rana pretiosa*) mortality following forest spraying of DDT. Herpetological Review 19:51–53.

Kirkland, A.H. 1897. The habits, food and economic value of the American toad, *Bufo lentiginosus americanus* (LeC.). Hatch Experiment Station of the Massachusetts Agricultural College, Bulletin No. 46.

Kirkland, A.H. 1904. Usefulness of the American toad. U.S. Department of Agriculture Farmer's Bulletin No. 196.

Kirkland, G.L., Jr., H.W. Snoddy, and T.L. Amsler. 1996. Impact of fire on small mammals and amphibians in a central Appalachian deciduous forest. American Midland Naturalist 135:253–260.

Kirkman, K.L., L.L. Smith, P.F. Quintana–Ascencio, M.J. Kaesner, S.W. Golladay, and A.L. Farmer. 2012. Is species richness congruent among taxa? Surrogacy, complementarity, and environmental correlates among three disparate taxa in geographically isolated wetlands. Ecological Indicators 18:131–139.

Kirton, M.P. 1974. Fall movements and hibernation of the wood frog, *Rana sylvatica*, in interior Alaska. M.S. thesis, University of Alaska, Fairbanks.

Kiviat, E. 1982. Geographic distribution. *Scaphiopus holbrookii holbrooki* (Eastern Spadefoot). Herpetological Review 13:51–52.

Kiviat, E. 2011. Frog call surveys in an urban wetland complex, the Hackensack Meadowlands, New Jersey, in 2006. Urban Habitats 6.

Kiviat, E., and J. Stapleton. 1983. *Bufo americanus* (American toad). Estuarine habitat. Herpetological Review 14:46.

Klaus, S.P. 2012. Correlates and temporal variation in call phenology of eastern Ontario frogs. M.S. thesis, Queen's University, Kingston, Ontario.

Kleinhenz, P., M.D. Boone, and G. Fellers. 2012. Effects of the amphibian chytrid fungus and four insecticides on Pacific Treefrogs (*Pseudacris regilla*). Journal of Herpetology 46:625–631.

Klemens, M.W. 1993. Amphibians and Reptiles of Connecticut and Adjacent Regions. State Geological and Natural History Survey of Connecticut, Bulletin 112.

Klemens, M.W. 2000. Amphibians & Reptiles in Connecticut. Connecticut Department of Environmental Protection, Bulletin 32.

Klemens, M.W., E. Kiviat, and R.E. Schmidt. 1987. Distribution of the northern leopard frog, *Rana pipiens*, in the lower Hudson and Housatonic River valleys. Northeastern Environmental Science 6:99–101.

Klemish, J.L., B.L. Johnson, E.R. Siddons, and E.R. Wild. 2012. Occurrence of *Batrachochytrium dendrobatidis* among populations of *Lithobates clamitans* and *L. pipiens* in Wisconsin, USA. Herpetological Review 43:282–288.

Kleopfer, J.D., and C.S. Hobson. 2011. A guide to the Frogs and Toads of Virginia. Bureau of Wildlife Special Publication No. 3, Virginia Department of Game & Inland Fisheries, Richmond. 44 pp.

Klimkiewicz, M.K. 1972. Amphibians of Mason Neck. Atlantic Naturalist 27:65-68.

Klimstra, W.D. 1949. Early bullfrog transformation. Copeia 1949:231.

Klimstra, W.D., and C.W. Myers. 1965. Foods of the toad, *Bufo woodhousei fowleri* Hinckley. Transactions of the Illinois State Academy of Science 58:11–26.

Kline, J. 1998. Monitoring amphibians in created and restored wetlands. Pp. 360–368 *In* M.J. Lannoo (ed.), Status & Distribution of Midwestern Amphibians. University of Iowa Press, Iowa City.

Klomberg, K.F., and C.A. Marler. 2000. The neuropeptide arginine vasotocin alters male call characteristics involved in social interactions in the grey treefrog, *Hyla versicolor*. Animal Behaviour 59:807–812.

Kluger, M.J. 1977. Fever in the frog *Hyla cinerea*. Journal of Thermal Biology 2:79–81.

Klugh, A.B. 1922. The economic value of the leopard frog. Copeia (102):14–15.

Klump, G.M., and H.C. Gerhardt. 1986. Rufstrategtien von Laubfroschmannchen (*Hyla* species) und Partnerwahl der Weibchen. Verhandlungen der Deutschen Zoologischen Gesellschaft 76:25–35.

Klump, G.M., and H.C. Gerhardt. 1987. Use of non–arbitrary acoustic criteria in mate choice by female gray tree frogs. Nature 326:286–288.

Klymus, K.E., S.C. Humfeld, V.T. Marshall, D. Cannatella, and H.C. Gerhardt. 2010. Molecular patterns of differentiation in canyon treefrogs (*Hyla arenicolor*): evidence for introgressive hybridization with the Arizona treefrog (*H. wrightorum*) and correlations with advertisement call differences. Journal of Evolutionary Biology 23:1425–1435.

Knapp, R.A. 2005. Effects of nonnative fish and habitat characteristics on lentic herpetofauna in Yosemite National Park, USA. Biological Conservation 121:265–279.

Knapp, R.A., and K.R. Matthews. 2000. Non–native fish introductions and the decline of the mountain yellow–legged frog from within protected areas. Conservation Biology 14:428–438.

Knapp, R.A., and J.A.T. Morgan. 2006. Tadpole mouthpart depigmentation as an accurate indicator of chytridiomycosis, an emerging disease of amphibians. Copeia 2006:188–197.

Knapp, R.A., K.R. Matthews, and O. Sarnelle. 2001. Resistance and resilience of alpine lake fauna to fish introductions. Ecological Monographs 71:401–421.

Knapp, R.A., K.R. Matthews, H.K. Preisler, and R. Jellison. 2003. Developing probabilistic models to predict amphibian site occupancy in a patchy landscape. Ecological Applications 13:1069–1082.

Knapp, R.A., D.M. Boiano, and V.T. Vredenburg. 2007. Removal of nonnative fish results in a population expansion of a declining amphibian (mountain yellow–legged frog, *Rana muscosa*). Biological Conservation 135:11–20.

Knapp, R.A., C.J. Briggs, T.C. Smith, and J.R. Maurer. 2011. Nowhere to hide: impact of a temperature–sensitive amphibian pathogen along an elevation gradient in the temperate zone. Ecosphere 2:1–26.

Knight, C.M., M.J. Parris, and W.H.N. Gutzke. 2009. Influence of priority effects and pond location on invaded larval amphibian communities. Biological Invasions 11:1033-1044.

Knight, E.L. 1972. Notes on cave–dwelling *Rana palustris* in Mississippi. Bulletin of the Maryland Herpetological Society 8:88.

Knight, M.T., G.J. Barbay, and E.O. Morrison. 1965. Incidence of infection by lung–fluke (*Haematoleochus*) of the bullfrog, *Rana catesbeiana*, in Jefferson County, Texas. Southwestern Naturalist 10:141–142.

Knowlton, G.F. 1944. Some insect food of *Rana pipiens*. Copeia 1944:119.

Knutson, M.G., J.R. Sauer, D.A. Olsen, M.J. Mossman, L.M. Hemesath, and M.J. Lannoo. 1999. Effects of landscape composition and wetland fragmentation on frog and toad abundance and species richness in Iowa and Wisconsin, U.S.A. Conservation Biology 13:1437-1446.

Knutson, M.G., J.R. Sauer, D.A. Olsen, M.J. Mossman, L.M. Hemesath, and M.J. Lannoo. 2000. Landscape associations of frogs and toad species in Iowa and Wisconsin, U.S.A. Journal of the Iowa Academy of Science 107:134–145.

Knutson, M.G., W.B. Richardson, D.M. Reineke, B.R. Gray, J.R. Parmelee, and S.E. Weick. 2004. Agricultural ponds support amphibian populations. Ecological Applications 14:669–684.

Koch, E.D., and C.R. Peterson. 1995. Amphibians & Reptiles of Yellowstone and Grand Teton National Parks. University of Utah Press, Salt Lake City.

Kocher, T.D., and R.D. Sage. 1986. Further genetic analyses of a hybrid zone between leopard frogs (*Rana pipiens* complex) in central Texas. Evolution 40:21–33.

Kofron, C.P. 1978. Foods and habitats of aquatic snakes (Reptilia, Serpentes) in a Louisiana swamp. Journal of Herpetology 12:543–554.

Koller, R.L., and A.J. Gaudin. 1977. An analysis of helminth infections in *Bufo boreas* (Amphibia: Bufonidae) and *Hyla regilla* (Amphibia: Hylidae) in southern California. Southwestern Naturalist 21:503–509.

Kolozsvary, M.B. 2003. Hydroperiod of wetlands and reproduction in wood frogs (*Rana sylvatica*) and spotted salamanders (*Ambystoma maculatum*). Ph.D. Dissertation, University of Maine, Orono.

Kolozsvary, M.B., and R.K. Swihart. 1999. Habitat fragmentation and the distribution of amphibians: patch and landscape correlates in farmland. Canadian Journal of Zoology 77:1288–1299.

Kolter, P.C. 1963. A hermaphroditic bullfrog. Turtox News 41(12):290–293.

Komarek, E.V., Sr. 1969. Fire and animal behavior. Proceedings of the Tall Timbers Conference on Fire Ecology 9:161–167.

Kopeny, M. 1988. White–tailed hawk. Pp. 97–104 *In* Proceedings of the Southwest Raptor Management Symposium, National Wildlife Federation, Washington, D.C.

Koprivnikar, J. 2010. Interactions of environmental stressors impact survival and development of parasitized larval amphibians. Ecological Applications 20:2263–2272.

Koprivnikar, J., R.L. Baker, and M.R. Forbes. 2006a. Environmental factors influencing trematode prevalence in grey tree frog (*Hyla versicolor*) tadpoles in southern Ontario. Journal of Parasitology 92:997–1001.

Koprivnikar, J., M. R. Forbes, and R. L. Baker. 2006b. On the efficacy of anti–parasite behaviour: a case study of tadpole susceptibility to cercariae of *Echinostoma trivolvis*. Canadian Journal of Zoology 84:1623–1629.

Koprivnikar, J., M. R. Forbes, and R. L. Baker. 2008. Larval amphibian growth and development under varying density: are parasitized individuals poor competitors? Oecologia 155:641– 649.

Koprivnikar, J., D.J. Marcogliese, J.R. Rohr, S.A. Orlofske, T.R. Raffel, and P.T.J. Johnson. 2012. Macroparasite infections of amphibians: What can they tell us? Eco-Health 9:342–360.

Korb, R.M. 2001. Wisconsin Frogs. Places to hear frogs and toads near our urban areas. Northeastern Wisconsin Audubon, Inc., Green Bay.

Korfel, C.A. 2012. Distribution and environmental correlates between amphibians and the fungal pathogen, *Batrachochytrium dendrobatidis*. Ph.D. Dissertation, Ohio State University, Columbus.

Korfel, C.A., W.J. Mitsch, T.E. Hetherington, and J.J. Mack. 2010. Hydrology, physiochemistry, and amphibians in natural and created vernal pool wetlands. Restoration Ecology 18:843–854.

Korky, J.D. 1978. Differentiation of the larvae of members of the *Rana pipiens* complex in Nebraska. Copeia 1978:455–459.

Korky, J.K., and J.A. Smallwood. 2011. Geographic variation in northern green frog larvae, *Lithobates clamitans melanotus*, in northwestern New Jersey. Bulletin of the Maryland Herpetological Society 47:1–10.

Kornilev, Y.V. 2007. *Bufo terrestris* (Southern toad). Arboreal behavior. Herpetological Review 38:319.

Korschgen, L.J. 1970. Soil–food–chain–pesticide wildlife relationships in aldrin–treated fields. Journal of Wildlife Management 34:186–199.

Korschgen, L.J., and D.L. Moyle. 1955. Food habits of the bullfrog in central Missouri farm ponds. American Midland Naturalist 54:332–341.

Korschgen, L.J., and T.S. Baskett. 1963. Foods of impound-ment– and stream–dwelling bullfrogs in Missouri. Herpetologica 19:89–99.

Koster, W.J. 1946. The robber frog in New Mexico. Copeia 1946:173.

Krakauer, T. 1968. The ecology of the Neotropical toad, *Bufo marinus*, in south Florida. Herpetologica 24:214–221.

Krakauer, T. 1970a. The invasion of the toads. Florida Naturalist 43(1):12–14.

Krakauer, T. 1970b. Tolerance limits of the toad, *Bufo marinus*, in south Florida. Comparative Biochemistry and Physiology 33:15–26.

Kramek, W.C. 1972. Food of the frog *Rana septentrionalis* in New York. Copeia 1972:390–392.

Kramek, W.C. 1976. Feeding behavior of *Rana septentrionalis* (Amphibia, Anura, Ranidae). Journal of Herpetology 10:251–252.

Kramek, W.C., and M.M. Stewart. 1980. Ontogenetic and sexual differences in the pattern of *Rana septentrionalis*. Journal of Herpetology 14:369–375.

Kramer, A.B. 2011. Effects of the aquatic to terrestrial habitat ratio on an amphibian predator and its prey. Ph.D. Dissertation, Washington University, St. Louis, Missouri.

Kramer, D.C. 1971. Seasonal movements of western chorus frogs (*Pseudacris triseriata triseriata*) tagged with radioactive cobalt. Ph.D. Dissertation, Ball State University, Muncie, Indiana.

Kramer, D.C. 1973. Movements of western chorus frogs *Pseudacris triseriata triseriata* tagged with Co[60]. Journal of Herpetology 7:231–235.

Kramer, D.C. 1974. Home range of the western chorus frog *Pseudacris triseriata triseriata*. Journal of Herpetology 8:245–246.

Kramer, D.C. 1978. Viability of the eggs of *Pseudacris triseriata* (Amphibia, Anura, Hylidae). Journal of Herpetology 12:119–120.

Kraus, F. 2009. Alien Reptiles and Amphibians. A Scientific Compendium and Analysis. Springer, Dordrecht, The Netherlands.

Kraus, F., and E.W. Campbell III. 2002. Human–mediated escalation of a formerly eradicable problem: the invasion of Caribbean frogs in the Hawaiian Islands. Biological Invasions 4:327–332.

Kraus, F., and F. Duvall. 2004. New records of alien reptiles and amphibians in Hawai'i. Bishop Museum Occasional Papers 79:62–64.

Kraus, F., and D.C. Duffy. 2010. A successful model from Hawaii for rapid response to invasive species. Journal for Nature Conservation 18:135–141.

Kraus, F., E.W. Campbell, A. Allison, and T. Pratt. 1999. *Eleutherodactylus* frog introductions to Hawaii. Herpetological Review 30:21–25.

Kress, S.W. 1965. Bullfrogs sing along with the jets. Audubon Magazine 67(2):93–96.

Kretzer, J.E., and J.F. Cully, Jr. 2001. Effects of black–tailed prairie dogs on reptiles and amphibians in Kansas short-grass prairie. Southwestern Naturalist 46:171–177.

Krivda, W. 1976. Movement of spring peepers. Blue Jay 34(1):17.

Kroll, A.J. 2009. Sources of uncertainty in stream–associated amphibian ecology and responses to forest management in the Pacific Northwest, USA: a review. Forest Ecology and Management 257:1188–1199.

Krupa, J.J. 1986a. Multiple egg clutch production in the Great Plains toad. Prairie Naturalist 18:151–152.

Krupa, J.J. 1986b. Distribution in Oklahoma of the bird–voiced treefrog (*Hyla avivoca*). Proceedings of the Oklahoma Academy of Science 66:37–38.

Krupa, J.J. 1986c. Anuran breeding dates in central Oklahoma. Proceedings of the Oklahoma Herpetological Society 11:10–13.

Krupa, J.J. 1987. Mate choice and mate location tactics in the Great Plains toad (*Bufo cognatus*). Ph.D. Dissertation, University of Oklahoma, Norman.

Krupa, J.J. 1988. Fertilization efficiency of the Great Plains toad (*Bufo cognatus*). Copeia 1988:800–802.

Krupa, J.J. 1989. Alternative mating tactics in the Great Plains toad. Animal Behaviour 37:1035–1043.

Krupa, J.J. 1990. Advertisement call variation in the Great Plains toad. Copeia 1990:884–886.

Krupa, J.J. 1994. Breeding biology of the Great Plains toad in Oklahoma. Journal of Herpetology 28:217–224.

Krupa, J.J. 1995. *Bufo woodhousii* (Woodhouse's Toad). Fecundity. Herpetological Review 26: 142, 144.

Krupa, J.J. 2002. Temporal shift in diet in a population of American bullfrog (*Rana catesbeiana*) in Carlsbad Caverns National Park. Southwestern Naturalist 47:461–467.

Kruse, K.C. 1973. Two morphotypes of the *Rana pipiens* complex in the central Plains states. M.S. thesis, Wayne State College, Detroit, Michigan.

Kruse, K.C. 1978. Causal factors limiting the distribution of Leopard Frogs in eastern Nebraska. Ph.D. Dissertation, University of Nebraska, Lincoln.

Kruse, K.C. 1981a. Mating success, fertilization potential, and male body size in the American toad *(Bufo americanus)*. Herpetologica 37:228–233.

Kruse, K.C. 1981b. Phonotactic responses of female northern leopard frogs (*Rana pipiens*) to *Rana blairi*, a presumed hybrid, and conspecific mating trills. Journal of Herpetology 15:145–150.

Kruse, K.C., and D.G. Dunlap. 1976. Serum albumins and hybridization in two species of the *Rana pipiens* complex in the north central United States. Copeia 1976:394–396.

Kruse, K.C., and M.G. Francis. 1977. A predation deterrent in larvae of the bullfrog, *Rana catesbeiana*. Transactions of the American Fisheries Society 106:248–252.

Kruse, K.C., and M. Mounce. 1982. The effects of multiple matings on fertilization capability in male American toads (*Bufo americanus*). Journal of Herpetology 16:410–412.

Kruse, K.C., and B.M. Stone. 1984. Largemouth bass (*Micropterus salmoides*) learn to avoid feeding on toad (*Bufo*) tadpoles. Animal Behaviour 32:1035–1039.

Krynak, T.J., T.L. Robison, and J.J. Scott. 2012. Detection of *Batrachochytrium dendrobatidis* in amphibian populations of northeast Ohio. Herpetological Review 43:87–89.

Krysko, K.L., and C.M. Sheehy III. 2005. Ecological status of the ocellated gecko, *Sphaerodactylus argus argus* Gosse 1850, in Florida, with additional herpetological notes from the Florida Keys. Caribbean Journal of Science 41:169–172.

Krysko, K.L., J.P. Burgess, M.R. Rochford, C.R. Gillette, D. Cueva, K.M. Enge, L.A. Somma, J.L. Stabile, D.C. Smith, J.A. Wasilewski, G.N. Kieckhefer III, M.C. Granatosky, and S.V. Nielsen. 2011. Verified non–indigenous amphibians and reptiles in Florida from 1863 through 2010: outlining the invasion process and identifying invasion pathways and stages. Zootaxa 3028:1–64.

Kubel, J.E., K.L. Derge, J. Williams, and R.H. Yahner. 2002. Amphibian and reptile inventory at Antietam National Battlefield, Maryland. Maryland Naturalist 45:59-74.

Kuchta, S.R. 2008. Amphibians. Pp. 198-220 *In* T.M. Haff, M.T. Brown, and W.B. Tyler (eds.), The Natural History of the UC Santa Cruz Campus. Second Edition. Environmental Studies Department, University of California, Santa Cruz, California.

Kuczynski, M., A. Vélez, J.J. Schwartz, and M.A. Bee. 2010. Sound transmission and the recognition of temporally degraded call structure in Cope's gray treefrog (*Hyla chrysoscelis*). Journal of Experimental Biology 213:2840–2850.

Kuntz, R.E. 1940. A study of anuran parasites of Commanche County, Oklahoma. M.S. thesis, University of Oklahoma, Norman.

Kuntz, R.E. 1941. The metazoan parasites of some Oklahoma anura. Proceedings of the Oklahoma Academy of Science 21:33–34.

Kuntz, R.E., and J.T. Self. 1943. An ecological study of the metazoan parasites of the Salientia of Commanche County, Oklahoma. Proceedings of the Oklahoma Academy of Science 24:35–38.

Kuperman, B.I., V.E. Matey, R.N. Fisher, E.L. Ervin, M.L. Warburton, L. Bakhireva, and C.A. Lehman. 2004. Parasites of the African clawed frog, *Xenopus laevis*, in Southern California, U.S.A. Comparative Parasitology 71:229–232.

Kupferberg, S.J. 1994. Exotic larval bullfrogs (*Rana catesbeiana*) as prey for native garter snakes: functional and conservation implications. Herpetological Review 25:95–97.

Kupferberg, S. J. 1996a. The ecology of native tadpoles (*Rana boylii* and *Hyla regilla*) and the impact of invading bullfrogs (*Rana catesbeiana*) in a northern California river. Ph.D. Dissertation, University of California, Berkeley.

Kupferberg, S.J. 1996b. Hydrologic and geomorphic factors affecting conservation of a river–breeding frog (*Rana boylii*). Ecological Applications 6:1332–1344.

Kupferberg, S.J. 1997a. Bullfrog (*Rana catesbeiana*) invasion of a California river: the role of larval competition. Ecology 78:1736–1751.

Kupferberg, S.J. 1997b. The role of larval diet in anuran metamorphosis. American Zoologist 37:146–159.

Kupferberg, S.J. 1997c. Facilitation of periphyton production by tadpole grazing: functional differences between species. Freshwater Biology 37:427–439.

Kupferberg, S.J. 1998. Predator mediated patch use by tadpoles (*Hyla regilla*): risk balancing or consequence of motionlessness? Journal of Herpetology 32:84–92.

Kupferberg, S.J., J.C. Marks, and M.E. Power. 1994. Effects of variation in natural algal and detrital diets on larval anuran (*Hyla regilla*) life–history traits. Copeia 1994:446–457.

Kupferberg, S.J., A. Catenazzi, K. Lunde, A.J. Lind, and W.J. Palen. 2009. Parasitic copepod (*Lernaea cyprinacea*) outbreaks in foothill yellow–legged frogs (*Rana boylii*) linked to unusually warm summers and amphibian malformations in northern California. Copeia 2009:529–537.

Kupferberg, S.J., A.J. Lind, V. Thill, and S.M. Yarnell. 2011. Water velocity tolerance in tadpoles of the foothill yellow–legged frog (*Rana boylii*): swimming performance, growth, and survival. Copeia 2011:141–152.

Kurzava, L.M., and P.J. Morin. 1998. Tests of functional equivalence: complementary roles of salamanders and fish in community organization. Ecology 79:477–489.

Kutka, F.J., and M.D. Bachmann. 1990. Acid sensitivity and water chemistry correlates of amphibian breeding ponds in northern Wisconsin, USA. Hydrobiologia 208:153–160.

Kuyt, E. 1991. A communal overwintering site for the Canadian toad, *Bufo americanus hemiophrys*, in the Northwest Territories. Canadian Field-Naturalist 105:119–121.

Kyser, P.D., D.J. Sausville, W.M. Ford, D.J. Schwab, and P.H. Brosew. 2004. Prescribed fire impacts to amphibians and reptiles in shelterwood–harvest oak–dominated forests. Virginia Journal of Science 55:159–168.

Labanick, G.M. 1976. Prey availability, consumption and selection in the cricket frog, *Acris crepitans* (Amphibia, Anura, Hylidae). Journal of Herpetology 10:293–298.

Labanick, G.M., and R.A. Schlueter. 1976. Growth rates of recently transformed *Bufo woodhousei fowleri*. Copeia 1976:824–826.

Lacan, I., K. Matthews, and K. Feldman. 2008. Interaction of an introduced predator with future effects of climate change in the recruitment dynamics of the imperiled Sierra Nevada yellow–legged frog (*Rana sierrae*). Herpetological Conservation and Biology 3:211–223.

Lacki, M.J., J.W. Hummer, and H.J. Webster. 1992. Mine–drainage treatment wetland as habitat for herpetofaunal wildlife. Environmental Management 16:513–520.

Lacoul, P., B. Freedman, and T. Clair. 2011. Effects of acidification on aquatic biota in Atlantic Canada. Environmental Reviews 19: 429–460.

Laerm, J., and A.S. Hopkins, Jr. 1997. Status of possible disjunct populations of the southern toad, *Bufo terrestris*, in the Piedmont and Blue Ridge of Georgia. Herpetological Review 28:162–163.

Laerm, J., N.K. Castleberry, M.A. Menzel, R.A. Moulis, G.K. Williamson, J.B. Jensen, B. Winn, and M.J. Harris. 2000. Biogeography of amphibians and reptiles of the Sea Islands of Georgia. Florida Scientist 63:193–231.

Lafferty, K.D., and C.J. Page. 1997. Predation on the endangered tidewater goby, *Eucyclogobius newberryi*, by the introduced African clawed frog, *Xenopus laevis*, with notes on the frog's parasites. Copeia 1997:589–592.

LaFiandra, E.M., and K.J. Babbitt. 2004. Predator induced phenotypic plasticity in the pinewoods tree frog, *Hyla femoralis*: necessary cues and the cost of development. Oecologia 138:350–359.

Lamb, A.C. III. 1980. Observations on the life history of the pig frog, *Rana grylio* Stejneger, in southwest Georgia. M.S. thesis. Auburn University, Auburn, Alabama.

Lamb, A.C. III. 1986. Hybridization and introgression between two species of treefrogs: mitochondrial DNA analysis. Ph.D. Dissertation, University of Georgia, Athens.

Lamb, T. 1984a. Amplexus displacement in the southern toad, *Bufo terrestris*. Copeia 1984:1023–1025.

Lamb, T. 1984b. The influence of sex and breeding condition on microhabitat selection and diet in the pig frog *Rana grylio*. American Midland Naturalist 111:311–318.

Lamb, T. 1987. Call site selection in a hybrid population of treefrogs. Animal Behaviour 35:1140–1144.

Lamb, T., and J.C. Avise. 1986. Directional introgression of mitochondrial DNA in a hybrid population of tree frogs: the influence of mating behavior. Proceedings of the National Academy of Sciences of the United States of America 83:2526–2530.

Lamb, T., and J.C. Avise. 1987. Morphological variability in genetically defined categories of anuran hybrids. Evolution 41:157–163.

Lamb, T., J.M. Novak, and D.L. Mahoney. 1990. Morphological asymmetry and interspecific hybridization: a case study using hylid frogs. Journal of Evolutionary Biology 3:295–309.

Lamb, T., R.W. Gaul, Jr., M.L. Tripp, J.M. Horton, and B.W. Grant. 1998. A herpetofaunal inventory of the lower Roanoke River floodplain. Journal of the Elisha Mitchell Scientific Society 114:43–55.

Lamb, T., B.K. Sullivan, and K. Malmos. 2000. Mitochondrial gene markers for the hybridizing toads *Bufo microscaphus* and *Bufo woodhousii* in Arizona. Copeia 2000:234–237.

Lamoureux, V.S., and D.M. Madison. 1999. Overwintering habitats of radio–implanted green frogs, *Rana clamitans*. Journal of Herpetology 33:430–435.

Lamoureux, V.S., J.C. Maerz, and D.M. Madison. 2002. Premigratory autumn foraging forays in the green frog, *Rana clamitans*. Journal of Herpetology 36:245–254.

Lampman, B.H. 1946. The Coming of the Pond Fishes. Binfords & Mort Publishers, Portland, Oregon.

Lance, S.L., and K.D. Wells. 1993. Are spring peeper satellite males physiologically inferior to calling males? Copeia 1993:1162–1166.

Lance, S.L., M. R. Erickson, R. W. Flynn, G. L. Mills, T. D. Tuberville, and D. E. Scott. 2012. Effects of chronic copper exposure on development and survival in the southern leopard frog (*Lithobates* [*Rana*] *sphenocephalus*). Environmental Toxicology and Chemistry 31:1587–1594.

Lanctot, C. 2012. The effects of glyphosate–based herbicides on the development of wood frogs, *Lithobates sylvaticus*. M.S. thesis, University of Ottawa, Ottawa, Ontario.

Landé, S.P., and S.I. Guttman. 1973. The effects of copper sulfate on the growth and mortality rate of *Rana pipiens* tadpoles. Herpetologica 29:22–27.

Landreth, H.F., and D.E. Ferguson. 1966a. Behavioral adaptations in the chorus frog, *Pseudacris triseriata*. Journal of the Mississippi Academy of Sciences 12:197–202.

Landreth, H.F., and D.E. Ferguson. 1966b. Celestial orientation of Fowler's toad, *Bufo fowleri*. Behaviour 26:105–123.

Landreth, H.F., and D.E. Ferguson. 1966c. Evidence of sun–compass orientation in the chorus frog, *Pseudacris triseriata*. Herpetologica 22:106–112.

Landreth, H.F., and D.E. Ferguson. 1967. Movements and orientation of the tailed frog, *Ascaphus truei*. Herpetologica 23:81–93.

Landreth, H.F., and D.E. Ferguson. 1968. The sun compass of Fowler's toad, *Bufo woodhousei fowleri*. Behaviour 30:27–43.

Landreth, H.F., and M.T. Christensen. 1971. Orientation of the Plains spadefoot toad, *Scaphiopus bombifrons*, to solar cues. Herpetologica 27:454–461.

Lang, E., C. Gerhardt, and C. Davidson. 2009. The Frogs and Toads of North America. A Comprehensive Guide to their Identification, Behavior, and Calls. Houghton Mifflin Harcourt, Boston, Massachusetts.

Langen, T. A., A. Machniak, E. K. Crowe, C. Mangan, D. F. Marker, N. Liddle, and B. Roden. 2007. Methodologies for surveying herpetofauna mortality on rural highways. Journal of Wildlife Management 71:1361–1368.

Langen, T. A., K. M. Ogden, and L. L. Schwarting. 2009. Predicting hot spots of herpetofauna road mortality along highway networks. Journal of Wildlife Management 73:104–114.

Langford, G.J., J.A. Borden, C.S. Major, and D.H. Nelson. 2007. Effects of prescribed fire on the herpetofauna of a southern Mississippi pine savanna. Herpetological Conservation and Biology 2:135–143.

Langlois, T.H. 1964. Amphibians and reptiles of the Erie islands. Ohio Journal of Science 64:11–25.

Langlois, V.S., A.C. Carew, B.D. Pauli, M.G. Wade, G.M. Cooke, and V.L. Trudeau. 2010. Low levels of the herbicide atrazine alters sex ratios and reduces metamorphic success in *Rana pipiens* tadpoles raised in outdoor mesocosms. Environmental Health Perspectives 118:552–557.

Lannoo, M. 1996. Okoboji Wetlands. A Lesson in Natural History. University of Iowa Press, Iowa City.

Lannoo, M. J. 1998. Amphibian conservation and wetland management in the upper Midwest: a catch–22 for the cricket frog? Pp. 330–339 *In* M.J. Lannoo (ed.), Status & Distribution of Midwestern Amphibians. University of Iowa Press, Iowa City.

Lannoo, M. J. (ed.). 2005. Amphibian Declines. The Conservation Status of United States Species. University of California Press, Berkeley.

Lannoo, M.J. 2008. Malformed Frogs. The Collapse of Aquatic Ecosystems. University of California Press, Berkeley.

Lannoo, M.J., K. Lang, T. Waltz, and G.S. Phillips. 1994. An altered amphibian assemblage: Dickinson County, Iowa, 70 years after Frank Blanchard's survey. American Midland Naturalist 131:311–319.

Lannoo, M.J., J.A. Holman, G.S. Casper, and E. Johnson. 1998. Mummification following winterkill of adult green frogs (Ranidae: *Rana clamitans*). Herpetological Review 29:82–84.

Lannoo, M.J., V.C. Kinney, J.L. Heemeyer, N.J. Engbrecht,

A.L. Gallant, and R.W. Klaver. 2009. Mine spoil prairies expand critical habitat for endangered and threatened amphibian and reptile species. Diversity 1:118–132.

Lannoo, M.J., C. Petersen, R.E. Lovich, P. Nanjappa, C. Phillips, J.C. Mitchell, and I. Macallister. 2011. Do frogs get their kicks from Route 66? Continental U.S. transect reveals spatial and temporal patterns of *Batrachochytrium dendrobatidis* infection. PLoS One 6(7):e22211.

LaPointe, J. 1953. Noteworthy amphibian records from Indiana Dunes State Park. Copeia 1953:129.

Laposata, M.M., and W.A. Dunson. 1998. Effects of boron and nitrate on hatching success of amphibian eggs. Archives of Environmental Contamination and Toxicology 35:615–619.

Lardie, R.L., and J.H. Black. 1981. The amphibians and reptiles of the Cimarron Gypsum Hills region in northwestern Oklahoma. Bulletin of the Oklahoma Herpetological Society 5:76–125.

La Rivers, I. 1948. Some Hawaiian ecological notes. The Wasmann Collector 7:85–110.

Larochelle, A. 1974. American toad as champion carabid beetle collector. Pan-Pacific Entomologist 50:203-204.

Larsen, K.W., and P.T. Gregory. 1988. Amphibians and reptiles in the Northwest Territories. Occasional Papers of the Prince of Wales Northern Heritage Centre No. 3:31–51.

Larson, K. 1997. Faunal diversity and richness of natural, restored, dam-created, and borrow pit wetlands in the Prairie Pothole Region of eastern South Dakota. M.S. thesis, South Dakota State University, Brookings.

Larson, K.A. 2004. Advertisement call complexity in northern leopard frogs, *Rana pipiens*. Copeia 2004:676–682.

Larson, M.D. 2012. Diet of the Cascades frog (*Rana cascadae*) as it relates to prey availability in the Klamath Mountains of northwest California. M.S. thesis, Humboldt State University, Arcata, California.

Latham, R. 1968a. Notes on the eating of May beetles by a Fowler's toad. Engelhardtia 1:29.

Latham, R. 1968b. Notes on some hibernating Fowler's toads. Engelhardtia 1:17.

Latham, R. 1970. The diet of shrikes on Long Island. Engelhardtia 3:29.

Latham, R. 1971. The leopard frog on eastern Long Island. Engelhardtia 4:58.

Laurin, M. 2010. How Vertebrates Left the Water. University of California Press, Berkeley.

Lawler, J.J., S.L. Shafer, B.A. Bancroft, and A.R. Blaustein. 2009. Projected climate impacts for the amphibians of the western hemisphere. Conservation Biology 24: 38–50.

Lawler, S.P. 1989. Behavioural responses to predators and predation risk in four species of larval anurans. Animal Behaviour 38:1039–1047.

Lawler, S.P., and P.J. Morin. 1993. Temporal overlap, competition, and priority effects in larval anurans. Ecology 74:174–182.

Lawler, S.P., D. Dritz, T. Strange, and M. Holyoak. 1999. Effects of introduced mosquitofish and bullfrogs on the threatened California red–legged frog. Conservation Biology 13:613–622.

Lawson, J. 1709. New Voyage to Carolina; Containing the Exact Description and Natural History of That Country: Together with the Present State Thereof. And a Journal of a Thousand Miles, Travel'd Thro' Several Nations of Indians. Giving a Particular Account of Their Customs, Manners, &c., London.

Layne, J.R., Jr. 1991. External ice triggers freezing in freeze–tolerant frogs at temperatures above their supercooling point. Journal of Herpetology 25:129–130.

Layne, J.R., Jr. 1995. Seasonal variation in the cryobiology of *Rana sylvatica* from Pennsylvania. Journal of Thermal Biology 20:349–353.

Layne, J.R., Jr., and M.A. Romano. 1985. Critical thermal minima of *Hyla chrysoscelis*, *H. cinerea*, *H. gratiosa* and natural hybrids (*H. cinerea* x *H. gratiosa*). Herpetologica 41:216–221.

Layne, J.R., Jr., and R.E. Lee Jr. 1987. Freeze tolerance and the dynamics of ice formation in wood frogs (*Rana sylvatica*) from southern Ohio. Canadian Journal of Zoology 65:2062–2065.

Layne, J.R., Jr., and R.E. Lee Jr. 1989. Seasonal variation in freeze tolerance and ice content of the tree frog *Hyla versicolor*. Journal of Experimental Biology 249:133–137.

Layne, J.R., Jr., and M.C. First. 1991. Resumption of physiological functions in the wood frog (*Rana sylvatica*) following freezing. American Journal of Physiology 261:R134–R137.

Layne, J.R., Jr., and R.E. Lee Jr. 1995. Adaptations of frogs to survive freezing. Climate Research 5:53–59.

Layne, J.R., Jr., and J. Kefauver. 1997. Freeze tolerance and postfreeze recovery in the frog *Pseudacris crucifer*. Copeia 1997:260–264.

Layne, J.R., and M.G. Stapleton. 2009. Annual variation in glycerol mobilization and effect of freeze rigor on post–thaw locomotion in the freeze–tolerant frog *Hyla versicolor*. Journal of Comparative Physiology 179B:215–221.

Layne, J. R., Jr., M. A. Romano, and S. I. Guttman. 1989. Responses to desiccation of the treefrogs *Hyla cinerea* and *H. gratiosa* and their natural hybrids. American Midland Naturalist 121: 61–67.`

Layne, J.R., Jr., R.E. Lee, Jr., and J.L. Huang. 1990. Inoculation triggers freezing at high subzero temperatures in a freeze–tolerant frog (*Rana sylvatica*) and insect (*Eurosta solidaginis*). Canadian Journal of Zoology 68:506–510.

Lazaroff, D.W., P.C. Rosen, and C.H. Lowe, Jr. 2006. Amphibians, Reptiles, and their Habitats at Sabino Canyon. University of Arizona Press, Tucson.

Lazell, J.D., Jr. 1968. Mr. Fowler's toad. Massachusetts Audubon 52(2):4 pp.

Lazell, J.D., Jr. 1969. Nantucket herpetology. Massachusetts Audubon 55(2):32–34.

Lazell, J.D., Jr. 1970. Nantucket gnome. Massachusetts Audubon 55(4): 4 pp.

Lazell, J.D., Jr. 1974. Reptiles & amphibians in Massachusetts. Rev. ed. Massachusetts Audubon Society, Lincoln.

Lazell, J.D., Jr. 1976. This Broken Archipelago. Cape Cod and the Islands, Amphibians and Reptiles. Quadrangle/New York Times Book Company, New York.

Lazell, J.D., Jr. 1989. Wildlife of the Florida Keys. Island Press, Washington, D.C.

Lazell, J.D., Jr., and T. Mann. 1991. Geographic distribution: *Bufo americanus charlesmithi* (Dwarf American Toad). Herpetological Review 22:62, 64.

Leach, D.J. 2011. Efficacy of small diameter pipe–trap refugia for capturing metamorph treefrogs. Herpetological Review 42:47–49.

Leary, C.J. 1996. *Bufo woodhousii* and *B. americanus* (Woodhouse's Toad and American Toad). Vocalization. Herpetological Review 27:139.

Leary, C.J. 1999. Comparison between release vocalizations emitted during artificial and conspecific amplexus in *Bufo americanus*. Copeia 1999:506–508.

Leary, C.J. 2001. Evidence of convergent character displacement in release vocalizations of *Bufo fowleri* and *Bufo terrestris* (Anura: Bufonidae). Animal Behaviour 61:431–438.

Leary, C.J. 2009. Hormones and acoustic communication in anuran amphibians. Integrative and Comparative Biology 49:452–470.

Leary, C.J., T.S. Jessop, A.M. Garcia, and R. Knapp. 2004. Steroid hormone profiles and relative body condition of calling and satellite toads: implications for proximate regulation of behavior in anurans. Behavioral Ecology 15:313–320.

Leary, C.J., D.J. Fox, D.B. Shepard, and A.M. Garcia. 2005. Body size, age, growth and alternative mating tactics in toads: satellite males are smaller but not younger than calling males. Animal Behaviour 70:663–671.

Leary, C.J., A. M. Garcia, and R. Knapp. 2006a. Elevated corticosterone levels elicit non–calling mating tactics in male toads independently of changes in circulating androgens. Hormones and Behavior 49:425–432.

Leary, C.J., A.M. Garcia, and R. Knapp. 2006b. Stress hormone is implicated in satellite–caller associations and sexual selection in the Great Plains toad. American Naturalist 168:431–440.

Leary, C.J., A.M. Garcia, and R. Knapp. 2008a. Density–dependent mating tactic expression is linked to stress hormone in Woodhouse's toad. Behavioral Ecology 19:1103–1110.

Leary, C.J., A.M. Garcia, R. Knapp, and D.L. Hawkins. 2008b. Relationships among steroid hormone levels, vocal effort and body condition in an explosive–breeding toad. Animal Behaviour 76:175–185.

Leavitt, D.J., and L.A. Fitzgerald. 2009. Diet of nonnative *Hyla cinerea* in a Chihuahuan Desert wetland. Journal of Herpetology 43:541–545.

Leclair, R., Jr., and L. Vallières. 1981. Régimes alimentaires de *Bufo americanus* (Holbrook) et *Rana sylvatica* LeConte (Amphibia: Anura) nouvellement metamorphoses. Le naturaliste canadien 108(4):325–329.

Leclair, R., Jr., and J. Castanet. 1987. A skeletochronological assessment of age and growth in the frog *Rana pipiens* Schreber (Amphibia, Anura) from southwestern Quebec. Copeia 1987:361–369.

Leclair, R., Jr., and G. Laurin. 1996. Growth and body size in populations of mink frogs *Rana septentrionalis* from two latitudes. Ecography 19:296–304.

Leclair, R., M.H. Leclair, J. Dubois, and J. Daoust. 2000. Age and size of wood frogs, *Rana sylvatica*, from Kuujjuarapik, northern Quebec. Canadian Field-Naturalist 114:381–387.

LeConte, J. 1825. Remarks on the American species of the genera *Hyla* and *Rana*. Annals of the Lyceum of Natural History of New York 1:278–282.

LeConte, J. 1855. Descriptive catalogue of the Ranina of the United States. Proceedings of the Academy of Natural Sciences of Philadelphia 7:423–431.

LeConte, J. 1856 (1857). Description of a new species of *Hyla* from Georgia. Proceedings of the Academy of Natural Sciences of Philadelphia 8:146.

Ledón–Rettig, C.C., D.W. Pfennig, and N. Nascone–Yoder. 2008. Ancestral variation and the potential for genetic accommodation in larval amphibians: implications for the evolution of novel feeding strategies. Evolution & Development 10:316–325.

Ledón–Rettig, C.C., D.W. Pfennig, and E.J. Crespi. 2009. Stress hormones and the fitness consequences associated with the transition to a novel diet in larval amphibians. Journal of Experimental Biology 212:3743–3750.

Ledón–Rettig, C.C., D.W. Pfennig, and E.J. Crespi. 2010. Diet and hormonal manipulation reveal cryptic genetic variation: implications for the evolution of novel feeding strategies. Proceedings of the Royal Society B 277:3569–3578.

Leduc, J.C., K.J. Kozlowicz, J.D. Litzgus, and D. Lesbarrères.

2012. Ecology of herpetofaunal populations in smelting tailings wetlands. Herpetology Notes 5:115–125.

Lee, D.S. 1968a. Observations on hybrid *Hyla gratiosa x cinerea* in central Florida. Bulletin of the Maryland Herpetological Society 4:76–78.

Lee, D.S. 1968b. Herpetofauna associated with central Florida mammals. Herpetologica 24:83-84.

Lee, D.S. 1969a. Floridian herpetofauna associated with cabbage palms. Herpetologica 25:71.

Lee, D.S. 1969b. Notes on the feeding behavior of cave–dwelling bullfrogs. Herpetologica 25:211–212.

Lee, D.S. 1969c. The treefrogs of Florida. Florida Naturalist 42(3):117–120.

Lee, D.S. 1972. List of amphibians and reptiles of Assateague Island. Bulletin of the Maryland Herpetological Society 8:90–95.

Lee, D.S. 1973a. Seasonal breeding distributions for selected Maryland and Delaware amphibians. Bulletin of the Maryland Herpetological Society 9:101–104.

Lee, D.S. 1973b. Additional reptiles and amphibians from Assateague Island. Bulletin of the Maryland Herpetological Society 9:110–111.

Lee, D.S. 1973c. Notes on a unique population of gopher frogs, *Rana areolata*, from central Florida. Bulletin of the Maryland Herpetological Society 9:1–5.

Lee, D.S., and R.A. Sanderson. 1970. Comments on the distribution of three species of frogs in Florida. Florida Naturalist 43(1):23.

Lee, J.C. 1986. Is the large–male mating advantage in anurans an epiphenonmenon? Oecologia 69:207–212.

Lee, J.C. 2001. Evolution of a secondary sexual dimorphism in the toad, *Bufo marinus*. Copeia 2001:928–935.

Lee, J.R. 2009. The herpetofauna of the Camp Shelby Joint Forces Training Center in the Gulf Coastal Plain of Mississippi. Southeastern Naturalist 8:639–652.

Lee, J.R., and C. Sabette. 2009. *Scaphiopus holbrookii holbrookii* (Eastern Spadefoot). Maximum size. Herpetological Review 40:211.

Lee, R.E., Jr., and J.P. Costanzo. 1993. Integrated physiological responses promoting anuran freeze tolerance. Pp. 501–510 *In* C. Carey, G.L. Florant, B.A. Wunder, and B. Horwitz (eds.), Life in the Cold. 3. Ecological, Physiological, and Molecular Mechanisms. Westview Press, Boulder, Colorado.

Lee–Yaw, J.A. 2007. The phylogeographic history of the wood frog (*Rana sylvatica*). M.S. thesis, McGill University, Montreal, Quebec.

Lee–Yaw, J.A., J.T. Irwin, and D.M. Green. 2008. Post–glacial range expansion from northern refugia by the wood frog, *Rana sylvatica*. Molecular Ecology 17:867–884.

Lee–Yaw, J., A. Davidson, B.H. McRae, and D.M. Green. 2009. Do landscape processes predict phylogeographic patterns in the wood frog. Molecular Ecology 18:1863–1874.

Lees, E. 1962. The incidence of helminth parasites in a particular frog population. Journal of Parasitology 52:95–102.

Lefcort, H. 1996. Adaptive, chemically mediated fright response in tadpoles of the southern leopard frog, *Rana utricularia*. Copeia 1996:455–459.

Lefcort, H. 1998. Chemically mediated fright response in southern toad (*Bufo terrestris*) tadpoles. Copeia 1998:445–450.

Lefcort, H., and S.M. Eiger. 1993. Antipredatory behaviour of feverish tadpoles: implications for pathogen transmission. Behaviour 126:13–27.

Lefcort, H., and A.R. Blaustein. 1995. Disease, predator avoidance, and vulnerability to predation in tadpoles. Oikos 74:469–474.

Lefcort, H., R.A. Meguire, L.H. Wilson, and W.F. Ettinger. 1998. Heavy metals alter the survival, growth, metamorphosis, and antipredatory behavior of Columbia spotted frog (*Rana luteiventris*) tadpoles. Archives of Environmental Contamination and Toxicology 35:447–456.

Lefcort, H., S. Thomson, E. Cowles, H. Harowicz, B. Livaudais, W. Roberts, and W. Ettinger. 1999. Ramification of predator avoidance: predator and heavy metal mediated competition between tadpoles and snails. Ecological Applications 9:1477-1489.

Leftwich, K.N., and P.D. Lilly. 1992. The effects of duration of exposure to acidic conditions on survival of *Bufo americanus* embryos. Journal of Herpetology 26:70–71.

LaHart, D.E. 1975. The freaky frogs of Florida. Florida Naturalist 48(5):2–5.

Laurin, G., and D.M. Green. 1990. Spring emergence and male breeding behaviour of Fowler's toads, (*Bufo woodhousei fowleri*), at Long Point, Ontario. Canadian Field-Naturalist 104:420–434.

Lehmann, D.L. 1960. Some parasites of central California amphibians. Journal of Parasitology 46:10.

Lehmann, D.L. 1965. Intestinal parasites of northwestern amphibians. Yearbook of the American Philosophical Society 1965:284–285.

Lehtinen, R.M. 2002. A historical study of the distribution of Blanchard's cricket frog (*Acris crepitans blanchardi*) in southeastern Michigan. Herpetological Review 33:194–197.

Lehtinen, R.M., and S.M. Galatowitsch. 2001. Colonization of restored wetlands by amphibians in Minnesota. American Midland Naturalist 145:388–396.

Lehtinen, R.M., and A.A. Skinner. 2006. The enigmatic decline of Blanchard's cricket frog (*Acris crepitans blanchar-*

di): a test of the habitat acidification hypothesis. Copeia 2006:159–167.

Lehtinen, R.M., and M.C. MacDonald. 2011. Live fast, die young? A six–year field study of longevity and survivorship in Blanchard's cricket frog (*Acris crepitans blanchardi*). Herpetological Review 42:504–507.

Lehtinen, R.M., S.M. Galatowitsch, and J.R. Tester. 1999. Consequences of habitat loss and fragmentation for wetland amphibian assemblages. Wetlands 19:1–12.

Leiden, Y.A., M.E. Dorcas, and J.W. Gibbons. 1999. Herpetofaunal diversity in coastal plain communities of South Carolina. Journal of the Elisha Mitchell Scientific Society 115:270–280.

Leips, J., and J. Travis. 1994. Metamorphic response to changing food levels in two species of hylid frogs. Ecology 75:1345–1356.

Leips, J., M.G. McManus, and J. Travis. 2000. Response of treefrog larvae to drying ponds: comparing temporary and permanent pond breeders. Ecology 81:2997–3008.

Lemm, J.M. 2006. Field Guide to Amphibians and Reptiles of the San Diego Region. University of California Press, Berkeley.

Lemmon, A.R., and E.M. Lemmon. 2008. A likelihood framework for estimating phylogeographic history on a continuous landscape. Systematic Biology 57:544–561.

Lemmon, E.M. 2007. Patterns and processes of speciation in North American chorus frogs. Ph.D. Dissertation, University of Texas, Austin.

Lemmon, E.M. 2009. Diversification of conspecific signals in sympatry: geographic overlap drives multidimensional reproductive character displacement in frogs. Evolution 63:1155–1170.

Lemmon, E.M., and A. R. Lemmon. 2010. Reinforcement in chorus frogs: lifetime fitness estimates including intrinsic natural selection and sexual selection against hybrids. Evolution 64:1748–1761.

Lemmon, E.M., A.R. Lemmon, and D.C. Cannatella. 2007a. Geological and climatic forces driving speciation in the continentally distributed trilling chorus frogs (*Pseudacris*). Evolution 61:2086–2103.

Lemmon, E.M., A.R. Lemmon, J.T. Collins, J.A. Lee–Yaw, and D.C. Cannatella. 2007b. Phylogeny–based delimitation of species boundaries and contact zones in the trilling chorus frogs (*Pseudacris*). Molecular Phylogenetics and Evolution 44:1068–1082.

Lemmon, E.M., A.R. Lemmon, J.T. Collins, and D.C. Cannatella. 2008. A new North American chorus frog species (Amphibia: Hylidae: *Pseudacris*) from the south–central United States. Zootaxa 1675:1–30.

Lemmon, E.M., M. Murphy, and T. E. Juenger. 2011. Identification and characterization of nuclear microsatellite loci for multiple species of chorus frogs (*Pseudacris*) for population genetic analyses. Conservation Genetics Resources 3:233–237.

Lemos Espinal, J.A., and H.M. Smith. 2007a. Anfibios y reptiles del estado de Chihuahua, México. Universidad Nacional Autónoma de México and Comisión Nacional para el Conocimiento y Uso de la Biodiversidad, México.

Lemos Espinal, J.A., and H.M. Smith. 2007b. Anfibios y reptiles del estado de Coahuila, México. Universidad Nacional Autónoma de México and Comisión Nacional para el Conocimiento y Uso de la Biodiversidad, México.

Lenaker, R.P., Jr. 1972. *Xenopus* in Orange County. Senior thesis, California State Polytechnic University, Kellogg–Voorhis Campus, Pomona.

Leney, J.L., K.G. Drouillard, and G.D. Haffner. 2006. Does metamorphosis increase the susceptibility of frogs to highly hydrophobic contaminants? Environmental Science & Technology 40:1491–1496.

Leonard, W.P., H.A. Brown, L.L.C. Jones, K.R. McAllister, and R.M. Storm. 1993. Amphibians of Washington and Oregon. Seattle Audubon Society, Seattle, Washington.

Leonard, W.P., N.P. Leonard, R.M. Storm, and P.E. Petzel. 1996. *Rana pretiosa* (Spotted Frog). Behavior and reproduction. Herpetological Review 27:195.

Leonard, W.P., L. Hallock, and K.R. McAllister. 1997a. *Rana pretiosa* (Oregon spotted frog). Behavior and reproduction. Herpetological Review 28:28.

Leonard, W.P., K.R. McAllister, and L.A. Hallock. 1997b. Autumn vocalizations by the red–legged frog (*Rana aurora*) and the Oregon spotted frog (*Rana pretiosa*). Northwestern Naturalist 78:73–74.

Leonard, W.P., K.R. McAllister, and R.C. Friesz. 1999. Survey and assessment of northern leopard frog (*Rana pipiens*) populations in Washington state. Northwestern Naturalist 80:51–60.

Lepage, M., R. Courtois, C. Daigle, and S. Matte. 1997. Surveying calling anurans in Québec using volunteers. SSAR Herpetological Conservation 1:128–140.

Lever, C. 2001. The Cane Toad. The History and Ecology of a Successful Colonist. Westbury Academic and Scientific Publishing, Otley, England.

Lever, C. 2003. Naturalized Reptiles and Amphibians of the World. Oxford University Press, Oxford, United Kingdom.

Levey, R., N. Shambaugh, D.J. Fort, and J. Andrews. 2003. Investigations into the causes of amphibian malformations in the Lake Champlain Basin of New England. Vermont Department of Environmental Conservation, Waterbury.

Levine, N.D., and R.R. Nye. 1977. A survey of blood and other tissue parasites of leopard frogs *Rana pipiens* in the United States. Journal of Wildlife Diseases 13:17–23.

Lewis, D. 2011. A Field Guide to the Amphibians and Reptiles of Wyoming. The Wyoming Naturalist, Douglas, Wyoming.

Lewis, D.L., G.T. Baxter, K.M. Johnson, and M.D. Stone. 1985. Possible extinction of the Wyoming toad, *Bufo hemiophrys baxteri*. Journal of Herpetology 19:166–168.

Lewis, R.J. 1962. Food habits of the leopard frog (*Rana pipiens sphenocephala*) in a minnow hatchery. Transactions of the Illinois State Academy of Science 55:78–79.

Lewis, T. 2009. New population of mountain yellow-legged frog (*Rana muscosa*) discovered. Herpetological Bulletin No. 108:1-2.

Lewis, T.H. 1946. Reptiles and amphibians of Smith Island, N.C. American Midland Naturalist 36:682–684.

Lewis, T.H. 1950. The herpetofauna of the Tularosa Basin and Organ Mountains of New Mexico with notes on some ecological features of the Chihuahuan Desert. Herpetologica 6:1–10.

Lewis, W.M., Jr. 1962. Stomach contents of bullfrogs (*Rana catesbeiana*) taken from a minnow hatchery. Transactions of the Illinois State Academy of Science 55:80–83.

Li, H., M.J. Vaughan, and R.K. Browne. 2009. A complex enrichment diet improves growth and health in the endangered Wyoming Toad (*Bufo baxteri*). Zoo Biology 28:197–213.

Liang, C.T. 2010. Habitat modeling and movements of the Yosemite toad (*Anaxyrus* (=*Bufo*) *canorus*) in the Sierra Nevada, California. Ph.D. Dissertation, University of California, Davis.

Liang, C.T., and T. J. Stohlgren. 2011. Habitat suitability of patch types: a case study of the Yosemite toad. Frontiers of Earth Science 5:217–228.

Licht, L.E. 1967. Growth inhibition in crowded tadpoles: intraspecific and interspecific effects. Ecology 48:736–745.

Licht, L.E. 1968. Unpalatability and toxicity of toad eggs. Herpetologica 24:93–98.

Licht, L.E. 1969a. Palatability of *Rana* and *Hyla* eggs. American Midland Naturalist 82:296–298.

Licht, L.E. 1969b. Unusual aspects of anuran sexual behavior as seen in the red–legged frog, *Rana aurora aurora*. Canadian Journal of Zoology 47:505–509.

Licht, L.E. 1969c. Comparative breeding behavior of the red–legged frog (*Rana aurora aurora*) and the western spotted frog (*Rana pretiosa pretiosa*) in southwestern British Columbia. Canadian Journal of Zoology 47:1287–1299.

Licht, L.E. 1971a. Breeding habits and embryonic thermal requirements of the frogs, *Rana aurora aurora* and *Rana pretiosa pretiosa*, in the Pacific Northwest. Ecology 52:116–124.

Licht, L.E. 1971b. The ecology of coexistence in two closely related species of frogs (*Rana*). Ph.D. Dissertation, University of British Columbia, Vancouver.

Licht, L.E. 1974. Survival of embryos, tadpoles, and adults of the frogs *Rana aurora aurora* and *Rana pretiosa pretiosa* sympatric in southwestern British Columbia. Canadian Journal of Zoology 52:613–627.

Licht, L.E. 1975. Comparative life history features of the western spotted frog, *Rana pretiosa*, from low– and high–elevation populations. Canadian Journal of Zoology 53:1254–1257.

Licht, L.E. 1976. Sexual selection in toads (*Bufo americanus*). Canadian Journal of Zoology 54:1277–1284.

Licht, L.E. 1986a. Comparative escape behavior of sympatric *Rana aurora* and *Rana pretiosa*. American Midland Naturalist 115:239–247.

Licht, L.E. 1986b. Food and feeding behavior of sympatric red–legged frogs, *Rana aurora*, and spotted frogs, *Rana pretiosa*, in southwestern British Columbia. Canadian Field-Naturalist 100:22–31.

Licht, L.E. 1991. Habitat selection of *Rana pipiens* and *Rana sylvatica* during exposure to warm and cold temperatures. American Midland Naturalist 125:259–268.

Licht, L.E. 2003. Shedding light on ultraviolet radiation and amphibian embryos. Bioscience 53:551–561.

Licht, L.E., and K.P. Grant. 1997. The effects of ultraviolet radiation on the biology of amphibians. American Zoologist 37:137–145.

Lichtenberg, J.S., S.L. King, J.B. Grace, and S.C. Walls. 2006. Habitat associations of chorusing anurans in the lower Mississippi River Alluvial Valley. Wetlands 26:736–744.

Ligas, F.J. 1960. The Everglades bullfrog life history and management. Proceedings of the 14th Annual Conference, Southeastern Association of Game and Fish Commissioners, pp. 9–14. [slightly modified version entitled "The Everglades Bullfrog" published in Florida Wildlife, 1963, 16(12):14–19]

Ligon, D.B., and P.A. Stone. 2003. *Kinosternon sonoriense* (Sonoran Mud Turtle) and *Bufo punctatus* (Red–spotted Toad). Predator–prey. Herpetological Review 34:141–142.

Ligon, N. F., and D. K. Skelly. 2009. Cryptic divergence: countergradient variation in the wood frog. Evolutionary Ecology Research 11:1099–1109.

Lillywhite, H.B. 1970. Behavioral temperature regulation in the bullfrog, *Rana catesbeiana*. Copeia 1970:158–168.

Lillywhite, H.B. 1971a. Temperature selection by the bullfrog, *Rana catesbeiana*. Comparative Biochemistry and Physiology 40A:213–227.

Lillywhite, H.B. 1971b. Thermal modulation of cutaneous mucus discharge as a determinant of evaporative water loss in the frog, *Rana catesbeiana*. Zeitschrift für vergleichende Physiologie 73:84–104.

Lillywhite, H.B. 1975. Physiological correlates of basking in amphibians. Comparative Biochemistry and Physiology 52A:323–330.

Lillywhite, H.B., and P. Licht. 1974. Movement of water over toad skin: functional role of epidermal sculpturing. Copeia 1974:165–171.

Lillywhite, H.B., and R.J. Wassersug. 1974. Comments on a postmetamorphic aggregation of *Bufo boreas*. Copeia 1974:984–986.

Limbaugh, B.A., and E.P. Volpe. 1957. Early development of the Gulf Coast Toad, *Bufo valliceps* Wiegmann. American Museum Novitates 1842:1–32.

Linck, M.H. 2000. Reduction in road mortality in a northern leopard frog population. Journal of the Iowa Academy of Science 107:209–211.

Lind, A.J. 2005. Reintroduction of a declining amphibian: determining an ecologically feasible approach for the foothill yellow–legged frog (*Rana boylii*) through analysis of decline factors, genetic structure, and habitat associations. Ph.D. Dissertation, University of California, Davis.

Lind, A.J., H.H. Welsh, Jr., and R.A. Wilson. 1996. The effects of a dam on breeding habitat and egg survival of the foothill yellow–legged frog (*Rana boylii*) in northwestern California. Herpetological Review 27:62–67.

Lind, A.J., J.B. Bettaso, and S.M. Yarnell. 2003. *Rana boylii* (Foothill Yellow–legged Frog) and *Rana catesbeiana* (Bullfrog). Reproductive behavior. Herpetological Review 34:234–235.

Lind, A.J., P.Q. Spinks, G.M. Fellers, and H.B. Shaffer. 2011. Rangewide phylogeography and landscape genetics of the western U.S. endemic frog *Rana boylii* (Ranidae): implications for the conservation of frogs and rivers. Conservation Genetics 12:269–284.

Linder, A.D., and E. Fichter. 1977. The Amphibians and Reptiles of Idaho. Idaho State University Press, Pocatello.

Lindholm, W.A. 1924. A forgotten description of a North American frog. Copeia (129):46–47.

Lindsay, H.L., Jr. 1958. Analysis of variation and factors affecting gene exchange in *Pseudacris clarki* and *Pseudacris nigrita* in Texas. Ph.D. Dissertation, University of Texas, Austin.

Liner, A.F., L.L. Smith, S.W. Golladay, S.B. Castleberry, and J.W. Gibbons. 2008. Amphibian distributions within three types of isolated wetlands in southwest Georgia. American Midland Naturalist 160:69-81.

Liner, A.E. 2006. Wetland predictors of amphibian distributions and diversity within the southeastern U.S. coastal plain. M.S. thesis, University of Georgia, Athens.

Liner, A.E., L.L. Smith, S.W. Golladay, S.B. Castleberry, and J.W. Gibbons. 2008. Amphibian distributions within three types of isolated wetlands in southwest Georgia.

American Midland Naturalist 160:69–81.

Liner, E.A. 1954. The herpetofauna of Lafayette, Terrebonne and Vermilion parishes, Louisiana. Proceedings of the Louisiana Academy of Sciences 17:65–85.

Liner, E.A. 1955. A herpetological consideration of the Bayou Tortue region of Lafayette Parish, Louisiana. Proceedings of the Louisiana Academy of Sciences 18:39–42.

Liner, E.A. 2007. Geographic distribution. *Euhyas planirostris*. Herpetological Review 38:214.

Ling, R.W., J.P. Van Amberg, and J.K. Werner. 1986. Pond acidity and its relationship to larval development of *Ambystoma maculatum* and *Rana sylvatica* in upper Michigan. Journal of Herpetology 20:230–236.

Link, E.E. 2012. A modeling and geospatial approach to predicting effects on biodiversity due to vineyard expansion in Napa county. M.S. thesis, Sacramento State University, Sacramento, California.

Linnaeus, C. 1758. Systema Naturae per Regna Tria Naturae, Secundum Classes, Ordines, Genera, Species cum Characteribus, Differentils, Synonymis, Locis. 10th ed., Vol. 1, L. Salvius, Stockholm, Sweden.

Linnenbach, M. 1985. Zum Feinbau der Haftscheiben von *Hyla cinerea* (Schneider, 1799) (Salientia: Hylidae). Salamandra 21:81–85.

Linsdale, J.M. 1927. Amphibians and reptiles of Doniphan County, Kansas. Copeia (164):75–81.

Linsdale, J.M. 1933a. A specimen of *Rana tarahumarae* from New Mexico. Copeia 1933:222.

Linsdale, J.M. 1933b. Records of *Ascaphus truei* in Idaho. Copeia 1933:223.

Linsdale, J.M. 1938. Environmental responses of vertebrates in the Great Basin. American Midland Naturalist 19:1–206.

Linsdale, J.M. 1940. Amphibians and reptiles in Nevada. Proceedings of the American Academy of Arts and Sciences 73:197–257.

Linzey, D.W. 1967. Food of the leopard frog, *Rana p. pipiens*, in central New York. Herpetologica 23:11–17.

Lipps, K.R. 1991. Vertebrates associated with tortoise (*Gopherus polyphemus*) burrows in four habitats in south–central Florida. Journal of Herpetology 25:477–481.

Little, E.E., R. Calfee, L. Cleveland, R. Skinker, A. Zaga–Parkhurst, and M.G. Barron. 2000. Photo–enhanced toxicity in amphibians: synergistic interactions of solar ultraviolet radiation and aquatic contaminants. Journal of the Iowa Academy of Science 107:67–71.

Little, E.L. 1940. Amphibians and reptiles of the Roosevelt Reservoir area, Arizona. Copeia 1940:260–265.

Little, E.L., and J. Keller. 1937. Amphibians and reptiles of the Jornada Experimental Range, New Mexico. Copeia 1937:215–222.

Little, J.B. 2007. A rare jewel. Audubon Magazine (March–April):69–75.

Little, M.L. 1983. The zoogeography of the *Hyla versicolor* complex in the central Appalachians, including physiological and morphological analyses. Ph.D. Dissertation, University of Louisville, Louisville, Kentucky.

Little, M.L., and T.K. Pauley. 1986. A new record of the diploid species of gray treefrog, *Hyla chrysoscelis*, in West Virginia. Proceedings of the West Virginia Academy of Science 58:57–58.

Little, M.L., B.L. Monroe, Jr., and J.E. Wiley. 1989. The distribution of the *Hyla versicolor* complex in the northern Appalachian highlands. Journal of Herpetology 23:299–303.

Littlejohn, M.J. 1958. Mating behavior in the treefrog *Hyla versicolor*. Copeia 1958:222–223.

Littlejohn, M.J. 1959. Artificial hybridization within the Pelobatidae and Microhylidae. Texas Journal of Science 9:57–59.

Littlejohn, M.J. 1960. Call discrimination and potential reproductive isolation in *Pseudacris triseriata* females from Oklahoma. Copeia 1960:370–371.

Littlejohn, M.J. 1961a. Artificial hybridization between some hylid frogs of the United States. Texas Journal of Science 13:176–184.

Littlejohn, M.J. 1961b. Mating call discrimination by females of the spotted chorus frog (*Pseudacris clarki*). Texas Journal of Science 13:49–50.

Littlejohn, M.J. 1971. A reappraisal of mating call differentiation in *Hyla cadaverina* (= *Hyla californiae*) and *Hyla regilla*. Evolution 25:98–102.

Littlejohn, M.J. 1977. Long–range acoustic communication in anurans: an integrated and evolutionary approach. Pp. 263–294 *In* D.H. Taylor and S.I. Guttman (eds.), The Reproductive Biology of Amphibians. Plenum Press, New York.

Littlejohn, M.J., and T.C. Michaud. 1959. Mating call discrimination by females of Strecker's chorus frog (*Pseudacris streckeri*). Texas Journal of Science 11:86–92.

Littlejohn, M.J., and M.J. Fouquette, Jr. 1960. Call discrimination by female frogs of the *Hyla versicolor* complex. Copeia 1960:47–49.

Littlejohn, M.J., and R.S. Oldham. 1968. *Rana pipiens* complex: mating call structure and taxonomy. Science 162:1003–1005.

Liu, H., R. J. Wassersug, and K. Kawachi. 1996. A computational fluid dynamics study of tadpole swimming. Journal of Experimental Biology 199:1245–1260.

Livezey, R.L. 1950. The eggs of *Acris gryllus crepitans* Baird. Herpetologica 6:139–140.

Livezey, R.L. 1952. Some observations on *Pseudacris nigrita triseriata* (Wied) in Texas. American Midland Naturalist 47:372–381.

Livezey, R.L. 1953. Late breeding of *Hyla regilla* Baird & Girard. Herpetologica 9:73.

Livezey, R.L. 1955. A northward range extension for *Bufo canorus* Camp. Herpetologica 11:212.

Livezey, R.L. 1960. Description of the eggs of *Bufo boreas exsul*. Herpetologica 16:48.

Livezey, R.L. 1962. Food of adult and juvenile *Bufo exsul*. Herpetologica 17:267–268.

Livezey, R.L., and A.H. Wright. 1945. Descriptions of four salientian eggs. American Midland Naturalist 34:701–706.

Livezey, R.L., and A.H. Wright. 1947. A synoptic key to the salientian eggs of the United States. American Midland Naturalist 37:179–222.

Livezey, R.L., and H.M. Johnson. 1948. *Rana grylio* in Texas. Herpetologica 4:164.

Livo, L. 1981a. Leopard frog (*Rana pipiens*) reproduction in Boulder County, Colorado. M.A. thesis, University of Colorado, Denver.

Livo, L. 1984. Dry years along with bullfrogs are creating problems for Colorado's two species of leopard frogs. Colorado Outdoors 33(4):16–18.

Livo, L.J. 1990. *Bufo cognatus* (Great Plains Toad). Microhabitat selection. Herpetological Review 21:58.

Livo, L. 1998. Identification guide to montane amphibians of the southern Rocky Mountains. Colorado Division of Wildlife, Denver.

Livo, L. 1999. The role of predation in the early life history of *Bufo boreas* in Colorado. Ph.D. Dissertation, University of Colorado, Boulder.

Livo, L.J. 2000. *Bufo punctatus* (Red–spotted Toad). Larval predation. Herpetological Review 31:99.

Livo, L., and D. Yeakley. 1997. Comparison of current with historical elevational range in the boreal toad, *Bufo boreas*. Herpetological Review 28:143–144.

Livo, L.J., D. Chiszar, and H.M. Smith. 1997. *Spea multiplicata* (New Mexico Spadefoot). Defensive posture. Herpetological Review 28:148.

Livo, L.J., and B.C. Kondratieff. 2000. *Bufo punctatus* (Red–spotted Toad). Predation. Herpetological Review 31:168–169.

Livo, L.J., and B.A. Lambert. 2001. *Bufo boreas* (Boreal Toad). Phoretic host. Herpetological Review 32:179.

Livo, L.J., G.A. Hammerson, and H.M. Smith. 1998. Summary of amphibians and reptiles introduced into Colorado. Northwestern Naturalist 79:1–11.

Livo, L.J., L. Harvey, and F. Bunch. 2012. Geographic distribution: *Anaxyrus* (=*Bufo*) *woodhousii* (Woodhouse's

Toad). Herpetological Review 43:612–613.

Lockley, T.C. 1990. Predation on the green treefrog by the star–bellied orb weaver, *Acanthepeira stellata* (Araneae, Araneidae). Journal of Arachnology 18:359.

Loda, J.L., and D.L. Otis. 2009. Low prevalence of amphibian chytrid fungus (*Batrachochytrium dendrobatidis*) in northern leopard frog (*Rana pipiens*) populations on north–central Iowa, USA. Herpetological Review 40:428–431.

Lodato, M.J., and M.H. Kerr. 1974. The western bird–voiced treefrog *Hyla. a. avivoca* Viosca in Henderson County, Kentucky. Journal of Herpetology 8:259–260.

Loding, H.P. 1922. A preliminary catalogue of the Alabama amphibians and reptiles. Geological Survey of Alabama, Museum Paper No. 5.

Loftin, C.S., A.J.K. Calhoun, S.J. Nelson, A.A. Elskus, and K. Simon. 2012. Mercury bioaccumulation in wood frogs developing in seasonal pools. Northeastern Naturalist 19:579–600.

Loftus–Hills, J.J. 1975. The evidence for reproductive character displacement between the toads *Bufo americanus* and *B. woodhousii fowleri*. Evolution 29:368–369.

Loftus–Hills, J.J., and M.J. Littlejohn. 1992. Reinforcement and reproductive character displacement in *Gastrophryne carolinensis* and *G. olivacea* (Anura: Microhylidae): a re-examination. Evolution 46:896–906.

Logier, E.B.S. 1928. The amphibians and reptiles of the Lake Nipigon region. Transactions of the Royal Canadian Institute 16:279–291.

Logier, E.B.S. 1930. A faunal investigation of King Township, York County, Ontario. IV. The amphibians and reptiles of King Township. Transactions of the Royal Canadian Institute 17:203–208.

Logier, E.B.S. 1931. The amphibians and reptiles of Long Point. Transactions of the Royal Canadian Institute 18:229–236.

Logier, E.B.S. 1932. Some account of the amphibians and reptiles of British Columbia. Transactions of the Royal Canadian Institute 18:311–336.

Logier, E.B.S. 1937. Amphibians of Ontario. Royal Ontario Museum, Zoology Handbook No. 3.

Logier, E.B.S. 1941. The amphibians and reptiles of Prince Edward County, Ontario. University of Toronto Studies, Biological Sciences No. 48:93–106.

Logier, E.B.S. 1942. Reptiles and amphibians of the Sault Ste. Marie region, Ontario. Transactions of the Royal Canadian Institute 24:154–163.

Logier, E.B.S. 1949. Effect of DDT on amphibians and reptiles. Ontario Department of Lands and Forests, Biological Bulletin 2:49–56.

Logier, E.B.S. 1952. The Frogs, Toads and Salamanders of Eastern Canada. Clarke, Irwin & Co., Toronto, Ontario.

Logier, E.B.S., and G.C. Toner. 1961. Check list of the amphibians & reptiles of Canada & Alaska. Royal Ontario Museum, Toronto, Ontario.

Lohoefener, R., and R. Altig.1983. Mississippi Herpetology. Mississippi State University Research Center at National Space Technology Laboratories Bulletin No. 1.

Lomolino, M.A., and G.A. Smith. 2004. Terrestrial vertebrate communities at black–tailed prairie dog (*Cynomys ludovicianus*) towns. Biological Conservation 115:89–100.

Londos, P.L., and R.J. Brooks. 1988. Effect of temperature acclimation on locomotory performance curves in the toad, *Bufo woodhousii fowleri*. Copeia 1988:26–32.

Londos, P.L., and R.J. Brooks. 1990. Time course of temperature acclimation of locomotory performance in the toad, *Bufo woodhousii woodhousii*. Copeia 1990:827–835.

Long, C.A. 1964. The badger as a natural enemy of *Ambystoma tigrinum* and *Bufo boreas*. Herpetologica 20:144.

Long, C.A. 1998. Notes on development and mortality in giant toads from Washington Island, Door County, Wisconsin. Bulletin of the Chicago Herpetological Society 33:161–164.

Long, C.A., and C.A. Long. 1976. Some amphibians and reptiles collected on islands in Green Bay, Lake Michigan. The Jack-Pine Warbler 54:54–58.

Long, L.E., L.S. Saylor, and M.E. Soulé. 1995. A pH/UV–B synergism in amphibians. Conservation Biology 9:1301–1303.

Long, M.C. 1970. Food habits of *Rana muscosa* (Anura: Ranidae). Herpeton, Journal of the Southwestern Herpetologists Society 5(1):1–8.

Long, Z. 2010. *Anaxyrus boreas* (Western Toad). Advertisement vocalization. Herpetological Review 41:332–333.

Long, Z.L., and E.E. Prepas. 2012. Scale and landscape perception: the case of refuge use by boreal toads (*Anaxyrus boreas boreas*). Canadian Journal of Zoology 90:1015–1022.

Longcore, J.R., J.E. Longcore, A.P. Pessier, and W.A. Halteman. 2007. Chytridiomycosis widespread in anurans of northeastern United States. Journal of Wildlife Management 71:435–444.

Loomis, R.B., and J.K. Jones, Jr. 1953. Records of the wood frog, *Rana sylvatica*, from western Canada and Alaska. Herpetologica 9:149–151.

López, L.O., G.A. Woolrich Piña, and J.A. Lemos Espinal. 2009. La familia Bufonidae en México. Universidad Nacional Autónomade de México, México.

Lopez, T.J., and L.R. Maxon. 1990. *Rana catesbeiana* (Bullfrog). Polymely. Herpetological Review 21:90.

Lord, R.D., Jr., and W.B. Davis. 1956. A taxonomic study

of the relationship between *Pseudacris nigrita triseriata* Wied and *Pseudacris clarki* Baird. Herpetologica 12:115–120.

Lorraine, R.K. 1984. *Hyla crucifer crucifer* (Northern Spring Peeper). Reproduction. Herpetological Review 15:16–17.

Lotshaw, D.P. 1977. Temperature adaptation and effects of thermal acclimation in *Rana sylvatica* and *Rana catesbeiana*. Comparative Biochemistry and Physiology 56B:287–294.

Lott, T. 2012. Observations on the distribution of the Rio Grande chirping frog, *Eleutherodactylus cystignathoides campi*, in the United States (Anura, Eleutherodactylidae). SWCHR Bulletin 2(1):8-12.

Lotz, A., and C.R. Allen. 2007. Observer bias in anuran call surveys. Journal of Wildlife Management 71:675–679.

Louisiana Department of Conservation. 1931. Frog industry in Louisiana. Division of Fisheries, Educational Pamphlet No. 2.

Louisiana Department of Conservation. 1938. Frog industry in Louisiana. Division of Fisheries, Bulletin No. 26.

Love, E.K., and M.A. Bee. 2010. An experimental test of noise–dependent voice amplitude regulation in Cope's grey treefrog, *Hyla chrysoscelis*. Animal Behaviour 80:509–515.

Love, W.B. 1995. *Osteopilus septentrionalis* (Cuban Treefrog). Predation. Herpetological Review 26:201–202.

Lovich, J.E., and T.R. Jaworski. 1988. Annotated list of amphibians and reptiles reported from Cedar Bog, Ohio. Ohio Journal of Science 88:139–143.

Lovich, R.E. 2009. Phylogeography and conservation of the Arroyo Toad (*Bufo californicus*). Ph.D. Dissertation, Loma Linda University, Loma Linda, California.

Lovich, R., M.J. Ryan, A.P. Pessier, and B. Claypool. 2008. Infection with the fungus *Batrachochytrium dendrobatidis* in a non–native *Lithobates berlandieri* below sea level in the Coachella Valley, California, USA. Herpetological Review 39:315–317.

Low, B.S. 1976. The evolution of amphibian life histories in the desert. Pp. 149-196 *In* D.W. Goodall (ed.), Evolution of Desert Biota. University of Texas Press, Austin.

Lowcock, L.A., T.F. Sharbel, J. Bonin, M. Ouellet, J. Rodrigue, and J.–L. DesGranges. 1997. Flow cytometric assay for in vivo genotoxic effects of pesticides in green frogs (*Rana clamitans*). Aquatic Toxicology 38:241–255.

Lowe, C.H., Jr. 1954. Isolating mechanisms in sympatric populations of southwestern anurans. Texas Journal of Science 6:265–270.

Lowe, J. 2009. Amphibian chytrid (*Batrachochytrium dendrobatidis*) in post–metamorphic *Rana boylii* in Inner Coast Ranges of California. Herpetological Review 40:180–182.

Lucas, E.A., and W.A. Reynolds. 1967. Temperature selection by amphibian larvae. Physiological Zoology 40:159–171.

Lucas, W. 1965. Bullfrog legs becoming big thing for farmers in Pender County Com. Rocky Mount (N.C.) Telegram, June 14, 1965, p. 9.

Luce, D., and J.J. Moriarty. 1999. *Rana sylvatica* (Wood Frog). Coloration. Herpetological Review 30:94.

Lucke, B. 1952. Kidney carcinoma in the leopard frog: a virus tumor. Annals of the New York Academy of Sciences 54:1093–1109.

Lucker, J.T. 1931. A new genus and a new species of trematode worms of the family Plagiorchiidae. Proceedings of the United States National Museum 79:1–8.

Luepschen, L.K. 1981. *Bufo punctatus* (Red–spotted Toad). Larval coloration. Herpetological Review 12:79.

Luhring, T.M. 2008. "Problem species" of the Savannah River Site, such as Brimley's chorus frog (*Pseudacris brimleyi*), demonstrate the hidden biodiversity concept on an intensively studied government reserve. Southeastern Naturalist 7:371–373.

Luhring, T.M., and Z.D. Ross. 2012. *Gastrophryne carolinensis* (Easten Narrow–mouthed Toad). Predation. Herpetological Review 43:118.

Luja, V.H., and R. Rodríguez–Estrella. 2010. The invasive bullfrog *Lithobates catesbeianus* in oases of Baja California Sur, Mexico: potential effects in a fragile ecosystem. Biological Invasions 12:2979–2983.

Lukose, R.L., and H.K. Reinert. 1998. Absence of cold acclimation response in gray treefrogs, *Hyla chrysoscelis* and *Hyla versicolor*. Journal of Herpetology 32:283–285.

Lund, E., M. Hayes, T. Curry, J. Marsten, and K. Young. 2008. Predation on the Coastal Tailed Frog (*Ascaphus truei*) in Washington State. Northwestern Naturalist 89:200–202.

Lunde, K.B. 2011. Investigations of altered aquatic ecosystems: biomonitoring, disease, and conservation. Ph.D. Dissertation, University of California, Berkeley.

Lunde, K.B., and P.T. Johnson. 2012. A practical guide for the study of malformed amphibians and their causes. Journal of Herpetology 46:429–441.

Lutterschmidt, W.I., G.A. Marvin, and V.H. Hutchison. 1996. *Rana catesbeiana* (Bullfrog). Record size. Herpetological Review 27:74–75.

Lyerla, T.A., and D.J. Jameson. 1968. Development of color in chimeras of Pacific tree frogs. Copeia 1968: 113–128.

Lykens, D.V., and D.C. Forester. 1987. Age structure in the spring peeper: do males advertise longevity? Herpetologica 43:216–223.

Lynch, J.D. 1964a. The toad *Bufo americanus charlesmithi* in

Indiana, with remarks on the range of the subspecies. Journal of the Ohio Herpetological Society 4:103–104.

Lynch, J.D. 1964b. Two additional predators of the spadefoot toad, *Scaphiopus holbrookii* (Harlan). Journal of the Ohio Herpetological Society 4:79.

Lynch, J.D. 1970. A taxonomic revision of the leptodactylid frog genus *Syrrhophus* Cope. University of Kansas Publications, Museum of Natural History 20:1–45.

Lynch, J.D. 1978. The distribution of leopard frogs (*Rana blairi* and *Rana pipiens*) (Amphibia, Anura, Ranidae) in Nebraska. Journal of Herpetology 12:157–162.

Lynch, J.D. 1985. Annotated checklist of the amphibians and reptiles of Nebraska. Transactions of the Nebraska Academy of Sciences 13:33–57.

Lynn, W.G., and A. Edelman. 1936. Crowding and metamorphosis in the tadpole. Ecology 17:104–109.

Mabie, R.W. 1936. American Frog Raising. American Frog Publishing Company, Ridgefield, New Jersey. 28 pp.

Mable, B.K., and L. Rye. 1994. Developmental abnormalities in triploid hybrids between tetraploid and diploid tree frogs (genus *Hyla*). Canadian Journal of Zoology 70:2072–2076.

Mabry, C.M. 1984. The distribution and reproduction of the plains spadefoot toad, *Scaphiopus bombifrons* in Iowa. M.A. Thesis. Drake University, Des Moines, Iowa.

Mabry, C.M., and J.L. Christiansen. 1991. The activity and breeding cycle of *Scaphiopus bombifrons* in Iowa. Journal of Herpetology 25:116–119.

MacArthur, D.L., and J.W.T. Dandy. 1982. Physiological aspects of overwintering in the boreal chorus frog (*Pseudacris triseriata maculata*). Comparative Biochemistry and Physiology 72A:137–141.

MaCartney, J.M, and P.T. Gregory. 1981. Differential susceptibility of sympatric garter snake species to amphibian skin secretions. American Midland Naturalist 106:271–281.

MacCoy, C.V. 1931. Key for the identification of New England amphibians and reptiles. Bulletin of the Boston Society of Natural History 59:25-33.

MacCracken, J.G., and J.L. Stebbings. 2012. Test of a body condition index with amphibians. Journal of Herpetology 46:346–350.

MacCulloch, R.D. 2002. The ROM Field Guide to Amphibians and Reptiles of Ontario. Royal Ontario Museum, Toronto.

MacCulloch, R.D., and J.R. Bider. 1975. New records of amphibians and garter snakes in the James Bay Area of Quebec. Canadian Field-Naturalist 89:80–82.

MacDonald, S.O., and J.A. Cook. 2007. Mammals and Amphibians of Southeast Alaska. Museum of Southwestern Biology, Special Publication No. 8.

Mace, T.F., and R.C. Anderson. 1975. Development of the giant kidney worm, *Dioctophyma renale* (Goeze, 1782) (Nematoda: Dioctophymatoidea). Canadian Journal of Zoology 53:1552–1568.

Macey, J.R., and T.J. Papenfuss. 1991. Amphibians. Pp. 277–290 *In* C.A. Hall (ed.), Natural History of the White–Inyo Range: Eastern California. California Natural History Guides No. 55., University of California Press, Berkeley.

Macey, J.R., J.L. Strasburg, J.A. Brisson, V.T. Vredenburg, M. Jennings, and A. Larson. 2001. Molecular phylogenetics of western North American frogs of the *Rana boylii* species group. Molecular Phylogenetics and Evolution 19:131–143.

Maciel, N.M., R.G. Collevatti, G.R. Colli, and E.F. Schwartz. 2010. Late Miocene diversification and phylogenetic relationships of the huge toads in the *Rhinella marina* (Linnaeus, 1758) species group (Anura: Bufonidae). Molecular Phylogenetics and Evolution 57:787–797.

MacKay, W.P., S.J. Loring, T.M. Frost, and W.G. Whitford. 1990. Population dynamics of a playa community in the Chihuahuan Desert. Southwestern Naturalist 35:393–402.

MacKenzie, D.I., J.D. Nichols, G.B. Lachman, S. Droege, J.A. Royle, and C.A. Langtimm. 2002. Estimating site occupancy rates when detection probabilities are less than one. Ecology 83:2248–2255.

Mackey, M.J., and M.D. Boone. 2009. Single and interactive effects of malathion, overwintered green frog tadpoles, and cyanobacteria on gray treefrog tadpoles. Environmental Toxicology and Chemistry 28:637–643.

MacTague, L., and P.T. Northen. 1993. Underwater vocalization by the foothill yellow–legged frog (*Rana boylii*). Transactions of the Western Section of the Wildlife Society 29:1–7.

Maerz, J.C., B. Blossey, and V. Nuzzo. 2005a. Green frogs show reduced foraging success in habitats invaded by Japanese knotweed. Biodiversity and Conservation 14:2901–2911.

Maerz, J.C., C.J. Brown, C.T. Chapin, and B. Blossey. 2005b. Can secondary compounds of an invasive plant affect larval amphibians? Functional Ecology 19:970–975.

Maerz, J.C., J.S. Cohen, and B. Blossey. 2010. Does detritus quality predict the effect of native and non-native plants on the performance of larval amphibians? Freshwater Biology 55: 1694–1704.

Mahan, J.T., and C.J. Biggers. 1977. Electrophoretic investigation of blood and parotoid venom proteins in *Bufo americanus* and *Bufo woodhousei fowleri*. Comparative Biochemistry and Physiology 57C:121–126.

Mahan, R.D., and J.R. Johnson. 2007. Diet of the gray treefrog (*Hyla versicolor*) in relation to foraging site location. Journal of Herpetology 41:16–23.

Mahaney, P.A. 1994. Effects of freshwater petroleum contamination on amphibian hatching and metamorphosis. Environmental Toxicology and Chemistry 13:259–265.

Mahoney, J.J., and V.H. Hutchison. 1969. Photoperiod acclimation and 24–hour variations in the Critical Thermal Maxima of a tropical and a temperate frog. Oecologia 2:143–161.

Mahrdt, C.R., and F.T. Knefler. 1972. Pet or pest? The African clawed frog. Environment Southwest (446):2–5.

Mahrdt, C.R., and F.T. Knefler. 1973. The clawed frog—again. Environment Southwest (450):1–3.

Mahrdt, C.R., R.E. Lovich, and S.J. Zimmitti. 2002. *Bufo californicus* (California Arroyo Toad). Habitat and population status. Herpetological Review 33:123–125.

Maisonneuve, C., and S. Rioux. 2001. Importance of riparian habitats for small mammal and herpetofaunal communities in agricultural landscapes of southern Québec. Agriculture, Ecosystems and Environment 83:165–175.

Majji, S., S. LaPatra, S.M. Long, R. Sample, L. Bryan, A. Sinning, and V.G. Chinchar. 2006. *Rana catesbeiana* virus Z (RCV–Z): a novel pathogenic ranavirus. Diseases of Aquatic Organisms 73:1–11.

Majka, C.G. 1974. A deletion in the known range of the gray treefrog (*Hyla versicolor*) in northern New Brunswick. Canadian Field-Naturalist 88:498.

Majka, C.G. 1981. Gray treefrog: an endangered species in New Brunswick. New Brunswick Naturalist 11:45-46.

Malaret, L. 1978. Herpetofauna of Lacreek National Wildlife Refuge. Transactions of the Kansas Academy of Science 80:145–150.

Malfatti, M. 1998. The relict leopard frog, *Rana onca*. Vivarium 9:36-38, 58-60.

Mallory, M.A., and S.J. Richardson. 2005. Complex interactions of light, nutrients and consumer density in a stream periphyton–grazer (tailed frog tadpoles) system. Journal of Animal Ecology 74:1020–1028.

Malmos, K.B., R. Reed, and B. Starret. 1995. Hybridization between *Bufo woodhousii* and *Bufo punctatus* from the Grand Canyon region of Arizona. Great Basin Naturalist 55:368–371.

Malmos, K.B., B.K. Sullivan, and T. Lamb. 2001. Calling behavior and directional hybridization between two toads (*Bufo microscaphus* x *B. woodhousii*) in Arizona. Evolution 55:626–630.

Malone, J.H. 1999. *Bufo speciosus* (Texas Toad). Diet. Herpetological Review 30:222–223.

Mancuso, J. 2003. The amphibians of Westchester County New York. Greenburgh Nature Center, Scarsdale, New York. 19 pp.

Maniero, G.D., and C. Carey. 1997. Changes in selected aspects of immune function in the leopard frog, *Rana pipiens*, associated with exposure to cold. Journal of Comparative Physiology 167B:256–263.

Manion, J.J. 1952. Comparative ecological studies on the amphibians of Cass County, Michigan, and vicinity. Ph.D. Dissertation, University of Notre Dame, South Bend, Indiana.

Manion, J.J., and L. Cory. 1952. Winter kill of *Rana pipiens* in shallow ponds. Herpetologica 8:32.

Manis, M.L., and D.L. Claussen. 1986. Environmental and genetic influences on the thermal physiology of *Rana sylvatica*. Journal of Thermal Biology 11:31–36.

Mannan, R.N. 2008. An assessment of survey methodology, calling activity, and habitat associations of wood frogs (*Rana sylvatica*) and boreal chorus frogs (*Pseudacris maculata*) in a tundra biome. M.S. thesis, Texas Tech University, Lubbock.

Mansueti, R. 1941. A descriptive catalogue of the amphibians and reptiles found in and around Baltimore city, Maryland within a radius of twenty miles. Proceedings of the Natural History Society of Maryland 7:1–53.

Mansueti, R. 1945. Note on a hibernating spring peeper. Maryland, A Journal of Natural History 15(3):59.

Mansueti, R. 1948. Winter expedition in Maryland. Maryland Naturalist 18:3-9.

Manter, H.W. 1938. A collection of trematodes from Florida amphibia. Transactions of the American Microscopical Society 57:26–37.

Manville, R.H. 1939. Notes on the herpetology of Mount Desert Island, Maine. Copeia 1939:174.

Mao, J., D.E. Green, G. Fellers, and V.G. Chinchar. 1999. Molecular characterization of iridoviruses isolated from sympatric amphibians and fish. Virus Research 63:45–52.

Marchisin, A., and J.D. Anderson. 1978. Strategies employed by frogs and toads (Amphibia, Anura) to avoid predation by snakes (Reptilia, Serpentes). Journal of Herpetology 12:151–155.

Marco, A., J.M. Kiesecker, D.P. Chivers, and A. Blaustein. 1998. Sex recognition and mate choice by male western toads, *Bufo boreas*. Animal Behaviour 55:1631–1635.

Marco, A., C. Quilchano, and A.R. Blaustein. 1999. Sensitivity of nitate and nitrite in pond–breeding amphibians from the Pacific Northwest. Environmental Toxicology and Chemistry 18:2836–2839.

Marcogliese, D.J., K.C. King, H.M. Salo, M. Fournier, P. Brousseau, P. Spear, L. Champoux, J.D. McLaughlin, and M. Boily. 2009. Combined effects of agricultural activity and parasites on biomarkers in the bullfrog, *Rana catesbeiana*. Aquatic Toxicology 91:126– 134.

Marcotte, L. 1907. Addition à notre faune batracienne. Naturaliste Canadien 34(10):155–157.

Marcotte, L. 1918. La "*Hyla pickeringii* Holbrook." Naturaliste Canadien 44:113.

Mares, M.A. 1972. Notes on *Bufo marinus* tadpole aggregations. Texas Journal of Science 23:433–435.

Maret, E. 1867. Frogs in Newfoundland. Proceedings of the Nova Scotian Institute of Science 1(3):6.

Marimon, S. 1923. Notes on the color changes of frogs. Pomona College Journal of Entomology and Zoology 15:27–31.

Marker, G.M., and R.E. Gatten, Jr. 1993. Individual variability in sprint performance, lactate production, and enzyme activity in frogs (*Rana pipiens*). Journal of Herpetology 27:294–299.

Markle, T.M., and D.M. Green. 2009. New distributional records of amphibians in Quebec and Labrador. Herpetological Review 40:240–241.

Marnell, L.F. 1997. Herpetofauna of Glacier National Park. Northwestern Naturalist 78:17–33.

Marr, S.K., S.A. Johnson, A.H. Hara, and M.E. McGarrity. 2010. Preliminary evaluation of the potential of the helminth parasite *Rhabdias elegans* as a biological control agent for invasive Puerto Rico coquis (*Eleutherodactylus coqui*) in Hawaii. Biological Control 54:69–74.

Marsh, D.M., and P.C. Trenham. 2001. Metapopulation dynamics and amphibian conservation. Conservation Biology 15:40-49.

Marsh, R.L., and H.B. John–Alder. 1994. Jumping performance of hylid frogs measured with high–speed cine film. Journal of Experimental Biology 188:131–141.

Marshall, J.L., and C.D. Camp. 1995. Aspects of the feeding ecology of the little grass frog, *Pseudacris ocularis* (Anura: Hylidae). Brimleyana 22:1–7.

Marshall, V.T., S.C. Humfeld, and M.A. Bee. 2003. Plasticity of aggressive signaling and its evolution in male spring peepers, *Pseudacris crucifer*. Animal Behaviour 65:1223–1234.

Marshall, V.T., J.J. Schwartz, and H.C. Gerhardt. 2006. Effects of heterospecific call overlap on the phonotactic behaviour of grey treefrogs. Animal Behaviour 72:449–459

Marshall, W.H., and M.F. Buell. 1955. A study of the occurrence of amphibians in relation to a bog succession, Itasca State Park, Minnesota. Ecology 36:381–387.

Martin, C.H. 1940. Life cycle of toads in Yosemite Valley. Yosemite Nature Notes 19:90–92.

Martin, D.L. 1991. Population census of a species of special concern: the Yosemite toad (*Bufo canorus*). Fourth Biennial Conference of Research in California's National Parks. University of California, Davis.

Martin, D.L. 2008. Decline, movement, and habitat utilization of the Yosemite toad (*Bufo canorus*): an endangered anuran endemic to the Sierra Nevada of California. Ph.D. Dissertation, University of California, Santa Barbara.

Martin, T.R., and D.B. Conn. 1990. The pathogenicity, localization, and cyst structure of echinostomatid metacercariae (Trematoda) infecting the kidneys of the frogs *Rana clamitans* and *Rana pipiens*. Journal of Parasitology 76:414–419.

Martinez, J.L. 1948. Cuban frog leg industry. U.S. Fish and Wildlife Service, Fishery Leaflet 284.

Martinez–Ortiz, M. 2004. Predicting habitat suitability and occurrence for Blanchard's cricket frogs (*Acris crepitans blanchardi*) in northwest Ohio. M.S. thesis, Bowling Green State University, Bowling Green, Ohio.

Martinez–Rivera, C.C. 2008. Call timing, aggressive behavior, and the role of acoustic cues in chorus formation of treefrogs. Ph.D. Dissertation, University of Missouri, Columbia.

Martínez–Rivera, C.C., and H.C. Gerhardt. 2008. Advertisement–call modification, male competition, and female preference in the bird–voiced treefrog *Hyla avivoca*. Behavioral Ecology and Sociobiology 63:195–208.

Martinez Rodriguez, E., T. Gamble, M.V. Hurt, and S. Cotner. 2009. Presence of *Batrachochytrium dendrobatidis* at the headwaters of the Mississippi River, Itasca State Park, Minnesota, USA. Herpetological Review 40:48–50.

Martof, B.S. 1951. An ecological study of the green frog, *Rana clamitans*, in southeastern Michigan. Ph.D. Dissertation, University of Michigan, Ann Arbor.

Martof, B.S. 1952. Early transformation of the greenfrog, *Rana clamitans* Latreille. Copeia 1952:115–116.

Martof, B.S. 1953a. Home range and movements of the green frog, *Rana clamitans*. Ecology 34:529–543.

Martof, B.S. 1953b. Territoriality in the green frog, *Rana clamitans*. Ecology 34:165–174.

Martof, B. 1954. The barking frogs, *Hyla gratiosa*, in the Cumberland Plateau of Georgia. Copeia 1954: 157.

Martof, B.S. 1955. Observations on the life history and ecology of the amphibians of the Athens area, Georgia. Copeia 1955:166–170.

Martof, B.S. 1956a. Growth and development of the green frog, *Rana clamitans*, under natural conditions. American Midland Naturalist 55:101–117.

Martof, B.S. 1956b. Factors influencing the size and composition of populations of *Rana clamitans*. American Midland Naturalist 56:224–245.

Martof, B.S. 1956c. Amphibians and Reptiles of Georgia. University of Georgia Press, Athens.

Martof, B.S. 1960. Autumnal breeding of *Hyla crucifer*. Copeia 1960:58–59.

Martof, B.S. 1961a. Vocalization as an isolating mechanism

in frogs. American Midland Naturalist 65:118–126.

Martof, B.S. 1961b. Aberrant reproductive behavior in frogs. Herpetologica 17:66–67.

Martof, B.S. 1962a. Some observations on the role of olfaction among salientian amphibia. Physiological Zoology 35:270–272.

Martof, B.S. 1962b. The behavior of Fowler's toad under various conditions of light and temperature. Physiological Zoology 35:38–46.

Martof, B.S. 1962c. An unusual color variant of *Rana pipiens*. Herpetologica 17:269–270.

Martof, B.S. 1963a. Some observations on the herpetofauna of Sapelo Island, Georgia. Herpetologica 19:70–72.

Martof, B.S. 1963b. Some observations on the feeding of Fowler's toad. Copeia 1963:439.

Martof, B.S., and R.L. Humphries. 1955. Observations on some amphibians from Georgia. Copeia 1955:245–248.

Martof, B.S., and E.F. Thompson. 1958. Reproductive behavior of the chorus frog, *Pseudacris nigrita*. Behaviour 13:243–257.

Martof, B.S., and R.L. Humphries. 1959. Geographic variation in the wood frog *Rana sylvatica*. American Midland Naturalist 61:350–389.

Martof, B.S., and E.F. Thompson, Jr. 1964. A behavioral analysis of the mating call of the chorus frog, *Pseudacris triseriata*. American Midland Naturalist 71:198–209.

Martof, B.S., W.M. Palmer, J.R. Bailey, and J.R. Harrison III. 1980. Amphibians and Reptiles of the Carolinas and Virginia. University of North Carolina Press, Chapel Hill.

Maskell, A.J., J.H. Waddle, and K.G. Rice. 2003. *Osteopilus septentrionalis* (Cuban Treefrog). Diet. Herpetological Review 34:137.

Maslin, T.P. 1947. *Rana sylvatica cantabrigensis* Baird in Colorado. Copeia 1947:158–162.

Maslin, T.P. 1959. An annotated check list of the amphibians and reptiles of Colorado. University of Colorado Studies, Series in Biology No. 6:1–98.

Masta, S.E., B.K. Sullivan, T. Lamb, and E.J. Routman. 2002. Molecular systematics, hybridization, and phylogeography of the *Bufo americanus* complex in eastern North America. Molecular Phylogenetics and Evolution 24:302–314.

Masta, S.E., N.M. Laurent, and E.J. Routman. 2003. Population genetic structure of the toad *Bufo woodhousii*: an empirical assessment of the effects of haplotype extinction on nested cladistic analysis. Molecular Ecology 12:1541–1554.

Mata–Silva, V., L.D. Wilson, J.D. Johnson, A. Rocha, and W.D. Lukefahr. 2012. *Anaxyrus punctatus* (Red–spotted Toad). Predation. Herpetological Review 43:629.

Materna, E.J., C.F. Rabini, and T.W. LaPoint. 1995. Effects of the synthetic pyrethroid insecticide, esfenvalerate, on larval leopard frogs (*Rana* spp.). Environmental Toxicology and Chemistry 14:613–622.

Mather, F. 1900. Modern Fish Culture in Fresh and Salt Waters. Forest and Stream Publishing Co., New York. [Chapter 40, Frog Culture, pp. 301-305].

Mathews, R.C., and A.C. Echternacht. 1984. Herpetofauna of the spruce–fir ecosystem in the southern Appalachian Mountains regions, with emphasis on the Great Smoky Mountains National Park. Pp. 155–167 *In* P.S. White (ed.), The Southern Appalachian Spruce–Fir Ecosystem. National Park Service, Research/Resources Management Report SER–71.

Mathewson, R.F. 1955. Reptiles and amphibians of Staten Island. Proceedings of the Staten Island Institute of Arts and Sciences 17:28-50.

Matson, T.O. 1990a. A morphometric comparison of gray treefrogs, *Hyla chrysoscelis* and *H. versicolor*, from Ohio. Ohio Journal of Science 90:98–101.

Matson, T.O. 1990b. Erythrocyte size as a taxonomic character in the identification of Ohio *Hyla chrysoscelis* and *H. versicolor*. Herpetologica 46:457–462.

Matsuda, B.M. 2001. The effects of clear–cut timber harvest on the movement patterns of tailed frogs (*Ascaphus truei*) in southwestern British Columbia. M.S. thesis, University of British Columbia, Vancouver.

Matsuda, B.M., and J.S. Richardson. 2005. Movement patterns and relative abundance of coastal tailed frogs in clearcuts and mature forest stands. Canadian Journal of Forest Resources 35:1131–1138.

Matsuda, B.M., D.M. Green, and P.T. Gregory. 2006. Amphibians and Reptiles of British Columbia. Royal BC Museum Handbook, Victoria, British Columbia.

Mattfeldt, S.D., L.L. Bailey, and E.H.C. Grant. 2009. Monitoring multiple species: estimating state variables and exploring the efficacy of a monitoring program. Biological Conservation 142:720–737.

Matthews, C.E., C.E. Moorman, C.H. Greenberg, and T.A. Waldrop. 2010. Response of reptiles and amphibians to repeated fuel reduction treatments. Journal of Wildlife Management 74:1301–1310.

Matthews, K.R. 2003. Response of mountain yellow–legged frogs, *Rana muscosa*, to short distance translocation. Journal of Herpetology 37:621–626.

Matthews, K.R., and K.L. Pope. 1999. A telemetric study of the movement patterns and habitat use of *Rana muscosa*, the mountain yellow–legged frog, in a high–elevation basin in Kings Canyon National Park, California. Journal of Herpetology 33:615–624.

Matthews, K.R., and C. Miaud. 2007. A skeletochronological study of the age structure, growth, and longevity of the Mountain yellow–legged frog, *Rana muscosa*, in the Sierra Nevada, California. Copeia 2007:986–993.

Matthews, K.R., and H.K. Preisler. 2010. Site fidelity of the declining amphibian *Rana sierrae* (Sierra Nevada yellow-legged frog). Canadian Journal of Fisheries and Aquatic Sciences 67: 243–255.

Matthews, K.R., K.L. Pope, H.K. Preisler, and R.A. Knapp. 2001. Effects of nonnative trout on Pacific treefrogs (*Hyla regilla*) in the Sierra Nevada. Copeia 2001:1130–1137.

Matthews, K.R., R.A. Knapp, and K.L. Pope. 2002. Garter snake distributions in high–elevation aquatic ecosystems: is there a link with declining amphibian populations and nonnative trout introductions? Journal of Herpetology 36:16–22.

Matthews, T.C. 1971. Genetic changes in a population of boreal chorus frogs (*Pseudacris triseriata*) polymorphic for color. American Midland Naturalist 85:208–221.

Matthews, T., and D. Pettus. 1966. Color inheritance in *Pseudacris triseriata*. Herpetologica 22:269–275.

Mattoon, A.T. 1996. Differential response to temperature by gray treefrog (*Hyla chrysoscelis*) larvae: a study of geographic variation. M.S. thesis, University of Michigan, Ann Arbor.

Matutte, B., K.B. Storey, F.C. Knoop, and J.M. Conlon. 2000. Induction of synthesis of an antimicrobial peptide in the skin of the freeze-tolerant frog, *Rana sylvatica*, in response to environmental stimuli. FEBS Letters 483:135-138.

Maunder, J.E. 1983. Amphibians of the Province of Newfoundland. Canadian Field-Naturalist 97:33–46.

Maunder, J.E. 1997. Amphibians of Newfoundland and Labrador: status changes since 1983. SSAR Herpetological Conservation 1:93–99.

Mautz, W.J., and M.R. Dohm. 2004. Respiratory and behavioral effects of ozone on a lizard and a frog. Comparative Biochemistry and Physiology 139A:371–377.

Maxell, B.A., K.J. Nelson, and S. Browder. 2002. Record clutch size and observations on breeding and development of the western toad (*Bufo boreas*) in Montana. Northwestern Naturalist 83:27-390.

Maxson, L.R., and D.L. Jameson. 1968. A comparison of the chromosomes of *Hyla regilla* and *Hyla californiae*. Copeia 1968:704–707.

Maxson, L.R., and A.C. Wilson. 1974. Convergent morphological evolution detected by studying proteins of tree frogs in the *Hyla exima* species group. Science 185:66–68.

Maxson, L.R., and A.C. Wilson. 1975. Albumin evolution and organismal evolution in tree frogs (Hylidae). Systematic Zoology 24:1–15.

Maxson, L., E. Pepper, and R.D. Maxon. 1977. Immunological resolution of a diploid–tetraploid species complex of tree frogs. Science 197:1012–1013.

Maxson, L.R., and C.H. Daugherty. 1980. Evolutionary relationships of the monotypic toad family Rhinophrynidae: a biochemical perspective. Herpetologica 36:275–280.

Maxson, R.D., and L.R. Maxson. 1978. Reply to Ralin. Science 202:336.

Maxwell, B.A., and D.G. Hokit. 1999. Amphibians and reptiles. Pp. 2.1–2.29 *In* G. Joslin and H. Youmans (coordinators), Effects of Recreation on Rocky Mountain Wildlife. A Review for Montana. Committee on Effects of Recreation on Wildlife, Montana Chapter of the Wildlife Society. http://joomla.wildlife.org/Montana/images/Documents/2hp1.pdf

Maxwell, B.A., J.K. Werner, P. Hendricks, and D.L. Flath. 2003. Herpetology in Montana. Northwest Fauna No. 5.

Maxwell, J. P. 1999. Effects of species, size, and density on the feeding behavior of three anuran tadpoles. M.S. thesis, Mississippi State University, Mississippi State.

Mayer, F.L., and M.R. Ellersieck. 1986. Manual of acute toxicity: interpretation and data base for 410 chemicals and 66 species of freshwater animals. U.S. Fish and Wildlife Service Resource Publication 160.

Mayhew, W.W. 1962. *Scaphiopus couchi* in California's Colorado Desert. Herpetologica 18:153–161.

Mayhew, W.W. 1965. Adaptations of the amphibian, *Scaphiopus couchi*, to desert conditions. American Midland Naturalist 74:95–109.

Maynard, E.A. 1934. The aquatic migration of the toad, *Bufo americanus* Le Conte. Copeia 1934:174–177.

Mazanti, L.E. 1999. The effects of atrazine, metolachlor and chlorpyrifos on the growth and survival of larval frogs under laboratory and field conditions. Ph.D. Dissertation, University of Maryland, College Park.

Mazerolle, M.J. 2001. Amphibian activity, movement patterns, and body size in fragmented peat bogs. Journal of Herpetology 35:13–20.

Mazerolle, M.J. 2003. Detrimental effects of peat mining on amphibian abundance and species richness in bogs. Biological Conservation 113:215–223.

Mazerolle, M.J. 2004. Amphibian road mortality in response to nightly variations in traffic intensity. Herpetologica 60:45–53.

Mazerolle, M.J. 2006. Choosing the safest route: frog orientation in an agricultural landscape. Journal of Herpetology 40:435–441.

Mazerolle, M.J., and A. Desrochers. 2005. Landscape resistance to frog movements. Canadian Journal of Zoology 83:455-464.

Mazerolle, M.J., M. Huot, and M. Gravel. 2005. Behavior of amphibians on the road in response to car traffic. Herpetologica 61:380–388.

Mazerolle, M.J., A. Desrochers, and L. Rochefort. 2005. Landscape characteristics influence pond occupancy by frogs after accounting for detectability. Ecological Applications 15:824-834.

Mazzoni, R., A.A. Cunningham, P. Daszak, A. Apolo, E. Perdomo, and G. Speranza. 2003. Emerging pathogen of wild amphibians in frogs (*Rana catesbeiana*) farmed for international trade. Emerging Infectious Diseases 9:995–998.

McAdam, S.H., and M. Nagel-Hisey. 1998. Winter northern leopard frogs at Cypress Hills Inter- provincial Park. Blue Jay 56:166-168.

McAlister, W. 1954. Natural history notes on the barking frog. Herpetologica 10:197–199.

McAlister, W.H. 1958. Species distribution in a mixed *Scaphiopus–Bufo* breeding chorus. Southwestern Naturalist 3:227–229.

McAlister, W.H. 1959. The vocal structures and method of call production in the genus *Scaphiopus* Holbrook. Texas Journal of Science 11:60–77.

McAlister, W.H. 1961a. The mechanics of sound production in the North American *Bufo*. Copeia 1961:86–95.

McAlister,W.H. 1961b. Artificial hybridization between *Rana a. areolata* and *R. p. pipiens* from Texas.Texas Journal of Science 13:423–426.

McAlister, W.H. 1962. Variation in *Rana pipiens* Schreber in Texas. American Midland Naturalist 67:334–363.

McAlister, W.H. 1963. A post–breeding concentration of the spring peeper. Herpetologica 19:293.

McAllister, C.T. 1987. Protozoan and metazoan parasites of Strecker's chorus frog, *Pseudacris streckeri streckeri* (Anura: Hylidae), from north–central Texas. Proceedings of the Helminthological Society of Washington 54:271–274.

McAllister, C.T. 1991. Protozoan, helminth, and arthropod parasites of the spotted chorus frog, *Pseudacris clarkii* (Anura: Hylidae), from north–central Texas. Journal of the Helminthological Society of Washington 58:51–56.

McAllister, C.T., and S.P. Tabor. 1985. *Gastrophryne olivacea* (Great Plains Narrowmouth Toad). Coexistence. Herpetological Review 16:109.

McAllister, C.T., and S. J. Upton. 1987. Parasites of the Great Plains narrowmouth toad (*Gastrophryne olivacea*) from northern Texas. Journal of Wildlife Diseases. 23: 686-688.

McAllister, C.T., and D.B. Conn. 1990. Occurrence of tetrathyridia of *Mesocestoides* sp. (Cestoidea: Cyclophyllidea) in North American anurans (Amphibia). Journal of Wildlife Diseases 26:540-543.

McAllister, C.T., and P.S. Freed. 1992. Larval *Abbreviata* sp. (Spirurida: Psysalopteridae) in introduced Rio Grande chirping frogs, *Syrrhopus cystignathoides campi* (Anura: Leptodactylidae), from Houston, Texas. Texas Journal of Science 44:359–361.

McAllister, C.T., and S.E. Trauth. 1995. New host records for *Myxidium serotinum* (Protozoa: Myxosporea) from North American amphibians. Journal of Parasitology 81:485–488.

McAllister, C.T., and S.E. Trauth. 1996. Ultrastructure of *Cedpedia virginiensis* (Protista: Haptophrynidae) from the gall bladder of the pickerel frog, *Rana palustris*, in Arkansas. Proceedings of the Arkansas Academy of Science 50:133–136.

McAllister, C.T., and C.R. Bursey. 2012. *Anaxyrus americanus charlesmithi* (Dwarf American Toad). Nematode parasite. Herpetological Review 43:117.

McAllister, C.T., S.J. Upton, and D.B. Conn. 1989. A comparative study of endoparasites in three species of sympatric *Bufo* (Anura: Bufonidae), from Texas. Proceedings of the Helminthological Society of Washington 56:162–167.

McAllister, C.T., S.J. Upton, S.E. Trauth, and C.R. Bursey. 1995a. Parasites of wood frogs, *Rana sylvatica* (Ranidae), from Arkansas, with a description of a new species of *Eimera* (Apicomplexa: Eimeriidae). Journal of the Helminthological Society of Washington 62:143–149.

McAllister, C.T., S.E. Trauth, and L.D. Gage. 1995b. Vertebrate fauna of abandoned mines at Gold Mine Springs, Independence County, Arkansas. Proceedings of the Arkansas Academy of Science 49:184–187.

McAllister, C.T., S.E. Trauth, and C.R. Bursey. 1995c. Parasites of the pickerel frog, *Rana palustris* (Anura: Ranidae), from the southern part of its range. Southwestern Naturalist 40:111–116.

McAllister, C.T., C.R. Bursey, and D.B. Conn. 2005. Endoparasites of Hurter's spadefoot, *Scaphiopus hurterii* and plains spadefoot, *Spea bombifrons* (Anura: Scaphiopodidae), from southern Oklahoma. Texas Journal of Science 57:383-389.

McAllister, C.T., C.R. Bursey, and S.E. Trauth. 2008a. New host and geographic records for some endoparasites (Myxosporea, Trematoda, Cestoidea, Nematoda) of amphibians and reptiles from Arkansas and Texas, U.S.A. Comparative Parasitology 75:241–254.

McAllister, C.T., F.L. Frye, and S.R. Goldberg. 2008b. *Hyla chrysoscelis* (Cope's Gray Treefrog). Ocular pathology. Herpetological Review 39:78–79.

McAllister, K.R. 1995. Distribution of amphibians and reptiles in Washington State. Northwest Fauna 3:81–112.

McAllister, K.R., and W.P. Leonard. 1997. Washington State

status report for the Oregon spotted frog. Washington Department of Fish and Wildlife, Olympia.

McAllister, K.R., W.P. Leonard, and R.M. Storm. 1993. Spotted frog (*Rana pretiosa*) surveys in the Puget Trough of Washington, 1989–1991. Northwestern Naturalist 74:10–15.

McAllister, K.R., J.W. Watson, K. Risenhoover, and T. McBride. 2004. Marking and radiotelemetry of Oregon spotted frogs (*Rana pretiosa*). Northwestern Naturalist 85:20–25.

McAlpine, D.F. 1996. Acanthocephala parasitic in North American amphibians: a review with new records. Alytes 14:115–121.

McAlpine, D.F. 1997a. Historical evidence does not suggest New Brunswick amphibians have declined. SSAR Herpetological Conservation 1:117–127.

McAlpine, D.F. 1997b. Helminth communities in bullfrogs (*Rana catesbeiana*), green frogs (*Rana clamitans*), and leopard frogs (*Rana pipiens*) from New Brunswick, Canada. Canadian Journal of Zoology 75:1883–1890.

McAlpine, D.F. 1997c. A simple transect technique for estimating abundance of aquatic ranid frogs. SSAR Herpetological Conservation 1:180–184.

McAlpine, D.F., and T.G. Dilworth. 1989. Microhabitat and prey size among three species of *Rana* (Anura: Ranidae) sympatric in eastern Canada. Canadian Journal of Zoology 67:2244–2252.

McAlpine, D.F., and D.A. Vail. 1997. Hyla Park: managing an amphibian conservation area in an eastern Canadian urban setting. Herpetological Bulletin 94:17-21.

McAlpine, D.F., and M.D.B. Burt. 1998. Helminth communities in bullfrogs (*Rana catesbeiana*), green frogs (*Rana clamitans*), and leopard frogs (*Rana pipiens*) from New Brunswick, Canada. Canadian Field-Naturalist 112:50–68.

McAlpine, D.F., S.W. Gorham, and A.D.B. Heward. 1980. Distributional status and aspects of the biology of the gray treefrog, *Hyla versicolor*, in New Brunswick. Journal of the New Brunswick Museum (1980):92-102.

McAlpine, D.F., T.J. Fletcher, S.W. Gorham, and I.T. Gorham. 1991. Distribution and habitat of the tetraploid gra treefrog, *Hyla versicolor*, in New Brunswick and eastern Maine. Canadian Field-Naturalist 105:526-529.

McAlpine, D.F., B. Cougle, and T.J. Fletcher. 2001. *Rana septentrionalis* (Mink Frog). Larval predation. Herpetological Review 32:183–184.

McAlpine, D.F., R.W. Harding, and R. Curley. 2006. Occurrence and biogeographic significance of the pickerel frog (*Rana palustris*) on Prince Edward Island, Canada. Herpetological Natural History 10:95–98.

McAlpine, D.F., J.D.H. Pratt, and J.M. Terhune. 2009. Apparent continuing expansion in the range of the gray treefrog, *Hyla versicolor*, in New Brunswick. Canadian Field-Naturalist 123:309-312.

McAlpine, S. 1993. Genetic heterozygosity and reproductive success in the green treefrog, *Hyla cinerea*. Heredity 70:553–558.

McAlpine, S., and M.H. Smith. 1995. Genetic correlates of fitness in the green treefrog, *Hyla cinerea*. Herpetologica 51:393–400.

McAtee, W.L. 1921. Homing and other habits of the bullfrog. Copeia (96):39–40.

McAuliffe, J.R. 1978. Biological survey and management of sport–hunted bullfrog populations in Nebraska. Nebraska Game and Parks Commission, Lincoln.

McCaffery, R.M.W. 2011. Population dynamics of the Columbia spotted frog (*Rana luteiventiris*): inference from long–term demography. Ph.D. Dissertation, University of Montana, Missoula.

McCaffery, R., A. Solonen, and E. Crone. 2012. Frog population viability under present and future climate conditions: a Bayesian state–space approach. Journal of Animal Ecology 81:978–985.

McCallum, M.L. 1999a. *Rana sphenocephala* (Southern Leopard Frog) malformities found in Illinois with behavioral notes. Transactions of the Illinois State Academy of Science 92:257–264.

McCallum, M.L. 1999b. *Acris crepitans* (Northern Cricket Frog). Death feigning. Herpetological Review 30:90.

McCallum, M.L. 2003. Reproductive ecology and taxonomic status of *Acris crepitans blanchardi* with additional investigations on the Hamilton and Zuk hypotheses. Ph.D. Dissertation, Arkansas State University, Jonesboro.

McCallum, M.L. 2010. Future climate change spells catastrophe for Blanchard's cricket frog, *Acris blanchardi* (Amphibia: Anura: Hylidae). Acta Herpetologica 5:119–130.

McCallum, M.L. 2011a. Road mortality of turtles and bullfrogs during a major flood. Herpetology Notes 4:183–186.

McCallum, M.L. 2011b. Orientation and directional escape by Blanchard's Cricket Frog (*Acris blanchardi*) in response to a human predator. Acta Herpetologica 6:161–168.

McCallum, M.L., and J.L. McCallum. 2012. Does body size reflect foraging ability in post–metamorphic marine toads? Herpetology Notes 5:15–18.

McCallum, M.L., and J.L. McCallum. 2005. *Hyla chrysoscelis* (Cope's Gray Treefrog). Tadpole over–wintering. Herpetological Review 3:54.

McCallum, M.L., and J. Rayburn. 2006. Herpetological Husbandry: A simple method for housing *Xenopus* during oviposition and obtaining eggs for use in FETAX. Herpetological Review 37:331.

McCallum, M.L., and S.E. Trauth. 2001a. Are tadpoles of the Illinois chorus frog (*Pseudacris streckeri illinoensis*) cannibalistic? Transactions of the Illinois State Academy of Science 94:171–178.

McCallum, M.L., and S.E. Trauth. 2001b. *Pseudacris streckeri illinoensis* (Illinois Chorus Frog). Terrestrial feeding. Herpetological Review 32:35.

McCallum, M.L., and S.E. Trauth. 2001c. *Hyla cinerea* (Green Treefrog) Pre–hibernation live body weights. Herpetological Review 32:182.

McCallum, M.L., and S.E. Trauth. 2002. Performance of wood frog (*Rana sylvatica*) tadpoles on three soybean meal – corn meal rations. Podarcis 3:78–85.

McCallum, M.L., and S.E. Trauth. 2003a. A forty–three year museum study of northern cricket frog (*Acris crepitans*) abnormalities in Arkansas: upward trends and distributions. Journal of Wildlife Diseases 39:522–528.

McCallum, M.L., and S.E. Trauth. 2003b. *Acris crepitans* (Northern Cricket Frog). Communal hibernaculum. Herpetological Review 34:228.

McCallum, M.L., and S.E. Trauth. 2004. Blanchard's cricket frog in Nebraska and South Dakota. Prairie Naturalist 36:129–135.

McCallum, M.L., and S.E. Trauth. 2006. An evaluation of the subspecies *Acris crepitans blanchardi* (Anura, Hylidae). Zootaxa 1104:1–21.

McCallum, M.L., and S.E. Trauth. 2007. Physiological trade–offs between immunity and reproduction in the northern cricket frog (*Acris crepitans*). Herpetologica 63:269–274.

McCallum, M.L., and S.E. Trauth. 2008. Egg mass fidelity suggests oophagy in the Illinois chorus frog (*Pseudacris streckeri illinoensis*). Journal of Kansas Herpetology 28:14.

McCallum, M.L., and S.E. Trauth. 2009. *Hyla cinerea* (Green Tree Frog) and *Pseudacris triseriata* (Western Chorus Frog). Calling in amplexus. Herpetological Review 40:204–205.

McCallum, M.L., B.A. Wheeler, and S.E. Trauth. 2001a. *Pseudacris streckeri illinoensis* (Illinois Chorus Frog). Dysfunctional vocal sac. Herpetological Review 32:248–249.

McCallum, M.L., B.A. Wheeler, and S.E. Trauth. 2001b. *Acris crepitans* (Northern Cricket Frog). Attempted cannibalism. Herpetological Review 32:99–100.

McCallum, M.L., B.A. Wheeler, and S.E. Trauth. 2001c. The "Frog Box:" A new bulk holding device for anurans. Herpetological Review 33:107.

McCallum, M.L., T.L. Klotz, and S.E. Trauth. 2003a. *Rana sylvatica* (Wood Frog). Death feigning. Herpetological Review 34:54–55.

McCallum, M.L., T.L. Klotz, and S.E. Trauth. 2003b. *Rana sylvatica* (Wood Frog). Phonotactic stalking. Herpetological Review 34:54.

McCallum, M.L., R.G. Neal, and S.E. Trauth. 2003c. *Pseudacris streckeri illinoensis* (Illinois Chorus Frog). Satellite behavior. Herpetological Review 34:53.

McCallum, M.L., S.E. Trauth, R.G. Neal, V. Hoffman. 2003d. A herpetofaunal inventory of Arkansas Post National Memorial, Arkansas County, Arkansas. Journal of the Arkansas Academy of Science 57:123–130.

McCallum, M.L., S.E. Trauth, M.N. Mary, C.R. McDowell, and B.A. Wheeler. 2004. Fall breeding of the southern leopard frog (*Rana sphenocephala*) in northeastern Arkansas. Southeastern Naturalist 3:401–408.

McCallum, M.L., S.E. Trauth, C.R. McDowell, R.G. Neal, and T.L. Klotz. 2006. Calling site characterization of the Illinois Chorus frog (*Pseudacris streckeri illinoensis*) with additional notes on population size and breeding choruses. Herpetological Natural History 9:183–185.

McCallum, M.L, V. Bogosian III, and E. Walsh. 2008. Mortality of bullfrogs (*Rana catesbeiana*) from an in–ground swimming pool. Journal of Kansas Herpetology 28:15.

McCallum, M.L., C. Brooks, R. Mason, and S.E. Trauth. 2011a. Growth, reproduction, and life span in Blanchard's cricket frog (*Acris blanchardi*) with notes on the growth of the northern cricket frog (*Acris crepitans*). Herpetology Notes 4:25–35.

McCallum, M.L., W.E. Moser, B.A. Wheeler, and S.E. Trauth. 2011b. Amphibian infestation and host size preference by the leech *Placobdella picta* (Verrill, 1872) (Hirudinida: Rhynchobdellida: Glossiphoniidae) from the eastern Ozarks, USA. Herpetology Notes 4:147–151.

McCarthy, K., and R.G. Lathrop. 2011. Stormwater basins of the New Jersey coastal plain: subsidies or sinks for frogs and toads? Urban Ecosystems 14:395-413.

McClanahan, L., Jr. 1964. Osmotic tolerance of the muscles of two desert–inhabiting toads, *Bufo cognatus* and *Scaphiopus couchi*. Comparative Biochemistry and Physiology 12:501–508.

McClanahan, L., Jr. 1967. Adaptations of the spadefoot toad, *Scaphiopus couchi*, to desert environments. Comparative Biochemistry and Physiology 20:73–99.

McClanahan, L., Jr. 1972. Changes in body fluid of burrowed spadefoot toads as a function of soil water potential. Copeia 1972:209–216.

McClanahan, L.L., R. Ruibal, and V.H. Shoemaker. 1994. Frogs and toads in deserts. Scientific American 270(3):82–88.

McClelland, B.E., and W. Wilczynski. 1989. Release call characteristics of male and female *Rana pipiens*. Copeia 1989:1045–1049.

McClelland, B.E., W. Wilczynski, and M.J. Ryan. 1996.

Correlations between call characteristics and morphology in male cricket frogs (*Acris crepitans*). Journal of Experimental Biology 199:1907–1919.

McClure, K.A. 1996. Ecology of *Pseudacris brachyphona*: a second look. M.S. thesis, Marshall University, Huntington, West Virginia.

McClure, K.V., J.W. Mora, and G.R. Smith. 2009. Effects of light and group size on the activity of wood frog tadpoles (*Rana sylvatica*) and their response to a shadow stimulus. Acta Herpetologica 4:103–107.

McCoid, M.J. 1985. An observation of reproductive behavior in a wild population of African clawed frogs, *Xenopus laevis*, in California. California Fish and Game 71:245–246.

McCoid, M.J. 2005. *Rana berlandieri* (Rio Grande Leopard Frog). Salinity tolerance. Herpetological Review 36:437–438.

McCoid, M.J., and T.H. Fritts. 1980a. Notes on the diet of a feral population of *Xenopus laevis* (Pipidae) in California. Southwestern Naturalist 25:272–275.

McCoid, M.J., and T.H. Fritts. 1980b. Observations of feral populations of *Xenopus laevis* (Pipidae) in Southern California. Bulletin of the Southern California Academy of Sciences 79:82–86.

McCoid, M.J., and T.H. Fritts. 1989. Growth and fatbody cycles in feral populations of the African clawed frog, *Xenopus laevis* (Pipidae), in California with comments on reproduction. Southwestern Naturalist 34:499–505.

McCoid, M.J., and T.H. Fritts. 1993. Speculations on colonizing success of the African clawed frog, *Xenopus laevis* (Pipidae), in California. South African Journal of Zoology 28:59–61.

McCoid, M.J., and T.H. Fritts. 1995. Female reproductive potential and winter clawed frogs (Pipidae: *Xenopus laevis*) in California. California Fish and Game 81:39–42.

McCoid, M.J., G.K. Pregill, and R.M. Sullivan. 1993. Possible decline of *Xenopus* populations in Southern California. Herpetological Review 24:29–30.

McCoid, M.J., F.S. Guthery, S. Kopp, and R. Howard.1999. *Scaphiopus holbrookii hurterii* (Hurter's Spadefoot). Predation. Herpetological Review 30:94.

McCollum, S.A. 1993. Ecological consequences of predator–induced polyphenism in larval hylid frogs. Ph.D. Dissertation, Duke University, Durham, North Carolina.

McCollum, S.A., and J. Van Buskirk. 1996. Costs and benefits of a predator–induced polyphenism in the gray treefrog *Hyla chrysoscelis*. Evolution 50:583–593.

McCollum, S.A., and J.D. Leimberger. 1997. Predator–induced morphological changes in an amphibian: predation by dragonflies affects tadpole shape and color. Oecologia 109:615–621.

McComb, W.C., and R.E. Noble. 1981. Herpetofaunal use of natural tree cavities and nest boxes. Wildlife Society Bulletin 9:261–267.

McConnell, A.P. 1966. The effect of temperature upon the larval development of amphibians. M.S. thesis, North Carolina State University, Raleigh.

McCormick, J. 1970. The Pine Barrens: a preliminary ecological inventory. Research Report No. 2, New Jersey State Museum, Trenton.

McCormick, S., and G.A. Polis. 1982. Invertebrates that prey on vertebrates. Biological Review 57:29–58.

McCoy, C.J. 1969. Diet of bullfrogs (*Rana catesbeiana*) in central Oklahoma farm ponds. Proceedings of the Oklahoma Academy of Science 48:44–45.

McCoy, C.J. 1982. Amphibians and reptiles in Pennsylvania. Carnegie Museum of Natural History Special Publication No. 6.

McCoy, C.J., and C.J. Durden. 1965. New distribution records of amphibians and reptiles in eastern Canada. Canadian Field-Naturalist 79:156–157.

McCoy, C.J., H.M. Smith, and J.A. Tihen. 1967. Natural hybrid toads, *Bufo punctatus* x *Bufo woodhousei*, from Colorado. Southwestern Naturalist 12:45–54.

McCoy, K.A. 2006. The reproductive and ecotoxicological effects of agricultural exposure on anurans. Ph.D. Dissertation, University of Florida, Gainesville.

McCoy, K.A., L.J. Bortnick, C.M. Campbell, H.J. Hamlin, L.J. Guillette, Jr., and C.M. St. Mary. 2008a. Agriculture alters gonadal form and function in the toad *Bufo marinus*. Environmental Health Perspectives 116:1526–1532.

McCoy, K.A., L.K. Hoang, L.J. Guillette, Jr., and C.M. St. Mary. 2008b. Renal pathologies in giant toads (*Bufo marinus*) vary with land use. Science of the Total Environment 407:348–357.

McCoy, M.W. 2003. *Bufo terrestris* (Southern Toad). Predation. Herpetological Review 34:135–136.

McCoy, M.W. 2006. Trait–mediated predator–prey interactions between treefrog tadpoles and aquatic invertebrates: integrating size–structure, density, and among–individual variation. Ph.D. Dissertation, University of Florida, Gainesville.

McCoy, M.W. 2007. Conspecific density determines the magnitude and character of predator–induced phenotype. Oecologia 153:871–878.

McCoy, M.W., and B.M. Bolker. 2008. Trait–mediated interactions: influence of prey size, density and experience. Journal of Animal Ecology 77:478–486.

McCreary, B., and C.A. Pearl. 2008. *Rana cascadae* (Cascades Frog). Albinism. Herpetological Review 39:79–80.

McCreary, B., and C.A. Pearl. 2010. *Pseudacris regilla* (Northern Pacific Treefrog). Cavity use. Herpetological Review 41:202–203.

McDaniel, V.R., and J.E. Gardner. 1977. Cave fauna of Arkansas: vertebrate taxa. Proceedings of the Arkansas Academy of Science 31:68–71.

McDiarmid, R.W. 1983. Large–scale operations management test of use of the White Amur for control of problem aquatic plants. Reports 2 and 3, Volume 5. The herpetofauna of Lake Conway, Florida: community analysis. U.S. Army Corps of Engineers, Technical Report A–78–2, Waterways Experiment Station, Vicksburg, Mississippi.

McDonald, D.G., J.L. Ozog, and B.P. Simons. 1984. The influence of low pH environments on ion regulation in the larval stages of the anuran amphibian, *Rana clamitans*. Canadian Journal of Zoology 62:2171–2177.

McGarrity, M.E., and S.A. Johnson. 2009. Geographic trend in sexual size dimorphism and body size of *Osteopilus septentrionalis* (Cuban treefrog): implications for invasion of the southeastern United States. Biological Invasions 11:1411–1420.

McGarrity, M.E., and S.A. Johnson. 2010. A radio telemetry study of invasive Cuban treefrogs. Florida Scientist 73:225–235.

McGehee, R.T., R. Reams, and M.E. Brown. 2001. *Bufo valliceps* (Gulf Coast Toad). Diet. Herpetological Review 32:101–102.

McGrath, E.A., and M.M. Alexander. 1979. Observations on the exposure of larval bullfrogs to fuel oil. Transactions of the Northeast Section of The Wildlife Society 36:45–51.

McHenry, D.J. 2010. Genetic variation and population structure in the endangered Houston Toad in contrast to its common sympatric relative, the coastal plain toad. Ph.D. Dissertation, University of Missouri, Columbia.

McHenry, D.J., M.A. Gaston, and M.R.J. Forstner. 2010. *Bufo houstonensis* (Houston Toad). Predation. Herpetological Review 41:193.

McIntyre, P.B., and S.A. McCollum. 2000. Responses of bullfrog tadpoles to hypoxia and predators. Oecologia 125:301–308.

McKamie, J.A., and G.A. Heidt. 1974. A comparison of spring food habits of the bullfrog, *Rana catesbeiana*, in three habitats of central Arkansas. Southwestern Naturalist 19:105–119.

McKay, A.H. 1896. Batrachia and reptilia of Nova Scotia. Transactions of the Nova Scotian Institute of Science 2:101-102.

McKeever, S. 1977. Observations of *Corethrella* feeding on tree frogs (*Hyla*). Mosquito News 37:522–523.

McKeever, S., and F.E. French. 1991. *Corethrella* (Diptera: Corethrellidae) of eastern North America: laboratory life history and field responses to anuran calls. Annals of the Entomological Society of America 84:493–497.

McKeown, J.P. 1968. The ontogenetic development of Y–axis orientation in four species of anurans. Ph.D. Dissertation, Mississippi State University, Mississippi State.

McKeown, S. 1996. A Field Guide to Reptiles and Amphibians in the Hawaiian Islands. Diamond Head Publishing, Los Osos, California.

McKibbina, R., W.T. Dushenkob, G. van Aggelenc, and C.A. Bishop. 2008. The influence of water quality on the embryonic survivorship of the Oregon spotted frog (*Rana pretiosa*) in British Columbia, Canada. Science of the Total Environment 395: 28–40.

McKinnell, R.G. 1984. Lucké tumor of frogs. Pp. 581–605 *In* G.L. Hoff, F.L. Frye, and E.R. Jacobson (eds.), Diseases of Amphibians and Reptiles. Plenum Press, New York.

McKinnell, R.G., and J. Zambernard. 1968. Virus particles in renal tumors obtained from spring *Rana pipiens* of known geographic origin. Cancer Research 28:684–688.

McKinnell, R.G., and V.L. Ellis. 1972. Herpesviruses in tumors of post spawning *Rana pipiens*. Cancer Research 32:1154-1159.

McKinnell, R.G., and D.L. Carlson. 2005. Lucké renal adenocarcinoma. Pp. 96–102 *In* M.J. Lannoo (ed.), Amphibian Declines. The Conservation Status of United States Species. University of California Press, Berkeley.

McKinnell, R.G., D.M. Hoppe, and B.K. McKinnell. 2005. Monitoring pigment pattern morphs of northern leopard frogs. Pp. 328–337 *In* M.J. Lannoo (ed.), Amphibian Declines. The Conservation Status of United States Species. University of California Press, Berkeley.

McKnight, D.T. 2012. *Hyla chrysoscelis* (Cope's Gray Treefrog). Calling/mate choice. Herpetological Review 43:120–121.

McLaren, I.A. 1965. Temperature and frog eggs. Journal of General Physiology 48:1071–1079.

McLeod, D. S. 1999. A re-survey of amphibian populations in Nebraska after twenty years: A test for declines. M.S. thesis, University of Nebraska-Lincoln.

McLeod, D.S. 2005. Nebraska's declining amphibians. Pp. 292–294 *In* M.J. Lannoo (ed.), Amphibian Declines. The Conservation Status of United States Species. University of California Press, Berkeley.

McLeod, R.F. 1995. The effects of timber harvest and prescribed burning on the distribution and abundance of reptiles and amphibians at Remington Farms, Maryland. M.S. thesis, Frostburg State University, Frostburg, Maryland.

McLeod, R.F., and J.E. Gates. 1998. Response of herpetofaunal communities to forest cutting and burning at Chesapeake Farms, Maryland. American Midland Naturalist

139:164–177.

McMahon T.A., N.T. Halstead, S. Johnson, T.R. Raffel, J.M. Romansic, P.W. Crumrine, R.K. Boughton, L.B. Martin, and J.R. Rohr. 2011. The fungicide Chlorothalonil is nonlinearly associated with corticosterone levels, immunity, and mortality in amphibians. Environmental Health Perspectives 119:1098–1103.

McMahon, T.A., L.A. Brannelly, M.W.H. Chatfield, P.T.J. Johnson, M.B. Joseph, V.J. McKenzie, C.L. Richards–Zawacki, M.D. Venesky, and J.R. Rohr. 2012. Chytrid fungus *Batrachochytrium dendrobatidis* has nonamphibian hosts and releases chemicals that cause pathology in the absence of infection. Proceedings of the National Academy of Sciences of the United States of America. 110:210–215.

McManus, M.I. 1924. The frogs and toads of the Columbia, South Carolina region. M.A. thesis, University of South Carolina, Columbia.

McMenamin, S.K., E.A. Hadly, and C.K. Wright. 2008. Climatic change and wetland desiccation cause amphibian decline in Yellowstone National Park. Proceedings of the National Academy of Sciences of the United States of America 105:16988–16993.

McMenamin, S.K., E.A. Hadly, and C.K. Wright. 2009. Reply to Patla et al.: amphibian habitat and populations in Yellowstone damaged by drought and global warming. Proceedings of the National Academy of Sciences of the United States of America 106:E23.

McMurry, S.T., L.M. Smith, K.D. Dupler, and M.B. Gutierrez. 2009. Influence of land use on body size and splenic cellularity in wetland breeding *Spea* spp. Journal of Herpetology 43:421–430.

McNicholl, M.K. 1972. An observation of apparent death–feigning by a toad. Blue Jay 30:54–55.

Meacham, W.R. 1958. Factors affecting gene exchange between two allopatric populations of the *Bufo woodhousei* complex. Ph.D. Dissertation, University of Texas, Austin.

Meacham, W.R. 1962. Factors affecting secondary intergradation between two allopatric populations in the *Bufo woodhousei* complex. American Midland Naturalist 67:282–304.

Means, D. B. 1977. Aspects of the significance to terrestrial vertebrates of the Apalachicola River Basin, Florida. Florida Marine Research Publications (26):37–67.

Means, D.B. 1983. The enigmatic Pine Barrens Treefrog. Florida Wildlife (May-Jun):16-19.

Means, D.B. 2004. *Rana capito* (Florida Gopher Frog). Defensive behavior. Herpetological Review 35:163–164.

Means, D.B., and C.J. Longden. 1976. Aspects of the biology and zoogeography of the Pine Barrens treefrog (*Hyla andersonii*) in northern Florida. Herpetologica 32:117–130.

Means, D.B., and D. Simberloff. 1987. The peninsula effect: habitat correlated species decline in Florida's herpetofauna. Journal of Biogeography 14:551–568.

Means, D.B., and R.C. Means. 2005. Effects of sand pine silviculture on pond–breeding amphibians in the Woodville Karst Plain of north Florida. Pp. 56–61 *In* W.E. Meshaka, Jr. and K.J. Babbitt (eds.), Amphibians and Reptiles. Status and Conservation in Florida. Krieger Publishing, Malabar, Florida.

Means, D.B., and S.C. Richter. 2007. Genetic verification of possible gigantism in a southern toad, *Bufo terrestris*. Herpetological Review 38:297–298.

Means, R.C., and R. Franz. 2005. Herpetofauna of impacted wetlands in east Florida: a pre–augmentation assessment. Pp. 23–31 *In* W.E. Meshaka, Jr. and K.J. Babbitt (eds.), Amphibians and Reptiles. Status and Conservation in Florida. Krieger Publishing, Malabar, Florida.

Mecham, J.S. 1954. Geographic variation in the green frog, *Rana clamitans* Latreille. Texas Journal of Science 6:1–24.

Mecham, J.S. 1957. Some hybrid combinations between Strecker's chorus frog, *Pseudacris streckeri*, and certain related forms. Texas Journal of Science 9:337–345.

Mecham, J.S. 1958. Some Pleistocene amphibians and reptiles from Friesenhahn Cave, Texas. Southwestern Naturalist 3:17–27.

Mecham, J.S. 1959. Experimental evidence of the relationship of two allopatric chorus frogs of the genus *Pseudacris*. Texas Journal of Science 11:343–347.

Mecham, J.S. 1960a. Introgressive hybridization between two southeastern treefrogs. Evolution 14:445–457.

Mecham, J.S. 1960b. Natural hybridization between the tree frogs *Hyla versicolor* and *Hyla avivoca*. Journal of the Elisha Mitchell Scientific Society 76:64–67.

Mecham, J.S. 1961. Isolating mechanisms in anuran amphibians. Pp. 24–61 *In* W.F. Blair (ed.), Vertebrate Speciation. University of Texas Press, Austin.

Mecham, J.S. 1964. Ecological and genetic relationships of the two cricket frogs, genus *Acris*, from Alabama. Herpetologica 20:84–91.

Mecham, J.S. 1965. Genetic relationships and reproductive isolation in southeastern frogs of the genera *Pseudacris* and *Hyla*. American Midland Naturalist 74:269–308.

Mecham, J.S. 1968. Evidence of reproductive isolation between two populations of the frog, *Rana pipiens*, in Arizona. Southwestern Naturalist 13:35–44.

Mecham, J.S. 1969. New information from experimental crosses on genetic relationships within the *Rana pipiens* species group. Journal of Experimental Zoology 170:169–180.

Mecham, J.S. 1971.Vocalizations of the leopard frog, *Rana pipiens*, and three related Mexican species. Copeia 1971: 505–516.

Mecham, J.S. 1979. The biogeographical relationships of the amphibians and reptiles of the Guadalupe Mountains. Pp. 169-180 *In* H.H. Genoways and R.J. Baker (eds.), Biological Investigations in the Guadalupe Mountains National Park, Texas. Proceedings and Transactions Series No. 4. National Park Service, Washington, D.C.

Mecham, J.S., M.J. Littlejohn, R.S. Oldham, L.E. Brown, and J.R. Brown. 1973. A new species of leopard frog (*Rana pipiens* complex) from the plains of the central United States. Occasional Papers. the Museum, Texas Tech University 18:1–11.

Meehan, W.E. 1905. Frog Culture. Transactions of the American Fisheries Society 34:257-264.

Meehan, W.E. 1906a. Frog–farming. Pennsylvania Department of Fisheries, Bulletin No. 4.

Meehan, W.E. 1906b. Frog culture. Pp.51-54 *In* Report of the Department of Fisheries of the Commonwealth of Pennsylvania. December 1, 1904 to November 30, 1905. Harrisburg, Pennsylvania.

Meehan, W.E. 1907. Frog culture. Pp. 72-75 *In* Report of the Department of Fisheries of the Commonwealth of Pennsylvania. December 4, 1905 to November 30, 1906. Harrisburg, Pennsylvania.

Meehan, W.E. 1908a. Frog farming an industry. Technical World Magazine 9(3):246–250.

Meehan, W.E. 1908b. Possibilities of frog farming. Country Life in America 13:614–615, 640.

Meehan, W.E., and E.A. Andrews. 1908. Frogs, *Rana* spp., Ranidae. Pp. 394–395 *In* L.H. Bailey (ed.), Cyclopedia of American Agriculture. A Popular Survey of Agricultural Conditions, Practices and Ideals in the United States and Canada. Macmillan, New York.

Meehan, W.E. 1913. Fish Culture in Ponds and Other Inland Waters. Sturgis & Walton, New York. [Chapter 17, Frog Culture, pp. 219-233]

Meek, S.E., and D.G. Elliot. 1899. Notes on a collection of cold–blooded vertebrates from the Olympic Mountains. Field Columbian Museum Publication 31, Zoological Series 1:225–236.

Meeks, D.E., and J.W. Nagel. 1973. Reproduction and development of the wood frog, *Rana sylvatica*, in eastern Tennessee. Herpetologica 29:188–191.

Meier, P.T. 2007. Fine spatial scale phenotypic divergence in wood frogs (*Lithobates sylvaticus*). Canadian Journal of Zoology 85: 873–882.

Meijden, A. van der, M. Vinces, S. Hoegg, R. Boistel, A. Channing, and A. Meyer. 2007. Nuclear gene phylogeny of narrow–mouthed toads (Family: Microhylidae) and a discussion of competing hypotheses concerning their biogeographical origins. Molecular Phylogenetics and Evolution 44:1017–1030.

Meik, J.M., E. Mociño–Deloya, and K. Setser. 2007. *Bufo punctatus* (Red–spotted Toad). Predation. Herpetological Review 38:180.

Meisler, J.A. 2005. *Pseudacris triseriata* (Western Chorus Frog). Reproduction. Herpetological Review 36:55.

Mendonça, M.T., P. Licht, M.J. Ryan, and R. Barnes. 1985. Changes in hormone levels in relation to breeding behavior in male bullfrogs (*Rana catesbeiana*) at the individual and population levels. General and Comparative Endocrinology 58:270–279.

Mengel, D.C. 2010. Amphibians as wetland restoration indicators on Wetlands Reserve Program sites in Lower Grand River Basin, Missouri. M.S. thesis, University of Missouri, Columbia, Missouri.

Mennell, L. 1997. Amphibians in southwestern Yukon and northwestern British Columbia. SSAR Herpetological Conservation 1:107–109.

Merkle, D.A. 1977. The occurrence of the eastern spadefoot, *Scaphiopus h. holbrooki*, in the central Piedmont of Virginia. Bulletin of the Maryland Herpetological Society 13:196–197.

Merovich, C.E., and J.H. Howard. 2000. Amphibian use of constructed ponds on Maryland's Eastern Shore. Journal of the Iowa Academy of Science 107:151–159.

Merrell, D.J. 1965. The distribution of the dominant Burnsi gene in the leopard frog, *Rana pipiens*. Evolution 19:69–85.

Merrell, D.J. 1968. A comparison of the estimated size and the "effective size" of breeding populations of the leopard frog, *Rana pipiens*. Evolution 22:274–283.

Merrell, D.J. 1969. Natural selection in a leopard frog population. Journal of the Minnesota Academy of Science 35:86–89.

Merrell, D.J. 1970. Migration and gene dispersal in *Rana pipiens*. American Zoologist 10:47–52.

Merrell, D.J. 1972. Laboratory studies bearing on pigment pattern polymorphisms in wild populations of *Rana pipiens*. Genetics 70:141–161.

Merrell, D.J. 1973. Ecological genetics of anurans as exemplified by *Rana pipiens*. Pp. 329–335 *In* J.L. Vial (ed.), Evolutionary Biology of the Anurans. University of Missouri Press, Columbia.

Merrell, D.J. 1977. Life history of the leopard frog, *Rana pipiens*, in Minnesota. University of Minnesota Bell Museum of Natural History, Occasional Papers No. 15.

Merrell, D.J., and C.F. Rodell. 1968. Seasonal selection in the leopard frog, *Rana pipiens*. Evolution 22:284–288.

Mertins, J.W., S.M. Torrence, and M.C. Sterner. 2011. Chiggers recently infesting *Spea* spp. in Texas, USA, were *Eu-*

trombicula affreddugesi, not *Hannemania* sp. Journal of Wildlife Diseases 47:612–617.

Meshaka, W.E., Jr. 1993. Hurricane Andrew and the colonization of five invading species in south Florida. Florida Scientist 56:193–201.

Meshaka, W.E., Jr. 1994a. Giant toad eaten by red–shouldered hawk. Florida Field Naturalist 22:54–55.

Meshaka, W.E., Jr. 1994b. Ecological correlates of successful colonization in the life history of the Cuban treefrog, *Osteopilus septentrionalis* (Anura: Hylidae). Ph.D. Dissertation, Florida International University, Miami.

Meshaka, W.E., Jr. 1996a. Diet and the colonization of buildings by the Cuban treefrog, *Osteopilus septentrionalis* (Anura: Hylidae). Caribbean Journal of Science 32:59–63.

Meshaka, W.E., Jr. 1996b. Vagility and the Florida distribution of the Cuban treefrog (*Osteopilus septentrionalis*). Herpetological Review 27:37–40.

Meshaka, W.E., Jr. 1996c. Retreat use by the Cuban treefrog (*Osteopilus septentrionalis*): implications for successful colonization in Florida. Journal of Herpetology 30:443–445.

Meshaka, W.E., Jr. 1996d. *Osteopilus septentrionalis* (Cuban Treefrog). Maximum size. Herpetological Review 27:74.

Meshaka, W.E., Jr. 2000. *Bufo terrestris* (Southern Toad). Maximum size. Herpetological Review 31:169.

Meshaka, W.E., Jr. 2001. The Cuban Treefrog in Florida. University Press of Florida, Gainesville.

Meshaka, W.E., Jr. 2009. The terrestrial ecology of an Allegheny amphibian community: implications for land management. Maryland Naturalist 50(1):30–56.

Meshaka, W.E., Jr. 2011a. A runaway train in the making: the exotic amphibians, reptiles, turtles, and crocodilians of Florida. Herpetological Conservation and Biology, Monograph 1:1–107.

Meshaka, W., Jr. 2011b. Overwintering by tadpoles of the green frog, *Lithobates clamitans melanota* (Rafinesque, 1820), in western Pennsylvania. Herpetology Notes 4:311–314.

Meshaka, W.E., Jr., and B. Ferster. 1995. Two species of snakes prey on Cuban treefrogs in southern Florida. Florida Field Naturalist 23:97–98.

Meshaka, W.E., Jr., and K.P. Jansen. 1997. *Osteopilus septentrionalis* (Cuban Treefrog). Predation. Herpetological Review 28:147–148.

Meshaka, W.E., Jr., and G.E. Woolfenden. 1999. Relation of temperature and rainfall to movements and reproduction of the eastern narrow–mouthed toad (*Gastrophryne carolinensis*) in south–central Florida. Florida Scientist 62:213–221.

Meshaka, W.E., Jr., and A.L. Mayer. 2005. Diet of the southern toad (*Bufo terrestris*) from the southern Everglades. Florida Scientist 68:261–266.

Meshaka, W.E., Jr., and J.T. Collins. 2010. A Pocket Guide to Pennsylvania Frogs & Toads. Mennonite Press Inc., Newton, Kansas.

Meshaka, W.E., Jr., and R. Powell. 2010. Diets of the native southern toad (*Anaxyrus terrestris*) and the exotic cane toad (*Rhinella marina*) from a single site in south–central Florida. Florida Scientist 73:175–179.

Meshaka, W.E., Jr., and S.D. Marshall. 2011a. Clutch characteristics of the southern leopard frog, *Lithobates sphenocephalus* (Cope, 1886), in Natchitoches, Louisiana. Bulletin of the Maryland Herpetological Society 47:36–37.

Meshaka, W.E., Jr., and S.D. Marshall. 2011b. Clutch characteristics of the pickerel frog, *Lithobates palustris* (LeConte, 1825), in Natchitoches, Louisiana. Bulletin of the Maryland Herpetological Society 47:45–46.

Meshaka, W.E., Jr., and S.D. Marshall. 2011c. Clutch characteristics of the spring peeper, *Pseudacris crucifer* (Wied–Newied, 1838), in Natchitoches, Louisiana. Journal of Kansas Herpetology 38:11.

Meshaka, W.E., Jr., and S.D. Marshall. 2011d. Aspects of a breeding aggregation of Hurter's Spadefoot (*Scaphiopus hurteri* Strecker, 1910) in Natchitoches, Louisiana. Journal of Kansas Herpetology, 38:12.

Meshaka, W.E., Jr., and S.D. Marshall. 2012. Seasonal activity, reproductive cycles, and growth of the bronze frog (*Lithobates clamitans clamitans*) in Florida. Florida Scientist 75:176–188.

Meshaka, W.E., Jr., W.F. Loftus, and T. Steiner. 2000. The herpetofauna of Everglades National Park. Florida Scientist 63:84–103.

Meshaka, W.E., Jr., B.P. Butterfield, and J.B. Hauge. 2004. The Exotic Amphibians and Reptiles of Florida. Krieger Publishing, Malabar, Florida.

Meshaka Jr, W.E., J.N. Huff, and R.C. Leberman. 2008. Amphibians and reptiles of Powdermill Nature Reserve in western Pennsylvania. Journal of Kansas Herpetology 25:12-18.

Meshaka, W.E., Jr., S.D. Marshall, L.R. Raymond, and L.M. Hardy. 2009a. Seasonal activity, reproductive cycles, and growth of the bronze frog (*Lithobates clamitans clamitans*) in northern Louisiana: the long and the short of it. Journal of Kansas Herpetology 29:12–20.

Meshaka, W.E., Jr., J. Boundy, S.D. Marshall, and J. Delahoussaye. 2009b. Seasonal activity, reproductive cycles, and growth of the bronze frog (*Lithobates clamitans clamitans*) in southern Louisiana: an endpoint in its geographic distribution and the variability of its geographic life history traits. Journal of Kansas Herpetology 31:12–17.

Meshaka, W.E., Jr., J. Boundy, and A. Williams. 2009c. The

dispersal of the greenhouse frog, *Eleutherodactylus planirostris* (Anura: Eleutherodactylidae), in Louisiana, with preliminary observations on several potential exotic colonizing species. Journal of Kansas Herpetology 32:13–16.

Meshaka, W.E., Jr., S.D. Marshall, and D. Heinicke. 2011a. Seasonal activity, reproductive cycles, and growth of the bronze frog (*Lithobates clamitans clamitans*) at the western edge of its geographic range. Bulletin of the Maryland Herpetological Society 47:11–22.

Meshaka, W.E., Jr., P.R. Delis, and S.A. Mortzfeldt. 2011b. Seasonal activity, reproductive cycles, and growth of the northern leopard frog, *Lithobates pipiens* (Schreber, 1782), from Pennsylvania. Bulletin of the Maryland Herpetological Society 47:23–35.

Meshaka, W.E., Jr., E. Wingert, R.W. Cassell, P.R. Delis, and S.A. Mortzfeldt. 2011c. Breeding episodes of the eastern spadefoot, *Scaphiopus holbrookii* (Harlan, 1835), in central Pennsylvania. Journal of Kansas Herpetology 40:10–12.

Meshaka, W.E., Jr., S. Bartle, P.R. Delis, and E. Wingert. 2012a. Adult body size and clutch characteristics of the spring peeper, *Pseudacris crucifer* (Wied–Neuwied, 1838), from a single site in south–central Pennsylvania. Collinsorum 1:7–8.

Meshaka, W.E., Jr., E. Wingert, R.W. Cassell, P.R. Delis, and N. Potter, Jr. 2012b. Status and distribution of the eastern spadefoot, *Scaphiopus holbrookii* (Harlan, 1835), in Pennsylvania: statewide conservation implications for an imperiled species. Collinsorum 1:20–24.

Meshaka, W.E., Jr., N. Edwards, and P. R. Delis. 2012c. Seasonal activity, reproductive cycles, and growth of the pickerel frog, *Lithobates palustris* (LeConte, 1825), from Pennsylvania. Herpetological Bulletin 119:1–8.

Meshaka, Jr., W.E., E. Wingert, P.R. Delis, and S.A. Mortzfeldt. 2012d. Three breeding episodes of the eastern spadefoot, *Scaphiopus holbrookii* (Harlan, 1835), in south-central Pennsylvania. Journal of Kansas Herpetology 40:10–12.

Meshaka, Jr., W.E., P.J. Warny, and A.N. Klippel). 2012e. Larval growth and transformation size of the green frog, *Lithobates clamitans melanota* (Rafinesque 1820), in southeastern New York. Herpetology Notes 5:59–62.

Metcalf, M.M. 1923. Opalinid ciliate infusorians. Bulletin of the United States National Museum (120):1–484.

Metcalf, M.M. 1928. The bell toads and their opalinid parasites. American Naturalist 62:5–21.

Metcalf, Z.P. 1921. The food capacity of the toad. Copeia (100):81–82.

Meteyer, C.U. 2000. Field guide to malformations of frogs and toads, with radiographic interpretations. U.S. Geological Survey, Biological Science Report USGS/BRD/BSR–2000–0005.

Meteyer, C.U., I.K. Loeffler, J.F. Fallon, K.A. Converse, E. Green, J.C. Helgen, S. Kersten, R. Levey, L. Eaton–Poole, and J.G. Burkhart. 2000a. Hind limb malformations in free–living northern leopard frogs (*Rana pipiens*) from Maine, Minnesota, and Vermont suggest multiple etiologies. Teratology 62:151–171.

Meteyer, C.U., R.A. Cole, K.A. Converse, D.E. Docherty, M. Wolcott, J.C. Helgren, R. Levey, L. Eaton–Poole, and J.G. Burkhart. 2000b. Defining anuran malformations in the context of a developmental problem. Journal of the Iowa Academy of Science 107:72–78.

Metter, D.E. 1960. The distribution of amphibians in eastern Washington. M.S. thesis, Washington State University, Pullman.

Metter, D.E. 1961. Water levels as an environmental factor in the breeding season of *Bufo boreas boreas* (Baird and Girard). Copeia 1961:488.

Metter, D.E. 1962. *Ascaphus truei* Stejneger in the Blue Mountains of southeastern Washington. Occasional Papers, Department of Biology, College of Puget Sound No. 21:205–206.

Metter, D.E. 1963. Stomach contents of Idaho larval *Dicamptodon*. Copeia 1963:435–436.

Metter, D.E. 1964a. A morphological and ecological comparison of two populations of the tailed frog, *Ascaphus truei* Stejneger. Copeia 1964:181–195.

Metter, D.E. 1964b. On breeding and sperm retention in *Ascaphus*. Copeia 1964:710–711.

Metter, D.E. 1966. Some temperature and salinity tolerances of *Ascaphus truei* Stejneger. Journal of the Idaho Academy of Science 4:44–47.

Metter, D.E. 1967. Variation in the ribbed frog *Ascaphus truei* Stejneger. Copeia 1967:634–649.

Metter, D.E. 1968. The influence of floods on population structure of *Ascaphus truei* Stejneger. Journal of Herpetology 1:105–106.

Metter, D.E., and R.J. Pauken. 1969. An analysis of the reduction of gene flow in *Ascaphus truei* in the northwest U.S. since the Pleistocene. Copeia 1969:307–310.

Metts, B.S., J.D. Lanham, and K.R. Russell. 2001. Evaluation of herpetofaunal communities on upland streams and beaver–impounded streams in the upper Piedmont of South Carolina. American Midland Naturalist 145:54–65.

Metts, B.S., K.A. Buhlmann, D.E. Scott, T.D. Tuberville, and W.A. Hopkins. 2012. Interactive effects of maternal and environmental exposure to coal combustion wastes decrease survival of larval southern toads (*Bufo terrestris*). Environmental Pollution 164:211–218.

Meyer, F.P. 1965. The experimental use of Guthion as a selective fish eradicator. Transactions of the American Fisheries Society 94:203–209.

Meyer, L.R., J. Geiger–Hayes, and P. Owen. 2007. Examination of the intestinal contents of *Rana clamitans* and *Rana catesbeiana* tadpoles for symbiotic, cellulose–digesting bacteria. Herpetological Review 38:393–395.

Meyers, J.M., and D.A. Pike. 2006. Herpetofaunal diversity of Alligator River National Wildlife Refuge, North Carolina. Southeastern Naturalist 5:235–252.

Meylan, P. 2005. Late Pliocene anurans from Inglis 1A, Citrus County, Florida. Bulletin of the Florida Museum of Natural History 45:171–178.

Micancin, J.P. 2008. Acoustic variation and species discrimination in southeastern sibling species, the cricket frogs *Acris crepitans* and *Acris gryllus*. Ph.D. Dissertation, University of North Carolina, Chapel Hill.

Micancin, J.P. 2010. *Acris gryllus* (Southern Cricket Frog). Dispersal. Herpetological Review 41:192.

Micancin, J.P., and J.T. Mette. 2009. Acoustic and morphological identification of the sympatric cricket frogs *Acris crepitans* and *A. gryllus* and the disappearance of *A. gryllus* near the edge of its range. Zootaxa 2076:1–36.

Micancin, J.P., and J.T. Mette. 2010. *Acris crepitans* (Northern Cricket Frog) and *Acris gryllus* (Southern Cricket Frog). Interspecific agonism. Herpetological Review 41:192.

Micancin, J.P., A. Tóth, R. Anderson, and J.T. Mette. 2012. Sympatry and syntopy of the cricket frogs *Acris crepitans* and *Acris gryllus* in southeastern Virginia, USA and decline of *A. gryllus* at the northern edge of its range. Herpetological Conservation and Biology 7:276–298.

Michaud, T.C. 1962. Call discrimination by females of the chorus frogs, *Pseudacris clarki* and *Pseudacris nigrita*. Copeia 1962:213–215.

Michaud, T.C. 1964. Vocal variation in two species of chorus frogs, *Pseudacris nigrita* and *Pseudacris clarki*, in Texas. Evolution 18:498–506.

Michelsen, A., M. Jorgensen, J. Christensen–Dalsgaard, and R.R. Capranica. 1986. Directional hearing of awake, unrestrained treefrogs. Naturwissenschaften 73:682–683.

Middendorf, L.J. 1957. Observations on the early spring activities of the western spotted frog (*Rana pretiosa pretiosa*) in Gallatin County, Montana. Proceedings of the Montana Academy of Science 17:55–56.

Middlemis Maher, J.E. 2011. Ecological and evolutionary context of the stress response in larval anurans. Ph.D. Dissertation, University of Michigan, Ann Arbor.

Miera, V. 1999. Simple introductions – major repercussions: The story of bullfrogs and crayfish in Arizona. Arizona Wildlife Views 42:25-27.

Mierzwa, K.S. 1998. Status of northeastern Illinois amphibians. Pp. 115–124 *In* M.J. Lannoo (ed.), Status & Distribution of Midwestern Amphibians. University of Iowa Press, Iowa City.

Mierzwa, K.S. 2000. Wetland mitigation and amphibians: preliminary observations at a southwestern Illinois bottomland hardwood forest restoration site. Journal of the Iowa Academy of Science 107:191–194.

Mifsud, D.A., and R. Mifsud. 2008. Golf courses as refugia for herpetofauna in an urban river floodplain. Pp. 303–310 *In* J.C. Mitchell, R.E. Jung Brown, and B. Bartholomew (eds.), Urban Herpetology. Herpetological Conservation 3. Society for the Study of Amphibians and Reptiles, Salt Lake City, Utah.

Mikaelian, I., M. Ouellet, B. Pauli, J. Rodrigue, J.C. Harshbarger, and D.M. Green. 2000. *Ichthyophonus*–like infection in wild amphibians from Québec, Canada. Diseases of Aquatic Organisms 40:195–201.

Mikulka, P., J. Highes, and G. Aggerup. 1980. The effect of pretraining procedures and discriminative stimuli on the development of food selection behaviors in the toad (*Bufo terrestris*). Behavioral and Neural Biology 29:52–62.

Milius, S. 1998. Fatal skin fungus found in U.S. frogs. Science News 154:7–8.

Miller, B.T., J.W. Lamb, and J.L. Miller. 2005. The herpetofauna of Arnold Air Force Base in the barrens of south–central Tennessee. Southeastern Naturalist 4:51–62.

Miller, C.E. 1968. Frogs with five legs. Carolina Tips No. 31(1).

Miller, D.A.W., C.S. Brehme, J.E. Hines, J.D. Nichols, and R.N. Fisher. 2012a. Joint estimation of habitat dynamics and species interactions: disturbance reduces co–occurrence of non–native predators with an endangered toad. Journal of Animal Ecology 81:1288-1297.

Miller, D.A.W., L. A. Weir, B.T. McClintock, E. H. Campbell Grant, L.L. Bailey, and T.R. Simons. 2012b. Experimental investigation of false positive errors in auditory species occurrence surveys. Ecological Applications 22:1665–1674.

Miller, D.G., Jr. 1899. A new treefrog from the District of Columbia. Proceedings of the Biological Society of Washington 13:75–78.

Miller, D.L., C.R. Bursey, M.J. Gray, and L.M. Smith. 2004. Metacercariae of *Clinostomum attenuatum* in *Ambystoma tigrinum marvortium*, *Bufo cognatus* and *Spea multiplicata* from west Texas. Journal of Helminthology 78:373–376.

Miller, D.L., S. Rajeev, M.J. Gray, and C.A. Baldwin. 2007. Frog Virus 3 infection, cultured American Bullfrogs. Emerging Infectious Diseases 13:342–343.

Miller, D. L., M. J. Gray, and A. Storfer. 2011. Ecopathology of ranaviruses infecting amphibians. Viruses 3:2351–2373.

Miller, D.M., R.A. Young, T.W. Gatlin, and J.A. Richardson. 1982. Amphibians and reptiles of the Grand Canyon. Grand Canyon Natural History Association, Monograph 4:i–144.

Miller, J.D. 1978. Observations on the diets of *Rana pretiosa, Rana pipiens*, and *Bufo boreas* from western Montana. Northwest Science 52:243–249.

Miller, K., and G.C. Packard. 1977. An altitudinal cline in critical thermal maxima of chorus frogs (*Pseudacris triseriata*). American Naturalist 111:267–277.

Miller, L., and A.H. Miller. 1936. The northward occurrence of *Bufo californicus* in California. Copeia 1936:176.

Miller, L.K., and P.J. Dehlinger. 1969. Neuromuscular function and low temperatures in frogs from cold and warm climates. Comparative Biochemistry and Physiology B 28:915–921.

Miller, N. 1909a. The American toad (*Bufo lentiginosus americanus*, LeConte). A study in dynamic biology. American Naturalist 43:641–668.

Miller, N. 1909b. The American toad (*Bufo lentiginosus americanus*, LeConte). 2. A study in dynamic biology. American Naturalist 43:730–745.

Miller, W.D.W. 1913. Late activity of Pickering's Hyla. Copeia (1):4.

Miller, W.D., and J. Chapin. 1910. The toads of the northeastern United States. Science 32:315–317.

Mills, N.E., and R.D. Semlitsch. 2004. Competition and predation mediate the indirect effects of an insecticide on southern leopard frogs. Ecological Applications 14:1041–1054.

Millzner, R. 1924. A larval acathocephalid, *Centrorhynchus californicus*, sp. nov., from the mesentery of *Hyla regilla*. University of California Publications in Zoology 26:225–230.

Milko, L.V. 2012. Integrating museum and GIS data to identify changes in species distributions driven by a disturbance–induced invasion. Copeia 2012:307–320.

Milstead, W.W., J.S. Mecham, and H. McClintock. 1950. The amphibians and reptiles of the Stockton Plateau in northern Terrell County, Texas. Texas Journal of Science 2:543–562.

Milstead, W.W., A.S. Rand, and M.M. Stewart. 1974. Polymorphism in cricket frogs: an hypothesis. Evolution 28:489–491.

Miner, B. G., S. E. Sultan, S. G. Morgan, D. K. Padilla, and R. A. Relyea. 2005. Ecological consequences of phenotypic plasticity. Trends in Ecology & Evolution 20:685–692.

Minton, J.E. 1949. Coral snake preyed upon by the bullfrog. Copeia 1949:288.

Minton, S.A., Jr. 1958. Observations on amphibians and reptiles of the Big Bend region of Texas. Southwestern Naturalist 3:28–54.

Minton, S.A., Jr. 1968. The fate of amphibians and reptiles in a suburban area. Journal of Herpetology 2:113–116.

Minton, S.A., Jr. 2001. Amphibians & Reptiles of Indiana. Indiana Academy of Science, Indianapolis.

Minton, S.A., J.C. List, and M.J. Lodato. 1982. Recent records and status of amphibians and reptiles in Indiana. Proceedings of the Indiana Academy of Science 92:489–498.

Mirarchi, R.E., M.A. Bailey, J.T. Garner, T.M. Haggerty T.L. Best, M.F. Mettee and P. O'Neil (eds.). 2004. Alabama Wildlife. Vol. 4. Conservation and Management Recommendations for Imperiled Wildlife. University of Alabama Press, Tuscaloosa.

Mitchell, J.C. 1982. A checklist of amphibians and reptiles of Back Bay National Wildlife Refuge and False Cape State Park, Virginia Beach, Virginia. Catesbeiana 2(2):13-15.

Mitchell, J.C. 1986. Life history patterns in a central Virginia frog community. Virginia Journal of Science 37:262–271.

Mitchell, J.C. 1990a. *Pseudacris feriarum* (Upland Chorus Frog). Predation. Herpetological Review 21:89–90.

Mitchell, J.C. 1990b. Contributions to the history of Virginia herpetology I: John B. Lewis' "List of amphibians observed in Amelia, Brunswick, and Norfolk Counties." Catesbeiana 10(2):36-42.

Mitchell, J.C. 1991a. Amphibians and reptiles. Pp. 411–476 *In* K. Terwilliger (ed.), Virginia's Endangered Species. McDonald Woodward Publishing Company, Blacksburg, Virginia.

Mitchell, J.C. 1991b. Contributions to the history of Virginia herpetology III: John B. Lewis' "Amphibia of the Seward Forest and vicinity." Catesbeiana 11(1):3-9.

Mitchell, J.C. 1992. Invertebrate prey of *Bufo woodhousii fowleri* (Anura: Bufonidae) from a Virginia barrier island. Banisteria 1:13-15.

Mitchell, J.C. 1995. Amphibians and reptiles of Sugarland Run, Fairfax and Loudon counties, Virginia: estimated numbers and commercial value. Catesbeiana 14(1):15-22.

Mitchell, J.C. 1999a. *Osteopilus septentrionalis* (Cuban Treefrog). VA: Fauquier Co., Warrenton. Catesbeiana 19:32.

Mitchell, J.C. 1999b. Checklist and keys to the amphibians and reptiles of the Eastern Shore of Virginia. Catesbeiana 19(1):3-18.

Mitchell, J.C. 2002. An overview of amphibian and reptile assemblages on Virginia's Eastern Shore, with comments on conservation. Banisteria 20:31–45.

Mitchell, J.C. 2004. Occurrence of intradermal mite, *Hannemania* sp. (Acarina: Trombiculidae), parasites in two species of amphibians in Virginia. Banisteria 23:50-51.

Mitchell, J.C. 2012. Amphibians and reptiles of the Eastern Shore of Virginia National Wildlife Refuge and Fisherman Island National Wildlife Refuge. Banisteria 39:21–33.

Mitchell, J.C., and J.M. Anderson. 1994. Amphibians and Reptiles of Assateague and Chincoteague Islands. Virginia Museum of Natural History, Special Publication No. 2.

Mitchell, J.C., and S.M. Roble. 1998. Annotated checklist of the amphibians and reptiles of Fort A.P. Hill, Virginia, and vicinity. Banisteria 11:19-32.

Mitchell, J.C., and K.A. Buhlmann. 1999. Amphibians and reptiles of the Shenandoah Valley sinkhole pond system in Virginia. Banisteria 13:129–142.

Mitchell, J.C., and K.K. Reay. 1999. Atlas of Amphibians and Reptiles in Virginia. Virginia Department of Game and Inland Fisheries, Wildlife Diversity Division, Special Publication No.1.

Mitchell, J.C., and D.E. Green. 2003. *Hyla gratiosa* (Barking Treefrog). Intestinal hernia. Herpetological Review 34:230–231.

Mitchell, J.C., and B. Burgess. 2004. A malformed Fowler's Toad (*Bufo fowleri*) from the Shenandoah Valley of Virginia. Banisteria 24:51–52.

Mitchell, J.C., and C.T. Georgel. 2005. Anopthalmia in an upland chorus frog (*Pseudacris feriarum feriarum*) from southeastern Virginia. Banisteria 25:53–54.

Mitchell, J.C., and C.T. Georgel. 2005. Bilateral ectromelia in a northern cricket frog (*Acris crepitans crepitans*) metamorph from Virginia. Banisteria 25:54-55.

Mitchell, J.C., and L. McGranaghan. 2005. Albinism in American Bullfrog (*Rana catesbeiana*) tadpoles in Virginia. Banisteria 25:51.

Mitchell, J.C., and T.K. Pauley. 2005. *Pseudacris brachyphona* (Cope, 1889). Mountain Chorus Frog. Pp. 465–466 *In* M.J. Lannoo (ed.), Amphibian Declines. The Conservation Status of United States Species. University of California Press, Berkeley.

Mitchell, J.C., and J. White. 2005. Leucistic wood frog (*Rana sylvatica*) tadpole from northern Virginia. Banisteria 25:52–53.

Mitchell, J.C., S.Y. Erdle, and J.F. Pagels. 1993. Evaluation of capture techniques for amphibian, reptile, and small mammal communities in saturated wetlands. Wetlands 13:130–136.

Mitchell, J.C., C.A. Pague, and D.J. Schwab. 2000. Herpetofauna of the Great Dismal Swamp. Pp. 155-174 *In* R.K. Rose (ed.), Natural History of the Great Dismal Swamp: Proceedings of the Third Dismal Swamp Symposium. Old Dominion University, Norfolk, Virginia.

Mitchell, J.C., A.R. Breisch, and K.A. Buhlmann. 2006. Habitat Management Guidelines for Amphibians and Reptiles of the Northeastern United States. Partners in Amphibian and Reptile Conservation, Technical Publication HMG–3, Montgomery, Alabama.

Mitchell, S.L., and G.L. Miller. 1991. Intermale spacing and calling site characteristics in a southern Mississippi chorus of *Hyla cinerea*. Copeia 1991:521–524.

Mitrovich, M.J., E.A. Gallegos, L.M. Lyren, R.E. Lovich, and R.N. Fisher. 2011. Habitat use and movement of the endangered arroyo toad (*Anaxyrus californicus*) in coastal Southern California. Journal of Herpetology 45:319–328.

Mittleman, M.B. 1945. The status of *Hyla phaeocrypta* with notes on its variation. Copeia 1945:31–37.

Mittleman, M.B. 1946. Nomenclatural notes on two southeastern frogs. Herpetologica 3:57–60.

Mittleman, M.B. 1950. Miscellaneous notes on some amphibians and reptiles from the southeastern United States. Herpetologica 6:20–24.

Mittleman, M.B., and G.S. Myers. 1949. Geographic variation in the ribbed frog, *Ascaphus truei*. Proceedings of the Biological Society of Washington 62:57–68.

Mittleman, M.B., and J.C. List. 1953. The generic differentiation of the swamp treefrogs. Copeia 1953:80–83.

Mizelle, J.D., D.C. Kritsky, and H.D. McDougal. 1969. Studies on monogenetic trematodes. 42. New species of *Gyrodactylus* from amphibia. Journal of Parasitology 55:740–741.

Modzelewski, E.H., Jr., and D.D. Culley, Jr. 1974a. Growth responses of the bullfrog, *Rana catesbeiana* fed various live foods. Herpetologica 30:396–405.

Modzelewski, E.H., Jr., and D.D. Culley, Jr. 1974b. Occurrence of the nematode *Eustrongylides wenrichi* in laboratory reared *Rana catesbeiana*. Copeia 1974:1000–1001.

Mohr, C.E. 1939. The amphibians of Berks County. Proceedings of the Pennsylvania Academy of Science 13:76–78.

Mohr, C.E. 1948. Unique animals inhabit subterranean Texas. National Speleological Society Bulletin 10:15–21, 88.

Mohr, J.R., and M.E. Dorcas. 1999. A comparison of anuran calling patterns at two Carolina Bays in South Carolina. Journal of the Elisha Mitchell Scientific Society 115:63–70.

Moler, P.E. 1981. Notes on *Hyla andersonii* in Florida and Alabama. Journal of Herpetology 15: 441–444.

Moler, P.E. 1983. Winter serenade. Florida Wildlife 36(5):18–23.

Moler, P.E. 1985. A new species of frog (Ranidae: *Rana*) from northwestern Florida. Copeia 1985:379–383.

Moler, P.E. 1992. Florida Bog Frog *Rana okaloosae* Moler. Pp. 30–33 *In* P.E. Moler (ed.), Rare and Endangered Biota of Florida. Vol. 3. Amphibians and Reptiles. University Press of Florida, Gainesville.

Monello, R.J., and R.G. Wright. 1999. Amphibian habitat preferences among artificial ponds in the Palouse Region of northern Idaho. Journal of Herpetology 33:298–303.

Monello, R.J., J.J. Dennehy, D.L. Murray, and A.J. Wirsing. 2006. Growth and behavioral responses of tadpoles of two native frogs to an exotic competitor, *Rana catesbeiana*. Journal of Herpetology 40:403–407.

Monroe, B.L., Jr., and R.W. Taylor. 1972. Occurrence of the barking treefrog, *Hyla gratiosa*, in Kentucky. Journal of Herpetology 6:78.

Monroe, B.L., Jr., and R.W. Giannini. 1977. Distribution of the barking treefrog in Kentucky. Transactions of the Kentucky Academy of Science 38:143–144.

Monsen, K.J., and M.S. Blouin. 2003. Genetic structure in a montane ranid frog: restricted gene flow and nuclear–mitochondrial discordance. Molecular Ecology 12:3275–3286.

Monsen, K.J., and M.S. Blouin. 2004. Extreme isolation by distance in a montane frog. Conservation Genetics 5:827–835.

Monsen–Collar, K., L. Hazard, and R. Dussa. 2010. Comparison of PCR and RT–PCR in the first report of *Batrachochytrium dendrobatidis* in amphibians in New Jersey, USA. Herpetological Review 41:460–462.

Monson, P.D., D.J. Call, D.A. Cox, K. Liber, and G.T. Ankley. 1999. Photoinduced toxicity of fluoranthene to northern leopard frogs (*Rana pipiens*). Environmental Toxicology and Chemistry 18: 308-312.

Montanucci, R. 2006. A review of the amphibians of the Jim Timmerman Natural Resources Area, Oconee and Pickens counties, South Carolina. Southeastern Naturalist 5 (Monograph 1):1–58.

Montgomery, F.A., Jr. 1936. Frog farming. Scientific American 155:280-281.

Moore, G.A. 1937. The spadefoot toad under drought conditions. Copeia 1937:225–226.

Moore, H. 1965. A mid–summer trip to Moosonee. The Wood Duck 19:163–165.

Moore, J., and B. Moore. 1939. Notes on the salientia of the Gaspé Peninsula. Copeia 1939:104.

Moore, J.A. 1939. Temperature tolerance and rates of development in eggs of Amphibia. Ecology 20:459–478.

Moore, J.A. 1940. Adaptive differences in the egg membranes of frogs. American Naturalist 74:89–93.

Moore, J.A. 1942a. Embryonic temperature tolerance and rates of development in *Rana catesbeiana*. Biological Bulletin 83:375–388.

Moore, J.A. 1942b. An embryological and genetical study of *Rana burnsi* Weed. Genetics 27:408–416.

Moore, J.A. 1944. Geographic variation in *Rana pipiens* Schreber of eastern North America. Bulletin of the American Museum of Natural History 82:345–370.

Moore, J.A. 1946a. Incipient intraspecific isolating mechanisms in *Rana pipiens*. Genetics 31:304–326.

Moore, J.A. 1946b. Hybridization between *Rana palustris* and different geographic forms of *Rana pipiens*. Proceedings of the National Academy of Sciences of the United States of America 32:209–212.

Moore, J.A. 1947. Hybridization between *Rana pipiens* from Vermont and eastern Mexico. Proceedings of the National Academy of Sciences of the United States of America 33:72–75.

Moore, J.A. 1949a. Patterns of evolution in the genus *Rana*. Pp. 315–338 *In* G.L. Jepsen, E. Mayr, and G.G. Simpson (eds.), Genetics, Paleontology and Evolution. Princeton University Press, Princeton, New Jersey.

Moore, J.A. 1949b. Geographic variation of adaptive characters in *Rana pipiens* Schreber. Evolution 3:1–24.

Moore, J.A. 1950. Further studies on *Rana pipiens* racial hybrids. American Naturalist 84:247–254.

Moore, J.A. 1952. An analytical study of the geographic distribution of *Rana septentrionalis*. American Naturalist 86:5–22.

Moore, J.A. 1955. Abnormal combinations of nuclear and cytoplasmic systems in frogs and toads. Advances in Genetics 7:139–182.

Moore, J.A. 1964. Diploid and haploid interracial hybrids in *Rana pipiens*. Pp. 431–435 *In* Genetics Today. Proceedings of the XI International Congress of Genetics. Pergamon Press. New York.

Moore, J.A. 1966. Hybridization experiments involving *Rana dunni*, *Rana megapoda*, and *Rana pipiens*. Copeia 1966:673–675.

Moore, J.A. 1975. *Rana pipiens*: the changing paradigm. American Zoologist 15:837–849.

Moore, J.E., and E.H. Strickland. 1954. Notes on the food of three species of Alberta amphibians. American Midland Naturalist 52:221–224.

Moore, J.E., and E.H. Strickland. 1955. Further notes on the food of Alberta amphibians. American Midland Naturalist 54:253–256.

Moore, M.K., and P.L. Klerks. 1998. Interactive effect of high temperature and low pH on sodium flux in tadpoles. Journal of Herpetology 33:588–592.

Moore, R.D. 1929. *Canis latrans lestes* Merriam feeding on tadpoles and frogs. Journal of Mammalogy 10:255.

Moore, R.G., and B.A. Moore. 1980. Observations on the body temperature and activity in the red spotted toad, *Bufo punctatus*. Copeia 1980:362–363.

Moore, R.H. 1976. Reproductive habits and growth of *Bufo speciosus* on Mustang Island, Texas, with notes on the ecology and reproduction of other anurans. Texas Journal of Science 27:173–178.

Moosman, D.L., and P.R. Moosman, Jr. 2005. *Bufo americanus* (American Toad) and *Rana catesbeiana* (American Bullfrog). Microhabitat. Herpetological Review 36:298.

Morafka, D.J. 1977. A biogeographical analysis of the Chihuahuan Desert through its herpetofauna. Biogeographica, Volume 9, Dr. W. Junk B.V. Publishers, The Hague, Netherlands.

Morafka, D.J., and B.H. Banta. 1972. The herpetozoogeography of the Gabilan Range, San Benito and Monterey counties, California. The Wasmann Journal of Biology 30:197–240.

Morafka, D.J., and B.H. Banta. 1976. Ecological relationships of the Recent herpetofauna of Pinnacles National Monument, Monterey and San Benito counties, California. The Wasmann Journal of Biology 34:304–324.

Moran, K., and C.E. Button. 2011. A GIS model for identifying eastern spadefoot toad (*Scaphiopus holbrookii*) habitat in eastern Connecticut. Applied Geography 31:980-989.

Morey, S.R. 1982. Evidence for predation as a selection factor in anuran color polymorphism. M.S. thesis, California State University, Fullerton.

Morey, S.R. 1990. Microhabitat selection and predation in the Pacific treefrog, *Pseudacris regilla*. Journal of Herpetology 24:292–296.

Morey, S.R. 1994. Age and size at metamorphosis in spadefoot toads: a comparative study of adaptation to uncertain environments. Ph.D. Dissertation, University of California, Riverside.

Morey, S.R. 1996. Pool duration influences size and body mass at metamorphosis in the western spadefoot toad: implications for vernal pool conservation. Pp. 86–91 *In* C.W. Whitman (ed.), Ecology, Conservation, and Management of Vernal Pool Eccosystems. California Native Plant Society, Sacramento.

Morey, S.R., and D.N. Janes. 1994. Variation in larval habitat duration influences metamorphosis in *Scaphiopus couchii*. Pp. 159–165 *In* P.R. Brown and J.W. Wright (eds.), Herpetology of the North American Deserts. Proceedings of a Symposium. Southwestern Herpetologists Society, Special Publication No. 5.

Morey, S., and D. Reznick. 2000. A comparative analysis of plasticity in larval development in three species of spadefoot toads. Ecology 81:1736–1749.

Morey, S., and D. Reznick. 2001. Effects of larval density on postmetamorphic spadefoot toads (*Spea hammondii*). Ecology 82:510–522.

Morey, S., and D. Reznick. 2004. The relationship between habitat permanence and larval development in California spadefoot toads: field and laboratory comparisons of developmental plasticity. Oikos 104:172–190.

Morgan, D. 2010. *Rana sylvatica* (Wood Frog). Egg mass survey. Herpetological Review 41:206.

Morgan, J.A.T., V.T. Vredenburg, L.J. Rachowicz, R.A. Knapp, M.J. Stice, T. Tunstall, R.E. Bingham, J.M. Parker, J.E. Longcore, C. Moritz, C.J. Briggs, and J.W. Taylor. 2007. Population genetics of the frog–killing fungus *Batrachochytrium dendrobatidis*. Proceedings of the National Academy of Sciences of the United States of America 104:13,845–13,850.

Moriarty, E.C., and D.C. Cannatella. 2004. Phylogenetic relationships of the North American chorus frogs (*Pseudacris*: Hylidae). Molecular Phylogenetics and Evolution 30:409–420.

Moriarty, J.J. 1995. Leaping leopards and other frogs, toads, and treefrogs. Minnesota Volunteer 58(3):20–33.

Moriarty, J.J. 2000. Status of amphibians in Minnesota. Pp. 166–168 *In* M.J. Lannoo (ed.), Status & Conservation of Midwestern Amphibians. University of Iowa Press, Iowa City.

Moriarty, J.J., A. Forbes and D. Jones. 1998. Geographic distribution: *Acris crepitans blanchardi* (Blanchard's cricket frog). Herpetological Review 29:172.

Morin, P.J. 1983. Predation, competition, and the composition of larval anuran guilds. Ecological Monographs 53:119–138.

Morin, P.J. 1985. Predation intensity, prey survival and injury frequency in an amphibian predator–prey interaction. Copeia 1985:638–644.

Morin, P.J. 1986. Interactions between intraspecific competition and predation in an amphibian predator–prey system. Ecology 67:713–720.

Morin, P.J. 1987. Predation, breeding asynchrony, and the outcome of competition among treefrog tadpoles. Ecology 68:675–683.

Morin, P.J., and E.A. Johnson. 1988. Experimental studies of asymmetric competition among anurans. Oikos 53:398–407.

Morin, P.J., S.P. Lawler, and E.A. Johnson. 1988. Competition between aquatic insects and vertebrates: interaction strength and higher order interactions. Ecology 69:1401–1409.

Morin, P.J., S.P. Lawler, and E.A. Johnson. 1990. Ecology and breeding phenology of larval *Hyla andersonii*: the disadvantages of breeding late. Ecology 71:1590–1598.

Morizot, D.C., and N.H. Douglas. 1970. Notes on *Pseudacris streckeri streckeri* in northwest Louisiana. Bulletin of the Maryland Herpetological Society 6:18.

Morris, M.A., and P.W. Smith. 1981. Endangered and threatened amphibians and reptiles. Pp.21–33 *In* Endangered and Threatened Vertebrate Animals and Vascular Plants of Illinois. Illinois Department of Conservation, Springfield.

Morris, M.R. 1989. Female choice of large males in the treefrog *Hyla chrysoscelis*: the importance of identifying the scale of choice. Behavioral Ecology and Sociobiology 25:275–281.

Morris, M.R., and S.L. Yoon. 1989. A mechanism for female choice of large males in the treefrog *Hyla chrysoscelis*. Behavioral Ecology and Sociobiology 25:65–71.

Morris, P.A. 1944. They Hop and Crawl. Jaques Cattell Press, Lancaster, Pennsylvania.

Morris, R.L. 1967. The ecology of the western spotted frog, *Rana pretiosa pretiosa*, Baird and Girard: a life history study. M.S. thesis, Brigham Young University, Provo, Utah.

Morris, R.L., and W.W. Tanner. 1969. The ecology of the western spotted frog, *Rana pretiosa pretiosa* Baird and Girard: a life history study. Great Basin Naturalist 29:45–81.

Morrison, E.O. 1967a. *Rhabdias ranae* infections in three species of Oklahoma anura. Southwestern Naturalist 12:335–336.

Morrison, E.O. 1967b. Mite and lungworm infections in *Acris crepitans* from southern Oklahoma. Texas Journal of Science 19:328–329.

Morrison, E.O. 1969. Erratum "Mite and lungworm infections in *Acris crepitans* from southern Oklahoma." Texas Journal of Science 20:303.

Morrissey, C.A., and R.J. Olenick. 2004. American Dipper, *Cinclus mexicanus*, preys upon larval tailed frogs, *Ascaphus truei*. Canadian Field-Naturalist 118:446–448.

Morse, M. 1904. Batrachians and reptiles of Ohio. Proceedings of the Ohio State Academy of Science 4: 93–144.

Morton, M.L., and K.N. Sokolski. 1978. Sympatry in *Bufo boreas* and *Bufo canorus* and additional evidence of natural hybridization. Bulletin of the Southern California Academy of Sciences 77:52–55.

Morton, M.L., and M.E. Pereyra. 2010. Habitat use by Yosemite toads: life history traits and implications for conservation. Herpetological Conservation and Biology 5:388–394.

Moseley, K.R., S.B. Castleberry, and S.H. Schweitzer. 2003a. Effects of prescribed fire on herpetofauna in bottomland hardwood forests. Southeastern Naturalist 2:475–486.

Moseley, K.R., S.B. Castleberry, J.L. Hanula, and W.M. Ford. 2003b. Diet of southern toads (*Bufo terrestris*) in loblolly pine (*Pinus taeda*) stands subject to coarse woody debris manipulations. American Midland Naturalist 153:327–337.

Mosher, D.D. 1982. Sun–compass orientation in the treefrogs, *Hyla crucifer* and *Pseudacris triseriata triseriata*. Ph.D. Dissertation, Ball State University, Muncie, Indiana.

Mosher, J.A., and P.F. Matray. 1974. Size dimorphism: a factor in energy savings for broad–winged hawks. Auk 91:325–341.

Mosimann, J.E., and G.B. Rabb. 1952. The herpetology of Tiber Reservoir area, Montana. Copeia 1952:23–27.

Mossman, M.J., L.M. Hartman, R. Hay, J.R. Sauer, and B.J. Dhuey. 1998. Monitoring long–term trends in Wisconsin frog and toad populations. Pp. 169–198 *In* M.J. Lannoo (ed.), Status & Conservation of Midwestern Amphibians. University of Iowa Press, Iowa City.

Moulton, C.A. 1996. Assessing effects of pesticides on frog populations (*Hyla squirella, Gastrophryne carolinensis, Scaphiopus holbrooki, Bufo terrestris, Acris gryllus, Hyla femoralis, Rana virgatipes*). Ph.D. Dissertation, North Carolina State University, Raleigh.

Moulton, C.A., W.J. Fleming, and B.R. Nerney. 1996. The use of PVC pipes to capture hylid frogs. Herpetological Review 27:186–187.

Moulton, J.M. 1954. A late August breeding of *Hyla cinerea* in Florida. Herpetologica 10:107.

Mount, R.H. 1975. The Reptiles & Amphibians of Alabama. Agricultural Experiment Station, Auburn University, Auburn, Alabama.

Mount, R.H. (ed.). 1986. Vertebrate Animals of Alabama in Need of Special Attention. Alabama Agricultural Experiment Station, Auburn University, Alabama.

Moyle, P.B. 1973. Effects of introduced bullfrogs, *Rana catesbeiana*, on the native frogs of the San Joaquin Valley, California. Copeia 1973:18–22.

Mueller, A.J. 1985. Vertebrate use of nontidal wetlands on Galveston Island, Texas. Texas Journal of Science 37:215–225.

Mueller, G.A., J. Carpenter, and D. Thornbrugh. 2006. Bullfrog tadpole (*Rana catesbeiana*) and red swamp crayfish (*Procambarus clarkii*) predation on early life stages of endangered razorback sucker (*Xryauchen texanus*). Southwestern Naturalist 51:258–261.

Muenz, T.K., S.W. Golladay, G. Vellidis, and L.L. Smith. 2006. Stream buffer effectiveness in an agriculturally influenced area, southwestern Georgia: responses of water quality, macroinvertebrates, and amphibians. Journal of Environmental Quality 35:1924–1938.

Muhse, E.F. 1909. The cutaneous glands of the common toads. American Journal of Anatomy 9:322–359.

Mulaik, S. 1937. Notes on *Leptodactylus labialis* (Cope). Copeia 1937:72–73.

Mulaik, S. 1945. New mites in the family Caeculidae. Bulletin of the University of Utah 35(17): 1–23.

Mulaik, S., and D. Sollberger. 1938. Notes on the eggs and habits of *Hypopachus cuneus* Cope. Copeia 1938:90.

Mulcahy, D.G., and J.R. Mendelsohn. 2000. Phylogeography and speciation of the morphologically variable, widespread species *Bufo valliceps*, based on molecular evidence from mtDNA. Molecular Phylogenetics and Evolution 17:173–189.

Mulcahy, D.G., M.R. Cummer, J.R. Mendelson III, B.L. Williams, and P.C. Ustach. 2002. Status and distribution of two species of *Bufo* in the northeastern Bonneville Basin of Idaho and Utah. Herpetological Review 33:287–289.

Mulcare, D.J. 1966. The problem of toxicity in *Rana palustris*. Proceedings of the Indiana Academy of Science 75:319–324.

Mulertt, H. 1896. An albino frog. The Oregon Naturalist 3(11):147.

Mulertt, H. 1897. An albino frog. The Aquarium (January):89-90.

Mulla, M.S. 1962. Frog and toad control with insecticides! Pest Control 30:20, 64.

Mulla, M.S. 1963. Toxicity of organochlorine insecticides to mosquitofish *Gambusia affinis* and the bullfrog *Rana catesbeiana*. Mosquito News 23:299–303.

Mullally, D.P. 1952. Habits and minimum temperatures of the toad *Bufo boreas halophilus*. Copeia 1952:274–276.

Mullally, D.P. 1953. Observations on the ecology of the toad *Bufo canorus*. Copeia 1953:182– 183.

Mullally, D.P. 1956. The relationships of the Yosemite and western toads. Herpetologica 12:133–135.

Mullally, D.P. 1958. Daily period of activity of the Western toad. Herpetologica 14:29–31.

Mullally, D.P. 1959. Notes on the natural history of *Rana muscosa* Camp in the San Bernardino Mountains. Herpetologica 15:78–80.

Mullally, D.P., and J.D. Cunningham. 1956a. Aspects of the thermal ecology of the Yosemite toad. Herpetologica 12:57–67.

Mullally, D.P., and J.D. Cunningham. 1956b. Ecological relations of *Rana muscosa* at high elevations in the Sierra Nevada. Herpetologica 12:189–198.

Mullally, D.P., and D.H. Powell. 1958. The Yosemite toad: northern range extension and possible hybridization with the Western toad. Herpetologica 14:31–33.

Munger, J.C., M. Gerber, K. Madrid, M.–A. Carroll, W. Petersen, and L. Heberger. 1998. U.S. National Wetlands Inventory classifications as predictors of the occurrence of Columbia spotted frogs (*Rana luteiventris*) and Pacific treefrogs (*Hyla regilla*). Conservation Biology 12:320–330.

Munsey, L.D. 1972. Salinity tolerance of the African clawed frog, *Xenopus laevis*. Copeia 1972:584–586.

Munz, P.A. 1920. A study of the food habits of the Ithacan species of anura during transformation. Pomona College Journal of Entomology and Zoology 12:33–56.

Murphy, C.G. 1994a. Chorus tenure of male barking treefrogs, *Hyla gratiosa*. Animal Behaviour 48:763–777.

Murphy, C.G. 1994b. Determinants of chorus tenure in barking treefrogs (*Hyla gratiosa*). Behavioral Ecology and Sociobiology 34:285–294.

Murphy, C.G. 1996. Evaluating the design of mate–choice experiments: the effect of amplexus on mate choice by female barking treefrogs, *Hyla gratiosa*. Animal Behaviour 51:881– 890.

Murphy, C.G. 1999. Nightly timing of chorusing by male barking treefrogs (*Hyla gratiosa*): the influence of female arrival and energy. Copeia 1999:333–347.

Murphy, C.G. 2003. The cause of correlations between nightly numbers of males and female barking treefrogs (*Hyla gratiosa*) attending choruses. Behavioral Ecology 14:271–281.

Murphy, C.G. 2012. Simultaneous mate–sampling by female barking treefrogs (*Hyla gratiosa*). Behavioral Ecology 23:1162–1169.

Murphy, C.G., and H.C. Gerhardt. 1996. Evaluating the design of mate–choice experiments: the effects of amplexus on mate choice by female barking treefrogs, *Hyla gratiosa*. Animal Behaviour 51:881–890.

Murphy, C.G., and H.C. Gerhardt. 2000. Mating preference functions of individual female barking treefrogs, *Hyla gratiosa*, for two properties of male advertisement calls. Evolution 54:660–669.

Murphy, C.G., and H.C. Gerhardt. 2002. Mate sampling by female barking treefrogs (*Hyla gratiosa*). Behavioral Ecology 13:472–480.

Murphy, C.G., and S.B. Floyd. 2005. The effect of call amplitude on male spacing in choruses of barking treefrogs, *Hyla gratiosa*. Animal Behaviour 69: 419–426.

Murphy, J.C. 1979. *Pseudacris streckeri*, a small frog with an interesting history. Bulletin of the Chicago Herpetological Society 14:120–129.

Murphy, J.F., E.T. Simandle, and D.E. Becker. 2003. Population status and conservation of the black toad, *Bufo exsul*. Southwestern Naturalist 48:54–60.

Murphy, M.A., R. Dezzani, D.S. Pilliod, and A. Storfer. 2010a. Landscape genetics of high mountain frog metapopulations: an application of gravity models. Molecular Ecology 19:3634–3649.

Murphy, M.A., J.S. Evans, and A. Storfer. 2010b. Quantifying ecological process at multiple spatial scales using landscape genetics: *Bufo boreas* connectivity in Yellowstone National Park. Ecology 91: 252–261.

Murphy, M.B., M. Hecker, K.K. Coady, A.R. Tompsett, P.D. Jones, L.H. DuPreez, G.J. Everson, K.R. Solomon, J.A. Carr, E.E. Smith, R.J. Kendall, G. Van Der Kraak, and J.P. Giesy. 2006. Atrazine concentrations, gonadal gross morphology and histology in ranid frogs collected in Michigan agricultural areas. Aquatic Toxicology 76:230–245.

Murphy, P.J., S. St.–Hilaire, S. Bruer, P.S. Corn, and C.R. Peterson. 2009. Distribution and pathogenicity of *Batrachochytrium dendrobatidis* in boreal toads from the Grand Teton area of western Wyoming. EcoHealth 6:109–120.

Murphy, R.C. 1917. Note on a one–legged frog. Copeia (50):100.

Murphy, R.K. 1987. Observations of the wood frog in northwestern North Dakota. Prairie Naturalist 19:262.

Murphy, R.W., and R.C. Drewes. 1976. Comments on the occurrence of *Smilisca baudini* (Dumeril and Bibron) (Amphibia: Hylidae) in Bexar County, Texas. Texas Journal of Science 27:406–407.

Murphy, T.D. 1963. Amphibian populations and movements at a small semi–permanent pond in Orange County, North Carolina. Ph.D. Dissertation, Duke University, Durham.

Murphy, T.D. 1965. High incidence of two parasitic infestations and two morphological abnormalities in a population of the frog, *Rana palustris* Le Conte. American Midland Naturalist 74:233–239.

Murray, D.L. 1990. The effects of food and density on growth and metamorphosis in larval wood frogs (*Rana sylvatica*) from central Labrador. Canadian Journal of Zoology 68:1221– 1226.

Murray, I., and C.W. Painter. 2003. Geographic distribution: *Eleutherodactylus augusti* (Barking Frog). Herpetological Review 34:161.

Musgrave, M.E. 1930. *Bufo alvarius*, a poisonous toad. Copeia 1930:96–98.

Mushet, D.M., N.H. Euliss, Jr., and C.A. Stockwell. 2012. Mapping anuran habitat suitability to estimate effects of grassland and wetland conservation programs. Copeia 2012:321–330.

Mushinsky, H.R. 1973. Ontogenetic development of habitat preference in amphibians: the influence of early experience. Ph.D. Dissertation, Clemson University, Clemson, South Carolina.

Muths, E. 2003. Home range and movements of boreal toads in undisturbed habitat. Copeia 2003:160–165.

Muths, E., and P.S. Corn. 1997. Basking by adult boreal toads (*Bufo boreas boreas*) during the breeding season. Journal of Herpetology 31:426–428.

Muths, E., and R.D. Scherer. 2011. Portrait of a small population of boreal toads (*Anaxyrus boreas*). Herpetologica 67:369–377.

Muths, E., T.L. Johnson, and P.S. Corn. 2001. Experimental repatriation of boreal toads (*Bufo boreas*) eggs, metamorphs, and adults in Rocky Mountain National Park. Southwestern Naturalist 46:106–113.

Muths, E., P.S. Corn, A.P. Pessier, and D.E. Green. 2003. Evidence for disease–related amphibian decline in Colorado. Biological Conservation 110:357–365.

Muths, E., S. Rittmann, J. Irwin, D. Keinath, and R. Scherer. 2005. Wood Frog (*Rana sylvatica*): a technical conservation assessment. USDA Forest Service, Rocky Mountain Region, Species Conservation Project. http//www.fs.fed.us/r2/projects/scp/assessments/woodfrog. pdf (accessed 20 June 2007).

Muths, E., R.D. Scherer, P.S. Corn, and B.A. Lambert. 2006. Estimation of temporary emigration in male toads. Ecology 87:1048–1056.

Muths, E., A.L. Gallant, E.H.C. Grant, W.A. Battaglin, D.E. Green, J.S. Staiger, S.C. Walls, M.S. Gunzburger, and R.F. Kearney. 2006. The Amphibian Research and Monitoring Initiative (ARMI): 5–year Report. U.S. Geological Survey Scientific Investigations Report 2006–5224.

Muths, E., D.S. Pilliod, and L.J. Livo. 2008. Distribution and environmental limitations of an amphibian pathogen in the Rocky Mountains, USA. Biological Conservation 141:1484–1492.

Muths, E., R.D. Scherer, and D.S. Pilliod. 2011. Compensatory effects of recruitment and survival when amphibian populations are perturbed by disease. Journal of Applied Ecology 48:873-879.

Muzzall, P. M. 1991. Helminth infracommunities of the frogs *Rana catesbeiana* and *Rana clamitans* from Turkey Marsh, Michigan. Journal of Parasitology 77: 366–371.

Muzzall, P.M., and C.R. Peebles. 1991. Helminths of the wood frog, *Rana sylvatica*, and spring peeper, *Pseudacris c. crucifer*, from southern Michigan. Journal of the Helminthological Society of Washington 58:263–265.

Muzzall, P.M., and E. Sonntag. 2012. Helminths and symbiotic protozoa of Blanchard's Cricket Frog, *Acris blanchardi* Harper, 1947 (Hylidae), from Michigan and Ohio, U.S.A. Comparative Parasitology 79:340–343.

Myers, C.W. 1958a. Subspecific identity of *Rana sylvatica* in Missouri. Copeia 1958:46.

Myers, C.W. 1958b. Amphibia in Missouri caves. Herpetologica 14:35–36.

Myers, C.W., and W.D. Klimstra. 1963. Amphibians and reptiles of an ecologically disturbed (strip–mined) area in southern Illinois. American Midland Naturalist 70:126–132.

Myers, G.S. 1926. A synopsis for the identification of the amphibians and reptiles of Indiana. Proceedings of the Indiana Academy of Science 35:337-340.

Myers, G.S. 1927. The differential characters of *Bufo americanus* and *Bufo fowleri*. Copeia (163):50–53.

Myers, G.S. 1930a. Notes on some amphibians in western North America. Proceedings of the Biological Society of Washington 43:55–64.

Myers, G.S. 1930b. The status of the Southern California toad, *Bufo californicus* (Camp). Proceedings of the Biological Society of Washington 43:73–78.

Myers, G. S. 1931a. *Ascaphus truei* in Humbolt County, California, with a note on the habits of the tadpole. Copeia 1931:56–57.

Myers, G. S. 1931b. The original descriptions of *Bufo fowleri* and *Bufo americanus*. Copeia 1931: 94–96.

Myers, G.S. 1942. The black toad of Deep Springs Valley, Inyo County, California. Occasional Papers of the Museum of Zoology, University of Michigan 460:1–13.

Myers, G. S. 1943. Notes on *Rhyacotriton olympicus* and *Ascaphus truei* in Humboldt County, California. Copeia 1943:125–126.

Myers, G. S. 1944. California records of the western spade-foot toad. Copeia 1944:58.

Myers, G. S. 1945. Possible introduction of Argentine toads into Florida. Copeia 1945:44.

Myers, G. S. 1950. The systematic status of *Hyla septentrionalis*, the large tree frog of the Florida Keys, the Bahamas and Cuba. Copeia 1950:204–214.

Myers, G. S. 1951. The most widely heard amphibian voice. Copeia 1951:179.

Myers, J.M., J.P. Ramsey, A.L. Blackman, A.E. Nichols, K.P. Minbiole, and R.N. Harris. 2012. Synergistic inhibition of the lethal fungal pathogen *Batrachochytrium dendrobatidis*: the combined effect of symbiotic bacterial metabolites and antimicrobial peptides of the frog *Rana muscosa*. Journal of Chemical Ecology 38:958–965.

Myers, T.K., C.H., L. Eigner, J.A. Harris, R. Hilman, M.D. Johnson, R. Kalinowski, J.J. Muri, M. Reyes, and L.E. Tucci. 2007. A comparison of ground–based and tree–based polyvinyl chloride pipe refugia for capturing *Pseudacris regilla* in northwestern California. Northwestern Naturalist 88:147–154.

Nace, G.W. 1974. Albino toad found in Michigan–continued. Herpetological Review 5:70.

Nagy, C., S. Aschen, R. Christie, and M. Weckel. 2011. Japanese stilt grass (*Microstegium vimineum*), a nonnative invasive grass, provides alternative habitat for native frogs in a suburban forest. Urban Habitats 6.

Najarian, H.H. 1955. Trematode parasites in the Salientia in the vicinity of Ann Arbor, Michigan. American Midland Naturalist 53:195–197.

Nash, C.W. 1905. Check list of the vertebrates of Ontario and catalogue of specimens in the Biological Section of the Provincial Museum. Department of Education, Toronto.

Nash, C.W. 1908. Vertebrates of Ontario. L.K. Cameron, Toronto.

Nash, R.F., G.G. Gallup, Jr., and M.K. McClure. 1970. The immobility reaction in leopard frogs (*Rana pipiens*) as a function of noise–induced fear. Psychonomic Science 21:155–156.

Naugle, D.E., T.D. Fischer, K.F. Higgins, and D.C. Backlund. 2005. Distribution of South Dakota anurans. Pp. 283–291 *In* M.J. Lannoo (ed.), Amphibian Declines. The Conservation Status of United States Species. University of California Press, Berkeley.

Nauman, R.S., and Y. Dettlaf. 1999. *Rana cascadae* (Cascades Frog). Predation. Herpetological Review 30:93.

Navarro–Martín, L., C. Lanctôt, C. Edge, J. Houlahan, and V.L. Trudeau. 2012. Expression profiles of metamorphosis–related genes during natural transformations in tadpoles of wild wood frogs (*Lithobates sylvaticus*). Canadian Journal of Zoology 90:1059–1071.

Nebeker, A.V., and G.S. Schuytema. 2000. Effects of ammonium sulfate on growth of larval Northwestern salamanders, red–legged and Pacific treefrog tadpoles, and juvenile fathead minnows. Bulletin of Environmental Contamination and Toxicology 64:271–278.

Nebeker, A.V., G.S. Schuytema, W.L. Griffis, and A. Cataldo. 1998. Impact of Guthion on survival and growth of the frog *Pseudacris regilla* and the salamanders *Ambystoma gracile* and *Ambystoma maculatum*. Archives of Environmental Contamination and Toxicology 35:48–51.

Neck, R. W. 1983. Origin of *Rana catesbeiana* populations in the Rio Grande delta of Texas. Herpetological Review 14:55.

Necker, W.L. 1939. Records of amphibians and reptiles of the Chicago region, 1935–1938. Bulletin of the Chicago Academy of Sciences 6:1–10.

Needham, J.G. 1905. The summer food of the bullfrog (*Rana catesbeiana* Shaw) at Saranac Inn. New York State Museum Bulletin 86:9–15.

Needham, J.G. 1924. Observations on the life of the ponds at the head of Laguna Canyon. Pomona College Journal of Entomology and Zoology 16:1–12.

Neill, W.E., and J.C. Grubb. 1971. Arboreal habits of *Bufo valliceps* in central Texas. Copeia 1971:347–348.

Neill, W.T. 1947a. Doubtful type localities in South Carolina. Herpetologica 4:75–76.

Neill, W.T. 1947b. A collection of amphibians from Georgia. Copeia 1947:271.

Neill, W.T. 1948a. Spiders preying on reptiles and amphibians. Herpetologica 4:158.

Neill, W.T. 1948b. Hibernation of amphibians and reptiles in Richmond County, Georgia. Herpetologica 4:107–114.

Neill, W.T. 1948c. A new subspecies of tree–frog from Georgia and South Carolina. Herpetologica 4:175–179.

Neill, W.T. 1949a. Hybrid toads from Georgia. Herpetologica 5:30–32.

Neill, W.T. 1949b. The status of Baird's chorus–frog. Copeia 1949:227–228.

Neill, W.T. 1949c. The status of *Hyla flavigula*. Copeia 1949:78.

Neill, W.T. 1950a. Reptiles and amphibians in urban areas of Georgia. Herpetologica 6:113–116.

Neill, W.T. 1950b. Taxonomy, nomenclature, and distribution of Southeastern cricket frogs, genus *Acris*. American Midland Naturalist 43:152–156.

Neill, W.T. 1951. A bromeliad herpetofauna in Florida. Ecology 32:140–143.

Neill, W.T. 1952a. New records of *Rana virgatipes* and *Rana grylio* in Georgia and South Carolina. Copeia 1952:194–195.

Neill, W.T. 1952b. Burrowing habits of *Hyla gratiosa*. Copeia 1952:196.

Neill, W.T. 1954. Ranges and taxonomic allocations of amphibians and reptiles in the Southeastern United States. Publications of the Research Division, Ross Allen's Reptile Institute 1:75–96.

Neill, W.T. 1957a. Notes on metamorphic and breeding aggregations of the eastern spadefoot, *Scaphiopus holbrooki* (Harlan). Herpetologica 13:185–188.

Neill, W.T. 1957b. The status of *Rana capito stertens* Schwartz and Harrison. Herpetologica 13:47–52.

Neill, W.T. 1957c. Homing by a squirrel treefrog, *Hyla squirella* Latreille. Herpetologica 13:217–218.

Neill, W.T. 1958a. The occurrence of amphibians and reptiles in saltwater areas, and a bibliography. Bulletin of Marine Science of the Gulf and Caribbean 8:1–97.

Neill, W.T. 1958b. The varied calls of the barking treefrog, *Hyla gratiosa* LeConte. Copeia1958:44–46.

Neill, W.T. 1961. River frog swallows eastern diamondback rattlesnake. Bulletin of the Philadelphia Herpetological Society 9:19.

Neill, W.T. 1964. Frogs introduced on islands. Quarterly Journal of the Florida Academy of Sciences 27:127–130.

Neils, A., and C. Bugbee. 2007. *Rana catesbeiana* (American Bullfrog). Diet. Herpetological Review 38:443.

Niemiller, M.L. 2005. The herpetofauna of the upper Duck River watershed in Coffee County, Tennessee. Journal of the Tennessee Academy of Science 80:6–12.

Niemiller, M.L., and B.T. Miller. 2005. *Rana clamitans melanota* (Northern Green Frog). Predation. Herpetological Review 36:440.

Nelson, C.E. 1971. Breakdown of ethological isolation between *Rana pipiens* and *Rana sylvatica*. Copeia 1971:344.

Nelson, C.E. 1972. Systematic studies of the North American microhylid genus *Gastrophryne*. Journal of Herpetology 6:111–137.

Nelson, C.E. 1973. Mating calls of the Microhylinae: descriptions and phylogenetic and ecological considerations. Herpetologica 29:163–176.

Nelson, C.E. 1974. Further studies on the systematics of *Hypopachus* (Anura: Microhylidae). Herpetologica 30:250–274.

Nelson, C.E. 1980. What determines the species composition of larval amphibian pond communities in south central Indiana, U.S.A.? Proceedings of the Indiana Academy of Science 89:149.

Nelson, C.E., and H.S. Cuellar. 1968. Anatomical comparison of tadpoles of the genera *Hypopachus* and *Gastrophryne* (Microhylidae). Copeia 1968:423–424.

Nelson, D.H. 1974a. Ecology of anuran populations inhabiting thermally stressed aquatic ecosystems, with an emphasis on larval *Rana pipiens* and *Bufo terrestris*. Ph.D. Dissertation, Michigan State University, East Lansing.

Nelson, D.H. 1974b. Growth and developmental responses of larval toad populations to heated effluent in a South Carolina reservoir. Pp. 264–276 *In* J.W. Gibbons and R.R. Sharitz (eds.), Thermal Ecology. Atomic Energy Commission Symposium Series, CONF–730505.

Nelson, G.L., and B.M. Graves. 2004. Anuran population monitoring: comparison of the North American Amphibian Monitoring Program's calling index with mark–recapture estimates for *Rana clamitans*. Journal of Herpetology 38:355–359.

Nelson, J. 1890. Descriptive Catalogue of the Vertebrates of New Jersey (A Revision of Dr. Abbott's Catalogue of 1868). Final Report of the State Geologist. Vol. II. Geological Survey of New Jersey, John L. Murphy, Trenton, New Jersey. [frogs on pp. 649-652]

Nelson, L., Jr., and M.W. Cummings. 1973. Bullfrog farming. Animal Briefs No. 26 (May):1-2. University of California, Agricultural Extension., Berkeley.

Nelson, R.T., B.J. Cochrane, P.R. Delis, and D.L. Testrake. 2002. Basidioboliasis in anurans in Florida. Journal of Wildlife Diseases 38:463–467.

Nero, R.W. 1967. A possible record of death–feigning in a toad. Blue Jay 25:193–194.

Nero, R.W. 1980. The Great Grey Owl, Phantom of the Northern Forest. Smithsonian Institution Press, Washington, D.C.

Nero, R.W., and F.R. Cook. 1964. A range extension for the wood frog in northeastern Saskatchewan. Canadian Field-Naturalist 78:268–269.

Netting, M.G. 1930 ("1929"). Further distinctions between

Bufo americanus Holbrook and *Bufo fowleri* Garman. Papers of the Academy of Science, Arts and Letters 11:437–443.

Netting, M.G. 1933. The amphibians of Pennsylvania. Proceedings of the Pennsylvania Academy of Science 7:1–11.

Netting, M.G. 1946. The amphibians and reptiles of Pennsylvania. Board of Fish Commissioners, Harrisburg.

Netting, M.G., and L.W. Wilson. 1940. Notes on amphibians from Rockingham County, Virginia. Annals of the Carnegie Museum 28:1–8.

Netting, M.G., and C.J. Goin. 1942. Additional notes on *Rana sevosa*. Copeia 1942:259.

Netting, M.G., and C.J. Goin. 1945. The occurrence of Fowler's toad, *Bufo woodhousii fowleri* Hinckley, in Florida. Proceedings of the Florida Academy of Sciences 7:181–184.

Netting, M.G., and C.J. Goin. 1946. *Acris* in Mexico and Trans–Pecos Texas. Copeia 1947:253.

Neufeldt, J.H. 2004. Terrestrial movements and habitat use of gopher frogs (*Rana capito*) at Fort Benning, Georgia. M.S. thesis, Columbus State University, Columbus, Georgia.

Nevo, E. 1973a. Adaptive color polymorphism in cricket frogs. Evolution 27:353–367.

Nevo, E. 1973b. Adaptive variation in size of cricket frogs. Ecology 54:1271–1281.

Nevo, E., and R.R. Capranica. 1985. Evolutionary origin of ethological reproductive isolation in cricket frogs, *Acris*. Pp. 147–214 *In* M.K. Hecht, B. Wallace, and G.T. Prance (ed.), Evolutionary Biology, Vol. 19. Plenum Press, New York.

Newman, C.E., and L.J. Rissler. 2011. Phylogeographic analyses of the southern leopard frog: the impact of geography and climate on the distribution of genetic lineages vs. subspecies. Molecular Ecology 20:5295–5312.

Newman, C.E., J.A. Feinberg, L.J. Rissler, J. Burger, and H.B. Shaffer. 2012. A new species of leopard frog (Anura: Ranidae) from the urban northeastern US. Molecular Phylogenetics and Evolution 63:445–455.

Newman, R.A. 1987. Effects of density and predation on *Scaphiopus couchi* tadpoles in desert ponds. Oecologia 71:301–307.

Newman, R.A. 1988a. Adaptive plasticity in development of *Scaphiopus couchii* tadpoles in desert ponds. Evolution 42:774–783.

Newman, R.A. 1988b. Genetic variation for larval anuran (*Scaphiopus couchii*) development time in an uncertain environment. Evolution 42:763–773.

Newman, R.A. 1989. Developmental plasticity of *Scaphiopus couchii* tadpoles in an unpredictable environment. Ecology 70:1775–1787.

Newman, R.A. 1992. Adaptive plasticity in amphibian metamorphosis. Bioscience 42:671–678.

Newman, R.A. 1994a. Genetic variation for phenotypic plasticity in the larval life history of spadefoot toads (*Scaphiopus couchii*). Evolution 48:1773–1785.

Newman, R.A. 1994b. Effects of changing density and food level on metamorphosis of a desert amphibian, *Scaphiopus couchii*. Ecology 75:1085–1096.

Newman, R.A. 1998. Ecological constraints on amphibian metamorphosis: interactions of temperature and larval density with responses to changing food level. Oecologia 115:9–16.

Newman, R.A. 1999. Body size and diet of recently metamorphosed spadefoot toads (*Scaphiopus couchii*). Herpetologica 55:507–515.

Newman, R.A., and A.E. Dunham. 1994. Size at metamorphosis and water loss in a desert anuran (*Scaphiopus couchii*). Copeia 1994:372–381.

Newman, R.A., and T. Squire. 2001. Microsatellite variation and fine–scale population structure in the wood frog (*Rana sylvatica*). Molecular Ecology 10:1087–1100.

Nichols, A. 1852. Occurrence of *Scaphiopus solitarius* in Essex County: with some notices of its history, habits, &c. Journal of the Essex County Natural History Society 1:113–117.

Nichols, R.J. 1937. Taxonomic studies on the mouth parts of larval anura. University of Illinois Bulletin, Illinois Biological Monographs 15:1–73.

Nicholson, D.J. 1961. Bullfrog captures warblers in Florida. Florida Naturalist 34(1):44.

Nicholson, F.E. 1932. Nature discloses more queer life. The Philadephia Ledger, 30 October. [republished in Literary Digest, 10 December 1932, pp. 31–32 and in the National Speleological Society Bulletin 10 (1948):22–26]

Nicholson, M.C. 1980. The effects of density and predation on the growth rate, larval period and survivorship of *Rana utricularia* and *Bufo terrestris* tadpoles. M.A. thesis, University of South Florida, Tampa.

Nickerson, M.A., and C.E. Mays. 1968. *Bufo retiformis* Sanders and Smith from the Santa Rosa Valley, Pinal County, Arizona. Journal of Herpetology 1:103.

Nickerson, M.A., and J.A. Hutchison. 1971. The distribution of the fungus *Basidiobolus ranarum* Eidam in fish, amphibians and reptiles. American Midland Naturalist 86:500–502.

Nickerson, M.A., and S.F. Celino. 2003. *Rana capito* (Gopher Frog). Drought shelter. Herpetological Review 34:137–138.

Nickum, J.G. 1961. Intra–specific variation in the chorus frog *Pseudacris nigrita* Le Conte. M.A. thesis, State University of South Dakota, Vermillion.

Nie, M., J.D. Crim, and G.R. Ultsch. 1999. Dissolved oxygen, temperature, and habitat selection by bullfrog (*Rana catesbeiana*) tadpoles. Copeia 1999:155–162.

Nielsen, H.I., and J. Dyck. 1978. Adaptation of the tree frog, *Hyla cinerea*, to colored backgrounds, and the role of the three chromatophore types. Journal of Experimental Zoology 205:79–94.

Nielsen, H. I. 1978. The effect of stress and adrenaline on the color of *Hyla cinerea* and *Hyla arborea*. General and Comparative Endocrinology 36: 543–552.

Nielson, M., K. Lohman, and J. Sullivan. 2001. Phylogeography of the tailed frog (*Ascaphus truei*): implications for the biogeography of the Pacific Northwest. Evolution 55:147–160.

Nielson, M., K. Lohman, C.H. Daugherty, F.W. Allendorf, K.I. Knudsen, and J. Sullivan. 2006. Allozyme and mitochondrial DNA variation in the tailed frog (Anura: *Ascaphus*): the influence of geography and gene flow. Herpetologica 62:235–258.

Niemiller, M.L. and R.G. Reynolds. 2011. The Amphibians of Tennessee. University of Tennessee Press, Knoxville.

Nieto, N.C., M.A. Camann, J.E. Foley, and J.O. Reiss. 2007. Disease associated with integumentary and cloacal parasites in tadpoles of northern red–legged frog *Rana aurora aurora*. Diseases of Aquatic Organisms 78:61–71.

Nigrelli, R.F. 1945. Trypanosomes from North American amphibians with a description of *Trypanosoma grylli* Nigrelli (1944) from *Acris gryllus* (LeConte). Zoologica 30:47-57.

Nityananda, V., and M.A. Bee M. 2011. Finding your mate at a cocktail party: frequency separation promotes auditory stream segregation of concurrent voices in multi–species frog choruses. PLoS One 6:e21191.

Nityananda, V., and M.A. Bee. 2012. Spatial release from masking in a free–field source identification task by gray treefrogs. Hearing Research 285:86–97.

Noble, G. K. 1923. The generic and genetic relations of *Pseudacris,* the swamp tree frogs. American Museum Novitates (70):1–6.

Noble, G.K., and R.C. Noble. 1923. The Anderson tree frog (*Hyla andersonii* Baird): observations on its habits and life history. Zoologica 11:413–455.

Noble, G.K., and E.J. Farris. 1929. The method of sex recognition in the wood frog, *Rana sylvatica* Le Conte. American Museum Novitates 363:1–17.

Noble, G.K., and P.G. Putnam. 1931. Observations on the life history of *Ascaphus truei* Stejneger. Copeia 1931:97–101.

Noble, G.K., and W.G. Hassler. 1936. Three salientians of geographic interest from southern Maryland. Copeia 1936:63–64.

Noble, G.K., and L.R. Aronson. 1942. The sexual behavior of Anura. 1. The normal mating pattern of *Rana pipiens*. Bulletin of the American Museum of Natural History 80:127–142.

Noland, R., and G.R. Ultsch. 1981. The roles of temperature and dissolved oxygen in microhabitat selection by the tadpoles of a frog (*Rana pipiens*) and a toad (*Bufo terrestris*). Copeia 1981:645–652.

Noles, P. 2010. Reconciling western toad phylogeography with Great Basin prehistory. M.S. thesis, University of Nevada, Reno.

Northen, P.T. 1970. The geographic and taxonomic relationships of the Great Basin spadefoot toad, *Scaphiopus intermontanus*, to other members of the subgenus *Spea*. Ph.D. Dissertation, University of Wisconsin, Madison.

Norton, V.M., and M.J. Harvey. 1975. Herpetofauna of Hardeman County, Tennessee. Journal of the Tennessee Academy of Science 50:131–136.

Nunn, R.B. 1932. Scientific Frog Farming. Nunn Frog Company, Houston, Texas. 16 pp.

Nunziata, S.O. 2011. Population and conservation genetics of crawfish frogs, *Lithobates areolatus*, at their northeastern range limits. M.S. thesis, Eastern Kentucky University, Richmond.

Nunziata, S.O., S.C. Richter, R.D. Denton, J.M. Yeiser, D.E. Wells, K.L. Jones, C.Hagen, and S.L. Lance. 2012. Fourteen novel microsatellite markers for the gopher frog, *Lithobates capito* (Amphibia: Ranidae). Conservation Genetics Resources 4:201–203.

Nussbaum, R.A., E.D. Brodie, Jr., and R.M. Storm. 1983. Amphibians & Reptiles of the Pacific Northwest. University Press of Idaho, Moscow.

NVDW (Nevada Division of Wildlife). 2003. Conservation agreement and conservation strategy. Columbia spotted frog (*Rana luteiventris*). Toiyabe Great Basin subpopulation Nevada. Nevada Division of Wildlife, Carson City.

Nyberg, D., and I. Lerner. 2000. Revitalization of ephemeral pools as frog breeding habitat in an Illinois forest preserve. Journal of the Iowa Academy of Science 107:187–190.

Nyman, S. 1985. An experimental study of two species of larval amphibians, *Rana sylvatica* and *Ambystoma maculatum*, as potential regulatory consumers in a temporary pond community. Ph.D. Dissertation, University of Rhode Island, Kingston.

Nyman, S. 1986. Mass mortality in larval *Rana sylvatica* attributable to the bacterium *Aeromonas hydrophila*. Journal of Herpetology 20:196–201.

Oberfoell, C.E.C., and J.L. Christiansen. 2001. Identification and distribution of the treefrogs *Hyla versicolor* and *Hyla chrysoscelis* in Iowa. Journal of the Iowa Academy of

Science 108:79–83.

O'Bryan, C.J., M.J. Gray and C.S. Brooks. 2012. Further presence of ranavirus infection in amphibian populations of Tennessee, USA. Herpetological Review 43:293–295.

O'Connor, M.P., and C.R. Tracy. 1992. Thermoregulation by juvenile toads of *Bufo woodhousei* in the field and in the laboratory. Copeia 1992:865–876.

Odlaug, T.O. 1954. Parasites of some Ohio batrachians. Ohio Journal of Science 54:126–128.

O'Donnell, S.P. 2012. Genetic diversity and genetic structuring at multiple spatial scales across the range of the Northern Leopard Frog, *Rana pipiens*. Ph.D. Dissertation, Utah State University, Logan.

Odum, R.A., and P.S. Corn. 2005. *Bufo baxteri* Porter, 1988. Wyoming Toad. Pp. 390–392 *In* M.J. Lannoo (ed.), Amphibian Declines. The Conservation Status of United States Species. University of California Press, Berkeley.

O'Hara, R.K. 1981. Habitat selection behavior in three species of anuran larvae: environmental cues, ontogeny and adaptive significance. Ph.D. Dissertation, Oregon State University, Corvallis.

O'Hara, R.K., and A.R. Blaustein. 1981. An investigation of sibling recognition in *Rana cascadae* tadpoles. Animal Behaviour 29:1121–1126.

O'Hara, R.K., and A.R. Blaustein. 1982. Kin preference behavior in *Bufo boreas* tadpoles. Behavioral Ecology and Sociobiology 11:43–49.

O'Hara, R.K., and A.R. Blaustein. 1985. *Rana cascadae* tadpoles aggregate with siblings: an experimental field study. Oecologia 67:44–51.

O'Hara, R.K., and A.R. Blaustein. 1988. *Hyla regilla* and *Rana pretiosa* tadpoles fail to display kin recognition behaviour. Animal Behaviour 36:946–948.

Okafor, J.I., D. Testrake, H.R. Mushinsky, and B.G. Yangco. 1984. A *Basidiobolus* sp. and its association with reptiles and amphibians in southern Florida. Sabouraudia, Journal of Medical and Veterinary Mycology 22:47–51.

Okonkwo, G.E. 2011. The use of small ephemeral wetlands and streams by amphibians in the mixedwood forest of boreal Alberta. M.S. thesis, University of Alberta, Edmonton.

Oláh–Hemmings, V., J.R. Jaeger, M.J. Sredl, M.A. Schlaepfer, R.D. Jennings, C.A. Drost, D.F. Bradford, and B.R. Riddle. 2009. Phylogeography of declining relict and lowland leopard frogs in the desert Southwest of North America. Journal of Zoology 280:343–354.

Oldfield, B., and J.J. Moriarty. 1994. Amphibians & Reptiles Native to Minnesota. University of Minnesota Press, Minneapolis.

Oldham, M.J. 1992. Declines in Blanchard's cricket frog in Ontario. Canadian Wildlife Service Occasional Paper No. 76:30–31.

Oldham, R.S. 1966. Spring movements in the American toad, *Bufo americanus*. Canadian Journal of Zoology 44:63–100.

Oldham, R.S. 1967. Orienting mechanisms of the green frog, *Rana clamitans*. Ecology 48:477–491.

Oldham, R.S. 1974. Mate attraction by vocalization in members of the *Rana pipiens* complex. Copeia 1974:982–984.

Oldham, R.S. 1975. Ovulation induced by vocalization in members of the *Rana pipiens* complex. Journal of Herpetology 9:248–249.

Oldham, R.S. 1976. Chorus maintenance in breeding populations of the *Rana pipiens* complex. Texas Journal of Science 27:323–325.

Oldham, R.S., and H.C. Gerhardt. 1975. Behavioral isolating mechanisms of the treefrogs *Hyla cinerea* and *H. gratiosa*. Copeia 1975:223–231.

Oliver, J.A., and J.R. Bailey. 1939. Amphibians and reptiles of New Hampshire, exclusive of marine forms. New Hampshire Fish and Game Commission, Biological Survey of Connecticut Watershed Report 4:195–217.

Oliver, J.A., and C.E. Shaw. 1953. The amphibians and reptiles of the Hawaiian Islands. Zoologica 38:65–95.

Oliver, J.H., Jr., M.P. Hayes, J.E. Keirans, and D.R. Lavender. 1993. Establishment of the foreign parthenogenetic tick *Amblyomma rotundatum* (Acari: Ixodidae) in Florida. Journal of Parasitology 79:786–790.

Olsen, O.W. 1937a. Description and life history of the trematode *Haplometrana utahensis* sp. nov. (Plagiorchiidae) from *Rana pretiosa*. Journal of Parasitology 23:13–28.

Olsen, O.W. 1937b. A new species of bladder fluke, *Gorgoderina tanneri* (Gorgoderidae: Trematoda) from *Rana pretiosa*. Journal of Parasitology 23:499–503.

Olsen, O.W. 1938. *Aplectana gigantica* (Cosmocercidae), a new species of nematode from *Rana pretiosa*. Transactions of the American Microscopical Society 57:200–203.

Olsen, O.W. 1962. Animal Parasites: Their Biology and Life Cycles. Burgess, Minneapolis, Minnesota.

Olson, C. A. 2011. Diet, density, and distribution of the introduced greenhouse frog, *Eleutherodactylus planirostris*, on the island of Hawaii. M.S. thesis. Utah State University, Logan.

Olson, C.A., and K.H. Beard. 2012. Diet of the introduced greenhouse frog in Hawaii. Copeia 2012:121–129.

Olson, C.A., K.H. Beard, D.N. Koons, and W.C. Pitt. 2012a. Detection probabilities of two introduced frogs in Hawaii:implications for assessing non–native species distributions. Biological Invasions 14:889–900.

Olson, C.A., K.H. Beard, and W.C. Pitt. 2012b. Biology and impacts of Pacific island invasive species. 8. *Eleutherodac-*

tylus planirostris, the greenhouse frog (Anura: Eleutherodactylidae). Pacific Science 66:255–270.

Olson, D.H. 1988. The ecological and behavioral dynamics of breeding in three sympatric anuran amphibians. Ph.D. Dissertation, Oregon State University, Corvallis.

Olson, D.H. 1989. Predation on breeding western toads (*Bufo boreas*). Copeia 1989:391–397.

Olson, D.H. 1992. Ecological susceptibility of amphibians to population declines. Pp. 55–62 *In* R.R. Harris, H.M. Kerner, and D.C. Erman (eds.), Proceedings of the Symposium on Biodiversity of Northwestern California. University of California, Berkeley.

Olson, D.H. 2001. Ecology and management of montane amphibians of the U.S. Pacific Northwest. Biota 2:51–74.

Olson, D.H., A.R. Blaustein, and R.K. O'Hara. 1986. Mating pattern variability among western toad (*Bufo boreas*) populations. Oecologia 70:351–356.

Olson, D. H., P. D. Anderson, C. A. Frissell, H.H. Welsh, and D.F. Bradford. 2007.Biodiversity management approaches for stream–riparian areas: perspectives for Pacific Northwest headwater forests, microclimates, and amphibians. Forest Ecology and Management 246: 81–107.

Olson, E.R. 2010. *Rana pipiens* (Northern Leopard Frog). Winter activity. Herpetological Review 41:206.

Olson, M.E., S. Gard, M. Brown, R. Hampton, and D.W. Morck. 1992. *Flavobacterium indologenes* infection in leopard frogs. Journal of the American Veterinary Medical Association 201:1766–1770.

Olson, R.E. 1959. Notes on some Texas herptiles. Herpetologica 15:48.

O'Neill, C.E. 2001. Determination of a terrestrial buffer zone for conservation of the cricket frog, *Acris crepitans.* M.S. thesis, Illinois State University, Normal.

O'Neill, E.D. 1995. Amphibian and reptile communities of temporary ponds in a managed pine flatwoods. M.S. thesis, University of Florida, Gainesville.

O'Neill, E.M., and K.H. Beard. 2011. Clinal variation in calls of native and introduced populations of *Eleutherodactylus coqui*. Copeia 2011:18–28.

Oplinger, C.S. 1963. The life history of the northern spring peeper, *Hyla crucifer crucifer* Wied at Ithaca, New York. Ph.D. Dissertation, Cornell University, Ithaca, New York.

Oplinger, C.S. 1966. Sex ratio, reproductive cycles, and time of ovulation in *Hyla crucifer crucifer* Wied. Herpetologica 22:276–283.

Oplinger, C.S. 1967. Food habits and feeding activity of recently transformed and adult *Hyla crucifer crucifer* Wied. Herpetologica 23:209–217.

Orchard, S.A. 1999. The American Bullfrog in British Columbia: the frog who came to dinner. Pp. 289–296 *In* R. Claudi and J.H. Leach (eds.), Nonindigenous Freshwater Organisms. Vectors, Biology, and Impacts. Lewis Publishers, Boca Raton, Florida.

Orchard, S.A. 2011. Removal of the American bullfrog *Rana (Lithobates) catesbeiana* from a pond and a lake on Vanvouver Island, British Columbia, Canada. Pp. 217–221 *In* C.R. Veitch, M.N. Clout and D.R. Towns (eds.), Island Invasives: Eradication and Management. IUCN, Gland, Switzerland.

Orlofske, S.A., and W.A. Hopkins. 2009. Energetics of metamorphic climax in the pickerel frog *(Rana palustris)*. Comparative Biochemistry and Physiology 154:191–196.

Orlofske, S.A., L.K. Belden, and W.A. Hopkins, W.A. 2009. Moderate *Echinostoma trivolvis* infection has no effects on physiology and fitness–related traits of larval pickerel frogs (*Rana palustris*). Journal of Parasitology 95:787–792.

Orlofske, S.A., R.C. Jadin, D.L. Preston, and P.T.J. Johnson. 2012. Parasite transmission in complex communities: predators and alternative hosts alter pathogenic infections in amphibians. Ecology 93:1247–1253.

O'Roke, E.C. 1924. The amphibians of South Dakota. Proceedings of the South Dakota Academy of Science 9:13-15.

Orr, J., and L. Mendoza. 2011. Herpetofaunal survey of Mason Neck State Park and Mason Neck National Wildlife Refuge. Catesbeiana 31:59-72.

Orr, L.P., J. Neumann, E. Vogt, and A. Collier. 1998. Status of northern leopard frogs in northeastern Ohio. Pp. 91–93 *In* M.J. Lannoo (ed.), Status & Conservation of Midwestern Amphibians. University of Iowa Press, Iowa City.

Ortenburger, A.I. 1924. Life history notes—*Scaphiopus*—the spadefoot toad. Proceedings of the Oklahoma Academy of Science 4:19–20.

Orton, G.L. 1939. Key to New Hampshire amphibian larvae. Pp. 218–221 *In* Biological Survey of the Connecticut Watershed. New Hampshire Fish and Game Department, Survey Report No. 4.

Orton, G.L. 1943. The tadpole of *Rhinophrynus dorsalis*. Occasional Papers of the Museum of Zoology, University of Michigan 472:1–7 + 1 plate.

Orton, G.L. 1946. Larval development of the eastern narrow–mouthed frog, *Microhyla carolinensis* (Holbrook), in Louisiana. Annals of the Carnegie Museum 30:241–248.

Orton, G.L. 1947. Notes on some hylid tadpoles in Louisiana. Annals of the Carnegie Museum 30:363–380 + 1 plate.

Orton, G.L. 1951a. An example of interspecific mating in toads. Copeia 1951:78.

Orton, G.L. 1951b. Notes on some tadpoles from southwestern Missouri. Copeia 1951:71–72.

Orton, G.L. 1952. Key to the genera of tadpoles in the United States and Canada. American Midland Naturalist 47:382–395.

Orton, G.L. 1954. Dimorphism in larval mouth–parts in spadefoot toads of the *Scaphiopus hammondii* group. Copeia 1954:97–100.

Osbourne, M.S. 2012. Initial juvenile movement of pond–breeding amphibians in altered forest habitat. Ph.D. Dissertation, University of Missouri, Columbia.

Oseen, K.L., and R.J. Wassersug. 2002. Environmental factors influencing calling in sympatric anurans. Oecologia 133:616–625.

Oseen, K.L., L.K.D. Newhook, and R.J. Wassersug. 2001. Turning bias in woodfrog (*Rana sylvatica*) tadpoles. Herpetologica 57:432–437.

Osgood, W.H. 1901. Natural history of the Cook Inlet region, Alaska. North American Fauna 21:51–81.

Ostergaard, E.C., K.O. Richter, and S.D. West. 2008. Amphibian use of stormwater ponds in the Puget lowlands of Washington, USA. Pp. 259–270 *In* J.C. Mitchell, R.E. Jung Brown, and B. Bartholomew (eds.),.Urban Herpetology. Herpetological Conservation 3. Society for the Study of Amphibians and Reptiles, Salt Lake City, Utah.

Otani, A., N. Palumbo, and G. Read. 1969. Pharmacodynamics and treatment of mammals poisoned by *Bufo marinus* toxin. American Journal of Veterinary Research 30:1865–1872.

Otto, C.V., J.W. Snodgrass, D.C. Forester, J.C. Mitchell, and R.W. Miller. 2007a. Climatic variation and the distribution of an amphibian polyploid complex. Journal of Animal Ecology 76:1053–1061.

Otto, C.V., D.C. Forester, and J.W. Snodgrass. 2007b. Influences of wetland and landscape characteristis on the distribution of carpenter frogs. Wetlands 27:261–269.

Ouellet, M., J. Bonin, J. Rodrigue, J.–L. DesGranges, and S. Lair. 1997. Hindlimb deformities (ectromelia, ectrodactyly) in free–living anurans from agricultural habitats. Journal of Wildlife Diseases 33:95–104.

Ouellet, M., I. Mikaelian, B.D. Pauli, J. Rodrigue, and D.M. Green. 2005a. Historical evidence of widespread chytrid infection in North American amphibian populations. Conservation Biology 19:1431–1440.

Ouellet, M., P. Galois, R. Pétel, and C. Fortin. 2005b. Les amphibians et les reptiles des collines montérégiennes: enjeux et conservation. Le naturaliste canadien 129:42–49.

Ouellet, M., C. Fortin, and M.–J. Grimard. 2009. Distribution and habitat use of the boreal chorus frog (*Pseudacris maculata*) at its extreme northeastern range limit. Herpetological Conservation and Biology 4:277–284.

Ovaska, K., T.M. Davis, and I.N. Flamarique. 1997. Hatching success and larval survival of the frogs *Hyla regilla* and *Rana aurora* under ambient and artificially enhanced solar ultraviolet radiation. Canadian Journal of Zoology 75:1081–1088.

Ovaska, K., and P. Govindarajulu. 2010. A guide to amphibians of British Columbia north of 50°. Brochure, BC FrogWatch and British Columbia Ministry of the Environment.

Over, W.H. 1923. Amphibians and reptiles of South Dakota. South Dakota Geological and Natural History Survey Series, Bulletin 12:1-34.

Over, W.H. 1943. Amphibians and reptiles of South Dakota. University of South Dakota Natural History Studies 6:1-31.

Overton, F. 1914a. Long Island flora and fauna. 3. The frogs and toads. The Museum of the Brooklyn Institute of Arts and Sciences Science Bulletin 2(3):21–40.

Overton, F. 1914b. Pickering's *Hyla* active in January. Copeia (4):1.

Overton, F. 1914c. The frogs and toads of Long Island. Brooklyn Museum Quarterly 1:30–38.

Overton, F. 1915. Annual occurrence of spade–foot toads. Copeia (20):1.

Owen, P.C. 2003. The structure, function, and evolution of aggressive signals in anuran amphibians. Ph.D. Dissertation, University of Connecticut, Storrs.

Owen, P.C., and S.A. Perrill. 1998. Habituation in the green frog, *Rana clamitans*. Behavioral Ecology and Sociobiology 44:209–213.

Owen, P.C., and S.A. Perrill. 2000. Habituation and neighbor–stranger discrimination in green frogs: a reply to Bee and Schachtman. Behavioral Ecology and Sociobiology 48:169–171.

Owen, P.C., and N.M. Gordon. 2005. The effect of perceived intruder proximity and resident body size on the aggressive responses of male green frogs, *Rana clamitans* (Anura: Ranidae). Behavioral Ecology and Sociobiology 58:446–455.

Owen, P.C., and J.K. Tucker. 2006. Courtship calls and behavior in two species of chorus frogs, genus *Pseudacris* (Anura: Hylidae). Copeia 2006:137–144.

Owen, R.D. 1996. Breeding phenology and microhabitat use among three chorus frog species (*Pseudacris*) in east–central Florida. M.S. thesis, University of Central Florida, Orlando.

Owen, R.D., and S.A. Johnson. 2005. *Pseudacris ocularis*

(Little Grass Frog). Predation. Herpetological Review 28:200.

Pace, A.E. 1972. Systematic and biological studies of the leopard frogs (*Rana pipiens* complex) of the United States. Ph.D. Dissertation, University of Michigan, Ann Arbor.

Pace, A.E. 1974. Systematics and biological studies of the leopard frogs (*Rana pipiens* complex) of the United States. Miscellaneous Publications, University of Michigan Museum of Zoology 148:1–140.

Pack, H.J. 1920. Eggs of the swamp tree frog. Copeia (77):7.

Pack, H.J. 1922. Toads in regulating insect outbreaks. Copeia (107):46–47.

Packard, A.S. 1866. List of vertebrates observed at Okak, Labrador, by Rev. Samuel Weiz, with annotations. Proceedings of the Boston Society of Natural History 10:264–277.

Packard, G.C., J.K. Tucker, and L.D. Lohmiller. 1998. Distribution of Strecker's chorus frogs (*Pseudacris streckeri*) in relation to their tolerance for freezing. Journal of Herpetology 32:437–440.

Padgett–Flohr, G.E. 2008. Pathogenicity of *Batrachochytrium dendrobatidis* in two threatened California amphibians: *Rana draytonii* and *Ambystoma californiense*. Herpetological Conservation and Biology 3:182–191.

Padgett–Flohr, G.E., and R.L. Hopkins III. 2009. *Batrachochytrium dendrobatidis*, a novel pathogen approaching endemism in central California. Diseases of Aquatic Organisms 83:1–9.

Padgett–Flohr, G.E., and M.P. Hayes. 2011. Assessment of the vulnerability of the Oregon spotted frog (*Rana pretiosa*) to the amphibian chytrid fungus (*Batrachochytrium dendrobatidis*). Herpetological Conservation and Biology 6:99–106.

Paetow, L. J. 2010. Effects of agricultural pesticides and the chytrid fungus *Batrachochytrium dendrobatidis* on the health of post–metamorphic northern leopard frogs (*Lithobates pipiens*). M.S. thesis. Concordia University, Montreal.

Paetow, L., D.K. Cone, T. Huyse, J.D. McLaughlin, and D.J. Marcogliese. 2009. Morphology and molecular taxonomy of *Gyrodactylus jennyae* n. sp. (Monogenea) from tadpoles of captive *Rana catesbeiana* Shaw (Anura), with a review of the species of *Gyrodactylus* Nordmann, 1832 parasitising amphibians. Systematic Parasitology 73:219–227.

Paetow, L.J., B.D. Pauli, J.D. McLaughlin, J. Bidulka, and D.J. Marcogliese. 2011. First detection of ranavirus in *Lithobates pipiens* in Quebec. Herpetological Review 42:211–214.

Paetow, L.J., J. D. McLaughlin, R. I. Cue, B. D. Pauli, and D. J. Marcogliese. 2012. Effects of herbicides and the chytrid fungus *Batrachochytrium dendrobatidis* on the health of post–metamorphic northern leopard frogs (*Lithobates pipiens*). Ecotoxicology and Environmental Safety 80: 372–380.

Pague, C.A., and J.C. Mitchell. 1987. The status of amphibians in Virginia. Virginia Journal of Science 38:304–318.

Pague, C.A., and J.C. Mitchell. 1991. The amphibians and reptiles of Back Bay, Virginia. Pp. 159–166 *In* H.G. Marshall and M.D. Norman (eds.), Proceedings of the Back Bay Ecological Symosium, Department of Biological Sciences, Old Dominion University, Norfolk, Virginia.

Pais, R.C., S.A. Bonney, and W.C. McComb. 1988. Herpetofaunal species richness and habitat associations in an eastern Kentucky forest. 1988 Proceedings of the Annual Conference of Southeastern Fish and Wildlife Agencies:448–455.

Palen, W.J. 2006. Freshwater vertebrates: American bullfrog (*Rana catesbeiana*). Pp. 146-147 *In* P.D. Boersma, S.H. Reichard, and A.N. Van Buren (eds.), Invasive Species in the Pacific Northwest. University of Washington Press, Seattle.

Palen, W.J., and D.E. Schindler. 2010. Water clarity, maternal behavior, and physiology combine to eliminate UV radiation risk to amphibians in a montane landscape. Proceedings of the National Academy of Sciences of the United States of America 107:9701–9706.

Palen, W.J., D.E. Schindler, M.J. Adams, C.A. Pearl, R.B. Bury, and S.A. Diamond. 2002. Optical characteristics of natural waters protect amphibians from UV–B in the U.S. Pacific Northwest. Ecology 83:2951–2957.

Palis, J. 1994a. Night of the spadefoots. Florida Wildlife 48(6):12–13.

Palis, J.G. 1994b. *Rana utricularia* (Southern Leopard Frog). Road mortality. Herpetological Review 25:119.

Palis, J.G. 1995. Winter chorus. Florida Wildlife 49(Jan/Feb):12–14.

Palis, J.G. 1997. Species profile: Gopher frog (*Rana capito* spp.) on military installations in the Southeastern United States. U.S. Army Corps of Engineers Waterways Experiment Station Technical Report SERDP–97–5.

Palis, J.G. 1998. Breeding biology of the gopher frog, *Rana capito*, in western Florida. Journal of Herpetology 32:217–223.

Palis, J. 2000a. *Scaphiopus holbrookii* (Eastern Spadefoot). Predation. Herpetological Review 31:42–43.

Palis, J. 2000b. *Rana sphenocephala* (Southern Leopard Frog). Subterranean vocalization. Herpetological Review 31:42.

Palis, J.G. 2005. *Pseudacris crucifer* (Spring Peeper). Predation. Herpetological Review 36:305.

Palis, J.G. 2007. If you build it, they will come: herpetofaunal colonization of constructed wetlands and adjacent terrestrial habitat in the Cache River drainage of southern Illinois. Transactions of the Illinois State Academy of Science 100:177–189.

Palis, J.G. 2009. Frog pond, fish pond: temporal co–existence of crawfish frog tadpoles and fishes. Proceedings of the Indiana Academy of Science 118:196–199.

Palis, J.G. 2010. *Hyla avivoca* (Bird–voiced Treefrog). Distance from breeding site. Herpetological Review 41:63–64.

Palis, J.G. 2012a. *Hyla avivoca* (Bird–voiced Treefrog). Constructed wetland colonization. Herpetological Review 43:119–120.

Palis, J.G. 2012b. *Pseudacris crucifer* (Spring Peeper). Predation. Herpetological Review 43:325.

Palis, J.G., and M.J. Aresco. 2007. Immigration orientation and migration distance of four pond–breeding amphibians in northwestern Florida. Florida Scientist 70:251–263.

Palmer, T.C. 1908. A familiar talk on Delaware County frogs. Proceedings of the Delaware County Institute of Science 4:12–22. [Delaware County, Pennsylvania]

Palmer, W.M., and A.L. Braswell. 1980. Additional records of albinistic amphibians and reptiles from North Carolina. Brimleyana 3:49–52.

Palmer, W.M., and A.L. Braswell. 1995. Reptiles of North Carolina. University of North Carolina Press, Chapel Hill.

Palmeri-Miles, A. 2012. Seasonal movement patterns and overwintering of western toads (*Anaxyrus boreas*) near Snoqualmie Pass, Washington. M.S. thesis, Central Washington University, Ellensburg.

Palmeri–Miles, A.F., K.A. Douville, J.A. Tyson, K.D. Ramsdell, and M.P. Hayes. 2010. Field observations of oviposition and early development of the coastal tailed frog (*Ascaphus truei*). Northwestern Naturalist 91:206–213.

Palorski, R.A. 2008. Relationship between lakeshore development and frog populations on central Wisconsin. Pp. 77–83 *In* J.C. Mitchell, R.E. Jung Brown, and B. Bartholomew (eds.), Urban Herpetology. Herpetological Conservation 3. Society for the Study of Amphibians and Reptiles, Salt Lake City, Utah.

Panik, H.R., and S. Barrett. 1994. Distribution of amphibians and reptiles along the Truckee River system. Northwest Science 68:197–204.

Panitz, E., and J.L. Briggs. 1968. *Rana cascadae*, new definitive host of *Megalodiscus microphagus* Ingles, 1936 (Trematoda, Paramphistomatidae) in Oregon. Bulletin of the Wildlife Disease Association 4:21.

Paoletti, D.J. 2009. Responses of foothill yellow–legged frog (*Rana boylii*) larvae to an introduced predator. M.S. thesis, Oregon State University, Corvallis.

Paoletti, D.J., D.H. Olson, and A.R. Blaustein. 2011. Responses of foothill yellow–legged frog (*Rana boylii*) larvae to an introduced predator. Copeia 2011:161–168.

Parker, H.W. 1934. A Monograph of the Frogs of the Family Microhylidae. The British Museum (Natural History), London.

Parker, J.M. 2000. Habitat use and movements of the Wyoming toad, *Bufo baxteri*: a study of wild juvenile, adult, and released captive–raised toads. M.S. thesis, University of Wyoming, Laramie.

Parker, J.M., and S.H. Anderson. 2003. Habitat use and movements of repatriated Wyoming toads. Journal of Wildlife Management 67:439–446.

Parker, J., S.H. Anderson, and F.J. Lindzey. 2000. *Bufo baxteri* (Wyoming Toad). Predation. Herpetological Review 31:167–168.

Parker, M.L., and M.I. Goldstein. 2004. Diet of the Rio Grande leopard frog (*Rana berlandieri*) in Texas. Journal of Herpetology 38:127–130.

Parker, M.V. 1937. Some amphibians and reptiles from Reelfoot Lake. Journal of the Tennessee Academy of Science 12:60–86.

Parker, M.V. 1939. The amphibians and reptiles of Reelfoot Lake and vicinity, with a key for the separation of the species and subspecies. Journal of the Tennessee Academy of Science 14:72–101.

Parker, M.V. 1941. The trematode parasites from a collection of amphibians and reptiles. Journal of the Tennessee Academy of Science 16:27–45.

Parker, M.V. 1951. Notes on the bird–voiced tree frog, *Hyla phaeocrypta*. Journal of the Tennessee Academy of Science 26:208–213.

Parker, W.S. 1973. Observations on a small, isolated population of *Bufo punctatus* in Arizona. Southwestern Naturalist 18:93–114.

Parmelee, J.R., M.G. Knutson, and J.E. Lyon. 2002. A field guide to amphibian larvae and eggs of Minnesota, Wisconsin, and Iowa. USGS Information and Technology Report 2002–0004, Washington, D.C.

Parmley, D.C., and J.D. Tyler. 1978. Gray treefrog in southwestern Oklahoma. Southwestern Naturalist 23:157–158.

Parris, M.J. 1998. Terrestrial burrowing ecology of newly metamorphosed frogs (*Rana pipiens* complex). Canadian Journal of Zoology 76:2124–2129.

Parris, M.J. 2000. Experimental analysis of hybridization in leopard frogs (Anura: Ranidae): larval performance in desiccating environments. Copeia 2000:11–19.

Parris, M.J., and R.D. Semlitsch. 1998. Asymmetric competition in larval amphibian communities: conservation implications for the northern crawfish frog, *Rana areola-*

ta circulosa. Oecologia 116:219–226.

Parris, M.J., and D.R. Baud. 2004. Interactive effects of a heavy metal and chytridiomycosis on gray treefrog larvae (*Hyla chrysoscelis*). Copeia 2004:344–350.

Parris, M.J., and J.G. Beaudoin. 2004. Chytridiomycosis impacts predator–prey interactions in larval amphibian communities. Oecologia 140:626–632.

Parris, M.J., R. D. Semlitsch, and R. D. Sage. 1999. Experimental analysis of the evolutionary potential of hybridization in leopard frogs (Anura: Ranidae). Journal of Evolutionary Biology 12: 662–671.

Parris, M.J., C.W. Laird, and R.D. Semlitsch. 2001. Differential predation on experimental populations of parental and hybrid leopard frogs (*Rana blairi* and *Rana sphenocephala*) larvae. Journal of Herpetology 35:479–485.

Parry, J.E., and A.W. Grundman. 1965. Species composition and distribution of the parasites of some common amphibians of Iron and Washington counties, Utah. Proceedings of the Utah Academy of Sciences, Arts, and Letters 42:271–279.

Pask, J.D., D.C. Woodhams, and L.A. Rollins–Smith. 2012. The ebb and flow of antimicrobial skin peptides defends northern leopard frogs (*Rana pipiens*) against chytridiomycosis. Global Change Biology 18:1231–1238.

Patch, C.L. 1918. The economic value of batrachians and reptiles. Ottawa Naturalist 32:29–30.

Patch, C.L. 1922. Some amphibians and reptiles from British Columbia. Copeia (111):74–79.

Patch, C.L. 1925. The frog eats the bird. Canadian Field-Naturalist 39:150.

Patch, C.L. 1929. Some amphibians of western North America. Canadian Field-Naturalist 43:137–138.

Patch, C.L. 1939. Northern records of the wood frog. Copeia 1939:235.

Patch, C.L. 1949. Further northern records of the wood–frog. Copeia 1949:233.

Patla, D.A. 1997. Changes in a population of spotted frogs in Yellowstone National Park between 1953 and 1995: the effects of habitat modification. M.S. thesis, Idaho State University, Pocatello.

Patla, D.A., and C.R. Peterson. 1999. Are amphibians declining in Yellowstone National Park? Yellowstone Science 7:2–11.

Patla, D.A., C.R. Peterson, and P.S. Corn. 2009. Amphibian decline in Yellowstone National Park. Proceedings of the National Academy of Sciences of the United States of America 106:E22.

Paton, P.W.C., and W.B. Crouch III. 2002. Using the phenology of pond-breeding amphibians to develop conservation strategies. Conservation Biology 16:194-204.

Paton, P., S. Stevens, and L. Longo. 2000. Seasonal phenology of amphibian breeding and recruitment at a pond in Rhode Island. Northeastern Naturalist 7:255–269.

Paton, P.W.C., R.S. Egan, J.E. Osenkowski, C.J. Raithel, and R.T. Brooks. 2003. *Rana sylvatica* (Wood Frog). Breeding behavior during drought. Herpetological Review 34:236–237.

Patrick, D.A., M.L. Hunter, and A.J. Calhoun. 2006. Effects of experimental forestry treatments on a Maine amphibian community. Forest Ecology and Management 234:323–332.

Patrick, D.A., A.J.K. Calhoun, and M.L. Hunter, Jr. 2007. Orientation of juvenile wood frogs, *Rana sylvatica*, leaving experimental ponds. Journal of Herpetology 41:158–163.

Patrick, D.A., E.B. Harper, M.L. Hunter, Jr., and A.J.K. Calhoun. 2008. Terrestrial habitat selection and strong density-dependent mortality in recently metamorphosed amphibians. Ecology 89:2563-2574.

Patrick, D.A., C.M. Schalk, J.P. Gibbs, and H.W. Woltz. 2010. Effective culvert placement and design facilitate passage of amphibians across roads. Journal of Herpetology 44:618–626.

Patrick, D.A., E.B. Harper, V.D. Popescu, Z. Bozic, A. Byrne, J. Daub, A. Lecheminant and J. Pierce. 2012a. The ecology of the mink frog, *Lithobates septentrionalis*, in the Adirondack Park, New York, with notes on conducting experimental research. Herpetological Review 43:396–398.

Patrick, D.A., J.P. Gibbs, V.D. Popescu, and D.A. Nelson. 2012b. Multi–scale habitat–resistance models for predicting road mortality "hotspots" for turtles and amphibians. Herpetological Conservation and Biology 7:407–426.

Patten, M.A., and S.J. Myers. 1992. Geographic distribution: *Bufo microscaphus californicus*. Herpetological Review 23:124.

Pauken, R.J., and D.E. Metter. 1971. Geographic representation of morphologic variation among populations of *Ascaphus truei* Stejneger. Systematic Zoology 20:434–441.

Paukstis, G.L., and L.E. Brown. 1987. Evolution of the intercalary cartilage in chorus frogs, genus *Pseudacris* (Salientia: Hylidae). Brimleyana 13:55–61.

Pauley, B.A., and T.K. Pauley. 2007. Survey of abandoned coal mines for amphibians and reptiles in New River Gorge National River, West Virginia. Proceedings of the West Virginia Academy of Science 79:22–30.

Pauley, T.K. 1980. Field notes on the distribution of terrestrial amphibians and reptiles of the West Virginia mountains above 970 meters. Proceedings of the West Virginia Academy of Science 52:84–92.

Pauley, T.K. 2000. Amphibians and reptiles in wetland habitats of West Virginia. Proceedings of the West Virginia

Academy of Science 72:78–88.

Pauley, T.K. 2001. Toads and frogs of West Virginia. West Virginia Division of Natural Resources, Elkins, West Virginia.

Pauley, T.K., J.C. Mitchell, R.R. Buech, and J.J. Moriarty. 2000. Ecology and management of riparian habitats for amphibians and reptiles. Pp. 169–192 *In* E.S. Verry, J.W. Hornbeck, and C.A. Dolloff (eds.), Riparian Management in Forests of the Continental Eastern United States. Lewis Publishers, Boca Raton, Florida.

Paulmier, F.C. 1902. Catalogue of New York reptiles and batrachians. Lizards, tortoises and batrachians of New York. New York State Museum Bulletin 51:389–409.

Paulson, B.K., and V.H. Hutchison. 1987. Origin of the stimulus for muscular spasms at the critical thermal maximum in anurans. Copeia 1987:810–813.

Pauly, G.B., D.M. Hillis, and D.C. Cannatella. 2004. The history of a Nearctic colonization: molecular phylogenetics and biogeography of the Nearctic toads (*Bufo*). Evolution 58:2517–2535.

Pauly, G.B., S.R. Ron, and L. Lerum. 2008. Molecular and ecological characterization of extralimital populations of red–legged frogs from western North America. Journal of Herpetology 42:669–679.

Peacock, M.M., K.H. Beard, E.M. O'Neill, V.S. Kirchoff, and M.B. Peters. 2009. Strong founder effects and low genetic diversity in introduced populatons of coqui frogs. Molecular Ecology 18:3603-3615.

Peacor, S.D. 2001. The role of phenotypic plasticity and indirect interactions on the structure of ecological communities. Ph.D. Dissertation, University of Michigan, Ann Arbor.

Peacor, S.D., and E.E. Werner. 2000. Predator effects on an assemblage of consumers through induced changes in consumer foraging behavior. Ecology 81:1998–2010.

Peacor, S.D., and E.E. Werner. 2001. The contribution of trait–mediated indirect effects to the net effects of a predator. Proceedings of the National Academy of Sciences of the United States of America 98: 3904–3908.

Pearl, C.A. 2000. *Bufo boreas* (Western Toad). Predation. Herpetological Review 31:233–234.

Pearl, C.A., and M.P. Hayes. 2002. Predation by Oregon spotted frogs (*Rana pretiosa*) on western toads (*Bufo boreas*) in Oregon. American Midland Naturalist 147:145–152.

Pearl, C.A., and D.E. Green. 2005. *Rana catesbeiana* (American Bullfrog). Chytridiomycosis. Herpetological Review 36:305–306.

Pearl, C.A., and J. Bowerman. 2006. Observations of rapid colonization of constructed ponds by western toads (*Bufo boreas*) in Oregon, USA. Western North American Naturalist 66:397–401.

Pearl, C.A., D.J. Major, and R.B. Bury. 2002. *Ascaphus truei* (Tailed Frog). Albinism. Herpetological Review 33:123.

Pearl, C.A., M.J. Adams, G.S. Schuytema, and A.V. Nebeker. 2003. Behavioral responses of anuran larvae to chemical cues of native and introduced predators in the Pacific Northwestern United States. Journal of Herpetology 37:572–576.

Pearl, C.A., M.J. Adams, R.B. Bury, and B. McCreary. 2004. Asymmetrical effects of introduced bullfrogs (*Rana catesbeiana*) on native ranid frogs in Oregon. Copeia 2004:11–20.

Pearl, C.A., M.P. Hayes, R. Haycock, J.D. Engler, and J. Bowerman. 2005a. Observations of interspecific amplexus between western North American ranid frogs and the introduced American Bullfrog (*Rana catesbeiana*) and an hypothesis concerning breeding interference. American Midland Naturalist 154:126–134.

Pearl, C.A., M.J. Adams, N. Leuthold, and R.B. Bury. 2005b. Amphibian occurrence and aquatic invaders in a changing landscape: implications for wetland mitigation in the Willamette Valley, Oregon, USA. Wetlands 25:76–88.

Pearl, C.A., J. Bowerman, and D. Knight. 2005c. Feeding behavior and aquatic habitat use by Oregon spotted frogs (*Rana pretiosa*) in central Oregon. Northwestern Naturalist 86:36–38.

Pearl, C.A., E.L. Bull, D.E. Green, J. Bowerman, M.J. Adams, A. Hyatt, and W.H. Wendt. 2007. Occurrence of the amphibian pathogen *Batrachochytrium dendrobatidis* in the Pacific Northwest. Journal of Herpetology 41:145–149.

Pearl, C.A., M.J. Adams, R.B. Bury, W.H. Wente, and B. McCreary. 2009a. Evaluating amphibian declines with site revisits and occupancy models: status of montane anurans in the Pacific Northwest USA. Diversity 1:166–181.

Pearl, C.A., M.J. Adams, and N. Leuthold. 2009b. Breeding habitat and local population size of the Oregon spotted frog (*Rana pretiosa*) in Oregon, USA. Northwestern Naturalist 90:136–147.

Pearl, C.A., J. Bowerman, M.J. Adams, and N.D. Chelgren. 2009c. Widespread occurrence of the chytrid fungus *Batrachochytrium dendrobatidis* on Oregon spotted frogs (*Rana pretiosa*). EcoHealth 6:209–218.

Pearman, P.B. 1991. The ecology of patchy habitats: effects of pond size on experimental tadpole populations. Ph.D. Dissertation, Duke Univesity, Durham, North Carolina.

Pearman, P.B. 1993. Effects of habitat size on tadpole populations. Ecology 74:1982–1991.

Pearman, P.B. 1995. Effects of pond size and consequent predator density on two species of tadpoles. Oecologia 102:1–8.

Pearse, A.S. 1910. The reactions of amphibians to light. Proceedings of the American Academy of Arts and Sciences 45:159–208.

Pearse, A.S. 1936. Estuarine animals at Beaufort, North Carolina. Journal of the Elisha Mitchell Scientific Society 52:174–222.

Pearson, H.A., R.R. Lohoefener and J.L. Wolfe. 1987. Amphibians and reptiles on longleaf–slash pine forests in Mississippi. Pp. 157–165 *In* Ecological, Physical, and Socioeconomic Relationships within Southern National Forests, Proceedings of the Southern Evaluation Project Workshop. USDA Forest Service, Southern Forest Experiment Station, New Orleans, Louisiana.

Pearson, K.J. 2009. Status of the Great Plains Toad (*Bufo* [*Anaxyrus*] *cognatus*) in Alberta: update 2009. Alberta Sustainable Resource Development, Wildlife Status Report No. 14 (update 2009), Edmonton.

Pearson, P.G. 1955. Population ecology of the spadefoot toad, *Scaphiopus h. holbrooki* (Harlan). Ecological Monographs 25:233–267.

Pearson, P.G. 1957. Further notes on the population ecology of the spadefoot toad. Ecology 38:580–586.

Pearson, P.G. 1958. Body measurements of *Scaphiopus holbrooki*. Copeia 1958:215–217.

Pearson, P.G. 1960. A description of a six–legged bullfrog, *Rana catesbeiana*. Copeia 1960:50–51.

Pease, K.M. 2011. Rapid evolution of anti–predator defenses in Pacific tree frog tadpoles exposed to invasive predatory crayfish. Ph.D. Dissertation, University of California, Los Angeles.

Pease, K.M., and R. K. Wayne. 2011. The response of Pacific tree frog (*Pseudacris regilla*) tadpoles to invasive predatory crayfish: a comparison of anti–predator traits between naive and exposed tadpoles. Integrative and Comparative Biology 51:E107–E107.

Pechmann, J.H.K., and R.D. Semlitsch. 1986. Diel activity patterns in the breeding migrations of winter–breeding anurans. Canadian Journal of Zoology 64:1116–1120.

Pechmann, J.H.K., D.E. Scott, J.W. Gibbons, and R.D. Semlitsch. 1989. Influence of wetland hydroperiod on diversity and abundance of metamorphosing juvenile amphibians. Wetlands Ecology and Management 1:3–11.

Pechmann, J.H.K., D.E. Scott, R.D. Semlitsch, J.P. Caldwell, L.J. Vitt, and J.W. Gibbons. 1991. Declining amphibian populations: the problem of separating human impacts from natural fluctuations. Science 253:892–895.

Pechmann, J.H.K., R.A. Estes, D.E. Scott, and J.W. Gibbons. 2001. Amphibian colonization and use of ponds created for trial mitigation of wetland loss. Wetlands 21:93–111.

Pehek, E.L. 1995. Competition, pH, and the ecology of larval *Hyla andersonii*. Ecology 76:1786–1793.

Pelgen, J.L. 1951. A *Rana catesbeiana* with six functional legs. Herpetologica 7:138–139.

Pemberton, C.E. 1933. Introductions to Hawaii of the tropical American toad *Bufo marinus*. Hawaiian Planters' Record 37:15–16.

Pemberton, C.E. 1934. Local investigations on the introduced tropical American toad *Bufo marinus*. Hawaiian Planters' Record 38:186–192.

Pemberton, C.E. 1949. Longevity of the tropical American toad, *Bufo marinus*. Science 110:512.

Penn, G.H. 1950. Utilization of crawfishes by cold–blooded vertebrates in the eastern United States. American Midland Naturalist 44:643–658.

Pennant, T. 1787. Arctic Zoology. Supplement, Robert Faulder, London. [frogs on pp. 80-82]

Penney, J.T. 1952. Distribution and bibliography of the amphibians and reptiles of South Carolina. University of South Carolina Publications, Biology. Series 3. 1(1):3-28.

Perkins, D.W. 2004. The effects of riparian timber management on amphibians in western Maine. Ph.D. Dissertation, University of Maine, Orono.

Perkins, D.W., and M.L. Hunter, Jr. 2006a. Effects of riparian timber management on amphibians in Maine. Journal of Wildlife Management 70:657–670.

Perkins, D.W., and M.L. Hunter, Jr. 2006b. Use of amphibians to define riparian zones of headwater streams. Canadian Journal of Forest Research 36:2124–2130.

Perrill, S.A. 1984. Male mating behavior in *Hyla regilla*. Copeia 1984:727–732.

Perrill, S.A., and R.E. Daniel. 1983. Multiple egg clutches in *Hyla regilla*, *H. cinerea*, and *H. gratiosa*. Copeia 1983:513–516.

Perrill, S.A., and M. Magier. 1988. Male mating behavior in *Acris crepitans*. Copeia 1988:245–248.

Perrill, S.A., and W.J. Shepherd. 1989. Spatial distribution and male–male communication in the northern cricket frog, *Acris crepitans blanchardi*. Journal of Herpetology 23:237–243.

Perrill, S.A., and L.C. Lower. 1994. Advertisement call discrimination by female cricket frogs (*Acris crepitans*). Journal of Herpetology 28:399–400.

Perrill, S.A., H.C. Gerhardt, and R. Daniel. 1978. Sexual parasitism in the green tree frog (*Hyla cinerea*). Science 200:1179–1180.

Perrill, S.A., H.C. Gerhardt, and R. Daniel. 1982. Mating strategy shifts in male green treefrogs (*Hyla cinerea*): an experimental study. Animal Behaviour 30:43–48.

Perry, R.W., D.C. Rudolph, and R.E. Thill. 2012. Effects of short–rotation controlled burning on amphibians and reptiles in pine woodlands. Forest Ecology and Management 271:124–131.

Peters, B., and F.S. Ruth. 1962. Albino bullfrog (*Rana catesbeiana*) population found in San Leandro Creek, Alameda County, California. Turtox News 40(1):7.

Peterson, C.R. 1974. A preliminary report on the amphibians and reptiles of the Black Hills of South Dakota and Wyoming. M.S. thesis, University of Illinois, Urbana–Champaign.

Peterson, H.W., R. Garrett, and J.P. Lantz. 1952. The mating period of the giant tree frog *Hyla dominicensis*. Herpetologica 8:63.

Peterson, J.A., and A.R. Blaustein. 1991. Unpalatability in anuran larvae as a defense against natural salamander predators. Ethology Ecology & Evolution 3:63–72.

Peterson, J.A., and A.R. Blaustein. 1992. Relative palatabilities of anuran larvae to natural aquatic insect predators. Copeia 1992:577–584.

Peterson, J.D. 2012. Physiological effects of chytridiomycosis, a cause of amphibian population declines. Ph.D. Dissertation, Auburn University, Auburn, Alabama.

Peterson, J.D., M.B. Wood, W.A. Hopkins, J.M. Unrine, and M.T. Mendonça. 2007. Prevalence of *Batrachochytrium dendrobatidis* in American bullfrog and southern leopard frog larvae from wetlands on the Savannah River Site, South Carolina. Journal of Wildlife Diseases 43:450–460.

Peterson, J.D., V.A. Peterson, and M.T. Mendonça. 2008. Growth and developmental effects of coal combustion residues on southern leopard frog (*Rana sphenocephala*) tadpoles exposed throughout metamorphosis. Copeia 2008:499–503.

Petranka, J.W. 1989a. Chemical interference competition in tadpoles: does it occur outside laboratory aquaria? Copeia 1989:921–930.

Petranka, J.W. 1989b. Response of toad tadpoles to conflicting chemical stimuli: predator avoidance versus "optimal" foraging. Herpetologica 45:283–292.

Petranka, J.W., and D.A.G. Thomas. 1995. Explosive breeding reduces egg and tadpole cannibalism in the wood frog, *Rana sylvatica*. Animal Behaviour 50:731–739.

Petranka, J.W., and L. Hayes. 1998. Chemically mediated avoidance of a predatory odonate (*Anax junius*) by American toad (*Bufo americanus*) and wood frog (*Rana sylvatica*) tadpoles. Behavioral Ecology and Sociobiology 42:263–271.

Petranka, J.W., and C.A. Kennedy. 1999. Pond tadpoles with generalized morphology: is is time to reconsider their functional roles in aquatic communities? Oecologia 120:621–631.

Petranka, J.W., L. B. Kats, and A. Sih. 1987. Predator–prey interactions among fish and larval amphibians: use of chemical cues to detect predatory fish. Animal Behaviour 35:420–425.

Petranka, J.W., M.E. Hopey, B.T. Jennings, S.D. Baird, and S.J. Boone. 1994. Breeding habitat segregation of wood frogs and American toads: the role of interspecific tadpole predation and adult choice. Copeia 1994:691–697.

Petranka, J.W., A.W. Rushlow, and M.E. Hopey. 1998. Predation by tadpoles of *Rana sylvatica* on embryos of *Ambystoma maculatum*: implications of ecological role reversals by *Rana* (predator) and *Ambystoma* (prey). Herpetologica 54:1–13.

Petranka, J.W., S.S. Murray, and C.A. Kennedy. 2003. Responses of amphibians to restoration of a Southern Appalachian wetland: perturbations confound post–restoration assessment. Wetlands 23:278–290.

Petranka, J.W., C.K. Smith, and A.F. Scott. 2004. Identifying the minimal demographic unit for monitoring pond–breeding amphibians. Ecological Applications 14:1065–1078.

Petrides, G.A. 1944. Life span of the spadefoot toad. Copeia 1944:122–123.

Petrovic, C.A. 1973. An "albino" frog, *Eleutherodactylus planirostris* Cope. Journal of Herpetology 7:49–51.

Pettus, D. 1955. Notes on the breeding behavior of the common treefrog (*Hyla versicolor*). Texas Journal of Science 7:345–346.

Pettus, D., and A.W. Spencer. 1964. Size and metabolic differences in *Pseudacris triseriata* (Anura) from different elevations. Southwestern Naturalist 9:20–26.

Pettus, D., and G.M. Angleton. 1967. Comparative reproductive biology of montane and Piedmont chorus frogs. Evolution 21:500–507.

Petzing, J.E., and C.A. Phillips. 1999. *Rana sphenocephala* (Southern Leopard Frog). Reproduction (fall breeding). Herpetological Review 30:93–94.

Pfeiffer, E.E. 1949. The toad: the gardener's assistant. Organic Farming Digest 1(12):18–23.

Pfennig, D.W. 1989. Evolution, development, and behavior of alternative amphibian morphologies. Ph.D. Dissertation, University of Texas, Austin.

Pfennig, D. 1990a. The adaptive significance of an environmentally–cued developmental swith in an anuran tadpole. Oecologia 85:101–107.

Pfennig, D.W. 1990b. "Kin recognition" among spadefoot toad tadpoles: a side–effect of habitat selection? Evolution 44:785–798.

Pfennig, D.W. 1992a. Polyphenism in spadefoot toad tadpoles as a locally adjusted evolutionarily stable strategy. Evolution 46:1408–1420.

Pfennig, D.W. 1992b. Proximate and functional causes of polyphenism in an an anuran tadpole. Functional Ecology 6:167–174.

Pfennig, D.W. 1999. Cannibalistic tadpoles that pose the greatest threat to kin are most likely to discriminate kin. Proceedings of the Royal Society B 266:57–61.

Pfennig, D.W., and P.J. Murphy. 2000. Character displacement in polyphenic tadpoles. Evolution 54:1738–1749.

Pfennig, D.W., and P.J. Murphy. 2002. How fluctuating competition and phenotypic plasticity mediate species divergence. Evolution 56:1217–1228.

Pfennig, D.W., and P.J. Murphy. 2003. A test of alternative hypotheses for character divergence between coexisting species. Ecology 84:1288–1297.

Pfennig, D.W., and R.A. Martin. 2009. A maternal effect mediates rapid population divergence and character displacement in spadefoot toads. Evolution 63:898–909.

Pfennig, D.W., A. Mabry, and D. Orange. 1991. Environmental causes of correlations between age and size at metamorphosis in *Scaphiopus multiplicatus*. Ecology 72:2240–2248.

Pfennig, K. S. 1998. The evolution of mate choice and the potential for conflict between species and mate–quality recognition. Proceedings of the Royal Society B 265:1743–1748.

Pfennig, K.S. 2000. Female spadefoot toads compromise on mate quality to ensure conspecific matings. Behavioral Ecology 11:220–227.

Pfennig, K.S. 2003. A test of alternative hypotheses for the evolution of reproductive isolation between spadefoot toads: support for the reinforcement hypothesis. Evolution 57:2842–2851.

Pfennig, K.S., and M.A. Simovich. 2002. Differential selection to avoid hybridization in two toad species. Evolution 56:1840–1848.

Pfennig, K.S., K. Rapa, and R. McNutt. 2000. Evolution of male mating behavior: male spadefoot toads preferentially associate with conspecific males. Behavioral Ecology and Sociobiology 48:69–74.

Pfingsten, R.A. 1998. Distribution of Ohio amphibians. Pp. 220–255 *In* M.J. Lannoo (ed.), Status & Distribution of Midwestern Amphibians. University of Iowa Press, Iowa City.

Pham, L., S. Boudreaux, S. Karhbet, B. Price, A.S. Ackleh, J. Carter, and N. Pal. 2007. Population estimates of *Hyla cinerea* (Schneider) (Green Tree Frog) in an urban environment. Southeastern Naturalist 6:203–216.

Phelps, J.P. 1993. The effect of clearcutting on herpetofauna of a South Carolina blackwater bottomland. M.S. thesis, North Carolina State University, Raleigh.

Phelps, J.P., and R. A. Lancia. 1995. Effects of a clearcut on the herpetofauna of a South Carolina bottomland swamp. Brimleyana 22:31–45.

Phillips, C.A., R.A. Brandon, and E.O. Moll. 1999. Field Guide to the Amphibians and Reptiles of Illinois. Illinois Natural History Survey, Champaign, Illinois.

Phillips, K.A., and G.C. Boone. 1976. Occurrence of yellow bullfrogs (*Rana catesbeiana* Shaw) in central Pennsylvania. Proceedings of the Pennsylvania Academy of Science 50:163–164.

Phillips, K.M. 1995. *Rana capito capito*, the Carolina gopher frog, in southeast Georgia: reproduction, early growth, adult movement patterns, and tadpole fright response. M.S. thesis, Georgia Southern University, Statesboro.

Phillips, P.C. 1998. Genetic constraints at the metamorphic boundary: morphological development in the wood frog, *Rana sylvatica*. Journal of Evolutionary Biology 11:453–463.

Phillips, P.C., and M.J. Wade. 1990. *Rana sylvatica* (Wood Frog). Reproductive mortality. Herpetological Review 21:59.

Phillipsen, I.C. 2010. Population genetics of ranid frogs: investigating effective population size and gene flow. Ph.D. Dissertation, Oregon State University, Corvallis.

Phillipsen, I.C., and A.E. Metcalf. 2009. Phylogeography of a stream–dwelling frog (*Pseudacris cadaverina*) in southern California. Molecular Phylogenetics and Evolution 53:152–170.

Phillipsen, I.C., J. Bowerman, and M.Blouin. 2010. Effective number of breeding adults in Oregon spotted frogs (*Rana pretiosa*): genetic estimates at two life stages. Conservation Genetics 11:737–745.

Phillipsen, I.C., W.C. Funk, E.A. Hoffman, K.J. Monsen, and M.S. Blouin. 2011. Comparative analyses of effective population size within and among species: ranid frogs as a case study. Evolution 65:2927–2945.

Piacenza, T. 2008. Population densities of the Cuban treefrog, *Osteopilus septentrionalis* and three native species of *Hyla* (Hylidae), in urban and natural habitats of southwest Florida. M.S. thesis, University of South Florida, Tampa.

Piatt, J. 1941. Observations on the breeding habits of *Bufo americanus*. Copeia 1941:264.

Pickens, A.L. 1927a. Intermediate between *Bufo fowleri* and *B. americanus*. Copeia (162):25–26.

Pickens, A.L. 1927b. Amphibians of the upper South Carolina. Copeia (165):106–110.

Pickwell, G. (ed.). 1930. Frogs Toads Salamanders. Western Nature Study 1:1–55.

Pickwell, G. 1947. Amphibians & Reptiles of the Pacific States. Stanford University Press, Stanford, California. [reissued by Dover Publications, 1972]

Pierce, B.A. 1993. The effects of acid precipitation on amphibians. Ecotoxicology 2:65–77.

Pierce, B.A., and N. Sikand. 1985. Intrapopulation variation in acid tolerance of Connecticut wood frogs: genetic and maternal effects. Canadian Journal of Zoology 63:1647–1651.

Pierce, B.A., and J.M. Harvey. 1987. Geographic variation in acid tolerance of wood frogs. Copeia 1987:94–103.

Pierce, B.A., and J. Montgomery. 1989. Effects of short–term acidification on growth rates of tadpoles. Journal of Herpetology 23:97–102.

Pierce, B.A., and D.K. Wooten. 1992. Genetic variation in tolerance of amphibians to low pH. Journal of Herpetology 26:422–429.

Pierce, B.A., and K.J. Gutzwiller. 2004. Auditory sampling of frogs: detection efficiency in relation to survey duration. Journal of Herpetology 38:495–500.

Pierce, B.A., J.B. Hoskins, and E. Epstein. 1984. Acid tolerance in Connecticut wood frogs (*Rana sylvatica*). Journal of Herpetology 18:159–167.

Pierce, B.A., M.A. Margolis, and L.J. Nirtaut. 1987. The relationship between egg size and acid tolerance in *Rana sylvatica*. Journal of Herpetology 21:178–184.

Pierce, J.R. 1975. Genetic compatibility of *Hyla arenicolor* with other species in the family, Hylidae. Texas Journal of Science 26:431–441.

Pierce, J.R. 1976. Distribution of two mating call types of the Plains spadefoot, *Scaphiopus bombifrons*, in southwestern United States. Southwestern Naturalist 20:578–582.

Pierce, J.R., and D.B. Ralin. 1972. Vocalizations and behavior of the males of three species in the *Hyla versicolor* complex. Herpetologica 28:329–337.

Piersol, W.H. 1913. Amphibia. Pp. 242–248 *In* J.H. Faull (ed.), The Natural History of the Toronto Region, Ontario, Canada. The Canadian Institute, Toronto.

Pieterson, C.B., L.M. Addison, J.N. Agobian, B. Brooks–Solveson, J. Cassani, and E.M. Everham III. 2006. Five years of the Southwest Florida Frog Monitoring Network: changes in frog communities as an indicator of landscape change. Florida Scientist 69 (Supplement 2):117–126.

Piha, H., M. Luoto, and J. Merilä. 2007. Amphibian occurrence is influenced by current and historic landscape characteristics. Ecological Applications 17:2298–2309.

Pike, N. 1886. Notes on the hermit spadefoot (*Scaphiopus holbrooki* Harlan; *S. solitarius* Holbr.). Bulletin of the American Museum of Natural History 1(14):213–221.

Pilliod, D.S. 1999. *Rana luteiventris* (Columbia Spotted Frog). Cannibalism. Herpetological Review 30:93.

Pilliod, D.S. 2002. Clark's Nutcracker (*Nucifraga columbiana*) predation on tadpoles of the Columbia spotted frog (*Rana luteiventris*). Northwestern Naturalist 83:59–61.

Pilliod, D.S., and C.R. Peterson. 2000. Evaluating effects of fish stocking on amphibian populations in wilderness lakes. Pp. 328–335 *In* Wilderness Science in a Time of Change Conference. Vol. 5. Wilderness Ecosystems, Threats, and Management. RMRS–P–15. http://www.fs.fed.us/rm/pubs/rmrs_p015_5/rmrs_p015_5_328_335.pdf

Pilliod, D.S., and C.R. Peterson. 2001. Local and landscape effects of introduced trout on amphibians in historically fishless watersheds. Ecosystems 4:322–333.

Pilliod, D.S., and E. Wind (eds.). 2008. Habitat Management Guidelines for Amphibians and Reptiles of the Northwestern United States and Western Canada. Partners in Amphibian and Reptile Conservation, Technical Publication HMG–4, Montgomery, Alabama.

Pilliod, D.S., C.R. Peterson, and P.I. Ritson. 2002. Seasonal migration of Columbia spotted frogs (*Rana luteiventris*) among complementary resources in a high mountain basin. Canadian Journal of Zoology 80:1849–1862.

Pilliod, D.S., R.B. Bury, E.J. Hyde, C.A. Pearl, and P.S. Corn. 2003. Fire and amphibians in North America. Forest Ecology and Management 178:163–181.

Pilliod, D.S., E. Muths, R.D. Scherer, P.E. Bartelt, P.S. Corn, B.R. Hossack, B.A. Lambert, R. McCaffery, and C. Gaughan. 2010. Effects of amphibian chytrid fungus on individual survival probability in wild boreal toads. Conservation Biology 24:1259–1267.

Pimentel, R.A. 1955. Habitat distribution and movements of *Bufo b. boreas*, Baird and Girard. Herpetologica 11:72.

Pinder, A., and S. Friet. 1994. Oxygen transport in egg masses of the amphibians *Rana sylvatica* and *Ambystoma maculatum*: convection, diffusion and oxygen production by algae. Journal of Experimental Biology 197:17–30.

Pinder, R. 2010. *Rana clamitans* (Green Frog). Coloration. Herpetological Review 41:66.

Pine, R.H. 1975. Star–nosed mole eaten by bull frog. Mammalia 39:713–714.

Piovia–Scott, J., K.L. Pope, S.P. Lawler, E.M. Cole, and J.E. Foley. 2011. Factors related to the distribution and prevalence of the fungal pathogen *Batrachochytrium dendrobatidis* in *Rana cascadae* and other amphibians in the Klamath Mountains. Biological Conservation 144:2913–2921.

Pittman, S.E., and M.E. Dorcas. 2006. Catawba River Corridor coverboard program: a citizen science approach to amphibian and reptile inventory. Journal of the North Carolina Academy of Science 122:142–151.

Pittman, S.E., A.L. Jendrek, S.J. Price and M.E. Dorcas. 2008. Habitat selection and site fidelity of Cope's gray treefrog (*Hyla chrysoscelis*) at the aquatic–terrestrial ecotone. Journal of Herpetology 42:378–385.

Pitt, A.L., C.M. Allard, J.E. Hawley, R.F. Baldwin, and B.L. Brown. 2011. *Lithobates sylvaticus*. (Wood Frog). Predation. Herpetological Review 42:263.

Placyk, J.S., Jr., M.J. Seider, and J.C. Gillingham. 2002. New herpetological records for High and Hog Islands of the Beaver Archipelago, Charlevoix County, Michigan. Herpetological Review 33:230.

Platin, T.J., and K.O. Richter. 1995. Amphibians as bioindicators of stress associated with watershed urbanization. Pp. 163–170 *In* Puget Sound Research 95 Proceedings, Vol. 1. Puget Sound Water Quality Authority, Olympia, Washington.

Platt, D.R. (Chairman of Conservation Committee). 1973. Rare, endangered and extirpated species in Kansas. 2. Amphibians and reptiles. Transactions of the Kansas Academy of Science 76:185–192.

Platt, S.G., K.R. Russell, W.E. Snyder, L.W. Fontenot, and S. Miller. 1999. Distribution and conservation status of selected amphibians and reptiles in the Piedmont of South Carolina. Journal of the Elisha Mitchell Scientific Society 115:8–19.

Platz, J.E. 1972. Sympatric interaction between two forms of leopard frog (*Rana pipiens* complex) in Texas. Copeia 1972:232–240.

Platz, J.E. 1976. Biochemical and morphological variation of leopard frogs in Arizona. Copeia 1976:660–672.

Platz, J.E. 1981. Suture zone dynamics: Texas populations of *Rana berlandieri* and *R. blairi*. Copeia 1981:733–734.

Platz, J.E. 1988a. Geographic variation in mating call among the four subspecies of the chorus frog: *Pseudacris triseriata* (Wied). Copeia 1988:1062–1066.

Platz, J.E. 1989. Speciation within the chorus frog *Pseudacris triseriata*: morphometric and mating call analyses of the boreal and western subspecies. Copeia 1989:704–712.

Platz, J.E. 1993. *Rana subaquavocalis*, a remarkable new species of leopard frog (*Rana pipiens* complex) from southeastern Arizona that calls under water. Journal of Herpetology 27: 154–162.

Platz, J.E., and A.L. Platz. 1973. *Rana pipiens* complex: hemoglobin phenotypes of sympatric and allopatric populations in Arizona. Science 179:1334–1336.

Platz, J.E., and J.S. Mecham. 1979. *Rana chiricahuensis*, a new species of leopard frog (*Rana pipiens* complex) from Arizona. Copeia 1979:383–390.

Platz, J.E., and J.S. Frost. 1984. *Rana yavapaiensis*, a new species of leopard frog (*Rana pipiens* complex). Copeia 1984:940–948.

Platz, J.E., and A. Lathrop. 1993. Body size and age assessment among advertising male chorus frogs. Journal of Herpetology 27:109–111.

Platz, J.E., and T.A. Grudzien. 2003. Limited genetic heterozygosity and status of two populations of the Ramsey Canyon leopard frog: *Rana subaquavocalis*. Journal of Herpetology 37:758–761.

Platz, J.E., R.W. Clarkson, J.C. Rorabaugh, and D.M. Hil-lis. 1990. *Rana berlandieri*: recently introduced populations in Arizona and southeastern California. Copeia 1990:324–333.

Platz, J.E., A. Lathrop, L. Hofbauer, and M. Vradenburg. 1997. Age distribution and longevity in the Ramsey Canyon leopard frog, *Rana subaquavocalis*. Journal of Herpetology 31:552–557.

Plotkin, M., and R. Atkinson. 1979. Geographic distribution. *Eleutherodactylus planirostris*. Herpetological Review 10:59.

Poile, M.E. 1982. The effects of genetic similarity and crowding on intraspecific competition in tadpoles. M.A. thesis, University of South Dakota, Vermillion.

Polis, G.A., and C.A. Myers. 1985. A survey of intraspecific predation among reptiles and amphibians. Journal of Herpetology 19:99–107.

Pollard, G.M., C.J. Biggers, and M.J. Harvey. 1973. Electrophoretic differentiation of the parotoid venoms of *Bufo americanus* and *Bufo woodhousei fowleri*. Herpetologica 29:251–253.

Pollet, I., and L.I. Bendell–Young. 2000. Amphibians as indicators of wetland qualiy in wetlands formed from oil sands effluent. Environmental Toxicology and Chemistry 19:2589–2597.

Pollio, C.A., and S.L. Kilpatrick. 2002. Status of *Pseudacris feriarum* in Prince William Forest Park, Prince William County, Virginia. Bulletin of the Maryland Herpetological Society 38:55–61.

Pollister, A.W., and J.A. Moore. 1937. Tables for the normal development of *Rana sylvatica*. Anatomical Record 68:489–496.

Pomeroy, L.V. 1981. Developmental polymorphism in the tadpoles of the spadefoot toad *Scaphiopus multiplicatus*. Ph.D. Dissertation, University of California, Riverside.

Pope, C.H. 1964. Amphibians and Reptiles of the Chicago Area. Chicago Natural History Museum, Chicago.

Pope, K.L. 1999. Mountain Yellow–legged Frog habitat use and movement patterns in a high elevation basin in Kings Canyon National Park. M.S. thesis, California Polytechnic State University, San Luis Obispo.

Pope, K.L. 1999b. *Rana muscosa* (Mountain Yellow–legged Frog). Diet. Herpetological Review 30:163–164.

Pope, K.L. 2008. Assessing changes in amphibian population dynamics following experimental manipulations of introduced fish. Conservation Biology 22:1572–1581.

Pope, K.L., and K.R. Matthews. 2001. Movement ecology and seasonal distribution of Mountain Yellow–Legged Frogs, *Rana muscosa*, in a high–elevation Sierra Nevada Basin. Copeia 2001:787–793.

Pope, K.L., and K.R. Matthews. 2002. Influence of anuran prey on the condition and distribution of *Rana muscosa* in the Sierra Nevada. Herpetologica 58:354–363.

Pope, K.L., J.M. Garwood, H.H. Welsh, Jr., and S.P. Lawler. 2008. Evidence of indirect impacts of introduced trout on native amphibians via facilitation of a shared predator. Biological Conservation 141:1321–1331.

Pope, P.H. 1915. The distribution of the northern frog, *Rana septentrionalis*, Baird, in Maine. Copeia (16):1–2.

Pope, P.H. 1919a. A note on the development of *Pseudacris feriarum* (Baird). Copeia (74):83–84.

Pope, P.H. 1919b. Some notes on the amphibians of Houston, Texas. Copeia (76):93–98.

Pope, S.E., L. Fahrig, and H.G. Merriam. 2000. Landscape complementation and metapopulation effects on leopard frog populations. Ecology 81:2498–2508.

Pope, T.E.B., and W.E. Dickinson. 1928. The amphibians and reptiles of Wisconsin. Bulletin of the Public Museum of the City of Milwaukee 8:1–138.

Popescu, V.D., and J.P. Gibbs. 2009. Interactions between climate, beaver activity, and pond occupancy by the cold–adapted mink frog in New York State, USA. Biological Conservation 142:2059–2068.

Popescu, V.D., and M.L. Hunter, Jr. 2011. Clear-cutting affects habitat connectivity for a forest amphibian by decreasing permeability to juvenile movements. Ecological Applications 21:1283-1295.

Popescu, V.D., B.S. Brodie, M.L. Hunter, and J.D. Zydlewski. 2012. Use of olfactory cues by newly metamorphosed wood frogs (*Lithobates sylvaticus*) during emigration. Copeia 2012:424–431.

Porej, D., and T.E. Hetherington. 2005. Designing wetlands for amphibians: the importance of predatory fish and shallow littoral zones in structuring of amphibian communities. Wetlands Ecology and Management 13:445–455.

Porej, D., M. Micacchion, and T.E. Hetherington. 2004. Core terrestrial habitat for conservation of local populations of salamanders and wood frogs in agricultural landscapes. Biological Conservation 120:399–409.

Porter, K.R. 1941. Diploid and androgenetic haploid hybridization between two forms of *Rana pipiens*, Schreber. Biological Bulletin 80:238-264.

Porter, K.R. 1961. Experimental crosses between *Rana aurora aurora* Baird and Girard and *Rana cascadae* Slater. Herpetologica 17:156–165.

Porter, K.R. 1964. Morphological and mating call comparisons in the *Bufo valliceps* complex. American Midland Naturalist 71:232–245.

Porter, K.R. 1968. Evolutionary status of a relict population of *Bufo hemiophrys* Cope. Evolution 22:583–594.

Porter, K.R. 1969a. Description of *Rana maslini*, a new species of wood frog. Herpetologica 25:212–215.

Porter, K.R. 1969b. Evolutionary status of the Rocky Mountain population of wood frogs. Evolution 23:163–170.

Porter, K.R., and W.F. Pyburn. 1967. Venom comparison and relationships of twenty species of New World toads (genus *Bufo*). Copeia 1967:298–307.

Porter, K.R., and D.E. Hakanson. 1976. Toxicity of mine drainage to embryonic and larval boreal toads (Bufonidae: *Bufo boreas*). Copeia 1976:327–331.

Post, D.D. 1972. Species differentiation in the *Rana pipiens* complex. Ph.D. Dissertation, Colorado State University, Fort Collins.

Post, D.D., and D. Pettus. 1966. Variation in *Rana pipiens* (Anura: Ranidae) of eastern Colorado. Southwestern Naturalist 11:476–482.

Post, D.D., and D. Pettus. 1967. Sympatry of two members of the *Rana pipiens* complex in Colorado. Herpetologica 23:323.

Post, T.J., and T. Uzzell. 1981. The relationships of *Rana sylvatica* and the monophyly of the *Rana boylii* group. Systematic Zoology 30:170–180.

Potthoff, T.L., and J.D. Lynch. 1986. Interpopulation variability in mouthparts of *Scaphiopus bombifrons* in Nebraska (Amphibia: Pelobatidae). Prairie Naturalist 18:15.

Pough, F.H., and S. Kamel. 1984. Post–metamorphic change in activity metabolism of anurans in relation to life history. Oecologia 65:138–144.

Pouliot, D., and J.-J. Frenette. 2010. Development and growth of northern leopard frog, *Lithobates pipiens*, tadpoles in North American Waterfowl Management Plan permanent basins and in natural wetlands. Canadian Field-Naturalist 124:159-168.

Powell, K.G. 2002. Watersheds and water quality as determinants of anuran distribution in western Newfoundland. B.Sc. Honour's thesis, Memorial University of Newfoundland, Corner Brook, Newfoundland.

Powell, R., J.T. Collins and E.D. Hooper, Jr. 2012. Key to the Herpetofauna of the Continental United States and Canada. Second edition, revised and updated. University Press of Kansas, Lawrence. [first edition, 1998]

Powell, R.L., and C.S. Lieb. 2003. *Hyla arenicolor* (Canyon Tree Frog). Toxic skin secretion. Herpetological Review 34:230.

Pramuk, J.B., T. Robertson, J.W. Sites, Jr., and B.P. Noonan. 2007. Around the world in 10 million years: biogeography of the nearly cosmopolitan true toads (Anura: Bufonidae). Global Ecology and Biogeography 17:72–83.

Prather, J.W., and J.T. Briggler. 2001. Use of small caves by anurans during a drought period in the Arkansas Ozarks. Journal of Herpetology 35:675–678.

Preble, E.A. 1902. A biological investigation of the Hudson Bay Region. North American Fauna 22:1–140.

Preble, E.A. 1908. Reptiles and batrachians of the Athabaska–MacKenzie region. North American Fauna 27:500–502.

Preston, D.L., J.S. Henderson, and P.T.J. Johnson. 2012. Community ecology of invasions: direct and indirect effects of multiple invasive species on aquatic communities. Ecology 93:1254–1261.

Preston, W.B. 1982. The Amphibians and Reptiles of Manitoba. Manitoba Museum of Man and Science, Winnipeg.

Preston, W.B. 2009. The distribution of the plains spadefoot, *Spea bombifrons*, in relation to soil type in southwestern Manitoba. Canadian Field-Naturalist 123:107-111.

Preston, W.B., and D.R.M. Hatch. 1986. The plains spadefoot, *Scaphiopus bombifrons*, in Manitoba. Canadian Field-Naturalist 100:123-125.

Prestwich, K.N., K.E. Brugger, and M. Topping. 1989. Energy and communication in three species of hylid frogs: power input, power output and efficiency. Journal of Experimental Biology 144:53–80.

Price, A.H. 2003. The Houston toad in Bastrop State Park 1990–2002: a narrative. Texas Parks and Wildlife Department, Open File Report 03–0401.

Price, R.M., and E.R. Meyer. 1979. An amplexus call made by the male American toad, *Bufo americanus americanus* (Amphibia, Anura, Bufonidae). Journal of Herpetology 13:506–509.

Price, S.J., D.R. Marks, R.W. Howe, J.M. Hanowski, and G.J. Niemi. 2004. The importance of spatial scale for conservation and assessment of anuran populations in coastal wetlands of the western Great Lakes, USA. Landscape Ecology 20:441–454.

Priddy, J.M., and D.D. Culley, Jr. 1971. The frog culture industry, past and present. Proceedings of the 25th Annual Conference of the Southeastern Association of Game and Fish Commissioners, pp. 597–601.

Priestley, A.S., T.A. Gorman, and C.A. Haas. 2010. Comparative morphology and identification of Florida bog frog and bronze frog tadpoles. Florida Scientist 73:20–26.

Provenzano, S.E., and M.D. Boone. 2009. Effects of density on metamorphosis of bullfrogs in a single season. Journal of Herpetology 43:49–54.

Pryor, G.S. 2003a. Growth rates and digestive abilities of bullfrog tadpoles (*Rana catesbeiana*) fed algal diets. Journal of Herpetology 37:560–566.

Pryor, G.S. 2003b. Roles of gastrointestinal symbionts in nutrition, digestion, and development of bullfrog tadpoles (*Rana catesbeiana*). Ph.D. Dissertation, University of Florida, Gainesville.

Pryor, G.S. 2008. Anaerobic bacteria isolated from the gastrointestinal tracts of bullfrog tadpoles (*Rana catesbeiana*). Herpetological Conservation and Biology 3:176–181.

Pryor, G.S., and E.C. Greiner. 2004. Expanded geographical range, new host accounts, and observations of the nematode *Gyrinicola batrachiensis* (Oxyuroidea: Pharyngodonidae) in tadpoles. Journal of Parasitology 90:189-191.

Pryor, G.S., and K.A. Bjorndal. 2005. Effects of the nematode *Gyrinicola batrachiensis* on development, gut morphology, and fermentation in bullfrog tadpoles (*Rana catesbeiana*): a novel mutualism. Journal of Experimental Zoology A 303:704-712.

Ptacek, M.B. 1992. Calling sites used by male gray treefrogs, *Hyla versicolor* and *Hyla chrysoscelis*, in sympatry and allopatry in Missouri. Herpetologica 48:373–382.

Ptacek, M.B. 1996. Interspecific similarity in life–history traits in symparic populations of gray treefrogs, *Hyla chrysoscelis* and *Hyla versicolor*. Herpetologica 52:323–332.

Ptacek, M. B., H. C. Gerhardt, and R. D. Sage. 1994. Speciation by polyploidy in treefrogs: multiple origins of the tetraploid, *Hyla versicolor*. Evolution 48: 898–908.

Puglis, H.J., and M.D. Boone. 2011. Effects of technical grade active ingredients vs. commercial formulation of seven pesticides in the presence or absence of UV radiation on survival of green frog tadpoles. Archives of Environmental Contamination and Toxicology 60:145–155.

Puglis, H.J., and M.D. Boone. 2012. Effects of terrestrial buffer zones on amphibians on golf courses. PLoS One 7:e39590.

Pulis, E.E., V.V. Tkach, and R.A. Neuman. 2011. Helminth parasites of the wood frog, *Lithobates sylvaticus*, in prairie pothole wetlands of the northern Great Plains. Wetlands 31:675–685.

Pullen, K.D., A.M. Best, and J.L. Ware. 2010. Amphibian pathogen *Batrachochytrium dendrobatidis* prevalence is correlated with season and not urbanization in central Virginia. Diseases of Aquatic Organisms 91:9–16.

Punzo, F. 1991a. Feeding ecology of spadefooted toads (*Scaphiopus couchi* and *Spea multiplicata*) in western Texas. Herpetological Review 22:79–80.

Punzo, F. 1991b. Group learning in tadpoles of *Rana heckscheri* (Anura: Ranidae). Journal of Herpetology 25:214–217.

Punzo, F. 1992a. Dietary overlap and activity patterns in sympatric populations of *Scaphiopus holbrooki* (Pelobatidae) and *Bufo terrestris* (Bufonidae). Florida Scientist 55:38–44.

Punzo, F. 1992b. Socially facilitated behavior in tadpoles of *Rana catesbeiana* and *Rana heckscheri*. Journal of Herpetology 26:219–222.

Punzo, F. 1993a. Effect of mercuric chloride on fertilization and larval development in the river frog, *Rana heckscheri* (Wright) (Anura: Ranidae). Bulletin of Environmental Contamination and Toxicology 51:575–581.

Punzo, F. 1993b. Ovarian effects of a sublethal concentration of mercuric chloride in the river frog, *Rana heckscheri* (Anura: Ranidae). Bulletin of Environmental Contamination and Toxicology 50:385–391.

Punzo, F. 1995. An analysis of feeding in the oak toad, *Bufo quercicus* (Holbrook) (Anura: Bufonidae). Florida Scientist 58:16–20.

Punzo, F., and L. Lindstrom. 2001. The toxicity of eggs of the giant toad, *Bufo marinus,* to aquatic predators in a Florida retention pond. Journal of Herpetology 35:693–697.

Punzo, F., J. Laveglia, D. Lohr, and P.A. Dahm. 1979. Organochlorine insecticide residues in amphibians and reptiles from Iowa and lizards from the Southwestern United States. Bulletin of Environmental Contamination and Toxicology 21:842–848.

Purcell, J.W. 1968. Embryonic temperature adaptations in southwestern populations of *Rana pipiens*. M.S. thesis, Texas Tech University, Lubbock.

Purgue, A.P. 1997. Tympanic sound radiation in the bullfrog, *Rana catesbeiana*. Journal of Comparative Physiology 181A:438–445.

Purrenhage, J.L. 2009. Importance of habitat structure for pond–breeding amphibians in multiple life stages. Ph.D. Dissertation, Miami University, Oxford, Ohio.

Purrenhage, J.L., and M.D. Boone. 2009. Amphibian community response to variation in habitat structure and competitor density. Herpetologica 65:14–30.

Putnam, R.F., and A.F. Bennett. 1981. Thermal dependence of behavioural performance of anuran amphibians. Animal Behaviour 29:502–509.

Putnam, R.W., and S.S. Hillman. 1977. Activity responses of anurans to dehydration. Copeia 1977:746–749.

Puttlitz, M.H., D.P. Chivers, J.M. Kiesecker, and A.R. Blaustein. 1999. Threat–sensitive predator avoidance by larval Pacific treefrogs (Amphibia, Hylidae). Ethology 105: 449–456.

Pyburn, W.F. 1958. Size and movements of a local population of cricket frogs (*Acris crepitans*). Texas Journal of Science 10:325–342.

Pyburn, W.F. 1960. Hybridization between *Hyla versicolor* and *H. femoralis*. Copeia 1960:55–56.

Pyburn, W.F. 1961a. The inheritance and distribution of vertebral stripe color in the cricket frog. Pp. 235–261 *In* W.F. Blair (ed.), Vertebrate Speciation, University of Texas Press, Austin.

Pyburn, W.F. 1961b. Inheritance of the green vertebral stripe in *Acris crepitans*. Southwestern Naturalist 6:164–167.

Pyburn, W.F., and J.P. Kennedy. 1960. Artificial hybridization of the Gray Treefrog, *Hyla versicolor* (Hylidae). American Midland Naturalist 64:216–223.

Pyburn, W.F., and J.P. Kennedy. 1961. Hybridization in U.S. treefrogs of the genus *Hyla*. Proceedings of the Biological Society of Washington 74:157–160.

Pytel, B.A. 1986. Biochemical systematics of the eastern North American frogs of the genus *Rana*. Herpetologica 42:273–282.

Quaintance, C.W. 1935. Reptiles and amphibians from Eagle Creek, Greenlee County, Arizona. Copeia 1935:183–185.

Quaranta, A., V. Bellantuono, G. Cassano, and C. Lippe. 2009. Why amphibians are more sensitive than mammals to xenobiotics. PLoS One 4(11):e7699.

Quinby, J.A. 1954. *Rana sylvatica sylvatica* Le Conte in South Carolina. Herpetologica 10:87–88.

Quinn, H. 1979. The Rio Grande chirping frog, *Syrrhophus cystignathoides campi* (Amphibia, Leptodactylidae), from Houston, Texas. Transactions of the Kansas Academy of Science 82:209–210.

Quinn, H. 1980. Captive propagation of endangered Houston toads. Herpetological Review 11:109.[reprinted from AAZPA Newsletter 21(5)]

Quinn, H.R., and G. Mengden. 1984. Reproduction and growth of *Bufo houstonensis* (Bufonidae). Southwestern Naturalist 29:189–195.

Quinn, H., K. Peterson, S. Mays, P. Freed, and K. Neitman. 1989. Captive propagation/ release and relocation program of the endangered Houston toad, *Bufo houstonensis*. Proceedings of the 1989 American Association of Zoological Parks and Aquariums National Conference, pp. 457–459.

Rabinowe, J.H., J.T. Serra, M.P. Hayes, and T. Quinn. 2002. *Rana aurora aurora* (Northern Red–legged Frog). Diet. Herpetological Review 33:128.

Rachowicz, L.J. 2002. Mouthpart pigmentation in *Rana muscosa* tadpoles: seasonal changes without chytridiomycosis. Herpetological Review 33:263–265.

Rachowicz, L.J., and V.T. Vredenburg. 2004. Transmission of *Batrachochytrium dendrobatidis* within and between amphibian life stages. Diseases of Aquatic Organisms 61:75–83.

Rachowicz, L.J., and C.J. Briggs. 2007. Quantifying the disease transmission function: effects of density on *Batrachochytrium dendrobatidis* transmission in the mountain yellow–legged frog *Rana muscosa*. Journal of Animal Ecology 76:711–721.

Rachowicz, L.J., R.A. Knapp, J.A.T. Morgan, M.J. Stice, V.T. Vredenburg, J.M. Parker, and C.J. Briggs. 2006. Emerging infectious disease as a proximate cause of amphibian mass mortality. Ecology 87:1671–1683.

Raffel, T.R., J.T. Hoverman, N.T. Halstead, P.J. Michel, and J.R. Rohr. 2010. Parasitism in a community context: trait-mediated interactions with competition and predation. Ecology 91:1900–1907.

Raffel, T.R., J.O. Lloyd–Smith, S.K. Sessions, P.J. Hudson, and J.R. Rohr. 2011. Does the early frog catch the worm? Disentangling potential drivers of a parasite age–intensity relationship in tadpoles. Oecologia 165:1031–1042.

Rafferty, K.A., Jr. 1964. Kidney tumors of the leopard frog: a review. Cancer Research 24:169–185.

Raimondo, S.M., C.L. Rowe, and J.D. Congdon. 1998. Exposure to coal ash impacts swimming performance and predator avoidance in larval bullfrogs (*Rana catesbeiana*). Journal of Herpetology 32:289–292.

Raithel, C.J. 2007. Environmental influences on pond-breeding larval amphibian abundance. M.S. thesis, University of Rhode Island, Kingston.

Raithel, C.J., P.W.C. Paton, P.S. Pooler, and F.C. Golet. 2011. Assessing long–term population trends of wood frogs using egg–mass counts. Journal of Herpetology 45:23–27.

Ralin, D.B. 1968. Ecological and reproductive differentiation in the cryptic species of the *Hyla versicolor* complex (Hylidae). Southwestern Naturalist 13:283–300.

Ralin, D.B. 1972. Genetic compatibility and a phylogeny of the temperate North American hylid fauna. Ph.D. Dissertation, University of Texas, Austin.

Ralin, D.B. 1976a. Comparative hybridization of a diploid–tetraploid cryptic species pair of treefrogs. Copeia 1976:191–196.

Ralin, D.B. 1976b. Behavioral and genetic differentiationin a diploid–tetraploid complex of treefrogs. Herpetological Review 7:97–98.

Ralin, D.B. 1977a. Evolutionary aspects of mating call variation in a diploid–tetraploid species complex of treefrogs (Anura). Evolution 31:721–736.

Ralin, D.B. 1977b. Hybridization of *Hyla cinerea* of the United States and *H. arborea savignyi* (Amphibia, Anura, Hylidae) of Israel. Journal of Herpetology 11:105–106.

Ralin, D.B. 1978. "Resolution" of the diploid–tetraploid tree frogs. Science 202:335–336.

Ralin, D.B., and J.S. Rogers. 1972. Aspects of tolerance to desiccation in *Acris crepitans* and *Pseudacris streckeri*. Copeia 1972:519–528.

Ralin, D.B., and J.S. Rogers. 1979. A morphological analysis of a North American diploid–tetraploid complex of treefrogs (Amphibia, Anura, Hylidae). Journal of Herpetology 13:261–269.

Ralin, D.B., and R.K. Selander. 1979. Evolutionary genetics of diploid–tetraploid species of treefrogs of the genus *Hyla*. Evolution 33:595–608.

Raloff, J. 2003. Hawaii's hated frogs: tiny invaders raise a big ruckus. Science News 163:11–13.

Ralph, C.L. 1978. Non–optic phototaxis of two species of ranid frogs (Amphibia, Anura, Ranidae) with special attention to the parapineal organ. Journal of Herpetology 12:197–201.

Ramer, J.D., T.A. Jenssen, and C.J. Hurst. 1983. Size–related variation in the advertisement call of *Rana clamitans* (Anura: Ranidae), and its effect on conspecific males. Copeia 1983:141–155.

Ramesh, R., K. Griffis–Kyle, G. Perry, and M. Farmer. 2012. Urban amphibians of the Texas Panhandle: baseline inventory and habitat associations in a drought year. Reptiles & Amphibians 19:243–253.

Ramirez, E.A., H. J. Puglis, A. Ritzenthaler, and M. Boone. 2012. Terrestrial movements and habitat preferences of male cricket frogs on a golf course. Copeia 2012:191–196.

Ramsey, J.P. 2011. Amphibian immune defenses against *Batrachochytrium dendrobatidis*: a war between host and pathogen. Ph.D. Dissertation, Vanderbilt University, Nashville, Tennessee.

Rand, A.S. 1950. Leopard frogs in caves in winter. Copeia 1950:324.

Randel, W.A. 1914. Frog Culture for Profit. Aqua Life Co., Seymour, Connecticut. 61 pp. [Note: Randel's name is not on the booklet]

Raney, E.C. 1940. Summer movements of the bullfrog, *Rana catesbeiana* Shaw, as determined by the jaw–tag method. American Midland Naturalist 23:733–745.

Raney, E.C., and W.M. Ingram. 1941. Growth of tagged frogs (*Rana catesbeiana* Shaw and *Rana clamitans* Daudin) under natural conditions. American Midland Naturalist 26:201–206.

Raney, E.C., and E.A. Lachner. 1947. Studies on the growth of tagged toads (*Bufo terrestris americanus* Holbrook). Copeia 1947:113–116.

Raphael, M.C. 1988. Long–term trends in abundance of amphibians, reptiles, and mammals in Douglas–fir forests of northwestern California. Pp. 23–31 *In* R.C. Szaro, K.E. Severson, and D.R. Patton (eds.), Management of Amphibians, Reptiles, and Small Mammals in North America. USDA Forest Service General and Technical Report RM–166.

Rathbun, G.B. 1998. *Rana aurora draytonii* (California Red–legged Frog). Egg predation. Herpetological Review 29:165.

Rathbun, G.B., and J. Schneider. 2001. Translocation of California red–legged frogs (*Rana aurora draytonii*). Wildlife Society Bulletin 29:1300–1303.

Rathbun, G.B., N.J. Scott, Jr., and T.G. Murphy. 1997. *Rana aurora draytonii* (California Red–legged Frog). Behavior. Herpetological Review 28:85–86.

Ratzlaff, K. 2012. Dynamics of chytrid fungus (*Batrachochytrium dendrobatidis*) infection in amphibians in the Rincon Mountains and Tucson, Arizona. M.S. thesis, University of Arizona, Tucson.

Raun, G.G. 1959. Terrestrial and aquatic vertebrates of a moist, relict area in central Texas. Texas Journal of Science 11:158–171.

Raun, G.G., and F.R. Gehlbach. 1972. Amphibians and reptiles in Texas. Taxonomic synopsis, bibliography, county distribution maps. Dallas Museum of Natural History, Bulletin 2.

Raveling, D.G. 1965. Variation in a sample of *Bufo americanus* from southwestern Illinois. Herpetologica 21:219–225.

Ray, J.E., D. Preston, and M.L. McCallum. 2006. *Bufo nebulifer* (Coastal Plains Toad). Urban road mortality. Herpetological Review 37:442.

Reagan, A.B. 1907. Animals, reptiles and amphibians of the Rosebud Indian Reservation, South Dakota. Transactions of the Kansas Academy of Science 21:163-164.

Reagan, D.P. 1974. Threatened native amphibians of Arkansas. Pp. 93–99 *In* Arkansas Natural Area Plan, Arkansas Department of Planning, Little Rock.

Reaser, J.K. 1996. *Rana pretiosa* (Spotted frog). Vagility. Herpetological Review 27:196–197.

Reaser, J.K. 2000. Demographic analysis of the Columbia spotted frog (*Rana luteiventris*): case study in spatiotemporal variation. Canadian Journal of Zoology 78:1158–1167.

Reaser, J.K., and R.E. Dexter. 1996. *Rana pretiosa* (Spotted Frog). Predation. Herpetological Review 27:75.

Reaser, J.K., and W.F. Knight. 2001. *Bufo marinus* (Giant Toad). Association with Oscars (*Astronotus oscellatus*). Herpetological Review 32:180.

Recuero, E., I. Martínez–Solano, G. Parra–Olea, and M. García–París. 2006a. Phylogeography of *Pseudacris regilla* (Anura: Hylidae) in western North America, with a proposal for a new taxonomic rearrangement. Molecular Phylogenetics and Evolution 39:293–304.

Recuero, E., I. Martínez–Solano, G. Parra–Olea, and M. García–París. 2006b. Corrigendum to "Phylogeography of *Pseudacris regilla* (Anura: Hylidae) in western North America, with a proposal for a new taxonomic rearrangement." Molecular Phylogenetics and Evolution 41:511.

Redmer, M. 1992. *Rana sphenocephala* (Southern Leopard Frog). Variation. Herpetological Review 23:58–59.

Redmer, M. 1996. Locality records of the northern leopard frog, *Rana pipiens*, in central and southwestern Illinois. Transactions of the Illinois State Academy of Science 89:215–219.

Redmer, M. 1998a. Status and distribution of two uncommon frogs, Pickerel Frogs and Wood Frogs, in Illinois. Pp. 83–90 *In* M.J. Lannoo (ed.), Status & Conservation of Midwestern Amphibians. University of Iowa Press, Iowa City.

Redmer, M. 1998b. *Hyla avivoca* (Bird–voiced Treefrog). Amplexus and oviposition. Herpetological Review 29:230–231.

Redmer, M. 2000. Demographic and reproductive characteristics of a southern Illinois population of the crayfish frog, *Rana areolata*. Journal of the Iowa Academy of Science 107:128–133.

Redmer, M. 2002. Natural history of the wood frog (*Rana sylvatica*) in the Shawnee National Forest, southern Illinois. Illinois Natural History Survey Bulletin 36:163–194.

Redmer, M., and S.E. Trauth. 2005. *Rana sylvatica* LeConte, 1825. Wood Frog. Pp. 590–593 *In* M.J. Lannoo (ed.), Amphibian Declines. The Conservation Status of United States Species. University of California Press, Berkeley.

Redmer, M., L.E. Brown, and R.A. Brandon. 1999a. Natural history of the bird–voiced treefrog (*Hyla avivoca*) and green treefrog (*Hyla cinerea*) in southern Illinois. Illinois Natural History Survey Bulletin 36:37–66.

Redmer, M., D.H. Jamieson, and S.E. Trauth. 1999b. Notes on the diet of female bird–voiced treefrogs (*Hyla avivoca*) in southern Illinois. Transactions of the Illinois State Academy of Science 92:271–275.

Redmond, W.H., and R.H. Mount. 1975. A biogeographic analysis of the herpetofauna of the Coosa Valley in Alabama. Journal of the Alabama Academy of Science 46:65–81.

Redmond, W.H., and A.F. Scott. 1996. Atlas of Amphibians in Tennessee. Center for Field Biology, Austin Peay State University, Miscellaneous Publication No. 12.

Reed, C.F. 1957. *Rana virgatipes* in southern Maryland, with notes upon its range from New Jersey to Georgia. Herpetologica 13:137–138.

Reed, C.F. 1960. New records for *Hyla cinerea* in Maryland, Delaware, Virginia, and North Carolina. Herpetologica 16:119–120.

Reeder, A.L., G.L. Foley, D.K. Nichols, L.G. Hansen, B. Wikoff, S. Faeh, J. Eisold, M.B. Wheeler, R. Warner, J.E. Murphy, and V.R. Beasley. 1998. Forms and prevalence of intersexuality and effects of environmental contaminants on sexuality in cricket frogs (*Acris crepitans*). Environmental Health Perspectives 106:261–266.

Reeder, A.L., M.O. Ruiz, A. Pessier, L.E. Brown, J.M. Levengood, C.A. Phillips, M.B. Wheeler, R.E. Warner, and V.R. Beasley. 2005. Intersexuality and the cricket frog decline: historic and geographic trends. Environmental Health Perspectives 113:261–265.

Reeder, N.M.M., A.P. Pessier, and V.T. Vredenburg. 2012. A reservoir species for the emerging amphibian pathogen *Batrachochytrium dendrobatidis* thrives in a landscape decimated by disease. PLoS One 7:e33567.

Reeves, M.K. 2008. *Batrachochytrium dendrobatidis* in wood frogs (*Rana sylvatica*) from three National Wildlife Refuges in Alaska, USA. Herpetological Review 39:68–70.

Reeves, M.K. 2010. Multiple stressors and the cause of amphibian abnormalities. Ph.D. Dissertation, University of California, Davis.

Reeves, M.K., and D.E. Green. 2006. *Rana sylvatica* (Wood Frog). Chytridiomycosis. Herpetological Review 37:450.

Reeves, M.K., C.L. Dolph, H. Zimmer, R.S. Tjeerdema, and K.A. Trust. 2008. Road proximity increases risk of skeletal abnormalities in wood frogs from National Wildlife Refuges in Alaska. Environmental Health Perspectives 116:1009–1015.

Reeves, M.K., P. Jensen, C. L. Dolph, M. Holyoak, and K. A. Trust. 2010. Multiple stressors and the cause of amphibian abnormalities. Ecological Monographs 80: 423–440.

Reeves, M.K., M. Perdue, G. D. Blakemore, D. J. Rinella, and M. Holyoak. 2011. Twice as easy to catch? A toxicant and a predator cue cause additive reductions in larval amphibian activity. Ecosphere 2:1–20.

Regan, G.T. 1972. Natural and man–made conditions determining the range of *Acris*. Ph.D. Dissertation, University of Kansas, Lawrence.

Regosin, J.V., B.S. Windmiller, and J.M. Reed. 2003. Terrestrial habitat use and winter densities of the wood frog (*Rana sylvatica*). Journal of Herpetology 37:390–394.

Regosin, J.V., B. S. Windmiller, R. N. Homan, and J. M. Reed. 2005. Variation in terrestrial habitat use by four pool–breeding amphibian species. Journal of Wildlife Management 69:1481–1493.

Reichard, S.M., and H.M. Stevenson. 1964. Records of *Eleutherodactylus ricordi* at Tallahassee. Florida Naturalist 37(3):96–B (inside back cover)

Reichert, M.S. 2011. Aggressive calling in Treefrogs. Ph.D. Dissertation, University of Missouri, Columbia.

Reichert, M.S., and H.C. Gerhardt. 2011. The role of body size on the outcome, escalation and duration of contests in the grey treefrog, *Hyla versicolor*. Animal Behaviour 82:1357– 1366.

Reichert, M.S., and H.C. Gerhardt. 2012. Trade–offs and upper limits to signal performance during close–range vocal competition in gray tree frogs *Hyla versicolor*. American Naturalist 180:425–437.

Reichling, S.B. 2008. Reptiles & Amphibians of the Southern Pine Woods. University Press of Florida, Gainesville.

Reilly, B.O., and P.T.K. Woo. 1982. The biology of *Trympanosoma andersoni* n. sp. and *Trympanosoma grylli* Nigrelli, 1944 (Kinoplastidia) from *Hyla versicolor* LeConte, 1825 (Anura). Canadian Journal of Zoology 60:116– 123.

Reimchen, T.E. 1990. Introduction and dispersal of the Pacific treefrog, *Hyla regilla*, on the Queen Charlotte Islands, British Columbia. Canadian Field-Naturalist 105:288–290.

Reintjes–Tolen, S. 2012. Geographic distribution of chytrid fungus (*Batrachochytrium dendrobatidis*) and *Ranavirus* spp. in amphibians in northern peninsular and panhandle Florida: with a case of a ranavirus die–off in Gold Head Branch State Park. Ph.D. Dissertation, University of Florida, Gainesville.

Relyea, R.A. 1998. Phenotypic plasticity in larval amphibians. Ph.D. Dissertation, University of Michigan, Ann Arbor.

Relyea, R.A. 2000. Trait–mediated indirect effects in larval anurans: reversing competition with the threat of predation. Ecology 81:2278–2289.

Relyea, R.A. 2001a. The relationship between predation risk and antipredator responses in larval anurans. Ecology 82:541–554.

Relyea, R.A. 2001b. Morphological and behavioral plasticity of larval anurans in response to different predators. Ecology 82:523–540.

Relyea, R.A. 2001c. The lasting effects of adaptive plasticity: predator–induced tadpoles become long–legged frogs. Ecology 82:1947–1953.

Relyea, R.A. 2002a. Competitor–induced plasticity in tadpoles: consequences, cues, and connections to predator– induced plasticity. Ecological Monographs 72:523–540.

Relyea, R.A. 2002b. Local population differences in phenotypic plasticity: predator–induced changes in wood frog populations. Ecological Monographs 72:77–93.

Relyea, R. A. 2002c. Costs of phenotypic plasticity. American Naturalist 159:272–282.

Relyea, R. A. 2002d. The many faces of predation: How selection, induction, and thinning combine to alter prey phenotypes. Ecology 83:1953–1964.

Relyea, R. A. 2003a. How prey respond to combined predators: A review and an empirical test. Ecology 84:1827– 1839.

Relyea, R. A. 2003b. Predator cues and pesticides: A double dose of danger for amphibians. Ecological Applications 13:1515–1521.

Relyea, R. A. 2003c. Predators come and predators go: The reversibility of predator–induced traits. Ecology 84:1840–1848.

Relyea, R.A. 2004a. Fine–tuned phenotypes: tadpole plasticity under 16 combinations of predators and competitors. Ecology 85:172–179.

Relyea, R.A. 2004b. Synergistic impacts of malathion and predatory stress on six species of North American tadpoles. Environmental Toxicology and Chemistry 23:1080–1084.

Relyea, R. A. 2004c. The growth and survival of five amphibian species exposed to combinations of pesticides. Environmental Toxicology and Chemistry 23:1737–1742.

Relyea, R.A. 2005a. The lethal impact of Roundup on aquatic and terrestrial amphibians. Ecological Applications 15:1118–1124.

Relyea, R.A. 2005b. The impact of insecticides and herbicides on the biodiversity and productivity of aquatic communities. Ecological Applications 15:618–627.

Relyea, R. A. 2005c. The heritability of inducible defenses in tadpoles. Journal of Evolutionary Biology 18:856–866.

Relyea, R. A. 2005d. The lethal impacts of Roundup and predatory stress on six species of North American tadpoles. Archives of Environmental Contamination and Toxicology 48:351–357.

Relyea, R. A. 2005e. Pesticides and amphibians: the importance of community context. Ecological Applications 15:1125–1134.

Relyea, R. A. 2006. The effects of pesticides, pH and predatory stress in amphibians under mesocosm conditions. Ecotoxicology 15:503–511.

Relyea, R.A. 2007. Getting out alive: how predators affect the decision to metamorphose. Oecologia 152:389–400.

Relyea, R.A. 2009. A cocktail of contaminants: how mixtures of pesticides at low concentrations affect aquatic communities. Oecologia 159:363–376.

Relyea, R. A. 2012. New effects of Roundup® on amphibians: Predators reduce herbicide mortality while herbicides induce anti–predator morphology. Ecological Applications 22:634–647.

Relyea, R.A., and E.E. Werner. 1999. Quantifying the relation between predator–induced behavior and growth performance in larval amphibians. Ecology 80:2117–2124.

Relyea, R. A., and E. E. Werner. 2000. Morphological plasticity of four larval anurans distributed along an environmental gradient. Copeia 2000:178–190.

Relyea, R.A., and N. Mills. 2001. Predator–induced stress makes the pesticide carbaryl more deadly to gray treefrog tadpoles (*Hyla versicolor*). Proceedings of the National Academy of Sciences of the United States of America 98:2491–2496.

Relyea, R. A., and K. L. Yurewicz. 2002. Predicting community outcomes from pairwise interactions: integrating density– and trait–mediated effects. Oecologia 131:569–579.

Relyea, R. A., and J. T. Hoverman. 2003. The impact of larval predators and competitors on the morphology and fitness of juvenile tree frogs. Oecologia 134:596–604.

Relyea, R.A., and J.R. Auld. 2004. Having the guts to compete: how intestinal plasticity explains costs of inducible defenses. Ecology Letters 7:869–875.

Relyea, R.A., and J.R. Auld. 2005. Predator– and competitor–induced plasticity: how changes in foraging morphology affect phenotypic trade–offs. Ecology 86:1723–1729.

Relyea, R.A., and N. Diecks. 2008. An unforeseen chain of events: Lethal effects of pesticides at sublethal concentrations. Ecological Applications 18:1728–1742.

Relyea, R.A., and J.T. Hoverman. 2008. Interactive effects of predators and a pesticide on aquatic communities. Oikos 117:1647–1658.

Relyea, R.A., and D. K. Jones. 2009. The toxicity of Roundup Original MAX® to 13 species of larval amphibians. Environmental Toxicology and Chemistry 28:2004–2008.

Relyea, R.A., and K. Edwards. 2010. What doesn't kill you makes you sluggish: how sublethal pesticides alter predator–prey interactions. Copeia 2010:558–567.

Relyea, R.A., N.M. Schoeppner, and J.T. Hoverman. 2005. Pesticides and amphibians: the importance of community context. Ecological Applications 15:1125–1134.

Renaud, M. 1977. Polymorphic and polytypic variation in the Arizona treefrog (*Hyla wrightorum*). Ph.D. Dissertation, Arizona State University, Tempe.

Resetarits, W.J., Jr. 1986. Ecology of cave use by the frog, *Rana palustris*. American Midland Naturalist 116: 256–266.

Resetarits, W.J., Jr. 1998. Differential vulnerability of *Hyla chrysoscelis* eggs and hatchlings to larval insect predators. Journal of Herpetology 32:440–443.

Resetarits, W.J., Jr. 2005. Habitat selection behaviour links local and regional scales in aquatic systems. Ecology Letters 8:480–486.

Resetarits, W.J., Jr., and R. D. Aldridge. 1988. Reproductive biology of a cave associated population of the frog, *Rana palustris*. Canadian Journal of Zoology 66:329–333.

Resetarits, W.J., Jr., and H.M. Wilbur. 1989. Choice of oviposition site by *Hyla chrysoscelis*: role of predators and competitors. Ecology 70:220–228.

Resetarits, W.J., Jr., and H.M. Wilbur. 1991. Calling site choice by *Hyla chrysoscelis*: effects of predators, competitors, and oviposition sites. Ecology 72:778–786.

Resetarits, W.J., Jr., and J.E. Fauth. 1998. From cattle tanks to Carolina Bays. The utility of model systems for understanding natural communities. Pp. 133–151 *In* W.J. Resetarits, Jr. and J. Bernardo (eds.), Experimental Ecology. Issues and Perspectives. Oxford University Press, New York.

Resetarits, W.J., Jr., and D. R. Chalcraft. 2007. Functional diversity within a morphologically conservative genus of predators: implications for functional equivalence and redundancy in ecological communities. Functional Ecology 21:793–804.

Resetarits, W.J., Jr., J. F. Rieger, and C.A. Binckley. 2004. Threat of predation negates density effects in larval gray treefrogs. Oecologia 138:532–538.

Resnick, L.E., and D.L. Jameson. 1963. Color polymorphism in Pacific tree frogs. Science 142:1081–1083.

Reynolds, T.D., and T.D. Stephens. 1984. Multiple ectopic limbs in a wild population of *Hyla regilla*. Great Basin Naturalist 44:166–169.

Rheinlaender, J., H.C. Gerhardt, D.D. Yager, and R.R. Capranica. 1979. Accuracy of phonotaxis by the green treefrog (*Hyla cinerea*). Journal of Comparative Physiology 133:247–255.

Rheinlaender, J., W. Walkowiak, and H.C. Gerhardt. 1981. Directional hearing in the green treefrog: a variable mechanism? Naturwissenschaften 68:430–431.

Rhoads, S.N. 1895. Contributions to the zoology of Tennessee. 1. Reptiles and amphibians. Proceedings of the Academy of Natural Sciences, Philadelphia 47:376–407.

Rhoden, R H., and M. G. Bolek. 2011. Distribution and reproductive strategies of *Gyrinicola batrachiensis* (Oxyuroidea: Pharyngodonidae) in larvae of eight species of amphibians from Nebraska. Journal of Parasitology 97:629–635.

Rhoden, R.H., and M. G. Bolek. 2012. Helminth and leech community structure in tadpoles and caudatan larvae of two amphibian species from western Nebraska. Journal of Parasitology 98: 236–244.

Rice, A.M., A.R. Leichty, and D.W. Pfennig. 2009. Parallel evolution and ecological selection: replicated character displacement in spadefoot toads. Proceedings of the Royal Society B 276:4189–4196.

Rice, A.N., T.L. Roberts IV, J.G. Pritchard, and M.E. Dorcas. 2001. Historical trends and perceptions of amphibian and reptile diversity in the western Piedmont of North Carolina. Journal of the Elisha Mitchell Scientific Society 117:264–273.

Rice, K.G., J.H. Waddle, M.W. Miller, M.E. Crockett, F.J. Mazzotti, and H.F. Percival. 2011. Recovery of native treefrogs after removal of nonindigenous Cuban treefrogs, *Osteopilus septentrionalis*. Herpetologica 67:105–117.

Rice, T.M., and D.H. Taylor. 1995. A simple test of prey discrimination that demonstrates learning in postlarval ranid frogs. Journal of Herpetology 29:320–322.

Richards, C.M. 1958. The inhibition of growth in crowded *Rana pipiens* tadpoles. Physiological Zoology 31:138–151.

Richards, C.M. 1962. The control of tadpole growth by alga–like cells. Physiological Zoology 35:285–296.

Richards, C.M., and G.C. Lehman. 1980. Photoperiodic stimulation of growth in postmetamorphic *Rana pipiens*. Copeia 1980:147–149.

Richards, C.M., and G.W. Nace. 1983. Dark pigment variants in anurans: classification, new descriptions, color changes and inheritance. Copeia 1983:979–990.

Richards, C.M., D.T. Tartof, and G.W. Nace. 1969. A melanoid variant in *Rana pipiens*. Copeia 1969:850–852.

Richards, L.P. 1958. Some locality records of Yosemite herps. Yosemite Nature Notes 37(9):118–126.

Richards–Hrdlicka, K.L., J.L. Richardson, and L. Mohabir. 2012. First survey for the amphibian chytrid fungus *Batrachochytrium dendrobatidis* in Connecticut (USA) finds widespread prevalence. Diseases of Aquatic Organisms 102:169–180.

Richardson, J. 2011. Movement, habitat use, and dispersal of juvenile Oregon spotted frogs (*Rana pretiosa*) on Joint Base Lewis McChord. M.E.S. thesis, Evergreen State University, Olympia, Washington.

Richardson, J.L. 2012a. The evolutionary ecology of pond–breeding amphibians: From local populations to regional landscapes. Ph.D. Dissertation, Yale University, New Haven, Connecticut.

Richardson, J.L. 2012b. Divergent landscape effects on population connectivity in two co– occurring amphibian species. Molecular Ecology 21: 4437–4451.

Richardson, J.M.L. 2001. A comparative study of activity levels in larval anurans and response to the presence of different predators. Behavioral Ecology 12:51–58.

Richardson, J.M.L. 2002a. A comparative study of phenotypic traits related to resource utilization in anuran communities. Evolutionary Ecology 16:101–122.

Richardson, J.M.L. 2002b. Burst swim speed in tadpoles inhabiting ponds with different top predators. Evolutionary Ecology Research 4:627–642.

Richmond, A.M., T. Tyning, and A.P. Summers. 1999. *Bufo americanus* (American Toad). Depth record. Herpetological Review 30:90–91.

Richmond, N.D. 1947. Life history of *Scaphiopus holbrookii holbrookii* (Harlan). 1. Larval development and behavior. Ecology 28:53–67.

Richmond, N.D. 1952. An addition to the herpetofauna of Nova Scotia, and other records of amphibians and reptiles on Cape Breton Island. Annals of the Carnegie Museum 32:331–332.

Richmond, N.D. 1964. The green frog (*Rana clamitans melanota*) developing in one season. Herpetologica 20:132.

Richmond, N.D., and C.J. Goin. 1938. Notes on a collection of amphibians and reptiles from New Kent County, Virginia. Annals of the Carnegie Museum 27:301–310.

Richter, J., L. Martin, and C.K. Beachy. 2009. Increased larval density induces accelerated metamorphosis independently of growth rate in the frog *Rana sphenocephala*. Journal of Herpetology 43:551–554.

Richter, K.O., and A.L. Azous. 1995. Amphibian occurrence

and wetland characteristics in the Puget Sound Basin. Wetlands 15:305–312.

Richter, S.C. 2000. Larval caddisfly predation on the eggs and embryos of *Rana capito* and *Rana sphenocephala*. Journal of Herpetology 34:590–593.

Richter, S.C., and R.A. Seigel. 2002. Annual variation in the population ecology of the endangered gopher frog, *Rana sevosa* Goin and Netting. Copeia 2002:962–972.

Richter, S.C., J.E. Young, R.A. Seigel, and G.N. Johnson. 2001. Postbreeding movements of the dark gopher frog, *Rana sevosa* Goin and Netting: implications for conservation and management. Journal of Herpetology 35:316–321.

Richter, S.C., J.E. Young, G.N. Johnson, and R.A. Seigel. 2003. Stochastic variation in reproductive success of a rare frog, *Rana sevosa*: implications for conservation and for monitoring amphibian populations. Biological Conservation 111:171–177.

Richter, S.C., B. I. Crother, and R.E. Broughton. 2009. Genetic consequences of population reduction and geographic isolation in the critically endangered frog, *Rana sevosa*. Copeia 2009:799–806.

Rickard, A. 2006. *Acris crepitans blanchardi* (Blanchard's Cricket Frog). Predation. Herpetological Review 37:439.

Rickard, A., and E. Sonntag. 2006. Blanchard's Cricket Frog (*Acris crepitans blanchardi*) morphological abnormalities in southeast Michigan. Endangered Species Update 23:131–134.

Rickard, A., E. Sonntag, and K. Zippel. 2004. Amphibian conservation strategies: translocating an entire population of Blanchard's Cricket frog (*Acris crepitans blanchardi*) in southeast Michigan. Endangered Species Update 21:128–131.

Ricker, W.E., and E.B.S. Logier. 1935. Notes on the occurrence of the ribbed toad (*Ascaphus truei* Stejneger) in Canada. Copeia 1935:46.

Rieger, J.F., C.A. Binckley, and W.J. Resetarits, Jr. 2004. Larval performance and oviposition site preference along a predation gradient. Ecology 85:2094–2099.

Riemer, W.J. 1958. Giant toads of Florida. Quarterly Journal of the Florida Academy of Sciences 21:207–211.

Rigley, L., and H. Hays. 1976. Sonic behavior of the green treefrog *Hyla cinerea*. Bulletin of the Maryland Herpetological Society 12:25–28.

Riha, V.F., and K.A. Berven. 1991. An analysis of latitudinal variation in the larval development of the wood frog (*Rana sylvatica*). Copeia 1991:209–221.

Riley, E.E., and M.R. Weil. 1986. The effects of thiosemicarbazide on development in the wood frog, *Rana sylvatica*. 1. Concentration effects. Ecotoxicology and Environmental Safety 12:154–160.

Riley, E.E., and M.R. Weil. 1987. The effects of thiosemicarbazide on development in the wood frog, *Rana sylvatica*. 2. Critical exposure length and age sensitivity. Ecotoxicology and Environmental Safety 13:202–207.

Riley, S.P.D., G.T. Busteed, L.B. Kats, T.L. Vandergon, L.E.S. Lee, R.G. Dagit, J.L. Kerby, R.N. Fisher, and R.M. Sauvajot. 2005. Effects of urbanization on the distribution and abundance of amphibians and invasive species in Southern California streams. Conservation Biology 19:1894-1907.

Rimer, R.L., and J.T. Briggler. 2010. Occurrence of the amphibian chytrid fungus (*Batrachochytrium dendrobatidis*) in Ozark caves, Missouri, USA. Herpetological Review 41:175–177.

Ripplinger, J.I., and R.S. Wagner. 2004. Phylogeography of northern populations of the Pacific treefrog, *Pseudacris regilla*. Northwestern Naturalist 85:118–125.

Ritchie, S.A. 1982. The green treefrog (*Hyla cinerea*) as a predator of mosquitoes in Florida. Mosquito News 42:619.

Ritchie, S.C., B.K. Rincon, and T.A. Gorman. 2008. *Rana clamitans* (Bronze Frog). Ranid aggression and interspecies amplexus. Herpetological Review 39:80.

Ritke, M. E., and J. G. Babb. 1991. Behavior of the gray treefrog (*Hyla chrysoscelis*) during the nonbreeding season. Herpetological Review 22:5–8.

Ritke, M.E., and M.L. Beck. 1991. An interspecific satellite pair association between *Hyla chrysoscelis* and *Hyla versicolor*. Herpetological Review 22:49–51.

Ritke, M.E., and R.D. Semlitsch. 1991. Mating behavior and determinants of male mating success in the gray treefrog, *Hyla chrysoscelis*. Canadian Journal of Zoology 69:246–250.

Ritke, M.E., and R.L. Mumme. 1993. Choice of callings sites and oviposition sites by gray treefrogs (*Hyla chrysoscelis*)-A comment. Ecology 74:623–626.

Ritke, M.E., and C.A. Lessman. 1994. Longitudinal study of ovarian dynamics in female gray treefrogs (*Hyla chrysoscelis*). Copeia 1994:1014–1022.

Ritke, M.E., J.G. Rabb, and M.K. Ritke. 1990. Life history of the gray treefrog (*Hyla chrysoscelis*) in western Tennessee. Journal of Herpetology 24:135–141.

Ritke, M.E., J.G. Rabb, and M.K. Ritke. 1991a. Breeding-site specificity in the Gray Treefrog (*Hyla chrysoscelis*). Journal of Herpetology 25:123–125.

Ritke, M.E., J.G. Babb, and M.K. Ritke. 1991b. Temporal patterns of reproductive activity in the gray treefrog (*Hyla chrysoscelis*). Journal of Herpetology 26:107–111.

Ritke, M.E., J.G. Babb, and M.K. Ritke. 1992. Annual growth rates of adult gray treefrogs (*Hyla chrysoscelis*). Journal of Herpetology 25:382–385.

Ritland, K., L.A. Dupuis, F.L. Bunnell, W.L.Y. Hung, and

J.E. Carlson. 2000. Phylogeography of the tailed frog (*Ascaphus truei*) in British Columbia. Canadian Journal of Zoology 78:1749–1758.

Rittenhouse, T.A.G. 2007. Behavioral choice and demographic consequences of wood frog habitat selection in response to land use. Ph.D. Dissertation, University of Missouri, Columbia.

Rittenhouse, T.A.G. 2011. Anuran larval habitat quality when reed canary grass is present in wetlands. Journal of Herpetology 45:491–496.

Rittenhouse, T.A.G., and R.D. Semlitsch. 2007a. Postbreeding habitat use of wood frogs in a Missouri oak–hickory forest. Journal of Herpetology 41:645–653.

Rittenhouse, T.A.G., and R.D. Semlitsch. 2007b. Distribution of amphibians in terrestrial habitat surrounding wetlands. Wetlands 27:153–161.

Rittenhouse, T.A.G., and R.D. Semlitsch. 2009. Behavioral response of migrating wood frogs to experimental timber harvest surrounding wetlands. Canadian Journal of Zoology 87:618–625.

Rittenhouse, T.A.G., E.B. Harper, L.R. Rehard, and R.D. Semlitsch. 2008. The role of microhabitats in the desiccation and survival of amphibians in recently harvested oak–hickory forest. Copeia 2008:807–814.

Rittenhouse, T.A.G., R.D. Semlitsch, and F.R. Thompson, III. 2009. Survival costs associated with wood frog breeding migrations: effects of timber harvest and drought. Ecology 90:1620–1630.

Rittmann, S.E., E. Muths, and D.E. Green. 2003. *Pseudacris triseriata* (Western Chorus Frog) and *Rana sylvatica* (Wood Frog). Chytridiomycosis. Herpetological Review 34:53.

Rittschof, D. 1975. Some aspects of the natural history and ecology of the leopard frog, *Rana pipiens*. Ph.D. Dissertation, University of Michigan, Ann Arbor.

Rizkalla, C.E. 2009. First reported detection of *Batrachochytrium dendrobatidis* in Florida, USA. Herpetological Review 40:189–190.

Rizkalla, C.E. 2010. Increasing detections of *Batrachochytrium dendrobatidis* in central Florida, USA. Herpetological Review 41:180–181.

Roberts, C.D., and T.E. Dickinson. 2012. *Ribeiroia ondatrae* causes limb abnormalities in a Canadian amphibian community. Canadian Journal of Zoology 90: 808–814.

Roberts, K.A., and R.B. Page. 2003. *Hyla cinerea* (Green Tree Frog). Reproduction. Herpetological Review 34:136.

Roberts, W.E. 1997. *Rana pretiosa* (Spotted Frog). Predation. Herpetological Review 28:86.

Roberts, W.R. 1998. The calliphorid fly (*Bufolucilia sylvarum*) parasitic on frogs in Alberta. Alberta Naturalist 28:48.

Roberts, W., and V. Lewin. 1979. Habitat utilization and population densities of the amphibians of northeastern Alberta. Canadian Field-Naturalist 93:144–154.

Robins, A., G. Lippolis, A. Bisazza, G. Vallortigara, and L.J. Rogers. 1998. Lateralized agonistic responses and hindlimb use in toads. Animal Behaviour 56:875–881.

Robinson, E.J., Jr. 1954. Notes on the occurrence and biology of filarial nematodes in southwestern Georgia. Journal of Parasitology 40:138–147.

Robinson, M., M.P. Donovan, and T.D. Schwaner. 1998. Western toad, *Bufo boreas*, in southern Utah: notes on a single population along the east fork of the Sevier River. Great Basin Naturalist 58:87–89.

Robinson, T.S. 1958. Notes on the development of a brood of Mississippi kites in Barber County, Kansas. Transactions of the Kansas Academy of Science 60:174–180.

Roble, S.M. 1979. Dispersal movements and plant associations of juvenile gray treefrogs, *Hyla versicolor*. Transactions of the Kansas Academy of Science 82:235–245.

Roble, S.M. 1985a. A study of the reproductive and population ecology of two hylid frogs in northeastern Kansas. Ph.D. Dissertation, University of Kansas, Lawrence.

Roble, S.M. 1985b. Observations on satellite males in *Hyla chrysoscelis, Hyla picta,* and *Pseudacris triseriata*. Journal of Herpetology 19:432–436.

Roble, S.M., A.C. Chazal, and A.K. Foster. 2000. A preliminary survey of the amphibians and reptiles of Savage Neck Dunes Natural Area Preserve, Northampton County, Virginia. Catesbeiana 20:63–74.

Robortella, J.M. 2006. Frogleg George: The legend no one really knew. Gates Historical Society, Rochester, New York. 36 pp.

Roby–Thomas, P.L. 2006. Amphibian colonization of new habitat : "if you build it, they will come". M.S. thesis, University of Louisville, Louisville, Kentucky.

Roche, L.M., B. Allen–Diaz, D.J. Eastburn, and K.W. Tate. 2012. Cattle grazing and Yosemite toad (*Bufo canorus* Camp) breeding habitat in Sierra Nevada meadows. Rangeland Ecology and Management 65:56–65.

Roche, L.M., A.M. Latimer, D.J. Eastburn, and K.W. Tate. 2012. Cattle grazing and conservation of a meadow–dependent amphibian species in the Sierra Nevada. PLoS One 7:e35734.

Rochester, C.J., C.S. Brehme, D.R. Clark, D.C. Stokes, S.A. Hathaway, and R.N. Fisher. 2010. Reptile and amphibian responses to large–scale wildfires in southern California. Journal of Herpetology 44:333–351.

Rödder, D. 2009. "Sleepless in Hawaii": does anthropogenic climate change enhance ecological and socioeconomic impacts of the alien invasive *Eleutherodactylus coqui* Thomas 1966 (Anura: Eleutherodactylidae)? North–Western Journal of Zoology 5:16–25.

Rödder, D., and F. Weinsheimer. 2009. Will future anthropogenic climate change increase the potential distribution of the alien invasive Cuban treefrog (Anura: Hylidae). Journal of Natural History 43:1207-1217.

Rodenhouse, N.L., L. M. Christenson, D. Parry, and L.E. Green. 2009. Climate change effects on native fauna of northeastern forests. Canadian Journal of Forest Research 39: 249–263.

Rodgers, L.O. 1941. *Diplorchis scaphiopi*, a new polystomatid monogenean fluke from the spadefoot toad. Journal of Parasitology 27:153–157.

Rodgers, L.O., and R.E. Kuntz. 1940. A new polystomatid monogenean fluke from a spadefoot. The Wasmann Collector 4:37–40.

Roe, J.H., W.A. Hopkins, and B.P. Jackson. 2005. Species- and stage–specific differences in trace element tissue concentrations in amphibians: implications for the disposal of coal–combustion wastes. Environmental Pollution 136:353–363.

Rogers, A.M. 1999. Ecology and natural history of *Rana clamitans melanota* in West Virginia. M.S. thesis, Marshall University, Huntington, West Virginia.

Rogers, C.D. 1996. *Rana catesbeiana* (Bullfrog). Predation. Herpetological Review 27:19.

Rogers, C.P. 1996. *Rana catesbeiana* (Bullfrog). Predation. Herpetological Review 27:75.

Rogers, J.S. 1972. Discriminant function analysis of morphological relationships within the *Bufo cognatus* species group. Copeia 1972:381–383.

Rogers, J.S. 1973a. Protein polymorphism, genic heterozygosity and divergence in the toads *Bufo cognatus* and *B. speciosus*. Copeia 1973:322–330.

Rogers, J.S. 1973b. Biochemical and morphological analysis of potential introgression between *Bufo cognatus* and *B. speciosus*. American Midland Naturalist 90:127–142.

Rogers, J.S. 1976. Species density and taxonomic diversity of Texas amphibians and reptiles. Systematic Zoology 25:26–40.

Rogers, K.L. 1987. Pleistocene high altitude amphibians and reptiles from Colorado (Alamosa Local Fauna; Pleistocene; Irvingtonian). Journal of Vertebrate Paleontology 7:82–95.

Rogers, K.L., and L. Harvey. 1994. A skeletochronological assessment of fossil and recent *Bufo cognatus* from south–central Colorado. Journal of Herpetology 28:133–140.

Rogers, K.L., C.A. Repenning, R.M. Forester, E.E. Larson, S.A. Hall, G.R. Smith, E. Anderson, and T.J. Brown. 1985. Middle Pleistocene (Late Irvingtonian: Nebraskan) climatic changes in south–central Colorado. National Geographic Research 1:535–563.

Rogers, M.W., Jr. 1975. Development and behavior of larvae of the western toad, *Bufo boreas*. M.A. thesis, University of Texas, Arlington, Texas.

Rogers, S.D., and M.M. Peacock. 2012. The disappearing northern leopard frog (*Lithobates pipiens*): conservation genetics and implications for remnant populations in western Nevada. Ecology and Evolution 2:2040–2056.

Rogers, T.N., and D.R. Chalcraft. 2008. Pond hydroperiod alters the effect of density–dependent processes on larval anurans. Canadian Journal of Fisheries and Aquatic Sciences 65: 2761–2768.

Rohr, J.R., and P.W. Crumrine. 2005. Effects of an herbicide and an insecticide on pond community structure and processes. Ecological Applications 15:1135–1147.

Rohr, J.R., and K.A. McCoy. 2010. A qualitative meta–analysis reveals consistent effects of atrazine on freshwater fish and amphibians. Environmental Health Perspectives 118:20–32.

Rohr, J.R., A.M. Schotthoefer, T.R. Raffel, H.J. Carrick, N. Halstead, J.T. Hoverman, C.M. Johnson, L.B. Johnson, C. Lieske, M.D. Piwoni, P.K. Schoff, and V.R. Beasley. 2008a. Agrochemicals increase trematode infections in a declining amphibian species. Nature 455:1235–1239.

Rohr, J. R., T. R. Raffel, S. K. Sessions, and P. J. Hudson. 2008b. Understanding the net effects of pesticides on amphibian trematode infections. Ecological Applications 18:1743–1753.

Rohr, J.R., A. Swan, T.R. Raffel, and P.J. Hudson. 2009. Parasites, info–disruption, and the ecology of fear. Oecologia 159:447–454.

Rollins–Smith, L.A., L.K. Reinert, V. Miera, and J.M. Conlon. 2002. Antimicrobial peptide defenses of the Tarahumara frog, *Rana tarahumarae*. Biochemical and Biophysical Research Communications 297:361–367.

Roloff, G.J., T.E. Grazia, K.F. Millenbah, and A.J. Kroll. 2011. Factors associated with amphibian occupancy in southern Michigan forests. Journal of Herpetology 45:15–22.

Romano, M.A., D.B. Ralin, S.I. Guttman, and J.H. Skillings. 1987. Parallel electromorph variation in the diploid–tetraploid gray treefrog complex. American Naturalist 130:864–878.

Romansic, J.M., K.A. Diez, E.M. Higashi, and A.R. Blaustein. 2006. Effects of nitrate and the pathogenic water mould *Saprolegnia* on survival of amphibian larvae. Diseases of Aquatic Organisms 68:235–243.

Romansic, J.M., E.M. Higashi, K.A. Diez, and A.R. Blaustein. 2007. Susceptibility of newly–metamorphosed frogs to a pathogenic water mould (*Saprolegnia* sp.). Herpetological Journal 17:161–166.

Romansic, J.M., A.A. Waggener, B.A. Bancroft, and A.R. Blaustein. 2009a. Influence of ultraviolet–B radiation on growth, prevalence of deformities, and susceptibility to predation in Cascades frog (*Rana cascadae*) larvae. Hydrobiologia 624:219–233.

Romansic, J.M., K.A. Diez, E.M. Higashi, J. Johnson, and A.R. Blaustein. 2009b. Effects of the pathogenic water mold *Saprolegnia ferax* on survival of amphibian larvae. Diseases of Aquatic Organisms 83: 187–193.

Romansic, J.M., P.T.J. Johnson, C.L. Searle, J.E. Johnson, T.S. Tunstall, B.A. Han, J.R. Rohr, and A.R. Blaustein. 2011. Individual and combined effect of multiple pathogens on Pacific treefrogs. Oecologia 166:1029–1041.

Rombough, C.J., and M.P. Hayes. 2005. *Rana boylii* (Foothill Yellow–legged Frog). Predation: eggs and hatchlings. Herpetological Review 36:163–164.

Rombough, C.J. 2010. *Rana catesbeiana* (American Bullfrog). Predation. Herpetological Review 41:204.

Rombough, C.J., and M.P. Hayes. 2005. Novel aspects of oviposition site preparation by foothill yellow–legged frogs (*Rana boylii*). Northwestern Naturalist 86:157–160.

Rombough, C.J., and C.A. Pearl. 2005. *Rana pretiosa* (Oregon Spotted Frog). Aggregation and habitat use. Herpetological Review 36:307–308.

Rombough, C.J., and A.M. Schwab. 2006. *Rana catesbeiana* (American Bullfrog). Mortality. Herpetological Review 37:448.

Rombough, C.J., and M.P. Hayes. 2007. *Rana boylii* (Foothill Yellow–legged Frog). Reproduction. Herpetological Review 38:70–71.

Rombough, C.J., and M.P. Hayes. 2008. *Rana pretiosa* (Oregon Spotted Frog). Reproduction. Herpetological Review 39:340–341.

Rombough, C.J., and M. Bradley. 2010. *Pseudacris regilla* (Northern Pacific Treefrog). Predation. Herpetological Review 41:203.

Rombough, C.J., D.J. Jordan, and C.A. Pearl. 2003. *Rana cascadae* (Cascades Frog). Cannibalism. Herpetological Review 34:138.

Rombough, C.J., J. Chastain, A.M. Schwab, and M.P. Hayes. 2005. *Rana boylii* (Foothill Yellow–legged Frog). Predation. Herpetological Review 36:438–439.

Rombough, C.J., M.P. Hayes, and J.D. Engler. 2006. *Rana pretiosa* (Oregon Spotted Frog). Maximum size. Herpetological Review 37:210.

Rondeau, S.L., and J.H. Gee. 2005. Larval amphibians adjust buoyancy in response to substrate ingestion. Copeia 2005:188–195.

Rorabaugh, J. 2004. Barking frogs (*Eleutherodactylus augusti*) of the Santa Rita Mountains. Sonoran Herpetologist 17:72–73.

Rorabaugh, J.C. 2005a. *Rana pipiens* Screber, 1782. Northern Leopard Frog. Pp. 570–577 *In* M.J. Lannoo (ed.), Amphibian Declines. The Conservation Status of United States Species. University of California Press, Berkeley.

Rorabaugh, J.C. 2005b. *Rana berlandieri* Baird. Rio Grande Leopard Frog. Pp. 530–532 *In* M.J. Lannoo (ed.), Amphibian Declines. The Conservation Status of United States Species. University of California Press, Berkeley.

Rorabaugh, J.C., and J. Humphrey. 2002. The Tarahumara frog: return of a native. Endangered Species Technical Bulletin 27(2):24–26.

Rorabaugh, J.C., and S.F. Hale. 2005. *Rana tarahumarae* Boulenger, 1917. Tarahumara Frog. Pp. 593–595 *In* M.J. Lannoo (ed.), Amphibian Declines. The Conservation Status of United States Species. University of California Press, Berkeley.

Rorabaugh, J.C., and L. Elliot. 2006. Tarahumara frog (*Rana tarahumarae*) call types and characteristics. Sonoran Herpetologist 19:134–136.

Rorabaugh, J.C., M.J. Sredl, V. Miera, and C.A. Drost. 2002. Continued invasion by an introduced frog (*Rana berlandieri*): southwestern Arizona, southeastern California, and Rio Colorado, México. Southwestern Naturalist 47:12-20.

Rorabaugh, J.C., J.M. Howland, and R.D. Babb. 2004. Distribution and habitat use of the Pacific treefrog (*Pseudacris regilla*) on the lower Colorado River and in Arizona. Southwestern Naturalist 49:94–99.

Rose, F.L. 1962. A case of albinism in *Rana pipiens* Schreber. Herpetologica 18:72.

Rose, F.L. 1963. Delayed melanism in tadpoles of *Rana pipiens* Schreber. Herpetologica 18:274–275.

Rose, F.L., T.R. Simpson, M.R.J. Forstner, D.J. McHenry, and J. Williams. 2006. Taxonomic status of *Acris gryllus paludicola*: in search of the pink frog. Journal of Herpetology 40:428–434.

Rose, G.J., and E.A. Brenowitz. 1991. Aggressive thresholds of male Pacific treefrogs for advertisement calls vary with amplitude of neighbors' calls. Ethology 89:244–252.

Rose, G.J., and E.A. Brenowitz. 1997. Plasticity of aggressive thresholds in *Hyla regilla*: discrete accommodation to encounter calls. Animal Behaviour 53:353–361.

Rose, G.J., and E.A. Brenowitz. 2002. Pacific treefrogs use temporal integration to differentiate advertisement from encounter calls. Animal Behaviour 63:1183–1190.

Rose, S.M. 1959. Failure of survival of slowly growing members of a population. Science 129:1026.

Rose, S.M. 1960. A feedback mechanism of growth control in tadpoles. Ecology 41:188–199.

Rose, S.M., and F.C. Rose. 1961. Growth–controlling exudates of tadpoles. Symposia of the Society for Experimental Biology 15:207–218.

Rosen, M., and L.E. Lemon. 1974. The vocal behavior of spring peepers, *Hyla crucifer*. Copeia 1974:940–950.

Rosen, P.C. 2007. *Rana yavapaiensis* (Lowland Leopard Frog). Larval cannibalism. Herpetological Review 38:195–196.

Rosen, P.C. 2011. *Rana yavapaiensis* (Lowland Leopard Frog). Larval cannibalism. Herpetological Review 42:589–590.

Rosen, P.C., and C.R. Schwalbe. 2002. Widespread effects of introduced species on reptiles and amphibians in the Sonoran Desert region. Pp. 220–240 *In* B. Tellman (ed.), Invasive Exotic Species in the Sonoran Region. University of Arizona Press and the Arizona–Sonora Desert Museum, Tucson, Arizona.

Rosen, P.C., C.R. Schwalbe, D.A. Parizek, Jr., P.A. Holm, and C.H. Lowe. 1995. Introduced aquatic vertebrates in the Chiricahua region: effects on declining native ranid frogs. Pp. 251–260 *In* L.F. DeBano, G.J. Gottfried, R.H. Hamre, C.B. Edminster, P.F. Ffolliott, and A. Ortega–Rubio (technical coordinators), Madrean Archipelago: the Sky Islands of the Southwestern United States and Northwestern Mexico. U.S. Forest Service Rocky Mountain Station General Technical Report RM–GTR–264.

Rosenberg, E.A. 1990. Effect of low pH on individual tadpole growth in *Pseudacris clarkii* and *Bufo valliceps*. M.S. thesis, Baylor University, Waco, Texas.

Rosenberg, E.A., and B.A. Pierce. 1995. Effect of initial mass on growth and mortality at low pH in tadpoles of *Pseudacris clarkii* and *Bufo valliceps*. Journal of Herpetology 29:181–185.

Rosenberry, D.O. 2001. Malformed frogs in Minnesota: an update. USGS Fact Sheet 043–01, Mounds View, Minnesota.

Rosenfield, R.N., M.W. Gratson, and L.B. Carson. 1984. Food brought by broad–winged hawks to a Wisconsin nest. Journal of Field Ornithology 55:246–247.

Rosenshield, M.L., M.B. Jofré, and W.H. Karasov. 1999. Effects of polychlorinated biphenyl 126 on green frog (*Rana clamitans*) and leopard frog (*R. pipiens*) hatching success, development, and metamorphosis. Environmental Toxicology and Chemistry 18:2478–2486.

Rosine, W.N. 1952. Notes on the occurrence of polydactylism in a second species of Amphibia in Muskee Lake, Colorado. Journal of the Colorado–Wyoming Academy of Science 4:100.

Rosine, W.N. 1955. Polydactylism in a second species of amphibia in Muskee Lake, Colorado. Copeia 1955:136.

Ross, D.A., and B.J. Richardson. 1996. *Rana pretiosa* (Spotted Frog). Basking behavior. Herpetological Review 26:203.

Ross, D.A., D.W. Kuehn, and M.C. Stanger. 1994. *Rana pretiosa* (Spotted Frog). Reproduction. Herpetological Review 25:118.

Ross, D.A., T.C. Esque, R.A. Fridell, and P. Hovingh. 1995. Historical distribution, current status, and a range extension of *Bufo boreas* in Utah. Herpetological Review 26:187–189.

Ross, D.A., J.K. Reaser, P. Kleeman, and D.L. Drake. 1999. *Rana luteiventris* (Columbia Spotted Frog). Mortality and site fidelity. Herpetological Review 30:163.

Rossi, J.V. 1981. *Bufo marinus* in Florida: some natural history and its impact on native vertebrates. M.A. thesis, University of South Florida, Tampa.

Rossi, J.V. 1983. The use of olfactory cues by *Bufo marinus*. Journal of Herpetology 17:72–73.

Rossman, D.A. 1959. Chorus frog (*Pseudacris nigrita* ssp.) intergradation in southwestern Illinois. Herpetologica 15:38–40.

Rossman, D.A. 1960. Herpetofaunal survey of the Pine Hills area of southern Illinois. Quarterly Journal of the Florida Academy of Sciences 22:207–225.

Roster, N.O., D.L. Clark, and J.C. Gillingham. 1995. Prey catching behavior in frogs and toads using video–simulated prey. Copeia 1995:496–498.

Roth, A.H., and J.F. Jackson. 1987. The effect of pool size on recruitment of predatory insects and on mortality in a larval anuran. Herpetologica 43:224–232.

Roth, E.D., P.G. May, and T.M. Farrell. 1999. Pigmy rattlesnakes use frog–derived chemical cues to select foraging sites. Copeia 1999:772–774.

Rothermel, B.B. 2004. Migratory success of juveniles: a potential constraint on connectivity for pond–breeding amphibians. Ecological Applications 14:1535–1546.

Rothermel, B.B., and R.D. Semlitsch. 2002. An experimental investigation of landscape resistance of forest versus old–field habitats to emigrating juvenile amphibians. Conservation Biology 16:1324–1332.

Rothermel, B.B., S.C. Walls, J.C. Mitchell, C.K. Dodd Jr., L.K. Irwin, D.E. Green, V.M. Vasquez, J.W. Petranka, and D.J. Stevenson. 2008. Widespread occurrence of the amphibian chytrid fungus (*Batrachochytrium dendrobatidis*) in the Southeastern United States. Diseases of Aquatic Organisms 82:3–18.

Rot–Nikcevic, I., C.N. Taylor, and R.J. Wassersug. 2006. The role of images of conspecifics as visual cues in the development and behavior of larval amphibians. Behavioral Ecology and Sociobiology 60:19–25.

Rouse, J.D., C.A. Bishop, and J. Struger. 1999. Nitrogen pollution: an assessment of its threat to amphibian survival. Environmental Health Perspectives 107:799–803.

Rowe, C.F.B. 1899. Report on Zoology. Batrachians. Bulletin of the Natural History Society of New Brunswick 17:169–170.

Rowe, C.L., and W.A. Dunson. 1993. Relationships among abiotic parameters and breeding effort by three amphibians in temporary wetlands of central Pennsylvania. Wetlands 13:237–246.

Rowe, C.L., and W.A. Dunson. 1995. Impacts of hydroperiod on growth and survival of larval amphibians in temporary ponds of central Pennsylvania, USA. Oecologia 102:397–403.

Rowe, C.L., and J. Freda. 2000. Effects of acidification on amphibians at multiple levels of biological organization. Pp. 545–571 *In* D.W. Sparling, G. Linder, and C.A. Bishop (eds.), Ecotoxicology of Amphibians and Reptiles. SETAC Press, Pensacola, Florida.

Rowe, C.L., and W.A. Hopkins. 2003. Anthropogenic activities producing sink habitats for amphibians in the local landscape: a case study of lethal and sublethal effects of coal combustion residues in the aquatic environment. Pp. 271–282 *In* G. Linder, S.K. Krest, and D.W. Sparling (eds.), Amphibian Decline: An Integrated Analysis of Multiple Stressor Effects. SETAC Press, Pensacola, Florida.

Rowe, C.L., W. J. Sadinski, and W.A. Dunson. 1994. Predation of larval and embryonic amphibians by acid–tolerant caddisfly larvae (*Ptilostomis postica*). Journal of Herpetology 28:357–364.

Rowe, C.L., O.M. Kinney, A.P. Fiori, and J.D. Congdon. 1996. Oral deformities in tadpoles (*Rana catesbeiana*) associated with coal ash deposition: effects on grazing ability and growth. Freshwater Biology 36:723–730.

Rowe, C.L., O.M. Kinney, R.D. Nagle, and J.D. Congdon. 1998a. Elevated maintenance costs in an anuran (*Rana catesbeiana*) exposed to a mixture of trace elements during the embryonic and early larval periods. Physiological Zoology 71:27–35.

Rowe, C.L., O.M. Kinney, and J.D. Congdon. 1998b. Oral deformities in tadpoles of the bullfrog (*Rana catesbeiana*) caused by conditions in a polluted habitat. Copeia 1998:244–246.

Rowe, C.L., W.A. Hopkins, and V.R. Coffman. 2001. Failed recruitment of southern toads (*Bufo terrestris*) in a trace element–contaminated breeding habitat: direct and indirect effects that may lead to a local population sink. Archives of Environmental Contamination and Toxicology 40:399–405.

Rowe, C.L., A. Heyes, and W.A. Hopkins. 2009. Effects of dietary vanadium on growth and lipid storage in a larval anuran: Results from studies employing ad libitum and rationed feeding. Aquatic Toxicology 91:179–186.

Rowe, C.L., A. Heyes, and J. Hilton. 2011. Differential patterns of accumulation and depuration of dietary selenium and vanadium during metamorphosis in the gray treefrog *(Hyla versicolor)*. Archives of Environmental Contamination and Toxicology 60:336–342.

Rowe, J.C., and T.S. Garcia. 2012. *Lithobates catesbeianus* (American Bullfrog). Diet. Herpetological Review 43:633–634.

Roy, J.–S. 2009. Structure and dynamics of a natural hybrid zone between the toads, *Anaxyrus americanus* and *Anaxyrus hemiophrys*, in southeastern Manitoba. M.S. thesis, McGill University, Montreal.

Rozdzial, M.M. 1980. Digestive assimilation and biomass production efficiencies and gut evacuation rate as a function of temperature in four anuran amphibians: a bioenergetic approach. M.S. thesis, California State University, Fullerton.

Roznik, E.A. 2007. Terrestrial ecology of juvenile and adult Gopher Frogs (*Rana capito*). M.S. thesis, University of Florida, Gainesville.

Roznik, E.A., and S.A. Johnson. 2007. *Rana capito* (Gopher Frog). Refuge during fire. Herpetological Review 38:442.

Roznik, E.A., and S.A. Johnson. 2009a. Canopy closure and emigration by juvenile gopher frogs. Journal of Wildlife Management 73:260–268.

Roznik, E.A., and S.A. Johnson. 2009b. Burrow use and survival of newly metamorphosed gopher frogs (*Rana capito*). Journal of Herpetology 43:431–437.

Roznik, E.A., and S.A. Johnson. 2009c. *Rana capito* (Gopher frog). Burrow cohabitation. Herpetological Review 40:209.

Roznik, E.A., S.A. Johnson, C.H. Greenberg, and G.W. Tanner. 2009. Terrestrial movements and habitat use of gopher frogs in longleaf pine forests: a comparative study of juveniles and adults. Forest Ecology and Management 259:187–194.

Rubbo, M.J., and J.M. Kiesecker. 2004. Leaf litter composition and community structure: translating regional species changes into local dynamics. Ecology 85:2519–2525.

Rubbo, M.J., and M.J. Kiesecker. 2005. Amphibian breeding distribution in an urbanized landscape. Conservation Biology 19:504-511.

Rubbo, M.J., L.K. Belden, and J.M. Kiesecker. 2008. Differential responses of aquatic consumers to variations in leaf–litter inputs. Hydrobiologia 605:37–44.

Rudolph, D.C., and J.G. Dickson. 1990. Streamside zone width and amphibian and reptile abundance. Southwestern Naturalist 35:472–476.

Ruffner, B.M. 1933. Practical Frog Raising. Southern Frog Farms, Jennings, Louisiana. 80 pp.

Rugh, R. 1934. The space factor in the growth rate of tadpoles. Ecology 15:407–411.

Rugh, R. 1941. Experimental studies on the reproductive physiology of the male spring peeper, *Hyla crucifer*. Proceedings of the American Philosophical Society 84:617–633.

Ruibal, R. 1957. An altitudinal and latitudinal cline in *Rana pipiens*. Copeia 1957:212–221.

Ruibal, R. 1959. The ecology of a brackish water population of *Rana pipiens*. Copeia 1959:315–322.

Ruibal, R. 1962. The ecology and genetics of a desert population of *Rana pipiens*. Copeia 1962:189–195.

Ruibal, R., and S. Hillman. 1981. Cocoon structure and function in the burrowing hylid frog, *Pternohyla fodiens*. Journal of Herpetology 15:403–408.

Ruibal, R., L. Tevis, Jr., and V. Roig. 1969. The terrestrial ecology of the spadefoot toad *Scaphiopus hammondii*. Copeia 1969:571–584.

Ruiz, A.M., J.C. Maerz, A.K. Davis, M.K. Keel, A.R. Ferreira, M.J. Conroy, L.A. Morris, and A.T. Fisk. 2010. Patterns of development and abnormalities among tadpoles in a constructed wetland receiving treated wastewater. Environmental Science & Technology 44:4862–4868.

Rundquist, E.M. 1978. The spring peeper, *Hyla crucifer* Wied (Anura, Hylidae) in Kansas. Transactions of the Kansas Academy of Science 80:155–158.

Runkle, L.S., K.D. Wells, C.C. Robb, and S.L. Lance. 1994. Individual, nightly, and seasonal variation in calling behavior of the gray tree frog, *Hyla versicolor*: implications for energy expenditure. Behavioral Ecology 5:318–325.

Rusling, W.J. 1951. Food habits of New Jersey owls. Proceedings of the Linnean Society of New York 58–62:38–45.

Russell, A.P., and A.M. Bauer. 2000. The Amphibians and Reptiles of Alberta. A Field Guide and Primer of Boreal Herpetology. 2nd ed. University of Calgary Press, Calgary, Alberta. [first edition, 1993]

Russell, D.M., C.S. Goldberg, L.P. Watts, and E.B. Rosenblum. 2010. *Batrachochytrium dendrobatidis* infection dynamics in the Columbia spotted frog, *Rana luteiventris*, in north Idaho. Diseases of Aquatic Organisms 92:223-230.

Russell, D.M., C.S. Goldberg, L. Sprague, L.P. Waits, D.E. Green, K.L. Schuler, and E.B. Rosenblum. 2011. Ranavirus outbreaks in amphibian populations of northern Idaho. Herpetological Review 42:223–225.

Russell, K.R., D.H. Van Lear, and D.C. Guynn, Jr. 1999. Prescribed fire effects on herpetofauna: review and management implications. Wildlife Society Bulletin 27:374–384.

Russell, K.R., D.C. Guynn, Jr., and H.G. Hanlin. 2002a. Importance of small isolated wetlands for herpetofaunal diversity in managed, young growth forests in the Coastal Plain of South Carolina. Forest Ecology and Management 163:43–59.

Russell, K.R., H.G. Hanlin, T.B. Wigley, and D.C. Guynn, Jr. 2002b. Responses of isolated wetland herpetofauna to upland forest management. Journal of Wildlife Management 66:603–617.

Russell, K.R., T.B. Wigley, W.M. Baughman, H.G. Hanlin, and W.M. Ford. 2004. Responses of southeastern amphibians and reptiles to forest management: a review. Pp. 319-334 *In* H.H. Rauscher, and K. Johnson (eds.), Southern Forest Science: Past, Present, and Future. USDA Forest Service General Technical Report SRS-75. Southern Research Station, Asheville, NC.

Russell, R.W. 2008. Spring peepers and pitcher plants: a case of commensalism? Herpetological Review 39:154–155.

Russell, R.W., S.J. Hecnar, and G.D. Haffner. 1995. Organochlorine pesticide residues in southern Ontario spring peepers. Environmental Toxicology and Chemistry 14:815–817.

Russell, R.W., K.A. Gillan, and G.D. Haffner.1997. Polychlorinated biphenyls and chlorinated pesticides in southern Ontario, Canada, green frogs. Environmental Toxicology and Chemistry 16:2258–2263.

Russell, R.W., G.J. Lipps, Jr., S.J. Hecnar, and G.D. Haffner. 2002. Persistent organic pollutants in Blanchard's cricket frogs (*Acris crepitans blanchardi*) from Ohio. Ohio Journal of Science 102:119-122.

Rutherford, P.L., D.L. McRuer, and M.R. Forbes. 2005. Condition and immune traits of frogs from Ontario baitshops: risks of practice not ameliorated by sale of healthy frogs. Herpetological Review 36:129–133.

Ruthig, G.R. 2008. The influence of temperature and spatial distribution on the susceptibility of southern leopard frog eggs to disease. Oecologia 156:895–903.

Ruthig, G.R. 2009. Water molds of the genera *Saprolegnia* and *Leptolegnia* are pathogenic to the North American frogs *Rana catesbeiana* and *Pseudacris crucifer*, respectively. Diseases of Aquatic Organisms 84:173–178.

Ruthig, G.R., and K.N. Provost–Javier. 2012. Multihost saprobes are facultative pathogens of bullfrog *Lithobates catesbeianus* eggs. Diseases of Aquatic Organisms 101:13–21.

Ruthven, A. G. 1907. A collection of reptiles and amphibians from southern New Mexico and Arizona. Bulletin of the American Museum of Natural History 23:483–604.

Ruthven, A.G. 1908. The cold–blooded vertebrates of Isle Royale. Ecology of Isle Royale, Michigan Survey 1908:329–333.

Ruthven, A.G. 1910. A contribution to the herpetology of Iowa. Proceedings of the Iowa Academy of Science for 1910:198–209.

[5]Ruthven, A.G. 1912. The herpetology of Michigan. Michigan Geological and Biological Survey Publication 10:11–166.

Ruthven, A. G. 1917. On the occurrence of *Bufo fowleri* in Michigan. Occasional Papers of the Museum of Zoolo-

---

[5]The amphibian sections of this monograph were actually written by Crystal Thompson and Helen Thompson.

gy, University of Michigan. (47):1–5.

Ruthven, A.G., and H.T. Gaige. 1915. The reptiles and amphibians collected in northeastern Nevada by the Walker–Newcomb expedition of the University of Michigan. Occasional Papers of the Museum of Zoology, University of Michigan 8:1–33.

Ruthven III, D.C., R.T. Kazmaier, J.F. Gallagher, and D.R. Synatzke. 2002. Seasonal variation in herpetofaunal abundance and diversity in the south Texas plains. Southwestern Naturalist 47:102–109.

Ruthven III, D.C., R.T. Kazmaier, and M.W. Janis. 2008. Short–term response of herpetofauna to various burning regimes in the south Texas plains. Southwestern Naturalist 53:280–287.

Ryan, M.J. 1978. A thermal property of the *Rana catesbeiana* (Amphibia, Anura, Ranidae) egg mass. Journal of Herpetology 12:247–248.

Ryan, M.J. 1980. The reproductive behavior of the bullfrog (*Rana catesbeiana*). Copeia 1980:108–114.

Ryan, M.J., and W. Wilczynski. 1988. Coevolution of sender and receiver: effect on local mate preference in cricket frogs. Science 240:1786–1788.

Ryan, M.J., and B.K. Sullivan. 1989. Transmission effects on the temporal structure of the advertisement call of two species of toads, *Bufo woodhousii* and *Bufo valliceps*. Ethology 80:182–185.

Ryan, M.J., and W. Wilczynski. 1990 (1991). Evolution of intraspecific variation in the advertisement call of a cricket frog (*Acris crepitans*, Hylidae). Biological Journal of the Linnean Society 44:249–271.

Ryan, M.J., R.B. Cocroft, and W. Wilczynski. 1991. The role of environmental selection in intraspecific divergence of mate recognition signals in the cricket frog, *Acris crepitans*. Evolution 44:1869–1872.

Ryan, M.J., S.A. Perrill, and W. Wilczynski. 1992. Auditory tuning and call frequency predict population–based mating preferences in the cricket frog, *Acris crepitans*. American Naturalist 139: 1370–1383.

Ryan, M.J., K.M. Warkentin, B.E. McClelland, and W. Wilcznyski. 1995. Fluctuating asymmetries and advertisement call variation in the cricket frog, *Acris crepitans*. Behavioral Ecology 6:124–131.

Ryan, R.A. 1953. Growth rates of some ranids under natural conditions. Copeia 1953:73–80.

Ryan, T.J., and C.T. Winne. 2001. Effects of hydroperiod on metamorphosis in *Rana sphenocephala*. American Midland Naturalist 145:46–53.

Ryberg, W.A., and G.H. Dayton. 2004. *Scaphiopus couchii* (Couch's Spadefoot). Predation. Herpetological Review 35:263.

Ryder, J.A. 1878. A monstrous frog. American Naturalist 12:751–752.

Ryder, J.A. 1891. Notes on the development of *Engystoma*. American Naturalist 25:838–840.

Saber, P.A., and W.A. Dunson. 1978. Toxicity of bog water to embryonic and larval anuran amphibians. Journal of Experimental Zoology 204:33–42.

Sacerdote, A.B. 2009. Reintroduction of extirpated flatwoods amphibians into restored forested wetlands of northern Illinois: feasibility assessment, implementation, habitat restoration, and conservation implications. Ph.D. Dissertation, Northern Illinois University, DeKalb.

Sadinski, W.J., and W.A. Dunson. 1992. A multilevel study of effects of low pH on amphibians of temporary ponds. Journal of Herpetology 26:413–422.

Sadinski, W.J., M. Roth, S. Treleven, J. Theyerl, and P. Dummer. 2010. Detection of the chytrid fungus, *Batrachochytrium dendrobatidis*, on recently metamorphosed amphibians in the north–central United States. Herpetological Review 41:170–175.

Saenz, D., J.B. Johnson, C.K. Adams, and G.H. Dayton. 2003. Accelerated hatching of southern leopard frog (*Rana sphenocephala*) eggs in response to the presence of a crayfish (*Procambarus nigrocinctus*) predator. Copeia 2003:646–649.

Saenz, D., L.A. Fitzgerald, K.A. Baum, and R.N. Connor. 2006. Abiotic correlates of anuran calling phenology: the importance of rain, temperature and season. Herpetological Monographs 20:64–82.

Saenz, D., B.T. Kavanagh, and M.A. Kwiatkowski. 2010. *Batrachochytrium dendrobatidis* detected in amphibians from National Forests in eastern Texas, USA. Herpetological Review 41:47–49.

Safford, W. E. 1917. Natural history of Paradise Key and the nearby Everglades of Florida. Annual Report of the Smithsonian Institution for 1917, pp. 377-434. Folding map, 64 plates.

Sage, R.D., and R.K. Selander. 1979. Hybridization between species of the *Rana pipiens* complex in central Texas. Evolution 33:1069–1088.

Sagor, E.S., M. Ouellet, E. Barten, and D.M. Green. 1998. Skeletochronology and geographic variation in age structure in the wood frog, *Rana sylvatica*. Journal of Herpetology 32:469–474.

Salice, C.J. 2012. Multiple stressors and amphibians: contributions of adverse health effects and altered hydroperiod to population decline and extinction. Journal of Herpetology 46:675–681.

Salice, C.J., C.L. Rowe, J.H.K. Pechmann, and W.A. Hopkins. 2011. Multiple stressors and complex life cycles: insights from a population–level assessment of breeding site contamination and terrestrial habitat loss in an amphibian. Environmental Toxicology and Chemistry

30:2874–2882.

Salinas, C. 2009. The influence of environmental factors on the activity and movement of *Bufo nebulifer* (Coastal Plain Toad) in a disturbed area. M.S. thesis, University of Texas–Pan American, Edinburg, Texas.

Salt, G.W., and R.C. Stebbins. 1948. Northward occurrence of the canyon tree–toad in California. Copeia 1948:131.

Salt, J.R. 1979. Some elements of amphibian distribution and biology in the Alberta Rockies. Alberta Naturalist 9:125–136.

Salthe, S.N. 1969. Geographic variation of the lactate dehydrogenase of *Rana pipiens* and *Rana palustris*. Biochemical Genetics 2:271–303.

Salthe, S.N., and E. Nevo. 1969. Geographic variation in lactate dehydrogenase in the cricket frog, *Acris crepitans*. Biochemical Genetics 3:335–341.

Salthe, S.N., and M.L. Crump. 1977. A Darwinian interpretation of hindlimb variability in frog populations. Evolution 31:737–749.

Samollow, P.B. 1980. Selective mortality and reproduction in a natural population of *Bufo boreas*. Evolution 34:18–39.

Samoray, S.T., and K.J. Regester. 2001. Geographic distribution: *Rana sylvatica* (Wood Frog). Herpetological Review 32:190.

Sams, E., and M.D. Boone. 2010. Interactions between recently metamorphosed green frogs and American toads under laboratory conditions. American Midland Naturalist 163:269–279.

Sanabria, E.A., and L.B. Quiroga. 2010. *Lithobates catesbeianus* (American Bullfrog). Diet. Herpetological Review 41:339.

Sanchiz, B. 1998. Handbuch der Paläoherpetologie. 4. Salientia. Verlag Dr. Friedrich Pfeil, Munich, Germany.

Sanders, H.O. 1970. Pesticide toxicities to tadpoles of the western chorus frog *Pseudacris triseriata* and Fowler's toad *Bufo woodhousii fowleri*. Copeia 1970:246–251.

Sanders, O. 1953. A new species of toad, with a discussion of morphology of the bufonid skull. Herpetologica 9:25–47.

Sanders, O. 1961. Indications for the hybrid origin of *Bufo terrestris* Bonnaterre. Herpetologica 17:145–156.

Sanders, O. 1978. *Bufo woodhousei* in central Texas. Bulletin of the Maryland Herpetological Society 14:55–66.

Sanders, O. 1986. The heritage of *Bufo woodhousei* Girard in Texas (Salientia: Bufonidae). Occasional Papers of the Strecker Museum (1):1–28.

Sanders, O. 1987. Evolutionary hybridization and speciation in North American indigenous bufonids. Privately published, Dallas, Texas.

Sanders, O., and H.M. Smith. 1951. Geographic variation in toads of the *debilis* group of *Bufo*. Field & Laboratory 19:141–160.

Sanders, O., and J.C. Cross. 1964. Relationships between certain North American toads as shown by cytological study. Herpetologica 19:248–255.

Sanders, O., and H.S. Smith. 1971. Skin tags and ventral melanism in the Rio Grande leopard frog. Journal of Herpetology 5:31–38.

Santana, F.E. 2012. Mountain Yellow–legged frog (*Rana muscosa*) conservation: multiple approaches. M.S. thesis, San Diego State University, San Diego, California.

Sanwald, W. 1915. Green frog active in December. Copeia (30):35.

Sanzo, D. 2005. Water chemistry: its effects on amphibians in northwestern Ontario, Canada. M.S. thesis, Lakehead University, Thunder Bay, Ontario.

Sanzo, D., and S.J. Hecnar. 2006. Effects of road de–icing salt (NaCl) on larval wood frogs (*Rana sylvatica*). Environmental Pollution 140:247–256.

Sargent, L.G. 2000. Frog and toad population monitoring in Michigan. Journal of the Iowa Academy of Science 107:195–199.

Sartorius, S.S., and P.C. Rosen. 2000. Breeding phenology of the lowland leopard frog (*Rana yavapaiensis*): implications for conservation and ecology. Southwestern Naturalist 45:267–273.

Sattler, P.W. 1980. Genetic relationships among selected species of North American *Scaphiopus*. Copeia 1980:605–610.

Sattler, P.W. 1985. Introgressive hybridization between spadefoot toads *Scaphiopus bombifrons* and *S. multiplicatus* (Salientia: Pelobatidae). Copeia 1985:324–332.

Saumure, R.A. 1993a. *Rana clamitans* (Green Frog). Albinism. Herpetological Review 24:31.

Saumure, R.A. 1993b. *Rana catesbeiana* (Bullfrog). Predation. Herpetological Review 24:30–31.

Saunders, H.O. 1970. Pesticide toxicities to tadpoles of the western chorus frog, *Pseudacris triseriata*, and Fowler's toad, *Bufo woodhousii fowleri*. Copeia 1970:246–251.

Saura–Mas, S., M.D. Boone, and C.M. Bridges. 2002. Evaluation of direct effects of an insecticide on gray treefrogs: laboratory and field trials. Journal of Herpetology 36:715–719.

Savage, A.E., and J.R. Jaeger. 2009. Isolation and characterization of microsatellite markers in the lowland leopard frog (*Rana yavapaiensis*) and the relict leopard frog (*Rana onca*), two declining frogs of the North American desert southwest. Molecular Ecology Resources 9:199–202.

Savage, A.E., M.J. Sredl, and K.R. Zamudio. 2011. Disease dynamics vary spatially and temporally in a North American amphibian. Biological Conservation 144:1910–

1915.

Savage, J.M. 1954. A revision of the toads of the *Bufo debilis* complex. Texas Journal of Science 6:83–112.

Savage, J.M. 1959. A preliminary biosystematic analysis of the toads of the *Bufo boreas* group in Nevada and California. Yearbook of the American Philosophical Society 1959:251–254.

Savage, J.M., and F.W. Schuierer. 1961. The eggs of toads of the *Bufo boreas* group, with descriptions of the eggs of *Bufo exsul* and *Bufo nelsoni*. Bulletin of the Southern California Academy of Sciences 60:93–99.

Say, T. 1822. *In* E. James, Account of an expedition from Pittsburgh to the Rocky Mountains, performed in the years 1819 and '20, by order of the Honorable J.C. Calhoun, Secretary of War, under the command of Major Stephen H. Long. Vol. 2. H.C. Carey and I. Lea, Philadelphia, Pennsylvania.

Scarlata, J.K., and C.G. Murphy. 2003. Timing of oviposition by female barking treefrogs (*Hyla gratiosa*). Journal of Herpetology 37:580–582.

Schaaf, R.T., and J.S. Garton. 1970. Raccoon predation on the American toad, *Bufo americanus*. Herpetologica 26:334–335.

Schaaf, R.T., Jr., and P.W. Smith. 1970. Geographic variation in the pickerel frog. Herpetologica 26:240–254.

Schalk, G. 1997. Single and interactive effects of low pH and leech predators on frog embryos and adults. M.S. thesis, Carleton University, Ottawa, Ontario.

Schalk, G., M.R. Forbes, and P.J. Weatherhead. 2002. Developmental plasticity and growth rates of green frog (*Rana clamitans*) embryos and tadpoles in relation to a leech (*Macrobdella decora*) predator. Copeia 2002:445–449.

Schank, C.M.M. 2008. Assessing the effects of trout stocking on native amphibian communities in small boreal foothills lakes of Alberta. M.S. thesis, University of Alberta, Edmonton.

Schank, C.M.M., C.A. Paszkowski, W.M. Tonn and G.J. Scrimgeourb. 2011. Stocked trout do not significantly affect wood frog populations in boreal foothills lakes. Canadian Journal of Fisheries and Aquatic Sciences 68:1790–1801.

Schardien, B.J., and J.A. Jackson. 1982. Killdeers feeding on frogs. Wilson Bulletin 94:85–87.

Schaub, D.L., and J.H. Larsen, Jr. 1978. The reproductive ecology of the Pacific treefrog (*Hyla regilla*). Herpetologica 34:409–416.

Schechtman, A.M., and J.B. Olson. 1941. Unusual temperature tolerance of an amphibian egg (*Hyla regilla*). Ecology 22:409–410.

Schell, S.C. 1964. *Bunoderella metteri* gen. and sp. n. (Trematoda: Allocreadiidae) and other trematode parasites of

Ascaphus truei Stejneger. Journal of Parasitology 50:652–655.

Scherer, R.D. 2008. Detection of wood frog egg masses and implications for monitoring amphibian populations. Copeia 2008:669–672.

Scherer, R.D. 2010. Monitoring amphibian populations and the status of wood frogs and boreal chorus frogs in the Kawuneeche Valley of Rocky Mountain National Park. Ph.D. Dissertation, Colorado State University, Fort Collins.

Scherer, R.D., and J.A. Tracey. 2011. A power analysis for the use of counts of egg masses to monitor wood frog (*Lithobates sylvaticus*) populations. Herpetological Conservation and Biology 6:81–90.

Scherer, R.D., E. Muths, B.R. Noon, and P.S. Corn. 2005. An evaluation of weather and disease as causes of decline in two populations of boreal toads. Ecological Applications 15:2150–2160.

Scherer, R.D., E. Muths, and B.A. Lambert. 2008. Effects of weather on survival in populations of boreal toads in Colorado. Journal of Herpetology 42:508–517.

Scherer, R.D., E. Muths, B.R. Noon, and S. J. Oyler–McCance. 2012a. The genetic structure of a relict population of wood frogs. Conservation Genetics 13:1521–1530.

Scherer, R.D., E. Muths, and B.R. Noon. 2012b. The importance of local and landscape-scale processes to the occupancy of wetlands by pond-breeding amphibians. Population Ecology 54:487-498.

Scherff–Norris, K.L., and L.J. Livo. 1999. *Bufo boreas* (Boreal Toad). Phoretic host. Herpetological Review 30:162.

Schiesari, L.C. 2004. Performance tradeoffs across resource gradients in anuran larvae. Ph.D. Dissertation, University of Michigan, Ann Arbor.

Schiesari, L. 2006. Pond canopy cover: a resource gradient for anuran larvae. Freshwater Biology 51:412–423.

Schiesari, L., S. D. Peacor, and E. E. Werner. 2006. The growth–mortality tradeoff: evidence from anuran larvae and consequences for species distributions. Oecologia 149:194–202.

Schlaepfer, M.A., M.J. Sredl, P.C. Rosen, and M.J. Ryan. 2007. High prevalence of *Batrachochytrium dendrobatidis* in wild populations of lowland leopard frogs *Rana yavapaiensis* in Arizona. EcoHealth 4:421–427.

Schlauch, F.C. 1971. The subspecific status of the leopard frogs in a region of the Pine Barrens of Long Island. Engelhardtia 4:47–49.

Schlauch, F.C. 1978. New methodologies for measuring species status and their application to the herpetofauna of a suburban region. Engelhardtia 6:30–41.

Schlefer, E.K., M.A. Romano, S.I. Guttman, and S.B. Ruth.

1986. Effects of twenty years of hybridization in a disturbed habitat on *Hyla cinerea* and *Hyla gratiosa*. Journal of Herpetology 20:210–221.

Schlichter, L.C. 1981. Low pH affects the fertilization and development of *Rana pipiens* eggs. Canadian Journal of Zoology 59:1693–1699.

Schloegel, L.M., A.M. Picco, A.M. Kilpatrick, A.J. Davies, and A.D. Hyatt. 2009. Magnitude of the US trade in amphibians and presence of *Batrachochytrium dendrobatidis* and ranavirus infection in imported North America bullfrogs (*Rana catesbeiana*). Biological Conservation 142:1420–1426.

Schloegel, L.M., L.F. Toledo, J.E. Longcore, S.E. Greenspan, C.A. Vieira, M. Lee, S. Zhao, C. Wamgen, C.M. Ferreira, M. Hipolito, A.J. Davies, C.A. Cuomo, P. Daszak, and T.Y. James. 2012. Novel, panzootic and hybrid genotypes of amphibian chytridiomycosis associated with the bullfrog trade. Molecular Ecology 21: 5162–5177.

Schmetterling, D.A., and M.K. Young. 2008. Summer movements of boreal toads (*Bufo boreas*) in two western Montana basins. Journal of Herpetology 42:111–123.

Schmid, W.D. 1965a. Some aspects of the water economies of nine species of amphibians. Ecology 46:261–269.

Schmid, W.D. 1965b. High temperaure tolerances of *Bufo hemiophrys* and *Bufo cognatus*. Ecology 46:559–560.

Schmid, W.D. 1982. Survival of frogs in low temperatures. Science 215:697–698.

Schmid, W.D. 1986. Winter ecology. Ekologiia (Soviet Journal of Ecology) 6:29–35.

Schmidt, K.P. 1924. A list of amphibians and reptiles collected near Charleston, S.C. Copeia (132):67–69.

Schmidt, K.P. 1929. The frogs and toads of the Chicago area. Field Museum of Natural History, Zoology Leaflet 11.

Schmidt, K.P. 1935. How to make money from frog-farming. Turtox News 13:75–76.

Schmidt, K.P. 1938. A geographic variation gradient in frogs. Field Museum of Natural History, Zoology Series 20(29):377–382.

Schmidt, K.P. 1946. How to make money from frog–farming. Turtox News 24:169–170.

Schmidt, K.P. 1953. A Check List of North American Amphibians and Reptiles. 6th ed. American Society of Ichthyologists and Herpetologists, University of Chicago Press, Chicago, Illinois.

Schmidt, K.P., and W.L. Necker. 1935. Amphibians and reptiles of the Chicago region. Bulletin of the Chicago Academy of Sciences 5:57–77.

Schmidt, K.P., and T.F. Smith. 1944. Amphibians and reptiles of the Big Bend region of Texas. Zoological Series of Field Museum of Natural History 29:75–96.

Schmidt, R.S. 1968. Chuckle calls of the leopard frog (*Rana pipiens*). Copeia 1968:561–569.

Schmidt, R.S. 1971a. A model of the central mechanisms of male anuran acoustic behaviour. Behaviour 39:288–317.

Schmidt, R.S. 1971b. Neural mechanisms of mating call orientation in female toads (*Bufo americanus*). Copeia 1971:545–548.

Schmidt, R.S. 1972a. Mechanisms of clasping and releasing (unclasping) in *Bufo americanus*. Behaviour 43:85–96.

Schmidt, R.S. 1972b. Release calling and inflating movements in anurans. Copeia 1972:240–245.

Schmidt, R.S. 1976. Release–call pulse differences between two members of the *Rana pipiens* complex. Copeia 1976:721–727.

Schmidt, R.S. 1978. Neural correlates of frog calling: Circum-metamorphic "calling" in leopard frog. Journal of Comparative Physiology 126A: 49-56.

Schmidt, R.S. 1985. Mating call phonotaxis in female American toad: induction by intracerebroventricular prostaglandin. Copeia 1985:490–492.

Schmutzer, A.C., M.J. Gray, E.C. Burton, and D.L. Miller. 2008. Impacts of cattle on amphibian larvae and the aquatic environment. Freshwater Biology 53:2613–2625.

Schneider, E.A. 2012. Population genetics of the Illinois Chorus Frog (*Pseudacris illinoensis*). M.S. thesis, University of Illinois, Urbana–Champaign.

Schneider, J.G. 1799. Historia Amphibiorum Naturalis et Literarariae. Fasciculus Primus. Continens Ranas, Calamitas, Bufones, Salamandras et Hydros in Genera et Species Descriptos Notisque suis Distinctos. Friederici Frommanni, Jena.

Schnell, J.H. 1988. Common black hawk. Pp. 388–389 *In* R.S. Palmer (ed.), Handbook of North American Birds. Volume 4, Diurnal Raptors (part 1). Yale University Press, New Haven, Connecticut.

Schock, D.M., T.K. Bollinger, V.G. Chinchar, J.K. Jancovich, and J.P. Collins. 2008. Experimental evidence that amphibian ranaviruses are multi–host pathogens. Copeia 2008:133–143.

Schoenherr, A.A. 1976. The herpetofauna of the San Gabriel Mountains, Los Angeles County, California, including distribution and biogeography. Special Publication of the Southwestern Herpetological Society, xiv + 95 pp.

Schoenherr, A.A., C.R. Feldmeth, and M.J. Emerson. 1999. Natural History of the Islands of California. University of California Press, Berkeley.

Schoeppner, N.M., and R.A. Relyea. 2005. Damage, digestion, and defense: The roles of alarm cues and kairomones for inducing prey defenses. Ecology Letters 8:505–512.

Schoeppner, N.M., and R.A. Relyea. 2008. Detecting small

environmental differences: Risk–response curves for predator–induced behavior and morphology. Oecologia 154:743–754.

Schoeppner, N.M., and R.A. Relyea. 2009a. When should prey respond to consumed heterospecifics? Testing hypotheses of perceived risk. Copeia 2009:190–194.

Schoeppner, N.M., and R.A. Relyea. 2009b. Interpreting the smells of predation: How alarm cues and kairomones induce different prey defenses. Functional Ecology 23:1114–1121.

Schoeppner, N.M., and R.A. Relyea. 2009c. Phenotypic plasticity in response to fine–grained environmental variation in predation. Functional Ecology 23:587–594.

Schoff, P.K., C.M. Johnson, A.M. Schotthoefer, J.E. Murphy, C. Lieske, R. A. Cole, L.B. Johnson, and V.R. Beasley. 2003. Prevalence of skeletal and eye malformations in frogs from north–central United States: estimations based on collections from randomly selected sites. Journal of Wildlife Diseases 39:510–521.

Schonberger, C.F. 1945. Food of some amphibians and reptiles of Oregon and Washington. Copeia 1945:120–121.

Schooley, J.D., M.R. Schwemm, and J.M. Barkstedt. 2008. *Rana catesbeiana* (American Bullfrog): tadpole gigantism and deformity. Herpetological Review 39:339–340.

Schorsch, I.G. 1933. Ranaculture. George H. Buchanan Co., Philadelphia, Pennsylvania. 87 pp.

Schotthoefer, A.M. 2003. Environmental, landscape, and host–related factors associated with parasitism and its effects on larvae and metamorphic frogs of *Rana pipiens*. Ph.D. Dissertation, University of Illinois, Urbana–Champaign.

Schotthoefer, A.M., A.V. Koehler, C.U. Meteyer, and R.A. Cole. 2003a. Influence of *Ribeiroia ondatrae* (Trematoda: Digenea) infection on limb development and survival of northern leopard frogs (*Rana pipiens*): effects of host stage and parasite–exposure level. Canadian Journal of Zoology 81:1144–1153.

Schotthoefer, A.M., R.A. Cole, and V.R. Beasley. 2003b. Relationship of tadpole stage to location of echinostome cercariae encystment and the consequences for tadpole survival. Journal of Parasitology 89:475–482.

Schotthoefer, A.M., M. G. Bolek, R. A. Cole, and V. R. Beasley. 2009. Parasites of the mink frog (*Rana septentrionalis*) from Minnesota, U.S.A. Comparative Parasitology 76:240–246.

Schotthoefer, A.M., J.R. Rohr, R.A. Cole, A.V. Koehler, C.M. Johnson, L.B. Johnson, and V.R. Beasley. 2011. Effects of wetland vs. landscape variables on parasite communities of *Rana pipiens*: links to anthropogenic factors. Ecological Applications 21:1257–1271.

Schoville, S.D., T.S. Tustall, V.T. Vredenburg, A.R. Backlin, E. Gallegos, D.A. Wood, and R.N. Fisher. 2011. Conservation genetics of evolutionary lineages of the endangered mountain yellow–legged frog, *Rana muscosa* (Amphibia: Ranidae), in southern California. Biological Conservation 144:2031–2044.

Schreber, H. 1782. Der Naturforscher. Vol. 18. Johann Jacob Gebaur, Halle.

Schreiman, E., and R. Rugh. 1949. Effects of DDT on functional development of larvae of *Rana pipiens* and *Fundulus heteroclitus*. Proceedings of the Society for Experimental Biology and Medicine 70:431–435.

Schriever, T.A. 2007. Salinity influences on larval *Hyla cinerea* (Green Treefrog) development and alteration to southeastern Louisiana wetlands and its effects on the herpetofauna. M.S. thesis, Southeastern Louisiana University, Hammond.

Schriever, T.A., J. Ramspott, B.I. Crother, and C.L. Fontenot, Jr. 2009. Effects of hurricanes Ivan, Katrina and Rita on a southeastern Louisiana herpetofauna. Wetlands 29:112–122.

Schrode, K.M., J.L. Ward, A. Vélez, and M.A. Bee. 2012. Female preferences for spectral call properties in the western genetic lineage of Cope's gray treefrog (*Hyla chrysoscelis*). Behavioral Ecology and Sociobiology 66:1595–1606.

Schroeder, E.E. 1966. Age determination and population structure in Missouri bullfrogs. Ph.D. Dissertation, University of Missouri, Columbia.

Schroeder, E.E. 1968. Aggressive behavior in *Rana clamitans*. Journal of Herpetology 1:95–96.

Schroeder, E.E. 1976. Dispersal and movement of newly transformed green frogs, *Rana clamitans*. American Midland Naturalist 95:471–474.

Schroeder, E.E., and T.S. Baskett. 1965. Frogs and toads of Missouri. Missouri Conservationist 26:15–18.

Schroeder, E.E., and T.S. Baskett. 1968. Age estimation, growth rates, and population structure in Missouri bullfrogs. Copeia 1968:583–592.

Schueler, F.W. 1973. Frogs of the Ontario coast of Hudson Bay and James Bay. Canadian Field-Naturalist 87:409–418.

Schueler, F.W. 1975. Geographic variation in the size of *Rana septentrionalis* in Quebec, Ontario, and Manitoba. Journal of Herpetology 9:177–185.

Schueler, F.W. 1979. Geographic variation in skin pigmentation and dermal glands in the northern leopard frog, *Rana pipiens*. Ph.D. Dissertation, University of Toronto, Toronto, Ontario.

Schueler, F.W. 1982a. Geographic variation in skin pigmentation and dermal glands in the northern leopard frog, *Rana pipiens*. National Museums of Canada, Publications in Zoology No. 16.

Schueler, F.W. 1982b. Sexual colour differences in Canadian

C. Kenneth Dodd, Jr.

western toads, *Bufo boreas*. Canadian Field-Naturalist 96:329–332.

Schueler, F.W. 1987. *Rana septentrionalis* (Mink Frog). Terrestrial activity. Herpetological Review 18:72.

Schueler, F.W., and F.R. Cook. 1980. Distribution of the middorsal stripe dimorphism in the wood frog, *Rana sylvatica*, in eastern North America. Canadian Journal of Zoology 58:1643–1651.

Schueler, F.W., and R.M. Rankin. 1982. Terrestrial amplexus in the wood frog, *Rana sylvatica*. Canadian Field-Naturalist 96:348–349.

Schuierer, F.W. 1962. Remarks upon the natural history of *Bufo exsul* Myers, the endemic toad of Deep Springs Valley, Inyo County, California. Herpetologica 17:260–266.

Schuierer, F.W. 1963. Notes on two populations of *Bufo exsul* Myers and a commentary on speciation within the *Bufo boreas* group. Herpetologica 18:262–267.

Schuierer, F.W. 1966. Environmental effects on the plasma proteins of *Bufo boreas*. Ph.D. Dissertation, Stanford University, Palo Alto, California.

Schuierer, F.W. 1972. The current status of the endangered species *Bufo exsul* Myers, Deep Springs Valley, Inyo County, California. Herpetological Review 4:81–82.

Schuierer, F.W., and S.C. Anderson. 1990. Population status of *Bufo exsul* Myers, 1942. Herpetological Review 21:57.

Schulte–Hostedde, A.I., and C.M.M. Schank. 2009. Secondary sexual traits and individual phenotype in male green frogs (*Rana clamitans*). Journal of Herpetology 43:89–95.

Schurbon, J.M. 2000. Effects of prescribed burning on amphibian diversity in the Francis Marion National Forest, South Carolina. M.S. thesis, University of Charleston, Charleston, South Carolina.

Schurbon, J.M., and J.E. Fauth. 2003. Effects of prescribed burning on amphibian diversity in a southeastern U.S. national forest. Conservation Biology 17:1338–1349.

Schutt, S.L. 1934. Big profits in back yard frog raising. Modern Mechanix and Invention 12(May):87, 128.

Schuytema, G.S., and A.V. Nebeker. 1999a. Effects of ammonium nitrate, sodium nitrate, and urea on red–legged frogs, Pacific treefrogs, and African clawed frogs. Bulletin of Environmental Contamination and Toxicology 63:357–364.

Schuytema, G.S., and A.V. Nebeker. 1999b. Comparative effects of ammonium and nitrate compounds on Pacific treefrog and African clawed frog embryos. Archives of Environmental Contamination and Toxicology 36:200–206.

Schuytema, G.S., A.V. Nebeker, and W.L. Griffis. 1995. Comparative toxicity of Guthion and Guthion 2S to *Xenopus laevis* and *Pseudacris regilla* tadpoles. Bulletin of Environmental Contamination and Toxicology 54:382–388.

Schwalbe, C.R., and P.C. Rosen. 1988. Preliminary report on effect of bullfrogs on wetland herpetofaunas in southeastern Arizona. Pp. 166–173 *In* R.C. Szaro, K.E. Severson, and D.R. Patton (eds.), Management of Amphibians, Reptiles, and Small Mammals in North America. USDA Forest Service General and Technical Report RM–166.

Schwalbe, C.R., and B. Alberti. 1998. Ground–truthing a troll. Studying the barking frog at Coronado National Memorial. Park Science 18:26–27.

Schwaner, T.D., and B.K. Sullivan. 2005. *Bufo microscaphus* Cope 1867 "1866." Arizona Toad. Pp. 422–424 *In* M.J. Lannoo (ed.), Amphibian Declines. The Conservation Status of United States Species. University of California Press, Berkeley.

Schwaner, T.D., and B.K. Sullivan. 2009. Fifty years of hybridization: introgression between the Arizona toad (*Bufo microscaphus*) and Woodhouse's toad (*B. woodhousii*) along Beaver Dam Wash in Utah. Herpetological Conservation and Biology 4:198–206.

Schwaner, T.D., D.R. Hadley, K.R. Jenkins, M.P. Donovan, and B. Al Tait. 1997. Population dynamics and life history studies of toads: short– and long–range needs for understanding amphibian populations in southern Utah. Pp. 197–202 *In* L.M. Hill (ed.), Learning from the Land. Grand Staircase–Escalante National Monument Science Symposium Proceedings, USDI Bureau of Land Management BLM/UT/GI–98/006+1220.

Schwartz, A. 1952. *Hyla septentrionalis* Dumeril and Bibron on the Florida mainland. Copeia 1952:117–118.

Schwartz, A. 1955. The chorus frog *Pseudacris brachyphona* in North Carolina. Copeia 1955:138.

Schwartz, A. 1957. Chorus frogs (*Pseudacris nigrita* LeConte) in South Carolina. American Museum Novitates 1838:1–12.

Schwartz, A., and J.R. Harrison III. 1956. A new subspecies of gopher frog (*Rana capito* Leconte). Proceedings of the Biological Society of Washington 69:135–144.

Schwartz, J.J. 1987. The function of call alternation in anuran amphibians: a test of three hypotheses. Evolution 41:461–471.

Schwartz, J.J. 1989. Graded aggressive calls of the spring peeper, *Pseudacris crucifer*. Herpetologica 45:172–181.

Schwartz, J.J., and H.C. Gerhardt. 1995. Directionality of the auditory system and call pattern recognition during acoustic interference in the gray tree frog, *Hyla versicolor*. Auditory Neuroscience 1:195–206.

Schwartz, J.J., and H.C. Gerhardt. 1998. The neuroethology of frequency preferences in the spring peeper. Animal Behaviour 56:55–69.

Schwartz, J.J., and V.T. Marshall. 2006. Forms of call overlap and their impact on advertisement call attractiveness to females of the gray treefrog, *Hyla versicolor*. Bioacoustics 16:39–56.

Schwartz, J.J., and K.M. Rahmeyer. 2006. Calling behavior and the capacity for sustained locomotory exercise in the gray treefrog (*Hyla versicolor*). Journal of Herpetology 40:164–171.

Schwartz, J.J., M.A. Bee, and S.D. Tanner. 2000. A behavioral and neurobiological study of the responses of gray treefrogs, *Hyla versicolor*, to the calls of a predator, *Rana catesbeiana*. Herpetologica 56:27–37.

Schwartz, J.J., B.W. Buchanan, and H.C. Gerhardt. 2001. Female mate choice in the gray treefrog (*Hyla versicolor*) in three experimental environments. Behavioral Ecology and Sociobiology 49:443–455.

Schwartz, J.J., B.W. Buchanan, and H.C. Gerhardt. 2002. Acoustic interactions among gray treefrogs, *Hyla versicolor*, in a chorus setting.Behavioral Ecology and Sociobiology 53:9–19.

Schwartz, J.J., K. Huth, and T. Hutchin. 2004. How long do females really listen? Assessment time for female mate choice in the grey treefrog, *Hyla versicolor*. Animal Behaviour 68:533–540.

Schwartz, J.J., R. Brown, S.Turner, K. Dushaj, and M. Castano. 2008. Interference risk and the function of dynamic shifts in calling in the gray treefrog (*Hyla versicolor*). Journal of Comparative Psychology 122:283–288.

Schwartz, V., and D.M. Golden. 2002. Field Guide to Reptiles and Amphibians of New Jersey. New Jersey Division of Fish and Wildlife, Woodbine.

Scott, A.F., and D.F. Harker. 1968. First records of the barking treefrog, *Hyla gratiosa* Le Conte, from Tennessee. Herpetologica 24:82–83.

Scott, D.E., E. D. Casey, M.F. Donovan, and T.K. Lynch. 2007. Amphibian lipid levels at metamorphosis correlate to post–metamorphic terrestrial survival. Oecologia 153:521–532.

Scott, D.E., B.S. Metts, and J.W. Gibbons. 2008. Enhancing amphibian biodiversity on golf courses with seasonal wetlands. Pp. 285–292 *In* J.C. Mitchell, R.E. Jung Brown, and B. Bartholomew (eds.), Urban Herpetology. Herpetological Conservation 3. Society for the Study of Amphibians and Reptiles, Salt Lake City, Utah.

Scott, G.W., and C.C. Carpenter. 1956. Body temperatures of *Bufo w. woodhousei*. Proceedings of the Oklahoma Academy of Science 36:84–85.

Scott, N.J., Jr., and R.D. Jennings. 1985. The tadpoles of five species of New Mexican leopard frogs. Occasional Papers, the Museum of Southwestern Biology 3:1–21.

Scott, N.J., Jr., and R.D. Jennings. 1989. Our enigmatic leopard frogs. New Mexico Wildlife (March–April):6–9.

Scudday, J.F. 1965. *Eleutherodacylus latrans* in Terrell County, Trans–Pecos, Texas. Southwestern Naturalist 10:78.

Seale, D.B. 1980. Influence of amphibian larvae on primary production, nutrient flux, and competition in a pond ecosystem. Ecology 61:1531–1550.

Seale, D.B. 1982. Physical factors influencing oviposition by the wood frog, *Rana sylvatica*, in Pennsylvania. Copeia 1982:627–635.

Seale, D.B. 1987. Amphibia. Pp. 467–552 *In* T.J. Pandian and F.J. Vernberg (eds.). Animal Energetics, Volume II. Bivalvia through Reptilia. Academic Press, New York.

Seale, D.B., and R.J. Wassersug. 1979. Suspension feeding dynamics of anuran larvae related to their functional morphology. Oecologia 39:259–272.

Seale, D.B., and N. Beckvar. 1980. The comparative ability of anuran larvae (genera: *Hyla, Bufo* and *Rana*) to ingest suspended blue–green algae. Copeia 1980:495–503.

Searle, C.L. 2011. Host–pathogen dynamics in a changing environment: susceptibility of amphibians to an emerging infectious disease. Ph.D. Dissertation, Oregon State University, Corvallis.

Searle, C.L., L.K. Belden, B.A. Bancroft, B.A. Han, L.M. Biga, and A.R. Blaustein. 2010. Experimental examination of the effects of ultraviolet–B radiation in combination with other stressors on frog larvae. Oecologia 162:237–245.

Searle, C.L., L.M. Biga, J.W. Spatafora, and A.R. Blaustein. 2011a. A dilution effect in the emerging amphibian pathogen *Batrachochytrium dendrobatidis*. Proceedings of the National Academy of Sciences of the United States of America 108: 16322–16326.

Searle, C.L., S.S. Gervasi, J. Hua, J.L. Hammond, R.A. Relyea, D.H. Olson, and A.R. Blaustein. 2011b. Differential host susceptibility to *Batrachochytrium dendrobatidis*, an emerging amphibian pathogen. Conservation Biology 25:965–974.

Seba, A. 1734. Locupletissimi Rerum Naturalium Thesauri Accurata Descriptio, et Iconibus Artificiosissimis Expressio, per Universam Physices Historiam. Janssonio–Waesbergios, & J. Wetstenium, & Gul. Smith, Amsterdam.

Seburn, C.N.L., D.C. Seburn, and C.A. Paszkowski. 1997. Northern leopard frog (*Rana pipiens*) dispersal in relation to habitat. SSAR Herpetological Conservation 1:64–72.

Seburn, D.C., and K. Gunson. 2011. Has the western chorus frog (*Pseudacris triseriata*) declined in western Ottawa, Ontario? Canadian Field-Naturalist 125:220–226.

Seburn, D.C., C.N.L. Seburn, and W.F. Weller. 2008. A localized decline in the western chorus frog, *Pseudacris triseriata*, in eastern Ontario. Canadian Field-Naturalist 122:158–161.

Secor, S.M. 1988. Perch sites of calling male bird–voiced treefrogs, *Hyla avivoca*, in Oklahoma. Proceedings of the

Oklahoma Academy of Science 68:71–73.

Seeba, F., J.J. Schwartz, and M.A. Bee. 2010. Testing an auditory illusion in frogs: perceptual restoration or rule–based sensory biases? Animal Behaviour 79:1317–1328.

Seehorn, M.E. 1982. Reptiles and amphibians of southeastern National Forests. U.S. Department of Agriculture, Atlanta, Georgia.

Seib, G.W. 1977. Probable case of death–feigning by wood frog. Blue Jay 35:148.

Seibel, R.V. 1968. Some variables affecting the critical thermal maximum of the leopard frog, *Rana pipiens* Schreber. M.S. thesis, California State University, Fullerton.

Seibel, R.V. 1970. Variables affecting the critical thermal maximum of the leopard frog, *Rana pipiens* Schreber. Herpetologica 26:208–213.

Seigel, R.A. 1983. Natural survival of eggs and tadpoles of the wood frog, *Rana sylvatica*. Copeia 1983:1096–1098.

Seigel, R.A., A. Dinsmore, and S.C. Richter. 2006. Using well water to increase hydroperiod as a management option for pond–breeding amphibians. Wildlife Society Bulletin 34:1022– 1027.

Seiter, S.A., 2011. Predator presence suppresses immune function in a larval amphibian. Evolutionary Ecology Research 13:283–293.

Semlitsch, R.D. 1990. Effects of body size, sibship, and tail injury on the susceptibility of tadpoles to dragonfly predation. Canadian Journal of Zoology 68:1027–1030.

Semlitsch, R.D. 2000. Principles for management of aquatic breeding amphibians. Journal of Wildlife Management 64:615–631.

Semlitsch, R.D. 2001. Critical elements for biologically based recovery plans of aquatic-breeding amphibians. Conservation Biology 16:619-629.

Semlitsch, R.D. (ed.) 2003. Amphibian Conservation. Smithsonian Books, Washington, D.C.

Semlitsch, R.D. 2008. Differentiating migration and dispersal processes for pond–breeding amphibians. Journal of Wildlife Management 72:260–267.

Semlitsch, R.D., and J.P. Caldwell. 1982. Effects of density on growth, metamorphosis, and survivorship in tadpoles of *Scaphiopus holbrooki*. Ecology 63:905–911.

Semlitsch, R.D., and J.W. Gibbons. 1988. Fish predation in size–structured populations of treefrog tadpoles. Oecologia 75:321–326.

Semlitsch, R.D., and J.B. Jensen. 2001. Core habitat, not buffer zone. National Wetlands Newsletter 23:5–6.

Semlitsch, R.D., and J.R. Bodie. 2003. Biological criteria for buffer zones around wetlands and riparian habitats for amphibians and reptiles. Conservation Biology 17:1219–1228.

Semlitsch, R.D., and D.K. Skelly. 2008. Ecology and conservation of pool–breeding amphibians. Pp. 127–147 *In* A.J.K. Calhoun and P. deMaynadier (eds.). Science and Conservation of Vernal Pools in Northeastern North America. CRC Press, Boca Raton, Florida.

Semlitsch, R.D., J.W. Gibbons, and T.D. Tuberville. 1995. Timing of reproduction and metamorphosis in the Carolina gopher frog (*Rana capito capito*) in South Carolina. Journal of Herpetology 29:612–614.

Semlitsch, R.D., D.E. Scott, J.H.K. Pechmann, and J.W. Gibbons. 1996. Structure and dynamics of an amphibian community: evidence from a 16–year study of a natural pond. Pp. 217–248 *In* M.L. Cody and J.A. Smallwood (eds.), Long–term Studies of Vertebrate Communities. Academic Press, San Diego, California.

Semlitsch, R.D., J. Pickle, M. J. Parris, and R. D. Sage. 1999. Jumping performance and short–term repeatability of newly metamorphosed hybrid and parental leopard frogs (*Rana sphenocephala* and *Rana blairi*). Canadian Journal of Zoology 77: 748–754.

Semlitsch, R.D., C.M. Bridges, and A.M. Welch. 2000. Genetic variation and a fitness tradeoff in the tolerance of gray treefrog (*Hyla versicolor*) tadpoles to the insecticide carbaryl. Oecologia 125:179–185.

Semlitsch, R.D., C.A. Conner, D.J. Hocking, T.A.G. Rittenhouse, and E.B. Harper. 2008. Effects of timber harvesting on pond–breeding amphibian persistence: testing the evacuation hypothesis. Ecological Applications 18:283–289.

Semlitsch, R.D., B.D. Todd, S.M. Blomquist, A.J.K. Calhoun, J.W. Gibbons, J.P. Gibbs, G.J. Graeter, E.B. Harper, D.J. Hocking, M.L. Hunter, D.A. Patrick, T.A.G. Rittenhouse, and B.B. Rothermel. 2009. Effects of timber harvest on amphibian populations: understanding mechanisms from forest experiments. Bioscience 59:853–862.

Semsar, K., K.F. Klomberg, and C. Marler. 1998. Arginine vasotocin increases calling–site acquisition by nonresident male grey treefrogs. Animal Behaviour 56:983–987.

Servoss, J.M., and K.M. Sharrocks. 2006. *Rana chiricahuensis* (Chiricahua Leopard Frog) and *Rana catesbeiana* (American Bullfrog). Reproduction. Herpetological Review 37:208.

Sessions, S.K., and S.B. Ruth. 1990. Explanation for naturally occurring supernumerary limbs in amphibians. Journal of Experimental Zoology 254:38–47.

Sever, D.M., E. C. Moriarty, L.C. Rania, and W.C. Hamlett. 2001. Sperm storage in the oviduct of an internally fertilizing frog, *Ascaphus truei*. Journal of Morphology 248:1–21.

Sexton, O.J., and K.R. Marion. 1974. Probable predation by Swainson's Hawk on swimming spadefoot toads. Wilson Bulletin 86:167–168.

Sexton, O.J., and C. Phillips. 1986. A qualitative study of fish–amphibian interactions in 3 Missouri ponds. Transactions, Missouri Academy of Science 20:25–35.

Sexton, O.J., C.A. Phillips, T.J. Bergman, E.B. Wattenberg, and R.E. Preston. 1998. Abandon not hope: status of repatriated populations of spotted salamanders and wood frogs at the Tyson Research Center, St. Louis County, Missouri. Pp. 340–353 *In* M.J. Lannoo (ed.), Status & Distribution of Midwestern Amphibians. University of Iowa Press, Iowa City.

Seyle, C.W., Jr., and S.E. Trauth. 1982. *Pseudacris ornata* (Ornate Chorus Frog). Reproduction. Herpetological Review 13:45.

Seymour, R.S. 1972a. Physiological correlates of activity and dormancy in spadefoot toads. Ph.D. Dissertation, University of California, Los Angeles.

Seymour, R.S. 1972b. Behavioral thermoregulation by juvenile green toads, *Bufo debilis*. Copeia 1972:572–575.

Seymour, R.S. 1973a. Physiological correlates of forced activity and burrowing in the spadefoot toad, *Scaphiopus hammondii*. Copeia 1973:103–115.

Seymour, R.S. 1973b. Energy metabolism of dormant spadefoot toads (*Scaphiopus*). Copeia 1973:435–445.

Seymour, R.S. 1995. Oxygen uptake by embryos in gelatinous egg masses of *Rana sylvatica*: the roles of diffusion and convection. Copeia 1995:626–635.

Shaffer, H.B., G.M. Fellers, A. Magee, and S.R. Voss. 2000. The genetics of amphibian declines: population substructure and molecular differentiation in the Yosemite toad, *Bufo canorus* (Anura, Bufonidae), based on single–stranded conformation polymorphism analysis (SSCP) and mitochondrial DNA sequence data. Molecular Ecology 9:245–257.

Shaffer, H.B., G.M. Fellers, S.R. Voss, J.C. Oliver, and G.B. Pauly. 2004. Species boundaries, phylogeography and conservation genetics of the red–legged frog (*Rana aurora/draytonii*) complex. Molecular Ecology 13:2667–2677.

Shah, A.A., M.J. Ryan, E. Bevilacqua, and M.L. Schlaepfer. 2010. Prior experience alters the behavioral responses of prey to a nonnative predator. Journal of Herpetology 44:185–192.

Shannon, F.A. 1949. A western subspecies of *Bufo woodhousii* hitherto erroneously associated with *Bufo compactilis*. Bulletin of the Chicago Academy of Sciences 8:301–312.

Shannon, F.A., and C.H. Lowe, Jr. 1955. A new subspecies of *Bufo woodhousei* from the inland Southwest. Herpetologica 11:185–190.

Shannon, P.R. 1988. The parasites of breeding male eastern gray treefrogs, *Hyla versicolor* LeConte, 1825, from the Ashland Wildlife Research Area, Boone County, Missouri. M.S. thesis, University of Missouri, Columbia.

Sharma, B., F. Hu, J.A. Carr, and R. Patiño. 2011 (2014). Water quality and amphibian health in the Big Bend region of the Rio Grande Basin. Texas Journal of Science 63:233-266.

Sharp, C.C.R. 2008. Effects of copper and light exposure on the development and survival of the wood frog tadpole (*Rana sylvatica*). M.S. thesis, University of Akron, Akron, Ohio.

Shaw, G. 1802. General Zoology, or Systematic Natural History. Vol 3, Part I. Amphibia. Thomas Davison, London.

Shedd, J.D. 2005. Amphibians and Reptiles of Bidwell Park. Quadco Publishing, Chico, California.

Sheldon, A.B. 2006. Amphibians & Reptiles of the North Woods. Kollath + Stensaas Publishing, Duluth, Minnesota.

Shelford, V.E. 1913. The reactions of certain animals to gradients of evaporating power of air: a study in experimental ecology. Biological Bulletin 25:79–120.

Shelford, V.E. 1914. Modification of the behavior of land animals by contact with air of high evaporating power. Journal of Animal Behavior (Boston) 4:31–49.

Shelford, V.E., and A.C. Twomey. 1941. Tundra animal communities in the vicinity of Churchill, Manitoba. Ecology 22:47–61.

Shepard, D. B. 2000. Aggressive behavior and potential of territoriality in the green frog *Rana clamitans*. M.S. thesis, Illinois State University, Normal.

Shepard, D.B. 2002. Spatial relationships of male green frogs (*Rana clamitans*) throughout the activity season. American Midland Naturalist 148:394–400.

Shepard, D.B. 2004. Seasonal differences in aggression and site tenacity in male green frogs, *Rana clamitans*. Copeia 2004:159–164.

Shepard, D.B., and A.R. Kuhns. 2000. *Pseudacris triseriata* (Western Chorus Frog): calling sites after drought. Herpetological Review 31:235–236.

Shepard, D.B., and L.E. Brown. 2005. *Bufo houstonensis* Sanders, 1953: Houston Toad. Pp. 415–417 *In* M.J. Lannoo (ed.), Amphibian Declines. The Conservation Status of United States Species. University of California Press, Berkeley.

Sheridan, C.D., and D.H. Olson. 2003. Amphibian assemblages in zero–order basins in the Oregon Coast Range. Canadian Journal of Forest Research 33:1452–1477.

Sherwood, W.J. 1898. The frogs and toads found in the vicinity of New York City. Proceedings of the Linnean Society of New York 10:9–24.

Shields, M.A. 1985. Selective use of pitfall traps by southern leopard frogs. Herpetological Review 16:14.

Shinn, E.A., and J.W. Dole. 1978. Evidence for a role for olfactory cues in the feeding response of leopard frogs,

Rana pipiens. Herpetologica 34:167–172.

Shinn, E.A., and J.W. Dole. 1979a. Lipid components of prey odors elicit feeding responses in Western toads (*Bufo boreas*). Copeia 1979:275–278.

Shinn, E.A., and J.W. Dole. 1979b. Evidence for a role for olfactory cues in the feeding response of Western toads (*Bufo boreas*). Copeia 1979:163–165.

Shirer, H.W., and H.S. Fitch. 1970. Comparison from radiotracking of movements and denning habits of the raccoon, striped skunk, and opossum in northeastern Kansas. Journal of Mammalogy 51:491–503.

Shirose, L. J., and R. J. Brooks. 1995a. Growth rate and age at maturity in syntopic populations of *Rana clamitans* and *Rana septentrionalis* in central Ontario. Canadian Journal of Zoology 73:1468–1473.

Shirose, L. J., and R. J. Brooks. 1995b. Age structure, mortality, and longevity in syntopic populations of three species of ranid frogs in central Ontario. Canadian Journal of Zoology 73:1878–1886.

Shirose, L.J., and R.J. Brooks. 1997. Fluctuations in abundance and age structure in three species of frogs (Anura: Ranidae) in Algonquin Park, Canada, from 1985 to 1993. SSAR Herpetological Conservation 1:16–26.

Shirose, L.J., R.J. Brooks, J.R. Barta, and S.S. Desser. 1993. Intersexual differences in growth, mortality, and size at maturity in bullfrogs in central Ontario. Canadian Journal of Zoology 71:2363–2369.

Shirose, L.J., C.A. Bishop, D.M. Green, C.J. MacDonald, R.J. Brooks, and N.J. Helferty. 1997. Validation tests of an amphibian call survey technique in Ontario, Canada. Herpetologica 53:312–320.

Shive, P.P., D.S. Pilliod., and C.R. Peterson. 2010. Hyperspectral analysis of Columbia spotted frog habitat. Journal of Wildlife Management 74:1387–1394.

Shively, J.N., J.G. Songer, S. Prchal, M.S. Keasey III, and C.O. Thoen. 1981. *Mycobacterium marinum* infection in Bufonidae. Journal of Wildlife Diseases 17:3–7.

Shoemaker, V.H., L. McClanahan, Jr., and R. Ruibal. 1969. Seasonal changes in body fluids in a field population of spadefoot toads. Copeia 1969:585–591.

Shoo, L. P., D.H. Olson, S.K. McMenamin, K.A. Murray, M. Van Sluys, M.A. Donnelly, D. Stratford, J. Terhivuo, A. Merino-Viteri, S.M. Herbert, and P.J. Bishop. 2011. Engineering a future for amphibians under climate change. Journal of Applied Ecology 48:487–492.

Shoop, C.R., and C. Ruckdeschel. 2003. Herpetological biogeography of the Georgia Barrier islands: an alternative interpretation. Florida Scientist 66:43–51.

Shoop, C.R., and C. Ruckdeschel. 2006. Amphibians and reptiles of Cumberland Island, Georgia. Occasional Publications of the Cumberland Island Museum No. 2, St. Marys, Georgia.

Shovlain, A.M. 2006. Oregon spotted frog (*Rana pretiosa*) habitat use and herbage (or biomass) removal from grazing at Jack Creek, Klamath County, Oregon. M.S. thesis, Oregon State University, Corvallis.

Shreve, B. 1945. Application of the name *Eleutherodacylus ricordii*. Copeia 1945:117.

Shubin, N.H., and F.A. Jenkins, Jr. 1995. An Early Jurassic jumping frog. Nature 377:49–52.

Shulse, C.D. 2011. Building better wetlands for amphibians: investigating the roles of engineered wetland features and mosquitofish (*Gambusia affinis*) on amphibian abundance and reproductive success. Ph.D. Dissertation, University of Missouri, Columbia.

Shulse, C.D., R.D. Semlitsch, and K.M. Trauth. 2009. Development of an amphibian biotic index to evaluate wetland health in northern Missouri. World Environmental and Water Resources Congress, ASCE. 342:2657–2667.

Shulse, C.D., R.D. Semlitsch, K.M. Trauth, and A.D. Williams. 2010. Influences of design and landscape placement parameters on amphibian abundance in constructed wetlands. Wetlands 30:915–928.

Shulse, C.D., R.D. Semlitsch, K.M. Trauth, and J.E. Gardner. 2012a. Testing wetland features to increase amphibian reproductive success and species richness for mitigation and restoration. Ecological Applications 22:1675–1688.

Shulse, C.D., R. D. Semlitsch, K.M. Trauth, and J. E. Gardner. 2012b. Amphibian reproductive success in wetlands. Bulletin of the Ecological Society of America 93:236–237.

Shutler, D., and D.J. Marcogliese. 2011. Leukocyte profiles of northern leopard frogs, *Lithobates pipiens*, exposed to pesticides and hematozoa in agricultural wetlands. Copeia 2011:301–307.

Shutler, D., T.G. Smith, and S.R. Robinson. 2009. Relationships between leucocytes and *Hepatozoon* spp. in green frogs, *Rana clamitans*. Journal of Wildlife Diseases 45:67–72.

Sias, J. 2006. Natural history and distribution of the upland chorus frog, *Pseudacris feriarum* Baird, in West Virginia. M.S. thesis, Marshall University, Huntington, West Virginia.

Sibley, C.L. 1912 The frog in Canadian diet. Maclean's Magazine, 1 September.

Siegel, D.S., D.M. Sever, T.A. Schriever, and R.E. Chabarria. 2008. Ultrastructure and histochemistry of the adhesive breeding glands in male *Gastrophryne carolinensis* (Amphibia: Anura: Microhylidae). Copeia 2008:877–881.

Siegler, H.R. 1962. New Hampshire Nature Notes. Equity Publishing Corporation, Orford, New Hampshire.

Siekmann, J.M. 1949. A survey of the tadpoles of Louisiana. M.S. thesis, Louisiana State University, Baton Rouge.

Sielecki, L.E. 2010. Wildlife Identification Field Guide. Red and Blue Listed Amphibians and Reptiles in British Columbia. Ministry of Transportation and Infrastructure, Victoria.

Sievert, G., and L. Sievert. 2005. A Field Guide to Oklahoma's Amphibians and Reptiles. Oklahoma Department of Wildlife Conservation, Oklahoma City. [second printing 2006]

Sievert, G., and L. Sievert. 2011. A Field Guide to Oklahoma's Amphibians and Reptiles. Third Edition. Oklahoma Department of Wildlife Conservation, Oklahoma City.

Sievert, L. 1991. Thermoregulatory behaviour in the toads *Bufo marinus* and *Bufo cognatus*. Journal of Thermal Biology 16:309–312.

Sievert, L.M., and J.L. Poore. 1995. Thermoregulatory behavior of *Bufo americanus*: effects of melatonin, chlorpromazine, and time of day. Copeia 1995:490–494.

Sievert, L.M., and J.K. Bailey. 2000. Specific dynamic action in the toad, *Bufo woodhousii*. Copeia 2000:1076–1078.

Sih, A., A.M. Bell, and J.L. Kirby. 2004. Two stressors are far deadlier than one. Trends in Ecology & Evolution 19:274–276.

Sim, R. 1895. Moths and toads. Natural Science News 1(12):45.

Simandle, E.T. 2006. Population structure and conservation of two rare toad species (*Bufo exsul* and *Bufo nelsoni*) in the Great Basin, USA. Ph.D. Dissertation, University of Nevada, Reno.

Simmons, A.M., and M. E. Bean. 2000. Perception of mistuned harmonics in complex sounds by the bullfrog (*Rana catesbeiana*). Journal of Comparative Psychology 114:167–173.

Simmons, R., and J.D. Hardy, Jr. 1959. The river–swamp frog, *Rana heckscheri* Wright, in North Carolina. Herpetologica 15:36–37.

Simon, J.A., J.W. Snodgrass, R.E. Casey, and D.W. Sparling. 2009. Spatial correlates of amphibian use of constructed wetlands in an urban landscape. Landscape Ecology 24:361–373.

Simon, M.P., I. Vatnick, H.A. Hopey, K. Butler, C. Korver, C. Hilton, R.S. Weimann, and M.A. Brodkin. 2002. Effects of acid exposure on natural resistance and mortality on adult *Rana pipiens*. Journal of Herpetology 36:697–699.

Simons, L.H. 1998. *Rana cascadae* (Cascade's Frog). Predation. Herpetological Review 29:232.

Simovich, M.A. 1985. Analysis of a hybrid zone between the spadefoot toads *Scaphiopus multiplicatus* and *Scaphiopus bombifrons*. Ph.D. Dissertation, University of California–Riverside.

Simovich, M.A. 1994. The dynamics of a spadefoot toad (*Spea multiplicata* and *S. bombifrons*) hybridization system. Pp. 167–182 *In* P.R. Brown and J.W. Wright (eds.), Herpetology of the North American Deserts. Proceedings of a Symposium. Southwestern Herpetologists Society, Special Publication No. 5.

Simovich, M.A., C.A. Sassaman, and A. Chovnick. 1991. Post–mating selection of hybrid toads (*Scaphiopus multiplicatus* and *Scaphiopus bombifrons*). Proceedings of the San Diego Society of Natural History No. 5:1–6.

Sin, H., K.H. Beard, and W.C. Pitt. 2008. An invasive frog, *Eleutherodacylus coqui*, increases new leaf production and leaf litter decomposition rates through nutrient cycling in Hawaii. Biological Invasions 10:335–345.

Singer, S.R., and S. Grismaijer. 2005. Panic in Paradise. Invasive Species Hysteria and the Hawaiian Coqui Frog War. ISCD Press, Pahoa, Hawaii.

Skehan, P., Jr. 1959. Early emergence of *Bufo americanus* from hibernation. Herpetologica 15:80.

Skelly, D.K. 1992a. Larval distributions of spring peepers and chorus frogs: regulating factors and the role of larval behavior. Ph.D. Dissertation, University of Michigan, Ann Arbor.

Skelly, D.K. 1992b. Field evidence for a cost of behavioral antipredator response in a larval amphibian. Ecology 73:704–708.

Skelly, D.K. 1994. Activity level and the susceptibility of anuran larvae to predation. Animal Behaviour 47:465–468.

Skelly, D.K. 1995a. A behavioral tradeoff and its consequences for the distribution of *Pseudacris* treefrog larvae. Ecology 76:150–164.

Skelly, D.K. 1995b. Competition and the distribution of spring peeper larvae. Oecologia 103:203–207.

Skelly, D.K. 1996. Pond drying, predators, and the distribution of *Pseudacris* tadpoles. Copeia 1996:599–605.

Skelly, D.K. 1997. Tadpole communities: pond permanence and predation are powerful forces shaping the structure of tadpole communities. American Scientist 85:36–45.

Skelly, D.K. 2001. Distributions of pond–breeding anurans: an overview of mechanisms. Israel Journal of Zoology 47:313–332.

Skelly, D.K. 2002. Experimental venue and estimation of interaction strength. Ecology 83:2097–2101.

Skelly, D.K. 2004. Microgeographic countergradient variation in the wood frog, *Rana sylvatica*. Evolution 58:160–165.

Skelly, D.K. 2005. Experimental venue and estimation of interaction strength: reply. Ecology 86:1068–1071.

Skelly, D.K., and E.E. Werner. 1990. Behavioral and life–historical responses of larval American toads to an odonate predator. Ecology 71:2313–2322.

Skelly, D.K., and L.K. Freidenburg. 2000. Effects of beaver on the thermal biology of an amphibian. Ecology Letters 3:483–486.

Skelly, D.K., and J.M. Kiesecker. 2001. Venue and outcome in ecological experiments: manipulations of larval anurans. Oikos 94:198–208.

Skelly, D.K., and J. Golon. 2003. Assimilation of natural benthic substrates by two species of tadpoles. Herpetologica 59:37–42.

Skelly, D.K., and M. F. Benard. 2010. Mystery unsolved: missing limbs in deformed amphibians. Journal of Experimental Zoology B: Molecular and Developmental Evolution 314B:179–181.

Skelly, D.K., E.E. Werner, and S.A, Cortwright. 1999. Long–term distributional dynamics of a Michigan amphibian assemblage. Ecology 80:2326–2337.

Skelly, D.K., L.K. Freidenburg, and J.M. Kiesecker. 2002. Forest canopy and the performance of larval amphibians. Ecology 83:983–992.

Skelly, D.K., K.L. Yurewicz, E.E. Werner, and R.A. Relyea. 2003. Estimating decline and distributional change in amphibians. Conservation Biology 17:744–751.

Skelly, D. K., M. A. Halverson, L. K. Freidenburg, and M. C. Urban. 2005. Canopy closure and amphibian diversity in forested wetlands. Wetlands Ecology and Management 13:261–268.

Skelly, D.K., S.R. Bolden, M.P. Holland, L.K. Freidenburg, N.A. Friedenfelds, and T.R. Malcolm. 2006. Urbanization and disease in amphibians. Pp. 153–167 *In* S.K. Collinge, and C. Ray (eds.), Disease Ecology: Community Structure and Pathogen Dynamics. Oxford University Press, Cary, North Carolina.

Skelly, D. K., S. R. Bolden, L. K. Freidenburg, N. A. Freidenfelds, and R. Levey. 2007. *Ribeiroia* infection is not responsible for Vermont amphibian deformities. EcoHealth 4:156–163.

Skelly, D.K., S.R. Bolden, and K.B. Dion. 2010. Intersex frogs concentrated in suburban and urban landscapes. EcoHealth 7:374–379.

Skidds, D.E., F.C. Golet, P.W.C. Paton, and J.C. Mitchell. 2007. Habitat correlates of reproductive effort in wood frogs and spotted salamanders in an urbanizing watershed. Journal of Herpetology 41:439–450.

Slater, J.R. 1931. The mating of *Ascaphus truei* Stejneger. Copeia 1931:62–63.

Slater, J.R. 1939a. Description and life–history of a new *Rana* from Washington. Herpetologica 1:145–149.

[6]Slater, J.R. 1939b. Some species of amphibians new to the state of Washington. Occasional Papers, Department of Biology, College of Puget Sound No. 2:4–5.

Slater, J.R. 1941a. Island records of amphibians and reptiles for Washington. Occasional Papers, Department of Biology, College of Puget Sound No. 13:74–77.

Slater, J.R. 1941b. The distribution of amphibians and reptiles in Idaho. Occasional Papers, Department of Biology, College of Puget Sound No. 14:78–109.

Slater, J.R. 1955. Distribution of Washington amphibians. Occasional Papers, Department of Biology, College of Puget Sound No. 16:120–154.

Slater, J.R. 1964a. A key to adult amphibians of Washington State. Occasional Papers, Department of Biology, College of Puget Sound No. 25:235–236.

Slater, J.R. 1964b. County records of amphibians for Washington. Occasional Papers, Department of Biology, College of Puget Sound No. 26:237–242.

Slevin, J.R. 1928. The amphibians of western North America. An account of the species known to inhabit California, Alaska, British Columbia, Washington, Oregon, Idaho, Utah, Nevada, Arizona, Sonora, and Lower California. Occasional Papers of the California Academy of Sciences 16:1–144.

Sloan, A.J. 1964. Amphibians of San Diego County. Occasional Papers of the San Diego Natural History Society No. 13.

Slough, B.G. 2009. Amphibian chytrid fungus in Western toads (*Anaxyrus boreas*) in British Columbia and Yukon, Canada. Herpetological Review 40:319–321.

Slough, B.G., and R.L. Mennell. 2006. Diversity and range of amphibians of the Yukon Territory. Canadian Field-Naturalist 120:87-92.

Smith, A.K. 1973. Feeding strategies and population dynamics of the bullfrog (*Rana catesbeiana*). Ph.D. Dissertation, University of Kansas, Lawrence.

Smith, A.K. 1977. Attraction of bullfrogs (Amphibia, Anura, Ranidae) to distress calls of immature frogs. Journal of Herpetology 11:234–235.

Smith, B.E. 2003. Conservation assessment for the Northern Leopard Frog in the Black Hills National Forest, South Dakota and Wyoming. USDA Forest Service, Black Hills National Forest, Custer, South Dakota. http://www.fs.fed.us/r2/blackhills/projects/planning/assessments/leopard_frog.pdf

Smith, B.E., and D.A. Keinath. 2007. Northern leopard frog (*Rana pipiens*): a technical conservation assessment. USDA Forest Service, Rocky Mountain Region, Species Conservation Project. http://www.fs.fed.us/ r2/projects/scp/assessments/northernleopardfrog.pdf

Smith, C.C., and A.N. Bragg. 1949. Observations on the ecology and natural history of anura, 7. Food and feeding habits of the common species of toads in Oklahoma.

---

[6]W.C. Brown's name was apparently omitted as a junior author. He appears as such in the table of contents.

Ecology 30:333–349.

Smith, D.A. 1953. Northern swamp tree frog, *Pseudacris nigrita septentrionalis* (Boulenger) from Churchill, Manitoba. Canadian Field-Naturalist 67:181–182.

Smith, D.C. 1983. Factors controlling tadpole populations of the chorus frog (*Pseudacris triseriata*) on Isle Royale, Michigan. Ecology 64:501–510.

Smith, D.C. 1987. Adult recruitment in chorus frogs: effects of size and date at metamorphosis. Ecology 68:344–350.

Smith, D.C. 1990. Population structure and competition among kin in the chorus frog (*Pseudacris triseriata*). Evolution 44:1529–1541.

Smith, D.D., and R. Powell. 1983. *Acris crepitans blanchardi* (Blanchard's Cricket Frog). Anomalies. Herpetological Review 14:118–119.

Smith, D.G., C.R. Wilson, and H.W. Frost. 1972. The biology of the American kestrel in central Utah. Southwestern Naturalist 17:73–83.

Smith, G.R. 1999a. Microhabitat preferences of bullfrog tadpoles (*Rana catesbeiana*) of different ages. Transactions of the Nebraska Academy of Sciences 25:73–76.

Smith, G.R. 1999b. Among family variation in tadpole (*Rana catesbeiana*) responses to density. Journal of Herpetology 33:167–169.

Smith, G.R. 2001. Effects of acute exposure to commercial formulation of plyphosate on the tadpoles of two species of anurans. Bulletin of Environmental Contamination and Toxicology 67:483–488.

Smith, G.R. 2002. *Pseudacris crucifer* (Spring Peeper). Predation. Herpetological Review 33:48.

Smith, G.R. 2007. Lack of effect of nitrate, nitrite, and phosphorus on wood frog (*Rana sylvatica*) tadpoles. Applied Herpetology 4:287–291.

Smith, G.R., and A.K. Jennings. 2004. Spacing of the tadpoles of *Hyla versicolor* and *Rana clamitans*. Journal of Herpetology 38:616–618.

Smith, G.R., and B.L. Doupnik. 2005. Habitat use and activity level of large American bullfrog tadpoles: choices and repeatability. Amphibia–Reptilia 26:549–552.

Smith, G.R., and J.J. Schulte. 2008. Conservation management implications of substrate choice in recently metamorphosed American toads (*Bufo americanus*). Applied Herpetology 5:87–92.

Smith, G.R., and A.R. Awan. 2009. The roles of predator identity and group size in the antipredator responses of American toad (*Bufo americanus*) and bullfrog (*Rana catesbeiana*) tadpoles. Behaviour 146:225–243.

Smith, G.R., and D.T. Fortune. 2009. Hatching plasticity of wood frog (*Rana sylvatica*) eggs in response to mosquitofish (*Gambusia affinis*) cues. Herpetological Conservation and Biology 4:43–47.

Smith, G.R., and A.A. Burgett. 2012a. Effects of nutrient enrichment and changes in the background tadpole community on American bullfrog tadpoles. Herpetological Journal 22:173–178.

Smith, G.R., and A.A. Burgett. 2012b. Interaction between two species of tadpoles mediated by nutrient enrichment. Herpetologica 68:174–183.

Smith, G.R., and C.J. Dibble. 2012. Effects of an invasive fish (*Gambusia affinis*) and anthropogenic nutrient enrichment on American toad (*Anaxyrus americanus*) tadpoles. Journal of Herpetology 46:198–202.

Smith, G.R., J.E. Rettig, G.G. Mittelbach, J.L. Valiulis, and S.R. Schaack. 1999. The effects of fish on assemblages of amphibians in ponds: a field experiment. Freshwater Biology 41:829–837.

Smith, G.R., M.A. Waters, and J.E. Rettig. 2000. Consequences of embryonic UV–B exposure for embryos and tadpoles of the plains leopard frog. Conservation Biology 14:1903–1907.

Smith, G.R., A. Todd, J.E. Rettig, and F. Nelson. 2003. Microhabitat selection by northern cricket frogs (*Acris crepitans*) along a west–central Missouri creek: field and experimental observations. Journal of Herpetology 37:383–385.

Smith, G.R., H.A. Dingfelder, and D.A. Vaala. 2004. Asymmetric competition between *Rana clamitans* and *Hyla versicolor* tadpoles. Oikos 105:626–632.

Smith, G.R., C.J. Dibble, A. Boyd, M.E. Ogle, A.J. Terlecky, and C.B. Dayer. 2007. *Bufo americanus* (American Toad). Hindlimb deformities. Herpetological Review 38:318–319.

Smith, G.R., A. Boyd, C.B. Dayer, and K.E. Winter. 2008. Behavioral responses of American toad and bullfrog tadpoles to the presence of cues from the invasive fish, *Gambusia affinis*. Biological Invasions 10:743–748.

Smith, G.R., A. Boyd, C.B. Dayer, M.E. Ogle, and A.J. Terlecky. 2009. Responses of gray treefrog and American toad tadpoles to the presence of cues from multiple predators. Herpetological Journal 19:79–83.

Smith, G.R., A. Boyd, C.B. Dayer, M.E. Ogle, A.J. Terlecky, and C.J. Dibble. 2010. Effects of sibship and the presence of multiple predators on the behavior of green frog (*Rana clamitans*) tadpoles. Ethology 116:217–225.

Smith, G.R., S.V. Krishnamurthy, A.C. Burger, and L.B. Mills. 2011a. Differential effects of malathion and nitrate exposure on American toad and wood frog tadpoles. Archives of Environmental Contamination and Toxicology 60:327–335.

Smith, G.R., A.J. Terlecky, C.B. Dayer, A. Boyd, M.E. Ogle and C.J. Dibble. 2011b. Effects of mosquitofish and ammonium nitrate on activity of green frog (*Lithobates clamitans*) tadpoles: a mesocosm experiment. Journal of

Freshwater Ecology 26:59–63.

Smith, G.R., A.A. Burgett, and J.E. Rettig. 2012. Effects of the anuran tadpole assemblage and nutrient enrichment on freshwater snail abundance (*Physella* sp.). American Midland Naturalist 168:341–351.

Smith, H.C. 1975. Reproductive cycles of *Bufo woodhousii* and *Acris gryllus* in northern Mississippi, with additional notes on *Pseudacris triseriata* and *Hyla chrysoscelis*. Ph.D. Dissertation, Mississippi State University, Mississippi State.

Smith, H.H., R.T. Zappalorti, A.R. Breisch, and D.L. McKinley. 1995. The type locality of the frog *Acris crepitans*. Herpetological Review 26:14.

Smith, H.M. 1932a. A report upon amphibians hitherto unknown from Kansas. Transactions of the Kansas Academy of Science 35:93–96.

Smith, H.M. 1932b. *Ascaphus truei* Stejneger in Montana. Copeia 1932:100.

Smith, H.M. 1933. On the proper name for the brevicipitid frog *Gastrophryne texensis* (Girard). Copeia 1933: 217.

Smith, H.M. 1934. The amphibians of Kansas. American Midland Naturalist 15:377–528.

Smith, H.M. 1937. Notes on *Scaphiopus hurterii* Strecker. Herpetologica 1:104–108.

Smith, H.M. 1947a. Subspecies of the Sonoran toad (*Bufo compactilis* Wiegmann). Herpetologica 4:7–13.

Smith, H.M. 1947b. *Pseudacris clarkii* and *P. n. triseriata* in Texas. Herpetologica 3:183–184.

Smith, H.M. 1950. Handbook of amphibians and reptiles of Kansas. Miscellaneous Publications of the Museum of Natural History, University of Kansas (2):1–336. [second edition, 1956]

Smith, H.M., and B.C. Brown. 1947. The Texan subspecies of the treefrog, *Hyla versicolor*. Proceedings of the Biological Society of Washington 60:47–50.

Smith, H.M., and H.K. Buechner. 1947. The influence of the Balcones Escarpment on the distribution of amphibians and reptiles in Texas. Bulletin of the Chicago Academy of Sciences 8:1–16.

Smith, H.M., and D.A. Langebartel. 1949. The toad *Bufo valliceps* in Arkansas. Copeia 1949:230.

Smith, H.M., and J.C. List. 1951. The occurrence of digital pads as an anomaly in the bullfrog. Turtox News 29:118–119.

Smith, H.M., and O. Sanders. 1952. Distributional data on Texan amphibians and reptiles. Texas Journal of Science 4:204–219.

Smith, H.M., C.W. Nixon, and P.E. Smith. 1948. A partial description of the tadpole of *Rana areolata circulosa* and notes on the natural history of the race. American Midland Naturalist 39:608–614.

Smith, H.M., T.P. Maslin, and R.L. Brown. 1965. Summary of the distribution of the herpetofauna of Colorado. University of Colorado Studies, Series in Biology No. 15:1–52.

Smith, H.M., R. Mixter, and T. Spandler. 1966. The bullfrog and other reptiles and amphibians in western South Dakota. Journal of the Ohio Herpetological Society 5:106–107.

Smith, H.M., D. Chiszar, J.T. Collins, and F. van Breukelen. 1998. The taxonomic status of the Wyoming toad, *Bufo baxteri* Porter. Contemporary Herpetology 1. http://www.contemporaryherpetology.org/ch/1998/1/CH_1998_1.pdf

Smith, K.G. 2004. *Osteopilus septentrionalis* (Cuban Treefrog). Reproductive behavior. Herpetological Review 35:374–375.

Smith, K.G. 2005a. Effects of nonindigenous tadpoles on native tadpoles in Florida: evidence of competition. Biological Conservation 123:433–441.

Smith, K.G. 2005b. An exploratory assessment of Cuban treefrog (*Osteopilus septentrionalis*) tadpoles as predators of native and nonindigenous tadpoles in Florida. Amphibia–Reptilia 26:571–575.

Smith, K.G. 2005c. Nonindigenous herpetofauna of Florida: patterns of richness and case studies of the impacts of the tadpoles of two invasive amphibians, *Osteopilus septentrionalis* and *Bufo marinus*. Ph.D. Dissertation, University of Tennessee, Knoxville.

Smith, L.L., and R. Franz. 1994. Use of Florida round-tailed muskrat houses by amphibians and reptiles. Florida Field Naturalist 22:69-74.

Smith, L.L., and C.K. Dodd, Jr. 2003. Wildlife mortality on U.S. Highway 441 across Paynes Prairie, Alachua County, Florida. Florida Scientist 66:128–140.

Smith, L.L., K.G. Smith, W.J. Barichivich, C.K. Dodd, Jr., and K. Sorensen. 2005. Roads and Florida's herpetofauna: a review and mitigation case study. Pp. 32–40 *In* W.E. Meshaka, Jr. and K.J. Babbitt (eds.), Amphibians and Reptiles. Status and Conservation in Florida. Krieger Publishing, Malabar, Florida.

Smith, L.L., W.J. Barichivich, J.S. Staiger, K.G. Smith, and C.K. Dodd, Jr. 2006a. Detection probabilities and site occupancy estimates for amphibians at Okefenokee National Wildlife Refuge. American Midland Naturalist 155:149–161.

Smith, L.L., D.A. Steen, J.M. Stober, M.C. Freeman, S.W. Golladay, L.M. Conner, and J. Cochrane. 2006b. The vertebrate fauna of Ichauway, Baker County, GA. Southeastern Naturalist 5:599–620.

Smith, L.M., M.J. Gray, and A. Quarles. 2004. Diets of newly metamorphosed amphibians in west Texas playas. Southwestern Naturalist 49:257–263.

Smith, M.A. 2002. *Pseudacris triseriata triseriata* (Western Chorus Frog). Reproduction. Herpetological Review 33:127.

Smith, M.A. 2003. Spatial ecology of *Bufo fowleri*. Ph.D. Dissertation, McGill University, Montreal, Quebec.

Smith, M.A., and D.M. Green. 2002. *Bufo fowleri* (Fowler's Toad). Predation. Herpetological Review 33:124.

Smith, M.A, and D.M. Green. 2004. Phylogeography of *Bufo fowleri* at its northern range limit. Molecular Ecology 13:3723–3733.

Smith, M.A., and D.M. Green. 2005a. *Bufo fowleri* (Fowler's Toad). Predation. Herpetological Review 36:159–160.

Smith, M.A., and D.M. Green. 2005b. Dispersal and the metapopulation paradigm in amphibian ecology and conservation: are all amphibian populations metapopulations? Ecography 28:110–128.

Smith, M.A, and D.M. Green. 2006. Sex, isolation and fidelity: unbiased long distance dispersal in a terrestrial amphibian. Ecography 29:649–658.

Smith, M.A., C.M. Kapron, and M. Berrill. 2000. Induction of photolyase activity in wood frog (*Rana sylvatica*) embryos. Photochemistry and Photobiology 72:575–578.

Smith, M.A., M. Berrill, and C.M. Kapron. 2002. Photolyase activity of the embryo and the ultraviolet absorbance of embryo jelly for several Ontario amphibian species. Canadian Journal of Zoology 80:1109–1116.

Smith, M.J., M.M. Drew, M. Peebles, and K. Summers. 2005. Predator cues during the egg stage affect larval development in the gray treefrog, *Hyla versicolor* (Anura: Hylidae). Copeia 2005:169–173.

Smith, P.W. 1947. The reptiles and amphibians of eastern central Illinois. Bulletin of the Chicago Academy of Sciences 8:21–40.

Smith, P.W. 1951. A new frog and a new turtle from the western Illinois sand prairies. Bulletin of the Chicago Academy of Sciences 9:189–199.

Smith, P.W. 1953. A reconsideration of the status of *Hyla phaeocrypta*. Herpetologica 9:169–173.

Smith, P.W. 1955a. An addition to the list of pallid animals occurring in White Sands National Monument. Copeia 1955:135.

Smith, P.W. 1955b. *Pseudacris streckeri illinoensis* in Missouri. Transactions of the Kansas Academy of Science 58:411.

Smith, P.W. 1956. The status, correct name, and geographic range of the boreal chorus frog. Proceedings of the Biological Society of Washington 69:169–176.

Smith, P.W. 1957. An analysis of post–Wisconsin biogeography of the Prairie Peninsula region based on distributional phenomena among terrestrial vertebrate populations. Ecology 38:205–218.

Smith, P.W. 1961. The amphibians and reptiles of Illinois. Illinois Natural History Survey Bulletin 28:1–298.

Smith, P.W., and D.M. Smith. 1952. The relationship of the chorus frogs, *Pseudacris nigrita feriarum* and *Pseudacris n. triseriata*. American Midland Naturalist 48:165–180.

Smith, R.E. 1940. Mating and oviposition in the Pacific Coast tree toad. Science 92:379–380.

Smith, W.H. 1882. Report on the reptiles and amphibians of Ohio. Geological Survey of Ohio, Zoology & Botany, Volume IV, Part 1, Section 3, pp. 629-734.

Smith, W.H. 2011. Ecological determinants of herpetofaunal assemblage structure in the longleaf pine ecosystem. Ph.D. Dissertation, University of Alabama, Tuscaloosa.

Smith, W.H., and L.J. Rissler. 2010. Quantifying disturbance in terrestrial communities: abundance–biomass comparisons of herpetofauna closely track forest succession. Restoration Ecology 18 (Supplement 1):195–204.

Smith–Gill, S.J., and K.A. Berven. 1980. In vitro fertilization and assessment of male reproductive potential using mammalian gonadotropin–releasing hormone to induce spermiation in *Rana sylvatica*. Copeia 1980:723–728.

Smith-Gill, S.J., C.M. Richards, and G.W. Nace. 1972. Genetic and metabolic bases of two "albino" phenotypes in the leopard frog, *Rana pipiens*. Journal of Experimental Zoology 180:157-167.

Smithberger, S.I., and C.W. Swarth. 1993. Reptiles and amphibians of the Jug Bay Wetlands Sanctuary. Maryland Naturalist 37(3–4):28–46.

Smits, A.W. 1984. Activity patterns and thermal biology of the toad *Bufo boreas halophilus*. Copeia 1984:689–696.

Smits, A.W., and D.L. Crawford. 1984. Emergence of toads to activity: a statistical analysis of contributing cues. Copeia 1984:696–701.

Smits, J.E.G., B.D. Hersikorn, R.F. Young, and P.M. Fedorak. 2012. Physiological effects and tissue residues from exposure of leopard frogs to commercial naphthenic acids. Science of the Total Environment 437:36–41.

Snider, A.T., and J.K. Bowler. 1992. Longevity of reptiles and amphibians in North American collections. 2nd ed. SSAR Herpetological Circular No. 21.

Snodgrass, J.W., W.A. Hopkins, and J.H. Roe. 2003. Relationships among developmental stage, metamorphic timing, and concentrations of elements in bullfrogs (*Rana catesbeiana*). Environmental Toxicology and Chemistry 22:1597–1604.

Snodgrass, J.W., W.A. Hopkins, J. Broughton, D. Gwinn, J.A. Baionno, and J. Burger. 2004. Species–specific responses of developing anurans to coal combustion waste. Aquatic Toxicology 66:171–182.

Snodgrass, J.W., W.A. Hopkins, B.P. Jackson, J.A. Baionno, and J. Broughton. 2005. Influence of larval period on responses of overwintering green frog (*Rana clamitans*) larvae exposed to contaminated sediments. Environmental Toxicology and Chemistry 24:1508–1514.

Snodgrass, J.W., R.E. Casey, D. Joseph, and J.A. Simon. 2008. Mesocosm investigations of stormwater pond sediment toxicity to embryonic and larval amphibians: variation in sensitivity among species. Environmental Pollution 154:291–297.

Snow, N.P., and G. Witmer. 2010. American bullfrogs as invasive species: a review of the introduction, subsequent problems, management options, and future directions. Pp. 86-89 *In* R.M. Timm and K.A. Fagerstone (eds.), Proceedings of the Vertebrate Pest Conference, University of California, Davis.

Snow, N.P., and G. Witmer. 2011. A field evaluation of a trap for invasive American bullfrogs. Pacific Conservation Biology 17:285-291.

Synder, D.F. 1972. Amphibians and Reptiles of Land Between the Lakes. Tennessee Valley Authority. [no place of publication given]

Snyder, G.K., and G.A. Hammerson. 1993. Interrelationships between water economy and thermoregulation in the canyon treefrog *Hyla arenicolor*. Journal of Arid Environments 25:321–329.

Snyder, J.O. 1920. *Scaphiopus* in northern Nevada. Copeia (86):83–84.

Snyder, W.F., and D.L. Jameson. 1965. Multivariate geographic variation of mating call in populations of the Pacific tree frog (*Hyla regilla*). Copeia 1965:129–142.

Sodhi, N.S., D. Bickford, A.C. Diesmos, T.M. Lee, L.P. Koh, B.W. Brook, C.H. Sekercioglu, and C.J.A. Bradshaw. 2008. Measuring the meltdown: drivers of global amphibian extinction and decline. PLoS One 2(2):e1636.

Sonnini, C.S., and P.A. Latreille. 1801. Histoire naturelle des reptiles, avec figures designées d'après nature, 2. Chez Deterville, Paris.

Sornborger, M.B. 1979. Population dynamics of the western toad, *Bufo boreas halophilus*, at Hidden Lake, Mt. San Jacinto State Park. M.S. thesis, California State Polytechnic University, Pomona.

Sosa, J.A., M.J. Ryan, and M.A. Schlaepfer. 2009. Induced morphological plasticity in lowland leopard frog larvae (*Rana yavapaiensis*) does not confer a survival advantage against green sunfish (*Lepomis cyanellus*). Journal of Herpetology 43:460–468.

Souder, W. 2000. A Plague of Frogs. The Horrifying True Story. Hyperion, New York.

Sours, G.N., and J.W. Petranka. 2007. Intraguild predation and competition mediate stage–structured interactions between wood frog (*Rana sylvatica*) and upland chorus frog (*Pseudacris feriarum*) larvae. Copeia 2007:131–139.

Southworth, G.C., G. Mason, and J.R. Seed. 1968. Studies on frog trypanosomiasis. I. A 24–hour cycle in the para-

sitemia level of *Trypanosoma rotatorium* in *Rana clamitans* from Louisiana. Journal of Parasitology 54:255–258.

Sowers, A.D. 2009. The effects of wastewater effluents on a fish and amphibian species. Ph.D. Dissertation, Clemson University, Clemson, South Carolina.

Sowers, A.D., and S.J. Kaine. 2008. The effects of triclosan on the development of *Rana palustris*. Proceedings of the 2008 South Carolina Water Resources Conference, Institute of Computational Ecology, Clemson University, Clemson, South Carolina. 2 pp.

Sowers, A.D., M.A. Mills, and S.J. Klaine. 2009. The developmental effects of a municipal wastewater effluent on the northern leopard frog, *Rana pipiens*. Aquatic Toxicology 94: 145–152.

Sparling, D.W. 2003. A review of the role of contaminants in amphibian declines. Pp. 1099–1128 *In* D.J. Hoffman, B.A. Rattner, G.A. Burton, Jr., and J. Cairns, Jr. (eds.), Handbook of Ecotoxicology. Lewis Publishers, Boca Raton, Florida.

Sparling, D.W., and T.P. Lowe. 1996. Metal concentrations of tadpoles in experimental ponds. Environmental Pollution 91:149–159.

Sparling, D.W., and G. Fellers. 2007. Comparative toxicity of chlorpyrifos, diazinon, malathion and their oxon derivatives to larval *Rana boylii*. Environmental Pollution 147:535–539.

Sparling, D.W., T.P. Lowe, D. Day, and K. Dolan. 1995. Responses of amphibian populations to water and soil factors in experimentally–treated aquatic mesocosms. Archives of Environmental Contamination and Toxicology 29:455–461.

Sparling, D.W., T.P. Lowe, and A.E. Pinkney. 1997. Toxicity of Abate™ to green frog tadpoles. Bulletin of Environmental Contamination and Toxicology 58:475–481.

Sparling, D.W., G.M. Fellers, and L.L. McConnell. 2001. Pesticides and amphibian population declines in California, USA. Environmental Toxicology and Chemistry 20:1591–1595.

Sparling, D.W., G. Linder, C.A. Bishop, and S. Krest. (eds.), 2010. Ecotoxicology of Amphibians and Reptiles. Second edition, SETAC Press, Pensacola, Florida.

Spear, P.A., M. Boily, I. Giroux, C. DeBlois, M.H. Leclair, M. Levasseur, and R. Leclair. 2009. Study design, water quality, morphometrics and age of the bullfrog, *Rana catesbeiana*, in sub–watersheds of the Yamaska River drainage basin, Québec, Canada. Aquatic Toxicology 91:110–117.

Spear, S.F., and A. Storfer. 2008. Landscape genetic structure of coastal tailed frogs (*Ascaphus truei*) in protected vs. managed forests. Molecular Ecology 17:4642–4656.

Spear, S.F., and A. Storfer. 2010. Anthropogenic and natural disturbance lead to differing patterns of gene flow in the

Rocky Mountain tailed frog, *Ascaphus montanus*. Biological Conservation. 143:778–786.

Spear, S.F., C.M. Chrisafulli, and A. Storfer. 2012. Genetic structure among coastal tailed frog populations at Mount St. Helens is moderated by post–disturbance management. Ecological Applications 22: 856–869.

Speck, F.G. 1925. *Bufo americanus* in the Labrador Peninsula. Copeia (138):5–6.

Spencer, A.W. 1964a. The relationship of dispersal and migration to gene flow in the boreal chorus frog. Ph.D. Dissertation, Colorado State University, Fort Collins.

Spencer, A.W. 1964b. An unusual phoretic host for the clam, *Pisidium*. Journal of the Colorado–Wyoming Academy of Science 5(5):43–44.

Spencer, A.W. 1964c. Movement in a population of *Pseudacris triseriata*. Journal of the Colorado–Wyoming Academy of Science 5(5):44.

Spencer, A.W. 1971. Boreal chorus frogs (*Pseudacris triseriata*) breeding in the alpine in southwestern Colorado. Arctic and Alpine Research 3:353.

Sprankle, T. 2008. Giving leopard frogs a head start. Endangered Species Research 33:15–17.

Spriggs, A.N. 2009. Distribution and status of the northern leopard frog, *Rana pipiens*, in West Virginia. M.S. thesis, Marshall University, Huntington, West Virginia.

Springer, S. 1938. On the size of *Rana sphenocephala*. Copeia 1938:49.

Squire, T., and R.A. Newman. 2002. Fine–scale population structure in the wood frog (*Rana sylvatica*) in a northern woodland. Herpetologica 58:119–130.

Squires, W.A. 1950. *Hyla versicolor* from northern New Brunswick. New Brunswick Museum Nature News 1950(3):2.

Sredl, M.J. 1998. Arizona leopard frogs: balanced on the brink? Pp. 573–574 *In* M.J. Mac, P.A. Opler, C.E. Puckett, and P.D. Doran (eds.), Status and Trends of the Nation's Biological Resources. Vol. 2. U.S. Geological Survey, Washington, D.C.

Sredl, M.J. 2005a. *Rana yavapaiensis* Platz and Frost, 1984: Lowland Leopard Frog. Pp. 596–599 *In* M.J. Lannoo (ed.), Amphibian Declines. The Conservation Status of United States Species. University of California Press, Berkeley.

Sredl, M.J. 2005b. *Pternohyla fodiens* Boulenger, 1882: Lowland Burrowing Treefrog. Pp. 488–489 *In* M.J. Lannoo (ed.), Amphibian Declines. The Conservation Status of United States Species. University of California Press, Berkeley.

Sredl, M.J., and J.P. Collins. 1991. The effect of ontogeny on interspecific interactions in larval amphibians. Ecology 72:2232–2239.

Sredl, M.J., and J.P. Collins. 1992. The interaction of predation, competition, and habitat complexity in structuring an amphibian community. Copeia 1992:607–614.

Sredl, M.J., and J.M. Howland. 1995. Conservation and management of Madrean populations of the Chiricahua leopard frog. Pp. 379–385 *In* L.F. DeBano, G.J. Gottfried, R.H. Hamre, C.B. Edminster, P.F. Ffoliot, and A. Ortega–Rubio (eds.), Madrean Archipelago. The Sky Islands of the Southwestern United States and Northwestern Mexico. U.S. Forest Service Rocky Mountain Station General Technical Report RM–GTR–264.

Sredl, M.J., and R.D. Jennings. 2005. *Rana chiricahuensis* Platz and Mecham, 1979. Chiricahua Leopard Frog. Pp. 546–549 *In* M.J. Lannoo (ed.), Amphibian Declines. The Conservation Status of United States Species. University of California Press, Berkeley.

Stabler, R.M. 1948. Prairie rattlesnake eats spadefoot toad. Herpetologica 4:168.

St. Amant, J.A., and F.G. Hoover. 1969. Addition of *Misgurnus anguillicaudatus* (Cantor) to the California fauna. California Fish and Game 55:330–331.

St. Amant, J.A., F.G. Hoover, and G.R. Stewart. 1973. African clawed frog, *Xenopus laevis* (Daudin), established in California. California Fish and Game 59:151–153.

St.–Amour, V., T.W.J. Garner, A.I. Schulte–Hostedde, and D. Lesbarrères. 2010. Effects of two amphibian pathogens on the developmental stability of green frogs. Conservation Biology 24:788–794.

Stanback, M. 2010. *Gambusia holbrooki* predation on *Pseudacris feriarum* tadpoles. Herpetological Conservation and Biology 5:486–489.

Standaert, W.F. 1967. Growth, maturation, and population ecology of the carpenter frog (*Rana virgatipes* Cope). Ph.D. Dissertation, Rutgers University, New Brunswick, New Jersey.

Stangel, P.W. 1983. Least sandpiper predation on *Bufo americanus* and *Ambystoma maculatum* larvae. Herpetological Review 14:112.

Stanley, W.F., and A.H. Benton. 1966. An anomalous *Rana* from western New York. Copeia 1966:363.

Stapleton, N.M.R. 2011. The distribution and spread on non–indigenous anuran species in western Newfoundland. B.Sc. thesis, Memorial University of Newfoundland, Corner Brook, Newfoundland.

Starnes, S.M., C.A. Kennedy, and J.W. Petranka. 2000. Sensitivity of embryos of Southern Appalachian amphibians to ambient solar UV–B radiation. Conservation Biology 14:277–282.

Stead, J.E., and K.L. Pope. 2010. Predatory leeches (Hirudinea) may contribute to amphibian declines in the Lassen region, California. Northwestern Naturalist 91:30–39.

Stebbins, R.C. 1951. Amphibians of Western North America. 1st ed. University of California Press, Berkeley.

Stebbins, R.C. 1954. Amphibians and Reptiles of Western North America. McGraw–Hill Book Company, New York.

Stebbins, R.C. 1962. Amphibians of Western North America. 2nd printing. University of California Press, Berkeley.

Stebbins, R.C. 1972. Amphibians and Reptiles of California. California Natural History Guides No. 31. University of California Press, Berkeley.

Stebbins, R.C. 2003. A Field Guide to Western Amphibians and Reptiles. 3rd ed. Houghton Mifflin Co., Boston. [first edition, 1966, second edition, 1985]

Stebbins, R.C., and N.W. Cohen. 1995. A Natural History of Amphibians. Princeton University Press, Princeton, New Jersey.

Stebbins, R.C., and S.M. McGuiness. 2012. Field Guide to Amphibians and Reptiles of California. University of California Press, Berkeley.

Steele, C.W., S. Strickler–Shaw, and D.H. Taylor. 1989. Behavior of tadpoles of the bullfrog, *Rana catesbeiana* in response to sublehal lead exposure. Aquatic Toxicology 14:331-344.

Steele, C.W., S. Strickler–Shaw, and D.H. Taylor. 1991. Failure of *Bufo americanus* tadpoles to avoid lead–enriched water. Journal of Herpetology 25:241–243.

Steele, C.W., S. Strickler–Shaw, and D.H. Taylor. 1999. Effects of sublethal lead exposure on the behaviours of green frog (*Rana clamitans*), bullfrog (*Rana catesbeiana*) and American toad (*Bufo americanus*) tadpoles. Marine and Freshwater Behavioural Physiology 32:1–16.

Steelman, C.K., and M.E. Dorcas. 2010. Anuran calling survey optimization: developing and testing predictive models of anuran calling activity. Journal of Herpetology 44:61–68.

Steen, D.A. 2011. Wildlife restoration via forest management in fire–suppressed longleaf pine sandhills. Ph.D. Dissertation, Auburn University, Auburn, Alabama.

Steen, D.A., A.E.R. McGee, S.M. Hermann, J.A. Stiles, S.H. Stiles, and C. Guyer. 2010. Effects of forest management on amphibians and reptiles: generalist species obscure trends among native forest associates. Open Environmental Sciences 4:24–30.

Stein, R.J. 1985. *Rana sylvatica* (Wood Frog). Predation. Herpetological Review 16:109–110.

Steiner, G. 1924. Some nemas from the alimentary tract of the Carolina green treefrog (*Hyla carolinensis* Pennant). Journal of Parasitology 11:1–32.

Steiner, S.L., and R.M. Lehtinen. 2008. Occurrence of the amphibian pathogen *Batrachochytrium dendrobatidis* in Blanchard's cricket frog (*Acris crepitans blanchardi*) in the U.S. Midwest. Herpetological Review 39:193–196.

Steinwascher, K. 1978a. The effect of coprophagy on the growth of *Rana catesbeiana* tadpoles. Copeia 1978:130–134.

Steinwascher, K. 1978b. Interference and exploitation competition among tadpoles of *Rana utricularia*. Ecology 59:1039–1046.

Steinwascher, K. 1981. Competition for two resources. Oecologia 49:415–418.

Steinwascher, K., and J. Travis. 1983. Influence of food quality and quantity on early larval growth of two anurans. Copeia 1983:238–242.

Stejneger, L. 1893. Annotated list of the reptiles and batrachians collected by the Death Valley Expedition in 1891, with descriptions of new species. North American Fauna 7:159–228.

Stejneger, L. 1899. Description of a new genus and species of discoglossid toad from North America. Proceedings of the United States National Museum 21:899–901 + 1 plate.

Stejneger, L. 1901. A new species of bullfrog from Florida and the Gulf Coast. Proceedings of the United States National Museum 24:211–215.

Stejneger, L. 1915. A new species of tailless batrachian from North America. Proceedings of the Biological Society of Washington 28:131–132.

Stejneger, L., and T. Barbour. 1923. A Check List of North American Amphibians and Reptiles. 2nd ed. Harvard University Press, Cambridge, Massachusetts. [first edition, 1917]

Stephens, M.R. 2001. Phylogeography of the *Bufo boreas* (Anura, Bufonidae) species complex and the biogeography of California. M.A. thesis, Sonoma State University, Sonoma, California.

Stephenson, B. 2001. Mating behavior patterns and sexual interactions in the tailed frog, *Ascaphus truei*. M.S. thesis, Washington State University, Pullman.

Stephenson, B., and P. Verrell. 2003. Courtship and mating of the tailed frog (*Ascaphus truei*). Journal of Zoology 259:15-22.

Stevens, C.E., and C.A. Paszkowski. 2004. Using chorus–size ranks from call surveys to estimate reproductive activity of the wood frog (*Rana sylvatica*). Journal of Herpetology 38:404–410.

Stevens, C.E., and C.A. Paszkowski. 2005. A comparison of two pitfall trap designs in sampling boreal anurans. Herpetological Review 36:147–149.

Stevens, C.E., A.W. Diamond, and T.S. Gabor. 2002. Anuran call surveys on small wetlands in Prince Edward Island, Canada, restored by dredging of sediments. Wetlands 22:90–99.

Stevens, C.E., C.A. Paszkowski, and G.J. Scrimgeour. 2006a. Older is better: beaver ponds on boreal streams as breeding habitat for the wood frog. Journal of Wildlife Management 70:1360–1371.

Stevens, C.E., C.A. Paszkowski, and D. Stringer. 2006b. Occurrence of the western toad and its use of "borrow pits" in west–central Alberta. Northwestern Naturalist 87:107–117.

Stevenson, D.J. 2007. *Rana catesbeiana* (American Bullfrog). Mortality. Herpetological Review 38:71.

Stevenson, D.J. 2009. Born of new water. South Carolina Wildlife 56(1):16–23.

Stevenson, D., and D. Crowe. 1992. Geographic distribution: *Pseudacris crucifer bartramiana* (Southern Spring Peeper). Herpetological Review 23:86.

Stevenson, D.J., and K.J. Dyer. 2002. *Rana capito capito* (Carolina Gopher Frog). Refugia. Herpetological Review 33:128–129.

Stevenson, D.J., and J. Stackhouse. 2012. The amphibians and reptiles of the Altamaha River, Georgia. IRCF Reptiles & Amphibians 19:170–186.

Stevenson, H.M. 1959. Some altitude records of reptiles and amphibians. Herpetologica 15:118.

Stevenson, H.M. 1969. Occurrence of the carpenter frog in Florida. Quarterly Journal of the Florida Academy of Sciences 32:233–235.

Stewart, M.M. 1984. Redescription of the type of *Rana clamitans*. Copeia 1984:210–213.

Stewart, M.M., and P. Sandison. 1972. Comparative food habits of sympatric mink frogs, bull frogs, and green frogs. Journal of Herpetology 6:241–244.

Stewart, M.M., and J. Rossi. 1981. The Albany Pine Bush: a northern outpost for southern species of amphibians and reptiles in New York. American Midland Naturalist 106:282–292.

Stewart, P.A., and E.L. Hart. 1967. Incidental capture of vertebrate wildlife in blacklight insect traps. American Midland Naturalist 78:235–240.

Stille, W.T. 1952. The nocturnal amphibian fauna of the southern Lake Michigan beach. Ecology 33:149–162.

Stille, W.T. 1958. The water absorption response in an anuran. Copeia 1958:217–218.

Stille, W.T., Jr., and R.A. Edgren, Jr. 1942. The spring peeper in the Palos Hills area, Cook County, Ill. The Chicago Naturalist 5(3):63.

Stinner, J., N. Zarlinga, and S. Orcutt. 1994. Overwintering behavior of adult bullfrogs, *Rana catesbeiana*, in northeastern Ohio. Ohio Journal of Science 94:8–13.

Stitt, E.W. 2006. *Rana yavapaiensis* (Lowland Leopard Frog). Egg and tadpole predation. Herpetological Review 37:450.

Stitt, E.W., and C.P. Seltenrich. 2010. *Rana draytonii* (California Red–legged Frog). Prey. Herpetological Review 41:206.

Stitt, E.W., and P.S. Balfour. 2011. *Spea hammondii* (Western Spadefoot). Predation. Herpetological Review 42:91–92.

Stockwell, S., and M.L. Hunter, Jr. 1989. Relative abundance of herpetofauna among eight types of Maine peatland vegetation. Journal of Herpetology 23:409–414.

Stoddard, M. A., and J. P. Hayes. 2005. The influence of forest management on headwater stream amphibians at multiple spatial scales. Ecological Applications 15:811–823.

Stokely, P.S., and J.F. Berberian. 1953. On the jumping ability of frogs. Copeia 1953:187.

Stoler, A.B., and R.A. Relyea. 2011. Living in the litter: the influence of tree leaf litter on wetland communities. Oikos 120: 862–872.

Stolzenburg, W. 1997. The naked frog. Nature Conservancy 47(4):24-27.

Stone, L.M. 1986. "New" frog in the bog. Animal Kingdom 89(4):40–43.

Stone, P.A., and D.B. Ligon. 2011. *Bufo punctatus* (Red–spotted Toad) and *Thamnophis cyrtopsis* (Black–necked Garter Snake). Prey–predator. Herpetological Review 42:82–83.

Stone, W. 1906. Notes on the reptiles and batrachians of Pennsylvania, New Jersey and Delaware. American Naturalist 40:159-170.

Stone, W. 1932. Terrestrial activity of spade–foot toads. Copeia 1932:35–36.

Stone, W.B., and R.D. Manwell. 1969. Toxoplasmosis in cold blooded hosts. Journal of Protozoology 16:99–102.

Stoner, E.A. 1939. Butcher–bird butchers toad. Condor 41:126.

Storer, T.I. 1914. The California toad, an economic asset. University of California Journal of Agriculture 2:89–91.

Storer, T.I. 1922. The eastern bullfrog in California. California Fish and Game 8:219–224.

Storer, T.I. 1925. A synopsis of the amphibia of California. University of California Publications in Zoology 27:1–342.

Storer, T.I. 1933. Frogs and their commercial use. California Fish and Game 19:203–213.

Storey, J.M., and K.B. Storey. 1985. Adaptations of metabolism for freeze tolerance in the gray tree frog, *Hyla versicolor*. Canadian Journal of Zoology 63:49–54.

Storey, K.B. 1987. Glycolysis and the regulation of cryoprotectant synthesis in liver of the freeze tolerant wood frog. Journal of Comparative Physiology B 157:373–380.

Storey, K.B., and J.M. Storey. 1984. Biochemical adaptation

for freezing tolerance in the wood frog, *Rana sylvatica*. Journal of Comparative Physiology B 155:29–36.

Storey, K.B., and J.M. Storey. 1986. Freeze tolerance and intolerance as strategies of winter survival in terrestrially hibernating amphibians. Comparative Biochemistry and Physiology 83A:613–617.

Storey, K.B., and J.M. Storey. 1987. Persistence of freeze tolerance in terrestrially hibernating frogs after spring emergence. Copeia 1987:720–726.

Storey, K.B., and J.M. Storey. 1989. Freeze tolerance and freeze avoidance in ectotherms. Pp. 51–82 *In* L.C.H. Wang (ed.), Advances in Comparative and Environmental Physiology. Springer–Verlag, Berlin.

Storey, K.B., and J.M. Storey. 1992. Natural freeze tolerance in ectothermic vertebrates. Annual Review of Physiology 54:R619–R637.

Storey, K.B., J. Bischof, and B. Rubinsky. 1992. Cryomicroscopic analysis of freezing in liver of the freeze–tolerant wood frog. American Journal of Physiology 263:R185–R194.

Storm, R.M. 1952. Interspecific mating behaviour in *Rana aurora* and *Rana catesbeiana*. Herpetologica 8:108.

Storm, R.M. 1960. Notes on the breeding biology of the red–legged frog (*Rana aurora aurora*). Herpetologica 16:251–259.

Storrs, S.I., and J.M. Kiesecker. 2004. Survivorship patterns of larval amphibians exposed to low concentrations of atrazine. Environmental Health Perspectives 112:1054–1057.

Storrs, S.I., and R.D. Semlitsch. 2008. Species variation in somatic and ovarian development: predicting susceptibility of amphibians to estrogenic contaminants. General and Comparative Endocrinology 156:524–530.

Storrs–Méndez, S.I., and R.D. Semlitsch. 2009. Intersex gonads in frogs: understanding the time course of natural development and the role of endocrine disruptors. Journal of Experimental Zoology 314B:57–66.

Storrs–Méndez, S.I., D.E. Tillitt, T.A.G. Rittenhouse, and R.D. Semlitsch. 2009. Behavioral response and kinetics of terrestrial atrazine exposure in American toads (*Bufo americanus*). Archives of Environmental Contamination and Toxicology 57:590–597.

Storz, B.L. 2004. Reassessment of the environmental mechanisms controlling developmental polyphenism in spadefoot toad tadpoles. Oecologia 141:402–410.

Storz, B.L., and J. Travis. 2007. Temporally dissociated, trait-specific modifications underlie phenotypic polyphenism in *Spea multiplicata* tadpoles, which suggests modularity. The Scientific World Journal 7:715-726.

Stoutamire, R. 1932. Bullfrog farming and frogging in Florida. State of Florida, Department of Agriculture Bulletin (new series) No. 56.

Straughan, I.R. 1975. An analysis of the mechanisms of mating call discrimination in the frogs *Hyla regilla* and *H. cadaverina*. Copeia 1975:415–424.

Straw, R.M. 1958. Experimental notes on the Deep Springs toad *Bufo exsul*. Ecology 39:552–553.

Strecker, J.K., Jr. 1908a. Notes on the life history of *Scaphiopus couchii* Baird. Proceedings of the Biological Society of Washington 21:199–206.

Strecker, J.K., Jr. 1908b. A preliminary annotated list of the Batrachia of Texas. Proceedings of the Biological Society of Washington 21:53–62.

Strecker, J.K., Jr. 1908c. The reptiles and batrachians of Victoria and Refugio counties, Texas. Proceedings of the Biological Society of Washington 2:47-52.

Strecker, J.K., Jr. 1909. Notes on the narrow–mouthed toads (*Engystoma*) and the description of a new species from southeastern Texas. Proceedings of the Biological Society of Washington 22:115–120.

Strecker, J.K., Jr. 1910a. Description of a new solitary spadefoot (*Scaphiopus hurterii*) from Texas, with other herpetological notes. Proceedings of the Biological Society of Washington 23:115–122.

Strecker, J.K., Jr. 1910b. Studies in North American batrachology. Notes on the robber frog (*Lithodytes latrans* Cope). Transactions of the Academy of Science of St. Louis 19:73–82.

Strecker, J.K., Jr. 1915. Reptiles and amphibians of Texas. Baylor University Bulletin 18:1–82.

Strecker, J.K., Jr. 1922. An annotated catalogue of the amphibians and reptiles of Bexar County, Texas. Bulletin of the Scientific Society of San Antonio 4:1.

Strecker, J.K. 1924. Notes on the herpetology of Hot Springs, Arkansas. The Baylor Bulletin, Zoological Papers 27:27–47.

Strecker, J.K., Jr. 1926a. Chapters from the life–histories of Texas reptiles and amphibians: part one. Contributions from Baylor University Museum 8:1–12.

Strecker, J.K., Jr. 1926b. On the habits and variations of *Pseudacris ornata* (Holbrook). Contributions from Baylor University Museum 7:8–11.

Strecker, J.K., Jr. 1927. Chapters from the life–histories of Texas reptiles and amphibians: part two. Contributions from Baylor University Museum 10:1–14.

Strecker, J.K., and L.S. Frierson, Jr. 1926. The herpetology of Caddo and DeSoto parishes, Louisiana. Contributions from Baylor University Museum No. 5.

Strecker, J.K., Jr., and W.J. Williams. 1927. Herpetological records from the vicinity of San Marcos, Texas, with distributional data on the amphibians and reptiles of the Edwards Plateau region and central Texas. Contributions from Baylor University Museum 12:1–16.

Streicher, J.W. 2008. *Bufo americanus* (American Toad). Reproduction. Herpetological Review 39:75.

Streicher, J.W., C.L. Cox, J.A. Campbell, E.N. Smith, and R.O. de Sá. 2012. Rapid range expansion in the Great Plains narrow–mouthed toad (*Gastrophryne olivacea*) and a revised taxonomy for North American microhylids. Molecular Phylogenetics and Evolution 64:645–653.

Strickland, D., and R.J. Rutter. 1986. Reptiles and Amphibians of Algonquin Provincial Park. Friends of Algonquin Park, Whitney, Ontario.

Strickler-Shaw, S. 1988. The effects of low lead concentrations on acquisition and retention of avoidance learning in tadpoles of the bullfrog, *Rana catesbeiana*. M.S. thesis, Miami University, Oxford, Ohio.

Strickler-Shaw, S., and D.H. Taylor. 1990. Sublethal exposure to lead inhibits acquisition and retention of discriminate avoidance learning in green frog (*Rana clamitans*) tadpoles. Environmental Toxicology and Chemistry 9:47-52.

Strickler-Shaw, S., and D.H. Taylor. 1991. Lead inhibits acquisition and retention learning in bullfrog tadpoles. Neurotoxicology and Teratology 13:167-173.

Strojny, C.A., and M. L. Hunter, Jr. 2010. Relative abundance of amphibians in forest canopy gaps of natural origin vs. timber harvest origin. Animal Biodiversity and Conservation 33:1–13.

Stroud, C.P. 1949. A white spade–foot toad from the New Mexico white sands. Copeia 1949:232.

Stuart, J.N. 1991. Partial albinism in a New Mexico population of *Bufo woodhousei*. Bulletin of the Maryland Herpetological Society 27:33.

Stuart, J.N. 1995. *Rana catesbeiana* (Bullfrog). Diet. Herpetological Review 26:33.

Stuart, J.N., and C.W. Painter. 1993. *Rana catesbeiana* (Bullfrog). Cannibalism. Herpetological Review 24:103.

Stuart, J.N., and C.W. Painter. 1994. A review of the distribution and status of the boreal toad, *Bufo boreas boreas*, in New Mexico. Bulletin of the Chicago Herpetological Society 29:113–116.

Stewart, J.P., W. Hiler, C. McDowell, R. Neal, B.A. Wheeler, and S.E. Trauth. 2005. *Bufo speciosus* (Texas Toad). Maximum size. Herpetological Review 36:299.

Stewart, P.A. 1969. Prey in two screech owl nests. Auk 86:141.

Stuart, S., J.S. Chanson, N.A. Cox, B.E. Young, A.S.L. Rodrigues, D.L. Fishman, and R.W. Waller. 2004. Status and trends of amphibian declines and extinctions worldwide. Science 306:1783–1786.

Suhre, D.O. 2010. Dispersal and demography of the American bullfrog (*Rana catesbeiana*) in a semi–arid grassland. M.S. thesis, University of Arizona, Tucson.

Sullivan, B.K. 1982a. Sexual selection in Woodhouse's toad (*Bufo woodhousei*). 1. Chorus organization. Animal Behaviour 30:680–686.

Sullivan, B.K. 1982b. Male mating behavior in the Great Plains toad (*Bufo cognatus*). Animal Behaviour 30:939–940.

Sullivan, B.K. 1982c. Significance of size, temperature and call attributes to sexual selection in *Bufo woodhousei australis*. Journal of Herpetology 16:103–106.

Sullivan, B.K. 1983a. Sexual selection in Woodhouse's toad (*Bufo woodhousei*). 2. Female choice. Animal Behaviour 31:1011–1017.

Sullivan, B.K. 1983b. Sexual selection and mating system variation in the Great Plains toad (*Bufo cognatus* Say) and Woodhouse's toad (*Bufo woodhousei australis* Shannon and Lowe). Ph.D. Dissertation, Arizona State University, Tempe, Arizona.

Sullivan, B.K. 1983c. Sexual selection in the Great Plains toad (*Bufo cognatus*). Behaviour 84:258–264.

Sullivan, B.K. 1984. Advertisement call variation and observations on breeding behavior of *Bufo debilis* and *B. punctatus*. Journal of Herpetology 18:406–411.

Sullivan, B.K. 1985a. Sexual selection and mating system variation in anuran amphibians of the Arizona–Sonora Desert. Great Basin Naturalist 45:688–696.

Sullivan, B.K. 1985b. Male calling behavior in response to playback of conspecific advertisement call in two bufonids. Journal of Herpetology 19:78–83.

Sullivan, B.K. 1986a. Hybridization between the toads *Bufo microscaphus* and *Bufo woodhousei* in Arizona: morphological variation. Journal of Herpetology 20:11–21.

Sullivan, B.K. 1986b. Intra–population variation in the intensity of sexual selection in breeding aggregations of Woodhouse's toad (*Bufo woodhousei*). Journal of Herpetology 20:88–90.

Sullivan, B.K. 1986c. Advertisement call variation in the Arizona tree frog, *Hyla wrightorum* Taylor, 1938. Great Basin Naturalist 46:378–381.

Sullivan, B.K. 1987. Sexual selection in Woodhouse's toad (*Bufo woodhousei*). 3. Seasonal variation in male mating success. Animal Behaviour 35:912–919.

Sullivan, B.K. 1989a. Desert environments and the structure of anuran mating systems. Journal of Arid Environments 17:175–183.

Sullivan, B.K. 1989b. Interpopulation variation in vocalizations of *Bufo woodhousii*. Journal of Herpetology 23:368–373.

Sullivan, B.K. 1990. Natural hybrid between the Great Plains toad (*Bufo cognatus*) and the red–spotted toad (*Bufo punctatus*) from central Arizona. Great Basin Naturalist 50:371–372.

Sullivan, B.K. 1992a. Sexual selection and calling behavior in the American toad (*Bufo americanus*). Copeia 1992:1–7.

Sullivan, B.K. 1992b. Calling behavior of the southwestern toad (*Bufo microscaphus*). Herpetologica 48:383–389.

Sullivan, B.K. 1993. Distribution of the southwestern toad (*Bufo microscaphus*) in Arizona. Great Basin Naturalist 53:402–406.

Sullivan, B.K. 1995. Temporal stability in hybridization between *Bufo microscaphus* and *Bufo woodhousii* (Anura:Bufonidae): behavior and morphology. Journal of Evolutionary Biology 8:233–247.

Sullivan, B.K. 2009. Mate recognition, species boundaries and the fallacy of "species recognition." The Open Zoology Journal 2:86–90.

Sullivan, B.K., and G.E. Walsberg. 1985. Call rate and aerobic capacity in Woodhouse's toad (*Bufo woodhousei*). Herpetologica 41:404–407.

Sullivan, B.K., and E.A. Sullivan. 1985. Size–related variation in advertisement calls and breeding behavior of spadefoot toads (*Scaphiopus bombifrons, S. couchi* and *S. multiplicatus*). Southwestern Naturalist 30:349–355.

Sullivan, B.K., and M.K. Leek. 1986. Acoustic communication in Woodhouse's toad (*Bufo woodhousei*). 1. Response of calling males to variation in spectral and temporal components of advertisement calls. Behaviour 98:305–319.

Sullivan, B.K., and M.R. Leek. 1987. Acoustic communication in Woodhouse's toad (*Bufo woodhousei*). II. Responses of females to variation in spectral and temporal components of advertisement calls. Behaviour 103:16–26.

Sullivan, B.K., and T. Lamb. 1988. Hybridization between the toads *Bufo microscaphus* and *Bufo woodhousii* in Arizona: variation in release calls and allozymes. Herpetologica 44:325–333.

Sullivan, B.K., and W.E. Wagner, Jr. 1988. Variation in advertisement and release calls, and social influences on calling behavior in the Gulf Coast toad (*Bufo valliceps*). Copeia 1988:1014–1020.

Sullivan, B.K., and S.H. Hinshaw. 1990. Variation in advertisement calls and male calling behavior in the spring peeper (*Pseudacris crucifer*). Copeia 1990:1146–1150.

Sullivan, B.K., and S.H. Hinshaw. 1992. Female choice and selection on male calling behaviour in the grey treefrog *Hyla versicolor*. Animal Behaviour 44:733–744.

Sullivan, B.K., and K.B. Malmos. 1994. Call variation in the Colorado River toad (*Bufo alvarius*): behavioral and phylogenetic implications. Herpetologica 50:146–156.

Sullivan, B.K., and P.J. Fernandez. 1999. Breeding activity, estimated age–structure, and growth in Sonoran Desert anurans. Herpetologica 55:334–343.

Sullivan, B.K., K.B. Malmos, and M.F. Given. 1996a. Systematics of the *Bufo woodhousii* complex (Anura: Bufonidae): advertisement call variation. Copeia 1996:274–280.

Sullivan, B.K., R.W. Bowker, K.B. Malmos, and E.W.A. Gergus. 1996b. Arizona distribution of three Sonoran Desert anurans: *Bufo retiformis, Gastrophryne olivacea*, and *Pternohyla fodiens*. Great Basin Naturalist 56:38–47.

Sullivan, B.K., C.R. Propper, M.J. Demlong, and L.A. Harvey. 1996c. Natural hermaphroditic toad (*Bufo microscaphus X Bufo woodhousii*). Copeia 1996:470–472.

Sullivan, B.K., K.B. Malmos, E.W.A. Gergus, and R.W. Bowker. 2000. Evolutionary implications of advertisement call variation in *Bufo debilis, B. punctatus*, and *B. retiformis*. Journal of Herpetology 34:368–374.

Sullivan, J.J., and E.E. Byrd. 1970. *Choledocystus pennsylvaniensis*: life history. Transactions of the American Microscopical Society 89:384–396.

Sullivan, S.R., P.E. Bartelt, and C.R. Peterson. 2008. Midsummer ground surface activity patterns of western toads (*Bufo boreas*) in southeastern Idaho. Herpetological Review 39:35–40.

Sultan, S.E. 2007. Development in context: the timely emergence of eco-devo. Trends in Ecology & Evolution 22:575–582.

Sun, L.–X., W. Wilczynski, A.S. Rand, and M.J. Ryan. 2000. Trade–off in short– and long–distance communication in túngara (*Physalaemus pustulosus*) and cricket (*Acris crepitans*) frogs. Behavioral Ecology 11:102–109.

Surdick, J.A., Jr. 2005. Amphibian and avian species composition of forested depressional wetlands and circumjacent habitat: the influence of land use type and intensity. Ph.D. Dissertation, University of Florida, Gainesville.

Surface, H.A. 1913. First report on the economic features of the amphibians of Pennsylvania. Zoological Bulletin, Division of Zoology, Pennsylvania Department of Agriculture 3:66–152.

Suring, L.H., W.L. Gaines, B.C. Wales, K. Mellen-McLean, J.S. Begley, and S. Mohoric. 2011. Maintaining populations of terrestrial wildlife through land management planning: a case study. Journal of Wildlife Management 75:945–958.

Sustare, B.D. 1977. Characterizing parameters of response to light intensity for six species of frogs. Behavior Processes 2:101–112.

Sutherland, D. 2005. Parasites of North American frogs. Pp. 109–123 *In* M.J. Lannoo (ed.), Amphibian Declines. The Conservation Status of United States Species. University of California Press, Berkeley.

Sutherland, G.D. 2000. Risk assessment for conservation under ecological uncertainty: a case study with a stream–dwelling amphibian in managed forests. Ph.D. Dissertation, University of British Columbia, Vancouver.

Sutherland, G.D., and F.L. Bunnell. 2001. Cross–scale classification trees for assessing risks of forest practices to headwater stream amphibians. Pp. 550–555 *In* D.H. Johnson and T.A. O'Neil (Managing Directors), Wildlife–Habitat Relationships in Oregon and Washington. Oregon State University Press, Corvallis.

Sutherland, M.A.B., G.M. Gouchie, and R.J. Wassersug. 2009. Can visual stimulation alone induce phenotypically plastic responses in *Rana sylvatica* tadpole oral structures? Journal of Herpetology 43:165–168.

Sutherland, R.W., P.R. Dunning, and W.M. Baker. 2010. Amphibian encounter rates on roads with different amounts of traffic and urbanization. Conservation Biology 24:1626–1635.

Sutton, W.B. 2004. The ecology and natural history of the Northern Leopard Frog, *Rana pipiens* Schreber, in West Virginia. M.S. thesis, Marshall University, Huntington, West Virginia.

Sutton, W.B. 2010. Herpetofaunal response to thinning and prescribed burning in pine–hardwood forests. Ph.D. Dissertation, Alabama A&M University, Huntsville.

Sutton, W.B., K.E. Rastall, and T.K. Pauley. 2006. Diet analysis and feeding strategies of *Rana pipiens* in a West Virginia wetland. Herpetological Review 37:152–153.

Suzuki, H.K. 1951. Recent additions to the records of the distribution of the amphibians in Wisconsin. Transactions of the Wisconsin Academy of Sciences, Arts and Letters 40:215–234.

Suzuki, H.K. 1957. A study of leg length variations in the wood frog, *Rana sylvatica* Le Conte. Transactions of the Wisconsin Academy of Sciences, Arts and Letters 46:299–303.

Svihla, A. 1935. Notes on the western spotted frog, *Rana pretiosa pretiosa*. Copeia 1935:119–122.

Svihla, A. 1936. *Rana rugosa* Schelegel. Notes on the life history of this interesting frog. Mid-Pacific Magazine (April–June):124–125.

Svihla, A. 1953. Diurnal retreats of the spadefoot toad *Scaphiopus hammondii*. Copeia 1953:186.

Svihla, A., and R.D. Svihla. 1933a. Notes on *Ascaphus truei* in Kittitas County, Washington. Copeia 1933:37–38.

Svihla, A., and R.D. Svihla. 1933b. Amphibians and reptiles of Whitman County, Washington. Copeia 1933:125–128.

Swann, D. 2005. Rock star: canyon treefrog (*Hyla arenicolor* Cope, 1866). Sonoran Herpetologist 18(4):39–42.

Swann, J.M., T.W. Schultz, and J.R. Kennedy. 1996. The effects of organophosphorus insecticides Dursban™ and Lorsban™ on the ciliated epithelium of the frog palate *in vitro*. Archives of Environmental Contamination and Toxicology 30:188–194.

Swannack, T.M. 2007. Modeling aspects of the ecological and evolutionary dynamics of the endangered Houston toad. Ph.D. Dissertation, Texas A&M University, College Station.

Swannack, T.M., and M.R.J. Forstner. 2007. Possible cause for the sex–ratio disparity of the endangered Houston Toad (*Bufo houstonensis*). Southwestern Naturalist 52:386–392.

[7]Swannack, T.M., J.T. Jackson, and M.R.J. Forstner. 2006a. *Bufo houstonensis* (Houston Toad). Juvenile dispersal. Herpetological Review 37:199–200.

[7]Swannack, T.M., J.T. Jackson, and M.R.J. Forstner. 2006b. *Bufo houstonensis* (Houston Toad). Juvenile dispersal. Herpetological Review 37:335.

Swanack [sic: Swannack], T.M., W.E. Grant, and M.R.J. Forstner. 2009. Projecting population trends of endangered amphibian species in the face of uncertainty: a pattern–oriented approach. Ecological Modelling 220:148–159.

Swanson, D.L., and B.M. Graves. 1995. Supercooling and freeze intolerance in overwintering juvenile spadefoot toads (*Scaphiopus bombifrons*). Journal of Herpetology 29:280–285.

Swanson, D.L., and S.L. Burdick. 2010. Overwintering physiology and hibernacula microclimates of Blanchard's cricket frogs at their northwestern range boundary. Copeia 2010:247–253.

Swanson, D.L., B.M. Graves, and K.L. Koster. 1996. Freezing tolerance/intolerance and cryoprotectant synthesis in terrestrially overwintering anurans in the Great Plains, USA. Journal of Comparative Physiology B 166:110–119.

Swanson, E.M., S.M. Tekmen, and M.A. Bee. 2007. Do female frogs exploit inadvertent social information to locate breeding aggregations? Canadian Journal of Zoology 85: 921–932.

Swanson, P.L. 1939. Herpetological notes from Indiana. American Midland Naturalist 22:684–695.

Swarth, H.S. 1936. Origins of the fauna of the Sitkan District, Alaska. Proceedings of the California Academy of Sciences 23:59–78.

Sweet, S.S., and B.K. Sullivan. 2005. *Bufo californicus* Camp, 1915. Arroyo Toad. Pp. 396–400 *In* M.J. Lannoo (ed.), Amphibian Declines. The Conservation Status of United States Species. University of California Press, Berkeley.

Sweetman, H.L. 1944. Food habits and molting of the common tree frog. American Midland Naturalist 32:499–501.

Swierk, L. 2012. *Rana sylvatica* (wood frog): Larval duration.

---

[7]Identical in content and must have been mistakenly printed twice.

Herpetological Bulletin (119):39.

Sype, W.E. 1975. Breeding habits, embryonic thermal requirements and embryonic and larval development of the Cascade frog, *Rana cascadae* (Slater). Ph.D. Dissertation, Oregon State University, Corvallis.

Szafoni, R.E., C.A. Phillips, and M. Redmer. 1999. Translocations of amphibian species outside their native range: a comment on Thurow (1994–1997). Transactions of the Illinois State Academy of Science 92:277–283.

Szuroczki, D., and J.M.L. Richardson. 2009. The role of trematode parasites in larval anuran communities: an aquatic ecologist's guide to the major players. Oecologia 161:371–385.

Szuroczki, D., and J.M.L. Richardson. 2011. Palatability of the larvae of three species of *Lithobates*. Herpetologica 67:213–221.

Taggart, T.W. 1997. Status of *Bufo debilis* (Anura: Bufonidae) in Kansas. Kansas Herpetological Society Newsletter (109):7-12.

Taigen, T.L., and F.H. Pough. 1981. Activity metabolism of the toad (*Bufo americanus*): ecological consequences of ontogenetic change. Journal of Comparative Physiology 144:247–252.

Taigen, T.L., and K.D. Wells. 1985. Energetics of vocalization by an anuran amphibian (*Hyla versicolor*). Journal of Comparative Physiology 155B:163–170.

Taigen, T.L., S.B. Emerson, and F.H. Pough. 1982. Ecological correlates of anuran exercise physiology. Oecologia 52: 49–56.

Talley, H.J. 1937. The annual ovarian cycle of *Acris gryllus*. M.S. thesis, University of Oklahoma, Norman.

Tamsitt, J.R. 1962. Notes on a population of the Manitoba toad (*Bufo hemiophrys*) in the Delta Marsh region of Lake Manitoba, Canada. Ecology 43:147–150.

Tanner, V.M. 1928. Distributional list of the amphibians and reptiles of Utah. Copeia (166):23–28.

Tanner, V.M. 1931. A synoptical study of Utah amphibia. Utah Academy of Sciences 8:159–198.

Tanner, V.M. 1933. *Bufo lamentor* Girard. Copeia 1933:42.

Tanner, V.M. 1939. A study of the genus *Scaphiopus*, the spade–foot toads. Great Basin Naturalist 1:3–20.

Tanner, W.W. 1941. The reptiles and amphibians of Idaho No. 1. Great Basin Naturalist 2:87–97.

Tanner, W.W. 1950. Notes on the habits of *Microhyla carolinensis olivacea* (Hallowell). Herpetologica 6:47–48.

Tanner, W.W. 1989. Status of *Spea stagnalis* Cope (1875), *Spea intermontanus* Cope (1889), and a systematic review of *Spea hammondii* Baird (1839) (Amphibia: Anura). Great Basin Naturalist 49:503–510.

Tardell, J.H., R.C. Yates, and D.H. Schiller. 1981. New re-

cords and habitat observations of *Hyla andersoni* Baird (Anura: Hylidae) in Chesterfield and Marlboro counties, South Carolina. Brimleyana 6:153–158.

Tatarian, P.J. 2008. Movement patterns of California red–legged frogs (*Rana draytonii*) in an inland California environment. Herpetological Conservation and Biology 3:155–169.

Tatarian, P., and G. Tatarian. 2010. Chytrid infection of *Rana draytonii* in the Sierra Nevada, California, USA. Herpetological Review 41:325–327.

Tattersall, G.J., and G.R. Ultsch. 2008. Physiological ecology of aquatic overwintering in Ranid frogs. Biological Review 83:119–140.

Tavares, K. 2008. Coqui control, monitoring and outreach program. 2008 Annual Report. Hawaii Volcanoes National Park. http://www.hawaiisfishes.com/coqui/2008%20HAVO%20Report/2008_Coqui_Season–lo-res.pdf

Tavera–Mendoza, L., S. Ruby, P. Brousseau, M. Fournier, D. Cyr, and D. Marcogliese. 2002. Response of the amphibian tadpole (*Xenopus laevis*) to atrazine during sexual differentiation of the testis. Environmental Toxicology and Chemistry 21:527–531.

Taylor, B., D. K. Skelly, L. K. Demarchis, M. D. Slade, D. Galusha, and P. M. Rabinowitz. 2005. Proximity to pollution sources and risk of amphibian limb malformation. Environmental Health Perspectives 113:1497–1501.

Taylor, C.L., R. Altig, and C.R. Boyle. 1995. Can anuran tadpoles choose among foods that vary in quality? Alytes 13:81–86.

Taylor, C.N., K.L. Oseen, and R.J. Wassersug. 2004. On the behavioral response of *Rana* and *Bufo* tadpoles to echinostomatoid cercariae: implications to synergistic factors influencing trematode infections in anurans. Canadian Journal of Zoology 82:701–706.

Taylor, D.H. 1970. Orientational mechanisms of the southern cricket frog, *Acris gryllus*. Ph.D. Dissertation, Mississippi State University, State College, Mississippi.

Taylor, D.H., and D.E. Ferguson. 1969. Solar cues and shoreline learning in the southern cricket frog, *Acris gryllus*. Herpetologica 25:147–149.

Taylor, D.H., and D.E. Ferguson. 1970. Extraoptic celestial orientation in the southern cricket frog *Acris gryllus*. Science 168:390–392.

Taylor, D.H., C.W. Steele, and S. Strickler-Shaw. 1990. Responses of green frog (*Rana clamitans*) tadpoles to lead-polluted water. Environmental Toxicology and Chemistry 9:87-93.

Taylor, E.H. 1929. List of reptiles and batrachians of Morton County, Kansas, reporting species new to the state fauna. University of Kansas Science Bulletin 19:63–65.

Taylor, E.H. 1939 ("1938"). Frogs of the *Hyla eximia* group in Mexico, with descriptions of two new species. University of Kansas Science Bulletin 25:421–445.

Taylor, E.H. 1940. Herpetological miscellany. University of Kansas Science Bulletin 26:489–571.

Taylor, E.H., and J.S. Wright. 1932. The toad *Bufo marinus* (Linnaeus) in Texas. University of Kansas Science Bulletin 20:247–249.

Taylor, J. 1993. The Amphibians & Reptiles of New Hampshire. New Hampshire Fish and Game Department, Concord.

Taylor, P. 2009. An extension of gray treefrog range in Manitoba and into Saskatchewan. Blue Jay 67:235–241.

Taylor, R.C., B.W. Buchanan, and J.L. Doherty. 2007. Sexual selection in the squirrel treefrog *Hyla squirella*: the role of multimodal cue assessment in female choice. Animal Behaviour 74:1753–1763.

Taylor, R.J., and E.D. Michael. 1971. Habitat effects on monthly foods of bullfrogs in eastern Texas. Proceedings of the Annual Conference of the Southeast Association of Game and Fish Commissioners 25:176–186.

Taylor, S.K. 1998. Investigation of mortality of Wyoming toads and the effect of malathion on amphibian disease susceptibility. Ph.D. Dissertation, University of Wyoming, Laramie.

Taylor, S.K., E.S. Williams, and K.W. Mills. 1999a. Effects of malathion on disease susceptibility in Woodhouse's toads. Journal of Wildlife Diseases 35:536–541.

Taylor, S.K., E.S. Williams, and K.W. Mills. 1999b. Mortality of captive Canadian toads from *Basidiobolus ranarum* mycotic dermatitis. Journal of Wildlife Diseases 35:64–69.

Taylor, S.K., E.S. Williams, E.T. Thorne, K.W. Mills, D.I. Withers, and A.C. Pier. 1999c. Causes of mortality of the Wyoming toad. Journal of Wildlife Diseases 35:49–57.

Taylor, S.K., E.S. Williams, A.C. Pier, K.W. Mills, and M.D. Bock. 1999d. Mucormycotic dermatitis in captive adult Wyoming toads. Journal of Wildlife Diseases 35:70–74.

Taylor, S.K., E.S. Williams, and K.W. Mills. 1999e. Experimental exposure of Canadian toads to *Basidiobolus ranarum*. Journal of Wildlife Diseases 35:58–63.

Teale, E.W. 1948. Our lowly friend the toad. Coronet (September):21–24.

Tekiela, S. 2003. Reptiles & Amphibians of Minnesota Field Guide. Adventure Publications, Cambridge, Minnesota.

Tekiela, S. 2004a. Reptiles & Amphibians of Michigan Field Guide. Adventure Publications, Cambridge, Minnesota.

Tekiela, S. 2004b. Reptiles & Amphibians of Wisconsin Field Guide. Adventure Publications, Cambridge, Minnesota.

Telford, H.H., and J.A. Munro. 1944. Toads feed upon sweet clover weevils. North Dakota Agricultural Experiment Station Bi–monthly Bulletin 6(4):35–37.

Telford, S.R., Jr. 1952. A herpetological survey in the vicinity of Lake Shipp, Polk County, Florida. Quarterly Journal of the Florida Academy of Sciences 15:175–185.

Temminck, C.J., and H. Schlegel. 1838. *In* P.F. Von Siebold (ed.), Fauna Japonica sive Descriptio animalium, quae in itinere per Japonianum, jussu et auspiciis superiorum, qui summum in India Batava Imperium tenent, suscepto, annis 1823–1830 colleget, notis observationibus et adumbrationibus illustratis. Vol. 3 (Chelonia, Ophidia, Sauria, Batrachia). J.G. Lalau, Leiden, Germany.

Tenneson, M.G. 1983. Behavioral ecology and population decline of the mink frog, *Rana septentrionalis*. M.S. thesis, University of North Dakota, Grand Forks.

Tennessen, J.A., D.C. Woodhams, P. Chaurand, L.K. Reinert, D. Billheimer, Y. Shyr, R.M. Caprioli, M.S. Blouin, and L.A. Rollins-Smith. 2009. Variations in the expressed antimicrobial peptide repertoire of northern leopard frog (*Rana pipiens*) populations suggest intraspecies differences in resistance to pathogens. Developmental and Comparative Immunology 33:1247-1257.

Terres, J.K. 1968. Kingfishers eating bullfrog tadpoles (*Megaceryle alcyon, Rana catesbeiana*). Auk 85:140.

Test, F.C. 1898. A contribution to the knowledge of the variations of the tree frog *Hyla regilla*. Proceedings of the United States National Museum 21:477–492.

Test, F.H. 1958. Butler's garter snake eats amphibian. Copeia 1958:151–152.

Test, F.H., and R.C. McCann. 1976. Foraging behavior of *Bufo americanus* tadpoles in response to high densities of micro–organisms. Copeia 1976:576–578.

Tester, J.R., and W.J. Breckenridge. 1964a. Population dynamics of the Manitoba toad, *Bufo hemiophrys*, in northwestern Minnesota. Ecology 45:592–601.

Tester, J.R., and W.J. Breckenridge. 1964b. Winter behavior patterns of the Manitoba toad, *Bufo hemiophrys*, in northwestern Minnesota. Annales Academiae Scientiarum Fennicae (Biologica) 71:421–431.

Tester, J.R., A. Parker, and D.B. Siniff. 1965. Experimental studies on habitat preferences and thermoregulation of *Bufo americanus, B. hemiophrys*, and *B. cognatus*. Journal of the Minnesota Academy of Science 33:27–32.

Tevis, L., Jr. 1966. Unsuccessful breeding by desert toads (*Bufo punctatus*) at the limit of their ecological tolerance. Ecology 47:766–775.

Theodorakis, C.W., J. Rinchard, J.A. Carr, J.-W. Park, L. McDaniel, F. Liu, and M. Wages. 2006. Thyroid endocrine disruption in stonerollers and cricket frogs from perchlorate–contaminated streams in east–central Texas. Ecotoxicology 15:31–50.

Thiemann, G.W., and R.J. Wassersug. 2000. Patterns and consequences of behavioural responses to predators and parasites in *Rana* tadpoles. Biological Journal of the Linnean Society 71:513–528.

Thomas, E.O., L. Tsang, and P. Licht. 1993. Comparative histochemistry of the sexually dimorphic skin glands of anuran amphibians. Copeia 1993:133–143.

Thomas, G. 1698. An Historical and Geographical Account of the Province and Country of Pensilvania and of West-New-Jersey in America. A. Baldwin, London.

Thomas, L.A., and J. Allen. 1997. *Bufo houstonensis* (Houston Toad). Behavior. Herpetological Review 28:40–41.

Thomas, L.A., and G.O.U. Wogan. 1999. *Rana catesbeiana* (Bullfrog). Record size. Herpetological Review 30:223–224.

Thomas, R. 1966. New species of Antillean *Eleutherodactylus*. Quarterly Journal of the Florida Academy of Sciences 28:375–391.

Thomas, R.A., S.A. Nadler, and W.L. Jagers. 1984. Helminth parasites of the endangered Houston toad, *Bufo houstonensis* Sanders, 1953 (Amphibia, Bufonidae). Journal of Parasitology 70:1012–1013.

Thompson, C. 1912. The status of *Rana palustris* LeConte in Michigan. Fourteenth Report of the Michigan Academy of Science, p. 190.

Thompson, C. 1915. Notes on the habits of *Rana areolata* Baird and Girard. Occasional Papers of the Museum of Zoology, University of Michigan 9:1–7 + 3 plates.

Thompson, D.G., B.F. Wojtaszek, B. Staznik, D.T. Chartrand, and G.R. Stephenson. 2004. Chemical and biomonitoring to assess potential acute effects of Vision™ herbicide on native amphibian larvae in forest wetlands. Environmental Toxicology and Chemistry 23:843–849.

Thompson, E.F., Jr. 1982. A Guide to the Amphibians, Reptiles and Mammals of South Carolina. Privately published, Columbia, South Carolina.

Thompson, E.F., Jr., and B.S. Martof. 1957. A comparison of the physical characteristics of frog calls (*Pseudacris*). Physiological Zoology 30:328–341.

Thompson, H.B. 1913. Description of a new subspecies of *Rana pretiosa* from Nevada. Proceedings of the Biological Society of Washington 26:53–56.

Thompson, J. N., Jr., and E. M. Grigsby. 1971. The pickerel frog in northeastern Oklahoma. Southwestern Naturalist 16:219–220.

Thompson, P.D. 2004. Observations of boreal toad (*Bufo boreas*) breeding populations in northwestern Utah. Herpetological Review 35:342–344.

Thompson, P.D., R.A. Fridell, K.K. Wheeler, and C.L. Bailey. 2004. Distribution of *Bufo boreas* in Utah. Herpetological Review 35:255–257.

Thompson, Z. 1842. History of Vermont, Natural, Civil and Statistical, in Three Parts with a New Map of the State, and 200 Engravings. Chauncey Goodrich, Burlington.

Thornton, C.S., and T.W. Shields. 1945. Five cases of atypical regeneration in the adult frog. Copeia 1945:40–42.

Thornton, W.A. 1955. Interspecific hybridization in *Bufo woodhousei* and *Bufo valliceps*. Evolution 9:455–468.

Thornton, W.A. 1960. Population dynamics in *Bufo woodhousei* and *Bufo valliceps*. Texas Journal of Science 12:176–200.

Thorson, T.B. 1955. The relationship of water economy to terrestrialism in amphibians. Ecology 36:100–116.

Thorson, T.B. 1956. Adjustment of water loss in response to desiccation in amphibians. Copeia 1956:230–237.

Thorson, T.B., and A. Svihla. 1943. Correlation of habitats of amphibians with their ability to survive loss of body water. Ecology 24:374–381.

Thrall, J. 1971. Excavation of pits by juvenile *Rana catesbeiana*. Copeia 1971:751–752.

Thrall, J. 1972. Food, feeding, and digestive physiology of the larval bullfrog, *Rana catesbeiana* Shaw. Ph.D. Dissertation, Illinois State University, Normal.

Thurgate, N.Y. 2006. The ecology of the endangered dusky gopher frog (*Rana sevosa*) and a common congener, the southern leopard frog (*Rana sphenocephala*). Ph.D. Dissertation, University of New Orleans, New Orleans, Louisiana.

Thurgate, N.Y., and J.H.K. Pechmann. 2007. Canopy closure, competition, and the endangered dusky gopher frog. Journal of Wildlife Management 71:1845–1852.

Thurow, G.R. 1994. Experimental return of wood frogs to west–central Illinois. Transactions of the Illinois State Academy of Science 87:83–97.

Thurow, G. R.1997. *Rana sylvatica* (Wood Frog). Egg mass features and larval behavior. Herpetological Review 28:148.

Thurston, R.V., T.A. Gilfoil, E.L. Meyn, R.K. Zajdel, T.I. Aoki, and G.D. Veith. 1985. Comparative toxicity of 10 organic chemicals to 10 common aquatic species. Water Research 19:1145–1155.

Tierney, D., and M.M. Stewart. 2001. Geographic distribution: *Scaphiopus holbrookii holbrookii* (Eastern Spadefoot). Herpetological Review 32:56.

Tietge, J.E., S.A. Diamond, G.T. Ankley, D.L. DeFoe, G.W. Holcombe, K.M. Jensen, S.J. Degitz, G.E. Elonen, and E. Hammer. 2001. Ambient solar UV radiation causes mortality in larvae of three species of *Rana* under controlled exposure conditions. Photochemistry and Photobiology 74:261–268.

Tihen, J.A. 1937. Additional distributional records of amphibians and reptiles in Kansas counties. Transactions of the Kansas Academy of Science 40:401–409.

Tihen, J.A. 1962a. Osteological observations on New World

Bufo. American Midland Naturalist 67:157–183.

Tihen, J.A. 1962b. A review of New World fossil bufonids. American Midland Naturalist 68:1–50.

Tilley, S.G., and J.E. Huheey. 2001. Reptiles & Amphibians of the Smokies. Great Smoky Mountain Natural History Association, Gatlinburg, Tennessee.

Timken, R.L., and D.G. Dunlap. 1965. Ecological distribution of two species of *Bufo* in southeastern South Dakota. Proceedings of the South Dakota Academy of Science 44:113–117.

Timm, A., and V. Meretsky. 2004. Anuran habitat use on abandoned and reclaimed mining areas of southwestern Indiana. Proceedings of the Indiana Academy of Science 113:140–146.

Timm, B.C., and K. McGarigal. 2010a. The diets of sub-adult Fowler's toads (*Bufo fowleri*) and eastern spadefoot toads (*Scaphiopus h. holbrookii*) at Cape Cod National Seashore. Herpetological Review 41:154–156.

Timm, B.C., and K. McGarigal. 2010b. *Scaphiopus holbrookii* (Eastern Spadefoot). Predation. Herpetological Review 41:207.

Timm, B.C., and K. McGarigal. 2012. *Scaphiopus holbrookii* (Eastern Spadefoot). Possible transmitter expulsion. Herpetological Review 43:636.

Timm, B.C., K. McGarigal, and L.R. Gamble. 2007a. Emigration timing of juvenile pond–breeding amphibians in western Massachusetts. Journal of Herpetology 41:243–250.

Timm, B.C., K. McGarigal, and B.W. Compton. 2007b. Timing of large movement events of pond–breeding amphibians in western Massachusetts, USA. Biological Conservation 136:442–454.

Timm, B.C., K. McGarigal, and C.L. Jenkins. 2007c. Emigration orientation of juvenile pond–breeding amphibians in western Massachusetts. Copeia 2007:685–698.

Timoney, K.P. 1996. Canadian toads near their northern limit in Canada: observations and recommendations. Alberta Naturalist 26:49–50.

Timpe, E.K., S.P. Graham, R.W. Gagliardo, R.T. Hill, and M.G. Levy. 2008. Occurrence of the fungal pathogen *Batrachochytrium dendrobatidis* in Georgia's amphibian populations. Herpetological Review 39:447–449.

Ting, H.–P. 1951. Duration of the tadpole stage of the greenfrog, *Rana clamitans*. Copeia 1951:82.

Tinkham, E.R. 1962. Notes on the occurrence of *Scaphiopus couchii* in California. Herpetologica 18:204.

Tinkle, D.W. 1959. Observations of reptiles and amphibians in a Louisiana swamp. American Midland Naturalist 62:189–205.

Tinsley, R.C. 1990. The influence of parasite infection on mating success in spadefoot toads, *Scaphiopus couchii.*

American Zoologist 30:313–324.

Tinsley, R.C., and K. Tocque. 1995. The population dynamics of a desert anuran, *Scaphiopus couchii.* Australian Journal of Ecology 20:376–384.

Tinsley, R.C., and H.R. Kobel (eds.). 1996. The Biology of *Xenopus.* Oxford Science Publications, Oxford, United Kingdom.

Tipton, B.L., T.L. Hibbitts, T.D. Hibbitts, T.J. Hibbitts, and T.J. LaDuc. 2012. Texas Amphibians. A Field Guide. University of Texas Press, Austin.

Toal, K.R., and J.T. Collins. 2003. *Gastrophryne carolinensis* (Eastern Narrrow–mouthed Toad). Maximum size. Herpetological Review 34:50.

Tobey, F.J. (ed.). 1979. Amphibians and reptiles. Pp. 375–414 *In* D.W. Linzey (ed.), Endangered and Threatened Plants and Animals of Virginia. Center for Environmental Studies, Virginia Polytechnic and State University, Blacksburg.

Tobey, F.J. 1985. Virginia's Amphibians and Reptiles. A Distributional Survey. Virginia Herpetological Society, Purcellville, Virginia.

Tobey, F.J., and O. Fanning. 1986. Reptiles and amphibians around the Beltway: a survey. Audubon Naturalist Society of the Central Atlantic States, Chevy Chase, Maryland. 4 pp.

Tobias, M.L., C. Barnard, R. O'Hagan, S.H. Horng, M. Rand, and D.B. Kelley. 2004. Vocal communication between male *Xenopus laevis.* Animal Behaviour 67:353–365.

Tocque, K. 1993. The relationship between parasite burden and host resources in the desert toad (*Scaphiopus couchii*), under natural environmental conditions. Journal of Animal Ecology 62:683–693.

Tocque, K., and R.C. Tinsley. 1994. The relationship between *Pseudodiplorchis americanus* (Monogenea) density and host resources under controlled environmental conditions. Parasitology 108:175–183.

Tocque, K., R. Tinsley, and T. Lamb. 1995. Ecological constraints on feeding and growth of *Scaphiopus couchii.* Herpetological Journal 5:257–265.

Todd, B.D., and B.B. Rothermel. 2006. Assessing the quality of clearcut habitats for amphibians: effects on abundances versus vital rates in the Southern toad (*Bufo terrestris*). Biological Conservation 133:178–185.

Todd, B.D., and C.T. Winne. 2006. Ontogenetic and inter-specific variation in timing of movement and responses to climatic factors during migrations by pond–breeding amphibians. Canadian Journal of Zoology 84:715–722.

Todd, B.D., T.M. Luhring, B.B. Rothermel, and J.W. Gibbons. 2009. Effects of forest removal on amphibian migrations: implications for habitat and landscape connectivity. Journal of Applied Ecology 46:554–561.

Todd, B.D., D.E. Scott, J.H.K. Pechmann, and J.W. Gibbons. 2011a. Climate change correlates with rapid delays and advancements in reproductive timing in an amphibian community. Proceedings of the Royal Society B 278:2191–2197.

Todd, B.D., C.M. Bergeron, M.J. Hepner, and W.A. Hopkins. 2011b. Aquatic and terrestrial stressors in amphibians: a test of the double jeopardy hypothesis using maternally and trophically derived mercury. Environmental Toxicology and Chemistry 30:2277–2284.

Todd, B.D., C.M. Bergeron, M.J. Hepner, J.N. Burke, and W.A. Hopkins. 2011c. Does maternal exposure to an environmental stressor affect offspring response to predators? Oecologia 166:283–290.

Todd, B.D., C.M. Bergeron, and W.A. Hopkins. 2012a. Use of toe clips as a nonlethal index of mercury accumulation and maternal transfer in amphibians. Ecotoxicology 21:882–887.

Todd, B.D., J.D. Willson, C.M. Bergeron, and W.A. Hopkins. 2012b. Do effects of mercury in larval amphibians persist after metamorphosis? Ecotoxicology 21:87–95.

Todd, M.J., R.R. Cocklin, and M.E. Dorcas. 2003. Temporal and spatial variation in anuran calling activity in the western Piedmont of North Carolina. Journal of the North Carolina Academy of Science 119:103–110.

Todd–Thompson, M. 2010. Seasonality, variation in species prevalence, and localized disease for ranavirus in Cades Cove (Great Smoky Mountains National Park) amphibians. M.S. thesis, University of Tennessee, Knoxville.

Todd–Thompson, M., D.L. Miller, P.E. Super, and M.J. Gray. 2009. Chytridiomycosis–associated mortality in a *Rana palustris* collected in Great Smoky Mountains National Park, Tennessee, USA. Herpetological Review 40:321–323.

Toledo, L.F. 2005. Predation of juvenile and adult anurans by invertebrates: current knowledge and perspectives. Herpetological Review 36:395–400.

Toledo, L.F., I. Sazima, and C.F.B. Haddad. 2011. Behavioural defences of anurans: an overview. Ethology Ecology & Evolution 23:1–25.

Toline, C.A. 1999. Mitochondrial and nuclear DNA variation within and among populations of Columbia spotted frogs (*Rana luteiventris*) in Utah. M.S. thesis, Utah State University, Logan.

Tomko, D.S. 1975. The reptiles and amphibians of the Grand Canyon. Plateau 47:161–166.

Tomko, D.S. 1976. *Rana pipiens* (Ranidae) in the Grand Canyon of the Colorado River, Arizona. Southwestern Naturalist 21:131.

Toner, G.C., and N. de St. Remy. 1941. Amphibians of eastern Ontario. Copeia 1941:10–13.

Tordoff, W. III. 1980. Selective predation of gray jays, *Periso-reus canadensis*, upon boreal chorus frogs, *Pseudacris triseriata*. Evolution 34:1004–1008.

Tordoff, W. III., and D. Pettus. 1977. Temporal stability of phenotypic frequencies in *Pseudacris triseriata* (Amphibia, Anura, Hylidae). Journal of Herpetology 11:161–168.

Tordoff, W. III, D. Pettus, and T.C. Matthews. 1976. Microgeographic variation in gene frequencies in *Pseudacris triseriata* (Amphibia, Anura, Hylidae). Journal of Herpetology 10:35–40.

Tóth, A. 2012. A tale of two species – species distribution models for two cryptic cricket frog (*Acris*) species in syntopy. B.S. Honors thesis, College of William and Mary, Williamsburg, Virginia.

Touchon, J.C., I. Gomez–Mestre, and K.M. Warkentin. 2006. Hatching plasticity in two temperate anurans: responses to a pathogen and predation cues. Canadian Journal of Zoology 84:556–563.

Touré, T.A., and G.A. Middendorf. 2002. Colonization of herpetofauna to a created wetland. Bulletin of the Maryland Herpetological Society 38:99–117.

Townes, G.F. 1956. Dispersal and minimum habitat of the leopard frog. Herpetologica 12:290.

Townsend, D.S., and M.M. Stewart. 1994. Reproductive ecology of the Puerto Rican frog *Eleutherodactylus coqui*. Journal of Herpetology 28:34–40.

Townsend, S.C. 1964. A life history study of the chorus frog, *Pseudacris triseriata*, in Marion County. M.A. thesis, Ball State University, Muncie, Indiana.

Tracy, C.R. 1971. Evidence for the use of celestial cues by disbursing immature California toads (*Bufo boreas*). Copeia 1971:145–147.

Tracy, C.R. 1973. Observations on social behavior in immature California toads (*Bufo boreas*) during feeding. Copeia 1973:342–345.

Tracy, C.R. 1976. Model of dynamic exchanges of water and energy between a terrestrial amphibian and its environment. Ecological Monographs 46:293–326.

Tracy, C.R., and J.W. Dole. 1969. Orientation of displaced California toads, *Bufo boreas*, to their breeding sites. Copeia 1969:693–700.

Tracy, C.R., K.A. Christian, M.P. O'Connor, and C.R. Tracy. 1993. Behavioral thermoregulation by *Bufo americanus*: the importance of the hydric environment. Herpetologica 49:375–382.

Trapido, H. 1940. On finding the mink frog in northern Vermont. New England Naturalist (7):11–14.

Trapido, H. 1947. Range extension of *Hyla septentrionalis* in Florida. Herpetologica 3:190.

Trapido, H., and R.T. Clausen. 1938. Amphibians and reptiles of eastern Quebec. Copeia 1938:117–125.

Trauth, J.B., S.E. Trauth, and R.L. Johnson. 2006. Best management practices and drought combine to silence the Illinois chorus frog in Arkansas. Wildlife Society Bulletin 34:514–518.

Trauth, J.B., R.L. Johnson, and S.E. Trauth. 2007. Conservation implications of a morphometric comparison between the Illinois chorus frog (*Pseudacris streckeri illinoensis*) and Strecker's chorus frog (*P. s. streckeri*) (Anura: Hylidae) from Arkansas, Illinois, Missouri, Oklahoma, and Texas. Zootaxa 1589:23–32.

Trauth, S.E. 1989. Female reproductive traits of the southern leopard frog, *Rana sphenocephala* (Anura: Ranidae), from Arkansas. Proceedings of the Arkansas Academy of Science 43:105–108.

Trauth, S.E. 1992. Distributional survey of the bird–voiced treefrog, *Hyla avivoca* (Anura: Hylidae), in Arkansas. Proceedings of the Arkansas Academy of Science 46:80–82.

Trauth, S.E., and C.T. McAllister. 1983. Geographic Distribution. *Rana clamitans melanota*. Herpetological Review 14:83.

Trauth, S.E., and J.W. Robinette. 1990. Notes on distribution, mating activity, and reproduction in the bird–voiced treefrog, *Hyla avivoca*, in Arkansas. Bulletin of the Chicago Herpetological Society 25:218–219.

Trauth, S.E., and A. Holt. 1993. Notes on the breeding biology of Hurter's spadefoot toad, *Scaphiopus holbrookii hurterii*, in Arkansas. Bulletin of the Chicago Herpetological Society 28:236–239.

Trauth, S.E., and R.G. Neal. 2004. Geographic range extension and feeding response by the leech *Macrobdella diplotertia* (Annelida: Hirudinea) to wood frog and spotted salamander egg masses. Journal of the Arkansas Academy of Science 58:139–141.

Trauth, S.E., M.E. Cartwright, and W.E. Meshaka. 1989. Reproduction in the wood frog, *Rana sylvatica* (Anura: Ranidae), from Arkansas. Proceedings of the Arkansas Academy of Science 43:114–116.

Trauth, S.E., R.L. Cox, B. Butterfield, D.A. Saugey, and W.E. Meshaka. 1990. Reproductive phenophases and clutch characteristics of selected Arkansas amphibians. Proceedings of the Arkansas Academy of Science 44:107–113.

Trauth, S.E., A. Holt, J. Davis, and P. Daniel. 1992. A new state record for the plains leopard frog, *Rana blairi*, in Arkansas. Bulletin of the Chicago Herpetological Society 27:255.

Trauth, S.E., W.E. Meshaka, Jr., and R.L. Cox. 1999. Post–metamorphic growth and reproduction in the eastern narrow–mouthed toad (*Gastrophryne carolinensis*) from northeastern Arkansas. Journal of the Arkansas Academy of Science 53:120–124.

Trauth, S.E., M.L. McCallum, and M.E. Cartwright. 2000.

Breeding mortality in the wood frog, *Rana sylvatica* (Anura: Ranidae), from northcentral Arkansas. Journal of the Arkansas Academy of Science 54:154–156.

Trauth, S.E., H.W. Robison, and M.V. Plummer. 2004. The Amphibians and Reptiles of Arkansas. University of Arkansas Press, Fayetteville.

Trauth, S.E., M.N. Mary, and J.A. Sawyer. 2005. *Bufo americanus charlesmithi* (Dwarf American Toad). Maximum size. Herpetological Review 36:298.

Travis, J. 1980. Phenotypic variation and the outcome of interspecific competition in hylid tadpoles. Evolution 34:40–50.

Travis, J. 1981a. Control of larval growth variation in a population of *Pseudacris triseriata* (Anura: Hylidae). Evolution 35:423–432.

Travis, J. 1981b. Keys to the tadpoles of North Carolina. Brimleyana 6:119–127.

Travis, J. 1983a. Variation in development patterns of larval anurans in temporary ponds. 1. Persistent variation within a *Hyla gratiosa* population. Evolution 37:496–512.

Travis, J. 1983b. Variation in growth and survival of *Hyla gratiosa* larvae in experimental enclosures. Copeia 1983:232–237.

Travis, J. 1984. Anuran size at metamorphosis: experimental test of a model based on intraspecific competition. Ecology 65:1155–1160.

Travis, J. 2006. Is it what we know or who we know? Choice of organism and robustness of inference in ecology and evolutionary biology. American Naturalist 167:303–314.

Travis, J., and J.C. Trexler. 1986. Interactions among factors affecting growth, development and survival in experimental populations of *Bufo terrestris* (Anura: Bufonidae). Oecologia 69:110–116.

Travis, J., W.H. Keen, and J. Juilianna. 1985a. The role of relative body size in a predator–prey relationship between dragonfly naiads and larval anurans. Oikos 45:59–65.

Travis, J., W.H. Keen, and J. Juilianna. 1985b. The effects of multiple factors on viability selection in *Hyla gratiosa* tadpoles. Evolution 39:1087–1099.

Travis, J., S.B. Emerson, and M. Blouin. 1987. A quantitative–genetic analysis of larval life–history traits in *Hyla crucifer*. Evolution 41:145–156.

Travis, J., and J.C. Trexler. 1984. Investigations on the control of the color polymorphsm in *Pseudacris ornata*. Herpetologica 40:252–257.

Treanor, R.R., and S.J. Nicola. 1972. A preliminary study of the commercial and sporting utilization of the bullfrog, *R. catesbeiana* Shaw in California. California Department of Fish and Game, Inland Fisheries Administrative Report 72–4.

Treat, D.A. 1948. Frogs and toads. Audubon Nature Bulletin, Series 18, No. 8.

Trenham, P.C., W.D. Koenig, M.J. Mossman, S.L. Stark, and L.A. Jagger. 2003. Regional dynamics of wetland-breeding frogs and toads: turnover and synchrony. Ecological Applications 13:1522–1532.

Trowbridge, A.H., and H.M. Hefley. 1934. Preliminary studies on the parasite fauna of Oklahoma anurans. Proceedings of the Oklahoma Academy of Science 14:16–19.

Trowbridge, A.H., and M.S. Trowbridge. 1937. Notes on the cleavage rate of *Scaphiopus bombifrons* Cope, with additional remarks on certain aspects of its life history. American Naturalist 71:460–480.

Trowbridge, M.S. 1941. Studies on the normal development of *Scaphiopus bombifrons* Cope. 1. The cleavage period. Transactions of the American Microscopical Society 60:508–526.

Troyer, D. R. 1968. Growth rate and movements of the bullfrog, *Rana catesbeiana*, in Kansas. M.A. thesis, University of Kansas, Lawrence.

True, F.W. 1883. A list of the vertebrate animals of South Carolina. Pp. 209-264 *In* South Carolina. Resources and Population. Institutions and Industries. State Board of Agriculture of South Carolina, Columbia.

True, F.W. 1884. The useful aquatic reptiles and batrachians of the United States. Pp. 137–162 *In* G.B. Goode (ed.), The Fisheries and Fishery Industries of the United States. Part II. United States Commission of Fish and Fisheries, U.S. Government Printing Office, Washington, D.C.

Trueb, L., and C. Gans. 1983. Feeding specializations of the Mexican burrowing toad, *Rhinophrynus dorsalis* (Anura: Rhinophrynidae). Journal of Zoology 199:189–208.

Truitt, J.O. 1964. Observations on the defensive attitude of a southern toad (*Bufo terrestris*). British Journal of Herpetology 3:167.

Trumbo, D.R., A.A. Burgett, and J.H. Knouft. 2011. Testing climate–based species distribution models with recent field surveys of pond–breeding amphibians in eastern Missouri. Canadian Journal of Zoology 89: 1074–1083.

Tuberville, T.D., J.D. Willson, M.E. Dorcas, and J.W. Gibbons. 2005. Herpetofaunal species richness of southeastern national parks. Southeastern Naturalist 4:537–569.

Tucker, J.K. 1977. Notes on the food habits of Kirtland's water snake, *Clonophis kirtlandi*. Bulletin of the Maryland Herpetological Society 13:193–195.

Tucker, J.K. 1995. Early post–transformational growth in the Illinois chorus frog (*Pseudacris streckeri illinoensis*). Journal of Herpetology 29:314–316.

Tucker, J.K. 1997a. Food habits of the fossorial frog *Pseudacris streckeri illinoensis*. Herpetological Natural History 5:83–87.

Tucker, J.K. 1997b. Fecundity in Illinois chorus frogs (*Pseudacris streckeri illinoensis*) from Madison County, Illinois. Transactions of the Illinois State Academy of Science 90:167–170.

Tucker, J.K. 1998. Status of Illinois chorus frogs in Madison County, Illinois. Pp. 94–101 *In* M.J. Lannoo (ed.), Status & Conservation of Midwestern Amphibians. University of Iowa Press, Iowa City.

Tucker, J.K. 2000a. Growth and survivorship in the Illinois chorus frog (*Pseudacris streckeri illinoensis*). Transactions of the Illinois State Academy of Science 93:63–68.

Tucker, J.K. 2000b. *Pseudacris streckeri illinoensis* (Illinois Chorus Frog). Frost injuries. Herpetological Review 31:41–42.

Tucker, J.K., and M.E. Sullivan. 1975. Unsuccessful attempts by bullfrogs to eat toads. Transactions of the Illinois State Academy of Science 68:167.

Tucker, J.K., D.W. Soergel, and J.B. Hatcher. 1995a. Flood—associated activities of some reptiles and amphibians at Carlyle Lake, Fayette County, Illinois. Transactions of the Illinois State Academy of Science 88:73–81.

Tucker, J.K., J.B. Camerer, and J.B. Hatcher. 1995b. *Pseudacris streckeri illinoensis* (Illinois Chorus Frog). Burrows. Herpetological Review 26:32–33.

Tucker, R.K., and D.G. Crabtree. 1969. Toxicity of Tectran insecticide to several wildlife species. Journal of Economic Entomology 62:1307–1310.

Tuckerman, F. 1886. Supernumerary leg in a male frog (*Rana palustris*). Journal of Anatomy and Physiology 20:516–519.

Tumlison, R., and S.E. Trauth. 2006. A novel facultative mutalistic relationship between bufonid tadpoles and flagellated green algae. Herpetological Conservation and Biology 1:51–54.

Tupper, T.A., and R.P. Cook. 2008. Habitat variables influencing breeding effort in northern clade *Bufo fowleri*: implications for conservation. Applied Herpetology 5:101–119.

Tupper, T.A., L.B. Adams, and B.C. Timm. 2009. *Bufo fowleri* (Fowler's Toad). Diet. Herpetological Review 40:200–201.

Tupper, T.A., J.W. Streicher, S.E. Greenspan, B.C. Timm, and R.P. Cook. 2011. Detection of *Batrachochytrium dendrobatidis* in anurans of Cape Cod National Seashore, Barnstable County, Massachusetts. Herpetological Review 42:62–65.

Tupy, J.A. 2012. Terrestrial habitat selection by the dusky gopher frog (*Rana sevosa*). M.S. thesis, Western Carolina University, Cullowhee, North Carolina.

Turner, D.S., P.A. Holm, E.B. Wirt, and C.R. Schwalbe. 2003. Amphibians and reptiles of the Whetstone Mountains, Arizona. Southwestern Naturalist 48:347–355.

Turner, F.B. 1952a. Peculiar aggregations of toadlets on Alum Creek. Yellowstone Nature Notes 26(5):57–58.

Turner, F.B. 1952b. The mouth parts of tadpoles of the spadefoot toad, *Scaphiopus hammondi*. Copeia 1952:172–175.

Turner, F.B. 1955. Reptiles and amphibians of Yellowstone National Park. Yellowstone Interpretive Series No. 5, Yellowstone National Park, Wyoming.

Turner, F.B. 1958a. Some parasites of the western spotted frog, *Rana p. pretiosa*, in Yellowstone National Park. Journal of Parasitology 44:182.

Turner, F.B. 1958b. Life–history of the western spotted frog in Yellowstone National Park. Herpetologica 14:96–100.

Turner, F.B. 1959a. Pigmentation of the western spotted frog, *Rana pretiosa pretiosa*, in Yellowstone Park, Wyoming. American Midland Naturalist 61:162–176.

Turner, F.B. 1959b. An analysis of the feeding habits of *Rana p. pretiosa* in Yellowstone Park, Wyoming. American Midland Naturalist 61:403–413.

Turner, F.B. 1959c. Some features of the ecology of *Bufo punctatus* in Death Valley, California. Ecology 40:175–181.

Turner, F.B. 1960a. Size and dispersion of a Louisiana population of the cricket frog, *Acris gryllus*. Ecology 41:258–268.

Turner, F.B. 1960b. Population structure and dynamics of the western spotted frog, *Rana p. pretiosa* Baird and Girard, in Yellowstone Park, Wyoming. Ecological Monographs 30:251–278.

Turner, F.B. 1960c. Postmetamorphic growth in amphibians. American Midland Naturalist 64:327–338.

Turner, F.B. 1962a. An analysis of geographic variation and distribution of *Rana pretiosa*. Yearbook of the American Philosophical Society 1962:325–328.

Turner, F.B. 1962b. The demography of frogs and toads. Quarterly Review of Biology 37:303-314.

Turner, F.B., and R.H. Wauer. 1963. A survey of the herpetofauna of the Death Valley area. Great Basin Naturalist 23:119–128.

Turner, L.J., and D.K. Fowler. 1981. Utilization of surface mine ponds in east Tennessee by breeding amphibians. U.S. Fish and Wildlife Service, Biological Services Program, FWS/OBS–81/08.

Turner, P.S. 2009. The Frog Scientist. Houghton–Mifflin Harcourt, Boston, Massachusetts.

Turnipseed, G., and R. Altig. 1975. Population density and age structure of three species of hylid tadpoles. Journal of Herpetology 9:287–291.

Tverdy, L.M., N.J. Meis, C.G. Wicher, and D.G. Hokit. 2005. Comet assay used to detect genotoxic effects of mining sediments in Western toad tadpoles (*Bufo boreas*). Herpetological Review 36:152–155.

Twedt, B. 1993. A comparative ecology of *Rana aurora* Baird and Girard and *Rana catesbeiana* Shaw at Freshwater Lagoon, Humboldt County, California. M.A. thesis, Humboldt State University, Arcata, California.

Tweedell, K.S. 1965. Renal tumors in a western population of *Rana pipiens*. American Midland Naturalist 73:285–292.

Twitty, V., D. Grant, and O. Anderson. 1967. Amphibian orientation: an unexpected observation. Science 155:352–353.

Tyler, J.D. 1969. Distribution and vertebrate associates of the black–tailed prairie dog in Oklahoma. Ph.D. Dissertation, University of Oklahoma, Norman.

Tyler, J.D. 1970. Vertebrates in a prairie dog town. Proceedings of the Oklahoma Academy of Science 50:110–113.

Tyler, J.D. 1982. Cyanistic bullfrog from southwestern Oklahoma. Southwestern Naturalist 27:218.

Tyler, J.D. 1983. Notes on burrowing owl (*Athene cunicularia*) food habits in Oklahoma, USA. Southwestern Naturalist 28:100–102.

Tyler, J.D., and R.D. Hoestenbach, Jr. 1979. Differences in food of bullfrogs (*Rana catesbeiana*) from pond and stream habitats in southwestern Oklahoma. Southwestern Naturalist 24:33–38.

Tyler, M.S. 1994. Stalking amphibians. Maine Naturalist 2:33–44.

Tyning, T. F. 1990. A Guide to Amphibians and Reptiles. Little, Brown & Co., Boston, Massachusetts.

Ubelaker, J.E., D.W. Duszynski, and D.L. Beaver. 1967. Occurrence of the trematode, *Glypthelmins pennsylvaniensis* Cheng, 1961, in chorus frogs, *Pseudacris triseriata*, in Colorado. Bulletin of the Wildlife Disease Association 3:177.

Ugarte, C.A. 2004. Human impacts on pig frog (*Rana grylio*) populations in south Florida wetlands: harvest, water management and mercury contamination. Ph.D. Dissertation, Florida International University, Miami.

Ugarte, C.A., K.G. Rice, and M.A. Donnelly. 2005. Variation of total mercury concentrations in pig frogs (*Rana grylio*) across the Florida Everglades, USA. Science of the Total Environment 345:51–59.

Ugarte, C.A., K.G. Rice, and M.A. Donnelly. 2007. Comparison of diet, reproductive biology, and growth of the pig frog (*Rana grylio*) from harvested and protected areas of the Florida Everglades. Copeia 2007:436–448.

Ulmer, M.J. 1970. Studies on the helminth fauna of Iowa. 1. Trematodes of amphibians. American Midland Naturalist 83:38–64.

Ulmer, M.J., and H.A. James. 1976a . Studies on the helminth fauna of Iowa. 2. Cestodes of amphibians. Proceedings of the Helminthological Society of Washington

43:191–200.

Ulmer, M.J., and H.A. James. 1976b. *Nematotaenoides ranae* gen. et sp. n. (Cyclophyllidea: Nematotaeniidae), from the leopard frog (*Rana pipiens*) in Iowa. Proceedings of the Helminthological Society of Washington 43:185–191.

Ultsch, G.R., S.A. Reese, M. Nie, J.D. Crim, W.H. Smith, and C.M. LeBerte. 1999. Influences of temperature and oxygen upon habitat selection by bullfrog tadpoles and three species of freshwater fishes in two Alabama strip mine ponds. Hydrobiologia 416:149–162.

Ultsch, G.R., T.E. Graham, and C.E. Crocker. 2000. An aggregation of overwintering leopard frogs, *Rana pipiens*, and common map turtles, *Graptemys geographica*, in northern Vermont. Canadian Field-Naturalist 114:314–315.

Ultsch, G.R., S.A. Reese, and E.R. Stewart. 2004. Physiology of hibernation in *Rana pipiens*: metabolic rate, critical oxygen tension, and the effects of hypoxia on several plasma variables. Journal of Experimental Zoology 301A:169–176.

Underhill, J.C. 1960. Breeding and growth in Woodhouse's toad. Herpetologica 16:237–242.

Underhill, J.C. 1961a. Intraspecific variation in the Dakota toad, *Bufo hemiophrys*, from northeastern South Dakota. Herpetologica 17:220–227.

Underhill, J.C. 1961b. Variation in Woodhouse's toad, *Bufo woodhousei* Girard in South Dakota. Copeia 1961:333–336.

Underhill, D.K. 1966. An incidence of spontaneous caudal scoliosis in tadpoles of *Rana pipiens* Schreber. Copeia 1966:582–583.

Unrine, J.M., C.H. Jagoe, W.A. Hopkins, and H.A. Brant. 2004. Adverse effects of ecologically relevant dietary mercury exposure in southern leopard frog (*Rana sphenocephala*) larvae. Environmental Toxicology and Chemistry 23:2964–2970.

Unrine, J.M., C.H. Jagoe, A.C. Briton, H.A. Brant, and N.T. Garvin. 2005. Dietary mercury exposure and bioaccumulation in amphibian larve inhabiting Carolina bay wetlands. Environmental Pollution 135:245–253.

Unrine, J.M., W.A. Hopkins, C.S. Romanek, and B.P. Jackson. 2007. Bioaccumulation of trace elements in omnivorous amphibian larvae: implications for amphibian health and contaminant transport. Environmental Pollution 149:182–192.

Upton, S.J., and C.T. McAllister. 1988. The Coccidia (Apicomplexa: Eimeriidae) of Anura, with descriptions of four new species. Canadian Journal of Zoology 66:1822–1830.

USDC (U.S. Department of Commerce). 1922. Frog culture. Bureau of Fisheries, Information Leaflet I-2, Washington, D.C. [identical copies dated May 1922, January 1923, July 1924]

USDC (U.S. Department of Commerce). 1924. The common bullfrog. Bureau of Fisheries, Information Leaflet I-80, Washington, D.C. 1 p. [May, 1924].

USDC. 1932a. Frog industry. Bureau of Fisheries Information Leaflet I-2. Washington, D.C. 4 pp.

USDC. 1932b. Commercial frog industry of the United States. Bureau of Fisheries, Supplement to Information Leaflet I-2, Washington, D.C. 3 pp.

USDC. 1936. Frog culture and the frog industry. Bureau of Fisheries, Washington, D.C. 8 pp. [October, 1936].

USDI (U.S. Department of Interior). 1944. Frog culture and the frog industry. Fishery Leaflet 102.

USDI. 1956. Commercial possibilities and limitations in frog raising. Fishery Leaflet 436. [later issue dated 1965]

USDI. 1973. Threatened Wildlife of the United States. U.S. Fish and Wildlife Service Resource Publication 114. U.S. Government Printing Office, Washington, D.C.

USFWS (U.S. Fish and Wildlife Service). 1977a. Proposed endangered status and critical habitat for the Pine Barrens treefrog in Florida. Federal Register 42(65):18109–18111.

USFWS. 1977b. Proposed determination of critical habitat for the Houston toad. Federal Register 42(102):27009–27011.

USFWS. 1977c. Review of status of 10 species of amphibians. Federal Register 42(148):39119–39120.

USFWS. 1977d. Final endangered status and critical habitat for the Florida population of the Pine Barrens treefrog. Federal Register 42(218):58754–58756.

USFWS. 1978. Determination of critical habitat for the Houston toad. Federal Register 43(21): 4022–4026.

USFWS. 1982. Proposal to remove the Florida population of the Pine Barrens treefrog (*Hyla andersonii*) from the list of Endangered and Threatened Wildlife and Plants and to rescind previously determined Critical Habitat. Federal Register 47(179):40673–40675.

USFWS. 1983a. Proposed Endangered status for *Bufo hemiophrys baxteri* (Wyoming toad). Federal Register 48(19):3794–3796.

USFWS. 1983b. Final rule to remove the Florida population of the Pine Barrens treefrog from the list of Endangered and Threatened Wildlife and to rescind previously determined Critical Habitat. Federal Register 48(226):52740–52743.

USFWS. 1984. Determination that *Bufo hemiophrys baxteri* (Wyoming toad) is an Endangered species. Federal Register 49(11):1992–1994.

USFWS. 1994a. Proposed Endangered status for the California red–legged frog. Federal Register 59(22):4888–4895.

USFWS. 1994b. Determination of Endangered status for the Arroyo Southwestern toad. Federal Register 59(241):64859–64866.

USFWS. 1996. Determination of Threatened status for the California red legged frog. Federal Register 61(101):25813–25833.

USFWS. 1999a. Arroyo southwestern toad (*Bufo microscaphus californicus*) recovery plan. U.S. Fish and Wildlife Service, Portland, Oregon.

USFWS. 1999b. Proposed Endangered status for the Southern California distinct vertebrate population segment of the mountain yellow–legged frog. Federal Register 64(245):71714– 71722.

USFWS. 2000a. Proposed rule to list the Mississippi gopher frog distinct population segment of dusky gopher frog. Federal Register 65(1000):33283–33291.

USFWS. 2000b. Proposal to list the Chiricahua leopard frog as Threatened with a special rule. Federal Register 65(115):37343–37357.

USFWS. 2000c. Proposed designation of Critical Habitat for the California red–legged frog (*Rana aurora draytonii*). Federal Register 65(176):54892–54932.

USFWS. 2000d. 90–day finding on a petition to list the mountain yellow–legged frog as Endangered. Federal Register 65(198):60603–60605.

USFWS. 2000e. 90–day finding on a petition to list the Yosemite toad as Endangered. Federal Register 65(198):60607–60609.

USFWS. 2001a. Final designation of Critical Habitat for the arroyo toad. Federal Register 66(26):9414–9474.

USFWS. 2001b. Final designation of Critical Habitat for the arroyo toad. Federal Register 66(45):13656–13671.

USFWS. 2001c. Final rule to list the Mississippi gopher frog distinct population segment of dusky gopher frog as Endangered. Federal Register 66(233):62993–63002.

USFWS. 2002a. Listing the Chiricahua leopard frog (*Rana chiricahuensis*). Federal Register 67(114):40790–40811.

USFWS. 2002b. Determination of Endangered status for the Southern California distinct vertebrate population segment of the mountain yellow–legged frog. Federal Register 67(127):44382–44392.

USFWS. 2002c. 12–Month finding for a petition to list the Wasatch Front Columbia spotted frog as Threatened throughout its range. Federal Register 67(169):55758–55767.

USFWS. 2003. 12–month finding for a petition to list the Sierra Nevada distinct population segment of the mountain yellow–legged frog (*Rana muscosa*). Federal Register 68(11):2283–2303.

USFWS. 2004. Proposed designation of Critical Habitat for the California red–legged frog (*Rana aurora draytonii*). Federal Register 69(71):19620–19642.

USFWS. 2005a. Proposed designation of Critical Habitat for the Southern California distinct vertebrate population segment of the mountain yellow–legged frog (*Rana muscosa*). Federal Register 70(176):54106–54143.

USFWS. 2005b. Revised 12–month finding for the southern Rocky Mountain distinct population segment of the boreal toad (*Bufo boreas boreas*). Federal Register 70(188):56880–56884.

USFWS. 2006a. Designation of Critical Habitat for the California red–legged frog, and special rule exemption associated with final listing for existing routine ranching activities. Federal Register 71(71):19244–19345.

USFWS. 2006b. Designation of Critical Habitat for the Southern California distinct population segment of the mountain yellow–legged frog (*Rana muscosa*). Federal Register 71(178):54344–54386.

USFWS. 2006c. Proposed designation of Critical Habitat for the Southern California distinct vertebrate population segment of the mountain yellow–legged frog (*Rana muscosa*). Federal Register 71(127):37881–37886.

USFWS. 2007. 12–Month finding on a petition to list the Sierra Nevada distinct population segment of the mountain yellow–legged frog (*Rana muscosa*). Federal Register 72(121):34657–34661.

USFWS. 2008. Revised Critical Habitat for the California red–legged frog (*Rana aurora draytonii*). Federal Register 73(180):53492–53680.

USFWS. 2009. 90–day finding on a petition to list the northern leopard frog (*Lithobates* [=*Rana*] *pipiens*) in the western United States as Threatened. Federal Register 74(125):31389–31401.

USFWS. 2010a. Revised designation of Critical Habitat for the California red–legged frog. Federal Register 75(51):12815–12959.

USFWS. 2010b. Designation of Critical Habitat for Mississippi gopher frog. Federal Register 75(106):31387–31411.

USFWS. 2011a. Listing and designation of Critical Habitat for the Chiricahua leopard frog. Federal Register 76(50):14125–14207.

USFWS. 2011b. Designation of Critical Habitat for Mississippi gopher frog. Federal Register 76(187):59733–59802.

USFWS. 2011c. 12–Month finding on a petition to list the northern leopard frog in the western United States as Threatened. Federal Register 76(193):61896–61931.

USFWS. 2012a. Listing and designation of Critical Habitat for the Chiricahua leopard frog. Federal Register 77(54):16324–16424.

USFWS. 2012b. Designation of Critical Habitat for dusky

gopher frog (previously Mississippi gopher frog). Federal Register 77(113):35117–35161.

Uyehara, I.K., T. Gamble, and S. Cotner. 2010. The presence of *Ranavirus* in anuran populations at Itasca State Park, Minnesota, USA. Herpetological Review 41:177–179.

Vaala, D.A. 2010. The use of mobile phones for conducting acoustic surveys of calling amphibians. M.S. thesis, Rochester Institute of Technology, Rochester, New York.

Vaala, D.A., G.R. Smith, K.G. Temple, and H.A. Dingfelder. 2004. No effect of nitrate on gray treefrog (*Hyla versicolor*) tadpoles. Applied Herpetology 1:265–269.

Valett, B.B., and D.L. Jameson. 1961. The embryology of *Eleutherodactylus augusti latrans*. Copeia 1961:103–109.

Vallinoto, M., F. Sequeira, D. Sodré, J.A.R. Bernardi, I. Sampaio, and H. Schneider. 2010. Phylogeny and biogeography of the *Rhinella marina* species complex (Amphibia, Bufonidae) revisited: implications for Neotropical diversification hypotheses. Zoologica Scripta 39:128–140.

Van Allen, B.G., V.S. Briggs, M.W. McCoy, and J.R. Vonesh. 2010. Carry–over effects of the larval environment on post–metamorphic performance in two hylid frogs. Oecologia 164:891–898.

Van Bergeyk, W.A. 1967. Anticipatory feeding behaviour in the bullfrog (*Rana catesbeiana*). Animal Behaviour 15:231–238.

Van Buskirk, J. 2000. The costs of an inducible defense in anuran larvae. Ecology 81:2813–2821.

Van Buskirk, J., and R.A. Relyea. 1998. Selection for phenotypic plasticity in *Rana sylvatica* tadpoles. Biological Journal of the Linnaean Society 65:301–328.

Van Buskirk, J., and S.A. McCollum. 2000a. Functional mechanisms of an inducible defence in tadpoles: morphology and behaviour influence mortality risk from predation. Journal of Evolutionary Biology 13: 336–347.

Van Buskirk, J., and S.A. McCollum. 2000b. Influence of tail shape on tadpole swimming performance. Journal of Experimental Biology 203:2149–2158.

Van Buskirk, J., S.A. McCollum, and E.E. Werner. 1997. Natural selection for environmentally induced phenotypes in tadpoles. Evolution 51:1983–1992.

Van Camp, L.F., and C.J. Henry. 1975. The screech owl: its life history and population ecology in northern Ohio. U.S. Fish and Wildlife Service, American Fauna 71:1–65.

Van Compernolle, S.E., R.J. Taylor, K. Oswald–Richter, J. Jiang, B.E. Youree, J.H. Bowie, M. J. Tyler, J.M. Conlon, D. Wade, C. Aiken, T.S. Dermody, V.N. Kewal Ramani, L.A. Rollins–Smith, and D. Unutmazl. 2005. Antimicrobial peptides from amphibian skin potently inhibit human immunodeficiency virus infection and transfer of virus from dendritic cells to T cells. Journal of Virology 79:11,598–11,606.

Van Denburgh, J. 1898. Herpetological notes. 1. *Bufo boreas* in Alaska. Proceedings of the American Philosophical Society 37:139.

Van Denburgh, J. 1905. The reptiles and amphibians of the islands of the Pacific Coast of North America from the Farallons to Cape San Lucas and the Revilla Gigedos. Proceedings of the California Academy of Sciences, Series 3, Zoology 4:1–40.

Van Denburgh, J. 1912. Notes on *Ascaphus*, the discoglossid toad of North America. Proceedings of the California Academy of Sciences, Series 4, 3:259–264.

Van Denburgh, J. 1913. A list of the amphibians and reptiles of Arizona, with notes on the species in the collection of the Academy. Proceedings of the California Academy of Sciences. 3: 391–454.

Van Denburgh, J. 1924. Notes on the herpetology of New Mexico, with a list of species known from that state. Proceedings of the California Academy of Sciences, Series 4, 13:189–230.

Van Denburgh, J., and J.R. Slevin. 1914. Reptiles and amphibians of the islands of the West Coast of North America. Proceedings of the California Academy of Sciences, Series 4, 4:129–152.

Van Denburgh, J., and J.R. Slevin. 1915. A list of the amphibians and reptiles of Utah, with notes on the species in the collection of the Academy. Proceedings of the California Academy of Sciences, Series 4, 5:99–110.

Van Denburgh, J., and J.R. Slevin. 1921a. A list of the amphibians and reptiles of Nevada, with notes on the species in the collection of the Academy. Proceedings of the California Academy of Sciences, Series 4, 11:27–38.

Van Denburgh, J., and J.R. Slevin. 1921b. A list of the amphibians and reptiles of Idaho, with notes on the species in the collection of the Academy. Proceedings of the California Academy of Sciences, Series 4, 11:39–47.

Van Devender, T.R. 1969. A record of albinism in the canyon tree frog, *Hyla arenicolor* Cope. Herpetologica 25:69.

Van Hyning, O.C. 1933. Batrachia and reptilia of Alachua County, Florida. Copeia 1933:3–7.

Van Meter, R.J., C. M. Swan, and C. A. Trossen. 2012. Effects of road deicer (NaCl) and amphibian grazers on detritus processing in pond mesocosms. Environmental Toxicology and Chemistry 31:2306–2310).

Van Schmidt, N.D., T. L. Cary, M. E. Ortiz–Santaliestra, and W.H. Karasov. 2012. Effects of chronic polybrominated diphenyl ether exposure on gonadal development in the northern leopard frog, *Rana pipiens*. Environmental Toxicology and Chemistry 31:347–354.

Van Winkle, K. 1922. Extension of the range of *Ascaphus truei* Stejneger. Copeia (102):4–6.

Vance, T., and G. Talpin. 1977. Another yellow albino amphibian. Herpetological Review 8:4–5.

Vandenlangenberg, S.M., J.T. Canfield, and J.A. Magner. 2003. A regional survey of malformed frogs in Minnesota (USA) (Minnesota Malformed Frogs). Environmental Monitoring and Assessment 82:45-61.

Vanderburgh, D.J., and R.C. Anderson. 1987a. The relationship between nematodes of the genus *Cosmocercoides* Wilkie, 1930 (Nematoda: Cosmocercoidea) in toads (*Bufo americanus*) and slugs (*Deroceras laeve*). Canadian Journal of Zoology 65:1650–1661.

Vanderburgh, D.J., and R.C. Anderson. 1987b. Preliminary observations on seasonal changes in prevalence and intensity of *Cosmocercoides variabilis* (Nematoda: Cosmocercoidea) in *Bufo americanus* (Amphibia). Canadian Journal of Zoology 65:1666–1667.

Vandewalle, T.J., K.K. Sutton, and J.L. Christiansen. 1996. *Pseudacris crucifer*: an Iowa case history study of an amphibian call survey. Herpetological Review 27:183–185.

Vandewege, M.W. 2011. Using pedigree construction to test head–starting efficiency for endangered amphibians: field tested in the Houston Toad (*Bufo houstonensis*). M.S. thesis, Texas State University, San Marcos, Texas.

Vandewege, M.W., D.J. Brown, and M.R.J. Forstner. 2012. *Bufo* (=*Anaxyrus*) *houstonensis* (Houston Toad). Head-start juvenile dispersal. Herpetological Review 43:117–118.

Varhegyi, G., S.M. Mavroidis, B.M. Walton, C.A. Conaway, and A.R. Gibson. 1998. Amphibian surveys in the Cuyahoga Valley National Recreation Area. Pp.137–154 *In* M.J. Lannoo (ed.), Status & Conservation of Midwestern Amphibians. University of Iowa Press, Iowa City.

Vasconselos, D. 2003. Amphibian and vegetation dynamics in a restored wetland in Maine. M.S. thesis, University of Maine, Orono.

Vasconcelos, D., and A.J.K. Calhoun. 2004. Movement patterns of adult and juvenile *Rana sylvatica* (LeConte) and *Ambystoma maculatum* (Shaw) in three restored seasonal pools in Maine. Journal of Herpetology 38:551–561.

Vasconcelos, D., and A.J.K. Calhoun. 2006. Monitoring created seasonal pools for functional success: a six–year study of amphibian responses, Sears Island, Maine, USA. Wetlands 26:992–1003.

Vatnick, I., M.A. Brodkin, M.P. Simon, B.W. Grant, C.R. Conte, M. Gleave, R. Myers, and M.M. Sadoff. 1999. The effects of exposure to mild acidic conditions on adult frogs (*Rana pipiens* and *Rana clamitans*): mortality rates and pH preferences. Journal of Herpetology 33:370–374.

Vélez A., and M.A. Bee. 2010. Signal recognition by frogs in the presence of temporally fluctuating chorus–shaped noise. Behavioral Ecology and Sociobiology 64:1695–1709.

Vélez, A., and M.A. Bee. 2011. Dip listening and the cocktail party problem in grey treefrogs: signal recognition in temporally fluctuating noise. Animal Behaviour 82:1319–1327.

Vélez, A., G. Höbel, N.M. Gordon, and M.A. Bee. 2012. Dip listening or modulation masking? Call recognition by green treefrogs (*Hyla cinerea*) in temporally fluctuating noise. Journal of Comparative Physiology 198A:891–904.

Velo–Antón, G., P.A. Burrowes, R.L. Joglar, I. Martínez–Solano, K.H. Beard, and G. Parra–Olea. 2007. Phylogenetic study of *Eleutherodactylus coqui* (Anura: Leptodactylidae) reveals deep genetic fragmentation in Puerto Rico and pinpoints origins of Hawaiian populations. Molecular Phylogenetics and Evolution 45:716–728.

Venable, N.J. 2001. Scales, skinks, scutes & newts. West Virginia University Extention Service, Morganton.

Venesky, M.D. 2011. Dynamics of an emerging infectious disease of amphibians: From individuals to communities. Ph.D. Dissertation, University of Memphis, Memphis, Tennessee.

Venesky, M.D., and F.M. Blem. 2008. Occurrence of *Batrachochytrium dendrobatidis* in southwestern Tennessee, USA. Herpetological Review 39:319–320.

Venesky, M.D., M. J. Parris, and A. Storfer. 2009. Impacts of *Batrachochytrium dendrobatidis* infection on tadpole foraging performance. EcoHealth 6:565–576.

Venesky, M.D., R.J. Wassersug, and M.J. Parris. 2010. The impact of variation in labial tooth number on the feeding kinematics of tadpoles of southern leopard frog (*Lithobates sphenocephalus*). Copeia 2010:481–486.

Venesky, M.D., T.E. Wilcoxen, M.A. Rensel, L. Rollins–Smith, J.L. Kerby, and M.J. Parris. 2012. Dietary protein restriction impairs growth, immunity, and disease resistance in southern leopard frog tadpoles. Oecologia 169:23–31.

Vergeer, T. 1948. Frog catches mouse in natural environment. Turtox News 26:91.

Verrill, A.E. 1865. Catalogue of the reptiles and batrachians found in the vicinity of Norway, Oxford Co., Maine. Proceedings of the Boston Society of Natural History 9:195–199.

Vertucci, F.A., and P.S. Corn. 1996. Evaluation of episodic acidification and amphibian declines in the Rocky Mountains. Ecological Applications 6:449–457.

Vestal, E.H. 1941. Defensive inflation of the body in *Bufo boreas halophilus*. Copeia 1941:183.

Veysey, J.S., S.D. Mattfeldt, and K.J. Babbitt. 2011. Comparative influence of isolation, landscape, and wetland characteristics on egg–mass abundance of two pool–breeding amphibian species. Landscape Ecology 26:661–672.

Vhora, M.S. 2012. A seasonal and comparative study of helminth parasites in some anurans from Oklahoma. M.S. thesis, Oklahoma State University, Stillwater.

Vickers, C.R. 1980. Effects of pond ditching on cypress pond herpetofauna in north Florida. M.S. thesis, University of Florida, Gainesville.

Vickers, C.R., L.D. Harris, and B.F. Swindel. 1985. Changes in herpetofauna resulting from ditching of cypress ponds in Coastal Plains flatwoods. Forest Ecology and Management 11:17–29.

Viernes, K.J.F. 1995. Bullfrog predation on an endangered common moorhen chick at Hanalei National Wildlife Refuge, Kaua'i. 'Elepaio 55:37.

Viers, J.H., and D.E. Rheinheimer. 2011. Freshwater conservation options for a changing environment in California's Sierra Nevada. Marine and Freshwater Research 62:266–278.

Viets, B.E. 1993. An annotated list of the herpetofauna of the F.B. and Rena G. Ross Natural History Reservation. Transactions of the Kansas Academy of Science 96:103–113.

Villard, M.–A., M.J. Mazerolle, and S. Haché. 2012. L'impact des routes, au–delà des collisions: le cas des oiseaux forestiers et des amphibiens. Le naturaliste canadien 136(2):61–65.

Villella, J., and A. Mims. n.d. A field guide to amphibians of Opal Creek. Opal Creek Ancient Forest Center, Jawbone Flats, Oregon.

Vinson, S.B., C.E. Boyd, and D.E. Ferguson. 1963. Aldrin toxicity and possible cross–resistance in cricket frogs. Herpetologica 19:77–80.

Viosca, P., Jr. 1923a. An ecological study of the cold blooded vertebrates of southeastern Louisiana. Copeia (115):35–44.

Viosca, P., Jr. 1923b. Notes on the status of *Hyla phaeocrypta* Cope. Copeia (122):96–99.

Viosca, P., Jr. 1926. Distributional problems of the cold–blooded vertebrates of the Gulf Coastal Plain. Ecology 7: 307–314.

Viosca, P., Jr. 1928. A new species of *Hyla* from Louisiana. Proceedings of the Biological Society of Washington 41:89–92.

Viosca, P., Jr. 1931. Principles of bullfrog (*Rana catesbeiana*) culture. Transactions of the American Fisheries Society 61:262–269.

Viosca, P., Jr. 1934. Principles of bullfrog culture. Southern Biological Supply Company, New Orleans, Louisiana.

Viosca, P., Jr. 1938. Notes on the winter frogs of Alabama. Copeia 1938:201.

Viosca, P., Jr. 1944. Distribution of certain cold–blooded animals in Louisiana in relationship to the geology and physiography of the state. Proceedings of the Louisiana Academy of Sciences 8:47–62.

Viosca, P., Jr. 1960. Frogs and toads of Louisiana. Louisiana Conservationist 12 (2–3): 8–11, 20.

Viparina, S., and J.J. Just. 1975. The life period, growth and differentiation of *Rana catesbeiana* larvae occurring in nature. Copeia 1975:103–109.

Visher, S.S. 1914. A preliminary report on the biology of Harding County northwestern South Dakota. VI. A preliminary list of the reptiles and amphibians of Harding County. South Dakota Geological Survey Bulletin No. 6:92–93.

Vitt, L.J., and R.D. Ohmart. 1978. Herpetofauna of the lower Colorado River: Davis Dam to the Mexican border. Proceedings of the Western Foundation of Vertebrate Zoology 2:35–71.

Vladykov, V.D. 1941. A preliminary list of Amphibia from the Laurentides Park in Province of Quebec. Canadian Field-Naturalist 55:83–84.

Vleminckz, T. 1988. Kweken met de Noordamerikaanse Groene Boomkikker (*Hyla cinerea*). Lacerta 46:130–132.

Voeltz, J.B., and T.R. Jones. 2007. *Rana yavapaiensis* (Lowland Leopard Frog). Reproduction. Herpetological Review 38:326.

Vogel, L.S. 2007. The decline of Fowler's toad (*Bufo fowleri*) in southern Louisiana: molecular genetics, field experiments and landscape studies. Ph.D. Dissertation, University of New Orleans, New Orleans, Louisiana.

Vogel, L.S., and S.G. Johnson. 2008. Estimation of hybridization and introgression frequency in toads (Genus: *Bufo*) using DNA sequence variation at mitochondrial and nuclear loci. Journal of Herpetology 42:61–75.

Vogel, L.S., and J.H.K. Pechmann. 2010. Response of Fowler's toad (*Anaxyrus fowleri*) to competition and hydroperiod in the presence of the invasive Coastal Plain toad (*Incilius nebulifer*). Journal of Herpetology 44:382–389.

Vogt, R.C. 1981. Natural History of Amphibians and Reptiles of Wisconsin. Milwaukee Public Museum, Milwaukee.

Vogt, T., and D.L. Jameson. 1970. Chronological correlation between change in weather and change in morphology of the Pacific tree frog in southern California. Copeia 1970:135–144.

Volpe, E.P. 1952. Physiological evidence for natural hybridization of *Bufo americanus* and *Bufo fowleri*. Evolution 6:393–406.

Volpe, E.P. 1954. Hybrid inviability between *Rana pipiens* from Wisconsin and Mexico. Tulane Studies in Zoology 1(9): 109–124.

Volpe, E.P. 1955a. Intensity of reproductive isolation between sympatric and allopatric populations of *Bufo americanus* and *Bufo fowleri*. American Naturalist 89:303–318.

Volpe, E.P. 1955b. A taxo–genetic analysis of the status of *Rana kandiyohi* Weed. Systematic Zoology 4:75–82.

Volpe, E.P. 1956a. Mutant color patterns in leopard frogs. Journal of Heredity 47:79–85.

Volpe, E.P. 1956b. Experimental F$_1$ hybrids between *Bufo valliceps* and *Bufo fowleri*. Tulane Studies in Zoology 4:61–75.

Volpe, E. P. 1956c. Reciprocal mis–matings between *Hyla squirella* and *Microhyla carolinensis*. Copeia 1956:261–262.

Volpe, E.P. 1957a. Embryonic temperature adaptations in highland *Rana pipiens*. American Naturalist 91:303–310.

Volpe, E.P. 1957b. The early development of *Rana capito sevosa*. Tulane Studies in Zoology 5:207–225.

Volpe, E.P. 1957c. Embryonic temperature tolerance and rate of development in *Bufo valliceps*. Physiological Zoology 30:164–176.

Volpe, E.P. 1957d. Genetic aspects of anuran populations. American Naturalist 91:355–372.

Volpe, E.P. 1958. Interspecific gene exchange between Fowler's and the Southern toad. Anatomical Record 131:605–606.

Volpe, E.P. 1959a. Hybridization of *Bufo valliceps* with *Bufo americanus* and *Bufo terrestris*. Texas Journal of Science 11:335–342.

Volpe, E.P. 1959b. Experimental and natural hybridization between *Bufo terrestris* and *Bufo fowleri*. American Midland Naturalist 61:295–312.

Volpe, E.P. 1960a. Evolutionary consequences of hybrid sterility and vigor in toads. Evolution 14:181–193.

Volpe, E. P. 1960b. Interaction of mutant genes in the leopard frog. Journal of Heredity 51:150–155.

Volpe, E.P. 1961. Polymorphism in anuran populations. Pp. 221–234 *In* W.F. Blair (ed.), Vertebrate Speciation. University of Texas Press, Austin.

Volpe, E.P., and J.L. Dobie. 1959. The larva of the oak toad, *Bufo quercicus* Holbrook. Tulane Studies in Zoology 7:145–152.

Volpe, E.P., M.A. Wilkens, and J.L. Dobie. 1961. Embryonic and larval development of *Hyla avivoca*. Copeia 1961:340–349.

Vonesh, J.R., and O.R. De la Cruz. 2002. Complex life cycles density dependence: assessing the contribution of egg mortality to amphibian declines. Oecologia 133:325–333.

Vonesh, J.R., and J.C. Buck. 2007. Pesticide alters oviposition site selection in gray treefrogs. Oecologia 154:219–226.

Vonesh, J.R., and J.M. Kraus. 2009. Pesticide alters habitat selection and aquatic community composition. Oecolo-gia 160:379–385.

Vonesh, J.R., J.M. Kraus, J.S. Rosenberg, and J.M. Chase. 2009. Predator effects on aquatic community assembly: disentangling the roles of habitat selection and post–colonization processes. Oikos 118:1219–1229.

Voris, H.K., and J.P. Bacon, Jr. 1966. Differential predation on tadpoles. Copeia 1966:594–598.

Voss, W.J. 1961. Rate of larval development and metamorphosis of the spadefoot toad, *Scaphiopus bombifrons*. Southwestern Naturalist 6:168–174.

Vredenburg, V. T. 2000. *Rana muscosa* (Mountain Yellow–legged Frog). Egg predation. Herpetological Review 31:170–171.

Vredenburg, V.T. 2004. Reversing introduced species effects: experimental removal of introduced fish leads to rapid recovery of a declining frog. Proceedings of the National Academy of Sciences of the United States of America 101:7646–7650.

Vredenburg, V.T., and A.P. Summers. 2001. Field identification of chytridiomycosis in *Rana muscosa* (Camp 1915). Herpetological Review 32:151–152.

Vredenburg, V., G.M. Fellers, and C. Davidson. 2005. *Rana muscosa* Camp, 1917. Mountain Yellow–legged Frog. Pp. 563–566 *In* M.J. Lannoo (ed.), Amphibian Declines. The Conservation Status of United States Species. University of California Press, Berkeley.

Vredenburg, V.T., R. Bingham, R. Knapp, J.A.T. Morgan, C. Moritz, and D. Wake. 2007. Concordant molecular and phenotypic data delineate new taxonomy and conservation priorities for the endangered mountain yellow–legged frog. Journal of Zoology 271:361– 374.

Vredenburg, V.T., J.M. Romansic, L.M. Chan, and T. Tunstall. 2010a. Does UV–B radiation affect embryos of three high elevation amphibian species in California? Copeia 2010:502–512.

Vredenburg, V.T., R.A. Knapp, T.S. Tunstall, and C.J. Briggs. 2010b. Dynamics of an emerging disease drive large–scale amphibian population declines. Proceedings of the National Academy of Sciences of the United States of America 107:9689–9694.

Wada, H., C.M. Bergeron, F.M.A. McNabb, B.D. Todd, and W.A. Hopkins. 2011. Dietary mercury has no observable effects on thyroid–mediated processes and fitness–related traits in wood frogs. Environmental Science & Technology 45:7915–7922.

Waddle, J.H. 2006. Use of amphibians as ecosystem indicator species. Ph.D. Dissertation, University of Florida, Gainesville.

Waddle, J.H., K.G. Rice, F.J. Mazzotti, and H.F. Percival. 2008. Modeling the effect of toe clipping on treefrog survival: beyond the return rate. Journal of Herpetology 42:467–473.

Waddle, J.H., R.M. Dorazio, S.C. Walls, K.G. Rice, J. Beauchamp, M.J. Schulman, and F.J. Mazzotti. 2010. A new parameterization for estimating co–occurrence of interacting species. Ecological Applications 20:1467–1475.

Wagner, G. 1997. Status of the northern leopard frog (*Rana pipiens*) in Alberta. Alberta Wildlife Status Report 9, Alberta Conservation Association, Edmonton.

Wagner, W.E., Jr. 1986. Tadpoles and pollen: observations on the feeding behavior of *Hyla regilla* larvae. Copeia 1986:802–804.

Wagner, W.E. 1989a. Social correlates of variation in male calling behavior in Blanchard's cricket frog, *Acris crepitans blanchardi*. Ethology 82:27–45.

Wagner, W.E. 1989b. Fighting, assessment, and frequency alteration in Blanchard's cricket frog. Behavioral Ecology and Sociobiology 25:429–436.

Wagner, W.E. 1989c. Graded aggressive signals in Blanchard's cricket frog: vocal responses to opponent proximity and size. Animal Behaviour 38:1025–1038.

Wagner, W.E. 1991. Social selection on male calling behavior in Blanchard's cricket frog. Ph.D. Dissertation, University of Texas, Austin.

Wagner, W.E. 1992. Deceptive or honest signalling of fighting ability? A test of alternative hypotheses for the function of changes in call dominant frequency by male cricket frogs. Animal Behaviour 44:449–462.

Wagner, W.E., Jr., and B.K. Sullivan. 1992. Chorus organization in the Gulf Coast toad (*Bufo valliceps*): male and female behavior and the opportunity for sexual selection. Copeia 1992:647–658.

Wagner, W.E., Jr., and B.K. Sullivan. 1995. Sexual selection in the Gulf Coast toad, *Bufo valliceps*: female choice based on variable characters. Animal Behaviour 49:305–319.

Wahbe, T.R. 1996. Tailed frogs (*Ascaphus truei* Stejneger) in natural and managed coastal temperate rainforests of southwestern British Columbia, Canada. M.S. thesis. University of British Columbia, Vancouver.

Wahbe, T.R. 2003. Aquatic and terrestrial movements of tailed frogs (*Ascaphus truei*) in relation to timber harvest in Coastal British Columbia. Ph.D. Dissertation, University of British Columbia, Vancouver.

Wahbe, T.R., and F.L. Bunnell. 2001. Preliminary observations on movements of tailed frog tadpoles (*Ascaphus truei*) in streams through harvested and natural forests. Northwest Science 75:77–83.

Wahbe, T.R., and F.L. Bunnell. 2003. Relations among larval tailed frogs, forest harvesting, stream microhabitat, and site parameters in southwestern British Columbia. Canadian Journal of Forest Resources 33:1256–1266.

Wahbe, T.R, F.L. Bunnell, and R.B. Bury. 2004. Terrestrial movements of juvenile and adult tailed frogs in relation to timber harvest in coastal British Columbia. Canadian Journal of Forest Resources 34:2455–2466.

Wahbe, T.R., C. Ritland, F.L. Bunnell, and K. Ritland. 2005. Population genetic structure of tailed frogs (*Ascaphus truei*) in clearcut and old–growth stream habitats in south coastal British Columbia. Canadian Journal of Zoology 83:1460–1468.

Waitz, J.A. 1961a. Parasites of Idaho amphibians. Journal of Parasitology 47:89.

Waitz, J.A. 1961b. Parasitic helminths as aids in studying the distribution of species of *Rana* in Idaho. Transactions of the Illinois State Academy of Science 54:152–156.

Wake, D.B., and V.T. Vredenburg. 2008. Are we in the midst of the sixth mass extinction? A view from the world of amphibians. Proceedings of the National Academy of Sciences of the United States of America 105:11466–11473.

Waldick, R.C. 1994. Implications of forestry-associated habitat conversion on amphibians in the vicinity of Fundy National Park, New Brunswick. M.S. thesis, Dalhousie University, Halifax, Nova Scotia.

Waldick, R.C., B. Freedman, and R.J. Wassersug. 1999. The consequences for amphibians of the conversion of natural, mixed–species forests to conifer plantations in southern New Brunswick. Canadian Field-Naturalist 113:408–418.

Waldman, B. 1981. Sibling recognition in toad tadpoles: the role of experience. Zeitschrift für Tierpsychologie 56:341–358.

Waldman, B. 1982a. Sibling association among schooling toad tadpoles: field evidence and implications. Animal Behaviour 30:700–713.

Waldman, B. 1982b. Adaptive significance of communal oviposition in wood frogs (*Rana sylvatica*). Behavioral Ecology and Sociobiology 10:169–174.

Waldman, B. 1984. Kin recognition and sibling association among wood frog (*Rana sylvatica*) tadpoles. Behavioral Ecology and Sociobiology 14:171–180.

Waldman, B. 1985a. Sibling recognition in toad tadpoles: are kinship labels transferred among individuals? Zeitschrift für Tierpsychologie 68:41–57.

Waldman, B. 1985b. Olfactory basis of kin recognition in toad tadpoles. Journal of Comparative Physiology A 156:565–577.

Waldman, B. 1986a. Preference for unfamiliar siblings over familiar non–siblings in American toad (*Bufo americanus*) tadpoles. Animal Behaviour 34:48–53.

Waldman, B. 1986b. Chemical ecology of kin recognition in anuran amphibians. Pp. 225–242 *In* D. Duvall, D. Müller–Schwarze, and D.M. Silverstein (eds.). Chemical Signals in Vertebrates. 4. Ecology, Evolution, and Comparative Biology. Plenum Press, New York.

Waldman, B. 1989. Do anuran larvae retain kin recognition abilities following metamorphosis? Animal Behaviour 37:1055–1058.

Waldman, B. 1991. Kin recognition in amphibians. Pp. 162–219 *In* P.G. Hepper (ed.). Kin Recognition. Cambridge University Press, Cambridge, United Kingdom.

Waldman, B. 2001. Kin recognition, sexual selection, and mate choice in toads. Pp. 232–244 *In* M.J. Ryan (ed.), Anuran Communication. Smithsonian Institution Press, Washington, D.C.

Waldman, B., and K. Adler. 1979. Toad tadpoles associate preferentially with siblings. Nature 282:611–613.

Waldman, B., and M.J. Ryan. 1983. Thermal advantages of communal egg mass deposition in wood frogs (*Rana sylvatica*). Journal of Herpetology 17:70–72.

Waldman, B., J.E. Rice, and R.L. Honeycutt. 1992. Kin recognition and incest avoidance in toads. American Zoologist 32:18–30.

Waldron, E. 1953. A carnival of frogs. The New Yorker, April 11, 1953. pp. 81-101.

Waldron, J.L., Z. Felix, W. J. Humphries, and T. K. Pauley. 2001. Herpetofauna of the Bluestone National Wild and Scenic River, West Virginia. Proceedings of the West Virginia Academy of Science 73:1–10.

Walker, C.F. 1931. The amphibians of Ohio. M.S. thesis, Ohio State University, Columbus.

Walker, C.F. 1932. *Pseudacris brachyphona* (Cope), a valid species. Ohio Journal of Science 32: 379–384.

Walker, C.F. 1935. The generic relations of the North American terrestrial Hylidae. Ph.D. Dissertation, University of Michigan, Ann Arbor.

Walker, C.F. 1946. The Amphibians of Ohio. 1. The frogs and toads (Order Salientia). Ohio State Museum Science Bulletin 1(3):1–109.

Walker, C.F. 1967. The Amphibians of Ohio. 1. Frogs and toads. Second edition. Ohio State Museum Science Bulletin 1(3):1–109. [reissue]

Walker, D., and S.D. Busack. 2000. *Rana catesbeiana* (Bullfrog). Tadpole depth record. Herpetological Review 31:236.

Walker, J.M. 1963. Amphibians and reptiles of Jackson Parish Louisiana. Proceedings of the Louisiana Academy of Sciences 26:91–107.

Walker, M.V. 1946. Reptiles and amphibians of Yosemite National Park. Yosemite Nature Notes 25:1–48. [an undated issue with a different cover also was published]

Walker, R.F., and W.G. Whitford. 1970. Soil water absorption capabilities in selected species of anurans. Herpetologica 26:411–418.

Wallace, J.E., R.J. Steidl, and D.E. Swann. 2010. Habitat characteristics of lowland leopard frogs in mountain canyons of southeastern Arizona. Journal of Wildlife Management 74:808-815.

Wallace, R.L., and L.V. Diller. 1998. Length of the larval cycle of *Ascaphus truei* in coastal streams of the Redwood region, northern California. Journal of Herpetology 32:404–409.

Walls, J.G. 1978. Leaping leopard frogs. Tropical Fish Hobbyist (Nov.):91, 93, 95–96, 99–102.

Walls, S.C., A.R. Blaustein, and J.J. Beatty. 1992. Amphibian biodiversity of the Pacific Northwest with special reference to old–growth stands. Northwest Environmental Journal 8:53–69.

Walls, S.C., D.G. Taylor, and C.M. Wilson. 2002. Interspecific differences in susceptibility to competition and predation in a species–pair of larval amphibians. Herpetologica 58:104–118.

Walpole, A.A., J. Bowman, D.C. Tozer, and D.S. Badzinski. 2012. Community–level response to climate change: shifts in anuran calling phenology. Herpetological Conservation and Biology 7:249–257.

Walston, L.J., and S.J. Mullin. 2007a. Population responses of wood frog (*Rana sylvatica*) tadpoles to overwintered bullfrog (*Rana catesbeiana*) tadpoles. Journal of Herpetology 41:24–31.

Walston, L.J., and S.J. Mullin. 2007b. Responses of a pond–breeding amphibian community to the experimental removal of predatory fish. American Midland Naturalist 157:63–73.

Walston, L.J., and S.J. Mullin. 2008. Variation in amount of surrounding forest habitat influences the initial orientation of juvenile amphibians emigrating from breeding ponds. Canadian Journal of Zoology 86:141–146.

Walters, B. 1975. Studies of interspecific predation within an amphibian community. Journal of Herpetology 9:267–279.

Walton, A.C. 1938. The Nematoda as parasites of Amphibia. 4. Transactions of the American Microscopical Society 57:38–53.

Walton, A.C. 1941a. Notes on some helminths from California amphibia. Transactions of the American Microscopical Society 60:53–57.

Walton, A.C. 1941b. Notes on amphibian parasites. Proceedings of the Helminthological Society of Washington 7:87–91.

Walton, A.C. 1947. Parasites of the Ranidae (Amphibia). Anatomical Record 99:684–685.

Walton, B.M., C.C. Peterson, and A.F. Bennett. 1994. Is walking costly for anurans? The energetic cost of walking in *Bufo boreas halophilus*. Journal of Experimental Zoology 197:165–178.

Walton, J. 2012. The effects of mine land reclamation on

herpetofaunal communities. M.S. thesis, University of Texas, Arlington.

Wang, I.J. 2009. Fine–scale population structure in a desert amphibian: landscape genetics of the black toad (*Bufo exsul*). Molecular Ecology 18:3847–3856.

Wang, I.J. 2012. Environmental and topographic variables shape genetic structure and effective population sizes in the endangered Yosemite toad. Diversity and Distributions 18:1033–1041.

Ward, R., C.J. Rutledge, and E.G. Zimmerman. 1987. Genetic variation and population subdivision in the cricket frog *Acris crepitans*. Biochemical Systematics and Ecology 15: 377–384.

Ward, W.T. 1947. Some effects of DDT on the leucocytes in *Rana pipiens*. M.S. thesis, Fordham University, New York, New York.

Warkentin, I.G., C.E. Campbell, K.G. Powell, and T.D. Leonard. 2003. First record of the mink frog (*Rana septentrionalis*) for insular Newfoundland. Canadian Field-Naturalist 117:477–478.

Warkentin, K.M. 1992a. Effects of temperature and illumination on feeding rates of green frog tadpoles (*Rana clamitans*). Copeia 1992:735–730.

Warkentin, K.M. 1992b. Microhabitat use and feeding rate variation in green frog tadpoles (*Rana clamitans*). Copeia 1992:731–740.

Warne, R.W., E.J. Crespi, and J.L. Brunner. 2011. Escape from the pond: stress and developmental responses to ranavirus infection in wood frog tadpoles. Functional Ecology 25:139-146.

Warner, S.C., W.A. Dunson, and J. Travis. 1991. Interaction of pH, density, and priority effects on the survivorship and growth of two species of hylid tadpoles. Oecologia 88:331–339.

Warner, S.C., J. Travis, and W.A. Dunson. 1993. Effect of pH variation on interspecific competition between two species of hylid tadpoles. Ecology 74:183–194.

Warny, P.R., W.E. Meshaka, Jr., and A.N. Klippel. 2012. Larval growth and transformation size of the green frog, *Lithobates clamitans melanota* (Rafinesque, 1820), in south–eastern New York. Herpetology Notes 5:59–62.

Washburn, A.M. 1899. A peculiar toad. American Naturalist 33:139–141.

Wasserman, A.O. 1957a. Hybridization in three species of spadefoot toads. Copeia 1957:144–145.

Wasserman, A.O. 1957b. Factors affecting interbreeding in sympatric species of spadefoots (genus *Scaphiopus*). Evolution 11:320–338.

Wasserman, A.O. 1958. Relationships of allopatric populations of spadefoots (genus *Scaphiopus*). Evolution 12:311–318.

Wasserman, A.O. 1963. Further studies of hybridization in spadefoot toads (Genus *Scaphiopus*). Copeia 1963:115–118.

Wasserman, A.O. 1964. Recent and summarized interspecific hybridization within the Pelobatidae. Texas Journal of Science 16:334–341.

Wasserman, A.O. 1966. It could be a "froad"; maybe a "trog." Natural History 75(4):18–23.

Wasserman, A.O. 1970. Chromosomal studies of the Pelobatidae (Salientia) and some instances of ploidy. Southwestern Naturalist 15:239–248.

Wasserman, A.O., and J.P. Bogart. 1968. Chromosomes of two species of spadefoot toads (genus *Scaphiopus*) and their hybrid. Copeia 1968:303–306.

Wassersug, R.J. 1973. Aspects of social behavior in anuran larvae. Pp. 273–297 *In* J.L. Vial (ed.), Evolutionary Biology of the Anurans. University of Missouri Press, Columbia.

Wassersug, R.J. 1975. The adaptive significance of the tadpole stage with comments on the maintenance of complex life cycles in anurans. American Zoologist 15:405–417.

Wassersug, R.J., and D.G. Sperry. 1977. The relationship of locomotion to differential predation on *Pseudacris triseriata* (Anura: Hylidae). Ecology 58:830–839.

Wassersug, R.J., and K. Hoff. 1979. A comparative study of the buccal pumping mechanisms of tadpoles. Biological Journal of the Linnaean Society 12:225–259.

Wassersug, R.J., and M.Yamashita. 2001. Plasticity and constraints on feeding kinematics in anuran larvae. Comparative Biochemistry and Physiology A 131:183–195.

Wassersug, R.J., T. Naitoh, and M. Yamashita. 1999. Turning bias in tadpoles. Journal of Herpetology 33:543–548.

Waters, D.L., T.J. Hassler, and B.R. Norman. 1998. On the establishment of the Pacific chorus frog, *Pseudacris regilla* (Amphibian, Anura, Hylidae), at Ketchikan, Alaska. Bulletin of the Chicago Herpetological Society 33:124–127.

Waterstrat, F.T., R.L. Crawford, and M.P. Hayes. 2007. *Ascaphus truei* (Coastal Tailed Frog). Spider prey. Herpetological Review 38:318.

Waterstrat, F.T., A.P. McIntyre, M.P. Hayes, K.M. Phillips, and T.R. Curry. 2008. *Ascaphus truei* (Coastal Tailed Frog). Atypical amplexus. Herpetological Review 39:458.

Watkins, T.B. 1996. Predator–mediated selection on burst swimming performance in tadpoles of the Pacific tree frog, *Pseudacris regilla*. Physiological Zoology 69:154–167.

Watkins, T.B. 1997. A microevolutionary study of locomotor performance in the Pacific tree frog, *Hyla regilla*. Ph.D. Dissertation, University of California, Berkeley.

Watkins, T.B. 2001. A quantitative genetic test of adaptive decoupling across metamorphosis for locomotor and life history traits in the Pacific tree frog, *Hyla regilla*. Evolution 55:1668–1677.

Watkins, T.B., and M.A. McPeek. 2006. Growth and predation risk in green frog tadpoles (*Rana clamitans*): a quantitative genetic analysis. Copeia 2006:478–488.

Watling, J.I., C.R. Hickman, and J.L. Orrock. 2011a. Predators and invasive plants affect performance of amphibian larvae. Oikos 120:735–739.

Watling, J.I., C.R. Hickman, and J.L. Orrock. 2011b. Invasive shrub alters native forest amphibian communities. Biological Conservation 144:2597–2601.

Watling, J.I., C.R. Hickman, E. Lee, K. Wand and J.L. Orrock. 2011c. Extracts of the invasive shrub *Lonicera maackii* increase mortality and alter behavior of amphibian larvae. Oecologia 165:153–159.

Watrous, W.L. 1967. A population and homing study of the chorus frog. M.S. thesis, Ball State University, Muncie, Indiana.

Watson, J.W., K.R. McAllister, and D.J. Pierce. 2003. Home ranges, movements, and habitat selection of Oregon spotted frogs (*Rana pretiosa*). Journal of Herpetology 37:292–300.

Wauer, R.H. 1964. Reptiles and amphibians of Zion National Park. Zion Natural History Association, Zion National Park, Utah.

Wauer, R.H. 1973. Naturalist's Big Bend. Texas A&M University Press, College Station. [updated version, 1980]

Waye, H.L. 2001. Teflon tubing as radio transmitter belt material for northern leopard frogs (*Rana pipiens*). Herpetological Review 31:88–89.

Waye, H.L., and C.H. Shewchuk. 1995. *Scaphiopus intermontanus* (Great Basin Spadefoot). Production of odor. Herpetological Review 26:98–99.

Weatherby, C.A. 1982. Introgression between the American toad *Bufo americanus* and the southern toad *B. terrestris* in Alabama. Ph.D. Dissertation, Auburn University, Auburn, Alabama.

Weaver, W.L. 1857. The Battle of the Frogs, at Windham, 1758, with Various Accounts and Three of the Most Popular Ballads on the Subject. James Walden, Willimantic, Connecticut. 31 pp.

Webb, R.G. 1963. The larva of the casque–headed frog, *Pternohyla fodiens* Boulenger. Texas Journal of Science 15:89–97.

Webb, R.G. 1965. Observations on the breeding habits of the squirrel treefrog, *Hyla squirella* Bosc *in* Daudin. American Midland Naturalist 74:500–501.

Webb, R.G., and J.K. Korky. 1977. Variation in tadpoles of frogs of the *Rana tarahumarae* group in western Mexico (Anura: Ranidae). Herpetologica 33:73–82.

Webber, L.A., and D.C. Cochran. 1984. Laboratory observations on some freshwater vertebrates and secerial saline fishes exposed to a monomolecular organic surface film (ISA–20E). Mosquito News 44:68–69.

Webber, N., M.D. Boone, and C.A. Distel. 2010. Effects of aquatic and terrestrial carbaryl exposure on feeding ability, growth, and survival of American toads. Environmental Toxicology and Chemistry 29:2323–2327.

Weber, J.A. 1928. Herpetological observations in the Adirondack Mountains, New York. Copeia (169):106–112.

Weed, A.C. 1922. New frogs from Minnesota. Proceedings of the Biological Society of Washington 34:107–110.

Wegner, W. 1958. Frogs in the diet of a hognose snake. Herpetologica 14:276.

Weick, D.L. 1979. Osmotic regulation and salinity tolerance in the Pacific treefrog, *Hyla regilla*. M.S. thesis, California State University, Fullerton.

Weil, A.T., and W. Davis. 1994. *Bufo alvarius*: a potent hallucinogen of animal origin. Journal of Ethnopharmacology 41:1-8.

Weintraub, J.D. 1967. Herpetological observations in Saskatchewan and the MacKenzie District. Canadian Field-Naturalist 81:106–109.

Weintraub, J.D. 1974. Movement patterns of the red–spotted toad, *Bufo punctatus*. Herpetologica 30:212–215.

Weintraub, J.D. 1980. Selection of daytime retreats by recently metamorphosed *Scaphiopus multiplicatus*. Journal of Herpetology 14:83–84.

Weir, L.A., J.A. Royle, P. Nanjappa, and R.E. Jung. 2006. Modeling anuran detection and site occupancy on North American Amphibian Monitoring Program (NAAMP) routes in Maryland. Journal of Herpetology 39:627–639.

Weir, L.A., I.J. Fiske, and J.A. Royle. 2009. Trends in anuran occupancy from northeastern states of the North American Amphibian Monitoring Program. Herpetological Conservation and Biology 4:389–402.

Weir, S.M., S. Yu, and C. J. Salice. 2012. Acute toxicity of herbicide formulations and chronic toxicity of technical–grade trifluralin to larval green frogs (*Lithobates clamitans*). Environmental Toxicology and Chemistry 31:2029–2034.

Weis, J.S. 1975. The effect of DDT on tail regeneration in *Rana pipiens* and *R. catesbeiana* tadpoles. Copeia 1975:765–767.

Weiss, B.A., B.H. Stuart, and W.F. Strother. 1973. Auditory sensitivity in the *Rana catesbeiana* tadpole. Journal of Herpetology 7:211–214.

Weitzel, N.H., and H.R. Panik. 1993. Long–term fluctuations of an isolated population of the Pacific chorus frog

(*Pseudacris regilla*) in northwestern Nevada. Great Basin Naturalist 53:379–384.

Welbourn, W.C., Jr., and R.B. Loomis. 1975. *Hannemania* (Acarina: Trombiculidae) and their anuran hosts at Four-tynine Palms Oasis, Joshua Tree National Monument, California. Bulletin of the Southern California Academy of Sciences 74:15–19.

Welch, A.M. 2003. Genetic benefits of a female mating preference in gray tree frogs are context–dependent. Evolution 57:883–893.

Welch, A.M., R.D. Semlitsch, and H.C. Gerhardt. 1998. Call duration as an indicator of genetic quality in male gray treefrogs. Science 280:1928–1930.

Welch, N.E., and J.A. MacMahon. 2005. Identifying habitat variables important to the rare Columbia spotted frog in Utah (U.S.A.): an information theoretic approach. Conservation Biology 19:473–481.

Weldon, C., L.H. du Preez, A.D. Hyatt, R. Muller, and R. Speare. 2004. Origin of the amphibian chytrid fungus. Emerging Infectious Diseases 10:2100–2105.

Weller, W.F. 2009. Extension of the known range of the gray treefrog, *Hyla versicolor*, in northwestern Ontario. Canadian Field-Naturalist 123:372-374.

Weller, W.F., and D.M. Green. 1997. Checklist and current status of Canadian amphibians. SSAR Herpetological Conservation 1:309–328.

Wells, K.D. 1976a. Territorial behavior of the green frog (*Rana clamitans*). Ph.D. Dissertation, Cornell University, Ithaca, New York.

Wells, K.D. 1976b. Multiple egg clutches in the green frog (*Rana clamitans*). Herpetologica 32:85–87.

Wells, K.D. 1977a. The social behaviour of anuran amphibians. Animal Behaviour 25:666–693.

Wells, K.D. 1977b. Territoriality and male mating success in the green frog (*Rana clamitans*). Ecology 58:750–762.

Wells, K.D. 1978. Territoriality in the green frog (*Rana clamitans*): vocalizations and agonistic behaviour. Animal Behaviour 26:1051–1063.

Wells, K.D. 2007. The Ecology & Behavior of Amphibians. University of Chicago Press, Chicago, Illinois.

Wells, K.D., and T.L. Taigen. 1984. Reproductive behavior and aerobic capacities of male American toads (*Bufo americanus*): is behavior constrained by physiology? Herpetologica 40:292–298.

Wells, K.D., and T.L. Taigen. 1986. The effect of social interactions on calling energetics in the gray treefrog (*Hyla versicolor*). Behavioral Ecology and Sociobiology 19:9–18.

Wells, K.D., and C.R. Bevier. 1997. Contrasting patterns of energy substrate use in two species of frogs that breed in cold weather. Herpetologica 53:70–80.

Wells, K.D., and J.J. Schwartz. 2007. The behavioral ecology of anuran communication. Pp. 44–86 *In* P. Narins, A.S. Feng, and R.R. Fay (eds.), Hearing and Sound Communication in Amphibians. Springer, New York.

Wells, K.D., T.L. Taigen, S.W. Rusch, and C.C. Robb. 1995. Seasonal and nightly variation in glycogen reserves of calling gray treefrogs (*Hyla versicolor*). Herpetologica 51:359–368.

Wells, K.D., T.L. Taigen, and J.A. O'Brien. 1996. The effect of temperature on calling energetics of the spring peeper (*Pseudacris crucifer*). Amphibia–Reptilia 17:149–158.

Welsh, H.H. 1988. An ecogeographic analysis of the herpetofauna of the Sierra San Pedro Mártir Region, Baja California, with a contribution to the biogeography of the Baja California herpetofauna. Proceedings of the California Academy of Sciences, 4th Series, 46:1–72.

Welsh, H.H. 1990. Relictual amphibians and old–growth forests. Conservation Biology 4:309–319.

Welsh, H.H. 1993. Hierarchical analysis of the niche relationships of four amphibians from forested habitats of northwestern California. Ph.D. Dissertation, University of California, Berkeley.

Welsh, H.H., Jr., and L.M. Ollivier. 1998. Stream amphibians as indicators of ecosystem stress: a case study from California's redwoods. Ecological Applications 8:1118–1132.

Welsh, H.H., Jr., K.L. Pope, and D. Boiano. 2006. Sub–alpine amphibian distributions related to species palatability to non–native salmonids in the Klamath mountains of northern California. Diversity and Distributions 12:298–309.

Welsh, H.H., Jr., G.M. Fellers, and A.J. Lind. 2007. Amphibian populations in the terrestrial environment: is there evidence of declines of terrestrial forest amphibians in northwestern California? Journal of Herpetology 42:469–482.

Welter, W.A., and K. Carr. 1939. Amphibians and reptiles of northeastern Kentucky. Copeia 1939:128–130.

Wendelken, P.W. 1968. Differential predation and behavior as factors in the maintenance of the vertebral stripe polymorphism in *Acris crepitans*. M.S. thesis, University of Texas, Austin.

Wente, W.H., and J.B. Phillips. 2003. Fixed green and brown color morphs and a novel color-changing morph of the Pacific tree frog *Hyla regilla*. American Naturalist 162:461–473.

Wente, W.H., and J.B. Phillips. 2005. Seasonal color change in a population of Pacific treefrogs (*Pseudacris regilla*). Journal of Herpetology 39:161–165.

Wente, W.H., M.J. Adams, and C.A. Pearl. 2005. Evidence of decline for *Bufo boreas* and *Rana luteiventris* in and around the northern Great Basin, western USA. Alytes

22:95–108.

Werler, J.E., and J. McCallion. 1951. Notes on a collection of reptiles and amphibians from Princess Anne County, Virginia. American Midland Naturalist 45:245–252.

Werner, E.E. 1986. Amphibian metamorphosis: growth rate, predation risk, and the optimal size at transformation. American Naturalist 128:319–341.

Werner, E.E. 1988. Size, scaling, and the evolution of complex life cycles. Pp. 60–81 *In* B. Ebenman and L. Persson (eds.), Size–structured Populations: Ecology and Evolution. Springer–Verlag, New York.

Werner, E.E. 1991. Nonlethal effects of a predator on competitive interactions between two anuran larvae. Ecology 75:1709–1720.

Werner, E.E. 1992. Competitive interactions between wood frog and northern leopard frog larvae: the influence of size and activity. Copeia 1992:26–35.

Werner, E.E. 1994. Ontogenetic scaling of competitive relations: size–dependent effects and responses in two anuran larvae. Ecology 75:197–213.

Werner, E.E. 1998. Ecological experiments and a research program in community ecology. Pp. 3–26 *In* W.J. Resetarits and J. Bernardo (eds.), Experimental Ecology: Issues and Perspectives. Oxford University Press, Oxford, United Kingdom.

Werner, E.E., and M.A. McPeek. 1994. Direct and indirect effects of predators on two anuran species along an environmental gradient. Ecology 75:1368–1382.

Werner, E.E., and B.R. Anholt. 1996. Predator–induced behavioral indirect effects: consequences to competitive interactions in anuran larvae. Ecology 77:157–169.

Werner, E.E., and K.S. Glennemeier. 1999. Influence of forest canopy cover on the breeding pond distributions of several amphibian species. Copeia 1999:1–12.

Werner, E.E., G.A. Wellborn, and M.A. McPeek. 1995. Diet composition in postmetamorphic bullfrogs and green frogs: implications for interspecific predation and composition. Journal of Herpetology 29:600–607.

Werner, E. E., D. K. Skelly, R.A. Relyea, and K. L. Yurewicz. 2007a. Amphibian species richness across environmental gradients. Oikos 116:1697–1712.

Werner, E. E., K. L. Yurewicz, D. K. Skelly, and R.A Relyea. 2007b. Turnover in a metacommunity: The role of local and regional factors. Oikos 116:1713–1725.

Werner, E. E., R. A. Relyea, K. L. Yurewicz, D. K. Skelly, and C. J. Davis. 2009. Comparative landscape dynamics of two anuran species: interaction of local and regional processes with an ENSO climate driver. Ecological Monographs 79:503–521.

Werner, J. K. 2003. Status of the northern leopard frog (*Rana pipiens*) in western Montana. Northwestern Naturalist 84:24–30.

Werner, J.K., B.A. Maxwell, P. Hendricks, and D.L. Flath. 2004. Amphibians and Reptiles of Montana. Mountain Press Publishing Co., Missoula, Montana.

Werner, J.K., G. Heinz, and J. Lichtenberg. 2006. The status of two northern leopard frog populations in western Montana. Herpetological Review 37:325–330.

Wernz, J.G. 1969. Spring mating of *Ascaphus*. Journal of Herpetology 3:167–169.

Wernz, J.G., and R.M. Storm. 1969. Pre–hatching stages of the tailed frog, *Ascaphus truei* Stejneger. Herpetologica 25:86–93.

West, L.B. 1960. The nature of growth inhibitory material from crowded *Rana pipiens* tadpoles. Physiological Zoology 33:232–239.

Westerman, A.G., A.J. Wigginton, D.J. Price, G. Linder, and W.J. Birge. 2003a. Integrating amphibians into ecological risk assessment strategies. Pp. 283–313 *In* G. Linder, S.K. Krest, and D.W. Sparling (eds.), Amphibian Decline: An Integrated Analysis of Multiple Stressor Effects. SETAC Press, Pensacola, Florida.

Westerman, A.G., W. van der Schalie, S.L. Levine, B. Palmer, D. Shank, and R.G. Stahl. 2003b. Linking stressors with potential effects on amphibian survival. Pp. 73–109 *In* G. Linder, S.K. Krest, and D.W. Sparling (eds.), Amphibian Decline: An Integrated Analysis of Multiple Stressor Effects. SETAC Press, Pensacola, Florida.

Westerveld, J. 2012. Tiny caller: the northern cricket frog. New York State Conservationist 66(4):24-27.

Westman, A.D.J., J. Elliott, K. Cheng, G. Van Aggelen, and C.A. Bishop. 2010. Effects of enviormentally relevant concentrations of endosulfan, azinphosmethyl, and diazinon on Great Basin spadefoot (*Spea intermontana*) and Pacific treefrog (*Pseudacris regilla*). Environmental Toxicology and Chemistry 29:1604–1612.

Wetzel, E.J., and G.W. Esch. 1996. Seasonal population dynamics of *Halipegus occidualis* and *Halipegus eccentricus* (Diagenea: Hemiuridae) in their amphibian host, *Rana clamitans*. Journal of Parasitology 82:414–422.

Weyrauch, S.L., and J.P. Amon. 2002. Relocation of amphibians to created seasonal ponds in southwestern Ohio. Restoration Ecology 20:31–36.

Weyrauch, S.L., and T.C. Grubb, Jr. 2006. Effects of the interaction between genetic diversity and UV-B radiation on wood frog fitness. Conservation Biology 20:802-810.

Wharton, J.C. 1989. Ecological and life history aspects of the San Francisco garter snake (*Thamnophis sirtalis tetrataenia*). M.S. thesis, San Francisco State University, San Francisco, California.

Wheeler, C.A. 2007. Temporal breeding patterns and mating strategy of the foothill yellow–legged frog (*Rana boylii*). M.S. thesis, Humboldt State University, Arcata, California.

Wheeler, C.A., H.H. Welsh, Jr., and L.L. Heise. 2003. *Rana boylii* (Foothill Yellow–legged Frog). Oviposition behavior. Herpetological Review 34:234.

Wheeler, C.A., J.M. Garwood, and H.H. Welsh, Jr. 2005. *Rana boylii* (Foothill Yellow–legged Frog). Physiological skin color transformation. Herpetological Review 36:164–165.

Wheeler, C.H., and H.H. Welsh, Jr. 2008. Mating strategy and breeding patterns of the foothill yellow–legged frog (*Rana boylii*). Herpetological Conservation and Biology 3:128–142.

Wheeler, G.C. 1947. The amphibians and reptiles of North Dakota. American Midland Naturalist 38:162–190.

Wheeler, G.C. 1954. The amphibians and reptiles on the North Dakota Badlands. Theodore Roosevelt Natural History Association, Medora, North Dakota.

Wheeler, G.C., and J. Wheeler. 1966. The Amphibians and Reptiles of North Dakota. University of North Dakota, Grand Forks.

Whitaker, J.O., Jr. 1961. Habitat and food of mousetrapped young *Rana pipiens* and *Rana clamitans*. Herpetologica 17:173–179.

Whitaker, J.O., Jr. 1971. A study of the western chorus frog, *Pseudacris triseriata*, in Vigo County, Indiana. Journal of Herpetology 5:127–150.

Whitaker, J.O., D. Rubin, and J.R. Munsee. 1977. Observations of food habits of four species of spadefoot toads, genus *Scaphiopus*. Herpetologica 33:468–475.

White, H.Q. 2002. Oviposition habitat enhancement and population estimates of Oregon spotted frogs (*Rana pretiosa*) at Beaver Creek, Washington. M.S. thesis, Evergreen State University, Olympia, Washington.

White, J.F. 1988. Amphibians of Delaware. Transactions of the Delaware Academy of Science 16:17–22.

White, J.F., Jr., and A.W. White. 2002. Amphibians and Reptiles of Delmarva. 1st ed. Tidewater Publishers, Centreville, Maryland.

White, J.F., Jr., and A.W. White. 2007. Amphibians and Reptiles of Delmarva. 2nd ed. Tidewater Publishers, Centreville, Maryland.

Whitehurst, P.H., and B.A. Pierce. 1991. The relationship between allozyme variation and life–history traits of the spotted chorus frog, *Pseudacris clarkii*. Copeia 1991:1032–1039.

Whiteman, H.H., and N. Buschhaus. 2002. Behavioral thermoregulation in field populations of amphibian larvae. Pp. 79–84 *In* B.J. Ploger and K. Yasukawa (eds.), Exploring Animal Behavior in Laboratory and Field, Academic Press: An Hypothesis–testing Approach to the Development, Causation, Function, and Evolution of Animal Behavior. Academic Press, San Diego, Californa.

Whitford, W.G. 1967. Observations on territoriality and aggressive behavior in the western spadefoot toad, *Scaphiopus hammondii*. Herpetologica 23:318.

Whiting, A. 2004. Population ecology of the western chorus frog, *Pseudacris triseriata*. M.S. thesis, McGill University, Montreal, Quebec.

Whiting, M.J., and A.H. Price. 1994. Not the last picture show: new collection records from Paris, Texas (and other places). Herpetological Review 25:130.

Whiting, R.M., Jr., R.R. Fleet, and V.A. Rakowitz. 1987. Herpetofauna in loblolly-shortleaf pine stands of east Texas. Pp. 31-38 *In* H.A. Pearson, F.E. Smeins, and R.E. Thrill (compilers). Ecological, Physical, and Socioeconomic Relationships within Southern National Forests. USDA Forest Service, General and Technical Report SO-68, New Orleans, Louisiana.

Whitman, J.S. 2009. Diet and prey consumption rates of nesting Boreal Owls, *Aegolius funereus*, in Alaska. Canadian Field-Naturalist 123:112–116.

Whitney, C.L. 1980. The role of the "encounter" call in spacing of Pacific tree frogs, *Hyla regilla*. Canadian Journal of Zoology 58:75–78.

Whitney, C.L. 1981. The monophasic call of *Hyla regilla* (Anura: Hylidae). Copeia 1981:230–233.

Whitney, C.L., and J.R. Krebs. 1975a. Mate selection in Pacific tree frogs. Nature 255:325–326.

Whitney, C.L., and C.J. Krebs. 1975b. Spacing and calling in Pacific tree frogs, *Hyla regilla*. Canadian Journal of Zoology 53:1519–1527.

Wickbom, T. 1950. The chromosomes of *Ascaphus truei* and the evolution of anuran karyotypes. Hereditas (Lund) 36:406–418.

Wicknick, J.A., C.D. Anthony, and J.S. Reblin. 2005. An amphibian survey of Killbuck Marsh Wildlife Area. Ohio Journal of Science 105:2–7.

Wied–Neuwied, M. Prinz zu. 1838. Reise in das innere Nord–Amerika in den Jahren 1832 bis 1834. Vol. 1. J. Hoelscher, Coblenz.

Wiens, J.A. 1970. Effects of early experience on substrate pattern selection in *Rana aurora* tadpoles. Copeia 1970:543–548.

Wiens, J.J., and T.A. Titus. 1991. A phylogenetic analysis of *Spea* (Anura: Pelobatidae). Herpetologica 47:21–28.

Wiese, R.J. 1985. Ecological aspects of the bullfrog in northeastern Colorado. M.S. thesis, Colorado State University, Fort Collins.

Wiese, R.J. 1990. Genetic structure of native and introduced populations of the bullfrog, a successful colonist. Ph.D. Dissertation, Colorado State University, Fort Collins.

Wiest, J.A., Jr. 1982. Anuran succession at temporary ponds

in a post oak–savanna region of Texas. Pp. 39–47 *In* N.J. Scott (ed.), Herpetological Communities. U.S. Fish and Wildlife Service, Wildlife Research Report 13.

Wiewandt, T.A. 1969. Vocalization, aggressive behavior, and territoriality in the bullfrog, *Rana catesbeiana*. Copeia 1969:276–285.

Wiggins, D.A. 1992. Foraging success of leopard frogs (*Rana pipiens*). Journal of Herpetology 26:87–88.

Wiggins, I.L. 1943. Additional note on the range of *Bufo canorus* Camp. Copeia 1943:197.

Wilbur, H.M. 1971. Competition, predation, and the structure of the *Ambystoma–Rana sylvatica* community. Ph.D. Dissertation, University of Michigan, Ann Arbor.

Wilbur, H.M. 1972. Competition, predation, and the structure of the *Ambystoma–Rana sylvatica* community. Ecology 53:3–21.

Wilbur, H.M. 1976. Density–dependent aspects of metamorphosis in *Ambystoma* and *Rana sylvatica*. Ecology 57:1289–1296.

Wilbur, H.M. 1977a. Interactions of food level and population density in *Rana sylvatica*. Ecology 58:206–209.

Wilbur, H.M. 1977b. Density–dependent aspects of growth and metamorphosis in *Bufo americanus*. Ecology 58:196–200.

Wilbur, H.M. 1980. Complex life cycles. Annual Review of Ecology and Systematics 11:67–93.

Wilbur, H.M. 1982. Competition between tadpoles of *Hyla femoralis* and *Hyla gratiosa* in laboratory experiments. Ecology 63:278–282.

Wilbur, H.M. 1984. Complex life cycles and community organization in amphibians. Pp. 114– 131 *In* P.W. Price, C.N. Slobodchikoff, and W.S. Gaud (eds.), A New Ecology: Novel Approaches to Interactive Systems. Wiley, New York.

Wilbur, H.M. 1987. Regulation of structure in complex systems: experimental temporary pond communities. Ecology 68:1437–1452.

Wilbur, H.M. 1988. Interactions between growing predators and growing prey. Pp. 157–172 *In* B. Ebenmann and L. Penson (eds.), Interactions in Size–structured Populations: From Individual Behavior to Ecosystem Dynamics. Springer–Verlag, Berlin, Germany.

Wilbur, H.M. 1990. Coping with chaos: toads in ephemeral ponds. Trends in Ecology and Systematics 5:37–39.

Wilbur, H.M. 1996. Multistage life cycles. Pp. 177–219 *In* O.E. Rhodes, Jr., R.K. Chesser, and M.H. Smith (eds.), Spatial and Temporal Aspects of Population Processes. University of Chicago Press, Chicago, Illinois.

Wilbur, H.M. 1997. Experimental ecology of food webs: complex systems in temporary ponds. Ecology 78:2279–2302.

Wilbur, H.M., and R.A. Alford. 1985. Priority effects in experimental pond communities: responses of *Hyla* to *Bufo* and *Rana*. Ecology 66:1106–1114.

Wilbur, H.M., and J.E. Fauth. 1990. Experimental aquatic food webs: interactions between two predators and two prey. American Naturalist 135:176–204.

Wilbur, H.M., and R.D. Semlitsch. 1990. Ecological consequences of tail injury in *Rana* tadpoles. Copeia 1990:18–24.

Wilbur, H.M., D.I. Rubenstein, and L. Fairchild. 1978. Sexual selection in toads: the roles of female choice and male body size. Evolution 32:264–270.

Wilbur, H.M., P.J. Morin, and R.N. Harris. 1983. Salamander predation and the structure of experimental communities: anuran responses. Ecology 64:1423–1429.

Wilcox, E.V. 1891. Notes on Ohio batrachians. Otterbein Aegis 1(9):133–135.

Wilcox, J.T. 2005. *Rana catesbeiana* (American Bullfrog). Diet. Herpetological Review 36:306.

Wilcox, J.T. 2006. *Rana catesbeiana* (American Bullfrog). Mortality. Herpetological Review 37:447–448.

Wilcox, J.T. 2011. *Rana draytonii* (California Red–legged Frog). Predation. Herpetological Review 42:414–415.

Wilczynski, W., H.H. Zakon, and E.A. Brenowitz. 1984. Acoustic communication in spring peepers: call characteristics and neurophysiological aspects. Journal of Comparative Physiology A 155:577–584.

Wiley, J.E. 1982. Chromosome banding patterns of treefrogs (Hylidae) of the eastern United States. Herpetologica 38:507–520.

Wiley, J.E. 1983. Chromosome polymorphism in *Hyla chrysoscelis*. Copeia 1983:273–275.

Wiley, J.E., and A.L. Braswell. 1986. A triploid male *Rana palustris*. Copeia 1986:531–533.

Wilgers, D.J. 2005. Impacts of land management practices on tallgrass prairie herpetofauna. M.S. thesis, Kansas State University, Manhattan, Kansas.

Wilgers, D.J., and E.A. Horne. 2006. Effects of different burn regimes on tallgrass prarie herpetofaunal species diversity and community composition in the Flint Hills, Kansas. Journal of Herpetology 40:73–84.

Wilgers, D.J., E.A. Horne, B.K. Sandercock, and A.W. Volkmann. 2006. Effects of rangeland management on community dynamics of the herpetofauna of the tallgrass prairie. Herpetologica 62:378–388.

Wilkerson, C.R. 2001. Effect of spatial configuration of isolated temporary wetlands on anuran metapopulations as mediated through predator–prey relations. M.S. thesis, University of Florida, Gainesville.

Wilkinson, J.A. 2006. *Rana aurora draytonii* (California Red–legged Frog). Defensive behavior. Herpetological

Review 37:207–208.

Wilks, B.J., and H.E. Laughlin. 1962. Artificial hybridization between the microhylid genera *Hypopachus* and *Gastrophryne*. Texas Journal of Science 14:183–187.

Willhite, C., and P.V. Cupp, Jr. 1982. Daily rhythms of thermal tolerance in *Rana clamitans* (Anura: Ranidae) tadpoles. Comparative Biochemistry and Physiology 72A:255–257.

Williams, B.K., and R.D. Semlitsch. 2009. Larval responses of three Midwestern anurans to chronic, low–dose exposures of four herbicides. Archives of Environmental Contamination and Toxicology 58:819–827.

Williams, B.K., T.A.G. Rittenhouse, and R.D. Semlitsch. 2008a. Leaf litter input mediates tadpole performance across forest canopy treatments. Oecologia 155:377–384.

Williams, B.K., T.A.G. Rittenhouse, and R.D. Semlitsch. 2008b. Effects of shade and pond substrate on larval fitness traits in three anurans. Oecologia 155:377–384.

Williams, D.D., and S.J. Taft. 1980. Helminths of anurans from NW Wisconsin. Proceedings of the Helminthological Society of Washington 47:278.

Williams, K.L., and P.S. Chrapliwy. 1958. Selected records of amphibians and reptiles from Arizona. Transactions of the Kansas Academy of Science 61:299–301.

Williams, K.L., and K. Mullin. 1987a. Amphibians and reptiles of loblolly-shortleaf pine stands in central Louisiana. Pp. 77-80 *In* H.A. Pearson, F.E. Smeins, and R.E. Thrill (compilers). Ecological, Physical, and Socioeconomic Relationships within Southern National Forests. USDA Forest Service, General and Technical Report SO-68, New Orleans, Louisiana.

Williams, K.L., and K. Mullin. 1987b. Amphibians and reptiles of longleaf-slash pine stands in central Louisiana. Pp. 116-120 *In* H.A. Pearson, F.E. Smeins, and R.E. Thrill (compilers). Ecological, Physical, and Socioeconomic Relationships within Southern National Forests. USDA Forest Service, General and Technical Report SO-68, New Orleans, Louisiana.

Williams, M.Y. 1920. Mink frog at Moose Factory, James Bay. Canadian Field-Naturalist 34:125.

Williams, P.J., J.R. Robb, R.H. Kappler, T.E. Piening, and D.R. Karns. 2012a. Intraspecific density dependence in larval development of the crawfish frog, *Lithobates areolatus*. Herpetological Review 43:36–38.

Williams, P.J., J.R. Robb, and D.R. Karns. 2012b. Habitat selection by crawfish frogs (*Lithobates areolatus*) in a large mixed grassland/forest habitat. Journal of Herpetology 46:682–688.

Williams, P.J., J.R. Robb, and D.R. Karns. 2012c. Occupancy dynamics of breeding crawfish frogs in southeastern Indiana. Journal of Wildlife Management 36:350–357.

Williams, P.K. 1969. Ecology of *Bufo hemiophrys* and *B. americanus* tadpoles in northwestern Minnesota. M.S. thesis, University of Minnesota, Minneapolis.

Williamson, G.K., and R.A. Moulis. 1994. Distribution of amphibians and reptiles in Georgia. Vol. 1, maps; Vol. 2, locality data. Special Publication No. 3, Savannah Science Museum, Savannah, Georgia.

Williamson, M.A., P.W. Hyder, and J.S. Applegarth. 1994. Snakes, Lizards, Turtles, Frogs, Toads & Salamanders of New Mexico. Sunstone Press, Santa Fe, New Mexico.

Willis, Y.L., D.L. Moyle, and T.S. Baskett. 1956. Emergence, breeding, hibernation, movements and transformation of the bullfrog, *Rana catesbeiana*, in Missouri. Copeia 1956:30–41.

Willson, J.D., W.A. Hopkins, C.M. Bergeron, and B.D. Todd. 2012. Making leaps in amphibian ecotoxicology: translating individual–level effects of contaminants to population viability. Ecological Applications 22:1791–1802.

Wilmhoff, C.D., L. Williams, R. MacDonald, S. Fisher, and J. Slothouber. 1999. Geographic distribution: *Pseudacris brachyphona* (Mountain Chorus Frog). Herpetological Review 30:173.

Wilson, A.K. 2000. Amphibian and reptile surveys in the Kaskaskia River drainage of Illinois during 1997 and 1998. Journal of the Iowa Academy of Science 107:203–205.

Wilson, A. 2001. Land management impacts upon the amphibian and reptile fauna of Lake Creek and its riparian areas, Alexander County, Illinois. M.S. thesis, Southern Illinois University, Carbondale.

Wilson, D.J., and H. Lefcort. 1993. The effect of predator diet on the alarm response of red–legged frog, *Rana aurora*, tadpoles. Animal Behaviour 46:1017–1019.

Wilson, G.A., T.L. Fulton, K. Kendell, G. Scrimgeour, C.A. Paszkowski, and D.W. Coltman. 2008a. Genetic diversity and structure in Canadian northern leopard frog (*Rana pipiens*) populations: implications for reintroduction programs. Canadian Journal of Zoology 86:863–874.

Wilson, G.A., T.L. Fulton, K. Kendell, D.M. Schock, C.A. Paszkowski, and D.W. Coltman. 2008b. Genetic evidence for single season polygyny in the northern leopard frog (*Rana pipiens*). Herpetological Review 39:46–50.

Wilson, L. 2010. Diet, critical thermal minimum, and occurrence of *Batrachochytrium dendrobatidis* in Cuban treefrogs (*Osteopilus septentrionalis*). M.S. thesis, Valdosta State University, Valdosta, Georgia.

Wilson, L.D., and L. Porras. 1983. The ecological impact of man on the south Florida herpetofauna. University of Kansas, Museum of Natural History Special Publication 9:1–89.

Wilson, L.W. 1945. Amphibians of Droop Mountain State Park. Proceedings of the West Virginia Academy of Science 16:39–41.

Wilson, V.V. 1975. The systematics and paleoecology of two Pleistocene herpetofaunas of the southeastern United States. Ph.D. Dissertation, Michigan State University, East Lansing.

Windes, K.M. 2010. Treefrog (*Hyla squirella*) responses to rangeland management in semi–tropical Florida, USA. M.S. thesis, University of Central Florida, Orlando.

Windmiller B.S. 1990. The limitations of Massachusetts regulatory protection for temporary pool breeding amphibians. M.S. thesis, Tufts University, Medford, Massachusetts.

Windmiller, B., R.N. Homan, J.V. Regosin, L.A. Willitts, D.L. Wells, and J.M. Reed. 2008. Breeding amphibian population declines following loss of upland forest habitat around vernal pools in Massachusetts, USA. Pp. 41–51 *In* J.C. Mitchell, R.E. Jung Brown, and B. Bartholomew (eds.), Urban Herpetology. Herpetological Conservation 3. Society for the Study of Amphibians and Reptiles, Salt Lake City, Utah.

Winn, B., J.B. Jensen, and S. Johnson. 1999. Geographic distribution. *Eleutherodactylus planirostris*. Herpetological Review 30:49.

Winterbotham, W. 1796. An Historical, Geographical, Commercial, and Philosophical View of the United States of America and of the European Settlements in America and the West Indies. 4 Volumes. John Reid, New York. [herpetology on pp. 402–408 + 3 unpaginated plates in Volume 4]

Wirsing, A.J., J.D. Roth, and D.L. Murray. 2005. Can prey use dietary cues to distinguish predators? A test involving three terrestrial amphibians. Herpetologica 61:104–110.

Wiseman, K.D., and J. Bettaso. 2007. *Rana boylii* (Foothill Yellow–legged Frog). Cannibalism. Herpetological Review 38:193.

Wiseman, K.D., K.R. Marlow, R.E. Jackson, and J.E. Drennan. 2005. *Rana boylii* (Foothill Yellow–legged Frog). Predation. Herpetological Review 36:162–163.

Withers, D.I. 1992. The Wyoming toad (*Bufo hemiophrys baxteri*): an analysis of habitat use and life history. M.S. thesis, University of Wyoming, Laramie.

Witmer, G.W., and J.C. Lewis. 2001. Introduced wildlife of Oregon and Washington. Pp. 423-443 *In* D. Johnson and T. O'Neil (eds.), Wildlife-Habitat Relationships in Oregon and Washington. Oregon State University Press, Corvallis.

Witschi, E. 1929. Studies on sex differentiation and sex determination in amphibians. II. Sex reversal in female tadpoles of *Rana sylvatica* following the application of high temperature. Journal of Experimental Zoology 52:267-291.

Witschi, E. 1953. The Cherokee frog, *Rana sylvatica cherokiana nom. nov.*, of the Appalachian Mountain region. Proceedings of the Iowa Academy of Science 60:764–769.

Witt, W.L. 1993. Annotated checklist of the amphibians and reptiles of Shenandoah National Park, Virginia. Catesbeiana 13:26–35.

Witte, C.L., M.J. Sredl, A.S. Kane, and L.L. Hungerford. 2008. Epidemiologic analysis of factors associated with local disappearances of native ranid frogs in Arizona. Conservation Biology 22:375–383.

Witte, K., K.–C. Chen, W. Wilczynski, and M.J. Ryan. 2000. Influence of amplexus on phonotaxis in the cricket frog *Acris crepitans blanchardi*. Copeia 2000:257–261.

Witte, K., M.J. Ryan, and W. Wilczynski. 2001. Changes in the frequency structure of a mating call decrease its attractiveness to females in the cricket frog *Acris crepitans blanchardi*. Ethology 107: 685–699.

Witte, K., H.E. Farris, M.J. Ryan, and W. Wilczynski. 2005. How cricket frog females deal with a noisy world: habitat–related differences in auditory tuning. Behavioral Ecology 16:571–579.

Witters, L.R., and L. Sievert. 2001. Feeding causes thermophily in the Woodhouse's toad (*Bufo woodhousii*). Journal of Thermal Biology 26:205–208.

Witz, B.W., D.S. Wilson, and M.D. Palmer. 1991. Distribution of *Gopherus polyphemus* and its vertebrate symbionts in three burrow categories. American Midland Naturalist 126:152–158.

Wojtaszek, B.F., B. Staznik, D.T. Chartrand, G.R. Stephenson, and D.G. Thompson. 2004. Effects of Vision™ herbicide on mortality, avoidance response, and growth of amphibian larvae in two forest wetlands. Environmental Toxicology and Chemistry 23:832–842.

Wolf, A.J. 1996. Systematics of the *Rana catesbeiana* species group and the phylogenetic informativeness of anuran vocalizations. Ph.D. Dissertation, University of Michigan, Ann Arbor.

Wolf, K., G.L. Bullock, C.E. Dunbar, and M.C. Quimby. 1968. Tadpole edema virus: a viscerotropic pathogen for anuran amphibians. Journal of Infectious Disease 118:253–262.

Wolf, K., G.L. Bullock, C.E. Dunbar, and M.C. Quimby. 1969. Tadpole edema virus: pathogenesis of virus infected bullfrog tadpoles. Pp. 327–336 *In* M. Mizell (ed.)., Biology of Amphibian Tumors. Springer–Verlag, New York.

Wolff, B.G., S.M. Conway, and C.J. Dabney III. 2012. *Batrachochytrium dendrobatidis* and ranavirus in anurans inhabiting decorative koi ponds near Minneapolis, Minnesota. Herpetological Review 43:427–429.

Wollmuth, L.P., and L.I. Crawshaw. 1988. The effect of development and season on temperature selection in bullfrog tadpoles. Physiological Zoology 61:461–469.

Wollmuth, L.P., L.I. Crawshaw, R.B. Forbes, and D.A. Grahn. 1987. Temperature selection during development in a montane anuran species, *Rana cascadae*. Physiological Zoology 60:472–480.

Woltz, H.W., J.P. Gibbs, and P.K. Ducey. 2008. Road crossing structures for amphibians and reptiles: informing design through behavioral analysis. Biological Conservation 141:2745–2750.

Woo, P.T.K., and J.P. Bogart. 1984. *Trypanosoma* sp. (Protozoa; Kinetoplastidia) in Hylidae (Anura) from eastern North America with notes on their distribution and prevalences. Canadian Journal of Zoology 62:820–824.

Wood, J.T. 1948. *Microhyla c. carolinensis* in an ant nest. Copeia 1948:226.

Wood, K.V., J.D. Nichols, H.F. Percival, and J.H. Hines. 1998. Size–sex variation in survival rates and abundance of pig frogs, *Rana grylio*, in northern Florida wetlands. Journal of Herpetology 32:527–535.

Wood, S., and J. Richardson. 2010. Evidence for ecosystem engineering in a lentic habitat by tadpoles of the Western toad. Aquatic Sciences – Research Across Boundaries 72:499–508.

Wood, T.S. 1977. Food habits of *Bufo canorus*. M.S. thesis, Occidental College, Los Angeles, California.

Wood, W.F. 1935. Encounters with the western spadefoot, *Scaphiopus hammondii*, with a note on a few albino larvae. Copeia 1935:100–102.

Woodbury, A.M. 1952. Amphibians and reptiles of the Great Salt Lake Valley. Herpetologica 8:42–50.

Woodford, J.E., and M.W. Meyer. 2003. Impact of lakeshore development on green frog abundance. Biological Conservation 110:277–284.

Woodhams, D.C., V.T. Vredenburg, M.–A. Simon, D. Billheimer, B. Shakhtour, Y. Shyr, C.J. Briggs, L.A. Rollins–Smith, and R.N. Harris. 2007. Symbiotic bacteria contribute to innate immune defenses of the threatened mountain yellow–legged frog, *Rana muscosa*. Biological Conservation 138:390–398.

Woodhams, D.C., R.A. Alford, C.J. Briggs, M. Johnson, and L.A. Rollins–Smith. 2008. Life–history trade–offs influence disease in changing climates: strategies of an amphibian pathogen. Ecology 89:1627–1639.

Woodhams, D.C., A.D. Hyatt, D.G. Boyle, and L.A. Rollins–Smith. 2008. The northern leopard frog *Rana pipiens* in a widespread reservoir species harboring *Batrachochytrium dendrobatidis* in North America. Herpetological Review 39:66–68.

Woodhouse, S.W. 1854. Report on the natural history of the country passed over by the exploring expedition under the command of Brev. Capt. L. Sitgreaves, United States Topographical Engineers, during the year 1851. Pp. 33–40 *In* Report of an Expedition down the Zuni and Colorado Rivers, by Captain L. Sitgreaves. Beverley Tucker, Senate Printer, Washington, D.C.

Woodliffe, P.A. 1989. Inventory, assessment, and ranking of natural areas of Walpole Island. Proceedings of the Eleventh North American Prairie Conference, pp. 47–52.

Woodward, B.D. 1982a. Sexual selection and nonrandom mating patterns in desert anurans (*Bufo woodhousei*, *Scaphiopus couchi*, *S. multiplicatus* and *S. bombifrons*). Copeia 1982:351–355.

Woodward, B.D. 1982b. Tadpole competition in a desert anuran community. Oecologia 54:96–100.

Woodward, B.D. 1982c. Tadpole interactions in the Chihuahuan Desert at two experimental densities. Southwestern Naturalist 27:119–121.

Woodward, B.D. 1982d. Male persistence and mating success in Woodhouse's toad (*Bufo woodhousei*). Ecology 63:583–585.

Woodward, B.D. 1983a. Tadpole size and predation in the Chihuahuan Desert. Southwestern Naturalist 28:470–471.

Woodward, B.D. 1983b. Predator–prey interactions and breeding–pond use of temporary–pond species in a desert anuran community. Ecology 64:1549–1555.

Woodward, B.D. 1984a. Operational sex ratios and sex biased mortality in *Scaphiopus* (Pelobatidae). Southwestern Naturalist 29:232–233.

Woodward, B.D. 1984b. Arrival to and location of *Bufo woodhousei* in the breeding pond: effect on the operational sex ratio. Oecologia 62:240–244.

Woodward, B.D. 1986. Paternal effects on juvenile growth in *Scaphiopus multiplicatus* (the New Mexico spadefoot toad). American Naturalist 128:58–65.

Woodward, B.D. 1987a. Clutch parameters and pond use in some Chihuahuan Desert anurans. Southwestern Naturalist 32:13–19.

Woodward, B.D. 1987b. Interactions between Woodhouse's toad tadpoles (*Bufo woodhousei*) of mixed sizes. Copeia 1987:380–386.

Woodward, B.D. 1987c. Intra– and interspecific variation in spadefoot toad (*Scaphiopus*) clutch parameters. Southwestern Naturalist 32:127–131.

Woodward, B.D., and P. Johnson. 1985. *Ambystoma tigrinum* (Ambystomatidae) predation on *Scaphiopus couchi* (Pelobatidae) tadpoles of different sizes. Southwestern Naturalist 39:460–461.

Woodward, B.D., and S. Mitchell. 1990. Predation on frogs in breeding choruses. Southwestern Naturalist 35:449–450.

Woodward, B.D., and J. Travis. 1991. Paternal effects on juvenile growth and survival in spring peepers (*Hyla crucifer*). Evolutionary Ecology 5:40–51.

Woodward, B.D., and S. Mitchell. 1992. Temperature variation and species interactions in aquatic systems in Grand Teton National Park. Pp. 140–145 *In* University of Wyoming/National Park Service Research Center, 16th Annual Report.

Woodward, B. D., J. Travis, and S. Mitchell. 1988. The effects of the mating system on progeny performance in *Hyla crucifer* (Anura: Hylidae). Evolution 42:784–794.

Woolbright, L.L., E.J. Greene, and G.C. Rapp. 1990. Density–dependent mate searching strategies of male wood frogs. Animal Behaviour 40:135–142.

Woolbright, L.L., A.H. Hara, C.M. Jacobsen, W.J. Mautz, and F.L. Benevides, Jr. 2006. Population densities of the coquí, *Eleutherodactylus coqui* (Anura: Leptodactyidae) in newly invaded Hawaii and in native Puerto Rico. Journal of Herpetology 40:122–126.

Wright, A.A., and A.H. Wright. 1933. Handbook of Frogs and Toads. 1st ed. Comstock Publishing Associates, Ithaca, New York.

Wright, A.A., and A.H. Wright. 1942. Handbook of Frogs and Toads. 2nd ed. Comstock Publishing Associates, Ithaca, New York.

Wright, A.H. 1914. Life–histories of the Anura of Ithaca, New York. Carnegie Institution of Washington, Publication 197.

Wright, A.H. 1915. The mink frog, *Rana septentrionalis* Baird, in Ontario. Copeia (23):46–48.

Wright, A.H. 1920. Frogs: Their Natural History and Utilization. Department of Commerce, Bureau of Fisheries, No. 88.

Wright, A.H. 1923. The salientia of the Okefinokee Swamp, Georgia. Copeia (115):34.

Wright, A.H. 1924. A new bullfrog (*Rana heckscheri*) from Georgia and Florida. Proceedings of the Biological Society of Washington 37:141–152.

Wright, A.H. 1926. The vertebrate life of Okefenokee Swamp in relation to the Atlantic Coastal Plain. Ecology 7:77–95.

Wright, A.H. 1929. Synopsis and description of North American tadpoles. Proceedings of the United States National Museum 74:1–70.

Wright, A.H. 1932. Life–Histories of the Frogs of Okefinokee Swamp, Georgia. Macmillan Co., New York. [reprinted 2002 by Cornell University Press, Ithaca, New York]

Wright, A. H., and A.A. Allen. 1908. Notes on the breeding habits of the swamp cricket frog, *Chorophilus triseriatus* Wied. American Naturalist 42:39–42.

Wright, A.H., and J. Moesel. 1919. The toads and frogs of Monroe and Wayne counties, N.Y. Copeia (74):81–83.

Wright, A.H., and A.A. Wright.1938. Amphibians of Texas. Transactions of the Texas Academy of Science 21:5–44.

Wright, A.H., and A.A. Wright.1949. Handbook of Frogs and Toads. 3rd ed. Comstock Publishing Associates, Ithaca, New York. [reprinted 1995 by Comstock Publishing Associates, Ithaca, New York]

Wright, H.P., and G.S. Myers. 1927. *Rana areolata* at Bloomington, Indiana. Copeia (159):173–175.

Wright, J.W. 1966. Predation on the Colorado River toad, *Bufo alvarius*. Herpetologica 22:127–128.

Wright, M.F., and S.I. Guttman. 1995. Lack of an association between heterozygosity and growth rate in the wood frog, *Rana sylvatica*. Canadian Journal of Zoology 73:569–575.

Wyatt, J.L., and E.A. Forys. 2004. Conservation implications of predation by Cuban treefrogs (*Osteopilus septentrionalis*) on native hylids in Florida. Florida Scientist 3:695–700.

Wygoda, M.L. 1984. Low cutaneous evaporative water loss in arboreal frogs. Physiological Zoology 57:329–337.

Wygoda, M.L. 1989. Body temperature in arboreal and nonarboreal anuran amphibians: differential effects of a small change in water vapor density. Journal of Thermal Biology 14:239–242.

Wygoda, M.L., and A.A. Williams. 1991. Body temperature in free–ranging green tree frogs (*Hyla cinerea*): a comparison with "typical" frogs. Herpetologica 47:328–335.

Wylie, G.D., M.L.Casazza, and M. Carpenter. 2003. Diet of bullfrogs in relation to predation on giant garter snakes at Colusa National Wildlife Refuge. California Fish and Game 89:139–145.

Wylie, K., M.–L. Gramby, and M. Kostalos. 2009. Inhibition of metamorphosis in *Bufo americanus* tadpoles in Nine Mile Run Restoration Area, Pittsburg [sic], Pennsylvania USA. Froglog 91:10–12.

Wylie, S.R. 1981. Effects of basking on the biology of the canyon treefrog, *Hyla arenicolor* Cope. Ph.D. Dissertation, Arizona State University, Tempe.

Wyman, R.L. 1988. Soil acidity and moisture and the distribution of amphibians in five forests of southcentral New York. Copeia 1988:394–399.

Yahner, R.H., W.C. Bramble, and W.R. Byrnes. 2001. Response of amphibian and reptile populations to vegetation maintenance of an electric transmission line right–of–way. Journal of Arboriculture 27:215–221.

Yamaguti, S. 1961. Systema helminthum. Volume 3, Part 1. Nematodes of Vertebrates. Interscience Publishing, Inc., New York.

Yang, X.B., and D.O. TeBeest. 1992. Green treefrogs as vectors of *Colletotrichum gloeosporioides*. Plant Diseases

76:1266–1269.

Yang, X.B., D.O. TeBeest, and E.L. Moore. 1992. Biological vectors for dispersal of *Colletotrichtum gloeosporioides*. Proceedings of the Arkansas Academy of Science 46:96–99.

Yarrow, H.C. 1882. Check-list of North American Reptilia and Batrachia, with catalogue of specimens in U.S. National Museum. Proceedings of the United States National Museum 24:3-249.

Yeager, D. 1926. Miscellaneous notes. Yellowstone Nature Notes 3(4):7.

Yeary, C.M. 1979. Support for a mutualistic relationship between the western narrow–mouthed frog, *Gastrophryne olivacea*, and the tarantula, *Dugesiella hentzii*. M.S. thesis, University of Tulsa, Tulsa, Oklahoma.

Yoder, H.R. 1998. Community ecology of helminth parasites infecting molluscan and amphibian hosts. Ph.D. Dissertation, University of Wisconsin–Milwaukee.

Yoder, H.R., and J.R. Coggins. 1996. Helminth communities in the northern spring peeper, *Pseudacris c. crucifer* Wied, and the wood frog, *Rana sylvatica* LeConte, from southeastern Wisconsin. Journal of the Helminthological Society of Washington 63:211–214.

Yoder, H.R., and J.R. Coggins. 2007. Helminth communities in five species of sympatric amphibians from three adjacent ephemeral ponds in southeastern Wisconsin. Journal of Parasitology 93:755–760.

Yoder, H.R., J.R. Coggins, and J. C. Reinbold. 2001. Helminth parasites of the green frog (*Rana clamitans*) from southeastern Wisconsin, U.S.A. Comparative Parasitology 68: 269–272.

Young, C.M., and C.G. Blome. 1975. Summer feeding habits of kestrels in northern Ontario. Ontario Field Biologist 29:44–49.

Young, F.N., and C.C. Goff. 1939. An annotated list of the arthropods found in the burrows of the Florida gopher tortoise, *Gopherus polyphemus* (Daudin). Florida Entomologist 22:53–62.

Young, J.E., and B.I. Crother. 2001. Allozyme evidence for the separation of *Rana areolata* and *Rana capito* and for the resurrection of *Rana sevosa*. Copeia 2001:382–388.

Young, M.K., and D.A. Schmetterling. 2009. Age–related seasonal variation in captures of stream–borne boreal toads (*Bufo boreas boreas*, Bufonidae) in western Montana. Copeia 2009:117–124.

Young, M.K., G.T. Allison, and K. Foster. 2007. Observations of boreal toads (*Bufo boreas boreas*) and *Batrachochytrium dendrobatidis* in south–central Wyoming and north–central Colorado. Herpetological Review 38:146–150.

Young, M.T. 2011. The Guide to Colorado Reptiles and Amphibians. Fulcrum Publishing, Golden, Colorado.

Youngstrom, K.A., and H.M. Smith. 1936. Description of the larvae of *Pseudacris triseriata* and *Bufo woodhousii woodhousii* (Anura). American Midland Naturalist 17:629–632.

Zacharow, M., W.J. Barichivich, and C.K. Dodd, Jr. 2003. Using ground–placed PVC pipes to monitor hylid treefrogs: capture biases. Southeastern Naturalist 2:575–590.

Zaga, A., E.E. Little, C.F. Raben, and M.R. Ellersieck. 1998. Photoenhanced toxicity of a carbamate insecticide to early life stage anuran amphibians. Environmental Toxicology and Chemistry 17:2543–2553.

Zamparo, D., and D.R. Brooks. 2005. Three rarely reported digeneans inhabiting amphibians from Vancouver Island, Bristish Columbia, Canada. Journal of Parasitology 91:1242–1244.

Zampella, R.A., and J.F. Bunnell. 2000. The distribution of anurans in two river systems of a coastal plain watershed. Journal of Herpetology 34:210–221.

Zeiber, R.A., T.M. Sutton, and B.E. Fisher. 2008. Western mosquitofish predation on native amphibian eggs and larvae. Journal of Freshwater Ecology 23:663-671.

Zell, G.A. 1986. The clawed frog: an exotic from South Africa invades Virginia. Virginia Wildlife 47(2):28–29.

Zellmer, A.J. 2010. Ecological and evolutionary consequences of population connectivity in an amphibian with local adaptation. Ph.D. Dissertation, University of Michigan, Ann Arbor.

Zellmer, A.J., C.L. Richards, and L.M. Martens. 2008. Low prevalence of *Batrachochytrium dendrobatidis* across *Rana sylvatica* populations in southeastern Michigan, USA. Herpetological Review 39:196–199.

Zelmer, D.A., E.J. Wetzel, and G.W. Esch. 1999. The role of habitat in structuring *Halipegus occidualis* metapopulations in the green frog. Journal of Parasitology 85:19–24.

Zenisek, C.J. 1963. A study of the natural history and ecology of the leopard frog, *Rana pipiens* Schreber. Ph.D. Dissertation, Ohio State University, Columbus.

Zettergren, L.D., B.W. Boldt, D.H. Petering, M.S. Goodrich, D.N. Weber, and J.G. Zettergren. 1991. Effects of prolonged low–level cadmium exposure on the tadpole immune system. Toxicology Letters 55:11–19.

Zetts, A.J. 1985. Frog culture. Zetts Fish Farm and Hatcheries, Drifting, Pennsylvania. 13 pp.

Zeyl, C. 1993. Allozyme variation and divergence among populations of *Rana sylvatica*. Journal of Herpetology 27:233–236.

Zielinski, W.J., and G.T. Barthalmus. 1989. African clawed frog skin compounds: antipredatory effects on African and North American water snakes. Animal Behaviour 38:1083–1086.

Ziesmer, T.C. 1997. Vocal behavior of the foothill and mountain yellow–legged frogs (*Rana boylii* and *Rana muscosa*). M.S. thesis, California State University, Sonoma, California.

Zippel, K.C. 2005. *Hyla chrysoscelis* (Gray Treefrog). Deprivation tolerance. Herpetological Review 36:301–302.

Zippel, K.C., and C. Tabaka. 2008. Amphibian chytridiomycosis in captive *Acris crepitans blanchardi* (Blanchard's cricket frog) collected from Ohio, Missouri and Michigan, USA. Herpetological Review 39: 192–193.

Zippel, K.C., A.T. Snider, L. Gaines, and D. Blanchard. 2005. *Eleutherodactylus planirostris* (Greenhouse Frog). Cold tolerance. Herpetological Review 36:299–300.

Zug, G.R. 1985. Anuran locomotion: fatigue and jumping performance. Herpetologica 41:188–194.

Zug, G.R., and P.B. Zug. 1979. The marine toad, *Bufo marinus*: a natural history resumé of native populations. Smithsonian Contributions to Zoology (284):1–58.

Zulich, M.O. 1966. A population study of four species of frogs. M.S. thesis, Ball State University, Muncie, Indiana.

Zweifel, R.G. 1955. Ecology, distribution, and systematics of frogs of the *Rana boylei* group. University of California Publications in Zoology 54:207–292.

Zweifel, R.G. 1956a. Two pelobatid frogs from the Tertiary of North America and their relationships to fossil and recent forms. American Museum Novitates 1762:1–45.

Zweifel, R.G. 1956b. A survey of the frogs of the *augusti* group, genus *Eleutherodactylus*. American Museum Novitates 1813:1–35.

Zweifel, R.G. 1961. Larval development of the tree frogs *Hyla arenicolor* and *Hyla wrightorum*. American Museum Novitates 2056:1–19.

Zweifel, R.G. 1968a. Effects of temperature, body size and hybridization on mating calls of toads *Bufo a. americanus* and *Bufo woodhousei fowleri*. Copeia 1968:269–285.

Zweifel, R.G. 1968b. Reproductive biology of anurans of the arid Southwest, with adaptation of embryos to temperature. Bulletin of the American Museum of Natural History 140:1–64.

Zweifel, R.G. 1970a. Distribution and mating call of the treefrog, *Hyla chrysoscelis*, at the northeastern edge of its range. Chesapeake Science 11:94–97.

Zweifel, R.G. 1970b. Descriptive notes on larvae of toads of the *debilis* group, genus *Bufo*. American Museum Novitates 2407:1–13.

Zweifel, R.G. 1989. Calling by the frog, *Rana sylvatica*, outside the breeding season. Journal of Herpetology 23:185–186.

Catalogue of American Amphibians and Reptiles

Altig, Ronald and Philip C. Dumas. 1971. *Rana cascadae*. Catalogue of American Amphibians and Reptiles. (105):1-2.

Altig, Ronald and Philip C. Dumas. 1972. *Rana aurora*. Catalogue of American Amphibians and Reptiles. (160):1-4.

Altig, Ronald and Ren Lohoefener. 1982. *Rana grylio*. Catalogue of American Amphibians and Reptiles. (286):1-2.

Altig, Ronald and Ren Lohoefener. 1983. *Rana areolata*. Catalogue of American Amphibians and Reptiles. (324):1-4.

Ashton, Ray E., Jr. and Richard Franz. 1979. *Bufo quercicus*. Catalogue of American Amphibians and Reptiles. (222):1-2.

Blem, Charles R. 1979. *Bufo terrestris*. Catalogue of American Amphibians and Reptiles. (223):1-4.

Brown, Lauren E. 1973. *Bufo houstonensis*. Catalogue of American Amphibians and Reptiles. (133):1-2.

Brown, Lauren E. 1992. *Rana blairi*. Catalogue of American Amphibians and Reptiles. (536):1-6.

Caldwell, Janalee P. 1982. *Hyla gratiosa*. Catalogue of American Amphibians and Reptiles. (298):1-2.

Duellman, William E. 1968. *Smilisca*. Catalogue of American Amphibians and Reptiles. (58):1-2.

Duellman, William E. 1968. *Smilisca baudinii*. Catalogue of American Amphibians and Reptiles. (59):1-2.

Duellman, William E. and Ronald I. Crombie. 1970. *Hyla septentrionalis*. Catalogue of American Amphibians and Reptiles. (92):1-4.

Easteal, Simon. 1986. *Bufo marinus*. Catalogue of American Amphibians and Reptiles. (395):1-4.

Fouquette, M. J., Jr. 1969. Rhinophrynidae, *Rhinophrynus, R. dorsalis*. Catalogue of American Amphibians and Reptiles. (78):1-2.

Fouquette, M. J., Jr. 1970. *Bufo alvarius*. Catalogue of American Amphibians and Reptiles. (93):1-4.

Franz, Richard and Charles J. Chantell. 1978. *Limnaoedus, L.*

ocularis. Catalogue of American Amphibians and Reptiles. (209):1-2.

Gates, William R. 1988. *Pseudacris nigrita*. Catalogue of American Amphibians and Reptiles. (416):1-3.

Gaudin, Anthony J. 1979. *Hyla cadaverina*. Catalogue of American Amphibians and Reptiles. (225):1-2.

Glorioso, B. M. 2010. *Pseudacris ornata*. Catalogue of American Amphibians and Reptiles. (866):1-8.

Gosner, K. L. and I. H. Black. 1967. *Hyla andersonii*. Catalogue of American Amphibians and Reptiles. (54):1-2.

Gosner, Kenneth L. and Irving H. Black. 1968. *Rana virgatipes*. Catalogue of American Amphibians and Reptiles. (67):1-2.

Hall, John A. 1998. *Scaphiopus intermontanus*. Catalogue of American Amphibians and Reptiles. (650):1-17.

Hedeen, Stanley E. 1977. *Rana septentrionalis*. Catalogue of American Amphibians and Reptiles. (202):1-2.

Heyer, W. Ronald. 1971. *Leptodactylus labialis*. Catalogue of American Amphibians and Reptiles. (104):1-3.

Heyer, M. M., W. R. Heyer and R. de Sá. 2006. *Leptodactylus fragilis*. Catalogue of American Amphibians and Reptiles. (830):1-26.

Hoffman, Richard L. 1980. *Pseudacris brachyphona*. Catalogue of American Amphibians and Reptiles. (234):1-2.

Hoffman, Richard L. 1983. *Pseudacris brimleyi*. Catalogue of American Amphibians and Reptiles. (311):1-2.

Hoffman, Richard L. 1988. *Hyla femoralis*. Catalogue of American Amphibians and Reptiles. (436):1-3.

Hulse, Arthur C. 1978. *Bufo retiformis*. Catalogue of American Amphibians and Reptiles. (207):1-2.

Jennings, Mark R. 1988. *Rana onca*. Catalogue of American Amphibians and Reptiles. (417):1-2.

Karlstrom, E. L. 1973. *Bufo canorus*. Catalogue of American Amphibians and Reptiles. (132):1-2.

Korky, John K. 1999. *Bufo punctatus*. Catalogue of American Amphibians and Reptiles. (689):1-5.

Krupa, James J. 1990. *Bufo cognatus*. Catalogue of American Amphibians and Reptiles. (457):1-8.

Martof, Bernard S. 1970. *Rana sylvatica*. Catalogue of American Amphibians and Reptiles. (86):1-4.

Martof, Bernard S. 1975. *Hyla squirella*. Catalogue of American Amphibians and Reptiles. (168):1-2.

Metter, Dean E. 1968. *Acaphus, A. truei*. Catalogue of American Amphibians and Reptiles. (69):1-2.

Moler, P. E. 1993. *Rana okaloosae*. Catalogue of American Amphibians and Reptiles. (561):1-3.

Nelson, Craig E. 1972. *Gastrophryne carolinensis*. Catalogue of American Amphibians and Reptiles. (120):1-4.

Nelson, Craig E. 1972. *Gastrophryne olivacea*. Catalogue of American Amphibians and Reptiles. (122):1-4.

Nelson, Craig E. 1973. *Gastrophryne*. Catalogue of American Amphibians and Reptiles. (134):1-2.

Pierce, Benjamin A. and Patricia H. Whitehurst. 1990. *Pseudacris clarkii*. Catalogue of American Amphibians and Reptiles. (458):1-3.

Platz, James E. and John S. Mecham. 1984. *Rana chiricahuensis*. Catalogue of American Amphibians and Reptiles. (347):1-2.

Platz, James E. 1988. *Rana yavapaiensis*. Catalogue of American Amphibians and Reptiles. (418):1-2.

Platz, James E. 1991. *Rana berlandieri*. Catalogue of American Amphibians and Reptiles. (508):1-4.

Porter, Kenneth R. 1970. *Bufo valliceps*. Catalogue of American Amphibians and Reptiles. (94):1-4.

Price, Andrew H. and Brian K. Sullivan. 1988. *Bufo microscaphus*. Catalogue of American Amphibians and Reptiles. (415):1-3.

Redmer, M. and R.A Brandon. 2003. *Hyla cinerea*. Catalogue of American Amphibians and Reptiles. 766:1-14.

Sanders, Albert E. 1984. *Rana heckscheri*. Catalogue of American Amphibians and Reptiles. (348):1-2.

Schaaf, Raymond T., Jr and Philip W. Smith. 1971. *Rana palustris*. Catalogue of American Amphibians and Reptiles. (117):1-3.

Schwartz, Albert. 1974. *Eleutherodactylus planirostris*. Catalogue of American Amphibians and Reptiles. (154):1-4.

Smith, Philip W. 1966. *Hyla avivoca*. Catalogue of American Amphibians and Reptiles. (28):1-2.

Smith, Philip W. 1966. *Pseudacris streckeri*. Catalogue of American Amphibians and Reptiles. (27):1-2.

Stewart, Margaret M. 1983. *Rana clamitans*. Catalogue of American Amphibians and Reptiles. (337):1-4.

Trueb, Linda. 1969. *Pternohyla, P. dentata, P. fodiens*. Catalogue of American Amphibians and Reptiles. (77):1-4.

Turner, Frederick B. and Philip C. Dumas. 1972. *Rana pretiosa*. Catalogue of American Amphibians and Reptiles. (119):1-4.

Wasserman, Aaron O. 1968. *Scaphiopus holbrookii*. Catalogue of American Amphibians and Reptiles. (85):1-4.

Wasserman, Aaron O. 1970. *Scaphiopus couchii*. Catalogue of American Amphibians and Reptiles. (85):1-4.

Zweifel, Richard G. 1967. *Eleutherodactylus augusti*. Catalogue of American Amphibians and Reptiles. (41):1-4.

Zweifel, Richard. 1968. *Rana boylii*. Catalogue of American Amphibians and Reptiles. (71):1-2.

Zweifel, Richard. 1968. *Rana muscosa*. Catalogue of American Amphibians and Reptiles. (65):1-2.

Zweifel, Richard. 1968. *Rana tarahumarae*. Catalogue of American Amphibians and Reptiles. (66):1-2.

Additional anuran—inclusive bibliographies

Banta, B.H. 1965. An annotated chronological bibliography of the herpetology of the state of Nevada. The Wasmann Journal of Biology 23:1–224.

Campbell, R.W., M.G. Shepard, B.M. Van Der Raay, and P.T. Gregory. 1982. A bibliography of Pacific Northwest herpetology. British Columbia Provincial Museum, Heritage Record No. 14.

Carpenter, C.C., and J.J. Krupa. 1989. Oklahoma Herpetology, An Annotated Bibliography. University of Oklahoma Press, Norman.

Collins, J.T., and J. Caldwell. 1977. A bibliography of the amphibians and reptiles of Kansas (1854–1976). Reports of the Biological Survey of Kansas No. 12, 56 pp.

Collins, J.T., and R.E. Kurtz. 1977. A bibliography of the amphibians and reptiles of Kentucky (1820–1976). Meseraull Printing, Lawrence, Kansas.

Corn, P.S., R.B. Bury, and H.H. Welsh. 1984. Selected bibliography of Wyoming amphibians and reptiles. Smithsonian Herpetological Information Service No. 59.

Dixon, J.R. 2013. Amphibians and Reptiles of Texas, with Keys, Taxonomic Synopses, Bibliography, and Distribution Maps. 3rd ed. Texas A & M University Press, College Station. [first edition, 1987; second edition, 2000]

Dlutkowski, L.A., P.A. Cochran, and M.J. Mossman. 1987. Bibliography of Wisconsin herpetology. Wisconsin Endangered Species Resources Report 28.

Enge, K.M. 2002. An updated, indexed bibliography of the herpetofauna of Florida. Florida Fish and Wildlife Conservation Commission, Technical Report No. 19. [earlier edition with C.K. Dodd, Jr. as coauthor dated 1992] [ongoing online version at: http://legacy.myfwc. com/herpbibl/]

Moriarty, J.J., and D.G. Jones. 1988. An annotated bibliography of Minnesota herpetology 1900–1985. J.F. Bell Museum of Natural History, Minneapolis.

Morris, M.A., R.S. Funk, and P.W. Smith. 1983. An annotated bibliography of the Illinois herpetological literature 1960–1980, and an updated checklist of species of the state. Illinois Natural History Survey Bulletin 33:123–137.

Redmond, W.H., A.C. Echternacht, and A.F. Scott. 1990. Annotated checklist and bibliography of amphibians and reptiles of Tennessee (1835 through 1989). Miscellaneous Publications of the Center for Field biology, Austin Peay State University, Clarksville, Tennessee.

Ross, D.A. 1989. Amphibians and reptiles in the diets of North American raptors. Wisconsin Endangered Resources Report 59.

Ross, D.A. 1991. Amphibians and reptiles in the diets of North American raptors. Wisconsin Endangered Resources Report 59. [updated version]

Stewart, M.M., and L.F. Biuso. 1982. A bibliography of the Green Frog, *Rana clamitans* Latreille 1801–1981. Smithsonian Herpetological Information Service No. 56.

Watermolen, D.J. 1992. Wisconsin herpetology: a bibliographical update with taxonomic, subject, and geographic indices. Wisconsin Endangered Resources Report 87.

Part 2: 2013–2021

Abercrombie, S. A., C. de Perre, Y.J. Choi, B.J. Tornabene, B. J., M.S. Sepúlveda, L.S. Lee, and J.T. Hoverman. 2019. Larval amphibians rapidly bioaccumulate poly- and perfluoroalkyl substances. Ecotoxicology and Environmental Safety 178:137-145.

Abercrombie, S. A., M. Iacchetta, C. De Perre, R.W. Flynn, M.S. Sepúlveda, L.S. Lee, and J.T. Hoverman. 2021. Sublethal effects of dermal exposure to poly- and perfluoroalkyl substances on postmetamorphic amphibians. Environmental Toxicology and Chemistry 40:717-726.

Adams, A.J., C. Dellith, and S.S. Sweet. 2016. *Rana aurora* (Northern Red-legged Frog). Transport. Herpetological Review 47:648.

Adams, A.J., A.P. Pessier, and C.J. Briggs. 2017a. Rapid extirpation of a North American frog coincides with an increase in fungal pathogen prevalence: historical analysis and implications for reintroduction. Ecology and Evolution 7:10216-10232.

Adams, A.J., S.J. Kupferberg, M.Q. Wilbur, A.P. Pressier, M. Grefsrud, S. Bobzien, V.T. Vredenburg, and C.J. Briggs. 2017b. Extreme drought, host density, sex, and bullfrogs influence fungal pathogen infection in a declining lotic amphibian. Ecosphere 8:e01740.

Adams, C. 2015. Rediscovering the Mexican white-lipped frog (*Leptodactylus* [*labialis*] *fragilis*) in south Texas (Anura: Leptodactylidae). SWCHR Bulletin 5(2):17-18.

Adams, M.J., D.A.W. Miller, E. Muths, P.S. Corn, E.H.C. Grant, L.L. Bailey, G.M. Fellers, R.N. Fisher, W.J. Sadinski, H. Waddle, and S.C. Walls. 2013. Trends in amphibian occupancy in the United States. PLoS One 8:e64347.

Albecker, M.A., A. M. M. Stuckert, C. N. Balakrishnan, and M. W. McCoy. 2021. Molecular mechanisms of local adaptation for salt-tolerance in a treefrog. Molecular Ecology 30:2065-2086.

Alix, D.M., C. Guyer, and C.J. Anderson. 2014a. Expansion of the range of the introduced greenhouse frog, *Eleutherodactylus planirostris*, in coastal Alabama. Southeastern Naturalist 13:N59-N62.

Alix, D.M., C.J. Anderson, J.B. Grand, and C. Guyer. 2014b. Evaluating the effects of land use on headwater wetland amphibian assemblages in coastal Alabama. Wetlands 34:917-926.

Altig, R., and R.W. McDiarmid. 2015. Handbook of Larval Amphibians of the United States and Canada. Cornell University Press, Ithaca, New York.

Alvarez, G., J.D. Johnson, and V. Mata-Silva. 2018. *Anaxyrus punctatus* (Red-spotted Toad). Diet. Herpetological Review 49:93.

Alvarez, J.A. 2013. *Rana draytonii* (California Red-legged Frog). Cannibalism. Herpetological Review 44:126-127.

Alvarez, J.A., and J.T. Wilcox. 2021a. Observations of atypical habitat use by foothill yellow-legged frogs (*Rana boylii*) in the Coast Range of California. Western North American Naturalist 81:293-299.

Alvarez, J.A., and J.T. Wilcox. 2021b. *Anaxyrus boreas* (Western Toad). Opportunistic scavenging. Herpetological Review 52:821.

Alvarez, J.A., and J.T. Wilcox. 2021c. Observations of nocturnal upland habitat use by *Rana draytonii* (California Red-Legged Frog), and implications for restoration and other activities. Ecological Restoration 39:155-157.

Alvarez, J.A., D.G. Cook, J.L. Yee, M.G. van Hattem, D.R. Fong, and R.N. Fisher. 2013. Comparative microhabitat characteristics at oviposition sites of the California red-legged frog (*Rana draytonii*). Herpetological Conservation and Biology 8:539-551.

Alvarez, J.A., E. Lopez, T. Collins, A. Herman, J. Jelincic, M.L. Olsen, B. Piontek, A.I. Ringstad, and J. T. Wilcox. 2021a. Misdirected amplexus between a Pacific treefrog (*Pseudacris regilla*) and a western toad (*Anaxyrus boreas*) in a northern California upland. Northwestern Naturalist 102:161-163.

Alvarez, J.A., K.M. Garten, J.T. Wilcox, Z.A. Cava, A. Engstrom, J. Litteral, M.L. Olson, M. Peterson, H. Sardiñas, and M. Stake. 2021b. *Rana draytonii* (California Red-legged Frog). Diet. Herpetological Review 52:618-619.

Alvarez, J.A., K. M Garten, and D.G Cook. 2021c. Limb malformation in a foothill yellow-legged frog (*Rana boylii*) from Sonoma County, California. Northwestern Naturalist 102:258-260.

Alvarez, J.A., M.A. Shea, and J.T. Wilcox. 2021d. Upwardly mobile: vertical movement of California red-legged frogs (*Rana draytonii*) and its management implications. Herpetology Notes 14:717-723.

Amburgey, S.M., L.L. Bailey, M. Murphy, E. Muths, and W.C. Funk. 2014. The effects of hydropattern and predator communities on amphibian occupancy. Canadian Journal of Zoology 92:927-937.

Anderson, R.B. 2017. *Rana draytonii* (California Red-legged Frog). Life history. Herpetological Review 48:169.

Anderson, R.B., J.P. Rose, and S.P. Lawler. 2019. Evolutionary experience with the invasive *Lithobates catesbeianus* predicts lower survival of larval *Rana draytonii*. Herpetological Conservation and Biology 14:349-359.

Andis, A.Z. 2018. Geographic distribution: *Anaxyrus boreas* (= *Bufo boreas*) (Western Toad). Herpetological Review 49:706.

Andrews, J.S. 2019. The Vermont Reptile and Amphibian Atlas, 2019 Update. Privately Published, Middlebury, Vermont.

Andrews, J.S., and E. Talmage. 2021. Phenological differences in wood frog and spotted salamander egg-mass onset and peak accumulation. Northeastern Naturalist 28: 456-461.

Annich, N.C., E.M. Bayne, and C.A. Paszkowski. 2019. Identifying Canadian toad (*Anaxyrus hemiophrys*) habitat in northeastern Alberta, Canada. Herpetological Conservation and Biology 14:503-514.

Araos, H.L., K.L. Kroft, R.M. Bogardus, Y-M. Chang, K.R. Donohue, D. Hanley, K.A. Hatch, and K.W. Wilson. 2017. The Columbia spotted frog (*Rana luteiventris*) — another species persisting with *Batrachochytrium dendrobatidis* infection. Herpetological Review 48:782-786.

Arietta, A.Z.A., L.K. Friedenburg, M.C. Urban, S.B. Rdrigues, A. Rubinstein, and D.K. Skelly. 2020a. Phenological delay despite warming in wood frog *Rana sylvatica* reproductive timing: a 20-year study. Ecography 43:1-11.

Arietta, A.Z.A., A. Rubinstein, L.K. Friedenburg, and P.N.K. Johnson. 2020b. Multiple cases of hypomelanism in wood frog larvae (*Rana sylvatica*) associated with developmental retardation and mortality. Northeastern Naturalist 27:641-648.

Arregui, L., A.J. Kouba, J.M. Germano, L. Barrios, M. Moore, and C.K. Kouba. 2021. Fertilization potential of cold-stored Fowler's toad (*Anaxyrus fowleri*) spermatozoa: temporal changes in sperm motility based on temperature and osmolality. Reproduction, Fertility and Development 34(5). doi:10.1071/RD21037.

Ashpole, S.L., C.A. Bishop, and J.E. Elliott. 2014. Clutch size in the Great Basin spadefoot (*Spea intermontana*), South Okanogan Valley, British Columbia, Canada. Northwestern Naturalist 95:35-40.

Ashpole, S.L., C.A. Bishop, and S.D. Murphy. 2018. Reconnecting amphibian habitat through small pond construction and enhancement, South Okanagan River Valley, British Columbia, Canada. Diversity 10:108.

Atkinson, M., E. Hoffman, E. Karwacki, and A. E. Savage. 2021. Invasive *Xenopus tropicalis* alter disease dynamics in Florida amphibian communities. Abstract Booklet, Southeast Partners in Amphibian and Reptile Conservation Annual Meeting, 25–27 February 2021, p. 14.

Backlin, A.R., C.J. Hitchcock, E.A. Gallegos, J.L. Yee, and R.N. Fisher. 2013. The precarious persistence of the endangered Sierra Madre yellow-legged frog *Rana muscosa* in southern California, USA. Oryx 49:157-164.

Badje, A.F., and A.I. Dyke. 2018. Geographic distribution: *Anaxyrus fowleri* (Fowler's Toad). Herpetological Review 49:706-707.

Badje, A.F., T.J. Brandt, T.L. Bergeson, R.A. Paloski, J.M. Kapfer, and G.W. Schuurman. 2016. Blanchard's cricket frog *Acris blanchardi* overwintering ecology in southwestern Wisconsin. Herpetological Conservation and Biology 11:101-111.

Badje, A.F., T.J. Brandt, T.L. Bergeson, R.A. Paloski, and J.M. Kapfer. 2021. Spring movement ecology of Blanchard's cricket frog (*Acris blanchardi*) in southwestern Wisconsin, USA. Herpetological Conservation and Biology 16:271–280.

Baecher, J.A., P.N. Vogrinc, J.C. Guzy, J.C. Neal, and J.D. Willson. 2014. *Lithobates areolatus* (Crawfish Frog). Predation. Herpetological Review 45:681-682.

Baecher, J.A., P.N. Vogrine, J.C. Guzy, C.S. Kross, and J.D. Willson. 2018. Herpetofaunal communities in restored and unrestored remnant tallgrass prairie and associated wetlands in northwest Arkansas, USA. Wetlands 38:157-168.

Bailey, L.L., P. Jones, K.G. Thompson, H.P. Foutz, J.M. Logan, F.B. Wright, and H.J. Crockett. 2019. Determining presence of rare amphibian species: testing and combining novel survey methods. Journal of Herpetology 53:115-124.

Baker, E.A. 2021. Evaluating a predator-induced phenotype in a mixed species context. M.S. thesis, East Carolina University, Greenville, North Carolina.

Banker, S.E., A.R. Lemmon, A.B. Hassinger, M. Dye, S.D.

Holland, M.L. Kortyna, O.E. Ospina, H. Ralicki, and E.M. Lemmon. 2020. Hierarchical hybrid enrichment: multi-tiered genomic data collection across evolutionary scales, with application to chorus frogs (*Pseudacris*). Systematic Biology 69:756-773.

Barragán-Ramírez, J.L., J. de Jesús Ascencio-Arrayga, F. Rodriguez-Ramírez, and J.L. Navarrete-Heredia. 2014. *Spea multiplicata* (Mexican Spadefoot). Foraging site selection. Herpetological Review 45:115-116.

Barrow, L.N., H.F. Ralicki, S.A. Emme, and E.M. Lemmon. 2014. Species tree estimation of North American chorus frogs (Hylidae: *Pseudacris*) with parallel tagged amplicon sequencing. Molecular Phylogenetics and Evolution 75:78-90.

Barrett, K., N.P. Nibbelink, and J.C. Maerz. 2014. Identifying priority species and conservation opportunities under future climate scenarios: amphibians in a biodiversity hotspot. Journal of Fish and Wildlife Management 5:282-297.

Barrett, K., C. Guyer, S.T. Samoray, and Y. Kanno. 2016. Stream and riparian habitat use by anurans along a forested gradient in western Georgia, USA. Copeia 104:570-576.

Barrett, K., J.A. Crawford, Z. Reinstein, and J.R. Milanovich. 2017. Detritus quality produces species-specific tadpole growth and survivorship responses in experimental wetlands. Journal of Herpetology 51:227-231.

Barry, S.J., and G.M. Fellers. 2013. History and status of the California red-legged frog (*Rana draytonii*) in the Sierra Nevada, California, USA. Herpetological Conservation and Biology 8:456-502.

Bartelt, P.E., and R.W. Klaver. 2017. Response of anurans to wetland restoration on a Midwestern agricultural landscape. Journal of Herpetology 51:504-514.

Bartlett, R.D., and P.P. Bartlett. 2013. New Mexico's Reptiles & Amphibians. A Field Guide. University of New Mexico Press, Albuquerque.

Barton, B.T., M.A. Lashley, and R. Altig. 2018. *Ascaphus montanus* (Rocky Mountain Tailed Frog). Breeding. Herpetological Review 49:299.

Bateman, P.W., and P.A. Fleming. 2014. Living on the edge: effects of body size, group density and microhabitat selection on escape behaviour of southern leopard frogs *Lithobates sphenocephalus*. Current Zoology 60:712-718.

Battaglin, W.A., K.L. Smalling, C. Anderson, D. Calhoun, T. Chestnut, and E. Muths. 2016. Potential interactions among disease, pesticides, water quality and adjacent land cover in amphibian habitats in the United States. Science of the Total Environment 566-567:320-332.

Baumberger, K.L., M.V. Eitzel, M.E. Kirby, and H.H. Horn. 2019. Movement and habitat selection of the western spadefoot (*Spea hammondii*) in southern California.

PLoS One 14(10):e0222532.

Baumgardt, J. A., M. L. Morrison, L. A. Brennan, M. Thornley, and T.A. Campbell. 2021. Variation in herpetofauna detection probabilities: implications for study design. Environmental Monitoring and Assessment. 193:658.

Beard, K.H., S.A. Johnson, and A.B. Shiels. 2018. Frogs (coqui frogs, greenhouse frogs, Cuban tree frogs, and cane toads). Pp. 163-192 *In* W.C. Pitt, J.C. Beasley, and G.W. Witmer (eds.), Ecology and Management of Terrestrial Vertebrate Invasive Species in the United States. CRC Press, Boca Ration, Florida.

Beaudry, P., and G. Höbel. 2014. *Hyla cinerea* (Green Treefrog). Dorsal spots. Herpetological Review 45:111-112.

Bedwell, M.E. K.V.S. Hopkins, C. Dillingham, and C.S. Goldberg. 2021. Evaluating Sierra Nevada yellow-legged frog distribution using environmental DNA. Journal of Wildlife Management 85:945-952.

Benard, M.F. 2015. Warmer winters reduce frog fecundity and shift breeding phenology, which consequently alters larval development and metamorphic timing. Global Change Biology 21:1058-1065.

Bennet, A.R., R. Rivera, R.A. Saumure, T. O'Toole, J.R. Jaeger, and P.R. Bean. 2020. *Rana onca* (Relict Leopard Frog). Diet and Mortality. Herpetological Review 51:302-303.

Bennett, A.M., and D.L. Murray. 2015. Carryover effects of phenotypic plasticity: embryonic environment and larval response to predation risk in wood frogs (*Lithobates sylvaticus*) and northern leopard frogs (*Lithobates pipiens*). Canadian Journal of Zoology 93:867-877.

Bevier, C.R., and A.M.G. Gelder. 2018. The effects of parasite infection on phonotactic response in the mink frog, *Lithobates septentrionalis*. Journal of Herpetology 52:34-39.

Bezy, R.L, and C.J. Cole. 2014. Amphibians and reptiles of the Madrean Archipelago of Arizona and New Mexico. American Museum Novitates No. 3810, 23 pp.

Bieri, E. 2021. *Lithobates septentrionalis* (Mink Frog) and *Pseudacris crucifer* (Spring Peeper). Interspecific amplexus. Herpetological Review 52:832-833.

Billerman, S.M., B.R. Jesmer, A.G. Watts, P.E. Schlichting, M.-J. Fortin, W.C. Funk, P. Hapeman, E. Muths, and M.A. Murphy. 2019. Testing theoretical metapopulation conditions with genotypic data from boreal chorus frogs (*Pseudacris maculata*). Canadian Journal of Zoology 97:1042-1053.

Billet, L.S., and J.T. Hoverman. 2020. Pesticide tolerance induced by a generalized stress response in wood frogs (*Rana sylvatica*). Ecotoxicology 29:1476-1485.

Billet, L.S., V.P. Wuerthner, J. Hua, R.A. Relyea, and J.T. Hoverman. 2020. Timing and order of exposure to two echinostome species affect patterns of infection in larval amphibians. Parasitology 147:1515-1523.

Billet, L.S., V.P. Wuerthner, J.Hua, R.A. Relyea, and J.T. Hoverman. 2021. Population-level variation in infection outcomes not influenced by pesticide exposure in larval wood frogs (*Rana sylvatica*). Freshwater Biology 66: 1169-1181.

Bingham, D.M., A.J. Sepulveda, and S. Painter. 2021. A small proportion of breeders drive American bullfrog invasion of the Yellowstone River floodplain, Montana. Northwest Science 94:231-242.

Birkhead, R.D., J.P. McGuire, R. Conley, and C.K. Ward. 2017. Geographic distribution: *Incilius nebulifer* (Gulf Coast Toad). Herpetological Review 48:120.

Bishir, S., and A. King. 2018. *Lithobates sylvaticus* (Wood Frog). Predation. Herpetological Review 49:517.

Bishop, M.R., R.C. Drewes, and V.T. Vredenburg. 2014. Food web linkages demonstrate importance of terrestrial prey for the threatened California red-legged frog. Journal of Herpetology 48:137-143.

Blackburn, D.C., L. Roberts, M.C. Vallejo-Pareja, and E.L. Stanley. 2019. First record of the anuran family Rhinophrynidae from the Oligocene of eastern North America. Journal of Herpetology 53:316-323.

Blais, B.R. 2016. *Anaxyrus americanus* (American Toad). Ectoparasites. Herpetological Review 47:435-436.

Blais, B.R. 2017. *Lithobates pipiens* (Northern Leopard Frog). Freeze intolerance. Herpetological Review 48:165.

Blais, B.R. 2019. *Lithobates catesbeianus* (American Bullfrog). Unusual malformation. Herpetological Review 50:339.

Blais, B., C. Bubac, and B. Smith. 2015. *Pseudacris maculata* (Boreal Chorus Frog). Calling phenology. Herpetological Review 46:416-417.

Bleich, V.C. 2020. Locality records for Woodhouse's toad: have wet washes in a dry desert led to extralimital occurrences of an adaptable anuran? California Fish and Wildlife 106:258-266.

Bleich, V.C. 2021. An endemic anuran and a horny toad: distributional histories, the potential for sympatry, and implications for conservation. California Fish and Wildlife 107:8-20.

Boeing, W.J., K.L. Griffis-Kyle, and J.M. Jungels. 2014. Anuran habitat associations in the northern Chiricahua Desert, USA. Journal of Herpetology 48:103-110.

Boes, M.W., and M.F. Benard. 2013. Carry-over effects in nature: effects of canopy cover and individual pond on size, shape, and locomotor performance of metamorphosing wood frogs. Copeia 2013:717-722.

Bogan, M.T., and D.E. Eppehimer. 2017. Attempted predation of western desert tarantula by Sonoran desert toad. Southwestern Naturalist 62:146-148.

Bohannon, J.S., D.R. Gay, M.P. Hayes, C.D. Danilson, and K.I. Warheit. 2016. Discovery of the Oregon spotted-frog in the northern Puget Sound Basin, Wasington State. Northwestern Naturalist 97:82-97.

Bomske, C.M., and N. Bickford. 2019. Overwintering anuran niche preferences in a series of interconnected ponds in northwestern Florida. Southeastern Naturalist 18:256-269.

Bondi, C.A., S.M. Yarnell, and A.J. Lind. 2013. Transferability of habitat suitability criteria for a stream breeding frog (*Rana boylii*) in the Sierra Nevada, California. Herpetological Conservation and Biology 8:88-103.

Boundy, J., and J.L. Carr. 2017. Amphibians & Reptiles of Louisiana. An Identification and Reference Guide. Louisiana State University Press, Baton Rouge.

Bourque, R.M., J.R. Bourque, and S.A. Bourque. 2018. *Anaxyrus boreas* (Western Toad). Habitat. Herpetological Review 49:514.

Bowerman, J. 2020. *Rana pretiosa* (Oregon Spotted Frog). Freeze mortality. Herpetological Review 51:102-103.

Bowerman, J., and C.A. Pearl. 2020. Oregon spotted frog (*Rana pretiosa*) migration from an aquatic overwintering site: timing, duration, and potential environmental cues. American Midland Naturalist 184:87-97.

Bowerman, J., C. Rombough, and J. Wilmoth. 2019. *Lithobates catesbeianus* (American Bullfrog). Diet. Herpetological Review 50:546-547.

Bradley, P.W., S.S. Gervasi, J. Hua, R.D. Cothran, R.A. Relyea, D.H. Olson, and A.R. Blaustein. 2015. Differences in sensitivity to the fungal pathogen *Batrachochytrium dendrobatidis* among amphibian populations. Conservation Biology 29:1347-1356.

Brady, S.P. 2013. Microgeographic maladaptive performance and deme depression in response to roads and runoff. PeerJ 1:e163.

Brady, S.P., and D. Goedert. 2017. Positive sire effects and adaptive genotype by environment interaction occur despite pattern of local maladaptation in roadside populations of an amphibian. Copeia 105:533-542.

Brand, M.D., R.D. Hill, R. Brenes, J.C. Chaney, R.P. Wilkes, L. Grayfer, D.L. Miller, and M.J. Gray. 2016. Water temperature affects susceptibility to ranavirus. EcoHealth 13:350-359.

Brannelly, L.A., H.I. McCallum, L.F. Grogan, C.J. Briggs, M.P. Ribas, M. Hollanders, T. Sasso, M.F. López, D.A. Newell, and A.M. Kilpatrick. 2021. Mechanisms underlying host persistence following amphibian disease emergence determine appropriate management strategies. Ecology Letters 24: 130-148.

Brattstrom, B.H. 2019. *Anaxyrus microscaphus* (Arizona Toad) and *Lithobates yavapaiensis* (Lowland Leopard Frog). Predation. Herpetological Review 50:757-758.

Braunagel, T.M. 2021. Applying ecological theory to amphibian populations to determine if wood frogs (*Lithobates sylvaticus*) are ideal and free when selecting breeding habitat. M.S. thesis, University of Massachusetts, Amherst.

Brenes, R., M.J. Gray, T.B. Waltzek, R.P. Wilkes, and D.L. Miller. 2014. Transmission of ranavirus between ectothermic vertebrate hosts. PLoS One 9:e92476.

Brennan, T.C., and R.D. Babb. 2020. *Heterodon kenneryli*. Mexican Hog-nosed Snake. pp. 159-167 *In* A.R. Holycross and J.C. Mitchell (eds.), Snakes of Arizona. ECO Publishing, Rodeo, New Mexico.

Briggler, J.T., and T.R. Johnson. 2021. The Amphibians and Reptiles of Missouri. Third edition. Missouri Department of Conservation, Jefferson City.

Brinley Buckley, E.M., B.L. Gottesman, A.J. Caven, M.J. Harner, and B.C. Pijanowski. 2021. Assessing ecological and environmental influences on boreal chorus frog (*Pseudacris maculata*) spring calling phenology using multimodal passive monitoring technologies. Ecological Indicators 121:107171.

Brown, C., L.R. Wilkinson, and K.B. Kiehl. 2014a. Comparing the status of two sympatric amphibians in the Sierra Nevada, California: insights on ecological risk and monitoring common species. Journal of Herpetology 48:74-83.

Brown, C., M.P. Hayes, G.A. Green, and D.C. Macfarlane. 2014b. Mountain yellow-legged frog conservation assessment for the Sierra Nevada Mountains of California, USA. USDA Forest Service, Pacific Southwest Region, RS-TP-038.

Brown, C., L.R. Wilkinson, K.K. Wilkinson, T. Tunstall, R. Foote, B.D. Todd, and V.T. Vredenburg. 2019. Demography, habitat, and movements of the Sierra Nevada yellow-legged frog (*Rana sierrae*) in streams. Copeia 107:661-675.

Brown, C., A. J. Nowakowski, N. C. Keung, S. P. Lawler, and B. D. Todd. 2021. Untangling multi-scale habitat relationships of an endangered frog in streams to inform reintroduction programs. Ecosphere 12(10):e03799.

Brown, C.T., J.M. Yahn, and W.H. Karasov. 2021. Warmer temperature increases toxicokinetic elimination of PCBs and PBDEs in northern leopard frog larvae (*Lithobates pipiens*). Aquatic Toxicology 234:105806.

Brown, D.J., M.C. Jones, J. Bell, and M.R.J. Forstner. 2012 (2015). Feral hog damage to endangered Houston toad (*Bufo houstonensis*) habitat in the Lost Pines of Texas. Texas Journal of Science 64:73-88.

Brown, D.J., T.M. Swannack, and M.R.J. Forstner. 2013a. Predictive models for calling and movement activity of the endangered Houston toad. American Midland Naturalist 169:303-321.

Brown, D.J., D.B. Preston, E. Ozel, and M.R.J. Forstner. 2013b. Wildfire impacts on red imported fire ant captures around forest ponds in the Lost Pines Ecoregion of Texas. Journal of Fish and Wildlife Management 4:129-133.

Brown, D.J., A. Duarte, I. Mali, M.C. Jones, and M.R.J. Forstner. 2014. Potential impacts of a high severity wildfire on abundance, movement, and diversity of herpetofauna in the Lost Pines Ecoregion of Texas. Herpetological Conservation and Biology 9:192-205.

Brown, D.J., T.M. Swannack, and M.R.J. Forstner. 2015. Using calling activity to predict calling activity: a case study with the endangered Houston toad (*Bufo* [*Anaxyrus*] *houstonensis*). Journal of North American Herpetology 2015:12-16.

Brown, J., and J. Kerby. 2013. *Batrachochytrium dendrobatidis* in South Dakota, USA amphibians. Herpetological Review 44:457-458.

Brown, J.R., T. Miiller, and J.L Kirby. 2013. The interactive effect of an emerging infectious disease and an emerging contaminant on Woodhouse's toad (*Anaxyrus woodhousii*) tadpoles. Environmental Toxicology and Chemistry 32:2003-2008.

Brown, M.E., and S.C. Walls. 2013. Variation in salinity tolerance among larval anurans: implications for community composition and the spread of an invasive, non-native species. Copeia 2013:543-551.

Brown, S.R., R.W. Flynn, and J.T. Hoverman. 2021. Perfluoroalkyl substances increase susceptibility of northern leopard frog tadpoles to tematode infection. Environmental Toxicology and Chemistry 40:689-694.

Brown, T.A., M.E. Fraker, and S.A. Ludsin. 2019. Space use of predatory larval dragonflies and tadpole prey in response to chemical cues of predation. American Midland Naturalist 181:53-62.

Browne, C.L., and C.A. Paszkowski. 2014. The influence of habitat composition, season and gender on habitat selection by western toads (*Anaxyrus boreas*). Herpetological Conservation and Biology 9:417-427.

Browne, C.L., and C.A. Paszkowski. 2018. Microhabitat selection by western toads (*Anaxyrus boreas*). Herpetological Conservation and Biology 13:317-330.

Brunet, J., B.A. Graham, J. Klemish, T.M. Burg, D. Johnson, and S. Wiseman. 2020. *Batrachochytrium dendrobatidis* from amphibian and eDNA samples in the Peace Region of British Columbia. Herpetological Review 51:732-737.

Brunner, J.L., L. Beaty, A. Guitard, and D. Russell. 2017. Heterogeneities in the infection process drive ranavirus transmission. Ecology 98:576-582.

Bryson, R.W., Jr., B.T. Smith, A. Nieto-Montes de Oca, U.O. García-Vázquez, and B.R. Riddle. 2014. The role of mitochondrial introgression in illuminating the evolutionary history of Nearctic treefrogs. Zoological Journal of the Linnean Society 172:103-116.

Bubac, C., B. Blais, and B. Smith. 2017. *Lithobates pipiens* (Northern Leopard Frog). Breeding behavior. Herpetological Review 48:165.

Bucciarelli, G.M., A.R. Blaustein, T.S. Garcia, and L.B. Kats. 2014. Invasion complexities: the diverse impacts of non-native species. Copeia 2014:611-632.

Buckley, E.M. Brinley, B.L. Gottesman, A.J. Caven, M.J. Harner, and B.C. Pijanowski. 2021. Assessing ecological and environmental influences on boreal chorus frog (*Pseudacris maculata*) spring calling phenology using multimodal passive monitoring technologies. Ecological Indicators 121: 107171.

Bunt, J., J.J. Webster, B. Jacobson, and F. Vilella. 2021. Predation of a brown bat (Vespertilionidae) by a green frog (*Lithobates clamitans*) in Ontario, Canada. Canadian Field-Naturalist 135:58-60.

Burrow, A.K. 2021. Plants matter: How human-driven changes to terrestrial and wetland vegetation may impact priority amphibian species in Southeastern pine savannas. Ph.D. Dissertation, University of Georgia, Athens.

Buxton, V.L., M.P. Ward, and J.H. Sperry. 2015. Use of chorus sounds for location of breeding habitat in 2 species of anuran amphibians. Behavioral Ecology 26:1111-1118.

Buxton, V.L., M.P. Ward, and J.H. Sperry. 2018. Evaluation of conspecific attraction as a management tool across several species of anurans. Diversity 10:6.

Cairns, N.A., A.S. Cicchino, K.A. Stewart, J.D. Austin, and S.C. Lougheed. 2021. Cytonuclear discordance, reticulations and cryptic diversity in one of North America's most common frogs. Molecular Phylogenetics and Evolution 156:107042.

Calatayud, N.E., T.T. Hammond, N.R. Gardner, M.J. Curtis, R.R. Swaisgood, and D.M. Shier. 2021. Benefits of overwintering in the conservation breeding and translocation of a critically endangered amphibian. Conservation Science and Practice 3:e341.

Calderon, V.S. 2021. Effects of climate change on the immune defenses and disease susceptibility of amphibian hosts. Ph.D. Dissertation, University of Pittsburgh, Pittsburgh, Pennsylvania.

Camp, C.D., and J.B. Jensen. 2021. Long-term observations of salamander abundance in twilight zones of caves in Georgia, USA. Herpetological Conservation and Biology 16:63-71.

Capps, K.A., K.A. Berven, and S.D. Tiegs. 2015. Modelling nutrient transport and transformation by pool-breeding amphibians in forested landscapes using a 21-year dataset. Freshwater Biology 60:500-511.

Carfagno, G.L.F., and P.P. Fong. 2014. Growth inhibition of tadpoles exposed to sertraline in the presence of conspecifics. Journal of Herpetology 48:571-576.

Carlson, B.E. 2014a. *Lithobates sylvaticus* (Wood Frog). Egg predation. Herpetological Review 45:479.

Carlson, B.E. 2014b. *Lithobates catesbeianus* (American Bullfrog). Habitat. Herpetological Review 45:479.

Carlson, B.E. 2016. *Lithobates sylvaticus* (Wood Frog). Melanistic tadpole. Herpetological Review 47:439.

Carlson, B.E., and T. Langkilde. 2013. Personality traits are expressed in bullfrog tadpoles during open-field trials. Journal of Herpetology 47:378-383.

Carlson, Z.A., and K. Geluso. 2018. Second sighting of Cope's gray treefrog (*Hyla chrysoscelis*) in Buffalo County, Nebraska. Collinsorum 7:18.

Carr, L.W., and L. Fahrig. 2021. Effect of road traffic on two amphibian species of differing vagility. Conservation Biology 15:1071-1078.

Carter, J., D. Johnson, and S. Merino. 2018. Exotic invasive *Pomacea maculata* (Giant Apple Snail) will depredate eggs of frog and toad species of the Southeastern US. Southeastern Naturalist 17:470-475.

Carter, J., D. Johnson, J. Boundy, and W. Vermillion. 2021. The Louisiana Amphibian Monitoring Program from 1997 to 2017: results, analyses, and lessons learned. PLoS One 16(9):e0257869.

Carter, S.K., D. Saenz, and V.H.W. Rudolf. 2018. Shifts in phenological distributions reshape interaction potential in natural communities. Ecology Letters 21:1143-1151.

Caseltine, J.S., S.L. Rumschlag, and M.D. Boone. 2016. Terrestrial growth in northern leopard frogs reared in the presence or absence of predators and exposed to the amphibian chytrid fungus at metamorphosis. Journal of Herpetology 50:404-408.

Casper, G.S., R.D. Rutherford, and T.G. Anton. 2015. Baseline distribution records for amphibians and reptiles in the Upper Peninsula of Michigan. Herpetological Review 46:391-406.

Casper, G.S., C.E. Smith, S.M. Nadeau, and A. Lewanski. 2017. Geographic distribution: *Acris blanchardi* (Blanchard's Cricket Frog). Herpetological Review 48:805.

Castaneda, E., V.R. Leavings, R.F. Noss, and M.K. Grace. 2020. The effects of traffic noise on tadpole behavior and development. Urban Ecosystems 23:245-253.

Catenazzi, A., and S.J. Kupferberg. 2013. The importance of thermal conditions to recruitment success in stream-breeding frog populations distributed across a productivity gradient. Biological Conservation 168:40-48.

Caudill, D., G. Caudill, B. Carpenter, K.M. Enge, and R.L. Engleman. 2014. *Scaphiopus holbrookii* (Eastern Spadefoot). Predation. Herpetological Review 45:481.

Cayuela, H., A. Valenzuela-Sánchez, L. Teulier, Í. Martínez-Solano, J.-P. Léna, J. Merilä, E. Muths, R. Shine, L. Quay, M. Denoël, J. Clobert, and B.R. Schmidt. 2020. Determinants and consequences of dispersal in vertebrates with complex life cycles: A review of pond-breeding amphibians. Quarterly Review of Biology 95:1-36.

Cayuela, H., Y. Dorant, B.R. Forester, D.L. Jeffries, R.M. Mccaffery, L.A. Eby, B.R. Hossack, J.M.W. Gippet, D.S. Pilliod, and W.C. Funk. 2021a. Genomic signatures of thermal adaptation are associated with clinal shifts of life history in a broadly distributed frog. Journal of Animal Ecology https://doi.org/10.1111/1365-2656.13545.

Cayuela, H., J.-F. Lemaître, E. Muths, R.M. McCaffery, T. Frétey, B. Le Garff, B.R. Schmidt, K. Grossenbacher, O. Lenzi, B.R. Hossack, L.A. Eby, B.A. Lambert, J. Elmberg, J. Merilä, J.M.W. Gippet, J.-M. Gaillard, and D.S. Pilliod. 2021b. Thermal conditions predict intraspecific variation in senescence rate in frogs and toads. Proceedings of the National Academy of Sciences of the United States of America 118: e2112235118.

Cha, E.S. M.T. Uhrin, S.J. McClelland, and S.K. Woodley. 2021. Brain plasticity in response to short-term exposure to corticosterone in larval amphibians. Canadian Journal of Zoology 99:839-844.

Chan, L.M., W. Eng, B. Hergert, A. Osbrink-McInroy, and C.N. Templeton. 2020. *Lithobates catesbeianus* (American Bullfrog). Diet. Herpetological Review 51:301.

Chandler, H.C., C.A. Haas, and T.A. Gorman. 2015a. The effects of habitat structure on winter aquatic invertebrate and amphibian communities in pine flatwoods wetlands. Wetlands 35:1201-1211.

Chandler, R.B., E. Muths, B.H. Sigafus, C.R. Schwalbe, C.J. Jarchow, and B.R. Hossack. 2015b. Spatial occupancy models for predicting metapopulation dynamics and viability following reintroduction. Journal of Applied Ecology 52:1325-1333.

Chatfield, M.W.H., L.A. Brannelly, M.J. Robak, L. Freeborn, S.P. Lailvaux, and C.L. Richards-Zawacki. 2013. Fitness consequences of infection by *Batrachochytrium dendrobatidis* in northern leopard frogs (*Lithobates pipiens*). EcoHealth 10:90-98.

Chelgren, N.D., and M.J. Adams. 2017. Inference of timber harvest effects on survival of stream amphibians is complicated by movement. Copeia 105:712-725.

Chen, Y., R.B.G. Clemente-Carvalho, and S.C. Lougheed. 2021. Seasonal shift in the age structure of calling males within a spring peeper (*Pseudacris crucifer*) chorus. Herpetological Review 52:499-506.

Chiari, Y., N. Moreno, J. Elmore, A. Hylton, A. Ray, R. Burkhardt, and S. Glaberman. 2017. Widespread occurrence of *Batrachochytrium dendrobatidis* in southern Alabama, USA. Herpetological Review 48:356-359.

Chinchar, V.G., A.L.J. Duffus, and J.L. Brunner. 2021. The impact of Ranavirus infections on amphibians. Pp. 405-447 *In* C.J. Hurst (ed.), Ecology of Viruses Infecting Ectothermic Vertebrates, Studies in Viral Ecology, Second Edition. John Wiley & Sons, New York.

Choquette, J.D., and E.A. Jolin. 2018. Checklist and status of the amphibians and reptiles of Essex County, Ontario: a 35 year update. Canadian Field-Naturalist 132:176-190.

Chuirazzi, C., M. Ocampo, and M.K. Takahashi. 2021. Influence of prey diet quality on predator-induced traits in wood frog tadpoles (*Lithobates sylvaticus*). Amphibia-Reptilia 42:331-341.

Claunch, N.M., R. Denton, M. Holding, E. Taylor, and S.J. Mullin. 2018. *Scaphiopus couchii* (Couch's Spadefoot Toad). Predation. Herpetological Review 49:312-313.

Cloyed, C.S., and P.K. Eason. 2017. Feeding limitations in temperate anurans and the niche vatiation hypothesis. Amphibia–Repilia 38:473-482.

Cole, E.M., and M.P. North. 2014. Environmental influences on amphibian assemblages across subalpine wet meadows in the Klamath Mountains, California. Herpetologica 70:135-148.

Cole, E.M., R. Hartman, and M.P. North. 2016. Hydroperiod and cattle use associated with lower recruitment in an r-selected amphibian with declining population trend in the Klamath Mountains, California. Journal of Herpetology 50:37-43.

Combs, J., M. Grisnik, K. Baldauf, M. Sigg, T. Swanson, R. Amos, C.M. White, D.M. Walker, T.D. Schwaner, A.D. Brown, T.R. Blincoe, and J.A. Wooten. 2015. A report of ranavirus infecting midland painted turtles with novel localities for frog infection in northwest Ohio. Herpetological Review 46:361-363.

Conlon, J.M., L.K. Reinert, M. Mechkarska, M. Prajeep, M.A. Meetani, L. Coquet, T. Jouenne, M.P. Hayes, G. Pagett-Flohr, and L.A. Rollins-Smith. 2013. Evaluation of the skin peptide defenses of the Oregon spotted frog against infection by the chytrid fungus *Batrachochytrium dendrobatidis*. Journal of Chemical Ecology 39:797-805.

Connior, M.B., C.T. McAllister, and C.R. Bursey. 2015a. *Scaphiopus hurterii* (Hurter's Spadefoot). Endoparasites. Herpetological Review 46:419-420.

Connior, M.B., T. Fulmer, C.T. McAllister, and S.E. Trauth. 2015b. *Scaphiopus hurterii* (Hurter's Spadefoot). Reproduction. Herpetological Review 46:420.

Connior, M.B., L.A. Durden, and C.T. McAllister. 2016. *Anaxyrus fowleri* (Fowler's Toad). Ectoparasites. Herpetological Review 47:104.

Cook, D.G., and A.F. Currylow. 2014. Seasonal spatial patterns of two sympatric frogs: California red-legged frog and American bullfrog. Western Wildlife 1:1-7.

Cook, M.T., S.S. Heppell, and T.S. Garcia. 2013. Invasive bullfrog larvae lack developmental plasticity to changing hydroperiod. Journal of Wildlife Management 77:655-662.

Corkran, C.C., and C. Thoms. 2020. Amphibians of Oregon, Washington and British Columbia. 3rd ed. Lone Pine Publishing, Edmonton, Alberta.

Coster, S.S., J.S. Veysey Powell, and K.J. Babbitt. 2014. Characterizing the width of amphibian movements during postbreeding migration. Conservation Biology 28:756-762.

Coster, S.S., K.J. Babbitt, and A.I. Kovach. 2015. High genetic connectivity in wood frogs (*Lithobates sylvaticus*) and spotted salamanders (*Ambystoma maculatum*) in a commercial forest. Herpetological Conservation and Biology 10:64-89.

Cothran, R.D., J.M. Brown, and R.A. Relyea. 2013. Proximity to agriculture is correlated with pesticide tolerance: evidence for the evolution of amphibian resistance to modern pesticides. Evolutionary Applications 6:832-841.

Coughlan, M.P., T.R. Waters, and J.C. Touchon. 2021. Salinity increases growth and pathogenicity of water mold to cause mortality and early hatching in *Rana sylvatica* embryos." FEMS Microbiology Ecology 97: fiaa257.

Coury, K.S., B.A. Neal, M.J. Shin, and E.D. Lindquist. 2016. Presence and prevalence of BD (*Batrachochytrium dendrobatidis*) in central Pennsylvanian woodland vernal pools. Journal of North American Herpetology 2016:11-14.

Crawford, J.A., C.A. Phillips, W.E. Peterman, I.E. MacAllister, N.A. Wesslund, A.R. Kuhns, and M.J. Dreslik. 2017. Chytrid infection dynamics in cricket frogs on military and public lands in the Midwestern United States. Journal of Fish and Wildlife Management 8:344-352.

Crockett, J.G., L.L. Bailey, and E. Muths. 2020. Highly variable rates of survival to metamorphosis in wild boreal toads (*Anaxyrus boreas boreas*). Population Ecology 62:258-268.

Crockett, J.G., W.E. Lanier, and L.R. Bailey. 2021. Few impacts of introduced cutthroat trout (*Oncorhynchus clarki*) on aquatic stages of boreal toads (*Anaxyrus boreas boreas*). Journal of Herpetology 55:310-317.

Crosby, J. 2014. Amphibian occurrence on south Okanagan roadways (2010-2012): investigating movement patterns, crossing hotspots and roadkill mitigation structure use at the landscape scales. M.E.S., University of Waterloo, Waterloo, Ontario.

Crother, B.I. (Committee Chair). 2017. Scientific and Standard English Names of Amphibians and Reptiles on North America North of Mexico, with Comments Regarding Confidence in Our Understanding. 8th ed. Society for the Study of Amphibians and Reptiles, Herpetological Circular No. 43.

Crump, M.L. 2015. Eye of Newt and Toe of Frog, Adder's Fork and Lizard's Leg. The Lore and Mythology of Amphibians and Reptiles. University of Chicago Press, Chicago.

Crump, P., K. Berven, T.E. Youker-Smith, D. Skelly, S. Thomas, and J. Houlahan. 2017. Predicting anuran abundance using an automated acoustics approach. Journal of Herpetology 51:582-589.

Cunningham, H.R., and N.H. Nazdrowicz (eds.). 2018. The Maryland Amphibian and Reptile Atlas. Johns Hopkins University Press, Baltimore.

Cunnington, G.M., and L. Fahrig. 2013. Mate attraction by male anurans in the presence of traffic noise. Animal Conservation 16:275-285.

Cunnington, G.M., E. Garrah, E. Eberhardt, and L. Fahrig. 2014. Culverts alone do not reduce road mortality in anurans. Ecoscience 21:69-78.

Cupp, P.V., Jr. 2017. *Anaxyrus americanus* (American Toad). Microhabitat. Herpetological Review 48:829.

Dananay, K.L., K.L. Krynak, T.J. Krynak, and M.F. Benard. 2015. Legacy of road salt: apparent positive larval effects counteracted by negative postmetamorphic effects in wood frogs. Environmental Toxicology and Chemistry 34:2417-2424.

D'Aoust-Messier, A.-M., and D. Lesbarrères. 2015. A peripheral view: post-glacial history and genetic diversity of an amphibian in northern landscapes. Journal of Biogeography 42:2078-2088.

D'Aoust-Messier, A.-M., P. Echaubard, V. Billy, and D. Lesbarrères. 2015. Amphibian pathogens at northern latitudes: presence of chytrid fungus and ranavirus in northeastern Canada. Diseases of Aquatic Organisms 113:149-155.

Davenport, J.M., P.A. Seiwert, L.A. Fishback, and W.B. Cash. 2013. The effects of two fish predators on wood frog (*Lithobates sylvaticus*) tadpoles in a subarctic wetland: Hudson Bay Lowlands, Canada. Canadian Journal of Zoology 91:866-871.

Davenport, J.M., P.A. Seiwert, L.A. Fishback, and W.B. Cash. 2016. The interactive effects of fish predation and conspecific density on survival and growth of tadpoles of *Rana sylvatica* in a subarctic wetland. Copeia 104:639-644.

Davenport, J.M., L.A. Fishback, and B.R. Hossack. 2020. Effects of experimental warming and nutrient enrichment on wetland communities at the Arctic's edge. Hydrobiologia 847:3677-3690.

Davis, C.L., D.A.W. Miller, S.C. Walls, W.J. Barichivich, J.W. Riley, and M.E. Brown. 2017. Species interactions and the effects of climate variability on a wetland amphibian metacommunity. Ecological Applications 27:285-296.

Davis, D.R., and J.L. Kerby. 2016. First detection of ranavirus in amphibians from Nebraska, USA. Herpetological Review 47:46-50.

Davis, D.R., and T.J. LaDuc. 2017. *Scaphiopus couchii* (Couch's Spadefoot). Aggregation. Herpetological Review 48:613.

Davis, D.R., and T.J. LaDuc. 2018. Amphibians and reptiles of C.E. Miller Ranch and the Sierra Vieja, Chihuahuan Desert, Texas, USA. ZooKeys 735:97-130.

Davis, D.R., L.M. Jackson, S.R. Siddons, and J.L. Kerby. 2016. *Rana blairi* (Plains Leopard Frog). Reproduction. Herpetological Review 47:648.

Davis, D.R., J.K. Farkas, T.R. Kruisselbrink, J.L. Watters, E.D. Ellsworth, J.L. Kerby, and C.D. Siler. 2019. Prevalence and distribution of ranavirus in amphibians from southeastern Oklahoma, USA. Herpetological Conservation and Biology 14:360-369.

DeBlieux, T. S., and J.T. Hoverman. 2019. Parasite-induced vulnerability to predation in larval anurans. Diseases of Aquatic Organisms 135:241-250.

DeGregorio, B.A., J.D. Willson, M.E. Dorcas, and J.W. Gibbons. 2014. Commercial value of amphibians produced from an isolated wetland. American Midland Naturalist 172:200-204.

DeGregorio, B.A., P. J. Wolff, and A. N. Rice. 2021. Evaluating hydrophones for detecting underwater-calling frogs. Herpetological Conservation and Biology 16:513–524.

De Jong, G.D., S.P. Canton, J.S. Lynch, and M. Murphy. 2015. Aquatic invertebrate and vertebrate communities of ephemeral stream eosystems in the arid southwestern United States. Southwestern Naturalist 60:349-359.

Delaney, K.S., G. Busteed, R.N. Fisher, and S.P.D. Riley. 2021. Reptile and amphibian diversity and abundance in an urban landscape: impacts of fragmentation and the conservation value of small patches. Ichthyology and Herpetology 109:424-435.

Delis, P., and W.E. Meshaka. 2019. The herpetofauna of Wallops Island, Accomack County, Virginia: a case study of a vulnerable community. Journal of the Pennsylvania Academy of Science 93:132-161.

Delis, P.R., E.D. McCoy, and H.R. Mushinsky. 2020. Observations of post-breeding migration of *Hyla gratiosa* (barking treefrog) adults. Herpetological Review 51:690-694.

DeMarchi, J.A., J.R. Gaston, A.N. Spadaro, C.A. Porterfield, and M.D. Venesky. 2015. Tadpole food consumption decreases with increasing *Batrachochytrium dendrobatidis* infection intensity. Journal of Herpetology 49:395-398.

Denton, R.D., and S.C. Richter. 2013. Amphibian communities in natural and constructed ridge top wetlands with implications for wetland construction. Journal of Wildlife Management 77:886-896.

Devan-Song, A., M.A. Walden, H.A. Moniz, J.M. Fox, M.-R. Low, E. Wilkinson, S.W. Buchanan, and N.E. Karraker. 2021. Confirmation bias perpetuates century-old ecological misconception: evidence against 'secretive' behavior of eastern spadefoots. Journal of Herpetology 55:137-150.

DeVore, J.L., and J.C. Maerz. 2014. Grass invasion increases top-down pressure on an amphibian via structurally mediated effects on an intraguild predator. Ecology 95:1724-1730.

Devos, T. 2020. *Lithobates sylvaticus* (Wood Frog). Necrophilia. Herpetological Review 51:821-822.

Deyrup, M., L. Deyrup, and J. Carrel. 2013. Ant species in the diet of a Florida population of eastern narrow-mouthed toads, *Gastrophryne carolinensis*. Southeastern Naturalist 12:367-378.

Dixon, J.R. 2013. Amphibians and Reptiles of Texas, with Keys, Taxonomic Synopses, Bibliography, and Distribution Maps. 3rd ed. Texas A&M University Press, College Station.

Dodd, C.E., and R. Buchholz. 2018. Apparent maladaptive oviposition site choice of Cope's gray treefrogs (*Hyla chrysoscelis*) when offered an array of pond conditions. Copeia 106:492-500.

Dodd, C.K., Jr. 2018. A bibliography of the anurans of the United States and Canada. Version 2, Updated and covering the period 1709-2012. Herpetological Conservation and Biology 13 (Monograph 7):1-328.

Dodd, C.K. Jr., and W.J. Barichivich. 2017. A survey of the amphibians of Savannah National Wildlife Refuge, South Carolina and Georgia. Southeastern Naturalist 16:529-545.

Dodd, C.K., Jr., and C.D. Anderson. 2018. Amphibian immigration and emigration at a temporary pond in the Florida sandhills, USA: implications for conservation. Herpetological Conservation and Biology 13:131-145.

Dodd, C.K., Jr., and M.R. Jennings. 2021. How to raise a bullfrog – The literature on frog farming in North America. Bibliotheca Herpetologica 15:77-100.

Dodd, C.K., Jr., W.J. Barichivich, S.A. Johnson, M.G. Aresco, and J.A. Staiger. 2017. Establishing a baseline: the amphibians of Lower Suwannee National Wildlife Refuge, Dixie and Levy counties, Florida. Florida Scientist 80:133-144.

Donald, D.B. 2021. Water quality limitations for tadpoles of the wood frog in the northern Great Plains, Canada. Environmental Monitoring and Assessment 193:1-13.

Donini, J., and J. S. Doody. 2021. Successful predation of invasive toxic cane toads (*Rhinella marina*) by the southern water snake (*Nerodia fasciata*). Journal of North American Herpetology 2021(1):12–18.

Douglas, A.J., L.A. Hug, and B.A. Katzenback. 2021. Composition of the North American wood frog (*Rana sylvat-*

ica) bacterial skin microbiome and seasonal variation in community structure. Microbial Ecology 81: 78-92.

Douglas, D.A., and A.N. Drayer. 2013. *Lithobates sylvaticus* (Wood Frog). Egg predator entanglement. Herpetological Review 44:497.

Drake, D.L., B.H. Ousterhout, J.R. Johnson, T.L. Anderson, W.E. Peterman, C.D. Shulse, D.J. Hocking, K.L. Lohraff, E.B. Harper, T.A.G. Rittenhouse, B.B. Rothermel, L.S. Eggert, and R.D. Semlitsch. 2015. Pond-breeding amphibian community composition in Missouri. American Midland Naturalist 174:180-187.

Drayer, A.N., and S.C. Richter. 2016. Physical wetland characteristics influence amphibian community composition differently in constructed wetlands and natural wetlands. Ecological Engineering 93:166-174.

Drayer, A.N., J.C. Guzy, R. Caro, and S.J. Price. 2020. Created wetlands managed for hydroperiod provide habitat for amphibians in western Kentucky, USA. Wetlands Ecology and Management 28:543-558.

Drost, C.A. 2020. *Thamnophis elegans*. Terrestrial Gartersnake. pp. 401-417 *In* A.R. Holycross and J.C. Mitchell (eds.), Snakes of Arizona. ECO Publishing, Rodeo, New Mexico.

Duarte, A., D.J. Brown, and M.R.J. Forstner. 2014. Documenting extinction in real time: decline of the Houston toad on a primary recovery site. Journal of Fish and Wildlife Management 5:363-371.

Duarte, A., J.T. Peterson, C.A. Pearl, J.C. Rowe, B. McCreary, S.K. Galvan, and M.J. Adams. 2020. Estimation of metademographic rates and landscape connectivity for a conservation-reliant anuran. Landscape Ecology 35:1459-1479.

Duarte, A., C. A. Pearl, B. McCreary, J.C. Rowe, and M. J. Adams. 2021. An updated assessment of status and trend in the distribution of the Cascades frog (*Rana cascadae*) in Oregon, USA. Herpetological Conservation and Biology 16:361–373.

Dubois-Gagnon, M.-P., L. Bernatchez, M. Bélisle, Y. Dubois, and M.J. Mazerolle. 2021. Distribution of the boreal chorus frog (*Pseudacris maculata*) in an urban environment using environmental DNA. Environmental DNA 2021:00:1-13.

Duellman, W.E., A.B. Marion, and S.B. Hedges. 2016. Phylogenetics, classification, and biogeography of the treefrogs (Amphibia: Anura: Arboranae). Zootaxa 4104:1-109.

Duran, C.M., and D.W. Hall. 2013. Geographic distribution: *Lithobates sphenocephalus* (Southern Leopard Frog). Herpetological Review 44:622.

Durso, A.M., and L. Smith. 2017. *Eleutherodactylus planirostris* (Greenhouse Frog). Predation. Herpetological Review 48:606.

Dyck, A., S.A. Robinson, S.D. Young, J.B. Renaud, L.Sabourin, D.R. Lapen, and F.R. Pick. 2021. The Effects of ditch management in agroecosystems on embryonic and tadpole survival, growth, and development of northern leopard frogs (*Lithobates pipiens*). Archives of Environmental Contamination and Toxicology 81:107-122.

Dykes, S., P. Wyatt, J. Sibley, K.D. Edwards, and A. Edwards. 2017. Geographic distribution: *Hyla cinerea* (Green Treefrog). Herpetological Review 48:119.

Dziadzio, M.C., and L.L. Smith. 2016. Vertebrate use of gopher tortoise burrows and aprons. Southeastern Naturalist 15:586-594.

Eakin, C.J., A.J.K. Calhoun, and M.L. Hunter, Jr. 2019a. Effects of suburbanizing landscapes on reproductive effort of vernal pool-breeding amphibians. Herpetological Conservation and Biology 14:515-532.

Eakin, C.J., M.L. Hunter, Jr., and A.J.K. Calhoun. 2019b. The influence of land cover and within-pond characteristics on larval, froglet, and adult wood frogs along a rural to suburban gradient. Urban Ecosystems 22:493-505.

Earl, J.E., and R.D. Semlitsch. 2013. Carryover effects in amphibians: are characteristics of the larval habitat needed to predict juvenile survival? Ecological Applications 23:1429-1442.

Earl, J.E., and M.J. Gray. 2014. Introduction of ranavirus to isolated wood frog populations could cause local extinctions. EcoHealth 11:581-592.

Earl, J.E., and R.D. Semlitsch. 2015. Importance of forestry practices relative to microhabitat and microclimate changes for juvenile pond-breeding amphibians. Forest Ecology and Management 357:151-160.

Earl, J.E., P.O. Castello, K.E. Cohagen, and R.D. Semlitsch. 2014. Effects of subsidy quality on reciprocal subsidies: how leaf litter species changes frog biomass export. Oecologia 175:209-218.

Earl, J.E., J.C. Chaney, W.B. Sutton, C.E. Lillard, A.J. Kouba, C. Langhorne, J. Krebs, R.P. Wilkes, R.D. Hill, D.L. Miller, and M.J. Gray. 2016. Ranavirus could facilitate local extinction of rare amphibian species. Oecologia 182:611-623.

Eaton, B.R., C.A. Paszkowski, and R. Chapman. 2003. *Rana sylvatica* (Wood Frog). Parasite. Herpetological Review 34:55.

Echaubard, P., B.D. Pauli, V.L. Trudeau, and D. Lesbarrères. 2016. Ranavirus infection in northern leopard frogs: the timing and number of exposures matter. Journal of Zoology 298:30-36.

Eck, B., A. Byrne, V.D. Popescu, E.A. Harper, and D.A. Patrick. 2014. Effects of water temperature on larval amphibian predator-prey dynamics. Herpetological Conservation and Biology 9:302-308.

Ecoclub Amphibian Group, K.L. Pope, G.M. Wengert, J.E. Foley, D.T. Ashton, and R.G. Botzler. 2016. Citizen scientists monitor a deadly fungus threatening amphibian communities in northern coastal California, USA. Journal of Wildlife Diseases 52:516-523.

Ecrement, S.M., and S.C. Richter. 2017. Amphibian use of wetlands created by military activity in Kisatchie National Forest, Louisiana, USA. Herpetological Conservation and Biology 12:321-333.

El Balaa, R., and G. Blouin-Demers. 2013. Does exposure to cues of fish predators fed different diets affect morphology and performance of northern leopard frog (*Lithobates pipiens*) larvae? Canadian Journal of Zoology 91:203-211.

Ellison, S., R. Knapp, and V. Vredenburg. 2021. Longitudinal patterns in the skin microbiome of wild, individually marked frogs from the Sierra Nevada, California. ISME Communications 1(45):1-11. https://doi.org/ 10.1038/ s43705-021-00047-7.

Engbrecht, N.J., M.J. Lannoo, P.J. Williams, J.R. Robb, T.A. Gerardot, D.R. Karns, and M.J. Lodato. 2013. Is there hope for the Hoosier frog? An update on the status of crawfish frogs (*Lithobates areolatus*) in Indiana, with recommendations for their conservation. Proceedings of the Indiana Academy of Science 121:147-157.

Engbrecht, N.J., J.J. Mirtl, and E.M. Johnson. 2018. Geographic distribution: *Hyla cinerea* (Green Treefrog). Herpetological Review 49:708.

Engbrecht, N.J., J.J. Mirtl, Z.T. Truelock, and J. Burris. 2020. Geographic distribution: *Lithobates blairi* (Plains Leopard Frog). Herpetological Review 51:70-71.

Environment Canada. 2015. Recovery Strategy for the Oregon Spotted Frog (*Rana pretiosa*) in Canada. Species at Risk Act Recovery Strategy Series. Environment Canada, Ottawa. https://www.canada.ca/en/environment-climate-change/services/species-risk-public-registry/recovery-strategies/oregon-spotted-frog-2015.html.

Environment and Climate Change Canada. 2016a. Management Plan for the Western Toad (*Anaxyrus boreas*) in Canada [Proposed]. Species at Risk Act Management Plan Series. Environment and Climate Change Canada, Ottawa. iv + 38 pp.

Environment and Climate Change Canada. 2016b. Management Plan for the Coastal Tailed Frog (*Ascaphus truei*) in Canada [Proposed]. Species at Risk Act Management Plan Series. Environment and Climate Change Canada, Ottawa. 2 parts, vii + 49 pp.

Erdmann, J.A., W.A. Estes-Zumpf, C. Snoberger, Z.J. Walker, and A. Pocewicz. 2018. Expanding knowledge of *Batrachochytrium dendrobatidis* in Wyoming, USA. Herpetological Review 49:37-41.

Ersan, J.S.M., B.J. Halstead, E.L.Wildy, M.L. Casazza, and G.D. Wylie. 2020. Giant gartersnakes (*Thamnophis gigas*) exploit abundant nonnative prey while maintaining their appetite for native anurans. Herpetologica 76:290-296.

Erwin, K.J., H.C. Chandler, J.G. Palis, T.A. Gorman, and C.A. Haas. 2016. Herpetofaunal communities in ephemeral wetlands embedded within longleaf pine flatwoods of the Gulf Coastal Plain. Southeastern Naturalist 15:431-447.

Ethier, J.P., A. Fayard, P. Soroye, D. Choi, M.J. Mazerolle, and V.L. Trudeau. 2021. Life history traits and reproductive ecology of North American chorus frogs of the genus *Pseudacris* (Hylidae). Frontiers in Zoology 18:1-18.

Everman, E., and P. Klawinski. 2013. Human-facilitated jump dispersal of a non-native frog species on Hawai'i Island. Journal of Biogeography 40:1961-1970.

Eversole, C.B., and A.L. Brenk. 2021. *Eleutherodactylus campi* (Rio Grande Chirping Frog). Herpetological Review 52:790.

Faccio, S.D., K.L. Buckman, J.D. Lloyd, and A.N. Curtis. 2019. Bioaccumulation of methylmercury in wood frogs and spotted salamanders in Vermont vernal pools. Ecotoxicology 28:717-731.

Farr, W.L. 2019. *Lithobates berlandieri* (Rio Grande Leopard Frog). Voluntary thermal maximum. Herpetological Review 50:338.

Farr, W.L, and M.R.J. Forstner. 2014. *Lithobates sphenocephalus* (Southern Leopard Frog). Diet. Herpetological Review 45:682.

Feinberg, J. A. 2015. An unexpected journey: Anuran decline research and the incidental elucidation of a new cryptic species endemic to the urban northeast and mid-Atlantic US. Ph.D. Dissertation, Rutgers, The State University of New Jersey, New Brunswick.

Feinberg, J.A., C.E. Newman, G.J. Watkins-Colwell, M.D. Schlesinger, B. Zarate, B.R. Curry, H.B. Shaffer, and J. Burger. 2014. Cryptic diversity in Metropolis: confirmation of a new leopard frog species (Anura: Ranidae) from New York City and surrounding Atlantic Coast Regions. PLoS One 9:e108213.

Fellers, G.M., P.M. Kleeman, D.A.W. Miller, B.J. Halstead, and W.A. Link. 2013. Population size, survival, growth, and movements of *Rana sierrae*. Herpetologica 69:147-162.

Fellers, G.M., P.M. Kleeman, and D.A.W. Miller. 2015. Wetland occupancy of pond-breeding amphibians in Yosemite National Park, USA. Journal of North American Herpetology 2015:22-33.

Fellers, G.M., P.M. Kleeman, D.A.W. Miller, and B.J. Halstead. 2017. Population trends, survival, and sampling methodologies for a population of *Rana draytonii*. Journal of Herpetology 51:567-573.

Ferreira, R.B., K.H. Beard, R.T. Choi, and W.C. Pitt. 2015. Diet of the nonnative greenhouse frog (*Eleutherodacty-*

lus planirostris) in Maui, Hawaii. Journal of Herpetology 49:586-593.

Feuka, A.B., K.E. Hoffman, M.L. Hunter, Jr., and A.J.K. Calhoun. 2017. Effects of light pollution on habitat selection in post-metamorphic wood frogs (*Rana sylvatica*) and unisexual blue-spotted salananders (*Ambystoma laterale x jeffersonianum*). Herpetological Conservation and Biology 12:470-476.

Fielder, C.M., D.K. Walkup, W.A. Ryberg, and T.J. Hibbitts. 2020. New county records for the Panhandle region of Texas, USA. Herpetological Review 51:87-89.

Fields, S.E. 2019. Amphibians of the central and southwestern Piedmont Province of South Carolina. Southeastern Naturalist 18:202-223.

Flaxington, W.C. 2021. Amphibians and Reptiles of California. Field Observations, Distribution, and Natural History. Fieldnotes Press, Anaheim, California.

Flynn, R. W., M.F. Chislock, M.E. Gannon, S.J. Bauer, B.J. Tornabene, J.T. Hoverman, and M.S. Sepúlveda. 2019. Acute and chronic effects of perfluoroalkyl substance mixtures on larval American bullfrogs (*Rana catesbeiana*). Chemosphere 236:124350.

Flynn, R. W., M. Iacchetta, C. De Perre, L.S. Lee, M.S. Sepúlveda, and J.T. Hoverman. 2021. Chronic per-/polyfluoroalkyl substance exposure under environmentally relevant conditions delays development in northern leopard frog (*Rana pipiens*) larvae. Environmental Toxicology and Chemistry 40:711-716.

Foguth, R.M., T.D. Hoskins, G.C. Clark, M. Nelson, R.W. Flynn, C. de Perre, J.T. Hoverman, L.S. Lee, M.S. Sepúlveda, and J.R. Cannon. 2020. Single and multiple per- and polyfluoroalkyl substances accumulate in developing northern leopard frog brains and produce complex neurotransmission alterations. Neurotoxicology and Teratology 81:106907.

Fonseca, T.N., L.M. Thompson, A.C. Wood, and H.J. Howell. 2020. *Eleutherodactylus planirostris* (Greenhouse Frog). Habitat. Herpetological Review 51:562-563.

Forbes, J. 2021. Diet and prey preference of the Illinois chorus frog (*Pseudacris streckeri illinoensis*). M.S. thesis, Southern Illinois University at Edwardsville, Edwardsville.

Ford, J., and D.M. Green. 2021. Captive rearing oligotrophic-adapted toad tadpoles in mesocosms. Herpetological Review 52:777-779.

Ford, N.B. 2020. *Thamnophis marcianus*. Checkered Gartersnake. pp. 433-439 *In* A.R. Holycross and J.C. Mitchell (eds.), Snakes of Arizona. ECO Publishing, Rodeo, New Mexico.

Forrest, M.J., and J.R. Jaeger. 2013. First reports of amphibian chytrid fungus infections in the Ash Meadows National Wildlife Refuge. Pp. 46-68 *In* M.J. Forrest, Ecological and Geochemical Aspects of Terrestrial Hydrothermal Systems. Ph.D. Dissertation, University of California, San Diego.

Forrest, M.J., K.D. Urquhart, J.F. Harter, O.J. Miano, and B.D. Todd. 2013. Habitat loss and the amphibian chytrid fungus *Batrachochytrium dendrobatidis* may threaten the Dixie Valley toad, a narrowly distributed endemic species. Pp. 69-101 *In* M.J. Forrest, Ecological and Geochemical Aspects of Terrestrial Hydrothermal Systems. Ph.D. Dissertation, University of California, San Diego.

Forrest, M.J., M.S. Edwards, R. Rivera, J.C. Sjöberg, and J.R. Jaeger. 2015. High prevalence and seasonal persistence of amphibian chytrid fungus infections in the desert-dwelling Amargosa toad, *Anaxyrus nelsoni*. Herpetological Conservation and Biology 10:917-925.

Forrest, M.J., J. Stiller, T.L. King, and G.W. Rouse. 2017. Between hot rocks and dry places: the status of the Dixie Valley toad. Western North American Naturalist 77:162-175.

Forzán, M.J., and J. Wood. 2013. Low detection of ranavirus DNA in wild postmetamorphic green frogs, *Rana* (*Lithobates*) *clamitans*, despite previous or concurrent tadpole mortality. Journal of Wildlife Diseases 49:879-886.

Fouquette, M.J. Jr., and A. Dubois. 2014. A Checklist of North American Amphibians and Reptiles. Volume 1-Amphibians. Privately published, 613 pp.

Fournier, A.M.V., K. Eulinger, T. McDaniel, and G. Calvert. 2019. *Lithobates catesbeianus* (American Bullfrog). Diet. Herpetological Review 50:338.

Fox, J. 2020. Field notes: *Hyla squirella* (Squirrel Treefrog). County record. Catesbeiana 40:66.

Franklin, C.J., and M.G. Oyer-Vides. 2017. *Spea bombifrons* (Plains Spadefoot). Anisocoria. Herpetological Review 48:833.

Freidenburg, L.K. 2017. Environmental drivers of carry-over effects in a pond-breeding amphibian, the wood frog (*Rana sylvatica*). Canadian Journal of Zoology 95:255-262.

Fuchs, L.D., T.A. Tupper, C.A. Bozarth, D. Ferrnandez, and R. Aguilar. 2018. An investigation of co-infection by *Batrachochytrium dendrobatidis* and *Ranavirus* (FV3) in anurans of two natural areas in Anne Arundal County, Maryland and Fairfax County, Virginia, USA. Catesbeiana 38:45-55.

Fulbright, M.C., J.P. Flaherty, C.M. Gienger, and J. MacGregor. 2014. Geographic distribution: *Lithobates blairi* (Plains Leopard Frog). Herpetological Review 45:87.

Fulmer, T., and M.B. Connior. 2013. Geographic distribution: *Hyla squirella* (Squirrel Treefrog). Herpetological Review 44:620-621.

Funk, W.C., M.A. Murphy, K.L. Hoke, E. Muths, S.M. Amburgey, E.M. Lemmon, and A.R. Lemmon. 2016.

Elevational speciation in action? Restricted gene flow associated with adaptive divergence across an altitudinal gradient. Journal of Evolutionary Biology 29:241-252.

Gabrielsen, C.G., A.I. Kovach, K.J. Babbitt, and W.H. McDowell. 2013. Limited effects of suburbanization on the genetic structure of an abundant vernal pool-breeding amphibian. Conservation Genetics 14:1083-1097.

Gahm, K., A.Z.A. Arietta, and D.K. Skelley. 2021. Temperature-mediated trade-off between development and performance in larval wood frogs (*Rana sylvatica*). Journal of Experimental Zoology, Ecological and Integrative Physiology 335:146-157.

Gallagher, S. J., B.J. Tornabene, T.S. DeBlieux, K.M. Pochini, M.F. Chislock, Z.A. Compton, L.K. Eiler, K.M. Verble, and J.T. Hoverman. 2019. Healthy but smaller herds: Predators reduce pathogen transmission in an amphibian assemblage. Journal of Animal Ecology 88:1613–1624.

Galt, N., M.S. Atkinson, B.M. Glorioso, J.H. Waddle, M. Litton, and A.E. Savage. 2021. Widespread ranavirus and perkinsea infections in Cuban treefrogs (*Osteopilus septentrionalis*) invading New Orleans, USA. Herpetological Conservation and Biology 16:17-29.

Garcia, N., and C.M. Schalk. 2020. *Incilius nebulifer* (Gulf Coast Toad). Polymelia. Herpetological Review 51:99.

Garcia, T.S., J.C. Urbina, E.M. Bredeweg, and M.C.O. Ferrari. 2017. Embryonic learning and developmental carry-over effects in an invasive anuran. Oecologia 184:623-631.

Garcia, T.S., E.M. Bredeweg, J. Urbina, and M.C.O. Ferrari. 2019. Evaluating adaptive, carry-over, and plastic anti-predator responses across a temporal gradient in Pacific chorus frogs. Ecology 100:e02825.

Gardner, S.T. 2021. Assessing stress physiology of cane toads in Florida: tradeoffs with dispersal. Ph.D. Dissertation, Auburn University, Auburn, Alabama.

Gardner, S.T., M. Kepas, C.R. Simons, L.M. Horne, A.H. Savitzky, and M.T. Mendonça. 2021. Differences in morphology and in composition and release of parotoid gland secretion in introduced cane toads (*Rhinella marina*) from established populations in Florida, USA. Ecology and Evolution 11:1013–1022.

Garner, J.L. 2013. Movement and habitat use of the Great Basin spadefoot (*Spea intermontana*) at its northern range limit. M.S. thesis, Thompson Rivers University, Kamloops, British Columbia.

Garner, T.W.J., B.R. Schmidt, A. Martel, F. Pasmans, E. Muths, A.A. Cunningham, C. Weldon, M.C. Fisher, and J. Bosch. 2016. Mitigating amphibian chytridiomycoses in nature. Philosophical Transactions of the Royal Society B 371:20160207.

Gavel, M.J., S.D. Young, R.L. Dalton, C. Soos, L. McPhee, M.R. Forbes, and S.A. Robinson. 2021. Effects of two pesticides on northern leopard frog (*Lithobates pipiens*) stress metrics: Blood cell profiles and corticosterone concentrations. Aquatic Toxicology 235:105820.

Geluso, K., and M.J. Harner. 2013. Reexamination of herpetofauna on Mormon Island, Hall County, Nebraska, with notes on natural history. Transactions of the Nebraska Academy of Sciences 33:7-20.

Geluso, K., B.T. Krohn, M.J. Harner, and M.J. Assenmacher. 2013. Whooping cranes consume plains leopard frogs at migratory stopover sites in Nebraska. Prairie Naturalist 45:91-93.

Ghioca-Robrecht, D.M., and L.M. Smith. 2013. Factors influencing premetamophic body mass of two polymorphic spadefoot species in cropland and grassland playas. Journal of Herpetology 47:299-307.

Gibbs, J.P., S. Rouhani, and L. Shams. 2017. Frog and toad habitat occupancy across a polychlorinated biphenyl (PCB) contamination gradient. Journal of Herpetology 51:209-214.

Gibson, J.D., and T. Anthony. 2019. Eastern spadefoots in Virginia: observations made from volunteer herpetologists around the state. Catesbeiana 39:70-81.

Gibson, J.D., and R.L. Hoffman. 2019. Prey of American bullfrogs (*Rana catesbeiana*) from Henry and Patrick counties, Virginia (Anura: Ranidae). Banisteria 52:37-41.

Gibson, J.D., and K. Ivanov. 2020. Field notes: *Anaxyrus americanus* (American Toad). Diet. Catesbeiana 40:65-66.

Gibson, J.D., and P. Sattler. 2020a. Natural history notes on the anurans of White Oak Mountain Wildlife Management Area. Catesbeiana 40:29-48.

Gibson, J.D., and P. W. Sattler. 2020b. The phenotype of the eastern American toad in the south-western piedmont of Virginia. Catesbeiana 40:109-123.

Gibson, J.D., E. Mazur, and A. Mazur. 2020. Field notes: *Hyla chrysoscelis/versicolor* (Cope's/Gray Treefrog). Color variant. Catesbeiana 40:130-132.

Gilhen, J., and R.W. Russell. 2015. Three records of rare blue American bullfrogs, *Lithobates catesbeianus*, in Nova Scotia, Canada. Canadian Field-Naturalist 129:395-398.

Gillingwater, S.D., and A.S. MacKenzie. 2015. Photo Field Guide to the Reptiles and Amphibians of Ontario. St. Thomas Field Naturalist Club, St. Thomas, Ontario.

Ginnan, N.A., J.R. Lawrence, M.E.T. Russell, D.L. Eggett, and K.A. Hatch. 2014. Toe clipping does not affect the survival of leopard frogs (*Rana pipiens*). Copeia 104:650-653.

Giovanetto, L.A. 2021. *Lithobates sylvaticus* (Wood Frog). Female mortality. Herpetological Review 52:833.

Glinski, D.A., R.J. Van Meter, S.T. Purucker, and W.M. Henderson. 2021. Route of exposure influences pesticide body burden and the hepatic metabolome in post-metamorphic leopard frogs. Science of The Total Environment 779:146358.

Glorioso, B.M., J.H. Waddle, J.A. Jenkins, H.M. Olivier, and R.R. Layton. 2015. *Hyla chrysoscelis* (Cope's Gray Treefrog) x *Hyla cinerea* (Green Treefrog). Putative natural hybrid. Herpetological Review 46:410-411.

Glorioso, B.M., J.H. Waddle, L.J. Muse, N.D. Jennings, M. Litton, J. Hamilton, S. Gergen, and D. Heckard. 2018. Establishment of the exotic invasive Cuban treefrog (*Osteopilus septentrionalis*) in Louisiana. Biological Invasions 20:2707-2713.

Glorioso, B.M., L.J. Muse, and J.H. Waddle. 2020. Egg counts of southern leopard frog, *Lithobates sphenocephalus*, egg masses from southern Louisiana, USA. Herpetology Notes 13:187-189.

Godwin, C.D., O. Ljustina, and J.A. Erdmann. 2018. *Gastrophryne carolinensis* (Eastern Narrow-mouthed Toad). Arboreal activity. Herpetological Review 49:95-96.

Godwin, J.C. 2017. *Lithobates capito* (Gopher Frog). Tadpole morphology. Herpetological Review 48:830-831.

Goedert, D., D. Clement, and R. Calsbeek. 2021. Evolutionary trade-offs may interact with physiological constraints to maintain color variation. Ecological Monographs 91: e01430.

Goetz, S.M., C.M. Romagosa, A.G. Appel, C. Guyer, and M.T. Mendonca. 2017. Reduced innate immunity of Cuban treefrogs at leading edge of range expansion. Journal of Experimental Zoology Part A 327:592-599.

Goetz, S., C. Guyer, S.M. Boback, and C.M. Romagosa. 2018. Toxic, invasive treefrog creates evolutionary trap for native gartersnakes. Biological Invasions 20:519-531.

Goldberg, S.R. 2016. Notes on reproduction of Red-spotted toads, *Anaxyrus punctatus* (Anura: Bufonidae), from Riverside County, California. Bulletin of the Chicago Herpetological Society 51:130-131.

Goldberg, S.R. 2017a. Reproduction in Gulf Coast toads, *Incilius nebulifer* (Anura: Bufonidae) from Texas. Bulletin of the Chicago Herpetological Society 52:159-161.

Goldberg, S.R. 2017b. Notes on reproduction of Woodhouse's Toads, *Anaxyrus woodhousii* (Anura: Bufonidae). Bulletin of the Chicago Herpetological Society 52:194-197.

Goldberg, S.R. 2017c. Reproduction of the Plains Spadefoot, *Spea bombifrons* (Anura: Scaphiopodidae) from New Mexico. Sonoran Herpetologist 30:65-67.

Goldberg, S.R. 2017d. Notes on reproduction of California Treefrogs, *Hyliola cadaverina* (Anura: Hylidae) from Riverside County, California. Sonoran Herpetologist 30:5-7.

Goldberg, S.R. 2018a. Notes on reproduction of Colorado River Toads, *Incilius alvarius* (Anura: Bufonidae) from Pima County, Arizona. Sonoran Herpetologist 31:38-39.

Goldberg, S.R. 2018b. Reproduction of the Texas Toad, *Anaxyrus speciosus* (Anura: Bufonidae) from Texas. Sonoran Herpetologist 31:2-4.

Goldberg, S.R. 2018c. Notes on reproduction of Eastern Spadefoot Toads, *Scaphiopus holbrookii* (Anura: Scaphiopodidae). Bulletin of the Chicago Herpetological Society 53:63-65.

Goldberg, S.R. 2018d. Reproduction of Couch's Spadefoot, *Scaphiopus couchii* (Anura: Scaphiopodidae) from Pima County, Arizona. Sonoran Herpetologist 31:28-29.

Goldberg, S.R. 2018e. Notes on reproduction of Great Plains Toads, *Anaxyrus cognatus* (Anura: Bufonidae) from southern Arizona. Bulletin of the Chicago Herpetological Society 53:131-134.

Goldberg, S.R. 2018f. Notes on reproduction of Western Narrow-mouthed Toads, *Gastrophryne olivacea* (Anura: Microhylidae) from Texas. Bulletin of the Chicago Herpetological Society 53:253-255.

Goldberg, S.R. 2018g. Notes on reproduction of the Arizona Treefrog, *Dryophytes wrightorum* (Anura: Hylidae) from Arizona. Sonoran Herpetologist 31:63-64.

Goldberg, S.R. 2019a. Notes on reproduction of Crawfish Frogs, *Lithobates areolatus* (Anura: Ranidae) from Oklahoma. Bulletin of the Chicago Herpetological Society 54:181-183.

Goldberg, S.R. 2019b. Reproduction of the Mexican Spadefoot Toad, *Spea multiplicata* (Anura: Scaphiopodidae), from New Mexico. Sonoran Herpetologist 32:17-19.

Goldberg, S.R. 2019c. Notes on reproduction of Great Basin Spadefoot Toads, *Spea intermontana* (Anura: Scaphiopodidae). Bulletin of the Chicago Herpetological Society 54:145-147.

Goldberg, S.R. 2019d. Notes on reproduction of the Foothill Yellow-legged frog, *Rana boylii* (Anura: Ranidae) from California. Sonoran Herpetologist 32:43-45.

Goldberg, S.R. 2019e. Notes on reproduction of Green Toads, *Anaxyrus debilis* (Anura: Bufonidae), from New Mexico. Sonoran Herpetologist 32:3-5.

Goldberg, S.R. 2019f. Notes on reproduction of Hurter's Spadefoot Toads, *Scaphiopus hurterii* (Anura: Scaphiopodidae), from Oklahoma. Bulletin of the Chicago Herpetological Society 54:9-11.

Goldberg, S.R. 2019g. Notes on reproduction of Columbia spotted frogs, *Rana luteiventris* (Ranidae: Anura). Bulletin of the Chicago Herpetological Society 54:228-230.

Goldberg, S.R. 2019h. Notes on reproduction of the lowland leopard frog, *Lithobates yavapaiensis* (Anura: Ranidae), from Arizona. Sonoran Herpetologist 32:62-64.

Goldberg, S.R. 2020a. Notes on reproduction of the Tarahumara frog, *Lithobates tarahumarae* (Anura: Ranidae). Sonoran Herpetologist 33:2-3.

Goldberg, S.R. 2020b. Notes on reproduction of Strecker's chorus frog, *Pseudacris streckeri* (Anura: Hylidae), from Oklahoma. Bulletin of the Chicago Herpetological Society 55:61- 63.

Goldberg, S.R. 2020c. Notes on reproduction of the Sierra Nevada yellow-legged frog from California. California Fish and Wildlife 106:7-10.

Goldberg, S.R. 2020d. Notes of reproduction of Rio Grande leopard frogs, *Lithobates berlandieri* (Anura: Ranidae), from Texas. Bulletin of the Chicago Herpetological Society 55:121-123.

Goldberg, S.R. 2020e. Notes on reproduction of the canyon treefrog, *Dryophytes arenicolor* (Anura: Hylidae) from Arizona. Sonoran Herpetologist 33:22-24.

Goldberg, S.R. 2020f. Notes on reproduction of Plains leopard frogs, *Lithobates blairi* (Anura: Ranidae) from Oklahoma. Bulletin of the Chicago Herpetological Society 55:163-165.

Goldberg, S.R. 2020g. Notes on reproduction of the Chiricahua leopard frog, *Lithobates chiricahuensis* (Anura: Ranidae), from Arizona. Sonoran Herpetologist 33:62-63.

Goldberg, S.R. 2020h. Notes on reproduction of the Sonoran green toad, *Anaxyrus retiformis* (Anura: Bufonidae) from Pima County, Arizona. Sonoran Herpetologist 33:97-98.

Goldberg, S.R. 2020i. Notes on reproduction of Cascades frogs from California. California Fish and Wildlife 106:215-219.

Goldberg, S.R. 2021a. Notes on reproduction of Cajun chorus frogs, *Pseudacris fouquettei* (Anura: Hylidae), from Oklahoma. Bulletin of the Chicago Herpetological Society 56:21- 22.

Goldberg, S.R. 2021b. Notes on reproduction of the Yosemite toad, *Anaxyrus canorus* (Anura: Bufonidae), from California. Sonoran Herpetologist 34:37-39.

Goldberg, S.R. 2021c. Notes on reproduction of black toads from California. California Fish and Wildlife Special CESA Issue:221-225.

Goldberg, S.R. 2021d. Notes on reproduction of pig frogs, *Lithobates grylio* (Anura: Ranidae), from Texas. Bulletin of the Chicago Herpetological Society 56:115-117.

Goldberg, S.R. 2021e. Notes on reproduction of the boreal chorus frog, *Pseudacris maculata* (Anura: Hylidae), from Colorado. Sonoran Herpetologist 34:66-69.

Goldberg, S.R. 2021f. Notes on reproduction of gopher frogs, *Lithobates capito* (Anura: Ranidae), from North and South Carolina. Bulletin of the Chicago Herpetological Society 56:156-158.

Goldberg, S.R., and C.R. Bursey. 2020. *Lithobates berlandieri* (Rio Grande Leopard Frog). Endoparasite. Herpetological Review 51:300.

Goldberg, S.R., and C.R. Bursey. 2021. *Lithobates capito* (Gopher Frog). Endoparasites. Herpetological Review 52:832.

Goldberg, S.R., C.R. Bursey, and J. Arreola. 2013. *Leptodactylus fragilis* (Mexican White-lipped Frog). Endoparasites. Herpetological Review 44:656.

Goldberg, S.R., C.R. Bursey, and F. Kraus. 2018. Helminths of the Green and Black Poison Frog, *Dendrobates auratus* (Anura, Dendrobatidae) from Hawaii. Alytes 35:29-42.

Goldstein, J.A., K. von Seckendorr Hoff, and S.D. Hillyard. 2017. The effect of temperature on development and behaviour of relict leopard frog tadpoles. Conservation Physiology 5(1):cow075.

Goodman, C.M. 2020. The potential for spread of a novel invader, the Tropical Clawed Frog in Florida. M.S. thesis, University of Florida, Gainesville.

Goodman, C.M., G.F.M. Jongsma, J.E. Hill, E.L. Stanley, Q.M. Tuckett, D.C. Blackburn, and C.M. Romagosa. 2021. A case of mistaken identity: genetic and anatomical evidence reveals the cryptic invasion of *Xenopus tropicalis* in central Florida. Journal of Herpetology 55:62-69.

Goodman, R.M., J.A. Tyler, D.M. Reinartz, and A.N. Wright. 2019. Survey of ranavirus and *Batrachochytrium dendrobatidis* in introduced frogs in Hawaii, USA. Journal of Wildlife Diseases 55:668-672.

Goodward, D.M., and M.D. Wilcox. 2019. The Rio Grande leopard frog (*Lithobates berlandieri*) and other introduced and native herpetofauna of the Coachella Valley, Riverside County, California. California Fish and Game 105:48-71.

Goodwin, K.A.J., A.M. Kissel, and W.J. Palen. 2019. Individual variation in thermal performance of a temperate, montane amphibian (*Rana cascadae*). Herpetological Conservation and Biology 14:420-428.

Gooley, A.C., and T.K. Pauley. 2013. Eastern spadefoots (*Scaphiopus holbrookii*) and other herpetofauna inhabiting an industrial fly-ash disposal site in southern Ohio. Ohio Biological Survey Notes 4:1-5.

Gordon, A.M., M.B. Youngquist, and M.D. Boone. 2016. The effects of pond drying and predation on Blanchard's cricket frogs (*Acris blanchardi*). Copeia 104:482-486.

Gordon, M.R., E.T. Simandle, and C.R. Tracy. 2017. A diamond in the rough desert shrublands of the Great Basin in the western United States: a new cryptic toad species (Amphibia: Bufonidae: *Bufo* (*Anaxyrus*)) discovered in northern Nevada. Zootaxa 4290:123-139.

Gordon, M.R., E.T. Simandle, F.C. Sandmeier, and C.R. Tracy. 2020. Two new cryptic endemic toads of *Bufo* discovered in central Nevada, western United States (Amphibia: Bufonidae: *Bufo* [*Anaxyrus*]). Copeia 108:166-183.

Gordon, N.M., and M. Hellman. 2015. Dispersal distance, gonadal steroid levels, and body condition in gray treefrogs (*Hyla versicolor*): seasonal and breeding night variation in females. Journal of Herpetology 49:655-661.

Gould, W.R., A.M. Ray, L.L. Bailey, D. Thoma, R. Daley, and K. Legg. 2019. Multistate occupancy modeling improves understanding of amphibian breeding dynamics in the Greater Yellowstone area. Ecological Applications 29:e01825.

Government of the Northwest Territories. 2017. Management Plan for Amphibians in the Northwest Territories. Species at Risk (NWT) Act Management Plan and Recovery Strategy Series. Environment and Natural Resources, Government of the Northwest Territories, Yellowknife.

Graham, A., and D.A. Wilcox. 2021. The impacts of Marcellus Shale gas drilling accidents on amphibians in a Pennsylvania fen. Wetlands Ecology and Management 29:155-167.

Graham, B.M., D.J. O'Hearn, I.E. MacAllister, and J.H. Sperry. 2020. Behavioral responses by adult northern leopard frogs to conspecific chemical cues. Journal of Herpetology 54:168-173.

Grant, E.H.C., E.F. Zipkin, J.D. Nichols, and J.P. Campbell. 2013. A strategy for monitoring and managing declines in an amphibian community. Conservation Biology 27:1245-1253.

Grant, E.H.C. , D.A.W. Miller, B.R. Schmidt, M.J. Adams, S.M. Amburgey, T. Chambert, S.S. Cruickshank, R.N. Fisher, D.M. Green, B.R. Hossack, P.T.J. Johnson, M.B. Joseph, T.A.G. Rittenhouse, M.E. Ryan, J.H. Waddle, S.C. Walls, L.L. Bailey, G.M. Fellers, T.A. Gorman, A.M. Ray, D.S. Pilliod, S.J. Price, D. Saenz, W. Sadinski, and E. Muths. 2016. Quantitative evidence for the effects of multiple drivers on continental-scale amphibian declines. Scientific Reports 6:25625.

Grant, E.H.C., D.A.W. Miller, and E. Muths. 2020. A synthesis of evidence of drivers of amphibian declines. Herpetologica 76:101-107.

Grant, T.J., D.L. Otis, and R.R. Koford. 2015. Short-term anuran community dynamics in the Missouri River floodplain following an historic flood. Ecosphere 6(10):1-16.

Grant, T.J., D.L. Otis, and R.R. Koford. 2018. Comparison between anuran call only and multiple life-stage occupancy designs in Missouri River floodplain wetlands following a catastrophic flood. Journal of Herpetology 52:371-380.

Gray, M.J., and V.G. Chinchar (eds.). 2015. Ranaviruses. Lethal Pathogens of Ectothermic Vertebrates. Springer Open, Cham, Switzerland.

Green, D.M. 2013. Sex ratio and breeding population size in Fowler's toad, *Anaxyrus* (=*Bufo*) *fowleri*. Copeia 2013:647-652.

Green, D.M. 2015. Implications of female body-size variation for the reproductive ecology of an anuran amphibian. Ethology Ecology & Evolution 27:173-184.

Green, D.M., and J. Middleton. 2013. Body size varies with abundance, not climate, in an amphibian population. Ecography 36:947-955.

Green, D.M., and K.T. Yagi. 2018. Ready for bed: pre-hibernation movements and habitat use by Fowler's toads (*Anaxyrus fowleri*). Canadian Field-Naturalist 132:46-52.

Green, D.M., L.A. Weir, G.S. Casper, and M.J. Lannoo. 2013. North American Amphibians. Distribution and Diversity. University of California Press, Berkeley.

Green, D.M., A.R. Yagi, and S.E. Hamill. 2019. Recovery Strategy for the Fowler's Toad (*Anaxyrus fowleri*) in Canada 2019. Ministry of the Environment and Natural Resources, Government of Canada. https://www.canada.ca/en/environment-climate-change/services/species-risk-public-registry/recovery-strategies/fowlers-toad-2019.html.

Green, J., P. Govindarajulu, and E. Higgs. 2021. Multiscale determinants of Pacific chorus frog occurrence in a developed landscape. Urban Ecosystems 24:587-600.

Green, T., E. Das, and D.M. Green. 2016. Springtime emergence of overwintering toads, *Anaxyrus fowleri*, in relation to environmental factors. Copeia 104:393-401.

Greenberg, C.H., S.J. Zarnoch, and J.D. Austin. 2017a. Weather, hydroregime, and breeding effort influence juvenile recruitment of anurans: implications for climate change. Ecosphere 8:e01789.

Greenberg, C.H., S.A. Johnson, R. Owen, and A. Storfer. 2017b. Amphibian breeding phenology and reproductive outcome: an examination using terrestrial and aquatic sampling. Canadian Journal of Zoology 95:673-684.

Greenberg, C.H., S.J. Zarnoch, and J.D. Austin. 2018a. Long term amphibian monitoring at wetlands lacks power to detect population trends. Biological Conservation 228:120-131.

Greenberg, C.H., C.E. Moorman, C. Matthews-Snoberger, T.A. Waldrop, D. Simon, A. Heh, and D. Hagan. 2018b. Long-term herpetofaunal response to repeated fuel reduction treatments. Journal of Wildlife Management 82:553-565.

Greenberg, D.A., and D.M. Green. 2013. Effects of an invasive plant on population dynamics in toads. Conservation Biology 27:1049-1057.

Greene, L., and T. Branston. 2013. Geographic distribution: *Anaxyrus woodhousii woodhousii* (Rocky Mountain

Toad). Herpetological Review 44:270.

Gregory, P.T. 2021. *Rana luteiventris* (Columbia Spotted Frog) and *Anaxyrus boreas* (Western Toad). Interspecific amplexus. Herpetological Review 52:837.

Griffis-Kyle, K.L., K. Mougey, M. Vanlandeghem, S. Swain, and J.C. Drake. 2018. Comparison of climate vulnerability among desert herpetofauna. Biological Conservation 225:164-175.

Groff, L.A., S.B. Marks, and M.P. Hayes. 2014. Using ecological niche models to direct rare amphibian surveys: a case study using the Oregon spotted frog (*Rana pretiosa*). Herpetological Conservation and Biology 9:354-368.

Groff, L.A., A.J.K. Calhoun, and C.S. Loftin. 2015. *Lithobates sylvaticus* (Wood Frog). Habitat use. Herpetological Review 46:234.

Groff, L.A., A.J.K. Calhoun, and C.S. Loftin. 2016. Hibernal habitat selection by wood frogs (*Lithobates sylvaticus*) in a northern New England montane landscape. Journal of Herpetology 50:559-569.

Groff, L.A., A.J.K. Calhoun, and C.S. Loftin. 2017a. Amphibian terrestrial habitat selection and movement patterns vary with annual life-history period. Canadian Journal of Zoology 95:433-442.

Groff, L.A., C.S. Loftin, and A.J.K. Calhoun. 2017b. Predictors of breeding site occupancy by amphibians in montane landscapes. Journal of Wildlife Management 81:269-278.

Groner, M.L., J.C. Buck, S. Gervasi, A.R. Blaustein, L.K. Reinert, L.A. Rollins-Smith, M.E. Bier, J. Hempel, and R.A. Relyea. 2013. Larval exposure to predator cues alters immune function and response to a fungal pathogen in post-metamorphic wood frogs. Ecological Applications 23:1443-1454.

Grossman, B., J. Roberts, and J.B. Jensen. 2014. *Lithobates sphenocephalus* (Southern Leopard Frog). Predation. Herpetological Review 45:478.

Guadiana, C.J., P.S. Robinson, M.J. Schalk, and D.R. Davis. 2020. New county records of amphibians and reptiles from south Texas, USA. Herpetological Review 51:799-803.

Gustafson, K.D., and R.A. Newman. 2016. Multiscale occupancy patterns of anurans in prairie wetlands. Herpetologica 72:293-302.

Gustafson, K.D., R.A. Newman, E.E. Pulis, and K.C. Cabarle. 2015. A skeletochronological assessment of age-parasitism relationships in wood frogs (*Lithobates sylvaticus*). Journal of Herpetology 49:122-130.

Gustafson, K.D., B.L. Bly, and R.A. Newman. 2019. Color and pigment polymorphisms of northern leopard frogs on a prairie landscape. Herpetological Conservation and Biology 14:223-234.

Gutiérrez-González, C.E., M.A. Gómez-Ramírez, D. Gutiérrez-Garcia, J. Valenzuela, J.C. Rorabaugh, and A.D. Flesch. 2016. Interactions between Sonoran Desert toads (*Incilius alvarius*) and mammalian predators at the Northern Jaguar Reserve, Sonora, Mexico. Sonoran Herpetologist 29:26.

Haggerty, C.J.E., T.L. Crisman. 2015. Pulse disturbance impacts from a rare freeze event in Tampa, Florida on the exotic invasive Cuban treefrog, *Osteopilus septentrionalis*, and native treefrogs. Biological Invasions 17:2103-2111.

Hales, J.-A., and K.W. Larsen. 2019. Habitat selection by newly metamorphosed Great Basin spadefoots (*Spea intermontana*): a microcosm study. Western Wildlife 6:61-68.

Hall, A.S. 2016. Acute artificial light diminishes central Texas anuran calling behavior. American Midland Naturalist 175:183-193.

Hall, E.M., S.P. Brady, N.M. Mattheus, R.L. Earley, M. Diamond, and E.J. Crespi. 2017. Physiological consequences of exposure to salinized roadside ponds on wood frog larvae and adults. Biological Conservation 209:98-106.

Hall, E.M., L.A. Rollins-Smith, and B.T. Miller. 2018. Axanthism in the southern leopard frog, *Lithobates sphenocephalus* (Cope, 1886), (Anura: Ranidae) from the state of Tennessee, USA. Herpetology Notes 11:601-602.

Hall, E.M., C.S. Goldberg, J.L. Brunner, and E.J. Crespi. 2018. Seasonal dynamics and potential drivers of ranavirus epidemics in wood frog populations. Oecologia 188:1253-1262.

Hall, E.M., J.L. Brunner, B. Hutzenbiler, and E.J. Crespi. 2020. Salinity stress increases the severity of ranavirus epidemics in amphibian populations. Proceedings of the Royal Society B 287:20200062.

Halliday, W.D., and G. Blouin-Demers. 2018. Habitat selection by common gartersnakes (*Thamnophis sirtalis*) is affected by vegetation but not by location of northern leopard frog (*Lithobates pipiens*) prey. Canadian Field-Naturalist 132:223-230.

Halstead, B.J., and P.M. Kleeman. 2017a. Occurrence of amphibians in northern California coastal dune drainages. Northwestern Naturalist 98:91-100.

Halstead, B.J., and P.M. Kleeman. 2017b. Frogs on the beach: ecology of California red-legged frogs (*Rana draytonii*) in coastal dune drainages. Herpetological Conservation and Biology 12:127-140.

Halstead, B.J., P.M. Kleeman, C.S. Goldberg, M. Bedwell, R.B. Douglas, and D.W. Ulrich. 2018a. Occurrence of California red-legged (*Rana draytonii*) and northern red-legged (*Rana aurora*) frogs in timberlands of Mendocino County, California, examined with environmental DNA. Northwestern Naturalist 99:9-20.

Halstead, B.J., P.M. Kleeman, and J.P. Rose. 2018b. Time-to-detection occupancy modeling: an efficient method for analyzing the occurrence of amphibians and reptiles. Journal of Herpetology 52:415-424.

Halstead, B.J., P.M. Kleeman, J.P. Rose, and K.J. Fouts. 2021. Water temperature and availability shape the spatial ecology of a hot springs endemic toad (*Anaxyrus williamsi*). Herpetologica 77:24-36.

Halstead, B.J., K.L. Baumberger, A.R. Backlin, P.M. Kleeman, M.N. Wong, E.A. Gallegos, J.P. Rose, and R.N. Fisher. 2021. Conservation implications of spatiotemporal variation in the terrestrial ecology of western spadefoots. Journal of Wildlife Management 85:1377-1393.

Hammond, T.T., M.J. Curtis, L.E. Jacobs, M.W. Tobler, R.R. Swaisgood, and D.M. Shier. 2021. Behavior and detection method influence detection probability of a translocated, endangered amphibian. Animal Conservation 35:401-411.

Hanlon, S.M., and R. Relyea. 2013. Sublethal effects of pesticides on predator-prey interactions in amphibians. Copeia 2013:691-698.

Hanlon, S.M., D. Smith, J.L. Kerby, E. Berg, W. Peterson, M.J. Parris, and J.E. Moore. 2014a. Occurrence of *Batrachochytrium dendrobatidis* in Wapanocca National Wildlife Refuge, Arkansas, USA. Herpetological Review 45:31-32.

Hanlon, S.M., D. Smith, J.L. Kerby, M.J. Parris, and J.E. Moore. 2014b. Detection of *Batrachochytrium dendrobatidis* infections in amphibian populations at Edward J. Meeman Biological Field Station, Tennessee, USA. Herpetological Review 45:32-34.

Hanlon, S.M., K.J. Lynch, J.L. Kerby, and M.J. Parris. 2015. Environmental substrates alter survival and foraging efficiencies in tadpoles. Herpetological Conservation and Biology 10:180-188.

Hanna, D.E.L., D.R. Wilson, G. Blouin-Demers, and D.J. Mennill. 2014. Spring peepers *Pseudacris crucifer* modify their call structure in response to noise. Current Zoology 60:438-448.

Hanson, K.M., and E.J. McElroy. 2015. Anthropogenic impacts and long-term changes in herpetofaunal diversity and community composition on a barrier island in the southeastern United States. Herpetological Conservation and Biology 10:765-780.

Harding, J.H., and D.A. Mifsud. 2017. Amphibians and Reptiles of the Great Lakes Region. Revised edition. University of Michigan Press, Ann Arbor.

Hardy, B.M., K. L. Pope, and E. K. Latch. 2021. Genomic signatures of demographic declines in an imperiled amphibian inform conservation action. Animal Conservation 24: 946-958.

Harings, N.M., and W.J. Boeing. 2014. Desert anuran occurrence and detection in artificial breeding habitats. Herpetologica 70:123-134.

Harner, M.J., J.N. Merlino, and G.D. Wright. 2013. Amphibian chytrid fungus in Woodhouse's toads, plains leopard frogs, and American bullfrogs along the Platte River, Nebraska, USA. Herpetological Review 44:459-461.

Harris, H.S., Jr. 2013. Cannibalism in the green frog, *Lithobates clamitans melanotus*, in Maryland. Bulletin of the Maryland Herpetological Society 49:47.

Harris, H.S., Jr. 2014. An isolating mechanism, between the cryptic species *Hyla chrysoscelis* Cope and *Hyla versicolor* Le Conte in a sympatric population, other than voice. Bulletin of the Maryland Herpetological Society 50:44-46.

Hartman, E.A.H., J.B. Belden, L.M. Smith, and S.T. McMurry. 2014. Chronic effects of strobilurin fungicides on development, growth, and mortality of larval Great Plains toads (*Bufo cognatus*). Ecotoxicology 23:396-403.

Hartman, R., K. Pope, and S. Lawler. 2013. Factors mediating co-occurrence of an economically valuable introduced fish and its native frog prey. Conservation Biology 28:763-772.

Hartzell, S.M. 2016. *Anayxrus americanus* (American Toad). Defensive behavior. Herpetological Review 47:102.

Hartzell, S.M. 2020. *Anayxrus americanus* (American Toad). Adult basking. Herpetological Review 51:95-96.

Hartzell, S.M. 2021. *Lithobates sylvaticus* (Wood Frog). Habitat. Herpetological Review 52:374.

Hayes, M.P., and K.M. Cassidy. 2013. *Rana luteiventris* (Columbia Spotted Frog). Maximum size. Herpetological Review 44:128.

Hecnar, S.J., and D.R. Hecnar. 2017. *Lithobates sylvaticus* (Wood Frog). Hibernation. Herpetological Review 48:165-166.

Hecnar, S.J., D.R. Hecnar, D.J. Brazeau, J. Prisciak, A. MacKenzie, T. Berkers, H. Brown, C. Lawrence, and T. Dobbie. 2018. Structure of coastal zone herpetofaunal communities in the southern Laurentian Great Lakes. Journal of Herpetology 52:19-27.

Heisler, L., A. Fortney, N.A. Cairns, A. Crosby, C. Sheffield, and R. Poulin. 2013. Herpetofauna observed during the Royal Saskatchewan Museum bioblitz of southwest Saskatchewan. The Canadian Herpetologist 3:10-13.

Hemenez, M.J., H.O. Clark, Jr., and R.K. Burton. 2013. Occurrence of a western spadefoot toad in the burrow system of the giant kangaroo rat, San Luis Obispo County, CA. Sonoran Herpetologist 26:24.

Herman, J.E. 2014. Road effects on amphibian populations: vehicles as a means of dispersal in treefrogs. Herpetological Review 45:203-205.

Herman, J.E., A. Benton, and I.N. Irizarry. 2015. *Osteopilus septentrionalis* (Cuban Treefrog). Cannibalism. Herpetological Review 46:75-76.

Hernández-Gómez, O., S.J. Kimble, J. Hua, V.P. Wuerthner, D.K. Jones, B.M. Mattes, R.D. Cothran, R.A. Relyea, G.A. Meindl, and J.T. Hoverman. 2019. Local adaptation of the MHC class II beta gene in populations of wood frogs (*Lithobates sylvaticus*) correlates with proximity to agriculture. Infection Genetics and Evolution 73:197-204.

Hernández-Salinas, U., A. Ramírez-Bautista, B.P. Stephenson, R. Cruz-Elizalde, C. Berriozabal-Islas, and C.J. Balderas-Valdivia. 2018. Amphibian life history in a temperate environment of the Mexican Plateau: dimorphism, phenology and trophic ecology of a hylid frog, *Hyla eximia* (=*Dryophytes eximius*). PeerJ 6:e5897.

Herrick, S.Z., K.D. Wells, T.E. Farkas, and E.T. Schultz. 2018. Noisy neighbors: acoustic interference and vocal interactions between two syntopic species of ranid frogs, *Rana clamitans* and *Rana catesbeiana*. Journal of Herpetology 52:176-184.

Herwig, B.R., L.W. Schroeder, K.D. Zimmer, M.A. Hanson, D.F. Staples, R.G. Wright, and J.A. Younk. 2013. Fish influences on amphibian presence and abundance in prairie and parkland landscapes of Minnesota, USA. Journal of Herpetology 47:489-497.

Heupel Greene, A. 2021. Rediscovering Florida's carpenter frogs. Abstract Booklet, Southeast Partners in Amphibian and Reptile Conservation Annual Meeting, 25–27 February 2021, p. 38.

Heuring, W.L., B.M. Poynter, S. Wells, and A.P. Pessier. 2021. Successful clearance of chytrid fungal infection in threatened Chiricahua leopard frog (*Rana chiricahuensis*) larvae and frogs using an elevated temperature treatment protocol. Salamandra 57:171-173.

Heyborne, W.H., C.E. Gardner, and I.W. Shipley. 2018. *Anaxyrus punctatus* (Red-spotted Toad). Cannibalism. Herpetological Review 49:728.

Hill, J., T.M. Terhune III, and J.A. Martin. 2019. *Lithobates catesbeianus* (American Bullfrog). Diet. Herpetological Review 50:761.

Hill, J.E., K.M. Lawson, and Q.M. Tuckett. 2017. First record of a reproducing population of the African clawed frog, *Xenopus laevis* Daudin, 1802 in Florida (USA). BioInvasions Records 6:87-94.

Hill, R.L., and M.G. Levy. 2014. Prevalence of *Batrachochytrium dendrobatidis* in pond-breeding amphibians of the Fall Line sandhills region of Georgia, USA. Herpetological Review 45:236-238.

Hinderer, R.K., A.R. Litt, and M. McCaffrey. 2017. Movement of imperiled Chiricahua leopard frogs during summer monsoons. Journal of Herpetology 51:497-503.

Höbel, G., D.S. Kim, and D. Neelon. 2014. Do green treefrogs (*Hyla cinerea*) eavesdrop on prey calls? Journal of Herpetology 48:389-393.

Hocking, D.J., and K.J. Babbitt. 2014. Amphibian contributions to ecosystem services. Herpetological Conservation and Biology 9:1-17.

Hoffmann, E.P., K. Williams, M.R. Hipsey, and N.J. Mitchell. 2021. Drying microclimates threaten persistence of natural and translocated populations of threatened frogs. Biodiversity and Conservation 30:15-34.

Holbrook, A.A. 2015. *Rhinella marina* (Cane Toad). Incidental translocation. Herpetological Review 46:237.

Holbrook, J.D., and N.J. Dorn. 2016. Fish reduce anuran abundance and decrease herpetofaunal species richness in wetlands. Freshwater Biology 61:100-109.

Holden, W.M., S.M. Hanlon, D.C. Woodhams, T.M. Chappell, H.L. Wells, S.M. Glisson, V.J. McKenzie, R. Knight, M.J. Parris, and L.A. Rollins-Smith. 2015. Skin bacteria provide early protection for newly metamorphosed southern leopard frogs (*Rana sphenocephala*) against the frog-killing fungus, *Batrachochytrium dendrobatidis*. Biological Conservation 187:91-102.

Holgerson, M.A., M.R. Lambert, L. Kealoha-Freidenburg, and D.K. Skelly. 2016. Suburbanization alters small pond ecosystems: shifts in nitrogen and food web dynamics. Canadian Journal of Fisheries and Aquatic Science 75:641-652.

Holt, B.D. 2015. Geographic distribution: *Lithobates areolatus* (Crawfish Frog). Herpetological Review 46:378-379.

Holycross, A.T., E.M. Nowak, B.L. Christman, and R.D. Jennings. 2020. *Thamnophis rufipunctatus*. Mogollon Narrow-headed Gartersnake. pp. 440-455 *In* A.R. Holycross and J.C. Mitchell (eds.), Snakes of Arizona. ECO Publishing, Rodeo, New Mexico.

Holycross, A.T., T. C. Brennan, and R. D. Babb. 2021. A Field Guide to Amphibians and Reptiles in Arizona. Second Edition. Arizona Game and Fish Department, Phoenix.

Holzer, K.A. 2014. Amphibian use of constructed and remnant wetlands in an urban landscape. Urban Ecosystems 17:955-968.

Holzer, K.A., and S.P. Lawler. 2015. Introduced reed canary grass attracts and supports a common native amphibian. Journal of Wildlife Management 79:1081-1090.

Homan, R.N., J.R. Bartling, R.J. Stenger, and J.L. Brunner. 2013. Detection of ranavirus in Ohio, USA. Herpetological Review 44:615-618.

Hoover, G.M., M.F. Chislock, B.J. Tornabene, S.C. Guffey, Y.J. Choi, C. De Perre, J.T. Hoverman, L.S. Lee, and M.S. Sepúlveda. 2017. Uptake and depuration of four per/polyfluoroalkyl substances (PFAAs) in northern leopard frog *Rana pipiens* tadpoles. Environmental Science & Technology Letters 4:399-403.

Horne, E.A., S. Foulks, and N.M. Bello. 2014. Visual display in Blanchard's cricket frogs (*Acris blanchardi*). Southwestern Naturalist 59:409-413.

Hossack, B.R., and R.K. Honeycutt. 2017. Declines revisited: long-term recovery and spatial population dynamics of tailed frog larvae after wildfire. Biological Conservation 212:274-278.

Hossack, B.R., W.H. Lowe, J.L. Ware, and P.S. Corn. 2013a. Disease in a dynamic landscape: host behavior and wildfire reduce amphibian chytrid infection. Biological Conservation 157:293-299.

Hossack, B.R., W.H. Lowe, and P.S. Corn. 2013b. Rapid increases and time–lagged declines in amphibian occupancy after wildfire. Conservation Biology 27:219–228.

Hossack, B.R., W.H. Lowe, R.K. Honeycutt, S.A. Parks, and P.S. Corn. 2013c. Interactive effects of wildfire, forest management, and isolation on amphibian and parasite abundance. Ecological Applications 23:479-492.

Hossack, B.R., M.J. Adams, C.A. Pearl, K.W. Wilson, E.L. Bull, K. Lohr, D. Patla, D.S. Pilliod, J.M. Jones, K.K. Wheeler, S.P. McKay, and P.S. Corn. 2013d. Roles of patch characteristics, drought frequency, and restoration in long-term trends of a widespread amphibian. Conservation Biology 27:1410-1420.

Hossack, B.R., W.R. Gould, D.A. Patla, E. Muths, R. Daley, K. Legg, and P.S. Corn. 2015. Trends in Rocky Mountain amphibians and the role of beaver as a keystone species. Biological Conservation 187:260-269.

Hossack, B.R., K.L. Smalling, C.W. Anderson, T.M. Preston, I.M. Cozzarelli, and R.K. Honeycutt. 2018. Effects of persistent energy-related brine contamination on amphibian abundance in national wildlife refuge wetlands. Biological Conservation 228:36-43.

Hossack, B.R., R.E. Russell, and R. McCaffrey. 2020. Contrasting demographic responses of toad populations to regionally synchronous pathogen (*Batrachochytrium dendrobatidis*) dynamics. Biological Conservation 241:108373.

Hoverman, J.T., Z. Olson, S. LaGrange, J. Grant, and R.N. Williams. 2015. A guide to larval amphibian identification in the field and laboratory. Purdue University Cooperative Extension Service, West Lafayette, Indiana. FNR-496-W.

Hoverman, J.T., M.E. Chislock, Z.A. Compton, and M.E. Gannon. 2019. Ranavirus reservoirs: assemblage of American bullfrog and green frog tadpoles maintains ranavirus infections across multiple seasons. Herpetological Review 50:275-278.

Howell, L.G., R. Frankham, J.C. Rodger, R.R. Witt, S. Clulow, R.M. Upton, and J. Clulow. 2021. Integrating biobanking minimises inbreeding and produces significant cost benefits for a threatened frog captive breeding programme. Conservation Letters 14:e12776.

Howell, P.E., B.R. Hossack, E. Muths, B.H. Sigafus, and R.B. Chandler. 2016. Survival estimates for reintroduced populations of the Chiricahua leopard frog (*Lithobates chiricahuensis*). Copeia 104:824-830.

Howell, P.E., E. Muths, B.R. Hossack, B.H. Sigafus, and R.B. Chandler. 2018. Increasing connectivity between metapopulation ecology and landscape ecology. Ecology 99:1119-1128.

Howell, P.E., B.H. Sigafus, B.R. Hossack, and E. Muths. 2019. Co-occurrence of Chiricahua leopard frogs (*Lithobates chiricahuensis*) with sunfish (*Lepomis*). Southwestern Naturalist 64:69-72.

Howell, P.E., B.R. Hossack, E. Muths, B.H. Sigafus, A. Chenevert-Steffler, and R.B. Chandler. 2020a. A statistical forecasting approach to metapopulation viability analysis. Ecological Applications 30:e02038.

Howell, P.E., B.R. Hossack, E. Muths, B.H. Sigafus, and R.B. Chandler. 2020b. Informing amphibian conservation efforts with abundance-based metapopulation models. Herpetologica 76:240-250.

Hromada, S.J., M.G. Iacchetta, B.J. Beas, J. Flaherty, M.C. Fulbright, K.H. Wild, A.F. Scott, and C.M. Gienger. 2021. Low-intensity agriculture shapes amphibian and reptile communities: insights from a 10-year monitoring study. Herpetologica 77:294-306.

Hua, J., and B.A. Pierce. 2013. Lethal and sublethal effects of salinity on three common Texas amphibians. Copeia 2013:562-566.

Hua, J., R. Cothran, A. Stoler, and R. Relyea. 2013. Cross-tolerance in amphibians: wood frog mortality when exposed to three insecticides with a common mode of action. Environmental Toxicology and Chemistry 32:932-936.

Hua, J., D.K. Jones, B.M. Mattes, R.D. Cothran, R.A. Relyea, and J.T. Hoverman. 2015a. The contribution of phenotypic plasticity to the evolution of insecticide tolerance in amphibian populations. Evolutionary Applications 8:586-596.

Hua, J., D.K. Jones, B.M. Mattes, R.D. Cothran, R.A. Relyea, and J.T. Hoverman. 2015b. Evolved pesticide tolerance in amphibians: Predicting mechanisms based on mode of action and pesticide novelty. Environmental Pollution 206:56-63.

Hua, J, V.P. Wuerthner, D.K. Jones, B. Mattes, R.D. Cothran, R.A. Relyea, and J.T. Hoverman. 2017. Evolved pesticide tolerance influences susceptibility to parasites in amphibians. Evolutionary Applications 10:802-812.

Huang, R., and L.A. Wilson. 2013. *Bratrachochytrium dendrobatidis* in amphibians of the Piedmont and Blue Ridge Provinces in northern Georgia, USA. Herpetological Review 44:95-98.

Hudgens, B, and M. Harbert. 2019. Amphipod predation on northern red-legged frog (*Rana aurora*) embryos. Northwestern Naturalist 100:126-131.

Hughes, D.F. 2014. *Lithobates clamitans melanota* (Northern Green Frog). Myiasis. Herpetological Review 45:305.

Hughes, D.F., and W.E. Meshaka, Jr. 2018. Life history of the Rio Grande leopard frog (*Lithobates berlandieri*) in Texas. Journal of Natural History 52:2221-2242.

Hughes, D.F., W.E. Meshaka, Jr., and P.R. Delis. 2017. Reproduction and growth of the southern leopard frog, *Lithobates sphenocephalus* (Cope, 1886), in Virginia: implications for seasonal shifts in response to global climate change. Basic and Applied Herpetology 31:17-31.

Hughes, D.F., P. Garcia-Bañuelos, R. Guevara, and W.E. Meshaka, Jr. 2019a. *Lithobates clamitans* (Green Frog). Partial leucism. Herpetological Review 50:115.

Hughes, D.F., M.L. Green, J.K. Warner, and P.C. Davidson. 2019b. There's a frog in my salad! A review of online media coverage for wild vertebrates found in prepackaged produce in the United States. Science of the Total Environment 675:1-12.

Hughes, D.F., M.L. Green, J.K. Warner, and P.C. Davidson. 2021. Evaluating exclusion barriers for treefrogs in agricultural landscapes. Wildlife Society Bulletin 45:305-311.

Hughey, M.C., M.H. Becker, J.B. Walke, M.C. Swartwout, and L.K. Belden. 2014. *Batrachochytrium dendrobatidis* in Virginia amphibians: within and among site infection variation. Herpetological Review 45:428-438.

Humfeld, S.C., and B. Grunert. 2015. Effects of temperature on spectral preferences of female gray treefrogs (*Hyla versicolor*). Herpetological Conservation and Biology 10:1013-1020.

Hunt, J.D. 2019. Improving monitoring and habitat assessment for gopher frog (*Rana [Lithobates] capito*) management in Georgia. Ph.D. Dissertation, University of Georgia, Athens.

Huss, M., L. Huntley, V. Vredenburg, J. Johns, and S. Green. 2014. Prevalence of *Batrachochytrium dendrobatidis* in archived specimens of *Lithobates catesbeianus* (American Bullfrog) collected in California, 1924-2007. EcoHealth 1:339-343.

Hutto Jr., D., and K. Barrett. 2021. Do urban open spaces provide refugia for frogs in urban environments? PLoS One 16(1):e0244932.

Hyman, O.J., and J.P. Collins. 2015. *Batrachochytrium dendrobatidis* dynamics in an isolated northern leopard frog (*Lithobates pipiens*) population in Arizona. Herpetological Review 46:535-537.

Igleski, M.J., and K.E. Nicholson. 2014. Spatial pattern of *Batrachochytrium dendrobatidis* infection in green frogs (*Lithobates clamitans*) in Michigan, USA. Herpetological Review 45:34-40.

Isidoro-Ayza, M., J.M. Lorch, D.A. Grear, M. Winzeler, D.L. Calhoun, and W.J. Barichivich. 2017. Pathogenic lineage of Perkinsea associated with mass mortality of frogs across the United States. Scientific Reports 7:10288.

Ivanov, K., and J.D. Gibson. 2020a. Field notes: *Gastrophryne carolinensis* (Eastern Narrow-mouthed Toad). Diet. Catesbeiana 40:128-130.

Ivanov, K., and J.D. Gibson. 2020b. Field notes: *Anaxyrus quercicus* (Oak Toad). Diet. Catesbeiana 40:126-127.

Jablonski, D., A. Alean, P. Vlček, and D. Jandzik. 2014. Axanthism in amphibians: A review and the first record in the widespread toad of the *Bufotes viridis* complex (Anura: Bufonidae). Belgian Journal of Zoology 144:93-101.

Jadin, R.C., S.A. Orlofske, T. Jezkova, and C. Blair. 2021. Single-locus species delimitation and ecological niche modelling provide insights into the evolution, historical distribution and taxonomy of the Pacific chorus frogs. Biological Journal of the Linnean Society 132:612-633.

Jaeger, J.R., A.W. Waddle, R. Rivera, D.T. Harrison, S. Ellison, M.J. Forrest, V.T. Vredenburg, and F. van Breukelen. 2017. *Batrachochytrium dendrobatidis* and the decline and survival of the relict leopard frog. EcoHealth 14:285-295.

James, T.Y., L.F. Toledo, R. Rödder, D. da Silva Leite, A.M. Belasen, C.M. Betancourt-Román, T.S. Jenkinson, C. Soto-Azat, C. Lambertini, A.V. Longo, J. Ruggeri, J.P. Collins, P.A. Burrowes, K.R. Lips, K.R. Zamudio, and J.E. Longcore. 2015. Disentangling host, pathogen, and environmental determinants of a recently emerged wildlife disease: lessons from the first 15 years of amphibian chytridiomycosis research. Ecology and Evolution 5:4079-4097.

Jancowski, K., and S. Orchard. 2013. Stomach contents from invasive American bullfrogs *Rana catesbeiana* (= *Lithobates catesbeianus*) on southern Vancouver Island, British Columbia. NeoBiota 16:17-37.

Jarboe, C.A., J.E. Colbert, B. Zulkiewicz, and D. Steen. 2019. Geographic distribution: *Osteopilus septentrionalis* (Cuban Treefrog). Herpetological Review 50:98.

Jarchow, C.J., B.R. Hossack, B.H. Sigafus, C.R. Schwalbe, and E. Muths. 2016. Modeling habitat connectivity to inform reintroductions: a case study with the Chiricahua leopard frog. Journal of Herpetology 50:63-69.

Jenkerson, J.T., D.M. Spontak, and J.D. Owen. 2017. Geographic distribution: *Incilius nebulifer* (Gulf Coast Toad). Herpetological Review 48:807.

Jennette, M.A., J.W. Snodgrass, and D.C. Forester. 2019. Variation in age, body size, and reproductive traits among urban and rural amphibian populations. Urban Ecosystems 22:137-147.

Jensen, J.B., and J. Wright. 2014. *Lithobates heckscheri* (River Frog). Mass metamorphosis. Herpetological Review 45:305-306.

Jeung, B. 2021. An examination of chemically-mediated

avoidance from captive-reared southern mountain yellow-legged frogs. M.S. thesis, California State University-East Bay, Hayward, California.

Johannsen, R.E. 2017. Updating the distributions of South Dakota amphibian and reptile species. B.S. thesis, University of South Dakota, Vermillion.

Johnson, M., S.M. Tekmen, and M.A. Bee. 2013. *Hyla chrysoscelis* (Cope's Gray Treefrog). Breeding activity. Herpetological Review 44:495.

Jones, B., J.W. Snodgrass, and D.R. Ownby. 2015. Relative toxicity of NaCl and road deicing salt to developing amphibians. Copeia 103:72-77.

Jones, C.A., and B.K. Sullivan. 2020. *Masticophis bilineatus.* Sonoran Whipsnake. pp. 240-251 *In* A.R. Holycross and J.C. Mitchell (eds.), Snakes of Arizona. ECO Publishing, Rodeo, New Mexico.

Jones, D., B. Forshee, and L. Fitzgerald. 2020. A herpetological survey of Edith L. Moore Nature Sanctuary. Check List 17(1):27-38.

Jones, D.K., and R.A. Relyea. 2015. Here today, gone tomorrow: short-term retention of pesticide-induced tolerance in amphibians. Environmental Toxicology and Chemistry 34:2295-2301.

Jones, D.K., T.D. Dang, J. Urbina, R.J. Bendis, J.C. Buck, R.D. Cothran, A.R. Blaustein, and R.A. Relyea. 2017. Effect of simultaneous amphibian exposure to pesticides and an emerging fungal pathogen, *Batrachochytrium dendrobatidis.* Environmental Science & Technology 51:671-679.

Jones, D.K., E.K. Yates, B.M. Mattes, W.D. Hintz, M.S. Schuler, and R.A. Relyea. 2018. Timing and frequency of sublethal exposure modifies the induction and retention of increased insecticide tolerance in wood frogs (*Lithobates sylvaticus*). Environmental Toxicology and Chemistry 37:2188-2197.

Jones, K.C., and G.W. Jones. 2017. *Scaphiopus holbrookii* (Eastern Spadefoot). Attempted predation by *Terrapene carolina.* Herpetological Review 48:171-172.

Jones, K.R., J.B. Walke, M.H. Becker, L.K. Belden, and M.C. Hughey. 2021. Time in the laboratory, but not exposure to chytrid fungus, results in rapid change in spring peeper (*Pseudacris crucifer*) skin bacterial communities. Ichthyology and Herpetology 109:75-83.

Jones, L.L.C., K.J. Halama, and R.E. Lovich (eds.). 2016. Habitat Management Guidelines for Amphibians and Reptiles of the Southwestern United States. Partners in Amphibian and Reptile Conservation, Technical Publication HMG-5, Birmingham, Alabama.

Jones, M.C., P. Crump, and M.R.J. Forstner. 2017. Sex ratios of captive raised Houston toads (*Bufo houstonensis*). Herpetological Review 48:82-84.

Jones, T.R., and F.R. Hensley. 2020. *Thamnophis cyrtopsis.* Black-necked Gartersnake. pp. 388-400 *In* A.R. Holycross and J.C. Mitchell (eds.), Snakes of Arizona. ECO Publishing, Rodeo, New Mexico.

Jones, T.R., M.J. Ryan, T.B. Cotton, and J.M. Servoss. 2020. *Thamnophis eques.* Mexican Gartersnake. pp. 418-432 *In* A.R. Holycross and J.C. Mitchell (eds.), Snakes of Arizona. ECO Publishing, Rodeo, New Mexico.

Jongsma, G.F.M., M.A. Empey, C.M. Smith, A.M. Bennett, and D.F. McAlpine. 2019. High prevalence of the amphibian pathogen *Batrachochytrium dendrobatidis* in plethodontid salamanders in protected areas in New Brunswick, Canada. Herpetological Conservation and Biology 14:91-96.

Joseph, M.B., and R.A. Knapp. 2018. Disease and climate effects on individuals drive post-reintroduction population dynamics of an endangered amphibian. Ecosphere 9:1-18.

Julian, J.T., R.P. Brooks, G.W. Glenney, and J.A. Coll. 2019. State-wide survey of amphibian pathogens in green frog (*Lithobates clamitans melanota*) reveals high chytrid infection intensities in constructed wetlands. Herpetological Conservation and Biology 14:199-211.

Kalnicky, E.A., K.H. Beard, and M.W. Brunson. 2013. Community-level response to habitat structure manipulations: an experimental case study in a tropical ecosystem. Forest Ecology and Management 307:313-321.

Kalnicky, E.A., M.W. Brunson, and K.H. Beard. 2014. A social-ecological systems approach to non-native species: habituation and its effect on management of coqui frogs in Hawaii. Biological Conservation 180:187-195.

Kamoroff, C., N. Daniele, R.L. Grasso, R. Rising, T. Espinoza, and C.S. Goldberg. 2020. Effective removal of the American bullfrog (*Lithobates catesbeianus*) on a landscape level: long term monitoring and removal efforts in Yosemite Valley, Yosemite National Park. Biological Invasions 22:617-626.

Karavlan, S.A., and M.D. Venesky. 2016. Thermoregulatory behavior of *Anaxyrus americanus* in response to infection with *Batrachochytrium dendrobatidis.* Copeia 104:746-751.

Karwacki, E.E., M.S. Atkinson, R.J. Ossiboff, and A.E. Savage. 2018. Novel quantitative PCR assay specific for the emerging Perkinsea amphibian pathogen reveals seasonal infection dynamics. Diseases of Aquatic Organisms 129:85-98.

Karwacki, E.E., K.R. Martin, and A.E. Savage. 2021. One hundred years of infection with three global pathogens in frog populations of Florida, USA. Biological Conservation 257:109088.

Kasper, S. 2014. *Anaxyrus debilis* (Green Toad). Predation. Herpetological Review 45:677.

Kasper, S. 2021. *Rana blairi* (Plains Leopard Frog). Defensive behavior and interspecific hibernaculum. Herpetological Review 52:373-374.

Kasper, S., and F.D. Yancey III. 2013. *Lithobates berlandieri* (Rio Grande Leopard Frog). Learned feeding behavior and diet. Herpetological Review 44:295.

Katzenberger, M., J. Hammond, M. Tejedo, and R. Relyea. 2018. Source of environmental data and warming tolerance estimation in six species of North American larval anurans. Journal of Thermal Biology 76:171-178.

Kean, C.M., and S.J.A. Kimble. 2021. *Lithobates sylvaticus* (Wood Frog) and *Ambystoma maculatum* (Spotted Salamander). Possibly fatal interspecific amplexus. Herpetological Review 52:614-615.

Keesing, F., and R.S. Ostfeld. 2021. Dilution effects in disease ecology. Ecology Letters 24: 2490-2505.

Kelehear, C., M. Seiler, and S.P. Graham. 2017. *Gastrophryne olivacea* (Western Narrow-mouthed Toad). Predation. Herpetological Review 48:606-607.

Kelly, D.O., R.J. Scott, C.E. Campbell, and I.G. Warkentin. 2017. Initial dispersal and breeding habitat use of newly introduced mink frogs in western Newfoundland, Canada. Copeia 105:389-398.

Kelly, P.W., D.W. Pfennig, and K.S. Pfennig. 2021. A condition-dependent male sexual signal predicts adaptive predator-induced plasticity in offspring. Behavioral Ecology and Sociobiology 75:28.

Kennedy, J.G., S.A. Johnson, J.S. Brewer, and C.J. Leary. 2021. The potential role of reproductive interference in the decline of native green treefrogs following Cuban treefrog invasions. Biological Invasions 23:553–568.

Keung, N.C. 2015. Longitudinal distribution and summer diurnal microhabitat use of California red-legged frogs (*Rana draytonii*) in coastal Waddell Creek. M.S. thesis, San Jose State University, San Jose, California.

Keung, N.C., S.P. Lawler, S.M. Yarnell, B.D. Todd, and C. Brown. 2021. Movement ecology study of stream-dwelling Sierra Nevada yellow-legged frogs (*Rana sierrae*) to inform reintroductions. Herpetological Conservation and Biology 16:72-85.

Kimble, S.J., R.N. Williams, and J.T. Hoverman. 2015. Ranavirus detected in *Lithobates clamitans* and *L. catesbeianus* in Indiana. Herpetological Review 46:532-534.

Kinsley, H., L. Cusack, and E. Stroh. 2020. Field notes: survey of Daniel Island. Catesbeiana 40:81-82.

Kirsch, D.R., S. Fix, J.M. Davenport, K.K. Cecala, and J.R. Ennen. 2021. Body size is related to temperature preference in *Hyla chrysoscelis* tadpoles. Journal of Herpetology 55:21-25.

Kirschman, L.J., J.G. Palis, K.A. Fritz, K. Althoff, and R.W. Warne. 2017. Two ranavirus-associated mass-mortality events among larval amphibians in Illinois, USA. Herpetological Review 48:779-782.

Kiesow, A.M., and D.R. Davis. 2020. Field Guide to Amphibians and Reptiles of South Dakota. Second Edition. South Dakota Department of Game, Fish and Parks, Pierre, South Dakota.

Kissel, A.M., W.J. Palen, P. Govindarajulu, and C.A. Bishop. 2014. Quantifying ecological life support: the biological efficacy of alternative supplementation strategies for imperiled amphibian populations. Conservation Letters 7:441-450.

Kissel, A.M., S. Tenan, and E. Muths. 2020. Density dependence and adult survival drive dynamics in two high elevation amphibian populations. Diversity 12:478.

Klaus, J.M., and R.F. Noss. 2016. Specialist and generalist amphibians respond to wetland restoration treatments. Journal of Wildlife Management 80:1106-1119.

Klawinski, P.D., B. Dalton, and A.B. Shiels. 2014. Coqui frog populations are negatively affected by canopy opening but not detritus deposition following an experimental hurricane in a tropical rainforest. Forest Ecology and Management 332:118-123.

Klemens, M.W., H.J. Gruner, D.P. Quinn, and E.R. Davison. 2021. Conservation of Amphibians and Reptiles in Connecticut. Department of Energy and Environmental Conservation, Hartford, Connecticut.

Knapp, R.A., G.M. Fellers, P.M. Kleeman, D.A.W. Miller, V.T. Vredenburg, E.B. Rosenblum, and C.J. Briggs. 2016. Large-scale recovery of an endangered amphibian despite ongoing exposure to multiple stressors. Proceedings of the National Academy of Sciences of the United States of America 113:11889-11894.

Kohl, K.D., T.L. Cary, W.H. Karasov, and M. D. Dearing. 2015. Larval exposure to polychlorinated biphenyl 126 (PCB-126) causes persistent alteration of the amphibian gut microbiota. Environmental Toxicology and Chemistry 34:1113-1118.

Kolby, J., and P. Daszak. 2016. The emerging amphibian fungal disease, Chytridiomycosis: A key example of the global phenonmenon of wildlife emerging infectious diseases. Microbiology Spectrum 4(3):El10-0004-2015.

Kolozsvary, M.B., and J.L. Brunner. 2016. Presence of *Ranavirus* in a created temporary pool complex in southeastern New York, USA. Herpetological Review 47:598-600.

Konvalina, J.D., and S.E. Trauth. 2015. *Hyla cinerea* (Green Treefrog). Predation. Herpetological Review 46:612-613.

Kraus, F. 2015. Impacts from invasive reptiles and amphibians. Annual Review of Ecology, Evolution, and Systematics 46:75–97.

Kreiser, B.R., E.Y. Kreiser, and J.Y. Lamb. 2016. *Anaxyrus terrestris* (Southern Toad). Forelimb malformation. Herpetological Review 47:104-105.

Kreitals, N., J.-M. Davies, D.W. Phillips, and I.D. Phillips. 2014. Overwintering observations of northern leopard frog (*Lithobates pipiens*) activity under ice. IRCF Reptiles & Amphibians 21:60-65.

Kross, C.S., and S.C. Richter. 2016. Species interactions in constructed wetlands result in population sinks for wood frogs (*Lithobates sylvaticus*) while benefitting eastern newts (*Notophthalmus viridescens*). Wetlands 36:385-393.

Kruger, A., and P.J. Morin. 2020. Predators induce morphological changes in tadpoles of *Hyla andersonii*. Copeia 108:316-325.

Krynak, K.L., and P.M. Dennis. 2014. Detection of ranavirus in northeast Ohio's wetlands. Herpetological Review 45:601-602.

Krynak, K.L., D.J. Burke, and M.F. Benard. 2015. Larval environment alters amphibian immune defenses differentially across life stages and populations. PLoS One 10(6):e013383.

Krynak, K.L., D.J. Burke, and M.F. Benard. 2016. Landscape and water characteristics correlate with immune defense traits across Blanchard's cricket frog (*Acris blanchardi*) populations. Biological Conservation 193:153-167.

Krynak, K.L., D.J. Burke, and M.F. Benard. 2017. Rodeo™ herbicide negatively affects Blanchard's cricket frogs (*Acris blanchardi*) survival and alters the skin-associated bacterial community. Journal of Herpetology 51:402-410.

Krysko, K.L., L.A. Somma, D.C. Smith, C.R. Gillette, D. Cueva, J.A. Wasilewski, K.M. Enge, S.A. Johnson, T.S. Campbell, J. R. Edwards, M.R. Rochford, R. Tompkins, J.L. Fobb, S. Mullin, C.J. Lechowicz, D. Hazelton, and A. Warren. 2016. New verified nonindigenous amphibians and reptiles in Florida through 2015: with a summary of over 152 years of introduction. IRCF Reptiles and Amphibians 23:110-143.

Krysko, K.L., K.M. Enge, and P.E. Moler. 2019. Amphibians and Reptiles of Florida. University of Florida Press, Gainesville.

Kuczynski, M.C., D. Bello-DeOcampo, and T. Getty. 2015. No evidence of terminal investment in the gray treefrog (*Hyla versicolor*): older males do not signal at greater effort. Copeia 103:530-535.

Kupferberg, S. 2021. *Rana boylii* (Foothill Yellow-legged Frog). Reproductive behavior and habitat use. Herpetological Review 52:835-836.

Kupferberg, S.J., H. Moidu, A.J. Adams, A. Catenazzi, M. Grefsrud, S. Bobzien, R. Leidy, and S.M. Carlson. 2021. Seasonal drought and its effects on frog population dynamics and amphibian disease in intermittent streams. Ecohydrology (2021):e2395.

Lambert, B.R., R.A. Schorr, S.C. Schneider, and E. Muths. 2016a. Influence of demography and environment on persistence in toad populations. Journal of Wildlife Management 80:1256-1266.

Lambert, M.R., and D.K. Skelly. 2016. Diverse sources for endocrine disruption in the wild. Endocrine Disruptors 4:e1148803.

Lambert, M.R., G.S.J. Giller, L.B. Barber, K.C. Fitzgerald, and D.K. Skelly. 2015. Suburbanization, estrogen contamination, and sex ratio in wild amphibian populations. Proceedings of the National Academy of Sciences of the United States of America 112:11881-11886.

Lambert, M.R., G.S.J. Giller, D.K. Skelly, and R.G. Bribiescas. 2016b. Septic systems, but not sanitary sewer lines, are associated with elevated estradiol in male frog metamorphs from suburban ponds. General and Comparative Endocrinology 323:109-114.

Lambert, M.R., B.E. Carlson, M.S. Smylie, and L. Swierk. 2017a. Ontogeny of sexual dicromatism in the explosively breeding wood frog. Herpetological Conservation and Biology 12:447-456.

Lambert, M.R., A.B. Stoler, M.S. Smylie, R.A. Relyea, and D.K. Skelly. 2017b. Interactive effects of road salt and leaf litter on wood frog sex ratios and sexual size dimorphism. Canadian Journal of Fisheries and Aquatic Sciences 74:141-146.

Lambert, M.R., M.S. Smylie, A.J. Roman, L. Kealoha-Freidenburg, and D.K. Skelly. 2018. Sexual and somatic development of wood frog tadpoles along a thermal gradient. Journal of Experimental Zoology 329:72-79.

Lanctôt, C., C. Robertson, L. Navarro-Martín, C. Edge, S.D. Melvin, J. Houlahan, and V.L. Trudeau. 2013. Effects of the glyphosate-based herbicide Roundup WeatherMax® on metamorphosis of wood frogs (*Lithobates sylvaticus*) in natural wetlands. Aquatic Toxicology 140-141:48-57.

Lanctôt, C., L. Navarro-Martín, C. Robertson, B. Park, P. Jackman, B.D. Pauli, and V.L. Trudeau. 2014. Effects of glyphosate-based herbicides on survival, development, growth and sex ratios of wood frog (*Lithobates sylvaticus*) tadpoles. II: Agriculturally relevant exposures to Roundup WeatherMax® and Vision® under laboratory conditions. Aquatic Toxicology 154:291-303.

Lannoo, M.J., and R.M. Stiles. 2017. Effects of short-term climate variation on a long-lived frog. Copeia 105:726-733.

Lannoo, M.J., and R.M. Stiles. 2020a. The Call of the Crawfish Frog. CRC Press, Boca Raton, Florida.

Lannoo, M.J., and R.M. Stiles. 2020b. Uncovering shifting amphibian ecological relationships in a world of environmental change. Herpetologica 76:144-152.

Lannoo, M.J., R.M. Stiles, M.A. Sisson, J.W. Swan, V.C.K. Terrell, and K.E. Robinson. 2017. Patch dynamics inform management decisions in a threatened frog species.

Copeia 105:53-63.

Lannoo, M.J., R.M. Stiles, D. Saenz, and T.J. Hibbitts. 2018. Comparative call characteristics in the anuran subgenus *Nenirana*. Copeia 106:575-579.

Lapp, S., T. Wu, C. Richards-Zawacki, J. Voyles, K.M. Rodriguez, H. Shamon, and J. Kitzes. 2021. Automated detection of frog calls and choruses by pulse repetition rate. Conservation Biology 35:1659-1668.

Larsen, A.S., J.H. Schmidt, H. Stapleton, H. Kristenson, D.Betchkal, and M.F. McKenna. 2021. Monitoring the phenology of the wood frog breeding season using bioacoustic methods. Ecological Indicators 131:108142.

Larson, D.J., L. Middle, H. Vu, W. Zhang, A.S. Serianni, J. Duman, and B.M. Barnes. 2014. Wood frog adaptations to overwintering in Alaska: new limits to freezing tolerance. Journal of Experimental Biology 217:2193-2200.

Laskow, S. 2017. The giant frog farms of the 1930s were a giant failure. Atlas Obscura, 25 October 2017, https://atlasobscura.com/articles/frog-farming-1930s-failure-ponds-canning-legs-conservation.

Laughlin, M.M., E.R. Olson, and J.G. Martin. 2017. Arboreal camera trapping expands *Hyla versicolor* complex (Hylidae) canopy use to new heights. Ecology 98:2221-2223.

LeClair, G., M.W.H. Chatfield, Z. Wood, J. Parmelee, and C.A. Frederick. 2021. Influence of the COVID-19 pandemic on amphibian road mortality. Conservation Science and Practice 3:e535.

Leduc, J., P. Echaubard, V. Trudeau, and D. Lesbarrères. 2016. Copper and nickel effects on survival and growth of northern leopard frog (*Lithobates pipiens*) tadpoles in field-collected smelting effluent water. Environmental Toxicology and Chemistry 35:687-694.

Lee, T.S., N.L. Kahal, H.L. Kinas, L.A. Randall, T.M. Baker, V.A. Carney, K. Kendell, K. Sanderson, and D. Duke. 2021. Advancing amphibian conservation through citizen science in urban municipalities. Diversity 13:211.

Leggett, S., J. Borrelli, D.K. Jones, and R. Relyea. 2021. The combined effects of road salt and biotic stressors on amphibian sex ratios. Environmental Toxicology and Chemistry 40:231-235.

LeGros, D.L., D. Lesbarrères, and B. Steinberg. 2021. Terrestrial dispersal of juvenile Mink Frog (*Lithobates septentrionalis*) in Algonquin Provincial Park, Ontario. Canadian Field-Naturalist 135:47-51.

Lehtinen, R.M., and J.R. Witter. 2014. Detecting frogs and detecting declines: an examination of occupancy and turnover patterns at the range edge of Blanchard's cricket frog (*Acris blanchardi*). Herpetological Conservation and Biology 9:502-515.

Lemm, J.M., and M.W. Tobler. 2021. Factors affecting the presence and abundance of amphibians, reptiles, and small mammals under artificial cover in Southern California. Herpetologica 77:307-319.

Lemmon, E.M., and T.E. Juenger. 2017. Geographic variation in hybridization across a reinforcement contact zone of chorus frogs (*Pseudacris*). Ecology and Evolution 7:9485-9502.

Lemos Espinal, J.A., H.M. Smith, and A. Cruz. 2013. Amphibians & Reptiles of the Sierra Tarahumara of Chihuahua, México. ECO Herpetological Publishing & Distribution, Rodeo, New Mexico.

Lent, E.M., and K.J. Babbitt. 2020. The effects of hydroperiod and predator density on growth, development, and morphology of wood frogs (*Rana sylvatica*). Aquatic Ecology 54:369-386.

Lentz, T.B., M.C. Allender, S.Y. Thi, A.S. Duncan, A.X. Miranda, J.C. Beane, D.S. Dombrowski, B.R. Forester, C.K. Akcali, N.A. Shepard, J.E. Corey, III, A.L. Braswell, L.A. Williams, C.R. Lawson, C. Jenkins, J.H.K. Pechmann, J. Blake, M. Hooper, K. Freitas, A.B. Somers, and B.L. Stuart. 2021. Prevalence of *Ranavirus*, *Batrachochytrium dendrobatidis*, *B. salamandrivorans*, and *Ophidiomyces ophiodiicola* in amphibians and reptiles of North Carolina, USA. Herpetological Review 52:285-293.

Le Sage, E. H., B. C. LaBumbard, L. K. Reinert, B. T. Miller, C. L. Richards-Zawacki, D. C. Woodhams, and L.A. Rollins-Smith. 2021a. Preparatory immunity: Seasonality of mucosal skin defences and *Batrachochytrium* infections in southern leopard frogs. Journal of Animal Ecology 90:542-554.

Le Sage, E.H., S.I. Duncan, T. Seaborn, J. Cundiff, L.J. Rissler, and E.J. Crespi. 2021b. Ecological adaptation drives wood frog population divergence in life history traits. Heredity 126:790-804.

Léveillé, A.N., E.G. Zeldenrust, and J.R. Barta. 2021. Multilocus genotyping of sympatric Hepatozoon species infecting the blood of Ontario ranid frogs reinforces species differentiation and identifies an unnamed Hepatozoon species. Journal of Parasitology 107:246-261.

Levis, N.A. 2013. *Lithobates sylvaticus* (Wood Frog). Unusual mortality. Herpetological Review 44:497-498.

Lewis, J.L., J.J. Borrelli, D.K. Jones, and R.A. Relyea. 2021. Effects of freshwater salinization and biotic stressors on amphibian morphology. Ichthyology and Herpetology 109:157-164.

Liang, C.T. 2013. Movements and habitat use of Yosemite toads (*Anaxyrus* (formerly *Bufo*) *canorus*) in the Sierra National Forest, California. Journal of Herpetology 47:555-564.

Liang, C.T., R.L. Grasso, J.J. Nelson-Paul, K.E. Vincent, and A.J. Lind. 2017. Fine-scale habitat characteristics related to occupancy of the Yosemite toad, *Anaxyrus canorus*. Copeia 105:120-127.

Lieto, M.J., and R.L. Burke. 2019. Tracking the spread of six invasive amphibians and reptiles using the geographic distribution records published in Herpetological Review. Herpetological Review 50:474-479.

Lillywhite, H.B., R. Shine, E. Jacobson, D.F. Denardo, M.S. Gordon, C.A. Navas, T. Wang, R.S. Seymour, K.B. Storey, H. Heatwole, D. Heard, B. Brattstrom, and G.M. Burghardt. 2017. Anesthesia and euthanasia of amphibians and reptiles used in scientific research: should hypothermia and freezing be prohibited? BioScience 67:53-61.

Lind, A.J., H.H. Welsh, Jr., and C.A. Wheeler. 2016. Foothill yellow-legged frog (*Rana boylii*) oviposition site choice at multiple spatial scales. Journal of Herpetology 50:263-270.

Lindauer, A.L., and J. Voyles. 2019. Out of the frying pan, into the fire? Yosemite toad (*Anaxyrus canorus*) susceptibility to *Batrachochytrium dendrobatidis* after development under drying conditions. Herpetological Conservation and Biology 14:185-198.

Lindemann, S.B., and A.M. O'Brien. 2019. Geographic distribution: *Hyla versicolor* (Gray Treefrog). Herpetological Review 50:97.

Lindemann, S.B., D.E. Putnam, M.L. Hunter, Jr., and T.B. Persons. 2019a. Spotless burnsi pattern in northern leopard frog (*Lithobates pipiens*) in Maine. Canadian Field-Naturalist 133:189-192.

Lindemann, S.B., A.M. O'Brien, T.B. Persons, and P.G. deMaynadier, 2019b. Axanthism in green frogs (*Lithobates clamitans*) and an American bullfrog (*Lithobates catesbeianus*) in Maine. Canadian Field-Naturalist 133:193-195.

Linhoff, L., and M. Donnelly. 2016. *Anaxyrus baxteri* (Wyoming Toad). Interactions with cattle. Herpetological Review 47:103-104.

Lingenfelter, A.R., K. Geluso, M.P. Nenneman, B.C. Peterson, and J.L. Kerby. 2014. Distribution, diet, and prevalence of amphibian chytrid fungus in non-native American bullfrogs (*Lithobates catesbeianus*) at the Valentine National Wildlife Refuge, Nebraska, USA. Journal of North American Herpetology 2014:81-86.

Litmer, A.R., and C.M. Murray. 2020. Critical thermal capacities of *Hyla chrysoscelis* in relation to season. Journal of Herpetology 54:413-417.

Lodato, M.J., N.M. Gordon, and N.J. Engbrecht. 2018. Geographic distribution: *Hyla cinerea* (Green Treefrog). Herpetological Review 49:707.

Lott, T. 2014. Update on the distribution of the Rio Grande chirping frog, *Eleutherodactylus cystignathoides campi*, in the United States (Anura: Eleutherodactylidae). SWCHR Bulletin 4(2):19-21.

Lott, T. 2016. A possible historical explanation for anomalous Texas locality records for the Mexican tree frog (*Smilisca baudinii*) (Amphibia: Anura: Hylidae). SWCHR Bulletin 6(3):43-45.

Loudon, A.H., A. Kurtz, E. Esposito, T.P. Umile, K.P.C. Minbiole, L.W. Parfrey, and B.A. Sheafor. 2020. Columbia spotted frogs (*Rana luteiventris*) have characteristic skin microbiota that may be shaped by cutaneous skin peptides and the environment. FEMS Microbiology Ecology 96(10):fiaa168.

Lovett, G.M. 2013. When do peepers peep? Climate and the date of first calling in the spring peeper (*Pseudacris crucifer*) in southeastern New York state. Northeastern Naturalist 20:333-340.

Lowe, W.H., T.E. Martin, D.K. Skelly, and H.A. Woods. 2021. Metamorphosis in an era of increasing climate variability. Trends in Ecology & Evolution 36:360-375.

Luccioni, M.D., and J.T. Wyman. 2021. *Incilius alvarius* (Sonoran Desert Toad). Ingestion of bullet casing. Herpetological Review 52:827.

Lucid, M.K., L. Robinson, A.C. Gygli, J. Neidler, A. Delima, and S. Ehlers. 2020. Status of pond-breeding amphibians in northern Idaho and northwestern Washington, USA. Herpetological Conservation and Biology 15:526-535.

Lucid, M.K., S. Ehlers, L. Robinson, and J. Sullivan. 2021. Genetic structure not detected in northern Idaho and northeast Washington western toad (*Anaxyrus boreas*) populations. Northwestern Naturalist 102:89-93.

Lukey, N. 2017. Management of introduced American bullfrogs (*Lithobates catesbeianus* Shaw 1802) in the South Okanagan, British Columbia. M.S. thesis, University of Waterloo, Waterloo, Ontario.

Mack, M., and L. Beaty. 2021. The influence of environmental and physiological factors on variation in American toad (*Anaxyrus americanus*) dorsal coloration. Journal of Herpetology 55:119-126.

MacLaren, A.R., S.F. McCracken, and M.R.J. Forstner. 2018. Automated monitoring techniques reveal new proximate cues of Houston toad chorusing behavior. Herpetological Review 49:622-626.

MacKnight, M., and A.K. Burrow. 2019. *Pseudacris ornata* (Ornate Chorus Frog). Albinism. Herpetological Review 50:549-550.

MacVean, S.R. 2021. *Rana pipiens* (Northern Leopard Frog). Overwintering. Herpetological Review 52:837-838.

Maier, P.A. 2018a. Evolutionary past, present, and future of the Yosemite toad (*Anaxyrus canorus*): a total evidence approach to delineating conservation units. Ph.D. Dissertation, University of California-Riverside and San Diego State University, California.

Maier, P.A. 2018b. *Anaxyrus canorus* (Yosemite Toad). Breeding behavior. Herpetological Review 49:727-728.

Maier, P.A. 2019. *Anaxyrus canorus* (Yosemite Toad). Larval diet. Herpetological Review 50:111-112.

Maldonado, B.R., B.M. Glorioso, and R.P. Kidder II. 2020. *Acris blanchardi* (Blanchard's Cricket Frog). Predation. Herpetological Review 51:296.

Mann, D.L., T. Mann, W. Lu, and N. Winstead. 2015. Geographic distribution: *Eleutherodactylus planirostris* (Greenhouse Frog). Herpetological Review 46:377.

Mannan, R.N., G. Perry, D.E. Anderson, and C.W. Boal. 2014. Call broadcasting and automated recorders as tools for anuran surveys in a subarctic tundra landscape. Journal of North American Herpetology 2014:47-52.

Marchetti, J.R., and K.H. Beard. 2021. Predator–prey reunion: Non-native coquí frogs avoid their native predators. Ichthyology & Herpetology 109:791-795.

Marcogliese, D.J., K.C. King, and K.A. Bates. 2021. Effects of multiple stressors on northern leopard frogs in agricultural wetlands. Parasitology 148:827-834.

Marhanka, E.C., J.L. Watters, N.A. Huron, A.L. McMillin, D.J. Curtis, D.R. Davis, J.K. Farkas, J.L. Kerby, and C.D. Siler. 2017. Detection of high prevalence of *Batrachochytrium dendrobatidis* in amphibians from southern Oklahoma, USA. Herpetological Review 48:70-74.

Marino, J.A., Jr. 2014. Interactions between echinostome parasites and larval anurans across ecological contexts and scales. Ph.D. Dissertation, University of Michigan, Ann Arbor.

Marino, J.A., Jr., and E.E. Werner. 2013. Synergistic effects of predators and trematode parasites on larval green frog (*Rana clamitans*) survival. Ecology 94:2697-2708.

Marlow, K.R., K.D. Wiseman, C.A. Wheeler, J.E. Drennan, and R.E. Jackman. 2016. Identification of individual foothill yellow-legged frogs (*Rana boylii*) using chin pattern photographs: a non-invasive and effective method for small population studies. Herpetological Review 47:193-198.

Marsh, D.L., B.J. Cosentino, K.S. Jones, J.J. Apodaca, K.H. Beard, J.M. Bell, C. Bozarth, D. Carper, J.F. Charbonnier, A. Dantas, E.A. Forys, M. Foster, J. General, K.S. Genet, M. Hanneken, K.R. Hess, S.A. Hill, F. Iqbal, N.E. Karraker, E.S. Kilpatrick, T.A. Langen, J. Langford, K. Lauer, A.J. McCarthy, J. Neale, S. Patel, A. Patton, C. Southwick, N. Stearrett, N. Steijn, M. Tasleem, J.M. Taylor, and J.R. Vonesh. 2017. Effects of roads and land use on frog distributions across spatial scales and regions in the eastern and central United States. Diversity and Distributions 23:158-170.

Martin, L.J., and B. Blossey. 2013. Intraspecific variation overrides origin effects in impacts of litter-derived secondary compounds on larval amphibians. Oecologia 173:449-459.

Matchett, J.R., P.B. Stark, S.M. Ostoja, R.A. Knapp, H.C.

McKenny, M.L. Brooks, W.T. Langford, L.N. Joppa, and E.L. Berlow. 2015. Detecting the influence of rare stressors on rare species in Yosemite National Park using a novel stratified permutation test. Scientific Reports 5:10702.

Mays, J.D. 2016. Geographic distribution: *Hyla* (sic) *feriarum* (Upland Chorus Frog). Herpetological Review 47:419.

McAllister, C.T. 2020. *Rana sphenocephala* (Southern Leopard Frog). Diet. Herpetological Review 51:303-304.

McAllister, C.T., and N.H. McAllister. 2013. *Lithobates clamitans melanota* (Green Frog). Unusual food item. Herpetological Review 44:295-296.

McAllister, C.T., and L.A. Durden. 2014. New records of chiggers and ticks from an Oklahoma amphibian and reptile. Proceedings of the Oklahoma Academy of Science 94:40-43.

McAllister, C.T., C.R. Bursey, M.B. Connior, and S.E. Trauth. 2013. Symbiotic protozoa and helminth parasites of the Cajun chorus frog, *Pseudacris fouquettei* (Anura: Hylidae), from southern Arkansas and northeastern Texas. Comparative Parasitology 80:96-104.

McAllister, C.T., R.S. Seville, C.R. Bursey, S.E. Trauth, M.B. Connior, and H.W. Robison. 2014a. A new species of *Eimeria* (Apicomplexa: Eimeriidae) from the green frog, *Lithobates clamitans* (Anura: Ranidae), from Arkansas, U.S.A. Comparative Parasitology 81:175-178.

McAllister, C.T., C.R. Bursey, M.B. Connior, and S.E. Trauth. 2014b. Myxozoan and helminth parasites of the dwarf American toad, *Anaxyrus americanus charlesmithi* (Anura: Bufonidae), from Arkansas and Oklahoma. Proceedings of the Oklahoma Academy of Science 94:51-58.

McAllister, C.T., C.R. Bursey, and D.M. Calhoun. 2015a. Commensal Protista, cnidaria and helminth parasites of the Cajun chorus frog, *Pseudacris fouquettei* (Anura: Hylidae), from Oklahoma. Proceedings of the Oklahoma Academy of Science 95:1-10.

McAllister, C.T., C.R. Bursey, M.B. Connior, S.E. Trauth, and L.A. Durden. 2015b. New host and geographic-distribution records for helminth and arthropod parasites of the southern toad *Anaxyrus terrestris* (Anura: Bufonidae) from Florida. Southeastern Naturalist 14:641-649.

McAllister, C.T., C.R. Bursey, and M.B. Connior. 2017. New host and distributional records for parasites (Apicomplexa, Trematoda, Nematoda, Acanthocephala, Acarina) of Texas herpetofauna. Comparative Parasitology 84:42-50.

McAllister, C.T., J.C. Mitchell, and L.A. Durden. 2018. *Acris blanchardi* (Blanchard's Cricket Frog). Ectoparasites. Herpetological Review 49:297-298.

McAlpine, D.F., and J. Gilhen. 2018. Erythrism in spring

peeper (*Pseudacris crucifer*) in Maritime Canada. Canadian Field-Naturalist 132:43-45.

McCaffrey, R.M., L.A. Eby, B.A. Maxwell, and P.S. Corn. 2014. Breeding site heterogeneity reduces variability in frog recruitment and population dynamics. Biological Conservation 170:169-176.

McCaffery, R., R.E. Russell, and B.R. Hossack. 2021. Enigmatic near-extirpation in a boreal toad metapopulation in northwestern Montana. Journal of Wildlife Management 85: 953-963.

McCallum, M.L., M. Mitchell, and Y. Tilahun. 2020. Can environmentally stressed amphibians alter foraging activity during the sudden onset of winter? Bulletin of the Maryland Herpetological Society 56:12-18.

McCallum, M.L., and S.E. Trauth. 2021. Complex associations of environmental factors may explain Blanchard's cricket frog, *Acris blanchardi*, declines and drive population recovery. Journal of North American Herpetology 2021:36-49.

McCartney-Melstad, E., M. Gidiş, and H. B. Shaffer. 2018. Population genomic data reveal extreme geographic subdivision and novel conservation actions for the declining foothill yellow-legged frog. Heredity 121:112-125.

McClelland, S.J., R.J. Bendis, R.A. Relyea, and S.K. Woodley. 2018. Insecticide-induced changes in amphibian brains: how sublethal concentrations of chlorpyrifos directly affect neurodevelopment. Environmental Toxicology and Chemistry 37:2692-2698.

McClure, R.R. 2021. *Gastrophryne olivacea* (Western Narrow-mouthed Toad). Habitat. Herpetological Review 52:608.

McConnell, R., T. McConnell, C. Guyer, and D. Laurencio. 2015. Geographic distribution: *Eleutherodacylus cystignathoides* (Rio Grande Chirping Frog). Herpetological Review 46:559.

McCoy, E.D., P.R. Delis, and H.R. Mushinsky. 2021. The importance of determining species sensitivity to environmental change: a tree frog example. Ecosphere 12:e03526.

McCoy, S.T.S., J. H. K. Pechmann, and L.A. Williams. 2021. Post-breeding movement, habitat selection, and natural history of Collinses' mountain chorus frog in North Carolina. Southeastern Naturalist 20:399-419.

McDonald, C. 2021. Too little, too late: Amphibian responses to chytrid fungus across time and space. Ph.D. Dissertation, Cornell University, Ithaca, New York.

McDonald, L., K.L. Grayson, H.A. Lin, and J.R. Vonesh. 2018. Stage specific effects of fire: Effects of prescribed burning on adult abundance, oviposition habitat selection, and larval performance of Cope's gray treefrog (*Hyla chrysoscelis*). Forest Ecology and Management 430:394-402.

McEwan, A.L., C.J. Johnson, M. Todd, and P. Govindarajulu. 2021. Resource selection and movement of the coastal tailed frog in response to forest harvesting. Forest Ecology and Management 497:119448.

McGinnis, S.M., and R.C. Stebbins. 2018. Peterson Field Guide to Western Reptiles and Amphibians. Houghton Mifflin Harcourt, Boston.

McGrath, S., A. Yirka, K. Frega, and A. Neighbors. 2020. Novel hylid survey technique: a clear alternative to traditional polyvinyl chloride pipe refugia. Herpetological Review 51:241– 244.

McGrath-Blaser, S., A. Neighbors, and O. Hyman. 2021. Novel, less invasive hylid survey device performs equally to traditional pipe shelters in a field-based comparison. Herpetological Conservation and Biology 16:355–360.

McKnight, D.T., and D.B. Ligon. 2013. *Scaphiopus hurterii* (Hurter's Spadefoot). Leucism. Herpetological Review 44:131-132.

McKnight, D.T., and H.J. Howell. 2015. A comparison of the flight initiation distances of male and female American bullfrogs (*Lithobates catesbeianus*) and green frogs (*Lithobates clamitans*). Herpetological Conservation and Biology 10:137-148.

McKnight, D.T., and D.B. Ligon. 2016. Chorusing patterns of a diverse anuran community, with an emphasis on southern crawfish frogs (*Lithobates areolatus areolatus*). Journal of North American Herpetology 2016:1-10.

McKnight, D.T., J.L. McKnight, T.L. Dean, and D.B. Ligon. 2013. *Acris blanchardi* (Blanchard's Cricket Frog). Feigning death. Herpetological Review 44:117.

McLean, R.P., G.D. Wright, and K. Geluso. 2015. Cope's gray treefrog (*Hyla chrysoscelis*) along the Platte River, Hall County, Nebraska. Collinsorum 4:2-4.

McMahon, T.A., R.K. Boughton, L.B. Martin, and J.R. Rohr. 2017. Exposure to the herbicide atrazine nonlinearly affects tadpole corticosterone levels. Journal of Herpetology 51:270-273.

McNamara, S.C. 2020. Habitat selection across a temperature and nutrient gradient by Cope's gray treefrog (*Hyla chrysoscelis*) and the mosquito, *Culex restuans*. M.S. thesis, University of Mississippi, Oxford.

McTaggart, A.L., T. Eberl, O. Keller, and M.L. Jameson. 2014. First report of *Batrachochytrium dendrobatidis* associated with amphibians in Kansas, USA. Herpetological Review 45:439-441.

Meadow, M., J.B. Tennessen, and L. Swierk. 2021. A novel counting method for wood frog mating calls that preserves accuracy. Herpetological Review 52:517-520.

Measey, J. 2017. Where do African clawed frogs come from? An analysis of trade in live *Xenopus laevis* imported into the USA. Salamandra 53:398-404.

Measey, G.J., D. Rödder, S.L. Green, R. Kobayashi, F. Lillo,

G. Lobos, R. Rebelo, and J.-M. Thirion. 2012. Ongoing invasions of the African clawed frog, *Xenopus laevis*: a global review. Biological Invasions 14:2255-2270.

Melvin, S.D., and J.E. Houlahan. 2014. Simulating selective mortality on tadpole populations in the lab yields improved estimates of effect size in nature. Journal of Herpetology 48:195-202.

Mendelson, J.R. III., and A.M. Kelly. 2017. Georgraphic distribution: *Eleutherodacylus planirostris* (Greenhouse Frog). Herpetological Review 48:382.

Mendelson, J.R., III, C.T. Kinsey, and J.B. Murphy. 2015. A review of the biology and literature of the Gulf Coast toad (*Incilius nebulifer*), native to Mexico and the United States. Zootaxa 3974:517-537.

Meshaka, W.E., Jr. 2013a. Larval period and transformation size in the wood frog (*Lithobates sylvaticus*) in a vernal pool at the Powdermill Nature Reserve in Pennsylvania. Collinsorum 2:12-15.

Meshaka, W.E., Jr. 2013b. Seasonal activity, reproduction, and growth to sexual maturity in the green frog (*Lithobates clamitans melanotus*) in south-central Pennsylvania: statewide and geographic comparisons. Bulletin of the Maryland Herpetological Society 49:8-19.

Meshaka, W.E., Jr., and D.F. Hughes. 2014. Adult body size and reproductive characteristics of the green frog, *Lithobates clamitans melanotus* (Rafinesque, 1820), from a single site in the northern Allegheny Mountains. Collinsorum 3:13-16.

Meshaka, W.E., Jr., and J.R. Layne. 2015. The herpetology of southern Florida. Herpetological Conservation and Biology 10(Monograph 5):1-353.

Meshaka, W.E., Jr., and E. Wingert. 2016. Reproduction and terrestrial movements of a herpetofaunal community on a Susquehanna River island in south-central Pennsylvania. Bulletin of the Maryland Herpetological Society 52:4-18.

Meshaka, W.E., Jr., and M.A. Morales. 2020. Environmental factors associated with breeding in four anuran species in south-central Pennsylvania. Collinsorum 9(3):17-24.

Meshaka, W.E., Jr., P.R. Delis, and K. Walters. 2015a. Geographic variation in selected life history traits of the eastern narrow-mouthed toad, *Gastrophryne carolinensis* (Holbrook, 1836), along the northeastern edge of its geographic range. Journal of North American Herpetology 2015:1-5.

Meshaka, W.E., V.R. Rep, P.R. Delis, and E. Wingert. 2015b. Selected life history traits of the American bullfrog, *Lithobates catesbeianus* (Shaw, 1802), and the green frog, *L. clamitans melanota* (Lareille, 1801), at a park in south-central Pennsylvania; geographic and interspecific comparisons. Collinsorum 4:13-18.

Meshaka, W.E., Jr., V.R. Rep, K. Relish, and M. Owen. 2015c. Colonization history and ecological aspects of the Cuban treefrog (*Osteopilus septentrionalis*) from Fakahatchee Strand Preserve State Park. Collinsorum 4:19-23.

Meshaka, W.E., Jr., E. Wingert, T. Frey, M. Roberts, C. Zwemer, S.R. Bartle, N.A. Davidson, P.R. Delis, and G.R. Cline. 2015d. Confirmation of species identity, body sizes, and clutch characteristics of the eastern gray treefrog (*Hyla versicolor* LeConte, 1825) from Cumberland County, Pennsylvania. Collinsorum 4:11-14.

Meshaka, W.E., Jr., L. Long, A. Tegeler, W.S. Humbert, and P.R. Delis. 2016. Reproductive phenophases of the American toad, *Anaxyrus americanus* (Holbrook, 1836), in the Ligonier Valley of southwestern Pennsylvania. Bulletin of the Maryland Herpetological Society 52:23-37.

Meshaka, W.E., Jr., M.L. McCallum, E. Wingert, P.R. Delis, and W. Humbert. 2017a. Gonadal cycles, clutch characteristics, and growth to sexual maturity in the American toad, *Anaxyrus americanus* (Holbrook), from an agricultural belt in south central Pennsylvania. Journal of the Pennsylvania Academy of Science 91:84-94.

Meshaka, W.E., Jr., S. Sambhare, and P.R. Delis. 2017b. Adult body size and reproductive characteristics of the pickerel frog, *Lithobates palustris* (LeConte, 1825), from Letterkenny Army Depot in south-central Pennsylvania. Bulletin of the Maryland Herpetological Society 53:12-16.

Meshaka, W.E., Jr., W.S. Humbert, M.L. McCallum, and P.R. Delis. 2020a. Loss of habitat leads to bigger toads and bigger eggs: natural area management predictions for the eastern American toad, *Anaxyrus americanus americanus* (Holbrook, 1836). Annals of the Carnegie Museum 86:77-88.

Meshaka, W.E., Jr., M.L. McCallum, C. Voirin, and J. Moore. 2020b. The Cuban treefrog (*Osteopilus septentrionalis*) dampens competitive superiority between two *Hemidactylus* gecko species on buildings. Urban Naturalist 34:1-10.

Meshaka, W.E., Jr., P.R. Delis, and D.F. Hughes. 2020c. Selected life history traits of the green treefrog, *Hyla cinerea* (Schneider, 1799), near the northeastern edge of its geographic range. Bulletin of the Maryland Herpetological Society 56:1-11.

Meshaka, W.E., Jr., R. Teffeteller, and E.Wingert. 2021. Colonization patterns of three artificial ponds by a Pennsylvania herpetofaunal community. Collinsorum 10:5-12.

Metts, B.S., K.A. Buhlmann, T.D. Tuberville, D.E. Scott, and W.A. Hopkins. 2013. Maternal transfer of contaminants and reduced reproductive success of southern toads (*Bufo [Anaxyrus] terrestris*) exposed to coal combustion waste. Environmental Science & Technology 47:2846-2853.

Middleton, J., and D.M. Green. 2015. Adult age-structure variability in an amphibian in relation to population decline. Herpetologica 71:190-195.

Mihaljevic, J.R., J.T. Hoverman, and P.T.J. Johnson. 2018. Co-exposure to multiple Ranavirus types enhances viral infectivity and replication in a larval amphibian system. Diseases of Aquatic Organisms. 132:23-35.

Miller, B.T., and J.L. Miller. 2021. *Anaxyrus americanus* (American Toad). Autumn amplexus. Herpetological Review 52:110-111.

Miller, G.L. 2021. The rain calls of frogs and the reigning paradigm of American herpetology. Archives of Natural History 48: 42-61.

Mims, M.C., I.C. Phillipsen, D.A. Lytle, E.E. Hartfield Kirk, and J.D. Olden. 2015. Ecological strategies predict associations between aquatic and genetic connectivity for dryland amphibians. Ecology 96:1371-1382.

Mirick, P.G., T. French, and J. Kubel. 2016. Field Guide to the Amphibians and Reptiles of Massachusetts. Massachusetts Division of Fisheries & Wildlife, Westborough, Massachusetts.

Mitchell, J.C. 2013. Body size and diet of *Amphiuma means* (Caudata: Amphiumidae) from southeastern Virginia. Journal of the North Carolina Academy of Science 129:66-68.

Mitchell, J.C. 2014. Natural history of an introduced population of squirrel treefrogs (*Hyla squirella*) on the Delmarva Peninsula. Journal of the North Carolina Academy of Science 130:6-10.

Mitchell, J.C. 2016. Restored wetlands in mid-Atlantic agricultural landscapes enhance species richness of amphibian assemblages. Journal of Fish and Wildlife Management 7:490-498.

Mitchell, J.C., and G.R. Johnston. 2013. *Osteopilus septentrionalis* (Cuban Treefrog). Predation. Herpetological Review 44:124.

Mitchell, J.C., and C.A. Pague. 2014. Filling gaps in life-history data: clutch sizes for 21 species of North American anurans. Herpetological Conservation and Biology 9:495-501.

Mitchell, J.C., R.T. Zappalorti, and R.F. Lukei, Jr. 2017. *Anaxyrus terrestris* (Southern Toad). Avian predation. Herpetological Review 48:829-830.

Moler, P.E., K.M. Enge, B. Tornwell, A.L. Farmer, and B.B. Harris. 2020. Status and current distribution of the Pine Barrens treefrog (*Hyla andersonii*) in Florida. Southeastern Naturalist 19:380-394.

Monsen-Collar, K., L. Hazard, and P. Dolcemascolo. 2013. A ranavirus-related mortality event and the first report of ranavirus in New Jersey. Herpetological Review 44:263-265.

Moore, M.N., and S.J. Hecnar. 2018. *Anaxyrus americanus* (American Toad). Electric fence mortality. Herpetological Review 49:299.

Moriarty, J.J., and C.D. Hall. 2014. Amphibians and Reptiles in Minnesota. University of Minnesota Press, Minneapolis.

Morrison, T.A., D. Keinath, W. Estes-Zumpf, J.P. Crall, and C.V. Stewart. 2016. Individual identification of the endangered Wyoming toad *Anaxyrus baxteri* and implications for monitoring species recovery. Journal of Herpetology 50:44-49.

Mosher, C.M., C.J. Johnson, B.W. Murray, and M. Todd. 2021. Environmental influences on the larval density and age-Class distribution of *Ascaphus truei* near the northern extent of its range. Ichthyology and Herpetology 109:1026-1035.

Muehlheim, R.L., and T.O. Matson. 2015. Ranavirus in three species of amphibian and unisexual *Ambystoma* in Ohio. Herpetological Review 46:200-202.

Muelleman, P.J., and C.E. Montgomery. 2013. *Batrachochytrium dendrobatidis* in amphibians of northern Calhoun County, Illinois, USA. Herpetological Review 44:614-615.

Mullaney, A.J., C.B. Eversole, D.R. Duffie, and S.E. Henke. 2021. *Gastrophryne carolinensis* (Eastern Narrow-mouthed Toad). Herpetological Review 52:791.

Muller, B.J., M.J. Schuman, and S.A. Johnson. 2020. Efficacy of acoustic traps for cane toad (*Rhinella marina*) management in Florida. Florida Scientist 83:42-53.

Murphy, C.M., L.L. Smith, J.J. O'Brien, and S.B. Castleberry. 2021. A comparison of vertebrate assemblages at gopher tortoise burrows and stump holes in the longleaf pine ecosystem. Forest Ecology and Management 482:118809.

Murphy, J.C. 2018. Arizona's Amphibians & Reptiles. Privately published by Book Services, Inc. [second edition, 2019]

Mushet, D.M., N.H. Euliss, Jr., Y. Chen, and C.A. Stockwell. 2013. Complex spatial dynamics maintain northern leopard frog (*Lithobates pipiens*) genetic diversity in a temporally varying landscape. Herpetological Conservation and Biology 8:163-175.

Mushet, D.M., J.L. Neau, and N.H. Euliss, Jr. 2014. Modeling effects of conservation grassland losses on amphibian habitat. Biological Conservation 174:93-100.

Muths, E., L.L. Bailey, and M.K. Watry. 2014. Animal reintroductions: an innovative assessment of survival. Biological Conservation 172:200-208.

Muths, E., R.D. Scherer, S.M. Amburgey, T. Matthews, A.W. Spencer, and P.S. Corn. 2016. First estimates of the probability of survival in a small-bodied, high-elevation frog (boreal chorus frog, *Pseudacris maculata*), or how historical data can be useful. Canadian Journal of

Zoology 94:599-606.

Muths, E., R.D. Scherer, S.M. Amburgey, and P.S. Corn. 2018. Twenty-nine years of population dynamics in a small-bodied montane amphibian. Ecosphere 9(12):e02522.

Muths, E., B.R. Hossack, E.H.C. Grant, D.S. Pilliod, and B.A. Mosher. 2020. Effects of snowpack, temperature, and disease on demography in a wild population of amphibians. Herpetologica 76:132-143.

Muzzall, P.M., and M.C. Kuczynski. 2017. Helminths of the eastern gray treefrog, *Hyla versicolor* (Hylidae), from a pond in southwestern lower Michigan, USA. Comparative Parasitology 84:55-59.

Nagel, L.D., S.A. McNulty, M.D. Schlesinger, and J.P. Gibbs. 2021. Breeding effort and hydroperiod indicate habitat quality of small, isolated wetlands for amphibians under climate extremes. Wetlands 41(2) (2021):1-11.

Navarro-Martín, L., C. Lanctôt, P. Jackman, B.J. Park, K. Doe, B.D. Pauli, and V.L. Trudeau. 2014. Effects of glyphosate-based herbicides on survival, development, growth and sex ratios of wood frog (*Lithobates sylvaticus*) tadpoles. I: Chronic laboratory exposures to Vision-Max®. Aquatic Toxicology 154:278-290.

Neal, K. M., B. B. Johnson, and H. B. Shaffer. 2018. Genetic structure and environmental niche modeling confirm two evolutionary and conservation units within the western spadefoot (*Spea hammondii*). Conservation Genetics 19:937-946

Neal, K.M., R.N. Fisher, M.J. Mitrovich, and H.B. Shaffer. 2020. Conservation genomics of the threatened western spadefoot, *Spea hammondii*, in urbanized Southern California. Journal of Heredity 111:613-627.

Neely, W.J., S.E. Greenspan, L.M. Stahl, S.D. Heraghty, V.M. Marshall, C.L. Atkinson, and C. G. Becker. 2021. Habitat disturbance linked with host microbiome dispersion and Bd dynamics in temperate amphibians. Microbial Ecology (2021):1-10.

Neff, M., and B. Balik. 2018. *Anaxyrus americanus americanus* (Eastern American Toad). Abnormal coloration. Herpetological Review 49:92-93.

Neto, J.G.D., T.A. Gorman, D.C. Bishop, and C.A. Haas. 2014. Population demographics of the Florida bog frog (*Lithobates okaloosae*). Southeastern Naturalist 13:128-137.

Niccoli, J.R. 2013. *Acris crepitans* (Northern Cricket Frog). Axanthism. Herpetological Review 44:117.

Nicholls, B., L.L. Manne, and R.R. Veit. 2017. Changes in distribution and abundance of anuran species of Staten Island, NY, over the last century. Northeastern Naturalist 24:65-81.

Nicoletto, P.F. 2013. Effects of Hurricane Rita on the herpetofauna of Village Creek State Park, Hardin County, Texas. Southwestern Naturalist 58:64-69.

Niebuhr, C.N., S.I. Jarvi, L. Kaluna, B.L.T. Fisher, A.R. Deane, I.L. Leinbach, and S.R. Siers. 2020. Occurrence of rat lungworm (*Angiostrongylus cantonensis*) in invasive coqui frogs (*Eleutherodactylus coqui*) and other hosts in Hawaii, USA. Journal of Wildlife Diseases 56:203-207.

Niemiller, M.L., K.S. Zigler, C.D.R. Stephen, E.T. Carter, A.T. Paterson, S.J. Taylor, and A.S. Engel. 2016. Vertebrate fauna in caves of eastern Tennessee within the Appalachians karst region, USA. Journal of Cave and Karst Studies 78:1-24.

Nikolakis, Z.L., A.K. Westfall, S.M. Goetz, D. Laurencio, and M.A. Miller. 2016. *Osteopilus septentrionalis* (Cuban Treefrog). Predation. Herpetological Review 47:439-440.

Nori, J., P. Lemes, N. Urbina-Cardona, D. Baldo, J. Lescano, and R. Loyola. 2015. Amphibian conservation, land-use changes and protected areas: A global overview. Biological Conservation 191:367-374.

Noss, C.F., and B.B. Rothermel. 2015. Juvenile recruitment of oak toads (*Anaxyrus quercicus*) varies with time-since-fire in seasonal ponds. Journal of Herpetology 49:364-370.

Nunziata, S.O., M.J. Lannoo, J.R. Robb, D.R. Karns, S.L. Lance, and S.C. Richter. 2013. Population and conservation genetics of crawfish frogs, *Lithobates areolatus*, at their northeastern range limit. Journal of Herpetology 47:361-368.

O'Connor, K.M., T.A.G. Rittenhouse, and J.L. Brunner. 2016. Ranavirus is common in wood frog (*Lithobates sylvaticus*) tadpoles throughout Connecticut, USA. Herpetological Review 47:394-397.

O'Hare, N.K., and M. Madden. 2018. Herpetofaunal inventory of Wormsloe State Historic Site, Savannah, Georgia. Southeastern Naturalist 17:1-18.

O'Loughlin, D., M. O'Loughlin, and C. Rombough. 2016. *Rana aurora* (Northern Red-legged Frog). Predation. Herpetological Review 47:647-648.

O'Neill, E.M., K.H. Beard, and C.W. Fox. 2018. Body size and life history traits in native and introduced populations of coqui frogs. Copeia 106:161-170.

O'Regan, S.M., W.J. Palen, and S.C. Anderson. 2014. Climate warming mediates negative impacts of rapid pond drying for three amphibian species. Ecology 95:845-855.

Ortiz-Santaliestra, M.E., T.A.G. Rittenhouse, T.L. Cary, and W.H. Karasov. 2013. Interspecific and postmetamorphic variation in susceptibility of three Norh American anurans to *Batrachochytrium dendrobatidis*. Journal of Herpetology 47:286-292.

Osland, M.J., P.W. Stevens, M.M. Lamont, R.C. Brusca, K.M. Hart, J.H. Waddle, C.A. Langtimm, C.M. Williams, B.D. Keim, A.J. Terando, E.A. Reyier, K.E. Marshall, M.E. Loik, R.E. Boucek, A.B. Lewis, and J.A. Seminoff. 2021. Tropicalization of temperate ecosystems in North America: the northward range expansion of tropical organisms in response to warming winter temperatures. Global Change Biology 27:3009-3034.

Ospina, O.E., L. Tieu, J.J. Apodaca, and E.M. Lemmon. 2020. Hidden diversity in the mountain chorus frog (*Pseudacris brachyphona*) and the diagnosis of a new species of chorus frog in the southeastern United States. Copeia 108:778-795.

Oswald, K.J., M.A. Roberts, P.E. Moler, R.G. Arndt, J.D. Camper, and J.M. Quattro. 2020. Wetlands, evolution, and conservation of the Pine Barrens treefrog (*Hyla andersonii*). Journal of Herpetology 54:206-215.

Owen, J.D., and L. Pustka. 2021. *Gastrophryne olivacea* (Western Narrow-mouthed Toad). Diet. Herpetological Review 52:116-117.

Owen, J.D., T.L. Marshall, and D.R. Davis. 2014. Geographic distribution: *Eleutherodactylus* (= *Syrrhophus*) *marnockii* (Cliff Chirping Frog). Herpetological Review 45:652.

Owens, A.K., M.J. Sredl, C.D. Mosley, M.J. Ryan, S.B. Riddle, J.P. Kraft, and A.H. McCall. 2019. Geographic distribution: *Rana sphenocephala* (Southern Leopard Frog). Herpetological Review 50:322-323.

Palis, J.G. 2013. Breeding frequency and success of eastern spadefoots, *Scaphiopus holbrookii*, in southern Illinois. Proceedings of the Indiana Academy of Science 121:158-162.

Palis, J.G. 2014a. *Pseudacris illinoensis* (Illinois Chorus Frog). Predation. Herpetological Review 45:480.

Palis, J.G. 2014b. Googling crawfish frogs: using satellite imagery and auditory surveys to locate breeding sites of a near-threatened species in southernmost Illinois. Bulletin of the Chicago Herpetological Society 49:57-60.

Palis, J.G. 2016. The eastern spadefoot, Illinois' storm frog. Illinois Audubon 335:19-21.

Palis, J.G. 2018a. Current distribution of crawfish frogs in southernmost Illinois. Transactions of the Illinois State Academy of Science 111:1-4.

Palis, J. G. 2018b. Small rural cities as habitat for amphibians and reptiles: the example of Jonesboro, Union County, Illinois. Bulletin of the Chicago Herpetological Society 53:33-40.

Palis, J.G. 2019. *Scaphiopus holbrookii* (Eastern Spadefoot) and *Hyla chrysoscelis* (Cope's Gray Treefrog). Interspecific amplexus. Herpetological Review 50:764-765.

Palis, J.G. 2020a. Solving the mystery of wood frogs at Snake Road. Bulletin of the Chicago Herpetological Society 55:117-120.

Palis, J.G. 2020b. The whistling "bird frogs" of southernmost Illinois. Illinois Audubon (353):18-20.

Palis, J.G. 2021. New records of bird-voiced treefrogs, *Hyla avivoca*, in southern Illinois. Transactions of the Illinois State Academy of Science 114:23-26.

Palis, J.G., and P.H. Mendenhall. 2021. *Anaxyrus americanus* (American Toad). Predation. Herpetological Review 52:603-604.

Palis, J.G., M.B. Boehler, and J.J. Vossler. 2019. An update and an addition to the anuran fauna of LaRue-Pine Hills, Union County, Illinois. Bulletin of the Chicago Herpetological Society 54:201-203.

Palis, J.G., S.A. Johnson, J.R. Marks, and M.B. Boehler. 2021. Cuban treefrogs (*Osteopilus septentrionalis*) in Illinois. Bulletin of the Chicago Herpetological Society 56:153-155.

Palmisano, J. N., C. Bockoven, S. M. McPherson, and T. M. Farrell. 2021. Infection experiment indicate common Florida anurans and lizards can serve as intermediate hosts for the invasive pentastome parasite, *Raillietiella orienatalis*. Abstract Booklet, Southeast Partners in Amphibian and Reptile Conservation Annual Meeting, 25–27 February 2021, p. 46.

Papoulias, D.M., M.S. Schwartz, and L. Mena. 2013. Gonadal abnormalities in frogs (*Lithobates* spp.) collected from managed wetlands in an agricultural region of Nebraska, USA. Environmental Pollution 172:1-8.

Parker, J.M., and S. Brito. 2013. Reptiles & Amphibians of the Mojave Desert. A Field Guide. Snell Press, Las Vegas, Nevada.

Parker, N. 2014. Recent distribution records for *Hyla cinerea* (Green Treefrog) from Tennessee, southern Kentucky, and northern Alabama, USA. Herpetological Review 45:298-299.

Parlin, A.F., S.A. Dinkelacker, A. McCall, M.S. Gosselin, C. Mettey, and R. Tibbett. 2019. Temporal and spatial assessment of the maritime forest herpetofauna diversity on a barrier island in North Carolina. Southeastern Naturalist 18:430-440.

Parmley, D., J.L. Clark, and A.J. Mead. 2020. Amphibians and squamates from the Late Pleistocene (Rancholabrean) Clark Quarry, coastal Georgia. Eastern Paleontologist 7:1-23.

Patla, D., S. St-Hilaire, A. Ray, B.R. Hossack, and C.R. Peterson. 2016. Amphibian mortality events and ranavirus outbreaks in the Greater Yellowstone Ecosystem. Herpetological Review 47:50-54.

Pauley, L.R., J.E. Earl, and R.D. Semlitsch. 2015. Ecological effects and human use of commercial mosquito insecticides in aquatic communities. Journal of Herpetology 49:28-35.

Pauley, N.M., M.P. Nenneman, and K. Geluso. 2018. *Lithobates catesbeianus* (American Bullfrog). Diet. Herpetological Review 49:517.

Pauly, G.B., M.C. Shaulsky, A.J. Barley, S. Kennedy-Gold, S.C. Stewart, S. Keeney, and R.C. Thomson. 2020a. Morphological change during rapid population expansion confounds leopard frog identifications in the southwestern United States. Copeia 108:299-308.

Pauly, G.B., L. Gordon, and R.E. Espinoza. 2020b. *Spea hammondii* (Western Spadefoot). Longevity. Herpetological Review 51:824-825.

Peace, A., S.M. O'Regan, J.A. Spatz, P.N. Reilly, R.D. Hill, E.D. Carter, R.P. Wilkes, T.B. Waltzek, D.L. Miller, and M.J. Gray. 2019. A highly invasive chimeric ranavirus can decimate tadpole populations rapidly through multiple transmission pathways. Ecological Modelling 410:108777.

Pearl, C.A., B. McCreary, J.C. Rowe, and M.J. Adams. 2018. Late-season movement and habitat use by Oregon spotted frog (*Rana pretiosa*) in Oregon, USA. Copeia 106:539-549.

Pease, K.M., and R.K. Wayne. 2014. Divergent responses of exposed and naive Pacific tree frog tadpoles to invasive predatory crayfish. Oecologia 174:241-252.

Peek, R.A., S.M. O'Rourke, and M.R. Miller. 2021. Flow modification associated with reduced genetic health of a river-breeding frog, *Rana boylii*. Ecosphere 12:e03496.

Pence, W. 2013. Survey of rare herpetofauna at the Fort Riley Military Reservation. M.S. thesis, Emporia State University, Emporia, Kansas.

Perata-García, A., J.H. Valdez-Villavicencio, B.D. Hollingsworth, and C.R. Mahrdt. 2020. *Rana draytonii* (California Red-legged Frog) and *Anaxyrus boreas* (Western Toad). Interspecific amplexus. Herpetological Review 51:568.

Perez, A., M.J. Mazerolle, and J. Brisson. 2013. Effects of exotic common reed (*Phragmites australis*) on wood frog (*Lithobates sylvaticus*) tadpole development and food availability. Journal of Freshwater Ecology 28:165-177.

Perez, L.K., J.D. Childress, M.A. Kwiatkowski, D. Saenz, and J.M. Gumm. 2021. Calling phenology and call structure of sympatric treefrogs in eastern Texas. Ichthyology and Herpetology 109:219-227.

Persons, T.B., and P.G. deMaynadier. 2016. Geographic distribution: *Lithobates septentrionalis* (Mink Frog). Herpetological Review 47:77.

Persons, T.B., P.G. deMaynadier, and D.T. Yorks. 2021. Geographic distribution: *Lithobates septentrionalis* (Mink Frog). Elevation. Herpetological Review 52:614.

Peterman, W.E., S.M. Feist, R.D. Semlitsch, and L.S. Eggert. 2013a. Conservation and management of peripheral populations: spatial and temporal influences on the genetic structure of wood frog (*Rana sylvatica*) populations. Biological Conservation 158:351-358.

Peterman, W.E., T.A.G. Rittenhouse, J.E. Earl, and R.D. Semlitsch. 2013b. Demographic network and multi-season occupancy modeling of *Rana sylvatica* reveal spatial and temporal patterns of population connectivity and persistence. Landscape Ecology 28:1601-1613.

Pfennig, K.S., and D.W. Pfennig. 2020. Dead spadefoot tadpoles adaptively modify development in future generations: a novel form of nongenetic inheritance? Copeia 108:116-121.

Phillips, C.A., N.A. Wesslund, and I.E. MacAllister. 2014. Occurrence of the chytrid fungus *Batrachochytrium dendrobatidis* in amphibians in Illinois, USA. Herpetological Review 45:238-240.

Picone, C. 2015. Effects of aquatic herbicides and housing density on abundance of pond-breeding frogs. Northeastern Naturalist 22:NENHC26-NENHC39.

Pierce, C.C., R.P. Shannon, and M.G. Bolek. 2018. Distribution and reproductive plasticity of *Gyrinicola batrachiensis* (Oxyuroidea: Pharyngodonidae) in tadpoles of five anuran species. Parasitology Research 117:461-470.

Pierce, M.T., J.T. Wilcox, A.L. Erway, V.L. Brunal, and J.A. Alvarez. 2021. *Rana draytonii* (California Red-legged Frog). Habitat use. Herpetological Review 52:619-620.

Pilliod, D.S., and R.D. Scherer. 2015. Managing habitat to slow or reverse population declines of the Columbia spotted frog in the northern Great Basin. Journal of Wildlife Management 79:579-590.

Pilliod, D.S., R.S. Arkle, J.M. Robertson, M.A. Murphy, and W.C. Funk. 2015. Effects of changing climate on aquatic habitat and connectivity for remnant populations of a wide-ranging frog species in an arid landscape. Ecology and Evolution 5:3979-3994.

Pilliod, D.S., M.B. Hausner, and R.D. Scherer. 2021. From satellites to frogs: quantifying ecohydrological change, drought mitigation, and population demography in desert meadows. Science of the Total Environment 758:143632.

Pintar, M.R., and W.J. Resetarits, Jr. 2017a. Larval development varies across pond age and larval density in Cope's gray treefrogs, *Hyla chrysoscelis*. Herpetologica 73:291-296.

Pintar, M.R., and W.E. Resetarits, Jr. 2017b. Out with the old, in with the new: oviposition preference matches larval success in Cope's gray treefrog, *Hyla chrysoscelis*. Journal of Herpetology 51:186-189.

Pintar, M.R., and W.E. Resetarits, Jr. 2018. Variation in pond hydroperiod affects larval growth in southern leopard frogs, *Lithobates sphenocephalus*. Copeia 106:70-76.

Piovia-Scott, J., K. Pope, S.J. Worth, E.B. Rosenblum, T. Poorten, J. Refsnider, L.A. Rollins-Smith, L.K. Reinert, H.L. Wells, D. Rejmanek, S. Lawler, and J. Foley. 2015. Correlates of virulence in a frog-killing fungal pathogen: evidence from a California amphibian decline. ISME

Journal 9:1570-1578.

Pisani, P.A., and G.R. Pisani. 2014. Late season chorusing by Blanchard's cricket frog. Collinsorum 3:9.

Pitt, A.L., M.M. Buckle, E.N. Wahlman, and V.E. Ditomo. 2015. *Anaxyrus americanus* (American Toad). Arboreal behavior. Herpetological Review 46:229-230.

Pitt, A.L., J.J. Tavano, R.F. Baldwin, and B.S. Stegenga. 2017. Movement ecology and habitat use of three sympatric anuran species. Herpetological Conservation and Biology 12:212-224.

Pochini, K.M., and J.T. Hoverman. 2017. Immediate and lag effects of pesticide exposure on parasite resistance in larval amphibians. Parasitology 144:817-822.

Polasik, J.S., M.A. Murphy, T. Abbott, and K. Vincent. 2016. Factors limiting early life stage survival and growth during endangered Wyoming toad reintroductions. Journal of Wildlife Management 80:540-552.

Pope, K.L., C.Brown, M. Hayes, G. Green, and D. Macfarlane. 2014. Cascades Frog Conservation Assessment. USDA Forest Service, Pacific Southwest Research Station, General Technical Report PSW-GTR-244.

Popescu, V.D., A.M. Kissel, M. Pearson, W.J. Palen, P. Govindarajulu, and C.A. Bishop. 2013. Defining conservation-relevant habitat selection by the highly imperiled Oregon spotted frog, *Rana pretiosa*. Herpetological Conservation and Biology 8:688-706.

Powell, R., R. Conant, and J.T. Collins. 2016. Peterson Field Guide to Reptiles and Amphibians of Eastern and Central North America. 4th ed., Houghton Mifflin Harcourt, Boston.

Powell, R., J.T. Collins, and E.D. Hooper, Jr. 2019. Key to the Herpetofauna of the Continental United States and Canada 3rd ed. University Press of Kansas, Lawrence.

Preston, D.B., and M.R.J. Forstner. 2015a. Aggregation status and cue type modify tadpole response to chemical cues. Journal of Fish and Wildlife Management 6:199-207.

Preston, D.B., and M.R.J. Forstner. 2015b. Houston toad (*Bufo* (*Anaxyrus*) *houstonensis*) tadpoles decrease their activity in response to chemical cues produced from the predation of conspecifics and congeneric (*Bufo* (*Incilius*) *nebulifer*) tadpoles. Journal of Herpetology 49:170-175.

Provincial Western Toad Working Group. 2014. Management Plan for the Western Toad (*Anaxyrus boreas*) in British Columbia. British Columbia Ministry for the Environment, Victoria, British Columbia.

Rabbani, M., B. Zacharczenko, and D.M. Green. 2015. Color pattern variation in a cryptic amphibian, *Anaxyrus fowleri*. Journal of Herpetology 49:649-654.

Rabe, A., M. Lannoo, and C.K. Beachy. 2013. *Lithobates pipiens* (Northern Leopard Frog). Malformation. Herpetological Review 44:296.

Raithel, C.J. 2019. Amphibians of Rhode Island. Their Status and Conservation. Rhode Island Division of Fish and Wildlife, Department of Environmental Management, West Kingston.

Ramsay, C. 2020. Using community ecology theory to enhance understanding of co-infections. Ph.D. Dissertation, University of Notre Dame, South Bend, Indiana.

Ramsay, C., and Jason R. Rohr. 2021. The application of community ecology theory to co-infections in wildlife hosts. Ecology 102: e03253.

Randall, L.A., L.D. Chalmers, A. Moehrenschlager, and A.P. Russell. 2014. Asynchronous breeding and variable embryonic development period in the threatened northern leopard frog (*Lithobates pipiens*) in the Cypress Hills, Alberta, Canada: conservation and management implications. Canadian Field-Naturalist 128:50-56.

Randall, L., L. Keating, R. Stanton, D. Stark, and A. Moehrenschlager. 2021. Feasibility assessment for recovery of the northern leopard frog (*Lithobates pipiens*) in the Kootenai Valley, Idaho. Wilder Institute/Calgary Zoo. Calgary, Alberta, Canada.

Rasche, E., M.O. Murphy, and M.D. Boone. 2020. Phonotactic response of juvenile Cope's gray treefrogs (*Hyla chrysoscelis*) exposed to conspecific and heterospecific acoustic signals. Herpetological Conservation and Biology 15:498-505.

Rashleigh, K.R., and M. Crowell. 2018. Spring peeper (*Pseudacris crucifer*) in Labrador, Canada: an update. Canadian Field-Naturalist 132:163-167.

Ray, A.M., C.R. Peterson, D.A. Patla, B.J. Thesing, D. Schneider, and J.J. Treanor. 2020. *Spea bombifrons* (Plains Spadefoot). Leucism. Herpetological Review 51:306-307.

Ray, J.D., M. Clack, and R.T. Kazmaier. 2017. *Anaxyrus woodhousii* (Woodhouse's Toad). Predation. Herpetological Review 48:830.

Ream, J.T., D. Zabriskie, and J. A. López. 2019. Herpetological inventory of the Stikine River region, Alaska, 2010-2018. Northwestern Naturalist 100:102-117.

Reeves, M.K., K.A. Medley, A.E. Pinckney, M. Holyoak, P.T.J. Johnson, and M.J. Lannoo. 2013. Localized hotspots drive continental geography of abnormal amphibians on US Wildlife Refuges. PLoS One 8:e77467.

Reeves, R.A., C.L. Pierce, M.W. Vandever, E. Muths, and K.L. Smalling. 2017. Amphibians, pesticides, and the amphibian chytrid fungus in restored wetlands in agricultural landscapes. Herpetological Conservation and Biology 12:68-77.

Reichert, M.S., and H.C. Gerhardt. 2013. Socially mediated plasticity in call timing in the gray treefrog, *Hyla versicolor*. Behavioral Ecology 24:393-401.

Reichert, M.S., and H.C. Gerhardt. 2014. Behavioral strategies and signaling in interspecific aggression interactions in gray tree frogs. Behavioral Ecology 25:520-530.

Relyea, R.A. 2018. The interactive effects of predator stress, predation, and the herbicide Roundup. Ecosphere 9:e02476.

Relyea, R.A., P.R. Stephens, L.N. Barrow, A.R. Blaustein, P.W. Bradley, J.C. Buck, A. Chang, J.P. Collins, B. Crother, J. Earl, S.S. Gervasi, J.T. Hoverman, O. Hyman, E. Moriarty Lemmon, T.M. Luhring, M. Michelson, C. Murray, S. Price, R.D. Semlitsch, A. Sih, A.B. Stoler, N. VandenBroek, A. Warwick, G. Wengert, and J.I. Hammond. 2018. Phylogenetic patterns of trait and trait plasticity evolution: insights from amphibian embryos. Evolution 72:663-678.

Renken, R.B., W.K. Gram, D.K. Fantz, S.C. Richter, T.J. Miller, K.B. Ricke, B. Russell, and X. Wang. 2004. Effects of forest management on amphibians and reptiles in Missouri Ozark forests. Conservation Biology 18:174-188.

Rep, V.R., E.Wingert, and W.E. Meshaka, Jr. 2015. Clutch size of a large bullfrog, *Lithobates catesbeianus* (Shaw, 1802), from south-central Pennsylvania. Collinsorum 4:5.

Resetarits, W.J. Jr., J.R. Bohenek, T.M. Breech, and M.R. Pintar. 2018. Predation risk and patch size jointly determine perceive [sic] patch quality in ovipositing treefrogs, *Hyla chrysoscelis*. Ecology 93:661–669.

Rettig, J.E., N.R. Teeters, and G.R. Smith. 2021. Effects of the interaction of bluegill and two species of tadpoles on experimental zooplankton communities. American Midland Naturalist 186:95-105.

Rhoden, H.R., and M.G. Bolek. 2015. Helminth community structure in tadpoles of northern leopard frogs (*Rana pipiens*) and Woodhouse's toads (*Bufo woodhousii*) from Nebraska. Parasitology Research 114:4685-4692.

Richmond, J.Q., K.R. Barr, A.R. Backlin, A.G. Vandergast, and R.N. Fisher. 2013. Evolutionary dynamics of a rapidly receding southern range boundary in the threatened California red-legged frog (*Rana draytonii*). Evolutionary Applications 6:808-822.

Richmond, J.Q., A.R. Backlin, P.J. Tatarian, B.G. Solvesky, and R.N. Fisher. 2014. Population declines lead to replicate patterns of internal range structure at the tips of the distribution of the California red-legged frog (*Rana draytonii*). Biological Conservation 172:128-137.

Richter, S.C., E.M. O'Neill, S.O. Nunziata, A. Rumments, E.S. Gustin, J.E. Young, and B.I. Crother. 2014. Cryptic diversity and conservation of gopher frogs across the southeastern United States. Copeia 2014:231-237.

Riddle, S.B., C.J. Herzog, H.L. Bateman, and M.J. Ryan. 2021a. Postfire breeding behavior in Couch's spadefoot toad (*Scaphiopus couchii*) along the San Pedro River, Arizona. Western North American Naturalist 80:531-535.

Riddle, S.B., C.J. Gill, E.J. Keister, M.J. Ryan, and S.L. Lashway. 2021b. *Lithobates catesbeianus* (American Bullfrog). Diet. Herpetological Review 52:613-614.

Rivera, B., K. Cook, K. Andrews, M.S. Atkinson, and A.E. Savage. 2019. Pathogen dynamics in an invasive frog compared to native species. EcoHealth 16:222-234.

RLFCT (Relict Leopard Frog Conservation Team). 2016. Conservation agreement and conservation assessment and strategy for the Relict Leopard Frog (*Rana onca* [=*Lithobates onca*]). Available at: http://www.ndow.org/uploadedFiles/ndoworg/Content/Our_Agency/Divisions/Fisheries/Relict-Leopard-Frog-Conservation Agreement.pdf.

Robertson, J.M., S.W. Fitzpatrick, B.B. Rothermel, and L.M. Chan. 2018. Fire does not strongly affect genetic diversity or structure of a common treefrog in the endangered Florida scrub. Journal of Heredity 2018:243-252.

Robinson, C.W., S.A. McNulty, and V.R. Titus. 2018. No safe space: prevalence and distribution of *Batrachochytrium dendrobatidis* in amphibians in a highly-protected landscape. Herpetological Conservation and Biology 13:373-382.

Robinson, K. 2021. Evaluating nutrient effects on trematode parasitism in larval frogs. M.S. thesis, Bradley University, Peoria, Illinois.

Rogic, A., N. Tessier, S. Noël, A. Gendron, A. Branchaud, and F.-J. Lapointe. 2015. A "Trilling" case of mistaken identity: call playbacks and mitochondrial DNA identify chorus frogs in southern Québec (Canada) as *Pseudacris maculata* and not *P. triseriata*. Herpetological Review 46:1-7.

Rogic, A., N. Tessier, and F.-J. Lapointe. 2019. Genetic characterization of imperiled boreal chorus frogs identifies populations for conservation. Journal of Herpetology 53:89-95.

Rohde, M.L., A.J. Forrester, M.J. Harner, C. Kruse, K. Geluso. 2021. *Lithobates pipiens* (Northern Leopard Frog). Egg predation. Herpetological Review 52:117-118.

Rollins-Smith, L.A. 2020. Global amphibian declines, disease, and the ongoing battle between *Batrachochytrium* fungi and the immune system. Herpetologica 76:178-188.

Rombough, C.J. 2016. *Rana luteiventris* (Columbia Spotted Frog). Cannibalism of embryos. Herpetological Review 47:118-119.

Rombough, C.J. 2020. *Pseudacris regilla* (Pacific Treefrog). Unusual mortality. Herpetological Review 51:101-102.

Rombough, C., and L. Trunk. 2017. *Rana draytonii* (California Red-legged Frog). Egg incubation period. Herpetological Review 48:831-832.

Rombough, C., and L. Trunk. 2019a. *Pseudacris regilla* (Pacific Treefrog). Predation. Herpetological Review

50:763-764.

Rombough, C., and L. Trunk. 2019b. *Pseudacris regilla* (Pacific Treefrog). Calling and temperature tolerance. Herpetological Review 50:762-763.

Rombough, C., and J. Bowerman. 2021. The structure of spotted frog (*Rana luteiventris* and *Rana pretiosa*) eggs: a physical description and historical perspective. Northwestern Naturalist 102:1-8.

Root, S.T., Y. Passamaneck, and J. Keele. 2017. Geographic distribution: *Rana sphenocephala* (Southern Leopard Frog). Herpetological Review 48:385.

Rorabaugh, J.C., and M.J. Sredl. 2014. Herpetofauna of the 100-mile circle. Chiricahua leopard frog (*Lithobates chiricahuensis*). Sonoran Herpetologist 27:61-68.

Rorabaugh, J.C., and J.A. Lemos-Espinal. 2016. A Field Guide to the Amphibians and Reptiles of Sonora, Mexico. ECO Herpetological Publishing and Distribution, Rodeo, New Mexico.

Rorabaugh, J.C., A.K. Owens, A. King, S.F. Hale, S. Poulin, M.J. Sredl, ad J.A. Lemos-Espinal. 2020. Reintroduction of the Tarahumara frog (*Rana tarahumarae*) in Arizona: lessons learned. Herpetological Conservation and Biology 15:372-389.

Rose, J.P., B.J. Halstead, and R.N. Fisher. 2020. Integrating multiple data sources and multi-scale land-cover data to model the distribution of a declining amphibian. Biological Conservation 241:108374.

Rose, J.P., S.J. Kupferberg, C.A. Wheeler, P.M. Kleeman, and B.J. Halstead. 2021. Estimating the survival of unobservable life stages for a declining frog with a complex life cycle. Ecosphere 12:e03381.

Ross, L., M. Wright, K. Wiskirchen, J. Grace, C. Lennon, J. Mantooth, D. Schneider, S.P. Hudman, M.I. Kelrick, and C.E. Montgomery. 2014. Prevalence of *Batrachochytrium dendrobatidis* in three frog species of the Bighorn National Forest, Wyoming USA. Herpetological Review 45:615-616.

Rothenberger, M.B., and A. Baranovic. 2021. Predator–prey relationships within natural, restored, and created vernal pools. Restoration Ecology 29:e13308.

Rothstein, A.P., R.A. Knapp, G.S. Bradburd, D.M. Boiano, C.J. Briggs, and E.B. Rosenblum. 2020. Stepping into the past to conserve the future: archived skin swabs from extant and extirpated populations inform genetic management of an endangered amphibian. Molecular Ecology 29:2598-2611.

Rowe, J.C., L. Wilson-Romine, and J. Romine. 2021a. Active season and pre-overwintering microhabitat use by Oregon spotted frogs (*Rana pretiosa*) and American bullfrogs (*Lithobates catesbeianus* [*Rana catesbeiana*]) at Conboy Lake National Wildlife Refuge, Washington. Northwestern Naturalist 102:55-75.

Rowe, J.C., A.Duarte, C.A. Pearl, B. McCreary, P.K. Haggerty, J.W. Jones, and M.J. Adams. 2021b. Demography of the Oregon spotted frog along a hydrologically modified river. Ecosphere 12:e03634.

Rowland, F.E., S.K. Tuttle, M.J. González, and M.J. Vanni. 2016. Canopy cover and anurans: nutrients are the most important predictor of growth and development. Canadian Journal of Zoology 94:225-232.

Roznik, E.A., and S.B. Reichling. 2021. Survival, movements and habitat use of captive-bred and reintroduced dusky gopher frogs. Animal Conservation 24:51-63.

Roznik, E. A., N. Cano, K. L. Surbaugh, C. T. Ramsay, and J. R. Rohr. 2021. Invasive Cuban treefrogs (*Osteopilus septentrionalis*) have more robust locomotor performance than two native treefrogs (*Hyla* spp.) in Florida, USA, in response to temperature and parasitic infections. Diversity 13(3):109.

Ruhe, B.M., and A.R. Ruhe. 2019. Geographic distribution: *Scaphiopus holbrookii* (Eastern Spadefoot). Herpetological Review 50:522.

Rumschlag, S.L., and M.D. Boone. 2015. How time of exposure to the amphibian chytrid fungus affects *Hyla chrysoscelis* in the presence of an insecticide. Herpetologica 71:169-176.

Rumschlag, S.L., and M.D. Boone. 2020a. Lethal and sublethal amphibian host responses to *Batrachochytrium dendrobatidis* exposure are determined by the additive influence of host resource availability. Journal of Wildlife Diseases 56:338-349.

Rumschlag, S.L., and M.D. Boone. 2020b. Amphibian infection risk changes with host life stage and across a landscape gradient. Journal of Herpetology 54:347-354.

Rumschlag, S.L., M.D. Boone, and G. Fellers. 2014. The effects of the amphibian chytrid fungus, insecticide exposure, and temperature on larval anuran development and survival. Environmental Toxicology and Chemistry 33:2545-2550.

Rumschlag, S. L., N.T. Halstead, J.T. Hoverman, T.R. Raffel, H.J. Carrick, P.J. Hudson, and J.R. Rohr. 2019. Effects of pesticides on exposure and susceptibility to parasites can be generalised to pesticide class and type in aquatic communities. Ecology Letters 22: 962-972.

Rumschlag, S.L., M.B. Mahon, J.T. Hoverman, T.R. Raffel, H.J. Carrick, P.J. Hudson, and J.R. Rohr. 2020. Consistent effects of pesticides on community structure and ecosystem function in freshwater systems. Nature Communications 11:1-9.

Ruso, G.E. N.S. Hogan, C. Sheedy, M.J. Gallant, T.D. Jardine. 2021. Effects of agricultural stressors on growth and an immune status indicator in wood frog (*Lithobates sylvaticus*) tadpoles and metamorphs. Environmental Toxicology and Chemistry 40:2269-2281.

Russell, R.E., B.J. Halstead, B.A. Mosher, E. Muths, M.J. Adams, E.H.C. Grant, R.N. Fisher, P.M. Kleeman, A.R. Backlin, C.A. Pearl, R.K. Honeycutt, and B.R. Hossack. 2019. Effect of amphibian chytrid fungus (*Batrachochytrium dendrobatidis*) on apparent survival of frogs and toads in the western USA. Biological Conservation 236:296-304.

Ruso, G.E., C.A. Morrissey, N.S. Hogan, C. Sheedy, M.J. Gallant, and T.D. Jardine. 2019. Detecting amphibians in agricultural landscapes using environmental DNA reveals the importance of wetland condition. Environmental Toxicology and Chemistry 38:2750-2763.

Ruthig, G.R. 2013. Temperature and water molds influence mortality of *Lithobates catesbeianus* eggs. Herpetological Conservation and Biology 8:707-714.

Ruthig, G.R., L.J. Bode, A. Cvetkovic, and P. Shrestha. 2020. Dead mink frogs (*Lithobates septentrionalis*) found in northern Minnesota were infected with both *Batrachochytrium dendrobatidis* and ranavirus. Herpetological Review 51:744-746.

Ryan, K.J., A.J.K. Calhoun, B.C. Timm, and J.D. Zydlewski. 2015. Monitoring eastern spadefoot (*Scaphiopus holbrookii*) response to weather with the use of a passive integrated transponder (PIT) system. Journal of Herpetology 49:257-263.

Ryan, Ma.J., I.M. Latella, C.W. Painter, J.T. Giermakowski, B.L. Christman, R.D. Jennings, and J.L. Voyles. 2014. First record of *Batrachochytrium dendrobatidis* in the Arizona toad (*Anaxyrus microscaphus*) in southwestern New Mexico, USA. Herpetological Review 45:616-618.

Ryan, M.J., C.T. McAllister, L.A. Durden, Y. Kiryu, J.H. Landsberg, I. Latella, J.T. Giermakowski, H. Snell, and J. Stabile. 2016a. *Anaxyrus microscaphus* (Arizona Toad). Ectoparasites. Herpetological Review 47:436-438.

Ryan, M.J., I.M. Latella, G. Gustafson, J.T. Giermakowski, and H. Snell. 2016b. *Anaxyrus microscaphus* (Arizona Toad). Diet, vertebrate prey. Herpetological Review 47:436.

Ryan, M.J., J.T. Giermakowski, I.M. Latella, and H.L. Snell. 2017. No evidence of hybridization between the Arizona toad (*Anaxyrus microscaphus*) and Woodhouse's toad (*A. woodhousii*) in New Mexico, USA. Herpetological Conservation and Biology 12:565-575.

Sacerdote, A.B., and R.B. King. 2014. Direct effects of an invasive European buckthorn metabolite on embryo survival and development in *Xenopus laevis* and *Pseudacris triseriata*. Journal of Herpetology 48:51-58.

Sacerdote-Velat, A., M.B. Manjerovic, and R. Santymire. 2016. Preliminary survey of *Batrachochytrium dendrobatidis* in the Chicago region of Illinois, USA. Herpetological Review 47:57-58.

Sadinski, W.J., A.L. Gallant, and J.E. Cleaver. 2020. Climate's cascading effects on disease, predation, and hatching success in *Anaxyrus canorus*, the threatened Yosemite toad. Global Ecology and Conservation 23:e01173.

Saenz, D., and C.K. Adams. 2020. Invasive plant leaf litter affects anuran embryo survival rates, time of hatching, and hatchling size. Herpetological Review 51:695-699.

Saenz, D., T.L. Hall, and M.A. Kwiatkowski. 2015. Effects of urbanization on the occurrence of *Batrachochytrium dendrobatidis*: do urban environments provide refuge from the amphibian chytrid fungus? Urban Ecosystems 18:333-340.

Sanders, D., E. Frago, R. Kehoe, C. Patterson, and K.J. Gaston. 2021. A meta-analysis of biological impacts of artificial light at night. Nature Ecology & Evolution 5:74-81.

Santana, F.E., R.R. Swaisgood, J.M. Lemm, R.N. Fisher, and R.W. Clark. 2015. Chilled frogs are hot: hibernation and reproduction of the endangered mountain yellow-legged frog *Rana muscosa*. Endangered Species Research 27:43-51.

Sasaki, K., D. Lesbarrères, G. Watson, and J. Litzgus. 2015. Mining-caused changes to habitat structure affect amphibian and reptile population ecology more than metal pollution. Ecological Applications 25:2240-2254.

Sauer, E. L., N. Trejo, J.T. Hoverman, and J.R. Rohr. 2019. Behavioural fever reduces ranaviral infection in toads. Functional Ecology 33:2172-2179.

Saumure, R.A. 2020. A Spring Without Water. Restoration of the Las Vegas Creek 2008 – 2020. Southern Nevada Water Authority, Environmental Resources Division, Las Vegas, Nevada.

Saumure, R.A., R. Rivera. J.R. Jaeger, T. O'Toole, A. Ambos, K. Guadelupe, A.R. Bennett, and Z. Marshall. 2021. Leaping from extinction: Rewilding the Relict Leopard Frog in Las Vegas, Nevada, USA. pp. 76–81 *In* P.S. Soorae (ed.), Global Conservation Translocation Perspectives: 2021. International Union for the Conservation of Nature / Species Survival Commission, Conservation Translocation Specialist Group, Gland, Switzerland.

Savage, A.E., B. Gratwicke, K. Hope, E. Bronikowski, and R.C. Fleischer. 2020. Sustained immune activation is associated with susceptibility to the amphibian chytrid fungus. Molecular Ecology 29:2889-2903.

Scheffers, B.R., and C.A. Paszkowski. 2016. Large body size for metamorphic wood frogs in urban stormwater wetlands. Urban Ecosystems 19:347-359.

Schiwitz, N.C., C.M. Schalk, and D. Saenz. 2020. Activity level-predation risk tradeoff in a tadpole guild: implications for community organization along the hydroperiod gradient. American Midland Naturalist 183:223-232.

Schlaepfer, M.A., D. Caldwell, J.C. Rorabaugh, M.J. Ryan, S. Stump, and M.J. Sredl. 2021. The use of evoked vocal responses to detect cryptic, low-density frogs in the field.

Journal of Herpetology 55:174-180.

Schlesinger, M.D., J.A. Feinberg, N.H. Nazdrowicz, J.D. Kloepfer, J. Beane, J.F. Bunnell, J. Burger, E. Corey, K. Gipe, J.W. Jaycox, E. Kiviat, J. Kubel, D. Quinn, C. Raithel, S. Wenner, E.L. White, B. Zarate, and H.B. Shaffer. 2017. Distribution, identification, landscape setting, and conservation of *Rana kauffeldi* in the Northeastern U.S. Final Report to the Wildlife Management Institute and the Northeastern Association of Fish and Wildlife Agencies. New York Natural Heritage Program, Albany.

Schriever, T.A., and D.D. Williams. 2013. Ontogenetic and individual diet variation in amphibian larvae across an environmental gradient. Freshwater Biology 58:223–236.

Schuett-Hames, J.P., D.E. Schuett-Hames, F.T. Waterstrat, and E.M. Lund. 2019. Amphibian movement across a new residential road in western Washington state. Northwestern Naturalist 100:186-197.

Schuett-Hames, J.P., and B.J. Blessing-Earle. 2021. Western toad (*Anaxyrus boreas*) breeding timing and larval development in Lake Cushman Reservoir, Washington State. Northwestern Naturalist 102:140-149.

Schuman, M.J., and I.A. Bartoszek. 2019a. *Rhinella marina* (Cane Toad). Diet. Herpetological Review 50:550-551.

Schuman, M.J., and I.A. Bartoszek. 2019b. *Anaxyrus terrestris* (Southern Toad) and *Rhinella marina* (Cane Toad). Interspecific amplexus. Herpetological Review 50:757.

Schuman, M.J., B.J. Muller, and S.A. Johnson. 2020. *Rhinella marina* (Cane Toad). Predation. Herpetological Review 51:304-305.

Schuman, M.J., I.A. Bartoszek, K.B. Worley, V.G. Booher, and J.R. Schmid. 2021. Spatial and temporal patterns for treefrog (Anura: Hylidae) communities in conservation areas of southwestern Florida. Herpetology Notes 14:913-922.

Schuman, M.J., S.L. Snyder, C.H. Smoak, and C.J. Dove. 2021. Mourning Dove (*Zenaida macroura*) and Eastern Bluebird (*Sialia sialis*) found in diet of the non-native cane toad (*Rhinella marina*) in Florida. Southeastern Naturalist 20(4):N115.

Schwartz, J.J., N.C. Crimarco, Y. Bregman, and K.R. Umeoji. 2013. An investigation of the functional significance of responses of the gray treefrog (*Hyla versicolor*) to chorus noise. Journal of Herpetology 47:354-360.

Seaborn, T., and C.S. Goldberg. 2020. Integrating genetics and metapopulation viability analysis to inform translocation efforts for the last northern leopard frog population in Washington State, USA. Journal of Herpetology 54:465-475.

Searcy, C.A., B. Gilbert, M. Krkošek, L. Rowe, and S.J. McCauley. 2018. Positive correlation between dispersal and body size in green frogs (*Rana clamitans*) naturally colonizing an experimental landscape. Canadian Journal of Zoology 96:1378-1384.

Seburn, D. 2013. How did the record hot spring of 2012 affect amphibian calling? The Canadian Herpetologist 3:8-10.

Seburn, D.C., K. Gunson, and F.W. Schueler. 2014. Apparent widespread decline of the boreal chorus frog (*Pseudacris maculata*) in eastern Ottawa. Canadian Field-Naturalist 128:151-157.

Seiler, M., S.P. Graham, and C. Kelehear. 2017. *Spea multiplicata* (New Mexico Spadefoot). Predation. Herpetological iReview 48:615.

Sekowska, J.M., M.C. Allender, and C.A. Phillips. 2014. Prevalence and spatial distribution of Frog Virus 3 in free-ranging populations of northern cricket frogs (*Acris crepitans*), American bullfrogs (*Lithobates catesbeianus*), and northern leopard frogs (*Lithobates pipiens*) in western Illinois and Wisconsin, USA. Herpetological Review 45:245-247.

Selman, W., and I.R. Garza. 2020. Geographic distribution: *Eleutherodactylus planirostris* (Greenhouse Frog). Herpetologcal Review 51:769.

Sepúlveda, A.J., and M. Layhee. 2015. Description of fall and winter movements of the introduced American bullfrog (*Lithobates catesbeianus*) in a Montana, USA, pond. Herpetological Conservation and Biology 10:978-984.

Sharma, B., F. Hu, J.A. Carr, and R. Patiño. 2011 (2014). Water quality and amphibian health in the Big Bend region of the Rio Grande Basin. Texas Journal of Science 63:233-266.

Shedd, J.D. 2016. Distribution of the western spadefoot (*Spea hammondii*) in the northern Sacramento Valley of California, with comments on status and survey methodology. Pp. 19-29 *In* R.A. Schlising, E.E. Gottschalk Fisher and C.M. Guilliams (eds.), Vernal Pools in Changing Landscapes. Studies from the Herbarium 2016(18).

Sheridan, J.A., N.M. Caruso, J.J. Apodaca, and L.J. Rissler. 2018. Shifts in frog size and phenology: testing predictions of climate change on a widespread anuran using data from prior to rapid climate warming. Ecology and Evolution 8:1316–1327.

Shier, D.M., T.T. Hammond, M.J. Curtis, L.E. Jacobs, N.E. Calatayud, and R.R. Swaisgood. 2021. Conservation breeding and reintroduction of the endangered mountain yellow-legged frog in Southern California, USA. pp. 65-69 *In* P.S. Soorae (ed.), Global Conservation Translocation Perspectives: 2021. Case Studies from Around the Globe. International Union for the Conservation of Nature / Species Survival Commission, Conservation Translocation Specialist Group, Gland, Switzerland.

Shine, R. 2018. Cane Toad Wars. University of California Press, Berkeley.

Shine, R., J.A. Lesku, and H.B. Lillywhite. 2019. Assessment of the cooling-then-freezing method for euthanasia of amphibians and reptiles. Journal of the American Veterinary Medical Association 255:48-50.

Shulse, C.D., R.D. Semlitsch, and K.M. Trauth. 2013. Mosquitofish dominate amphibian and invertebrate community development in experimental wetlands. Journal of Applied Ecology 50:1244-1256.

Shutler, D., A.D. Gendron, M. Rondeau, and D.J. Marcogliese. 2015. Nematode parasites and leukocyte profiles of northern leopard frogs, *Rana pipiens*: location, location, location. Canadian Journal of Zoology 93:41-49.

Siddons, S.R., and C.L. Searle. 2021. Exposure to a fungal pathogen increases the critical thermal minimum of two frog species. Ecology and Evolution 11(14):9589-9598.

Siddons, S.R., M.C. Bray, and C.L. Searle. 2020. Higher infection prevalence in amphibians inhabiting human-made compared to natural wetlands. Journal of Wildlife Diseases 56:823-836.

Sievert, G., and L. Sievert. 2021. A Field Guide to Oklahoma's Amphibians and Reptiles. Fourth Edition. Oklahoma Department of Wildlife Conservation, Oklahoma City.

Sigafus, B.H., C.R. Schwalbe, B.R. Hossack, and E. Muths. 2014. Prevalence of the amphibian chytrid fungus (*Batrachochytrium dendrobatidis*) at Buenos Aires National Wildlife Refuge, Arizona, USA. Herpetological Review 45:41-42.

Simpson, S.E., P.S. Crump, and T.J. Hibbitts. 2019. Geographic distribution: *Eleutherodacylus planirostris* (Greenhouse Frog). Herpetological Review 50:96.

Sinclair, B.J., J.R. Stinziano, C.M. Williams, H.A. MacMillan, K.E. Marshall, and K.B. Storey. 2013. Real-time measurement of metabolic rate during freezing and thawing of the wood frog, *Rana sylvatica*: implications for overwinter energy use. Journal of Experimental Biology 216:292-302.

Sirsi, S., M.J. Marsh, and M.R.J. Forstner. 2020. Evaluating the effects of red imported fore ants (*Solenopsis invicta*) on juvenile Houston toads (*Bufo* [= *Anaxyrus*] *houstonensis*) in Colorado County, Texas. PeerJ 8:e8480.

Skadsen, D.R., and D.R. Davis. 2020. Wood frogs (*Rana sylvatica*) in southwestern Roberts County and western Grant County, South Dakota. Prairie Naturalist 52:31-32.

Skibbe, J.R., J. Farrar, K. Watson, and S.C. Richter. 2021. Population genetics of wood frogs (*Lithobates sylvaticus*) in an altered forested ridgetop wetland ecosystem in Appalachia. Herpetological Conservation and Biology 16:1-10.

Slough, B.G. 2013. Occurrence of amphibians in British Columbia north of 57°N. Northwestern Naturalist 94:180-186.

Slough, B.G., and A. deBruyn. 2018. The observed decline of western toads (*Anaxyrus boreas*) over several decades at a novel winter breeding site. Canadian Field-Naturalist 132:53-57.

Smalling, K.L., G.M. Fellers, P.M. Kleeman, and K.M. Kuivila. 2013. Accumulation of pesticides in Pacific chorus frogs (*Pseudacris regilla*) from California's Sierra Nevada Mountains, USA. Environmental Toxicology and Chemistry 32:2026-2034.

Smalling, K.L., R. Reeves, E. Muths, M. Vandever, W.A. Battaglin, M.L. Hladik, and C.L. Pierce. 2015. Pesticide concentrations in frog tissue and wetland habitats in a landscape dominated by agriculture. Science of the Total Environment 502:80-90.

Smalling, K.L., J.C. Rowe, C.A. Pearl, L.R. Iwanowicz, C.E. Givens, C.W. Anderson, B. McCreary, and M.J. Adams. 2021. Monitoring wetland water quality related to livestock grazing in amphibian habitats. Environmental Monitoring and Assessment 193:58.

Smith, C.E. 2018. *Acris blanchardi* (Blanchard's Cricket Frog). Recolonization. Herpetological Review 49:298.

Smith, D.D., and P.A. Bredehoft. 2020. *Hyla chrysoscelis/Hyla versicolor* (Gray Treefrog/Cope's Gray Treefrog [sic]). Predation. Herpetological Review 51:299.

Smith, D.H.V., B. Jones, L. Randall, and D.R.C. Prescott. 2014. Difference in detection and occupancy between two anurans: the importance of species-specific monitoring. Herpetological Conservation and Biology 9:267-277.

Smith, G.R. 2014. Leucistic wood frog (*Rana sylvatica*) tadpole from central Ohio. Bulletin of the Maryland Herpetological Society 50:74-75.

Smith, G.R., and L.E. Smith. 2015. Effects of western mosquitofish (*Gambusia affinis*) on tadpole production of gray treefrogs (*Hyla versicolor*). Herpetological Conservation and Biology 10:723-727.

Smith, G.R., and J.B. Iverson. 2016. Observations of oviposition and calling in Cuban treefrogs, *Osteopilus septentrionalis*, from the Bahamas. Herpetological Bulletin 135:26-27.

Smith, G.R., C.J. Dibble, A.J. Terlecky, C.B. Dayer, A.B. Burner, and M.E. Ogle. 2013. Effects of invasive western mosquitofish and ammonium nitrate on green frog tadpoles. Copeia 2013:248-253.

Smith, G.R., J.E. Rettig, M. Smyk, M. Jones, G. Eng-Surowiec, D. Mirshavili, and J. Hollis. 2019. Consumption of the eggs, hatchlings, and tadpoles of green frogs (*Lithobates clamitans*) by native and non-native predators. Amphibia-Reptilia 40:383-387.

Smith, K.P.W., M.R. Parker, and W.F. Bien. 2013. *Anaxyrus fowleri* (Fowler's Toad). Interspecific nest use. Herpetological Review 44:492-493.

Smith, L.L., J.M. Howze, J.S. Staiger, E.R. Sievers, D. Burr, and K.M. Enge. 2021. Added value: Gopher tortoise surveys provide estimates of gopher frog abundance in tortoise burrows. Journal of Fish and Wildlife Management 12:3-11.

Smith, R.L., K.H. Beard, and A.B. Shiels. 2017. Different prey resources suggest little competition between non-native frogs and insectivorous birds despite isotopic niche overlap. Biological Invasions 9:1001-1013.

Smith, T.C., A.M. Picco, and R. Knapp. 2017. Ranaviruses infect mountain yellow-legged frogs (*Rana muscosa* and *Rana sierrae*) threatened by *Batrachochytrium dendrobatidis*. Herpetological Conservation and Biology 12:149-159.

Smith, W.H. 2013. Amphibians and large, infrequent forest disturbances: an extreme wind event facilitates habitat creation and anuran breeding. Herpetological Conservation and Biology 8:732-740.

Smits, A.P., D.K. Skelly, and S.R. Bolden. 2014. Amphibian intersex in suburban landscapes. Ecosphere 5(1):11.

Smyers, S.D., M.T. Jones, L.L. Willey, T. Tadevosyan, J. Martinez, K. Cormier, and D.B. Kemmett. 2021. Calling phenology in *Rana sylvatica* (Wood Frog) at high-elevation ponds in the White Mountains, New Hampshire. Northeastern Naturalist 28(sp 11):156-179.

Snyder, D.H., A.F. Scott, E.J. Zimmerer, and D.F. Frymire. 2016. Amphibians and Reptiles of Land Between the Lakes. University Press of Kentucky, Lexington.

Somma, L.A. 2021. Geographic distribution: *Eleutherodacylus coqui* (Common Coqui). Herpetological Review 52:76-77.

Sommers, M.D., L.A. Randall, and R.M.R. Barclay. 2018. Effects of environmental variables on the calling behaviour of northern leopard frogs (*Lithobates pipiens*) in Alberta, Canada. Canadian Journal of Zoology 96:163-169.

Sonn, J.M., R.M. Utz, and C.L. Richards-Zawacki. 2019. Effects of latitudinal, seasonal, and daily temperature variations on chytrid fungal infections in a North American frog. Ecosphere 10:e02892.

Sonn, J.M., W.P. Porter, P.D. Mathewson, and C.L. Richards-Zawacki. 2020. Predictions of disease risk in space and time based on the thermal physiology of an amphibian host-pathogen interaction. Frontiers in Ecology and Evolution 8:576065.

Stackhouse, A., and C.D. Foster. 2020. *Anaxyrus woodhousii* (Woodhouse's Toad). Hypomelanism/Leucism. Herpetological Review 51:296-297.

Standish, I., E. Leis, L. Zinnel, N. Schmitz, and J. Creedico. 2018. The identification of *Janthinobacterium lividum* on Wisconsin, USA amphibians. Herpetological Review 49:473-479.

Steen, D.A., C.J.W. McClure, and S.P. Graham. 2013. Relative influence of weather and season on anuran calling activity. Canadian Journal of Zoology 91:462-467.

Stephens, J.P., K.A. Berven, and S.D. Tiegs. 2013. Anthropogenic changes to leaf litter input affect the fitness of a larval amphibian. Freshwater Biology 58:1631-1646.

Stephens, J.P., K.A. Berven, S.D. Tiegs, and T.R. Raffel. 2015. Ecological stoichiometry quantitatively predicts responses of tadpoles to a food quality gradient. Ecology 96:2070-2076.

Stevenson, D.J., and H.C. Chandler. 2017. The herpetofauna of conservation lands along the Altamaha River, Georgia. Southeastern Naturalist 16:261-282.

Stewart, K.A. 2013. Contact zone dynamics and the evolution of reproductive isolation in a North American treefrog, the spring peeper (*Pseudacris crucifer*). Ph.D. Dissertation, Queen's University, Kingston, Ontario.

Stewart, K.A., R. Wang, and R. Montgomerie. 2016. Extensive variation in sperm morphology in a frog with no sperm competition. BMC Evolutionary Biology 16:29.

Stewart, K.A., J.D. Austin, K.R. Zamudio, and S.C. Lougheed. 2016. Contact zone dynamics during early stages of speciation in a chorus frog (*Pseudacris crucifer*). Heredity 116:239-247.

Stiles, R.M., and M.J. Lannoo. 2015. *Lithobates sphenocephalus* (Southern Leopard Frog). Fall breeding. Herpetological Review 46:414.

Stiles, R.M., M.J. Sieggreen, A. Preston, A.P. Pessier, S.J. Lannoo, and M.J. Lannoo. 2016. First report of ranavirus-associated mortality in crawfish frogs (*Lithobates areolatus*), a species of conservation concern, in Indiana, USA. Herpetological Review 47:389-391.

Stiles, R.M., J.W. Swan, J.L. Klemish, and M.J. Lannoo. 2017a. Amphibian habitat creation on postindustrial landscapes: a case study in a reclaimed coal strip-mine area. Canadian Journal of Zoology 95:67-73.

Stiles, R.M., T.R. Halliday, N.J. Engbrecht, J.W. Swan, and M.J. Lannoo. 2017b. Wildlife cameras reveal high resolution activity patterns in threatened crawfish frogs (*Lithobates areolatus*). Herpetological Conservation and Biology 12:160-170.

Stiles, R.M., V.C.K. Terrell, J.C. Maerz, and M.J. Lannoo. 2020. Density-dependent fitness attributes and carry-over effects in crawfish frogs (*Rana areolata*), a species of conservation concern. Copeia 108:443-452.

Stoler, A.B., and R.A. Relyea. 2013. Leaf litter quality induces morphological and developmental changes in larval amphibians. Ecology 94:1594-1603.

Strasburg, M., and M.D. Boone. 2021. Effects of trematode parasites on snails and northern leopard frogs (*Lithobates pipiens*) in pesticide-exposed mesocosm communities.

Journal of Herpetology 55:229-236.

Streicher, J.W., and M.K. Fujita. 2014. Observations on the captive maintenance and reproduction of the Balcones barking frog, *Craugastor augusti latrans*. Herpetological Review 45:49-51.

Sullivan, B.K. 2019. Book review: Arizona's Amphibians & Reptiles: A Natural History and Field Guide. Herpetological Review 50:173-175.

Sullivan, B.K., J. Wooten, T.D. Schwaner, K.O. Sullivan, and M. Takahashi. 2015. Thirty years of hybridization between toads along the Agua Fria River in Arizona: I. Evidence from morphology and mtDNA. Journal of Herpetology 49:150-156.

Suriyamongkol, T., A. Villamizar-Gomez, M.R.J. Forstner, and I. Mali. 2019. Detection of *Batrachochytrium dendrobatidis* in eastern New Mexico, USA. Herpetological Review 50:300-303.

Suriyamongkol, T., K. Forks, A. Villamizar-Gomez, H.-H. Wang, W.E. Grant, M. R. J. Forstner, and I. Mali. 2021. A simple conservation tool to aid restoration of amphibians following high-severity wildfires: use of PVC pipes by green tree frogs (*Hyla cinerea*) in central Texas, USA. Diversity 13:649.

Sutton, W.B., M.J. Gray, R.H. Hardman, R.P. Wilkes, A.J. Kouba, and D.L. Miller. 2014. High susceptibility of the endangered dusky gopher frog to ranavirus. Diseases of Aquatic Organisms 112:9-16.

Swan, K.D., V.C. Hawkes, and P.T. Gregory. 2015. Breeding phenology and habitat use of amphibians in the drawdown zone of a hydroelectric reservoir. Herpetological Conservation and Biology 10:864-873.

Swanson, C.L., and R.L. Swanson. 2017. Geographic distribution: *Eleutherodactylus cystignathoides* (Rio Grande Chirping Frog). Herpetological Review 48:382.

Swanson, J.E., E. Muths, C.L. Pierce, S.J. Dinsmore, M.W. Vandever, M.L. Hladik, and K.L. Smalling. 2018. Exploring the amphibian exposome in an agricultural landscape using telemetry and passive sampling. Scientific Reports 8:10045.

Swanson, J.E., C.L. Pierce, S.J. Dinsmore, K.L. Smalling, M.W. Vandever, T.W. Stewart, and E. Muths. 2019. Factors influencing anuran wetland occupancy in an agricultural landscape. Herpetologica 75:47-56.

Swartz, L.K., C.R. Faurot-Daniels, B.R. Hossack, and E. Muths. 2014. *Anaxyrus boreas* (Western Toad). Predation. Herpetological Review 45:303.

Swartz, L.K., W.H. Lowe, E.L. Muths, and B.R. Hossack. 2020. Species-specific responses to wetland mitigation among amphibians in the Greater Yellowstone Ecosystem. Restoration Ecology 28:206-214.

Swartz, T. M., E. K. Wilson, and J. R. Miller. 2019. Lithobates blairi (Plains Leopard Frog). Ectoparasites. Herpe-

tological Review 50:114.

Sweeney, M.R., C.M. Thompson, and V.D. Popescu. 2021. Sublethal, behavioral, and developmental effects of the neonicotinoid pesticide imidacloprid on larval wood frogs (*Rana sylvatica*). Environmental Toxicology and Chemistry 40:1838-1847.

Talbott, K., T.M. Wolf, P. Sebastian, M. Abraham, I. Bueno, M. McLaughlin, T. Harris, R. Thompson, A. Pessier, and D. Travis. 2018. Factors influencing detection and co-detection of ranavirus and *Batrachochytrium dendrobatidis* in midwestern North American anuran populations. Diseases of Aquatic Organisms 128:93-103.

Talley, B.L., C.R. Muletz, V.T. Vredenburg, R.C. Fleischer, and K.R. Lips. 2015. A century of *Batrachochytrium dendrobatidis* in Illinois amphibians (1888-1989). Biological Conservation 182:254-261.

Taus, M. 2018. Field Guide to Reptiles & Amphibians of Fort Ord Natural Reserve. UCSC Natural Reserves, Santa Cruz, California.

Taylor, M.E.D., and C.A. Paszkowski. 2018. Postbreeding movement patterns and habitat use of adult wood frogs (*Lithobates sylvaticus*) at urban wetlands. Canadian Journal of Zoology 96:521-532.

Terrell, V.C.K., J.L. Klemish, N.J. Engbrecht, J.A. May, P.J. Lannoo, R.M. Stiles, and M.J. Lannoo. 2014a. Amphibian and reptile colonization of reclaimed coal spoil grasslands. Journal of North American Herpetology 2014:59-68.

Terrell, V.C.K., N.J. Engbrecht, A.P. Pessier, and M.J. Lannoo. 2014b. Drought reduces chytrid fungus (*Batrachochytrium dendrobatidis*) infection intensity and mortality but not prevalence in adult crawfish frogs (*Lithobates areolatus*). Journal of Wildlife Diseases 50:56-62.

Tetzlaff, S.J., and K.D. Tetzlaff. 2019. *Lithobates catesbeianus* (American Bullfrog). Predation. Herpetological Review 50:547.

Thomas, M.A., and S.C. Follum. 2016. *Anaxyrus americanus* (American Toad). Partial leucism. Herpetological Review 47:102-103.

Thompson, C.M., and V.D. Popescu. 2021. Complex hydroperiod induced carryover responses for survival, growth, and endurance of a pond-breeding amphibian. Oecologia 195:1071-1081.

Thompson, M., and R.V. Rea. 2013. *Rana sylvatica* (Wood Frog). Leucism. Herpetological Review 44:128-129.

Thompson, M.D., M.G. Bolek, and R.V. Rhea. 2021. Calliphorid parasitism causing myiasis in amphibians: a new record for *Anaxyrus boreas* (Western Toad) in British Columbia, Canada. Herpetological Review 52:303-307.

Thompson, M.E., A.J. Nowakowski, and M.A. Donnelly. 2015. The importance of defining focal assemblages when evaluating amphibian and reptiles responses to land use. Conservation Biology 30:249-258.

Thurman, L.L., and T.S. Garcia. 2017. Differential plasticity in response to simulated climate warming in a high-elevation amphibian assemblage. Journal of Herpetology 51:232-239.

Tice, A.K., M.W. Brown, and R. Altig. 2016. *Hyla chrysoscelis* (Cope's Gray Treefrog), *H. avivoca* (Bird-voiced Treefrog), and *H. cinerea* (Green Treefrog). Egg and tadpole mortality. Herpetological Review 47:438-439.

Tidwell, K.S. 2017. Quantifying the impacts of a novel predator: the distinctive case of the Oregon spotted frog (*Rana pretiosa*) and the invasive American bullfrog (*Rana* (*Aquarana*) *catesbeiana*). Ph.D. Dissertation, Portland State University, Portland, Oregon.

Tidwell, K.S., and M.P. Hayes. 2013. Difference in flight initiation distance between recently metamorphosed Oregon spotted frogs (*Rana pretiosa*) and American bullfrogs (*Lithobates catesbeianus*). Herpetological Conservation and Biology 8:426-434.

Tidwell, K.S., D.J. Shepherdson, and M.P. Hayes. 2013. Interpopulation variability in evasive behavior in the Oregon spotted frog (*Rana pretiosa*). Journal of Herpetology 47:93-96.

Timm, B.C., and K. McGarigal. 2014. Fowler's toad (*Anaxyrus fowleri*) activity patterns on a roadway in Cape Cod National Seashore. Journal of Herpetology 48:111-116.

Timm, B.C., K. McGarigal, and R.C. Cook. 2014. Upland movement patterns and habitat selection of adult eastern spadefoots (*Scaphiopus holbrookii*) at Cape Cod National Seashore. Journal of Herpetology 48:84-97.

Titus, V.R., and T.M. Green. 2013. Presence of *Ranavirus* in green frogs and eastern tiger salamanders on Long Island, New York. Herpetological Review 44:266-267.

Todd, H., J.M. Fritzler, R.T. Kazmaier, and J.B. Johnson. 2019. Prevalence of *Batrachochytrium dendrobatidis* and ranavirus in western Texas, USA. Herpetological Review 50:505-507.

Todoroff, R.J. 2021. *Rana aurora* (Northern Red-legged Frog). Predation. Herpetological Review 52:120-121.

Tornabene, B.J. 2021. Ecotoxicological and physiological effects of salinity on amphibians. Ph.D. Dissertation, University of Montana, Missoula.

Tornabene, B.J., C.W. Breuner, and B.R. Hossack. 2021a. Comparative effects of energy-related saline wastewaters and sodium chloride on hatching, survival, and fitness-associated traits of two amphibian species. Environmental Toxicology and Chemistry 40:3137-3147.

Tornabene, B.J., M.F. Chislock, M.E. Gannon, M.S. Sepúlveda, and J.T. Hoverman. 2021b. Relative acute toxicity of three per- and polyfluoroalkyl substances on nine species of larval amphibians. Integrated Environmental Assessment and Management 2021:1-6.

Trites, M.J., L.V. Ferguson, C.T. Ogbuah, C.M. Dickson, and T.G. Smith. 2013. Factors determining the in vitro emergence of sexual stages of *Hepatozoon clamatae* from erythrocytes of the green frog (*Rana clamitans*). Canadian Journal of Zoology 91:219-226.

Tupper, T.A., C.A. Bozarth, K.S. Jones , and R.P. Cook. 2014. Detection of *Batrachochytrium dendrobatidis* in the eastern spadefoot, *Scaphiopus holbrookii*, at Cape Cod National Seashore, Barnstable County, Massachusetts, USA. Herpetological Review 45:445-447.

Tupper, T.A., L.D. Fuchs, C. O'Connor-Love, R. Aguilar, C. Bozarth, and D. Fernandez. 2017. Detection of the pathogenic fungus, *Batrachochytrium dendrobatidis*, in anurans of Huntley Meadows Park, Fairfax County, Virginia. Catesbeiana 37:109-120.

Tye, S.P., K. Geluso, and M.J. Harner. 2017. Geographic distribution: *Hyla chrysoscelis* (Cope's Gray Treefrog). Herpetological Review 48:382-383.

Underwood, J.G., and K.J. Letchworth. 2016. Improving bullfrog capture methods in areas managed for Hawaii's endangered endemic waterbirds. Pp. 380-383 *In* R.M. Timm and R.A. Baldwin (eds.), Proceedings of the 27th Vertebrate Pest Conference, University of California, Davis.

Urban, M.C., J.L. Richardson, N.A. Freidenfelds, D.L. Drake, J.F. Fischer, and P.P. Saunders. 2017. Microgeographic adaptation of wood frog tadpoles to an apex predator. Copeia 105:451-461.

USFWS. 2014. Endangered Species status for Sierra Nevada Yellow Legged Frog and Northern Distinct Population Segment of the Mountain Yellow-Legged Frog, and Threatened Species status for Yosemite Toad. Final Rule. Federal Register 79(82):24256-24310.

USFWS. 2015. Withdrawal of proposed rule to reclassify the Arroyo Toad as Threatened. Federal Register 80(246):79805-79816.

USFWS. 2016. Designation of Critical Habitat for the Sierra Nevada Yellow-Legged Frog, the Northern DPS of the Mountain Yellow-Legged Frog, and the Yosemite Toad; Final Rule. Federal Register 81(166):59046-59119.

Vallegos, J.G. 2021. *Rana sylvatica* (Wood Frog). Leucism. Herpetological Review 52:121.

Vandewege, M.W., T.M. Swannack, K.L. Greuter, D.J. Brown, and M.R.J. Forstner. 2013. Breeding site fidelity and terrestrial movement of an endangered amphibian, the Houston toad (*Bufo houstonensis*). Herpetological Conservation and Biology 8:435-446.

Van Kleek, M.J., and B.S. Holland. 2018. Gut check: predatory ecology of the invasive wrinkled frog (*Glandirana rugosa*) in Hawai'i. Pacific Science 72:199-208.

Vargas-Salinas, F., G.M. Cunnington, A. Amézquita, and L. Fahrig. 2014. Does traffic noise alter calling time in frogs and toads? A case study of anurans in eastern Ontario, Canada. Urban Ecosystems 17:945-953.

Vélez, A., and A.S. Guajardo. 2021. Individual variation in two types of advertisement calls of Pacific tree frogs, *Hyliola* (= *Pseudacris*) *regilla*, and the implications for sexual selection and species recognition. Bioacoustics 30:437-457.

Vélez, A., B.J. Linehan-Skillings, Y. Gu, Y. Sun, and M.A. Bee. 2013. Pulse-number discrimination by Cope's gray treefrog (*Hyla chrysoscelis*) in modulated and unmodulated noise. Journal of the Acoustical Society of America 134:3079-3089.

Vemulapally, S., A. Villamizar, T. Guerra, M.E. Tocidlowski, M. Spradley, S. Mays, M.R.J. Forstner, and D. Hahn. 2021. Mycobacteria skin lesions and the habitat of the endangered Houston toad (*Anaxyrus houstonensis*). Journal of Wildlife Diseases 57:503-514.

Vhora, M.S., and M.G. Bolek. 2013. New host and distribution records for *Aplectana hamatospicula* (Ascaridida: Cosmocercidae) in *Gastrophryne olivacea* (Anura: Microhylidae) from the Great Plains USA. Journal of Parasitology 99:417-420.

Vieira de Andrade, D., C.R. Bevier, and J. Eduardo de Carvalho (eds.). 2016. Amphibian and Reptile Adaptations to the Environment. Interplay between Physiology and Behavior. CRC Press, Boca Raton, Florida.

Villa, R.A., S. Riplog-Peterson, R.A. Repp, and K.C. Reeves. 2021. *Incilius alvarius* (Sonoran Desert Toad). Injury and Recovery. Herpetological Review 52:609.

Villamizar-Gomez, A., M.R.J. Forstner, T. Suriyamongkol, K.N. Forks, W.E. Grant, H.-H. Wang, and I. Mali. 2016. Prevalence of *Batrachochytrium dendrobatidis* in two sympatric treefrog species, *Hyla cinerea* and *Hyla versicolor*. Herpetological Review 47:601-5605.

Villena, O.C., J.A. Royle, L.A. Weir, T.M. Foreman, K.D. Gazenski, and E.H.C. Grant. 2016. Southeast regional and state trends in anuran occupancy from calling survey data (2001-2013) from the North American Amphibian Monitoring Program. Herpetological Conservation and Biology 11:373-385.

Waddle, A.W. 2017. Emerging infectious disease and the imperiled relict leopard frog. M.S. thesis, University of Nevada, Las Vegas.

Wallis, A.C., R.L. Smith, and K.H. Beard. 2016. Temporal foraging patterns of nonnative coqui frogs (*Eleutherodactylus coqui*) in Hawaii. Journal of Herpetology 50:582-588.

Walls, S.C., J.H. Waddle, and S.P. Faulkner. 2014a. Wetland Reserve Program enhances site occupancy and species richness in assemblages of anuran amphibians in the Mississippi alluvial valley, USA. Wetlands 34:197-207.

Walls, S.C., J.H. Waddle, W.J. Barichivich, I.A. Bartoszek, M.E. Brown, J.M. Hefner, and M.J. Schuman. 2014b. Anuran site occupancy and species richness as tools for evaluating restoration of a hydrologically-modified landscape. Wetlands Ecology and Management 22:625-639.

Ward, J.L., E.K. Love, A. Vélez, N.P. Buerkle, L.R. O'Bryan, and M.A. Bee. 2013. Multitasking males and multiplicative females: dynamic signaling and receiver preferences in Cope's gray treefrog. Animal Behaviour 86:231-243.

Warwick, A.R., J. Travis, and E. Moriarty Lemmon. 2015. Geographic variation in the Pine Barrens treefrog (*Hyla andersonii*): concordance of genetic, morphometric and acoustic signal data. Molecular Ecology 24:3281-3298.

Warwick, A.R., L.N. Barrow, M.L. Smith, D.B. Means, A.R. Lemmon, and E.M. Lemmon. 2021. Signatures of north-eastern expansion and multiple refugia: genomic phylogeography of the Pine Barrens tree frog, *Hyla andersonii* (Anura: Hylidae). Biological Journal of the Linnean Society 133:120–134.

Watermolen, D.J. 2017. Identification keys for the eggs of South Dakota amphibians. Bulletin of the Chicago Herpetological Society 52:212-215.

Watermolen, D.J. 2019. Crustacean ectoparasites of amphibians. Bulletin of the Chicago Herpetological Society 54:85-91.

Watters, J.L., R.L. Flanagan, D.R. Davis, J.K. Farkas, J.L. Kerby, M.J. Labonte, M.L. Penrod, and C.D. Siler. 2016. Screening natural history collections for historical presence of *Batrachochytrium dendrobatidis* in anurans from Oklahoma, USA. Herpetological Review 47:214-220.

Watters, J.L., D.R. Davis, T. Yuri, and C.D. Siler. 2018. Concurrent infection of *Batrachochytrium dendrobatidis* and ranavirus among native amphibians from northeastern Oklahoma, USA. Journal of Aquatic Animal Health 30:291-301.

Watters, J.L., S.L. McMillin, E.C. Marhanka, D.R. Davis, J.K. Farkas, J.L. Kerby, and C.D. Siler. 2019. Seasonality in *Batrachochytrium dendrobatidis* detection in amphibians in central Oklahoma, USA. Journal of Zoo and Wildlife Medicine 50:492-497.

Watts, A.G., P.E. Schlichting, S.M. Billerman, B.R. Jesmer, S. Micheletti, M.-J. Fortin, W.C. Funk, P. Haperman, E. Muths, and M.A. Murphy. 2015. How spatio-temporal habitat connectivity affects amphibian genetic structure. Frontiers in Genetics 6:275.

Weeks, D.M., and M.J. Parris. 2020. A *Bacillus thuringiensis kurstaki* biopesticide does not reduce hatching success or tadpole survival at environmentally relevant concentrations in southern leopard frogs (*Lithobates sphenocephalus*). Environmental Toxicology and Chemistry 39:155-161.

Weick, D.L., and B.H. Brattstrom. 2020. Salinity tolerance and osmoregulation in the wide-spread Pacific treefrog,

Pseudacris regilla. Bulletin of the Southern California Academy of Sciences 119(2):1-9.

Weir, L.A., J.A. Royle, K.D. Gazenski, and O. Villena. 2014. Northeast regional and state trends in anuran occupancy from calling survey data (2001-2011) from the North American Amphibian Monitoring Program. Herpetological Conservation and Biology 9:223-245.

Weir, S.M., R.W. Flynn, D.E. Scott, S. Yu, and S.L. Lance. 2016a. Environmental levels of zinc do not protect embryos from Cu toxicity in three species of amphibians. Environmental Pollution 214:161-168.

Weir, S.M., D.E. Scott, C.J. Salice, and S.L. Lance. 2016b. Integrating copper toxicity and climate change to understand extinction risk to two species of pond-breeding anurans. Ecological Applications 26:1721-1732.

Wen, A. 2015. Association between habitat characteristics, human activities, and anuran species in a wetland agricultural landscape. Journal of Herpetology 49:594-601.

Whalen, T.O. 2021. Geographic distribution: *Anaxyrus cognatus* (Great Plains Toad). Herpetological Review 52:76.

Wheeler, C.A., A.J. Lind, H.H. Welsh, Jr., and A.K. Cummings. 2018. Factors that influence the timing of calling and oviposition of a lotic frog in northwestern California. Journal of Herpetology 52:289-298.

Wheelwright, N.T., M.J. Gray, R.D. Hill, and D.L. Miller. 2014. Sudden mass die-off of a large population of wood frog (*Lithobates sylvaticus*) tadpoles in Maine, USA, likely due to a ranavirus. Herpetological Review 45:240-242.

Whitworth, T.L., M.G. Bolek, and G. Arias-Robledo. 2021. *Lucilia bufonivora*, not *Lucilia silvarum* (Diptera: Calliphoridae), causes myiasis in anurans in North America with notes about *Lucilia elongata* and *Lucilia thatuna*. Journal of Medical Entomology 58:88–92.

Wiesehan, K. 2021. Conservation status of the state-threatened Illinois chorus frog (*Pseudacris illinoensis*) in southwestern Illinois. M.S. thesis, Southern Illinois University at Edwardsville, Edwardsville.

Wilcox, J.T. 2017. *Rana boylii* (Foothill Yellow-legged Frog). Predation. Herpetological Review 48:612-613.

Wilder, A.E., and A.M. Welch. 2014. Effects of salinity and pesticide on sperm activity and oviposition site selection in green treefrogs, *Hyla cinerea*. Copeia 2014:659-667.

Williams, J.M., D.J. Brown, and P.B. Wood. 2017. Responses of terrestrial herpetofauna to persistent, novel ecosystems resulting from mountaintop removal mining. Journal of Fish and Wildlife Management 8:387-400.

Williams, P.J., N.J. Engbrecht, J.R. Robb, V.C.K. Terrell, and M.J. Lannoo. 2013. Surveying a threatened amphibian species through a narrow detection window. Copeia 2013:552-561.

Williams, R.N., B.J. MacGowan, Z. Walker, J. Hoverman, and N. Burgmeier. 2017. Frogs and Toads of Indiana. Purdue University Extension, West Lafayette, Indiana.

Wilson, A.C. 2016. Distribution of Cane Toads (*Rhinella marina*) in Florida and their status in natural areas. M.S. thesis, University of Florida, Gainesville

Wilson, T.P., J.M. Barbosa, E.A. Carver, B.R. Reynolds, D. Richards, Team Salamander, and T.M. Wilson. 2015. *Batrachochytrium dendrobatidis* prevalence in two ranid frogs on a former United States Department of Defense installation in southeastern Tennessee. Herpetological Review 46:37-41.

Wimsatt, J., S.H. Feldman, M. Heffron, M. Hammond, M.P. Roth Ruehling, K.L. Grayson, and J.C. Mitchell. 2014. Detection of pathogenic *Batrachochytrium dendrobatidis* using water filtration, animal and bait testing. Zoo Biology 33:577-585.

Windstam, S.T., and J.C. Olori. 2014. Proportion of hosts carrying *Batrachochytrium dendrobatidis*, causal agent of amphibian chytridiomycosis, in Oswego County, NY in 2012. Northeastern Naturalist 21:NENHC25-NENHC34.

Winzeler, M.E., R.N. Williams, and S.J.A. Kimble. 2016. Surveying for ranavirus in green frogs (*Lithobates clamitans*) at five locations in Indiana. Journal of North American Herpetology 2016:23-26.

Wise, R.S., S.L. Rumschlag, and M.D. Boone. 2014. Effects of amphibian chytrid fungus exposure on American toads in the presense of an insecticide. Environmental Toxicology and Chemistry 33:2541-2544.

Wolff, B.G., E. Wurm, and L. Tracey. 2013. *Pseudacris crucifer* (Sping Peeper). Myiasis. Herpetological Review 44:498-499.

Wolff, B.G., E. Wurm, S. Conway, and K. Kinzer. 2014. *Batrachochytrium dendrobatidis* infection rates differ over short distances between natural lakes and artificial ponds in Minnesota, USA. Herpetological Review 45:447-449.

Wooten, J.A., B.K. Sullivan, M.R. Klooster, T.D. Schwaner, K.O. Sullivan, A.D. Brown, M. Takahashi, and P.R. Bradford. 2019. Thirty years of hybridization between toads along the Agua Fria River in Arizona: Part II: Fine-scale assessment of genetic changes over time using microsatellites. Journal of Herpetology 53:104-114.

Wuerthner, V.P., J. Hua, and J.T. Hoverman. 2017. The benefits of coinfection: trematodes alter disease outcomes associated with virus infection. Journal of Animal Ecology 86:921-931.

Wyman, J.T., and Q.I. Agnew. 2021. *Craugastor augusti cactorum* (Western Barking Frog). Behavior. Herpetological Review 52:606-607.

Yagi, K.T., and D.M. Green. 2016. Mechanisms of density-dependent growth and survival in tadpoles of Fowler's toad, *Anaxyrus fowleri*: volume vs. abundance. Copeia 104:942-951.

Yagi, K.T., and D.M. Green. 2018. Post-metamorphic carry-over effects in a complex life history: behavior and growth at two life stages in an amphibian, *Anaxyrus fowleri*. Copeia 106:77-85.

Yahn, J.M., and W.H. Karasov. 2021. The effects of dietary polybrominated diphenyl ether exposure and rearing temperature on tadpole growth, development, and their underlying processes. Environmental Toxicology and Chemistry 40:3181-3192.

Yarnell, S.M., R.A. Peek, N. Keung, B.D. Todd, S. Lawler, and C. Brown. 2019. A lentic breeder in lotic waters: Sierra Nevada yellow-legged frog (*Rana sierrae*) habitat suitability in northern Sierra Nevada streams. Copeia 107:676-693.

Youker-Smith, T.E., C.M. Whipps, and S.J. Ryan. 2016. Detection of an FV3-like ranavirus in wood frogs (*Lithobates sylvaticus*) and green frogs (*Lithobates clamitans*) in a constructed vernal pool network in central New York state. Herpetological Review 47:595-598.

Youngquist, M.B., and M.D. Boone. 2014. Movement of amphibians through agricultural landscapes: the role of habitat on edge permeability. Biological Conservation 175:148-155.

Youngquist, M.B., and M.D. Boone. 2021a. Making the connection: combining habitat suitability and landscape connectivity to understand species distribution in an agricultural landscape. Landscape Ecology 36:2795-2809.

Youngquist, M.B., and M.D. Boone. 2021b. Larval development and survival of pond-breeding anurans in an agricultural landscape impacted more by phytoplankton than surrounding habitat. PLoS One 16:e0255058.

Youngquist, M.B., K. Downard, and M.D. Boone. 2015. Competitive interactions between cricket frogs (*Acris blanchardi*) and other anurans. Herpetologica 71:260-267.

Yuan, Z.-H., W.-W. Zhou, X. Chen, N.A. Poyarkay, H.-M. Chen, N.-H. Jang-Liaw, W.-H. Chou, N.J. Matzke, K. Iizuka, M.-S. Min, S.L. Kuzmin, Y.-P. Zhang, D.C. Cannatella, D.M. Hillis, and J. Che. 2016. Spatiotemporal diversification of the True Frogs (Genus *Rana*): a historical framework for a widely studied group of model organisms. Systematic Biology 65:824-842.

Zeitler, E.F., M. Kochinski, and K.K. Cecala. 2021. Evaluating potential mechanisms for altered amphibian performance in treated wastewater. Herpetological Conservation and Biology 16:412–424.

Zigler, K.S., M.L. Niemiller, C.D.R. Stephen, B.N. Ayala, M.A. Milne, N.S. Gladstone, A.S. Engel, J.B. Jensen, C.D. Camp, J.C. Ozier, and A. Cressler. 2020. Biodiversity from caves and other subterranean habitats of Georgia, USA. Journal of Cave and Karst Studies 82:125-167.

Zughaiyir, F.E., C. Rivera, S. Sirsi, and M.R.J. Forstner. 2021. *Anaxyrus houstonensis* (Houston Toad). Predation. Herpetological Review 52:111-112.

Zylstra, E.R., and M.K. Ward. 2013. *Hyla arenicolor* (Canyon Treefrog). Habitat use. Herpetological Review 44:494-495.

Zylstra, E.R., R.J. Steidl, D.E. Swann, and K. Ratzlaff. 2015. Hydrologic variability governs population dynamics of a vulnerable amphibian in an arid environment. PLoS One 10:e0125670.

Zylstra, E.R., D.E. Swan, B.R. Hossack, E. Muths, and R.J. Steidl. 2019a. Drought-mediated extinction of an arid-land amphibian: insights from a spatially explicit dynamic occupancy model. Ecological Applications 29:e01859.

Zylstra, E.R., D.E. Swann, and R.J. Steidl. 2019b. Surface-water availability governs survival of an amphibian in arid mountain streams. Freshwater Biology 64:164-174.

International Society for the History and Bibliography of Herpetology

Mission. The ISHBH aims to promote research related to historical herpetology. The Society is devoted to stimulating and advancing the study of the history of the herpetological discipline and its bibliography. We bring together individuals for whom the history and bibliography of herpetology are appealing. We promote the knowledge of these and related topics among members and the general public. The Society was established in 1998. Membership is open to anyone who shares the aims of the Society.

Activities. The society members meet every year, usually in connection with other general herpetological meetings with international participation. Activities include visits to private and public libraries, museums and other places with historical links to the discipline. We organize workshops and sessions in the fields that form parts of larger national or international meetings on herpetology. Our shared lunches adjacent to the yearly business meetings have become popular among members and guests. The Society works to facilitate both formal and informal contacts among members.

Journal. The ISHBH publishes the journal *Bibliotheca Herpetologica*, which is the central pillar of the Society. It contains articles, essays, bibliographies and news of people and events in our field and is a peer-reviewed. The many important contributions make the publication a vital source for bibliographers, historians and taxonomists alike but at the same time the papers are appealing to the layperson in the field. All issues of *Bibliotheca Herpetologica* are available Open Access on the society website, www.ISHBH.com. A print copy of each volume of *Bibliotheca Herpetologica*, containing all articles published that year, is distributed to current members.

The name of the journal, up to volume 5(1), was: *International Society for the History and Bibliography of Herpetology Newsletter and Bulletin*. All of these issues are also available, Open Access on the society's website.

Wahlgreniana. Beginning 2022, ISHBH began publishing *Wahlgreniana* — A series of book-length works complementing the ISHBH journal, *Bibliotheca Herpetologica*. Books in this series are published on an irregular basis and are sold separately from ISHBH subscriptions, discounted for society members. *Wahlgreniana* is named in honor of Richard Wahlgren (1946–2019) A founding member and first Chairman of the International Society for the History and Bibliography of Herpetology. Without Richard's tireless dedication to ISHBH, the society could not have made it through its early years.

Volume 1—May 2022

Bour, Roger and Josef F. Schmidtler. 2022. *Nikolaus Michael Oppel's Drawings, Watercolors, and Engravings 3. Crocodiles (1807–1817): A comparative study of some historical and recent crocodile illustrations*. ISHBH, Salt Lake City, x, 184 p. Hardcover, ISBN 978-0-578-29399-8. Includes a complete facsimile of Tiedemann, F., M. Oppel & J. Liboschitz 1817. *Naturgeschichte der Amphibien. Erstes Heft. Gattung Krokodil*. Joseph Engelmann, Heidelberg. v–vi, 1–88, vii–viii. Auf Kosten der Verfasser, München, 15 pls. Retail: $65.00; ISHBH members $39.00.

Volume 2—October 2022

Dodd, C. Kenneth, Jr. 2022. *Bibliography of the Anurans of the United States and Canada. Version 3. Part 1: 1698–2012. Part 2: 2013-2021*. ISHBH, Salt Lake City, x, 282 p. Hardcover, ISBN: 979-8-218-06245-3; eBook ISBH: 979-8-218-06246-0. Retail: $35.00; ISHBH members $21.00.

Membership. ISHBH membership categories are: Corresponding Membership US$25/year (open access only plus all other membership benefits), Regular Membership US$50/year, Institutional Memberships US$75/year, and Sponsoring US$75/year. All members, except Corresponding Memebers, receive a printed copy of *Bibliotheca Herpetologica*. Join online at www.ISHBH.com.